PROTEINS

Second Edition

SB
Sterling Book House
181, Dr. D. N. Road, Fort,
Mumbai-400 001, INDIA.
Tel. 261 2521 • Fax 262 3551

PROTEINS

Structures and Molecular Properties

Second Edition

Thomas E. Creighton

European Molecular Biology Laboratory
Heidelberg, Germany

W. H. Freeman and Company • New York

Cover image provided by Martin Noble.

Library of Congress Cataloging-in-Publication Data
Creighton, Thomas E., 1940–
    Proteins : structures and molecular properties / Thomas E.
Creighton. — 2nd ed.
      p.   cm.
    Includes bibliographical references and index.
    ISBN 0-7167-7030-X
    1. Proteins—Structure. 2. Proteins—Chemistry. I. Title.
QP551.C737   1993
574.19′245—dc20   92-6664
                                         CIP

Printed in the United States of America

Fourth printing 1996, KP

Contents

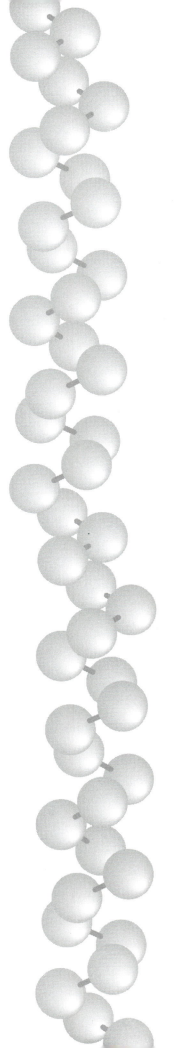

Preface   xi

1 *Chemical Properties of Polypeptides*   1

  1.1 The Polymeric Nature of Proteins   2

  1.2 The Polypeptide Backbone   5

  1.3 Amino Acid Residues   6

    1.3.1   Gly   7

    1.3.2   The Aliphatic Residues: Ala, Val, Leu, Ile   7

    1.3.3   The Cyclic Imino Acid: Pro   7

    1.3.4   The Hydroxyl Residues: Ser and Thr   8

    1.3.5   The Acidic Residues: Asp and Glu   8

    1.3.6   The Amide Residues: Asn and Gln   9

    1.3.7   The Basic Residues: Lys and Arg   10

    1.3.8   His   13

    1.3.9   The Aromatic Residues: Phe, Tyr, and Trp   14

    1.3.10 The Sulfur-Containing Residues: Met and Cys   17

  1.4 Detection of Amino Acids, Peptides, and Proteins   20

1.5 Determination of the Sizes of Proteins   23

   1.5.1 Sedimentation Analysis   24

   1.5.2 Gel Filtration   24

   1.5.3 SDS Polyacrylamide Gel Electrophoresis   26

1.6 Determination of the Covalent Strutures of Proteins   28

   1.6.1 Amino Acid Composition   28

   1.6.2 The Amino Acid Sequence   31

1.7 Nature of Amino Acid Sequences   43

1.8 Peptide Synthesis   43

Exercises   47

2 *Biosynthesis of Proteins*   49

2.1 Assembly of the Primary Structure   50

   2.1.1 Gene Structure   50

   2.1.2 Transcription   51

   2.1.3 Translation   53

2.2 Protein Engineering   59

   2.2.1 Isolating the Gene for a Protein   59

   2.2.2 Protein Sequences from Gene Sequences   59

   2.2.3 Identifying a Protein Specified by a Gene of Known Sequence   60

   2.2.4 Expressing a Cloned or Synthetic Gene   62

   2.2.5 Site-Directed Mutagenesis   63

2.3 Topogenesis   64

   2.3.1 Secreted, Lysosomal, and Membrane Proteins, Via the Endoplasmic Reticulum   65

   2.3.2 Import of Proteins into Other Organelles   71

   2.3.3 Nuclear Proteins   75

   2.3.4 Membrane Proteins   75

   2.3.5 Bacterial Secreted and Membrane Proteins   77

2.4 Posttranslational Covalent Modifications of Polypeptide Chains   78

   2.4.1 Proteolytic Processing   78

   2.4.2 Alteration of the Chain Termini   86

   2.4.3 Glycosylation   91

   2.4.4 Lipid Attachment   94

   2.4.5 Sulfation   95

   2.4.6 γ-Carboxy-Glu Residues   95

   2.4.7 Hydroxylation   96

   2.4.8 Phosphorylation   96

   2.4.9 ADP-Ribosylation   98

   2.4.10 Disulfide Bond Formation   98

   2.4.11 Common Nonenzymatic, Chemical Modifications   99

2.5 Nonribosomal Biosynthesis of Unusual Peptides   100

Exercises   102

3 *Evolutionary and Genetic Origins of Protein Sequences*   105

3.1 Primoridal Origins of Proteins   107

3.2 Evolutionary Divergence of Proteins   108

   3.2.1 Homologous Genes and Proteins   108

   3.2.2 Mutations and Protein Structure   111

   3.2.3 Genetic Divergence During Evolution   113

3.3 Reconstructing Evolution from Contemporary Sequences   114

   3.3.1 Variation among Species   114

   3.3.2 Variation within Species   125

3.4 Gene Rearrangements and the Evolution of Protein Complexity   127

   3.4.1 Products of Gene Duplications   127

3.4.2 Evolution of Metabolic
Pathways   131

3.4.3 Protein Elongation by
Intragene Duplication   132

3.4.4 Gene Fusion and Division   132

3.4.5 Genetic Consequences of
Duplicated Genes   134

3.5 Using Genetics to Probe Protein
Structure   134

3.5.1 Selecting for Functional
Mutations   135

3.5.2 Simulation of Evolution   136

Exercises   137

4 Physical Interactions That
Determine the Properties
of Proteins   139

4.1 The Physical Nature of Noncovalent Interactions   140

4.1.1 Short-Range Repulsions   140

4.1.2 Electrostatic Forces   142

4.1.3 Van der Waals Interactions   146

4.1.4 Hydrogen Bonds   147

4.2 The Proteins of Liquid Water and
the Characteristics of Noncovalent Interactions in This Solvent   148

4.2.1 Liquids   149

4.2.2 Water   150

4.2.3 Aqueous Solutions   153

4.3 The Hydrophobic Interaction   157

4.3.1 The Hydrophobic Interaction in Model Systems   157

4.3.2 Hydrophobicities of
Amino Acid Residues   160

4.4 Intramolecular Interactions   162

4.4.1 Effective Concentrations   163

4.4.2 Multiple Interactions   165

4.4.3 Cooperativity of Multiple
Interactions   165

Exercises   167

5 Conformational Properties
of Polypeptide Chains   171

5.1 Three-Dimensional Conformations   172

5.2 Polypeptides as Random
Polymers   173

5.2.1 Local Restrictions on Flexibility: The Ramachandran Plot   173

5.2.2 Statistical Properties of
Random Polypeptides   176

5.2.3 Rates of Conformational
Change   180

5.3 Regular Conformations of
Polypeptides   182

5.3.1 The α-Helix   182

5.3.2 β-Sheets   186

5.3.3 Other Regular Conformations   187

5.4 Experimental Characterization of
Polypeptides in Solution   189

5.4.1 Hydrodynamic Properties   189

5.4.2 Spectral Properties   190

5.5 Fibrous Proteins   193

5.5.1 Silk Fibroin   193

5.5.2 Coiled Coils   193

5.5.3 Collagen Triple Helix   196

Exercises   198

6 The Folded Conformations
of Globular Proteins   201

6.1 Three-Dimensional Structures by
X-ray Diffraction   202

6.1.1 Crystallizing Proteins   202

6.1.2 Basic Principles of
Diffraction   203

6.1.3 Phase Determination   208

6.1.4 Calculation of the Electron
Density Map   210

6.1.5 Interpretation of the
Electron Density Map   213

6.1.6 Refinement of the Model   214

6.1.7 Rapid Diffraction Measurements   215

6.1.8 Neutron Diffraction   216

6.2 The General Properties of Protein Structures   217

   6.2.1 The Tertiary Structure   217

   6.2.2 Secondary Structure   221

   6.2.3 Reverse Turns   225

   6.2.4 Supersecondary Structures   227

   6.2.5 Interiors and Exteriors   227

   6.2.6 Quaternary Structure   232

   6.2.7 Flexibility Detected Crystallographically   236

   6.2.8 The Solvent   238

6.3 Protein Structure Determination by Nuclear Magnetic Resonance Spectroscopy   238

   6.3.1 Nuclear Magnetic Resonance Spectra of Proteins   238

   6.3.2 Determining a Protein Structure by Nuclear Magnetic Resonance   243

6.4 Proteins with Similar Folded Conformations   244

   6.4.1 Evolutionarily Related Proteins   244

   6.4.2 Conformational Similarity without Apparent Sequence Homology   249

   6.4.3 Structural Homology within a Polypeptide Chain   251

6.5 Rationalization and Prediction of Protein Structure   253

   6.5.1 Predicting Secondary Structure   255

   6.5.2 Modeling Homologous Protein Structures   257

   6.5.3 De Novo Protein Design   258

Exercises   259

7 Proteins in Solution and in Membranes   261

7.1 Physical and Chemical Properties of Soluble Proteins   261

   7.1.1 Aqueous Solubility   262

   7.1.2 Hydrodynamic Properties in Aqueous Solution   264

   7.1.3 Spectral Properties   270

   7.1.4 Ionization   271

   7.1.5 Chemical Properties   272

7.2 Proteins in Membranes   276

   7.2.1 Association with Membranes   277

   7.2.2 Structures of Integral Membrane Proteins   278

   7.2.3 Identifying Amino Acid Sequences Likely to Traverse Membranes   280

   7.2.4 Dynamic Behavior in Membranes   280

7.3 Flexibility of Protein Structure   281

   7.3.1 Hydrogen Exchange   282

   7.3.2 Fluorescence Quenching and Depolarization   286

   7.3.3 Rotations of Side Chains   286

7.4 Stability of the Folded Conformation   287

   7.4.1 Reversible Unfolding Transitions   287

   7.4.2 Nature of the Unfolded State   291

   7.4.3 Physical Basis For Protein Denaturation   292

   7.4.4 Effects on Stability of Variation of the Primary Structure   303

7.5 Mechanism of Protein Folding   309

   7.5.1 Kinetic Analysis of Complex Reactions   309

   7.5.2 Kinetics of Unfolding   310

7.5.3 Kinetics of Refolding   311

7.5.4 Folding Pathways   316

7.5.5 Folding of Large Proteins   321

7.5.6 Biosynthetic Folding   323

Exercises   325

8 Interactions with Other
   Molecules   329

8.1 Structures of Protein – Ligand
     Complexes   330

8.1.1 The Difference Fourier
Crystallographic Technique   330

8.1.2 NMR   333

8.1.3 Chemical Methods of
Determining Binding Sites   333

8.1.4 General Properties of
Ligand Binding Interactions   334

8.2 Energetics and Dynamics of
     Binding   337

8.2.1 Binding Affinities   338

8.2.2 Accounting for Relative
Affinities   340

8.2.3 Rates of Binding and
Dissociation   344

8.2.4 Affinity Chromatography   346

8.3 Relationship between Protein
     Conformation and Binding   348

8.3.1 Immunoglobulins   348

8.3.2 DNA-Binding Proteins   355

8.3.3 Nucleotide Binding   360

8.3.4 Very Small Ligands   361

8.4 Allostery: Interactions between
     Binding Sites   367

8.4.1 Multiple Binding Sites and
Interactions between Them   368

8.4.2 Allosteric Models   372

8.4.3 The Allosteric Properties
of Hemoglobin   374

8.4.4 Other Allosteric O_2-
Binding Proteins   380

8.4.5 Negative Cooperativity   381

Exercises   382

9 Enzyme Catalysis   385

9.1 The Kinetics of Enzyme Action   386

9.1.1 Steady-State Kinetics   386

9.1.2 Reactions on the Enzyme   392

9.2 Theories of Enzyme Catalysis   396

9.2.1 Rate Enhancements   396

9.2.2 Transition-State
Stabilization   402

9.2.3 Transition-State
Analogues   403

9.2.4 Catalytic Antibodies   406

9.2.5 Substrate Specificity and
Induced Fit   406

9.2.6 Testing Theories of
Catalysis   410

9.3 Examples of Enzyme
     Mechanisms   413

9.3.1 Tyrosyl tRNA Synthetase   413

9.3.2 Proteases   417

9.3.3 Lysozyme   437

9.4 Regulation of Enzyme Activity   441

9.4.1 Enzyme Function in Vivo   442

9.4.2 Allosteric Regulation   444

9.4.3 Reversible Covalent
Modification   452

Exercises   459

10 Degradation   463

10.1 Chemical Aging   464

10.2 Protein Turnover in Vivo   465

10.2.1 Factors That Deter-
mine the Rate of Degradation   466

10.3 Mechanisms of Protein
     Degradation   468

10.3.1 Proteases Involved in
Protein Turnover   468

10.3.2 Lysosomes   469

10.3.3 Ubiquitin-Mediated
Pathway   470

Exercises   472

Appendix 1 *Major Protein
and DNA Sequence Data
Banks*   475

Appendix 2 *References to
Protein Structures Deter-
mined Crystallographically
to High Resolution*   477

Index   491

Preface

Because our understanding of protein structure and function has increased remarkably in the nine years since the first edition of this volume, most of this edition needed to be entirely rewritten. The structures of the twenty amino acid residues used in proteins, at least, have not changed. Also, there has been no marked change in our understanding of their intrinsic chemical properties (Chapter 1), although this probably reflects the decreased importance of chemical modification for studying protein function (Sec. 1.3) due to the advent of site-directed mutagenesis and protein engineering (Sec. 2.2). Procedures for the analysis, sequencing, and synthesis of peptides and proteins (Secs. 1.6 and 1.8) have been improved and have become routine and automated. A thorough knowledge of the chemical basis of such techniques is no longer absolutely necessary, but any serious scientist needs to understand them so as not to be misled by results obtained with automated procedures. The classical technique of mass spectrometry previously seemed of only limited use with nonvolatile samples like proteins, but new developments have made it of major importance in studying protein structure (Sec. 1.6.2.d).

Many new posttranslational modifications of proteins have been found, a number of physiological importance, and those best characterized are described in Section 2.4. The twenty-first amino acid to be incorporated directly into proteins during biosynthesis, selenocysteine, has been discovered (Sec. 2.1.3.d). The overall process of protein biosynthesis is described in Section 2.1.

One of the greatest technological advances has been the introduction of protein engineering (Sec. 2.2), whereby a protein of any desired amino acid sequence can be produced in substantial quantities by expressing a cloned or synthetic gene. The primary structure can be altered with remarkable specificity by site-directed mutagenesis of the gene. This has had a major impact on the study of protein structure, stability, and function, as described in Chapters 6 through 9. The relative simplicity of gene cloning and sequencing has resulted in an explosion of primary structures of proteins determined in this way, although the variety of posttranslational modifications possible requires the covalent structure of the final active protein to be determined. Even proteins that occur naturally in only minute quantities, and therefore are often those of greatest biological potency, can be identified and sequenced from their genes. Such proteins can then be produced in quantities sufficient for industrial or pharmaceutical use by expressing these genes.

The ease of sequencing proteins via their genes has resulted in great pressure to obtain further insight into the function of a protein from just its primary structure. This is best accomplished by scanning the sequence data banks (Appendix 1) for homologous proteins of known function or structure. It is too frequently assumed, however, that homologous proteins share the same functions, and an understanding of the relationship between protein structure and function, as explained in Chapters 6 through 9, is necessary to minimize the many opportunities to be misled.

After a protein polypeptide chain is synthesized, it must also be directed to its appropriate place in the cell, and great advances have been made in this area (Sec. 2.3). The roles of the very many proteins that are present in a cell continue to be studied in increasing numbers, and the cell biology of proteins will become increasingly important. As larger and more complex proteins are studied, the challenge of extrapolating our understanding of the fundamental properties of simpler proteins to the more complex cases will become greater. In many cases, the fundamental knowledge is available but overlooked, as in the rediscovery of coiled coils (Sec. 5.5.2) as so-called leucine zippers.

One striking aspect of the study of protein structure is how limited our understanding is of the fundamental physical chemistry that is the basis for protein structure and function (Chapter 4). For example, our understanding of the hydrophobic interaction has changed substantially in the past few years (Sec. 4.3). How to analyze the simplest physical interaction, that between charged groups, in complex systems like proteins, is only now becoming clear. Entropic considerations for protein structure and function (Sec. 4.4) are still largely ignored, and it is possible to calculate free-energy changes in proteins only when very small changes take place (Sec. 7.4). Although much has been accomplished, much remains to be learned or applied.

The technique of X-ray crystallography has been developed and refined further (Sec. 6.1), and crystallographers determine increasing numbers of protein structures, each more beautiful, awesome, and accurate than the last. There are now approximately 500 entries in the list of known protein structures (Appendix 2), whereas the first edition listed only 148. A major advance in determining protein structure has been the development of nuclear magnetic resonance spectroscopy (NMR) as a tool for determining the polypeptide chain fold in small proteins (Sec. 6.3). With technical advances and isotopic labeling, larger and larger proteins are being studied in this way. NMR has the advantages of allowing proteins to be studied in solution, thus eliminating the need for crystalline samples, and of giving dynamic information, but X-ray crystallography is also becoming a much more rapid and dynamic technique (Sec. 6.1.7).

With the availability of many more protein structures, the rules of protein architecture are slowly becoming understood. The power of protein engineering techniques for studying protein structure (Sec. 6.5), stability (Sec. 7.4), folding (Sec. 7.5), and function (Chapters 8 and 9) has resulted in a great deal of experimental data, but relatively few simple conclusions. Progress in being able to predict protein structure and function from just the amino acid sequence is painfully slow and is predominantly an empirical process, due largely to the complexity of the problem. The protein folding problem may be solved when we know most of the possible protein tertiary structures, which may number no more than 500 to 1000, so that every protein is found from its amino acid sequence to be homologous to a protein of known structure and function.

Membrane proteins were largely ignored in the first edition, but the high-resolution structures of the bacterial photosynthetic reaction center and porin have made it possible to describe this important class of proteins in molecular terms (Sec. 7.2). These two membrane proteins have turned out to have structures remarkably like water-soluble proteins but with nonpolar surfaces where they are interacting with the nonpolar parts of the lipid bilayer. The interactions of globular proteins with the aqueous solvent are now understood to a great extent, and it is now possible to discuss the aqueous solubilities of proteins in considerable detail (Sec. 7.1).

Virtually all proteins function by interacting specifically with other molecules (Chapter 8), and the greatest advance in this area has been with DNA-binding proteins (Sec. 8.3.2). These proteins play crucial roles in the replication and expression of genetic information and are central to understanding cell function. It is now clear that a number of protein structural motifs and types of interaction with nucleic acids are involved, but there is no simple code relating the sequence of the protein to the sequence of the DNA that it recognizes. Consequently, a comprehensive understanding is required of all the various types of DNA-binding proteins, so that all the many proteins involved in gene regulation and expression can be identified and characterized from just their amino acid sequences.

The interactions of antibodies with antigens have been characterized in detail in a number of instances, including cases where the antigen is another protein molecule (Sec. 8.3.1). This interaction was predicted to have some special properties, but all the observations so far indicate that it is typical of other protein–ligand interactions, described in Chapter 8. The phenomenon of allostery involving positive cooperativity of binding is now fairly clear (Secs. 8.4 and 9.4.2), but that concerning negative cooperativity largely remains a mystery (Sec. 8.4.5).

The study of enzyme catalysis has been revolutionized by the technology of protein engineering (Chapter 9), and it is now possible to measure quantitatively the contribution to binding and catalysis of each group of the side chains of the enzyme. Consequently, very detailed enzyme mechanisms are now known in a few instances (Sec. 9.3). It is clear that the most important property of enzymes is their complementarity to the transition state of the reaction (Sec. 9.2). This probably explains why increased binding interactions between a substrate and enzyme are so often manifested as increases in catalytic rate rather than just binding affinity, although a detailed explanation is not yet possible. The importance of complementarity to the transition state is also demonstrated by the generation of catalytic antibodies using transition-state analogues as immunogens (Sec. 9.2.4).

The importance of enzyme regulation by reversible covalent modification, especially phosphorylation, is increasingly evident, and the molecular basis of this phenomenon is now known in detail in two instances, glycogen phosphorylase and isocitrate dehydrogenase (Sec. 9.4.3).

Finally, the complex phenomenon of protein degradation is now known to be of importance physiologically, and the mechanism of the process is understood to a remarkable extent. Little was known nine years ago.

With all this new knowledge, the description of proteins presented here is much more complete than was possible with the first edition. As a consequence, the volume has grown somewhat larger, although this was kept to a minimum. Including the necessary recent references while minimizing the book's length required that many of the earlier references be omitted from this edition. It is recommended, therefore, that the first edition also be consulted for a more complete listing of references. I hope that the second edition has fewer errors than the first, and I wish to thank those who pointed out deficiencies and errors in the first edition. Various sections of this edition have been read by some of those most active in the field, and I would like to thank especially R. L. Baldwin, P. R. Evans, J. Ewbank, A. R. Fersht, L. N. Johnson, J. R. Knowles, W. Lipscomb, B. W. Matthews, E. Meyer, and H. K. Schachman for their valuable assistance. Of course, perfection would have required an undue delay in publication, and I remain responsible for any errors that might remain.

An innovation in this edition is the exercises at the end of each chapter. They are intended as informative and instructive exercises by which students can expand their understanding of proteins and of how scientific research progresses, primarily by referring to the literature. Most of the examples are of instances where errors or alternative interpretations were reported in the literature, because it is felt that such instances illustrate most clearly the adventure of scientific research. Most papers in the literature are highly polished, condensed, and edited reports of a selected aspect of a scientific investigation and do not report the often convoluted and erratic process by which the end product was attained. One can learn from the literature how science progresses in this way only when such instances are brought out into the open. It is in this spirit that the exercises are offered as interesting, informative, and sometimes amusing glimpses into how proteins are studied.

Of course, there is no substitute for participating in the adventure, and it is hoped that this volume may inspire young people with the appropriate interest and motivation to contribute to future research on the fascinating subject of proteins.

Thomas E. Creighton
May 1992

PROTEINS

Second Edition

Chemical Properties
of Polypeptides

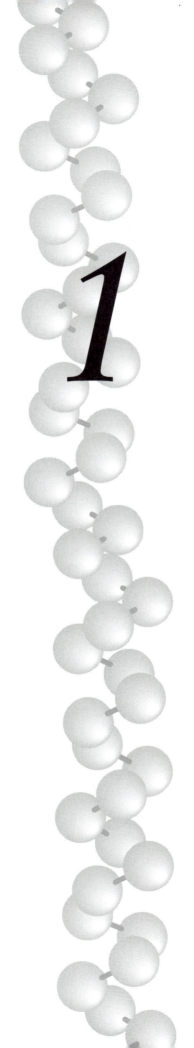

Virtually every property that characterizes a living organism is affected by proteins. Nucleic acids, also essential for life, encode genetic information —mostly specifications for the structures of proteins—and the expression of that information depends almost entirely on proteins (though some RNA molecules with catalytic activity have been discovered recently).

Life forms make use of many chemical reactions to supply themselves continually with chemical energy and to use it efficiently, but by themselves these reactions could not occur fast enough under physiological conditions (aqueous solution, 37 °C, pH 7, atmospheric pressure) to sustain life. The rates of these reactions are increased, by many orders of magnitude, in organisms by the presence of enzymes, which also are proteins. The subject of biochemistry is primarily a study of the roles of enzymes in living systems.

Proteins store and transport a variety of particles ranging from macromolecules to electrons. They guide the flow of electrons in the vital process of photosynthesis; as hormones, they transmit information between specific cells and organs in complex organisms; some proteins control the passage of molecules across the membranes that compartmentalize cells and organelles; proteins function in the immune systems of complex organisms to defend against intruders (the best known are the antibodies); and proteins control gene expression by binding to specific sequences of nucleic acids, thereby turning genes on and off. Proteins are the crucial components of muscles and other systems for converting chemical energy into mechanical energy. They also are necessary for sight, hearing, and the other senses. And many proteins are simply structural, providing the filamentous architecture within cells and the materials that are used in hair, nails, tendons, and bones of animals.

In spite of these diverse biological functions, proteins are a relatively homogeneous class of molecules. All are the same type of linear polymer, built of various combinations of the same 20 amino acids. They differ only in the sequence in which the amino acids are assembled into polymeric chains. The secret to their functional diversity lies partly in the chemical diversity of the amino acids but primarily in the diversity of the three-dimensional structures that these building blocks can form, simply by being linked in different sequences. The awesome functional properties of proteins can be understood only in terms of their relationship to the three-dimensional structures of proteins. That relationship is the topic of this volume.

1.1 The Polymeric Nature of Proteins

Proteins are more complex than most linear polymers in that they can incorporate 20 different monomers in their construction instead of only one or two. In another sense, proteins are structurally less complex. Most chemical polymers are synthesized by chemically polymerizing a mixture of monomers, thereby producing a distribution of chain lengths and, if more than one type of monomer is present, an approximately random sequence of monomers. Natural proteins, however, are linear and unbranched and have precise lengths and exact sequences of amino acids. Indeed, it is only the differences in length and sequence that distinguish one protein from any other and that make possible a diversity of structures and functions. Most importantly, the linear polymeric chain of almost every natural protein has the crucial property of being able to assume a specific three-dimensional folded conformation, as will be described in later chapters.

Of the 20 amino acids usually found in proteins, 19 have the general structure

$$\begin{array}{c} \mathbf{R} \\ | \\ H_2N - CH - CO_2H \end{array} \qquad (1.1)$$

and differ only in the chemical structures of the side chain $\mathbf{R}$. The 20th natural amino acid, proline, is similar, but its side chain is bonded to the nitrogen atom to give the imino acid:

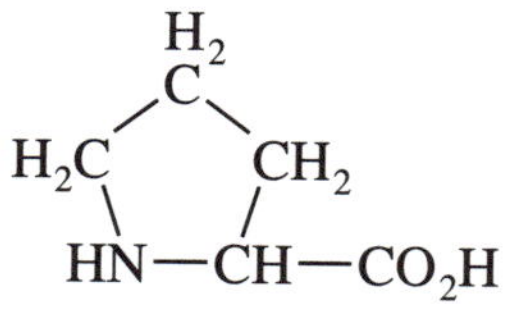

$$\begin{array}{c} H_2 \\ C \\ H_2C \qquad CH_2 \\ | \qquad | \\ HN - CH - CO_2H \end{array} \qquad (1.2)$$

Except in glycine, where the side chain is only a hydrogen atom, the central carbon atom is asymmetric and is always the L isomer:

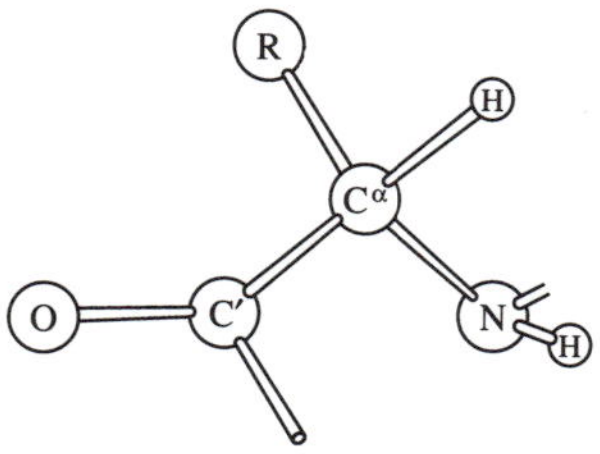

$$(1.3)$$

Unless indicated otherwise, all amino acids in this volume are L isomers.

The structures of the side chains of the 20 amino acids are illustrated in Figure 1.1. The central carbon atom depicted in Equation (1.1) is designated as α, and the atoms of the side chains are commonly designated β, γ, δ, ϵ, and ζ, in order away from the α carbon atom. Chemical groups, however, are usually designated by the carbon atom to which they are bonded; hence, the ζN atom of a Lys residue is part of the ϵ-amino group. Table 1.1 and Figure 1.1 also give the three- and one-letter abbreviations commonly used for designating the amino acids when they are incorporated into proteins. Three-letter abbreviations are used throughout this volume because their designations are obvious. The relationships of the one-letter codes to the amino acids are somewhat less obvious, but they are commonly used in compilations of long sequences because they save space and are less likely to be confused (e.g., Gln, Glu, and Gly can easily be confused but not Q, E, and G).

FIGURE 1.1

Side chains of the 20 amino acids that occur naturally in proteins. Small unlabeled spheres are hydrogen atoms, and large unlabeled spheres are carbon atoms; other atoms are labeled. Double bonds are black, and partial double bonds are shaded. In the case of Pro, the bonds of the polypeptide backbone are included and are black. Below the name of the amino acid are the three-letter and the one-letter abbreviations commonly used. Note that isoleucine and threonine have asymmetric centers in their side chains, and only the isomer illustrated is used biologically.

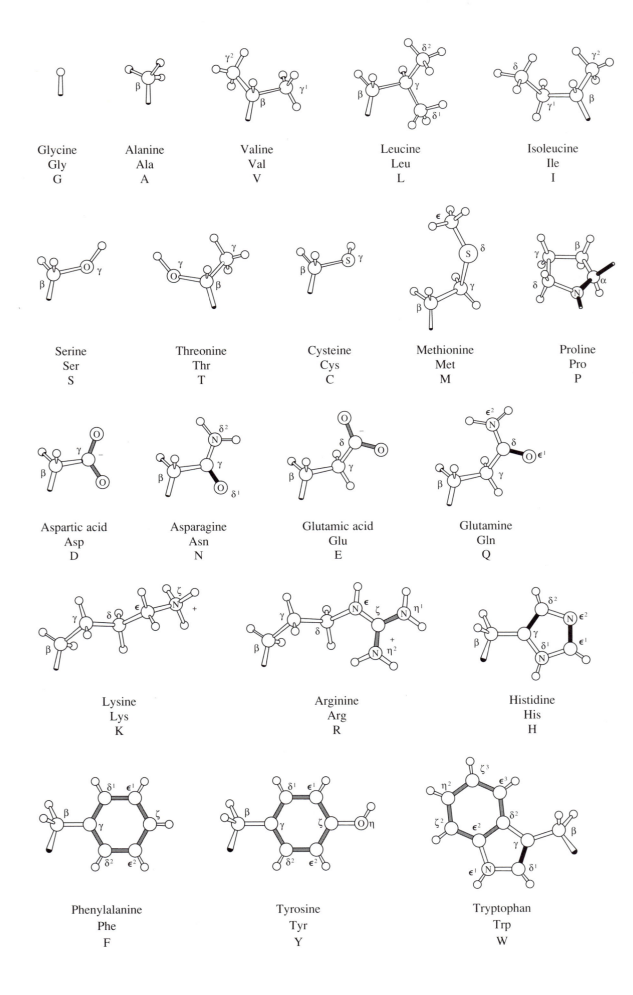

Glycine
Gly
G

Alanine
Ala
A

Valine
Val
V

Leucine
Leu
L

Isoleucine
Ile
I

Serine
Ser
S

Threonine
Thr
T

Cysteine
Cys
C

Methionine
Met
M

Proline
Pro
P

Aspartic acid
Asp
D

Asparagine
Asn
N

Glutamic acid
Glu
E

Glutamine
Gln
Q

Lysine
Lys
K

Arginine
Arg
R

Histidine
His
H

Phenylalanine
Phe
F

Tyrosine
Tyr
Y

Tryptophan
Trp
W

Table 1.1   *Properties of Individual Amino Acid Residues*

Residue	Residue mass[a] (daltons)	Van der Waals volume[b] (Å³)	Frequency in proteins[c] (%)
Ala (A)	71.09	67	8.3
Arg (R)	156.19	148	5.7
Asn (N)	114.11	96	4.4
Asp (D)	115.09	91	5.3
Cys (C)	103.15	86	1.7
Gln (Q)	128.14	114	4.0
Glu (E)	129.12	109	6.2
Gly (G)	57.05	48	7.2
His (H)	137.14	118	2.2
Ile (I)	113.16	124	5.2
Leu (L)	113.16	124	9.0
Lys (K)	128.17	135	5.7
Met (M)	131.19	124	2.4
Phe (F)	147.18	135	3.9
Pro (P)	97.12	90	5.1
Ser (S)	87.08	73	6.9
Thr (T)	101.11	93	5.8
Trp (W)	186.21	163	1.3
Tyr (Y)	163.18	141	3.2
Val (V)	99.14	105	6.6
Weighted average[d]	119.40	161	

[a] Molecular weight of nonionized amino acid minus that of water.

[b] Volume enclosed by van der Waals radii, calculated with volumes of atoms given by F. M. Richards, *J. Mol. Biol.* 82:1–14 (1974).

[c] Frequency of occurrence of amino acid residue in primary structures of 1021 unrelated proteins of known sequence (P. McCaldon and P. Argos, *Proteins* 4:99–122, 1988).

[d] Weighted by frequency of occurrence in proteins, to give the value for average residue in globular proteins.

The 20 amino acids are linked into proteins by the **peptide bond,** as illustrated here by the condensation of two amino acids:

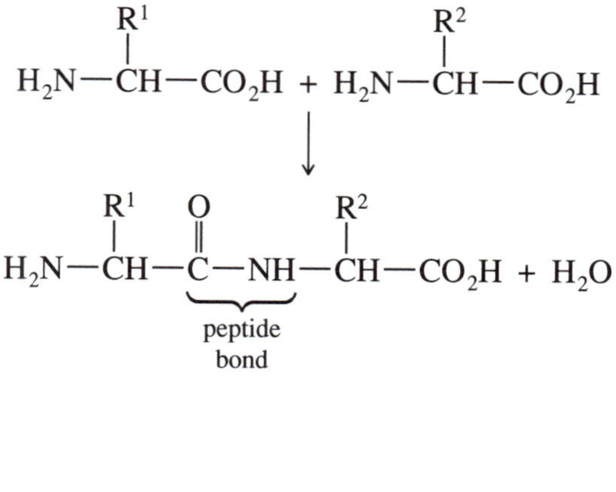

$$H_2N-\underset{\underset{R^1}{|}}{CH}-CO_2H + H_2N-\underset{\underset{R^2}{|}}{CH}-CO_2H$$

$$\downarrow$$

$$H_2N-\underset{\underset{R^1}{|}}{CH}-\overset{\overset{O}{\|}}{C}-NH-\underset{\underset{R^2}{|}}{CH}-CO_2H + H_2O \qquad (1.4)$$

peptide bond

Generally, between 50 and 3000 such amino acids are linked in this way to form a typical linear **polypeptide chain.** The polypeptide backbone is a repetition of the basic unit common to all amino acids. When the side chain is included, this unit is described as an amino acid **residue:**

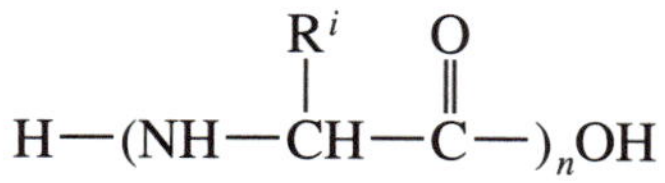

$$H-(NH-\underset{\underset{R^i}{|}}{CH}-\overset{\overset{O}{\|}}{C}-)_n OH \qquad (1.5)$$

Some of the most important properties of the amino acid residues are listed in Table 1.1.

All proteins and polypeptides have this basic structure. Proteins differ only in the number of amino acids linked together (n in Eq. 1.5) and the sequence in which the various amino acids occur. The sequence of amino acids in a polypeptide chain generally identifies a protein unambiguously.

Sequences of amino acids in proteins are usually written with either the three-letter or the one-letter abbreviations, starting with the N-terminal residue, which is at the left and is considered the first residue of the polypeptide chain. Hyphens are inserted between residues to indicate known peptide bonds, and other punctuation is used if the sequence is not known entirely: parentheses around segments of uncertain sequence, dots separating residues whose positions are almost certain, and commas between amino acid residues of unknown sequence. If it is not known whether a residue is Glu or Gln, the appropriate abbreviation is Glx or Z; for Asp or Asn, Asx or B is used.

Amino acid residues in proteins are properly referred to by adding *-yl* to the ends of their amino acid names (e.g., glycyl or alanyl residues). This complication is avoided by using the one- or three-letter abbreviations.

The following terms are used to describe the various types of polymerized amino acids:

- **Peptide**   A short chain of residues with a defined sequence. There is no maximum number of residues in a peptide, but the term is appropriate to a chain if its physical properties are those expected from the sum of its amino acid residues and if there is no fixed three-dimensional conformation.

- **Polypeptide**   A longer chain, usually of defined sequence and length.

- **Polyamino acids**   Random sequences of varying lengths generally resulting from nonspecific polymerization of one or a few amino acids.

- **Protein**   Usually reserved for those polypeptides that occur naturally and have a definite three-dimensional structure under physiological conditions. This class of polypeptides is the subject of this volume, though reference is also made to the others.

References

Proteins, Amino Acids and Peptides. E. J. Cohn and J. T. Edsall. Van Nostrand-Reinhold, Princeton, N.J., 1943.

X-ray studies of amino acids and peptides. R. B. Corey. *Adv. Protein Chem.* 4:385–406 (1948).

Crystal structure studies of amino acids and peptides. R. E. Marsh and J. Donohue. *Adv. Protein Chem.* 22:235–256 (1967).

The Structure and Action of Proteins. R. E. Dickerson and I. Geis. Harper & Row, New York, 1969.

Nomenclature and symbolism for amino acids and peptides. IUPAC-IUB Joint Commission on Biochemical Nomenclature. *J. Biol. Chem.* 260:14–42 (1985).

Protein Structure: A Practical Approach. T. E. Creighton, ed. IRL Press, Oxford, 1989.

1.2 The Polypeptide Backbone

The backbone of the linear polypeptide chain consists of a repeated sequence of three atoms of each residue in the chain — the amide N, the C^α, and the carbonyl C:

$$-N-CH-C- \quad (1.6)$$

These atoms are generally represented as N_i, C_i^α, and C_i', respectively, where i is the number of the residue, starting from the amino end of the chain.

The dimensions of the peptide group of a residue, given in Figure 1.2, have been derived from three-dimensional crystal-structure analyses of small peptides. The maximum distance between corresponding atoms of adjacent residues is 3.80 Å when the peptide bond is *trans*. In a fully extended chain, the residues are staggered, so the maximum linear dimension of a polypeptide with n residues is $n \times 3.63$ Å.

The presence of an asymmetric center at the C^α carbon atom, and only L amino acid residues, results in an inherent asymmetry of the polypeptide chain. Chapters 5 and 6 show this to be important for the spectral and conformational properties of polypeptides and proteins.

In principle, rotation could occur about any of the three bonds of each residue of the polypeptide backbone, but the peptide bond appears to have partial double-bonded character due to resonance:

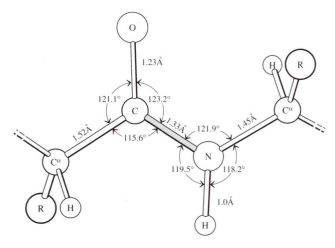

FIGURE 1.2
The geometry of the peptide backbone, with a *trans* peptide bond, showing all the atoms between two C^α atoms of adjacent residues. The peptide bond is stippled. The dimensions given are the averages observed crystallographically in amino acids and small peptides. (G. N. Ramachandran et al., *Biochim. Biophys. Acta* 359:298–302, 1974.)

bone, but the peptide bond appears to have partial double-bonded character due to resonance:

$$(1.7)$$

Consequently, the peptide bond length is only 1.33 Å, shorter than the usual C—N bond length of 1.45 Å, as in the C^α—N bond. It is, however, longer than the value of 1.25 Å for the average C=N double bond. The peptide bond appears to have approximately 40 percent double-bonded character. As a result, rotation of this bond is restricted, and the six atoms depicted in Equation (1.7) have a strong tendency to be coplanar. Two configurations of the planar peptide bond are possible, one in which the C^α atoms are *trans*, the other in which they are *cis*:

$$\textit{trans} \qquad \textit{cis} \quad (1.8)$$

The *trans* form is intrinsically favored energetically, probably owing to fewer repulsions between nonbonded atoms. If the residue that follows the peptide bond is Pro, however, its cyclic side chain diminishes the repulsions between atoms, and the intrinsic stability of the *cis* isomer is comparable to that of the *trans* isomer (Sec. 5.2.1).

Table 1.2 *Intrinsic* pK_a *Values of Ionizable Groups Found in Proteins*

Group	Observed pK_a[a]
α-Amino	6.8–8.0
α-Carboxyl	3.5–4.3
β-Carboxyl (Asp)	3.9–4.0
γ-Carboxyl (Glu)	4.3–4.5
δ-Guanido (Arg)	12.0
ϵ-Amino (Lys)	10.4–11.1
Imidazole (His)	6.0–7.0
Thiol (Cys)	9.0–9.5
Phenolic hydroxyl (Tyr)	10.0–10.3

[a] The ranges of values are given by different model compounds used to represent an isolated amino acid residue. The values for the terminal α-amino and α-carboxyl groups especially depend on the identity of the terminal residue. Values from C. Tanford, *Adv. Protein Chem.* 17:69–165 (1962); A. Bundi and K. Wüthrich, *Biopolymers* 18:285–297 (1979); J. B. Matthew et al., *CRC Crit. Rev. Biochem.* 18:91–197 (1985).

Resonance of the peptide bond (Eq. 1.7) tends to redistribute its electrons, and the polypeptide backbone is correspondingly polar. The H and N atoms appear to have, respectively, positive and negative equivalent charges of 0.20 electron, whereas C and O, respectively, have positive and negative equivalent charges of 0.42 electron. This gives the peptide bond a substantial permanent dipole moment of about 3.5 Debye units (corresponding to one unit charge separated by 0.73 Å; see Sec. 4.1.2.b). Note that the polypeptide backbone of each residue contains one potent hydrogen bond donor, —NH—, and a hydrogen bond acceptor, carbonyl —CO—. Chapters 5 and 6 will show this to be crucial for the three-dimensional architectures of proteins.

The other two types of bonds of the peptide backbone behave as normal C—C and C—N single bonds.

The peptide backbone is not very reactive chemically. The only groups usually ionized are the terminal α-amino and carboxyl groups, which normally have pK_a values of about 7.4 and 3.9, respectively (Table 1.2), depending on the nature of the terminal amino acid residue. A proton is added or lost to internal peptide bonds only at extremes of pH. The apparent pK_a value of the amide NH for deprotonation is between 15 and 18 and is in the region of -8 to -12 for protonation. The oxygen atom of the carbonyl group is protonated more readily, with an apparent pK_a of about -1. These properties facilitate the exchange of hydrogen isotopes between the backbone and aqueous solvents, which is important to the study of protein conformational fluctuations in solution (Chap. 7).

At pH values for which protonation or deprotonation of the peptide bond becomes significant, the polypeptide chain is usually hydrolyzed to the substituent amino acids. For example, the standard method for determining the amino acid composition of proteins or peptides is to maintain them at about 105°C for about 24 h in 6 *M* HCl. The peptide bond is hydrolyzed very slowly in more physiological conditions, its half-life having been measured to be approximately 7 years at neutral pH and room temperature; but this decreases to about 1 min at 250°C.

Other chemical alterations of the peptide chain require either drastic conditions or the very close proximity of certain reactive groups attached to the amino acid side chains. Such reactions are useful in the laboratory for selective cleavage of the chain at certain residues (see Sec. 1.6.2.e).

References

Hydrogen exchange. S. W. Englander et al. *Ann. Rev. Biochem.* 41:903–924 (1972).

The α-helix as an electric macro-dipole. A. Wada. *Adv. Biophys.* 9:1–63 (1976).

Coordinating properties of the amide bond. Stability and structure of metal ion complexes of peptides and related ligands. H. Sigel and R. B. Martin. *Chem. Rev.* 82:385–426 (1982).

Hydrolytic stability of biomolecules at high temperatures and its implication for life at 250°C. R. H. White. *Nature* 310:430–432 (1984).

The role of the α-helix dipole in protein function and structure. W. G. J. Hol. *Prog. Biophys. Mol. Biol.* 45:149–195 (1985).

Hydrolysis of a peptide bond in neutral water. D. Kahne and W. C. Still. *J. Amer. Chem. Soc.* 110:7529–7534 (1988).

1.3 Amino Acid Residues

The 20 different amino acid side chains possess a variety of chemical properties. This variety is greatly enhanced when the various groups are combined in various sequences in a single molecule, which gives a protein properties far beyond those of simpler molecules. The chemical properties of a protein molecule are far more complex than the sum of the properties of its constituent amino acids, but the properties of the 20 side chains are the starting point for understanding proteins. For example, if the thiol group of a Cys residue is important in the function of a protein, that function will be abolished by a reagent that reacts chemically with thiol groups (see Sec. 1.3.10). For such a test to be meaningful, of course, one must be certain that the reagent does not also react with other groups on the protein.

Each of the side chains will now be discussed briefly and their chemical properties described. Remember, however, that residues in biologically active proteins may have chemical or physical properties very different from those described here. The chemical reactions undergone by the various side chains that are described here are those that are at least moderately selective and sufficiently specific to be useful in determining the roles of the residues in proteins.

Some pertinent chemical and physical properties of the amino acid residues are summarized in Tables 1.1 and 1.2.

References

Chemistry of the Amino Acids. J. P. Greenstein and M. Winitz. John Wiley & Sons, New York, 1961.

Chemical Modification of Proteins. G. E. Means and R. E. Feeney. Holden-Day, San Francisco, 1971.

The chemical modification of proteins by group-specific and site-specific reagents. A. N. Glazer. In *The Proteins,* 3rd ed., H. Neurath and R. L. Hill, eds., vol. 2, pp. 1–103. Academic Press, New York, 1976.

Chemistry and Biochemistry of Amino Acids. G. C. Barrett, ed. Chapman and Hall, New York, 1985.

Amino acid side chain parameters for correlation studies in biology and pharmacology. J. L. Fauchere et al. *Intl. J. Peptide Protein Res.* 32:278–279 (1988).

Chemical modification. T. Imoto and H. Yamada. In *Protein Function: A Practical Approach,* T. E. Creighton, ed., pp. 247–277. IRL Press, Oxford, 1989.

1.3.1 Gly

Glycine is the simplest amino acid, with only a hydrogen atom for a side chain. The absence of a larger side chain gives the polypeptide backbone at Gly residues much greater conformation flexibility than other residues (discussed further in Chap. 5).

The alpha carbon atom of Gly is not asymmetric because it has two H atoms, so this amino acid does not occur as D or L isomers.

1.3.2 The Aliphatic Residues: Ala, Val, Leu, Ile

These amino acid residues have no reactive groups on their side chains, only inert methylene ($-CH_2-$) and methyl ($-CH_3$) groups. They have the important property, however, of *not* interacting favorably with water.

Ala   Val

Leu   Ile   (1.9)

Instead, they interact much more favorably with each other and with other nonpolar atoms than with water, which is one of the main factors in stabilizing the folded conformations of proteins (Sec. 4.3). Consequently, the aliphatic residues can be considered the "bricks" around which the functional parts of a protein molecule are assembled. The variety of their shapes is important for this structural role.

Note that the Ile side chain has an extra center of asymmetry, at C^β, and that only the one isomer occurs naturally and is incorporated into proteins. It is curious that there are amino acids in proteins with one, three, or four carbon atoms in their side chains but not two. The corresponding amino acid, α-amino butyric acid, is not used in protein biosynthesis.

1.3.3 The Cyclic Imino Acid: Pro

(1.10)

The side chain of proline is aliphatic like those of the preceding amino acids, with no functional groups. It is unique, however, in that it is bonded covalently to the nitrogen atom of the peptide backbone, which is indicated by the solid bonds in Equation (1.10). Therefore, the peptide backbone at Pro residues has no amide hydrogen for use as a donor in hydrogen bonding or in resonance stabilization of the peptide bond of which it is part. The cyclic five-membered ring also imposes rigid constraints on rotation about the $N-C^\alpha$ bond of the backbone, and the peptide bond preceding a Pro residue is more likely to adopt the *cis* configuration (see Secs. 1.2, 5.2.1). Pro residues, consequently, have significant effects on the conformation of the polypeptide backbone (discussed further in Chaps. 5 and 6).

The five-membered pyrrolidine ring of the Pro residue is invariably puckered, with the C^α, C^β, C^δ, and N atoms approximately coplanar but with the C^γ atom displaced from the plane by about 0.5 Å.

1.3.4 The Hydroxyl Residues: Ser and Thr

Ser   Thr   (1.11)

The side chains of Ser and Thr are small and aliphatic, except for the presence of a polar hydroxyl group on each. These hydroxyl groups are normally no more reactive chemically than ethanol, so there are few chemical reactions in which they participate that are useful with proteins. The only reaction that occurs readily is acetylation with acetyl chloride in aqueous trifluoroacetic acid.

Note that the side chain of Thr, like that of Ile, has a center of asymmetry at C^β and that only the one isomer occurs naturally.

1.3.5 The Acidic Residues: Asp and Glu

Asp   Glu   (1.12)

The side chains of Asp and Glu differ only in having one and two methylene groups, respectively. It might be thought, therefore, that they would be very similar chemically and functionally in proteins, but this is not the case. The slight difference in length of the side chains causes them to have different tendencies in their chemical interactions with the peptide backbone. Consequently, Asp and Glu residues have markedly different effects on the conformation and chemical reactivity of the peptide backbone.

The carboxyl groups of Asp and Glu are normally no more reactive than are those of corresponding organic molecules such as acetic acid. They can be chemi-

cally esterified, coupled with amino or other nucleophiles, and reduced to alcohols, but these reactions are somewhat harsh chemically. The most useful chemical modification of the carboxyl groups is their coupling to amines, which is catalyzed by carbodiimides:

$$
\begin{array}{cc}
\overset{O}{\overset{\|}{R^1-COH}} + R^2N{=}C{=}NR^3 \\
\text{carboxyl} \quad \text{carbodiimide}
\end{array}
$$

$$\downarrow H^+$$

$$
\begin{array}{c}
O \quad \overset{+}{HNR^2} \\
\overset{\|}{R^1-CO-}\overset{\|}{C} \\
| \\
HNR^3
\end{array}
\quad (1.13)
$$

$$
R^4-NH_2 \Big| \text{ amine} \downarrow
$$

$$
\overset{O}{\overset{\|}{R^1-C}}-NH-R^4 + R^2NH-\overset{O}{\overset{\|}{C}}-NHR^3
$$
$$
\text{amide} \qquad\qquad \text{urea derivative} \quad (1.14)
$$

The carboxyl groups are first activated by reaction with the carbodiimide (Eq. 1.13); then they react with an added amine to generate the amide (Eq. 1.14). A competing reaction is hydrolysis of the activated carbodiimide to regenerate the carboxyl and to produce the urea derivative of the carbodiimide:

$$
\begin{array}{c}
O \quad \overset{+}{HNR^2} \\
\overset{\|}{R^1-CO-}\overset{\|}{C} \\
| \\
HNR^3
\end{array}
$$

$$\downarrow H_2O$$

$$
\overset{O}{\overset{\|}{R^1-COH}} + R^2NH-\overset{O}{\overset{\|}{C}}-NHR^3
$$
$$
\text{carboxyl} \qquad \text{urea derivative} \quad (1.15)
$$

The carboxyl groups that are regenerated can react with another molecule of the carbodiimide. The carbodiimide reagents most useful with proteins have R_2 and R_3 groups that make the reagent soluble in water.

The carboxyl groups of Asp and Glu side chains ionize with intrinsic pK_a values of 3.9 and 4.3, respectively (Table 1.2), so these residues are ionized and very polar under physiological conditions. They can also be effective chelators of certain metal ions when held in appropriate proximity to the ions (Sec. 8.3.4.a).

The ionization state of a carboxyl can sometimes be determined by its susceptibility to modification by different reagents. Diazo compounds, such as diazoacetate esters and amides, react with the nonionized form:

$$-CO_2H + N_2CH-\overset{\overset{\displaystyle O}{\|}}{C}-NH-R$$

diazo amide

$$\downarrow$$

$$-\overset{\overset{\displaystyle O}{\|}}{C}-O-CH_2-\overset{\overset{\displaystyle O}{\|}}{C}-NH-R + N_2 \quad (1.16)$$

In contrast, epoxides react with the ionized form:

$$-CO_2^- + \overset{\displaystyle O}{CH_2-CHR}$$

epoxide

$$H^+ \downarrow$$

$$-\overset{\overset{\displaystyle O}{\|}}{C}-O-CH_2-\overset{\overset{\displaystyle OH}{|}}{CH}-R \quad (1.17)$$

The carboxyl groups of Glu and Asp residues differ only marginally in their physical and chemical properties from the terminal α-carboxyl group of the polypeptide chain.

References

Esterification. P. E. Wilcox. *Methods Enzymol.* 25:596–615 (1972).

Carbodiimide modification of proteins. K. L. Carraway and D. E. Koshland. *Methods Enzymol.* 25:616–623 (1972).

1.3.6 The Amide Residues: Asn and Gln

Asn   Gln   (1.18)

These residues are the amide forms of Asp and Glu, but the corresponding amino acids occur naturally and are incorporated directly into proteins; Asn and Gln residues do not occur as a result of amidation of Asp and Glu residues in proteins.

The Asn and Gln amide side chains do not ionize and are not very reactive chemically, but they are polar,

being both hydrogen-bond donors and acceptors. The amide groups are labile at extremes of pH and at high temperatures, and these residues can deamidate to Asp and Glu residues. At alkaline pH, the Asn residue is much more labile than the Gln residue because its side chain is sterically suited to interact with the —NH— group of the following residue to form transiently a cyclic succinimidyl derivative (Fig. 1.3). This derivative can undergo racemization and hydrolysis to cleave the peptide chain or to produce a mixture of D and L isomers of Asp and isoAsp residues. In an isoAsp residue, the peptide bond of the backbone is through the side-chain carboxyl group rather than the usual α-carboxyl. The deamidation reaction of Asn occurs 30–50 times more rapidly if the following residue is Gly because the absence of a side chain favors succinimide formation. The rate of this reaction also depends markedly on the conformation of the polypeptide backbone because only some conformations permit succinimide formation. This reaction may be a limiting factor in the longevities of some proteins (Chap. 10).

At acidic pH values, deamidation occurs by other mechanisms, and Gln is more reactive than Asn.

When Gln residues are at the N-terminus of the peptide chain, they spontaneously cyclize:

$$\begin{array}{cc} \overset{\displaystyle O}{\overset{\|}{NH_2C}} & CH_2 \\ | & / \quad \backslash \\ CH_2 & O{=}C \quad CH_2 \quad \overset{\displaystyle O}{\|} \\ | \quad \overset{\displaystyle O}{\|} & | \quad | \\ CH_2 & HN-CH-C- \\ | \\ H_2N-CH-C- \longrightarrow \end{array} \quad (1.19)$$

The resulting residue of pyrrolidone carboxylic acid makes the amino terminus unreactive in most procedures for determining amino acid sequences (Sec. 1.6.2); but it can be removed by the enzyme pyroglutamyl amino peptidase.

References

Rates of nonenzymatic deamidation of glutaminyl and asparaginyl residues in pentapeptides. A. B. Robinson et al. *J. Amer. Chem. Soc.* 95:8156–8159 (1973).

Deamidation, isomerization, and racemization at asparaginyl and aspartyl residues in peptides. Succinimide-linked reactions that contribute to protein degradation. T. Geiger and S. Clarke. *J. Biol. Chem.* 262:785–794 (1987).

Effect of protein conformation on rate of deamidation: ribonuclease A. S. J. Wearne and T. E. Creighton. *Proteins: Struct. Funct. Genet.* 5:8–12 (1989).

Succinimide formation from aspartyl and asparaginyl peptides as a model for the spontaneous degradation of pro-

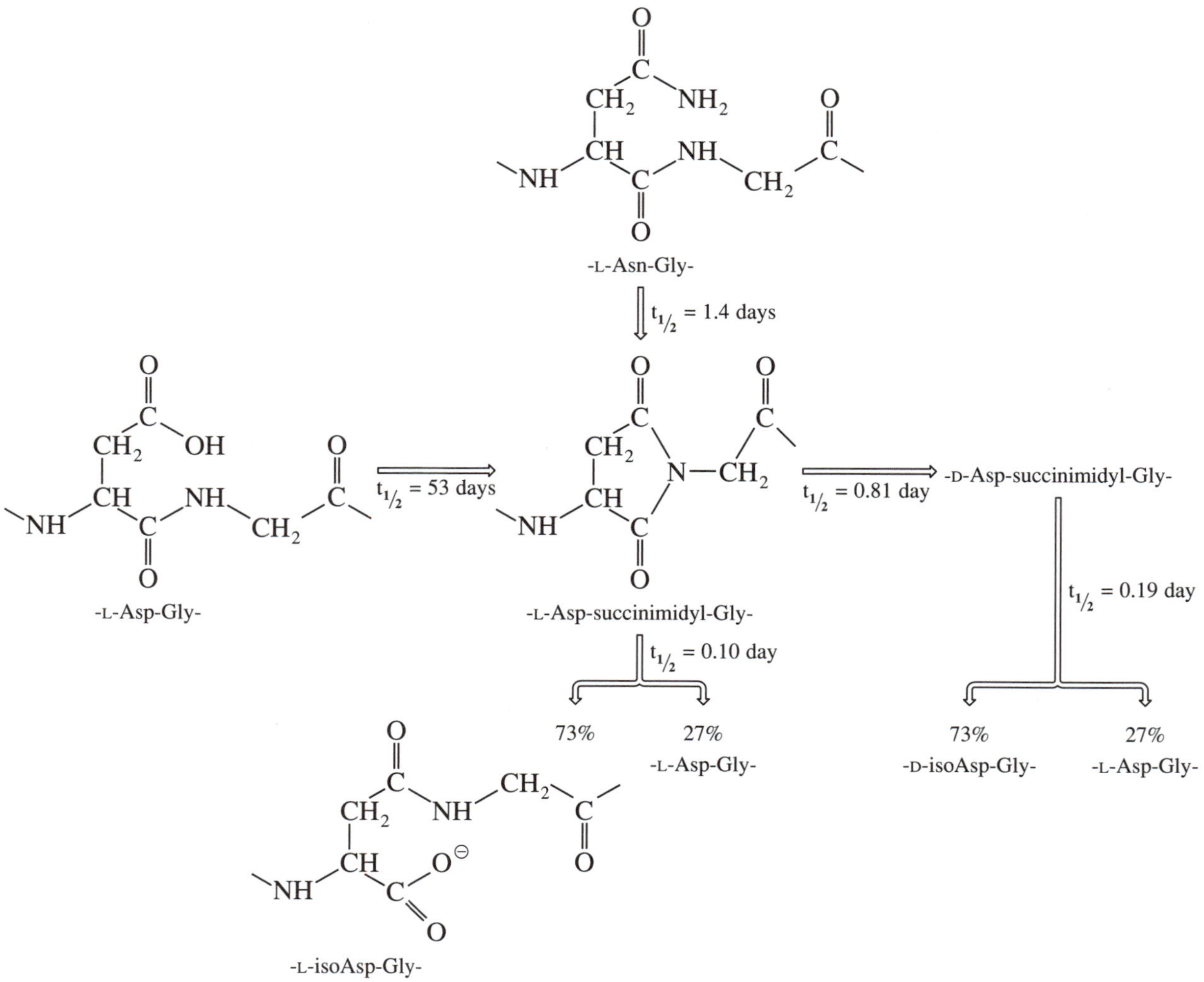

FIGURE 1.3
Spontaneous formation of peptide succinimides and the products of hydrolysis from Asp and Asn residues. The half-times shown for the various reactions are for the model peptides Val-Tyr-Pro-Asn-Gly-Ala and Val-Tyr-Pro-Asp-Gly-Ala at 37°C and pH 7.4. When the Gly residue is replaced by Leu, the rate of succinimide formation is 50 times slower. (From S. Clarke, *Intl. J. Peptide Protein Res.* 30:808–821, 1987.)

teins. R. C. Stephenson and S. Clarke. *J. Biol. Chem.* 264:6164–6170 (1989).

Identification of an isoaspartyl linkage formed upon deamidation of bovine calbindin D_{9k} and structural characterization by 2D ^1H NMR. W. J. Chazin et al. *Biochemistry* 28:8646–8653 (1989).

Lys   (1.20)

1.3.7 The Basic Residues: Lys and Arg

a. Lys

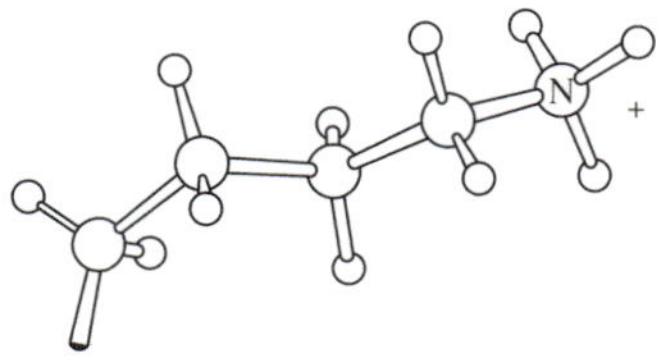

The side chain of Lys is a hydrophobic chain of four methylene groups capped by an amino group that ionizes with an intrinsic pK_a value of 11.1 in the absence of perturbing factors, so it is ionized under most physiological conditions. There is always a finite fraction of nonionized amino groups, however, which are potent nucleophiles. Consequently, the amino groups of Lys residues readily undergo a variety of acylation, alkylation, arylation, and amidination reactions. The rates of such reactions usually increase with increasing pH due to the increasing proportion of nonionized amino groups.

The number of amino groups present is frequently determined by arylation with 2,4,6-trinitrobenzene sulfonate (TNBS):

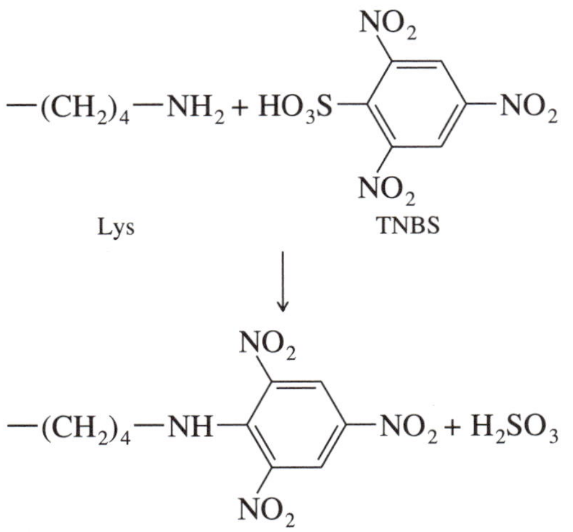

$$-(CH_2)_4-NH_2 + HO_3S-\overset{NO_2}{\underset{NO_2}{\bigcirc}}-NO_2$$

$$\text{Lys} \qquad\qquad \text{TNBS}$$

$$\downarrow$$

$$-(CH_2)_4-NH-\overset{NO_2}{\underset{NO_2}{\bigcirc}}-NO_2 + H_2SO_3 \qquad (1.21)$$

The modified amino group absorbs strongly at 367 nm, so the reaction can be followed spectrophotometrically.

Acetylation of amino groups with a variety of anhydrides—acetic, succinic, maleic, citraconic (i.e., methylmaleic), and 3,4,5,6-tetrahydrophthaloyl—occurs rapidly:

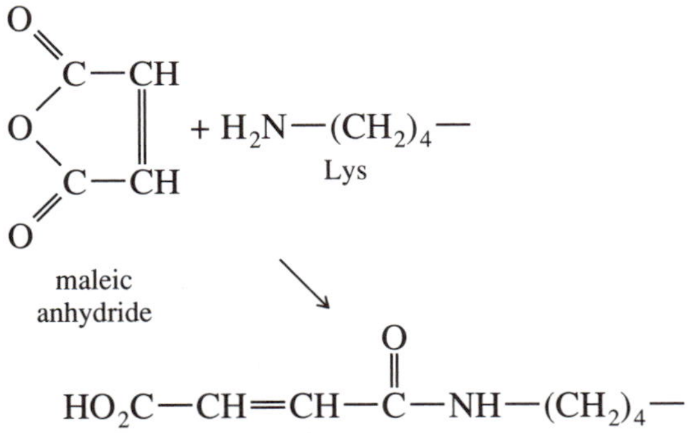

$$\begin{matrix} O \\ \| \\ O=C-CH \\ \quad | \\ C-CH \\ \| \\ O \end{matrix} + H_2N-(CH_2)_4-$$

maleic anhydride

Lys

$$\searrow$$

$$HO_2C-CH=CH-\overset{O}{\overset{\|}{C}}-NH-(CH_2)_4- \qquad (1.22)$$

The maleic, citraconic, and tetrahydrophthaloyl anhydrides are especially useful for temporary modifications because the reaction reverses readily in acidic conditions.

The amino group can be guanylated with O-methyl isourea to convert the Lys side chain to that of homoarginine:

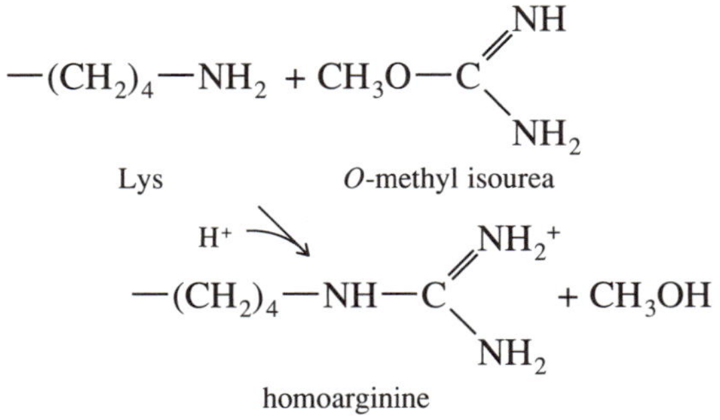

$$-(CH_2)_4-NH_2 + CH_3O-\overset{NH}{\underset{NH_2}{C}}$$

$$\text{Lys} \qquad\qquad \text{O-methyl isourea}$$

$$\overset{H^+}{\searrow}$$

$$-(CH_2)_4-NH-\overset{NH_2^+}{\underset{NH_2}{C}} + CH_3OH$$

homoarginine $\qquad\qquad (1.23)$

A related reaction is amidination:

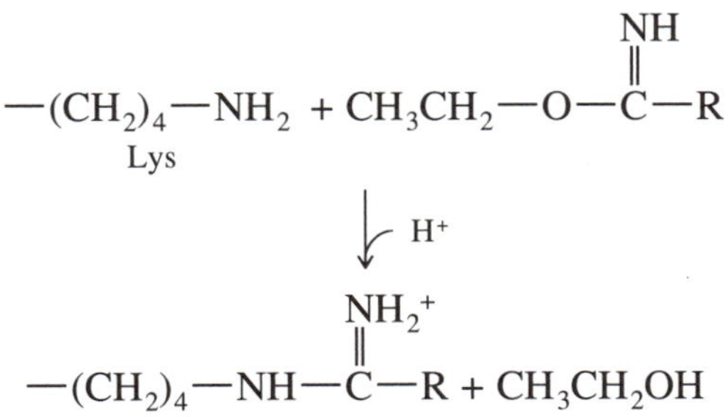

$$-(CH_2)_4-NH_2 + CH_3CH_2-O-\overset{NH}{\overset{\|}{C}}-R$$

Lys

$$\downarrow H^+$$

$$-(CH_2)_4-NH-\overset{NH_2^+}{\overset{\|}{C}}-R + CH_3CH_2OH \qquad (1.24)$$

In both of these reactions, the Lys side chains remain basic and are usually positively charged.

Lys residues also reversibly form Schiff bases with aldehydes, such as that of the natural cofactor pyridoxal phosphate. The Schiff base can be stabilized in the laboratory by reduction; for example, by sodium borohydride (NaBH$_4$):

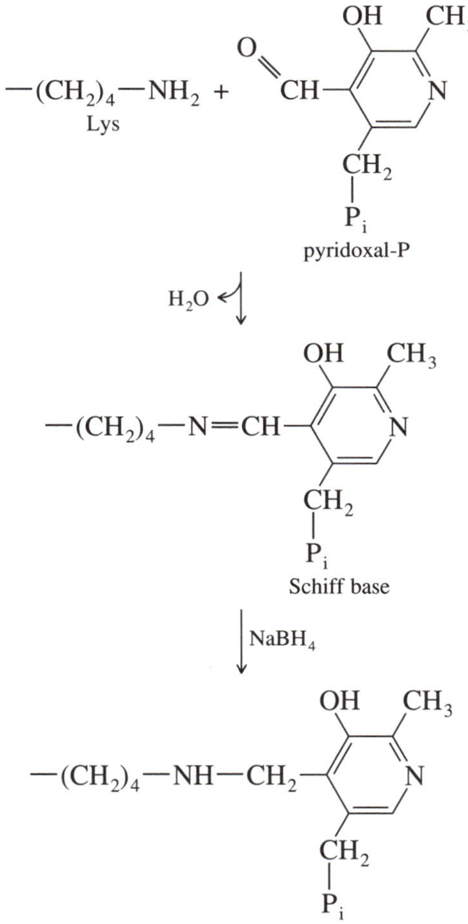

pyridoxal-P

$$H_2O \leftharpoondown \downarrow$$

Schiff base

$$\downarrow NaBH_4$$

$$(1.25)$$

A widely used modification is that of carbamylation by cyanate to form a homocitrulline residue:

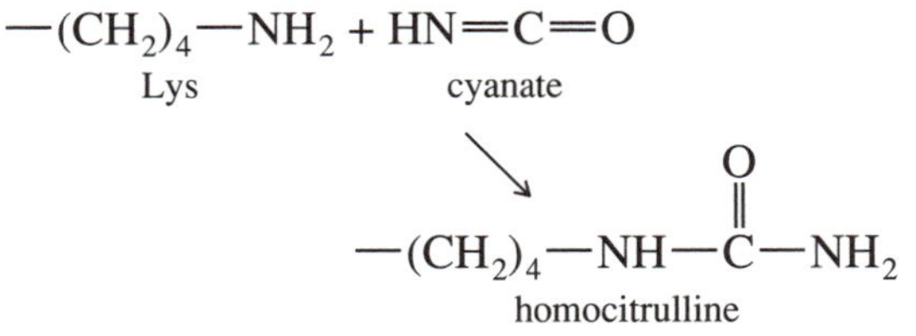

$$-(CH_2)_4-NH_2 + HN=C=O$$

$$\text{Lys} \qquad\qquad \text{cyanate}$$

$$\searrow$$

$$-(CH_2)_4-NH-\overset{O}{\overset{\|}{C}}-NH_2$$

homocitrulline $\qquad\qquad (1.26)$

These examples illustrate that the amino groups of Lys residues participate in a multitude of reactions. This reactivity makes it possible to convert Lys side chains to a variety of analogues that have positive, negative, or no charge under physiological conditions. The α-amino group of the polypeptide chain, however, tends to participate in the same reactions, differing from the ϵ-amino groups of Lys side chains primarily in its lower pK_a value: 7.4 versus 11.1 (Table 1.2); consequently, the α-amino groups have a lesser tendency to be ionized than those of Lys side chains.

References

Maleylation of amino groups. P. J. G. Butler and B. S. Hartley. *Methods Enzymol.* 25:191–199 (1972).

Acetylation. J. F. Riordan and B. L. Vallee. *Methods Enzymol.* 25:494–499 (1972).

Amidination. M. J. Hunter and M. L. Ludwig. *Methods Enzymol.* 25:585–596 (1972).

Carbonyl-amine reactions in protein chemistry. R. E. Feeney et al. *Adv. Protein Chem.* 29:135–203 (1975).

A method for the separation of hybrids of chromatographically identical oligomeric proteins. Use of 3,4,5,6-tetrahydrophthaloyl groups as a reversible "chromatographic handle." I. Gibbons and H. K. Schachman. *Biochemistry* 15:52–60 (1976).

b. Arg

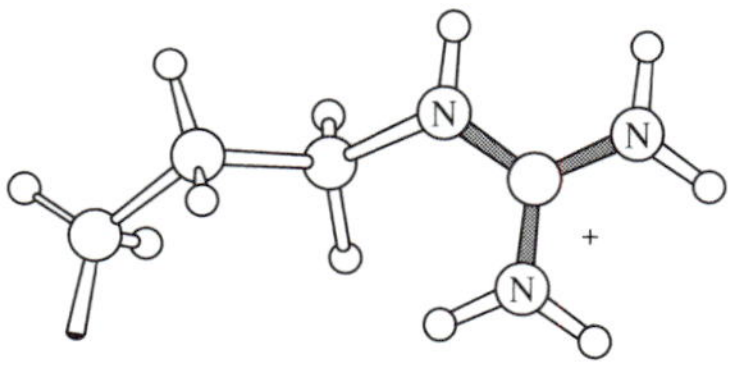

Arg   (1.27)

The Arg side chain consists of three nonpolar methylene groups and the strongly basic δ-guanido group. With a pK_a value usually of about 12, the guanido group is ionized over the entire pH range in which proteins exist naturally. The ionized guanido group is planar as a result of resonance:

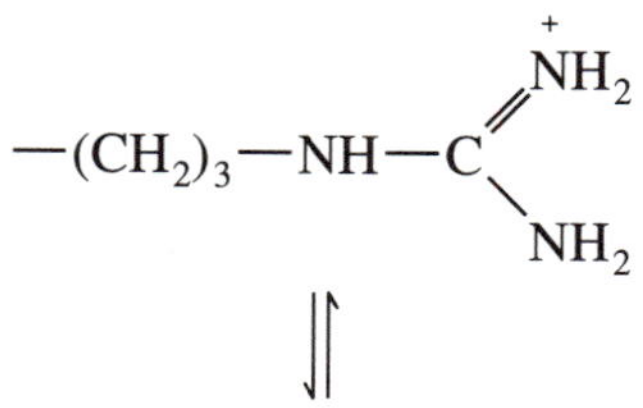

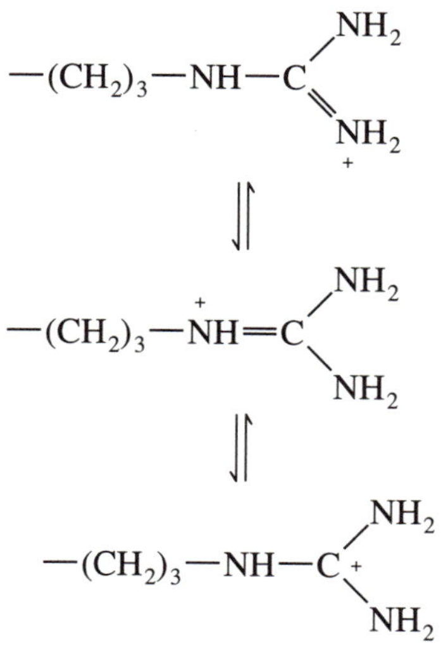

and the positive charge is effectively distributed over the entire group. In the protonated form, the guanido group is unreactive, and only very small fractions of the nonionized form are present at physiological pH values.

One reaction that is useful in the laboratory is the formation of heterocyclic condensation products with 1,2- and 1,3-dicarbonyl compounds, such as phenylglyoxal, 2,3-butanedione, and 1,2-cyclohexanedione:

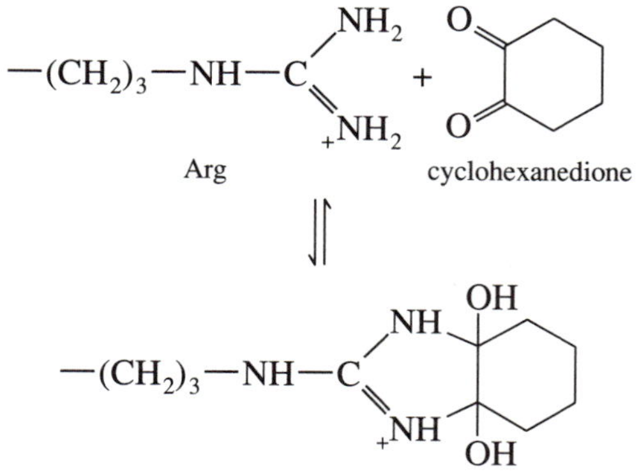

These reactions are favored because the distance between the two carbonyl groups of the reagents closely matches that between the two unsubstituted nitrogen atoms of the guanido group. The adduct formed can be further stabilized by the presence of borate, which complexes with the adjacent hydroxyl groups.

The guanido group can be cleaved by hydrazine (H_2NNH_2) to produce the side chain of ornithine (Orn):

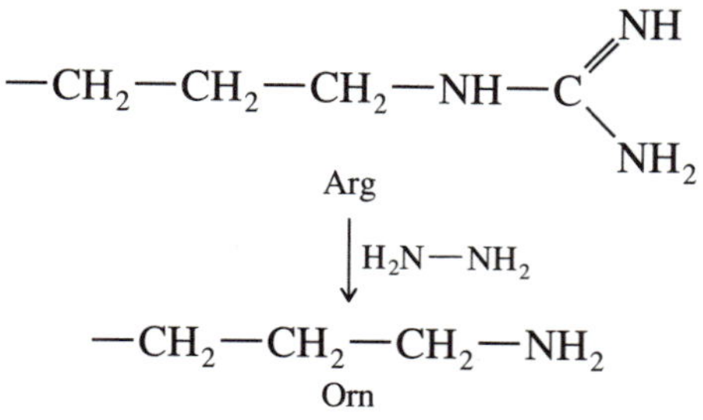

This reaction, however, is often accompanied by cleavage of the polypeptide backbone (Sec. 1.6.2.a).

References

Reversible blocking at arginine by cyclohexanedione. E. L. Smith. *Methods Enzymol.* 47:156–161 (1977).

Chemical modification of peptides by hydrazine. A. Honegger et al. *Biochem. J.* 199:53–59 (1981).

Modification of available arginine residues in proteins by *p*-hydroxyphenylglyoxal. R. B. Yamasaki et al. *Anal. Biochem.* 109:32–40 (1980).

1.3.8 His

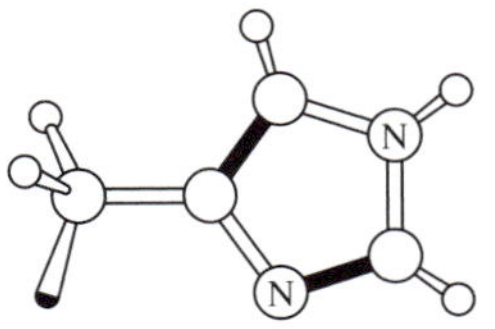

His   (1.31)

The imidazole side chain of His residues possesses several special properties that make it extremely effective as a nucleophilic catalyst. It is an amine, which is much more reactive than hydroxide ion in terms of its basicity. Furthermore, it is a tertiary amine, which is intrinsically more nucleophilic than primary or secondary amines. The enhanced reactivity of tertiary amines is usually canceled by their greater steric hindrance, but in imidazole the atoms bonded to the nitrogens are held back in a five-membered ring and cause relatively little steric hindrance. Imidazole has a pK_a value near 7, so it is one of the strongest bases that can exist at neutral pH. A weaker base would have a lower nucleophilic reactivity, whereas a stronger base would be protonated to a greater extent at neutral pH and would be correspondingly less reactive.

In the nonionized form of the imidazole ring, the nitrogen with the hydrogen atom is an electrophile and donor for hydrogen bonding, and the other nitrogen atom is a nucleophile and acceptor for hydrogen bonding. Consequently, this one side chain is extremely versatile—almost the chemical equivalent of being ambidextrous.

The imidazole group is, in principle, capable of undergoing numerous reactions, but most of these reactions occur more readily with amino and thiol groups; very few are suitable for modifying His residues specifically. The classical approach to modifying His residues was oxidation by light in the presence of dye sensitizers such as methylene blue or rose bengal. The reaction products are probably an aspartic acid residue and urea,

but the reaction is not well characterized. The most specific reaction is that with diethylpyrocarbonate, which can be reversed by hydroxylamine:

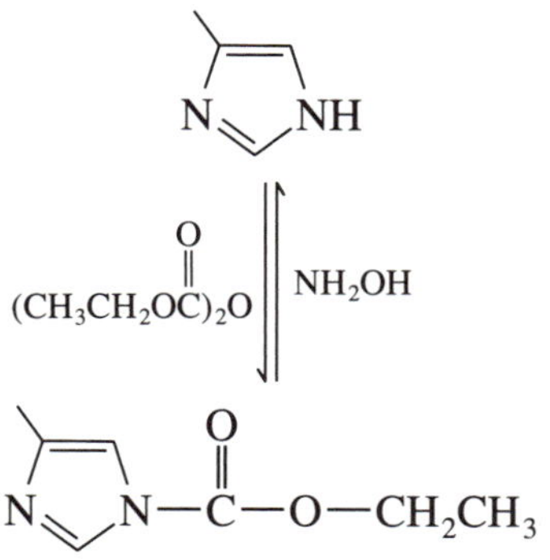

(1.32)

Reactions with amino groups can be minimized by carrying out the reaction at acidic pH.

The two nitrogen atoms of the His side chain are designated here as $\delta 1$ and $\epsilon 2$, but they are also known, respectively, as π and τ or as N-1 and N-3. The last designation is often ambiguous because biochemists usually assign the number 1 to the nitrogen atom adjacent to the side chain whereas organic chemists tend to designate this atom as 3.

The nonionized imidazole ring can exist as two tautomers, with the hydrogen atom on either the $\delta 1$ or the $\epsilon 2$ nitrogen atom:

(1.33)

[13]Carbon-NMR studies have shown that in model peptides the hydrogen atom is usually predominantly on the $\epsilon 2$ nitrogen atom, which has a pK_a value about 0.6 pH unit higher than that of the $\delta 1$ atom. The position of the hydrogen atom depends on the relative affinities of the two nitrogen atoms for protons, however, and can vary with conditions in the local environment; both forms are found in proteins.

The His side chain is readily protonated, with a pK_a value near 7 at the second N atom, which destroys its nucleophilicity. The positive charge is shared by the two N atoms by resonance:

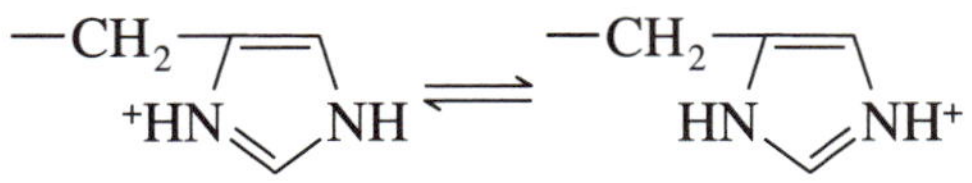

(1.34)

Both of the imidazole nitrogen atoms can also be deprotonated simultaneously, with an apparent pK_a value of about 14.4, to give the aromatic anion:

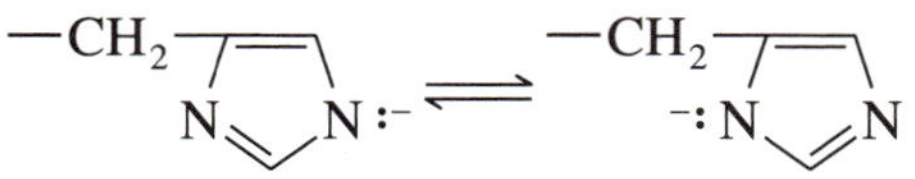

(1.35)

Table 1.3 *Spectroscopic Properties of the Aromatic Amino Acids at Neutral* pH

	Absorbance[a]		Fluorescence Emission[b]	
	λ_{max} (nm)	Molar absorbance ($M^{-1}cm^{-1}$)	λ_{max} (nm)	Quantum yield
Phenylalanine	257.4	197	282	0.04
Tyrosine	274.6	1420	303	0.21
Tryptophan	279.8	5600	348	0.20

[a] From J. E. Bailey, Ph.D. thesis, London University, 1966.

[b] From F. W. J. Teale and G. Weber, *Biochem. J.* 65:476–482 (1957).

This anion would be expected to be a potent nucleophile but is rarely present in substantial quantities.

The $C^{\epsilon 1}$ atom (often designated as C-2) is observed to exchange its hydrogen atom slowly with the solvent, indicating that it has a very small probability of being deprotonated:

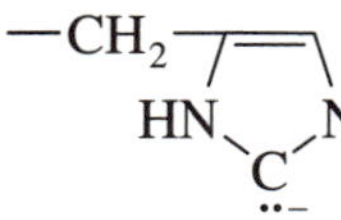

$$-CH_2 \underset{HN \underset{C}{} NH}{\overset{}{\boxed{}}} \qquad (1.36)$$

This exchange reaction provides a useful probe of the environments of His residues in proteins (Chap. 7).

His residues are especially useful in ^{1}H-NMR studies of proteins because the hydrogen atom on the $C^{\epsilon 1}$ atom is usually well resolved from the multitude of resonances of the other hydrogen atoms in proteins in one-dimensional spectra. Its resonance is also usually shifted by about 1 ppm to a lower field strength upon protonation of the side chain, often making it relatively easy to determine the pK_a values of individual His residues even in large proteins.

References

The roles of imidazole in biological systems. E. A. Barnard and W. D. Stein. *Adv. Enzymol.* 20:51–110 (1959).

Titration behaviour and tautomeric states of individual histidine residues of myoglobin. D. J. Wilbur and A. Allerhand. *J. Biol. Chem.* 252:4968–4975 (1977).

Modification of histidyl residues in proteins by diethylpyrocarbonate. E. W. Miles. *Methods Enzymol.* 47:431–442 (1977).

Tautomeric states of the histidine residues of bovine pancreatic ribonuclease A. D. E. Walters and A. Allerhand. *J. Biol. Chem.* 255:6200–6204 (1980).

^{1}H-NMR study on the tautomerism of the imidazole ring of histidine residues. I. Microscopic pK values and molar ratios of tautomers in histidine-containing peptides. M. Tanokura. *Biochim. Biophys. Acta* 742:576–585 (1983).

Chemical modification of pig kidney 3,4-dihydroxyphenylalanine decarboxylase with diethyl pyrocarbonate. P. Dominici et al. *J. Biol. Chem.* 260:10583–10589 (1985).

1.3.9 The Aromatic Residues: Phe, Tyr, and Trp

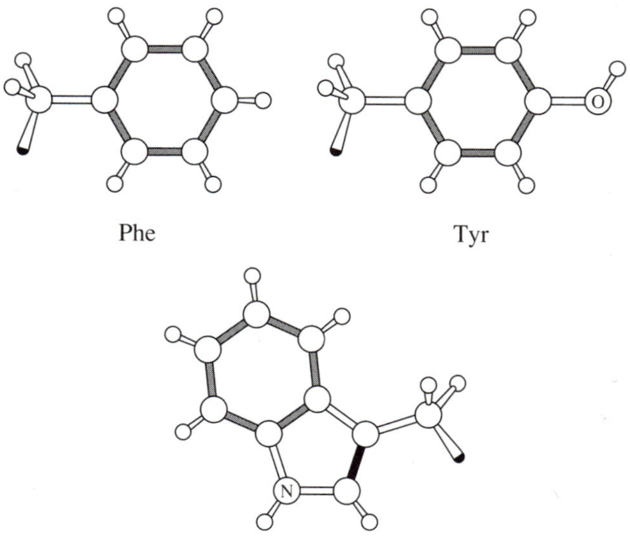

Phe   Tyr

Trp   (1.37)

These aromatic side chains are responsible for most of the ultraviolet absorbance and fluorescence properties of proteins. The spectral properties of the side chains (Table 1.3 and Fig. 1.4) are very sensitive to the immedi-

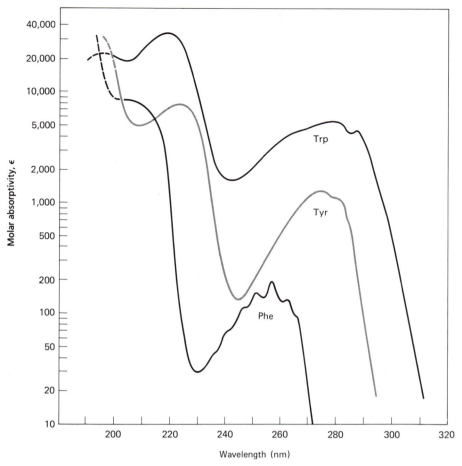

FIGURE 1.4

Ultraviolet absorbance spectra of the aromatic amino acids at pH 6. Incorporation of these amino acids into peptides has little direct effect on the absorbance properties of their side chains, unless they are placed into different environments. (From D. B. Wetlaufer, *Adv. Protein Chem.* 17:303–390, 1962.)

ate environment of the side chains, which makes them useful probes of protein structure (Chaps. 5 and 7).

a. Phe

The aromatic ring of Phe residues is chemically comparable to that of benzene or toluene. Consequently, it is nonpolar and is chemically reactive only under extreme conditions that are not applicable to proteins.

b. Tyr

The hydroxyl group of the phenolic ring of Tyr residues makes this aromatic ring relatively reactive in electrophilic substitution reactions. These usually occur at the positions designated here as $\epsilon 1$ and $\epsilon 2$, which elsewhere are often numbered as 3 and 5, respectively. Conse-

quently, Tyr side chains can be readily nitrated and iodinated to form the following derivatives:

$$(1.38)$$

Table 1.4 *Spectroscopic Properties of Tyrosine and Several Derivatives*

	pK_{app} of —OH	Nonionized OH		Ionized Hydroxyl	
		λ_{max} (nm)	Molar absorbance ($M^{-1}cm^{-1}$)	λ_{max} (nm)	Molar absorbance $M^{-1}cm^{-1}$
Tyrosine	10.1	274.5	1400	293	2400
ϵ-Iodotyrosine	8.2	283	2750	305	4100
$\epsilon1,\epsilon2$-Diiodotyrosine	6.5	287	2750	311	6250
ϵ-Nitrotyrosine	7.2	360	2790	428	4200
ϵ-Aminotyrosine	10.0^a	275	1600	320	4200
O-Acetyltyrosine	—	262	262	—	—

^{a} The pK_{app} of the aromatic amino group is approximately 4.8.

From A. N. Glazer, in H. Neurath and R. L. Hill (eds.), *The Proteins*, 3rd ed. vol. 2, pp. 1–103. New York, Academic Press, 1976.

The spectrophotometric properties of these groups are altered substantially, making spectrophotometric analysis particularly useful (Table 1.4).

The hydroxyl group of the Tyr side chain ionizes at alkaline pH values, with an intrinsic pK_a of 11.1 and with a change in its spectral properties (Table 1.4). The hydroxyl group can participate in hydrogen bonding and also can be acetylated by acetic anhydride:

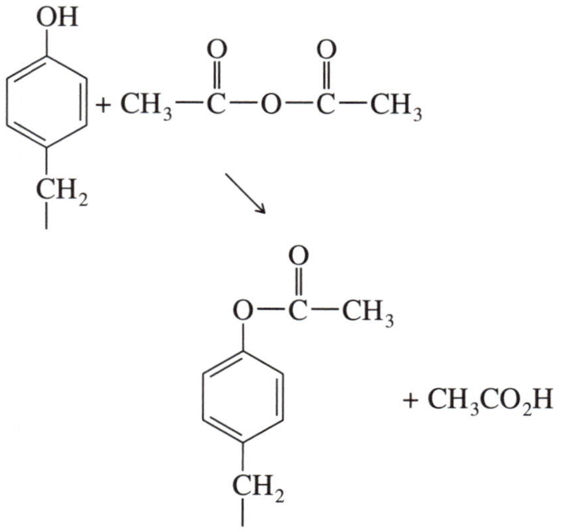

(1.39)

c. Trp

The indole side chain of Trp residues is the largest and the most fluorescent of the side chains of proteins. This amino acid also occurs least frequently, so proteins often have only one or a few Trp residues. Their spectral properties are often useful probes of protein structure (Chap. 7). Their fluorescence properties are especially sensitive to the environment of the side chain, but in largely unpredictable ways.

The indole ring is susceptible to irreversible oxidation. In particular, iodine and N-bromosuccinimide oxidize the Trp indole ring to that of oxindolealanine:

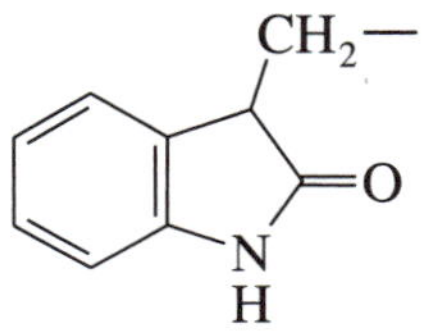

(1.40)

Ozone opens the indole ring to that of N-formylkynurenine:

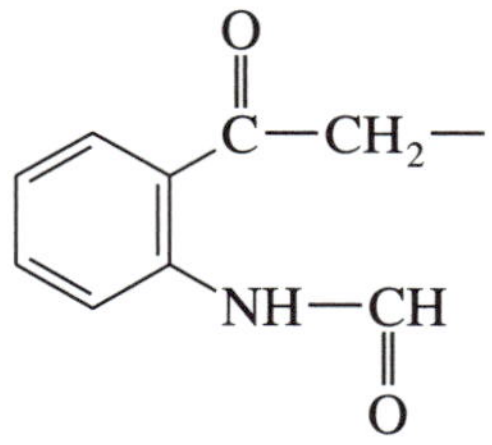

(1.41)

These reagents, however, also rapidly oxidize Cys thiol groups and react more slowly with Tyr and His residues.

The nitrogen atom of the indole ring can be reversibly formylated by anhydrous formic acid containing HCl to yield

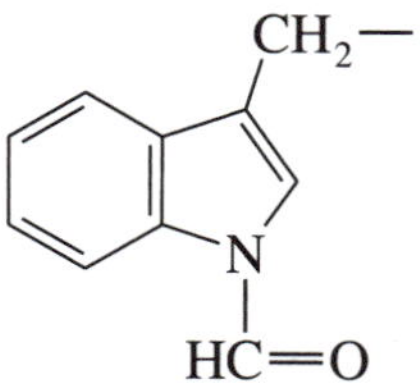

(1.42)

Certain reactive benzyl and sulfenyl halides have been designed that alkylate the $C^{\delta 1}$ atom of the indole ring:

$$\text{(1.43)}$$

A variety of groups with other spectral properties can be introduced into Trp residues in this way.

The pyrrole nitrogen atom of the indole ring may be a hydrogen bond donor. This is also the only side chain capable of participating as a donor in charge-transfer complexes with pyridinium compounds and other electrophiles.

References

Ultraviolet fluorescence of the aromatic amino acids. F. J. W. Teale and G. Weber. *Biochem. J.* 65:476–482 (1957).

Ultraviolet spectra of proteins and amino acids. D. B. Wetlaufer. *Adv. Protein Chem.* 17:303–390 (1962).

Nitration with tetranitromethane. J. F. Riordan and B. L. Vallee. *Methods Enzymol.* 25:515–521 (1972).

Fluorescence of aminotyrosyl residues in peptides and helical proteins. R. L. Seagle and R. W. Cowgill. *Biochim. Biophys. Acta* 439:461–469 (1976).

1.3.10 The Sulfur-Containing Residues: Met and Cys

Met   Cys   (1.44)

a. Met

The long side chain of Met residues is nonpolar and relatively unreactive and is the only unbranched nonpolar side chain of all the natural amino acids. The sulfur atom is somewhat nucleophilic, but unlike other nucleophiles in proteins it cannot be protonated. Consequently, it is the most potent nucleophile in proteins at acidic pH. It can be selectively modified under such conditions, readily forming sulfonium salts with alkylating agents such as methyl iodide:

$$-(CH_2)_2-S-CH_3$$

$$-(CH_2)_2-\overset{CH_3}{\underset{+}{S}}-CH_3 + I^- \quad (1.45)$$

This reaction can be reversed by thiols. The methyl group removed is equally likely to be the original one or that introduced by the methyl iodide, so this reaction offers the possibility of introducing an isotopic label in 50% of the residues by using labeled methyl iodide. The sulfur atom also has a tendency to interact noncovalently with the platinum derivative $PtCl_4^{2-}$, which is a useful heavy-atom derivative in protein crystallography (Chap. 6).

The sulfur atom of Met residues is also susceptible to oxidation by air and more potent oxidants such as peroxides. The sulfoxide is formed first, followed by the sulfone:

$$\text{(1.46)}$$

The first step, but not the second, can be reversed by sulfides and thiols. Either oxidation state makes the Met residue unreactive with alkylating reagents (Eq. 1.45).

b. Cys

The thiol group of Cys residues is the most reactive of any amino acid side chain, so only a few of the most important and useful examples can be given here. The Cys thiol usually ionizes at slightly alkaline pH values with an intrinsic pK_a in the region of 9.0 to 9.5. The thiolate anion formed is the reactive species in most instances. It reacts rapidly with alkyl halides, such as iodoacetate, iodoacetamide, and methyl iodide, to give the corresponding stable alkyl derivatives:

$$-CH_2-S^- \quad ICH_2CO_2^-$$
Cys   iodoacetate

$$-CH_2-S-CH_2-CO_2^- + I^-$$
carboxymethyl-Cys   (1.47)

The thiol group can also add across double bonds, such as those of N-ethylmaleimide or maleic anhydride:

$$
\begin{array}{c}
-CH_2-S^- + \\
\text{Cys}
\end{array}
\quad
\begin{array}{c}
O \\
\| \\
HC-C \\
\| \qquad N-CH_2CH_3 \\
HC-C \\
\| \\
O \\
\text{N-ethylmaleimide}
\end{array}
$$

$$H^+ \downarrow$$

$$
\begin{array}{c}
O \\
\| \\
-CH_2-S-CH-C \\
\qquad\quad N-CH_2CH_3 \\
H_2C-C \\
\| \\
O
\end{array}
\qquad (1.48)
$$

It can also open the ring of ethyleneimine:

$$
-CH_2-S^- +
\begin{array}{c}
CH_2 \\
| \quad NH \\
CH_2
\end{array}
$$

$$
2H^+ \searrow
$$

$$-CH_2-S-CH_2-CH_2-NH_3^+ \qquad (1.49)$$

The resulting positively charged side chain provides a new site for cleavage of polypeptides by trypsin, in addition to those of Lys and Arg, which is useful for amino acid sequence determination (Sec. 1.6.2).

The thiols of Cys residues form complexes of varying stability with a variety of metal ions. The most stable are those with divalent mercury, Hg^{2+}, but complexes with a variety of stoichiometries are formed. Consequently, univalent organic mercurials of the type $R—Hg^+$ tend to be used instead because they more reproducibly form one-to-one complexes with thiols. The best known is p-mercuribenzoic acid:

$$
-CH_2-S^- + {}^+Hg-\left\langle\text{benzene}\right\rangle-CO_2H
$$

$$\searrow$$

$$
-CH_2-S-Hg-\left\langle\text{benzene}\right\rangle-CO_2H \qquad (1.50)
$$

It has been used to titrate thiol groups by measuring the spectral change that takes place when binding occurs. Such reactions with mercurial compounds are also the most obvious and useful ways to make heavy-atom derivatives for protein crystallography (Chap. 6).

Thiol complexes with silver are less stable than those with mercury, but univalent Ag^+ reacts stoichiometrically and can be used to titrate thiols. Copper, iron, zinc, cobalt, molybdenum, manganese, and cadmium ions all form various complexes with the thiol groups of Cys residues.

Thiols are readily oxidized by oxygen, especially in the presence of trace amounts of metal ions such as Cu^{2+}, Fe^{2+}, Co^{2+}, and Mn^{2+}; it is likely that the metal complexes are the actual reactants with oxygen.

The sulfur atom of Cys residues can exist in a variety of oxidation states, but some of them are unstable. Besides the thiol form, only two oxidation states are generally encountered, the disulfide and the sulfonic acid. The disulfide is usually the end product of air oxidation:

$$2 -CH_2SH + \tfrac{1}{2}O_2 \longrightarrow$$

$$-CH_2S-SCH_2- + H_2O \qquad (1.51)$$

The sulfonic acids are produced by more potent oxidizing agents. For example, performic acid oxidizes both thiol and disulfide forms of Cys residues to cysteic acid residues (abbreviated $CysO_3H$ or Cya):

$$
\begin{array}{c}
SO_3H \\
| \\
CH_2 \quad O \\
| \qquad \| \\
-NH-CH-C- \\
\text{CysO}_3\text{H}
\end{array}
\qquad (1.52)
$$

Disulfide bonds between Cys residues occur in some proteins; two such residues linked by a disulfide bond are often designated as a **cystine** residue, after the amino acid cystine:

$$
\begin{array}{c}
NH_2 \qquad\qquad\qquad\qquad NH_2 \\
| \qquad\qquad\qquad\qquad\qquad | \\
CH-CH_2-S-S-CH_2-CH \\
| \qquad\qquad\qquad\qquad\qquad\quad | \\
CO_2H \qquad\qquad\qquad\qquad CO_2H
\end{array}
\quad (1.53)
$$

and individual Cys residues are often designated as $\tfrac{1}{2}$-cystines. It is now clear, however, that cystine is not incorporated into proteins as such (Chap. 2). Instead, the thiol form, cysteine, is used in protein biosynthesis, and disulfide bonds between Cys residues can be added later. The designation of cystine as an amino acid of proteins is therefore somewhat misleading, but references to it in the literature are still common.

Disulfide bonds are covalent and can be kept intact under appropriate conditions. The angle of rotation (see Sec. 5.1) about the disulfide bond in its preferred (most stable) geometry has a value close to either $+90°$ or $-90°$. Other angles are unstable by as much as 10 kcal/

mol. Interconversion of these two favored conformers is rapid and has an activation barrier of about 7 kcal/mol.

Thiols and disulfides undergo very rapid exchange at neutral to alkaline pH values:

$$R_1-S^- + R_2-S-S-R_2$$
$$\Updownarrow$$
$$R_2-S^- + R_1-S-S-R_2 \qquad (1.54)$$

The thiolate anion is the reactive species, so the rate is greatest at alkaline pH values. The equilibrium constant between simple aliphatic thiols and disulfides is close to unity.

Thiol–disulfide exchange with aromatic disulfides provides the most convenient method of assaying thiol groups because the liberated aromatic thiol is brightly colored. This is the basis of the popular Ellman assay, using dithiobisnitrobenzoic acid (DTNB):

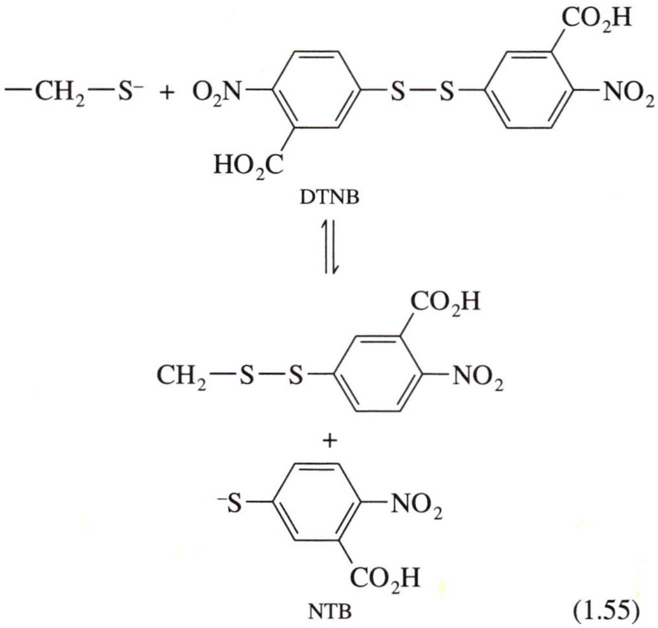

$$-CH_2-S^- + O_2N-\underset{HO_2C}{\overset{CO_2H}{\bigcirc}}-S-S-\bigcirc-NO_2$$

DTNB

$$\Updownarrow$$

$$CH_2-S-S-\overset{CO_2H}{\bigcirc}-NO_2$$

$$+$$

$$^-S-\underset{CO_2H}{\overset{NO_2}{\bigcirc}}$$

NTB   (1.55)

Adding excess DTNB results in formation of a stoichiometric amount of the yellow aromatic thiol, nitrothiobenzoate (NTB).

Disulfides in proteins can be effectively reduced by thiol–disulfide exchange with thiol reagents (RSH), such as mercaptoethanol:

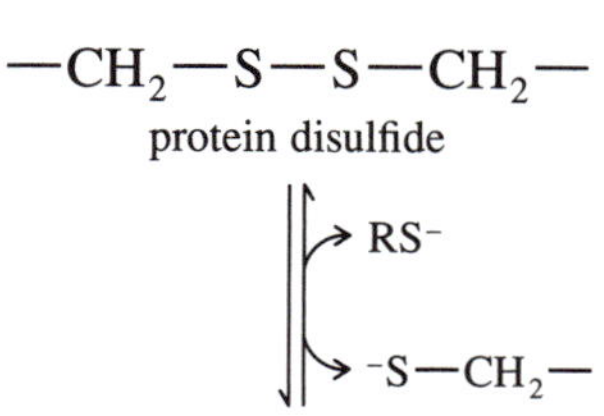

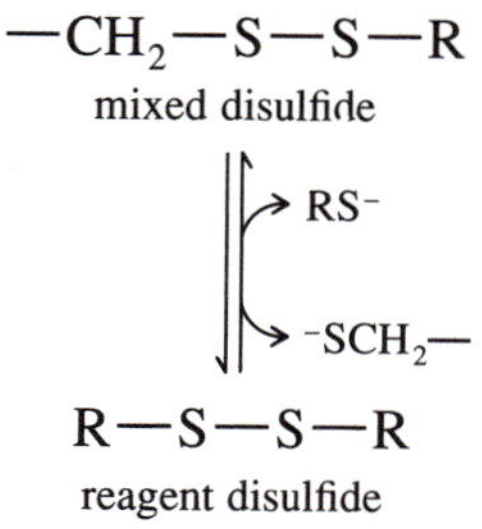

$$-CH_2-S-S-CH_2-$$
protein disulfide

$$\Updownarrow \begin{array}{l} \nearrow RS^- \\ \searrow {}^-S-CH_2- \end{array}$$

$$-CH_2-S-S-R$$
mixed disulfide

$$\Updownarrow \begin{array}{l} \nearrow RS^- \\ \searrow {}^-SCH_2- \end{array}$$

$$R-S-S-R$$
reagent disulfide   (1.56)

Two sequential thiol–disulfide exchanges are involved, proceeding via the mixed-disulfide intermediate. Most natural disulfide bonds in proteins are stabilized by the protein conformation, so a large excess of thiol compound is required to break them. The stabilities of protein disulfide bonds can easily be increased or decreased by controlling the relative concentrations of the thiol and disulfide forms of a reagent.

A most useful reagent for reducing protein disulfides or those of other molecules (RSSR) is dithiothreitol (Cleland's reagent or its isomer dithioerythritol):

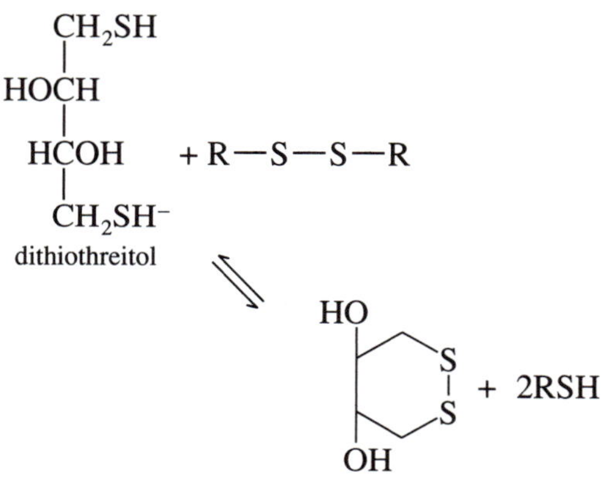

$$\begin{array}{l} CH_2SH \\ HOCH \\ HCOH \qquad + R-S-S-R \\ CH_2SH^- \end{array}$$
dithiothreitol

$$\Longleftarrow \overset{HO}{\underset{OH}{\bigcirc}}\begin{array}{l}S \\ | \\ S\end{array} + 2RSH$$

(1.57)

This reagent forms an intramolecular disulfide bond in a stable six-membered ring, so the equilibrium for Equation (1.57) lies far to the right (equilibrium constant of approximately $10^3 \, M$). The reaction proceeds in two steps (as in Eq. 1.56) through the mixed-disulfide intermediate, but this intermediate is unstable because the intramolecular ring of the disulfide form of the reagent is formed very rapidly. Disulfide bonds can also be reduced directly by electrodes or chemically with reducing reagents such as sodium borohydride.

Disulfides can also be cleaved by nucleophiles such as cyanide, sulfite, or hydroxide ion:

$$RS^- + RSCN \underset{CN^-}{\overset{}{\rightleftharpoons}} R-S-S-R \overset{OH^-}{\rightleftharpoons} RSOH + RS^-$$

$$\Updownarrow SO_3^-$$

$$R-S-SO_3^-$$
$$+$$
$$RS^- \qquad (1.58)$$

The equilibria for these reactions are such that it is difficult to drive the reactions to completion by simply adding reagent. Complete reaction can be accomplished, however, by adding a second reagent that reacts with the thiol generated. When this is nitrothiosulfobenzoate (NTSB),

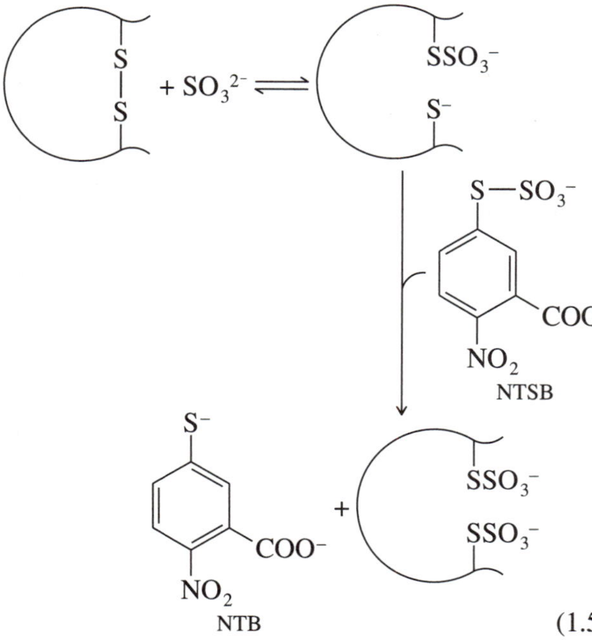

(1.59)

the reaction can go to completion, and 1 mol of the colored NTB is generated for each mole of protein disulfide originally present. This is the most convenient assay for protein disulfides.

References

Dithiothreitol, a new protective reagent for SH groups. W. W. Cleland. *Biochemistry* 3:480–482 (1964).

Biochemistry of the SH Group. P. C. Jocelyn. Academic Press, New York, 1972.

Disulfide bond formation in proteins. T. E. Creighton. *Methods Enzymol.* 107:305–329 (1984).

Selective oxidation and reduction of methionine residues in peptides and proteins by oxygen exchange between sulfoxide and sulfide. Y. Schechter. *J. Biol. Chem.* 261:66–70 (1986).

Detection and quantitation of biological sulfhydryls. A. Russo and E. A. Bump. *Methods Biochem. Anal.* 33:165–241 (1988).

1.4 Detection of Amino Acids, Peptides, and Proteins

The experimental study of proteins requires first that they be detectable. Assays are required to determine the quantity of protein present in a solution and to detect all proteins present after separation by techniques such as chromatography and electrophoresis.

Purified proteins are most easily detected and quantified by their ultraviolet (UV) absorbance in the aromatic region around 280 nm (Table 1.3), so long as no interfering substances that significantly absorb UV at the same wavelength are present. The absorbance of each protein depends on the number and positions of its aromatic amino acid residues Phe, Tyr, and Trp and on any disulfide bonds between Cys residues; so the molar absorbance coefficient of each protein must be known for this UV absorbance procedure to measure the amount of protein present quantitatively. Fortunately, a fully unfolded protein has absorbance properties in the aromatic region that are close to those of its constituent amino acids, making it possible to calculate the molar absorbance coefficient of a protein under unfolding conditions if the content of aromatic amino acids and disulfides is known. Conversely, such a spectrum can be used to determine the amino acid composition of a protein.

In the ideal detection procedure, all proteins would react to the same extent, and thus the assay would give an accurate measure of the amount of every protein present. Such an ideal assay would make use of some property common to all protein molecules, such as the peptide bonds of the backbone. Few of the procedures commonly used are ideal, however; that is, the response of various proteins depends to some extent on their amino acid compositions. Moreover, the assays usually make use of known quantities of one particular protein, often bovine serum albumin (BSA), to produce a standard curve. The accuracy of the results obtained consequently depends very much on how closely the standard protein corresponds to the protein being measured.

One procedure that does measure primarily peptide bonds, and is therefore specific to proteins, is the biuret reaction. A dilute solution of cupric sulfate in strongly alkaline tartrate is added to the protein solution. A purplish-violet compound with maximum absorbance at 540 nm forms. The nature of this colored compound is uncertain, but its color is probably due to the formation of a cupric ion complex coordinated with adjacent peptide groups:

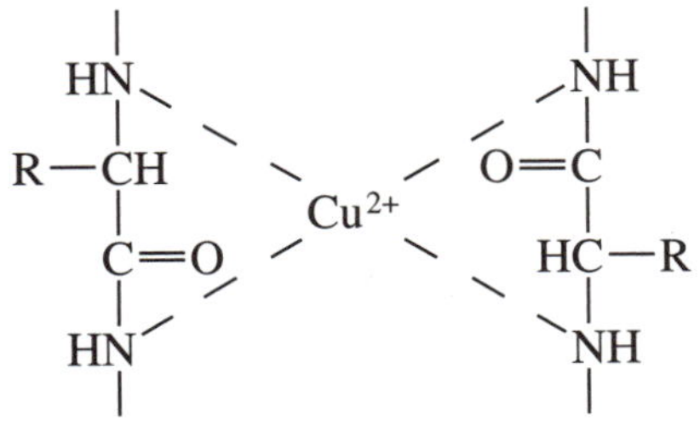

(1.60)

More reactions appear to take place under the alkaline conditions, and the cupric ion (Cu^{2+}) tends to oxidize the peptide backbone and to be reduced to Cu^+. The involvement of the NH group of the peptide backbone suggests that Pro residues do not participate in this reaction, and proteins with many Pro residues give low responses. Therefore, even this procedure, though it is based on the peptide backbone, is not applicable to all proteins. The greatest shortcoming of the biuret reaction, however, is its low sensitivity.

The classical assay for proteins has been the Folin phenol method devised by Lowry, Rosebrough, Farr, and Randall. It is based on the biuret reaction but greatly enhances the reaction. Copper ions are added to the protein-containing solution, followed by addition of the Folin-Ciocalteau reagent, the active constituents of which are the phosphomolybdic-tungstic mixed acids:

$$3\,H_2O \cdot P_2O_5 \cdot 13\,WO_3 \cdot 5\,MoO_3 \cdot 10\,H_2O$$
$$3\,H_2O \cdot P_2O_5 \cdot 14\,WO_3 \cdot 4\,MoO_3 \cdot 10\,H_2O \quad (1.61)$$

These mixed acids are reduced by the protein, probably through the Cu^+ generated, thereby losing one, two, or three oxygen atoms from the tungstates and molybdates and producing a number of reduced species with a characteristic blue color (maximum absorbance 720–750 nm). Pro residues prevent the adjacent peptide backbone from reacting, and the side chains of Tyr, Trp, His, and Asn residues appear to participate in the reaction and to produce a blue color. Consequently, the color response of different proteins in this assay is more variable than in the biuret reaction. The greatest problem with the assay, however, is the variety of nonprotein substances that interfere, either producing a blue color themselves or inhibiting color development by the protein.

Ninhydrin, one of the most widely used reagents, detects most amino acids, peptides, and proteins. It reacts with the amino groups that are almost invariably present or can be generated by hydrolysis of the polypeptide chain. The reaction is complex, but a major step appears to be the oxidative deamination of an amino acid to CO_2, NH_3, and an aldehyde containing one less carbon atom than the original amino acid. The ninhydrin is simultaneously reduced to the hydrindantin:

ninhydrin   amino acid

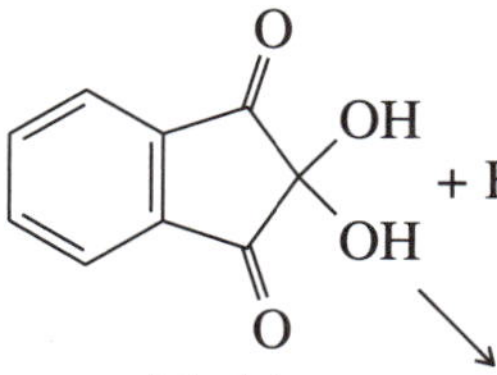

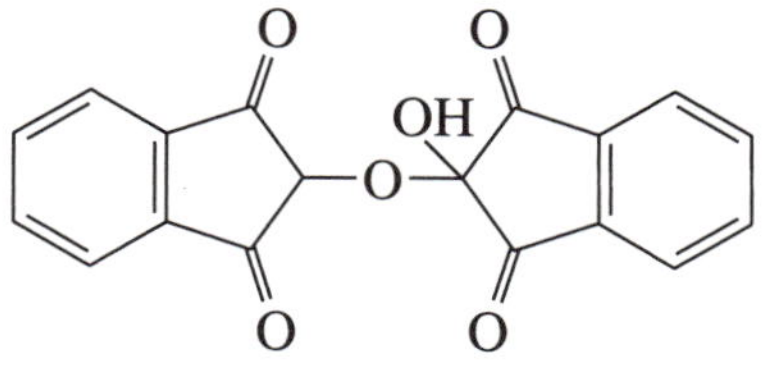

hydrindantin

$$+ \; RCH + CO_2 + NH_3 \quad (1.62)$$

The hydrindantin then reacts with the liberated NH_3 to give the purple product:

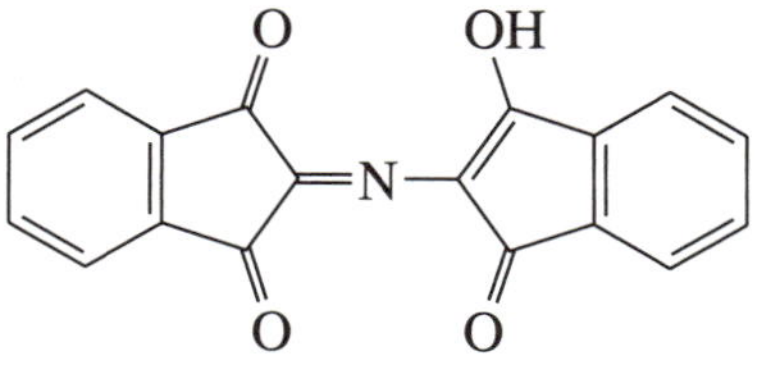

$$(1.63)$$

This product is intensely colored, with a maximum absorbance at 570 nm, whereas the ninhydrin is virtually colorless. The imino acid proline does not react in this way, but forms a yellow product of uncertain structure with maximum absorbance at 440 nm.

Ninhydrin is most often used to detect amino acids and peptides, either in solution (as in amino acid analyzers) or on paper or thin-layer plates. It has the disadvantage of destroying the amino acid or peptide so that the material detected cannot be characterized further. Excess ninhydrin reacts further if a treated protein or peptide is hydrolyzed.

Fluorescamine is a similar reagent that also reacts with amino groups but is much more sensitive and does not have the disadvantages of ninhydrin:

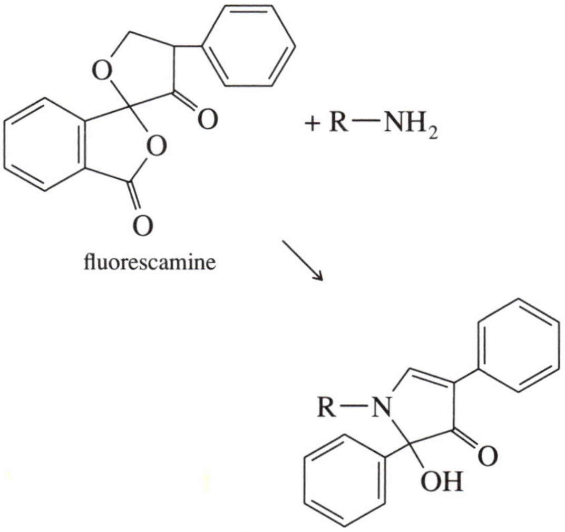

fluorescamine $+ \; R{-}NH_2$

$$(1.64)$$

In this case, the derivative is fluorescent, whereas the reagent is not. The amino groups that have reacted can be recovered in high yield by acid hydrolysis and can be identified by further analysis. Excess fluorescamine is unstable in the presence of water, being hydrolyzed to nonfluorescent and nonreactive products, and thus is destroyed. Consequently, peptides detected with fluorescamine can be analyzed directly. The other advantage of this reagent is its great sensitivity; it can be used to detect much smaller quantities of protein than can ninhydrin.

A related reagent is *ortho*-phthalaldehyde, which reacts with amino groups in the presence of mercaptoethanol and forms a fluorescent reaction product:

$$(1.65)$$

This sensitive detection procedure can detect picomole (10^{-12} mol) quantities of amino acids.

Ninhydrin, fluorescamine, and o-phthalaldehyde are not specific for proteins because they react with most amino groups.

A common reagent for staining proteins is **Coomassie Brilliant Blue,** in either the R250 or the G250 form:

R250: R=H
G250: R=CH_3

$$(1.66)$$

These dyes do not react chemically with proteins but merely form noncovalent complexes. The interaction is believed to be primarily ionic, involving the basic groups on proteins, but nonpolar van der Waals forces are probably also involved. Consequently, the dyes do not bind equally to all proteins. Although the two dyes are structurally similar, they require different physical and staining procedures and are not interchangeable in any particular protocol.

The Coomassie blue dyes, particularly G250, can be used to quantify the amount of protein in solution because complex formation changes their absorbance properties; under certain acidic conditions, their absorbance maximum shifts from 465 to 595 nm. Even though different proteins give somewhat different responses in this assay, its simplicity makes it widely used.

The most common use of the Coomassie blue dyes, particularly R250, is in staining proteins in gels. This procedure relies on the proteins being made insoluble and binding the dye tightly. Most, but not all, of the excess dye is removed by washing the gel, so proteins are apparent by the blue dye bound to them. Approximately 0.1 microgram (μg) of protein can be detected on a polyacrylamide gel. A number of other dyes also bind tightly to proteins, but most are not as sensitive as Coomassie blue.

The most sensitive staining procedure for fixed proteins on gels or blots (see Sec. 1.5.3) involves **silver** staining, which can detect less than a nanogram (10^{-9} g) of protein. The chemical basis for the silver-staining procedure is not known, but it appears to be related to the processes that occur in photography. Indeed, many of the procedures were adapted from photographic protocols. Silver nitrate is added to the protein-containing material under acidic conditions. The silver ion is believed to bind to various sites on the protein, which seem to involve basic groups but are probably composed of more than just individual chemical groups. This reaction is followed by the addition of formaldehyde under alkaline conditions to reduce selectively the ionic silver to metallic silver. This reduction process is critical for the type of staining that is obtained. The silver that is bound to sites on the protein nucleates the growth of globular metallic silver grains. If a protein site has a higher reducing potential than the bulk environment, silver grains grow on the protein more than elsewhere and the protein is positively stained. If the protein site has a potential that is less reducing, however, the protein is negatively stained. Under some conditions, proteins stain positively with different colors, which are believed to arise from the different sizes of the silver grains.

Although the silver-staining process is usually used, other metal salts probably interact in similar ways with proteins. They can be used to detect proteins if the unbound salts can be precipitated in the gel at areas where proteins are not present. Other sensitive staining procedures use the affinity of colloidal gold particles for proteins, which is probably due to both electrostatic and van der Waals interactions.

The chemical properties of many of the amino acid side chains, particularly those of His, Cys, Arg, Trp, and Tyr, can be used in staining procedures that are specific for each amino acid. They can then be used to detect the peptides and proteins that contain residues of the particular amino acid. Such staining procedures are too numerous to describe here, but most of them rely on the chemical properties of the amino acid side chains described in the previous section.

Individual proteins are usually detected specifically by their biological activities, which vary so greatly that no simple guide can be given (some of the biological activities of proteins are described primarily in Chaps. 8 and 9). Immunological techniques are useful for detecting specific proteins and are described in Section 8.3.1.

References

Fluorescamine: a reagent for assay of amino acids, peptides, proteins, and primary amines in the picomole range. S. Udenfriend et al. *Science* 178:871–872 (1972).

A rapid and sensitive method for the quantitation of microgram quantities of proteins utilizing the principle of protein dye binding. M. M. Bradford. *Anal. Biochem.* 72:248–254 (1976).

Review of the Folin phenol protein quantitation method of Lowry, Rosebrough, Farr, and Randall. G. L. Peterson. *Anal. Biochem.* 100:201–220 (1979).

o-Phthalaldehyde: fluorogenic detection of primary amines in the picomole range. Comparison with fluorescamine and ninhydrin. J. R. Benson and P. E. Hare. *Proc. Natl. Acad. Sci. USA* 72:619–622 (1979).

Why does Coomassie Brilliant Blue R interact differently with different proteins? A partial answer. M. Tal et al. *J. Biol. Chem.* 260:9976–9980 (1980).

Measurement of protein using bicinchonic acid. P. K. Smith et al. *Anal. Biochem.* 150:76–85 (1985).

On the chemical basis of the Lowry protein determination. G. Legler et al. *Anal. Biochem.* 150:278–287 (1985).

Determination of the concentration of protein by dry weight —a comparison with spectrophotometric methods. Y. Nozaki. *Arch. Biochem. Biophys.* 249:437–446 (1986).

Coloration of silver-stained protein bands in polyacrylamide gels is caused by light scattering from silver grains of characteristic sizes. C. R. Merril et al. *Proc. Natl. Acad. Sci. USA* 85:453–457 (1988).

Formation and instability of o-phthalaldehyde derivatives of amino acids. M. C. G. Alvarez-Coque et al. *Anal. Biochem.* 178:1–7 (1989).

Calculation of protein extinction coefficients from amino acid sequence data. S. C. Gill and P. H. von Hippel. *Anal. Biochem.* 182:319–326 (1989).

Determination of tryptophan, tyrosine, and phenylalanine by second derivative spectrophotometry. Y. Nozaki. *Arch. Biochem. Biophys.* 277:324–333 (1990).

1.5 Determination of the Sizes of Proteins

Proteins vary tremendously in size. Small proteins may consist of only 50 amino acid residues whereas polypeptide chains with more than 25,000 residues have been found; the average is approximately 250. Some proteins consist of a single polypeptide chain; others exist as aggregates of many copies of the same polypeptide chain or of one or more copies of different polypeptide chains.

The size of a protein strongly influences what can be determined about it (small proteins are much easier to characterize than large ones) and the types of procedures that can be used to study them. The prospects for determining the amino acid sequence or the three-dimensional structure of a protein of 50–100 amino acid residues are much greater than for one consisting of 1000 or more residues. Similarly, determining the amino acid sequence of a protein with a molecular weight of 40,000,000 is much more feasible if it is an aggregate of 2130 copies of one polypeptide chain that consists of 158 amino acid residues (as for the tobacco mosaic virus) than if it is a single polypeptide chain consisting of 336,500 residues. Therefore, the number of polypeptide chains present in a protein and the number of residues present in each polypeptide chain are pertinent to the characterization of a protein.

Extremely large proteins can be visualized in the electron microscope, which gives an estimate of size, but this is unusual. Most procedures for determining the size of a protein or of a polypeptide involve determining its molecular weight. Molecular weight is more properly designated as *molecular mass,* which is frequently expressed in units of daltons (though purists argue about whether this is appropriate); but the more common term *molecular weight* (M_W) is used here as a dimensionless number. Dividing the molecular weight of a protein by the weight of an average amino acid residue (Table 1.1) gives the approximate number of residues present. The molecular weight of a protein is determined solely by its amino acid composition only if no substantial additions have been made to its polypeptide chain, which often occurs after its biosynthesis (Chap. 2). The modification most pertinent to molecular weight measurements is addition of carbohydrate chains to amino acid side chains; carbohydrate chains can be large and can contribute significantly to the molecular weight of a protein. Consequently, it is important to measure the masses of the polypeptide part and of the added components separately.

To determine the size of the polypeptide chains that make up a protein, it is simplest to place the protein in conditions that dissociate it into its individual polypeptides. Such conditions include a strong denaturant, such as 6 M guanidinium chloride (see Sec. 7.4.3.b) or a detergent (Sec. 1.5.3); it is usual also to reduce any disulfide bonds that might be linking the polypeptides (Sec. 1.3.10). The identities and the molecular weights of the individual polypeptide chains can then be determined. Alternatively, the number and identities of the N-terminal residues can be determined by amino acid sequence analysis (Sec. 1.6.2).

The two techniques used most frequently to determine the molecular weights of native proteins in solution, without dissociating them into individual polypeptide chains, are sedimentation analysis and gel filtration (Secs. 1.5.1 and 1.5.2). The molecular weights of individual polypeptide chains are usually determined by gel filtration in a strong denaturant or by sodium dodecyl sulfate (SDS) gel electrophoresis (Sec. 1.5.3). In addition, the molecular weights of pure polypeptide chains can be obtained very accurately by the new technique of electrospray ionization mass spectrometry (Sec. 1.6.2.d).

1.5.1 Sedimentation Analysis

The classical procedure for estimating the approximate size of a protein is by its sedimentation coefficient, determined either in the analytical ultracentrifuge or by sucrose gradient centrifugation. Under a given set of conditions, the rate of sedimentation of a protein depends not only on its molecular weight but also on its density, overall shape, and interactions with the solvent (see Sec. 7.1.2). A protein that sediments rapidly is likely to be large, but it might just be aggregating reversibly. A protein that migrates more slowly might be smaller, or it might have an unusual shape. Consequently, it is dangerous to estimate the molecular weight of a protein from its sedimentation coefficient alone.

The uncertainties about the shape and hydration of a protein can be overcome by using **sedimentation equilibrium.** At lower centrifugal force, the tendency of a protein molecule to sediment is balanced by its tendency to diffuse. Consequently, a concentration gradient is established at equilibrium. The shape and hydration factors affect sedimentation and diffusion in the same way, so they do not influence the equilibrium concentration gradient. The steepness of this gradient depends only on the molecular weight and density of the protein. The density of the protein can be measured or estimated from its amino acid composition (Sec. 8.1.2) and is usually expressed as the reciprocal, the partial specific volume $\bar{v}$. Knowing this parameter, one can determine the molecular weight of the protein from the equilibrium concentration gradient.

The equation relating the concentration gradient dc/dr to the molecular weight M_W and other parameters is

$$\frac{d \ln c}{d r^2} = \frac{M_W(1 - \bar{v}\rho)\omega^2}{2RT} \qquad (1.67)$$

where c is the protein concentration at various radii r from the axis of rotation, ω is the radial velocity of the rotor in radians per second, ρ is the density of the solvent, R is the gas constant, and T is the absolute temperature.

Sedimentation equilibrium is most frequently used to measure the molecular weights of native proteins, to determine how many polypeptide chains are present, and to study the interactions between them.

References

Protein volume in solution. A. A. Zamyatnin. *Prog. Biophys. Mol. Biol.* 24:109–123 (1972).

Characterization of proteins by sedimentation equilibrium in the analytical ultracentrifuge. D. C. Teller. *Methods Enzymol.* 27:346–441 (1973).

The calculation of partial specific volumes of proteins in guanidine hydrochloride. J. C. Lee and S. N. Timasheff. *Arch. Biochem. Biophys.* 165:268–273 (1974).

Macromolecular characterization by sedimentation equilibrium in the preparative ultracentrifuge. R. J. Pollet et al. *J. Biol. Chem.* 254:30–33 (1979).

The calculation of partial specific volumes of proteins in 8M urea solution. V. Prakash and S. N. Timasheff. *Anal. Biochem.* 117:330–335 (1981).

1.5.2 Gel Filtration

The size of a protein molecule determines the rate of its passage through a molecular sieve, so gel filtration is a convenient method for estimating this parameter. Molecular sieves consist of small particles of materials with a network of pores into which molecules of less than some maximum size can penetrate (Fig. 1.5). Protein molecules in solution partition between the volume of the liquid external to the sieve particles and any internal volume of the pores of the sieving material accessible to them. The smaller the protein, the greater the probability that it will enter the internal volume of the particles.

In gel filtration a sample of proteins is applied to a column of a molecular sieve, which is then washed with an appropriate buffer (Fig. 1.5). The first proteins to

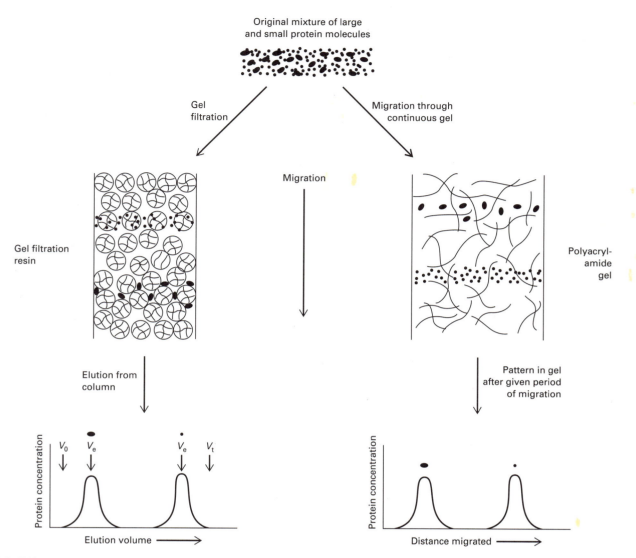

FIGURE 1.5

Schematic illustration of gel filtration and molecular sieving to determine the relative sizes of proteins. At the top is a mixture of two proteins, one large and the other small. When the mixture is subjected to gel filtration *(left)*, the large protein tends to be excluded from the pores of the beads of the gel filtration resin, so it moves through the column much faster than the small protein, which tends to enter the beads. The elution pattern at the bottom shows the larger protein eluting earlier than the smaller protein (see Fig. 1.6). Proteins that are totally excluded from the beads would elute at the void volume (V_0), whereas those that are free to enter all the pores of the beads would elute at the total volume (V_t). When migrating though a continous porous medium *(right)*, as in SDS electrophoresis through polyacrylamide gels, small molecules tend to move much faster through the pores than do larger molecules. If the molecules have the same shapes and charge densities, their rates of movement will be inversely proportional to their molecular weights (Fig. 1.7).

emerge from the column are those that are too large to enter the pores of the sieve. The volume of buffer required to elute them is known as the **void volume** (generally abbreviated as V_0) because it gives the volume of the column outside the particles. Other proteins are eluted in decreasing order of their molecular size. Molecules under some limiting size are not excluded from the molecular sieve particles and elute at a volume of buffer that represents the total buffer volume of the column (V_t). The particular volume at which a protein is observed to elute (V_e) is often used to define a distribution coefficient K_d:

$$K_d = \frac{V_e - V_0}{V_t - V_0}$$

(1.68)

K_d is essentially an equilibrium constant for partitioning between the interior pores and the external volume.

The relationship between K_d or V_e and the size of a protein molecule cannot be determined from first principles because there is no rigorous theory that accounts for gel filtration of protein molecules. Instead, a gel filtration column is calibrated by passing a set of proteins of known sizes through it; given this standard curve (Fig. 1.6), the elution volume on the column of some unknown protein gives an estimation of its size. The size of a protein measured by gel filtration is the average volume it occupies in solution, commonly expressed as the **Stokes radius**. This parameter is especially sensitive to the shape and physical properties of a protein, so gel filtration is subject to uncertainties similar to those of sedimentation velocity. The molecular weight of a protein can be estimated only if the column is calibrated with proteins that have the same shapes and physical properties. For example, the proteins of Figure 1.6A elute from the gel filtration column in the order expected from their molecular weights only because they are all globular proteins with similar shapes (see Chap. 6).

Molecular weight measurements by gel filtration are more certain with unfolded polypeptide chains, under denaturing conditions with all disulfide bonds reduced, when they all approximate random coils (see Chap. 5). Under these conditions, linear standard curves relating the molecular weight of a polypeptide chain to its elution volume or K_d can usually be obtained (Fig. 1.6B). The value of K_d is believed to be proportional to the effective hydrodynamic volume of the unfolded polypeptide chain; that is, to the cube of its average radius. The effective radius of an unfolded polypeptide chain is observed to be approximately proportional to $M_W^{0.555}$ (Sec. 5.2.2), so a plot of $K_d^{1/3}$ versus $M_W^{0.555}$ is linear for unfolded proteins.

1.5.3 SDS Polyacrylamide Gel Electrophoresis

The most popular method for estimating the size of a polypeptide chain is by its mobility in polyacrylamide gel electrophoresis (PAGE) in the presence of the detergent sodium dodecyl sulfate (SDS):

$$CH_3-(CH_2)_{11}-SO_4^- Na^+$$

(1.69)

This method is related to gel filtration in that the size of a protein is estimated by its migration through the small pores of a gel matrix (Fig. 1.5). In this case, however, the gel matrix is continuous rather than particulate, so smaller proteins migrate more rapidly than large proteins. The relationship between mobility and size is determined not by first principles but by relating the observed mobility to that of standard proteins of known size.

The electrophoretic mobility of a protein is determined not only by its molecular weight but also by its net charge and its shape. The last two parameters are standardized for all proteins by SDS electrophoresis; SDS binds to proteins, disrupts their structure and shape, dissociates them into polypeptide chains, and imposes comparable shapes and net charge densities on these chains. The physical basis for these effects of SDS on proteins is not known, in spite of its importance, and the approach is based entirely on the empirical observation that it usually works.

When SDS is present at a free monomer concentration of at least 8×10^{-4} M, it unfolds most proteins and binds to them at a constant 1.4 g per gram of protein; this is an average of one molecule of SDS for every two amino acid residues. The binding of SDS to polypeptides seems to depend on ionic interactions because nonionized polypeptides are not affected, even though nonpolar interactions are probably also important. This high level of binding of the charged detergent swamps the intrinsic charge of most proteins. The constant binding ratio results in different protein–SDS complexes that have virtually the same constant density of negative charge along the polypeptide chain. Only proteins that are intrinsically very basic or very acidic seem to produce complexes with SDS that have somewhat atypical charge densities, reflecting their intrinsic charges.

Proteins that interact in the typical manner with SDS also appear to produce complexes with similar shapes. The original conformations of the protein are disrupted by the detergent, but the nature of the conformation they adopt in SDS is not known. SDS-unfolded polypeptide chains appear to be somewhat elongated and flexible particles with a constant diameter and a length proportional to the number of amino acid residues in the polypeptide chain.

As a consequence of these properties, protein–SDS complexes are observed to have electrophoretic mobilities through polyacrylamide gels that are inversely proportional to the logarithm of the length of the polypeptide chain (Fig. 1.7). Comparison of the mobility of an unknown protein with that of a set of standard marker proteins usually makes it possible to determine the molecular weight of the unknown polypeptide chain to within 10% of the true value. Incorrect values result when there are abnormalities in SDS binding or protein conformation, large differences in intrinsic protein

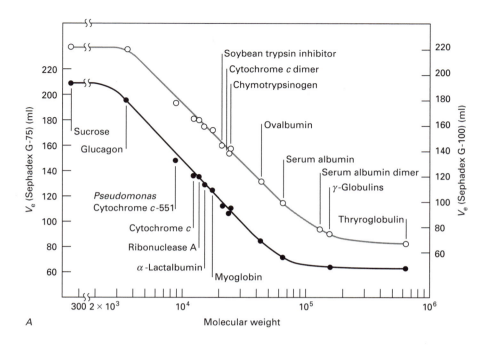

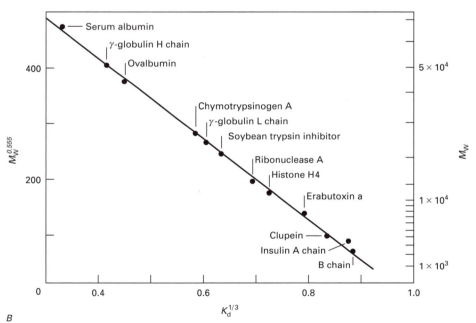

FIGURE 1.6

Gel filtration to determine the sizes of native globular proteins *(A)* and unfolded polypeptide chains *(B)*. A variety of proteins of known molecular weights were subjected to gel filtration under native conditions *(A)* on Sephadex G75 (solid circle) or G100 (open circle) or in 6 *M* guanidinium chloride under reducing conditions in which all disulfide bonds are broken *(B)*.

The data in *A* are plotted as the elution volume (V_e) versus the logarithm of the molecular weight of the protein or other small molecules. The plateaus at the left and right give the total (V_t) and void (V_0) volumes of the column, respectively. The observed correlation of elution volume with normal molecular weight of the protein is observed in this case because all of the proteins are globular, with similar three-dimensional shapes overall. (From P. Andrews, *Biochem J.* 91:222–233, 1964.)

The elution volumes of the unfolded polypeptides in *B* are expressed as distribution coefficients, K_d (Eq. 1.68). The value of $K_d^{1/3}$ is linearly related to the 0.555 power of the molecular weight M_w. (From N. Ui, *Anal. Biochem.* 97:65–71, 1979.)

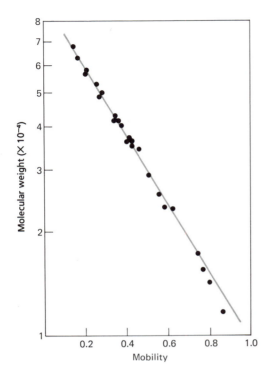

FIGURE 1.7
Correlation between mobility in SDS polyacrylamide gel electrophoresis and molecular weight of the polypeptide chain. Proteins of known molecular weight within the range 14,000–70,000 were used. (From K. Weber and M. Osborne, in *The Proteins*, 3rd ed., H. Neurath and R. L. Hill (eds.), vol. 1, pp. 179–223, Academic Press, New York, 1975.)

charge, or covalently attached nonprotein moieties, especially carbohydrates. Caution is advisable, therefore, in the use of this technique.

Nevertheless, the remarkable resolution and ease of SDS electrophoresis has made it one of the most widely used techniques, not only for determining molecular weights of proteins but also for studying biological processes in which the apparent molecular weights of proteins are altered. The remarkable resolving power of the technique makes it possible to analyze heterogeneous mixtures of proteins, including entire cell extracts. It is only necessary to be able to identify the band corresponding to the protein of interest. Even proteins that are present in extremely small amounts can be identified by using blotting techniques, such as Western blotting (Sec. 8.3.1.b). Of course, SDS electrophoresis gives information only about one aspect of a polypeptide chain, its length, because the SDS disrupts and masks its other properties.

References

Molecular weight estimation of polypeptide chains by electrophoresis in SDS-polyacrylamide gels. A. L. Shapiro et al. *Biochem. Biophys. Res. Commun.* 28:815–820 (1967).

Molecular weight determinations and the influence of gel density, protein charges, and protein shape in polyacrylamide gel electrophoresis. D. B. Blattler and F. J. Reithel. *J. Chromatogr.* 46:286–292 (1970).

Molecular characterization of proteins in detergent solutions. C. Tanford et al. *Biochemistry* 13:2369–2376 (1974).

Proteins and sodium dodecyl sulfate: molecular weight determination on polyacrylamide gels and related procedures. K. Weber and M. Osborne. In *The Proteins*, 3rd ed., H. Neurath and R. L. Hill, eds., vol. 1, pp. 179–223. Academic Press, New York, 1975.

Conformational properties of the complexes formed by proteins and sodium dodecyl sulfate. W. Mattice et al. *Biochemistry* 15:4264–4272 (1976).

Behavior of glycopolypeptides with empirical molecular weight estimation methods. 1. In sodium dodecyl sulfate. B. S. Leach et al. *Biochemistry* 19:5734–5741 (1980).

1.6 Determination of the Covalent Structures of Proteins

A typical protein of between 100 and 1000 amino acid residues consists of 1500–15,000 atoms, making it orders of magnitude more complex than a typical molecule whose structure might be determined by a chemist. To determine the covalent structure of a protein directly would be an impossible task were it not known that proteins are linear polymers of 20 amino acids of known structure, plus some variations produced by covalent modification after biosynthesis of the polypeptide chain (Sec. 2.4). The covalent structure is then defined simply by the sequence of the amino acid residues in the one-dimensional polypeptide chain and by the known structures of the polypeptide backbone and of the amino acid side chains.

1.6.1 Amino Acid Composition

The amino acid composition of a protein is routinely determined by completely hydrolyzing the peptide bonds of the polypeptide chain, then determining quantitatively the constituent amino acids that were released. The traditional method of hydrolyzing the polypeptide chain has been to incubate it anaerobically in 6 M HCl at approximately 110°C for 24–72 h, but recent methods use other acids, higher temperatures, and shorter periods of time. Most peptide bonds hydrolyze at similar rates, but the peptide bonds between the large nonpolar amino acid residues, particularly Val, Leu, and Ile, require longer hydrolysis times or the addition of organic acids such as trifluoroacetic acid. Hydrolysis is presumably hindered sterically by the bulky side chains.

Trp residues are usually destroyed completely by acid hydrolysis, probably as a result of reaction with chlorine produced by oxidation of the HCl. They can be protected, however, by the addition of thiol or sulfonic acid compounds, or of phenol to scavenge the chlorine. Alternatively, alkaline hydrolysis can be used to hydrolyze the polypeptide chain. Tyr residues also are susceptible to chlorination but are usually lost only partially. Cysteine is oxidized and partially destroyed by acid hydrolysis; it is best analyzed after performic acid oxidation of the protein to convert all the Cys residues to cysteic acid, usually abbreviated as Cya.

The amide side chains of Asn and Gln residues are quantitatively hydrolyzed to produce the amino acids aspartic acid and glutamic acid, respectively, by any chemical procedure used to hydrolyze the peptide bonds of the backbone. The number of Asn and Gln residues can be measured by the amount of ammonia released during hydrolysis, but otherwise it is not possible to distinguish between Asn and Asp and between Gln and Glu residues after hydrolysis of the polypeptide chain.

Some of the problems with acid hydrolysis can be overcome by using other procedures such as hydrolysis by alkali or by proteolytic enzymes (Sec. 1.6.2.b). Other amino acids, notably Ser and Thr, are destroyed by alkali, however, and total protease digestion is not straightforward. For most purposes, acid hydrolysis is adequate.

The identities and quantities of the various amino acids present in hydrolyzates are determined by automated amino acid analyzers. The amino acids are separated chromatographically and quantified as they emerge from the column. Traditional methods used ion-exchange chromatography, followed by detection with ninhydrin or fluorescent reagents (Sec. 1.4). Proline does not react in the usual manner with such reagents due to the absence of an amino group, so special procedures are required to measure it. Recent procedures are more rapid and sensitive, however, and involve reacting the amino acids with suitable reagents before the chromatographic separation rather than after. The favored method at present is to react the amino acids with phenylisothiocyanate (PITC, see Sec. 1.6.2.b) and then to separate the colored derivatives by reverse phase chromatography (Fig. 1.8). With this procedure, complete quantitative amino acid analyses can be carried out in 12 min with only picomole (10^{-12} mol) quantities of amino acids.

The numbers of aromatic residues (Phe, Tyr, and Trp) in proteins and peptides can usually be determined from the UV absorbance spectrum under conditions in which the polypeptide chain is fully unfolded so that its spectrum is the sum of its constituent residues.

Amino acid analysis does not directly give the number of residues of each amino acid per polypeptide chain. The most accurate result is the molar ratios of amino acids. The true molecular weight of the polypeptide chain must be known for the amino acid analysis to be converted to the number of residues of each amino acid per chain. Of course, for a homogeneous polypeptide this number should be an integer, but this is rare because of the lack of precision in the determination of both the amino acid composition and the molecular weight.

A totally different approach can give integral values for the number of residues of a few of the 20 amino acids, independent of any other property of the polypeptide chain, including its molecular weight. To determine the integral value of the number of residues n of a particular amino acid, the side chains of all its residues are modified in a specific way that alters some electrophoretic or chromatographic property of the polypeptide chain; modifications that change the charge on the side chain are particularly useful. Varying extents of the modification are carried out on the unfolded polypeptide chain; thus, a complete spectrum of molecules with $0, 1, 2, \ldots, n-1, n$ residues modified is generated and is separated by a procedure that is sensitive only to the number of modifications. The number of species is one more than the number of residues.

The procedure has been used for counting the integral number of Cys and Lys residues in proteins (Fig. 1.9). Here the number n of Cys residues was determined from the $n+1$ electrophoretic bands generated by using competition between the neutral and acidic thiol reagents, iodoacetamide and iodoacetate, respectively (Eq. 1.47). Amino groups can be counted by gradually modifying them with succinic anhydride (Eq. 1.22) to convert them from basic to acidic groups.

After the number of residues of one or more amino acids has been determined in this way, the relative amounts of all the amino acids determined by standard amino acid analysis can be converted directly to numbers of residues per polypeptide chain. These values are not necessarily integers, due to experimental errors, but they may be sufficiently close. The molecular weight of the polypeptide chain can then be calculated directly from the amino acid composition.

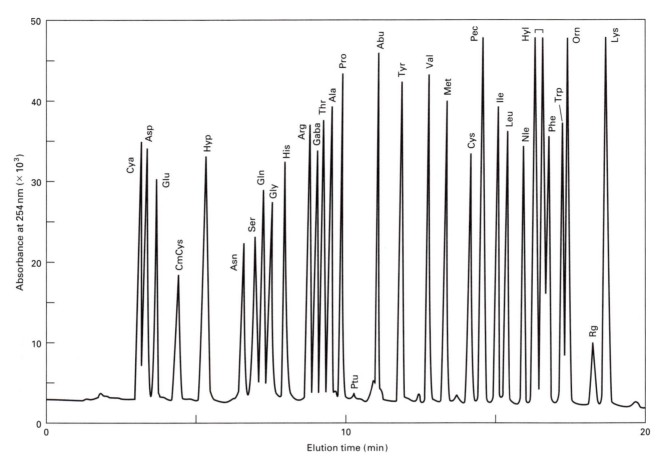

FIGURE 1.8

Chromatographic separation of amino acids after derivatization with phenylisothiocyanate. A sample containing 250 pmol of each amino acid was separated on a 3.9 × 300-mm Picotag reverse phase column. Elution was with a convex gradient from eluent A (0.14 M sodium acetate, 0.05% triethylamine, 6% acetonitrile, pH 6.4) to 54% A, 46% eluent B (60% acetonitrile in water).

The standard three-letter abbreviations are used for the usual amino acids, except that in this case Cys was the disulfide-bonded cystine. The nonstandard amino acids are CmCys, carboxymethyl cysteine; Hyp, hydroxyproline; Gaba, γ-amino butyric acid; Ptu, phenylthiourea (generated from ammonia by derivatization reaction); Abu, α-amino butyric acid; Nle, norleucine; Hyl, hydroxylsine; Orn, ornithine; Rg, contaminant from reagents. (From S. A. Cohen and D. J. Strydom, *Anal. Biochem.* 174:1–16, 1988.)

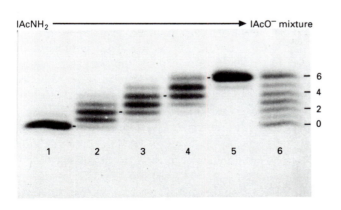

FIGURE 1.9

Counting integral numbers of Cys residues. Reduced bovine pancreatic trypsin inhibitor (BPTI) was reacted completely with iodoacetamide (IAcNH₂) in lane 1, with iodoacetate (IAcO⁻) in lane 5, and with varying ratios of the two reagents in lanes 2–4. A mixture of all these samples is in lane 6. The protein was subjected to electrophoresis under conditions in which its mobility is determined only by its net charge, which varies here according to the number of acidic carboxymethyl groups introduced by reaction of the Cys thiols with IAcO⁻. Varying extents of reaction with IAcO⁻ generated six new electrophoretic bands, the number of Cys residues present in the protein. (From T. E. Creighton, *Nature* 284:487–489, 1980.)

References

Hydrolysis of proteins. R. L. Hill. *Adv. Protein Chem.* 20:37–107 (1965).

Spectroscopic determination of tryptophan and tyrosine in proteins. H. Edelhoch. *Biochemistry* 6:1948–1954 (1967).

Counting integral numbers of amino acid residues per polypeptide chain. T. E. Creighton. *Nature* 284:487–489 (1980).

Counting integral numbers of amino groups per polypeptide chain. M. Hollecker and T. E. Creighton. *FEBS Letters* 119:187–189 (1980).

Hydrochloric acid hydrolysis of proteins and determination of tryptophan by reversed-phase high-performance liquid chromatography. L. T. Ng et al. *Anal. Biochem.* 167:47–52 (1987).

Amino acid analysis utilizing phenylisothiocyanate derivatives. S. A. Cohen and D. J. Strydom. *Anal. Biochem.* 174:1–16 (1988).

1.6.2 The Amino Acid Sequence

The amino acid sequence, appropriately called the **primary structure,** identifies a protein unambiguously, determines all its chemical and biological properties, and specifies (indirectly) the higher levels of protein structure (secondary, tertiary, and quaternary), which will be encountered in later chapters. The most basic step in characterizing a protein, therefore, is to determine its amino acid sequence.

The procedures described here are capable, in various combinations, of determining the entire amino acid sequences of most proteins. This is a substantial undertaking, however, and at the present time it is easier to determine the sequences of large, insoluble, or rare proteins from their gene sequences (Sec. 2.2.2). Even so, protein sequencing techniques are still required to determine the sequence of part of a protein, which is then used to design an oligonucleotide that is complementary to the part of the gene sequence that codes for that part of the protein. The gene for a protein is usually identified and cloned by its hybridization to such an oligonucleotide probe. Protein sequencing techniques are also necessary to determine what modifications occur to a polypeptide chain after it is generated from its gene sequence (Sec. 2.4).

References

The arrangement of amino acids in proteins. F. Sanger. *Adv. Protein Chem.* 7:1–67 (1952).

Recent developments in chemical modification and sequential degradation of proteins. G. R. Stark. *Adv. Protein Chem.* 24:261–308 (1970).

Strategy and methods of sequence analysis. W. H. Konigsberg and H. H. Steinmann. In *The Proteins,* 3rd ed., H. Neurath and R. L. Hill, eds., vol. 3, pp. 1–178. Academic Press, New York, 1977.

Advances in protein sequencing. K. A. Walsh et al. *Ann. Rev. Biochem.* 50:261–284 (1981).

Contemporary methodology for protein structure determination. M. W. Hunkapiller et al. *Science* 226:304–311 (1984).

Protein Sequencing: A Practical Approach. J. B. C. Findlay and M. J. Geisow, eds. IRL Press, Oxford (1989).

a. Amino-Terminal and Carboxyl-Terminal Residues

The usual methods for identifying N- and C-terminal residues require the terminal amino or carboxyl group to have normal chemical properties and to not be blocked covalently. In other cases, special techniques must be used.

Several chemical methods are available for identifying the N-terminal residue of a protein or peptide. They involve labeling chemically all the amino groups, hydrolyzing the peptide to its constituent amino acids, then determining which amino acid is labeled on the α-amino group. The two most widely used procedures employ **fluoro-2,4-dinitrobenzene** and **dansyl chloride:**

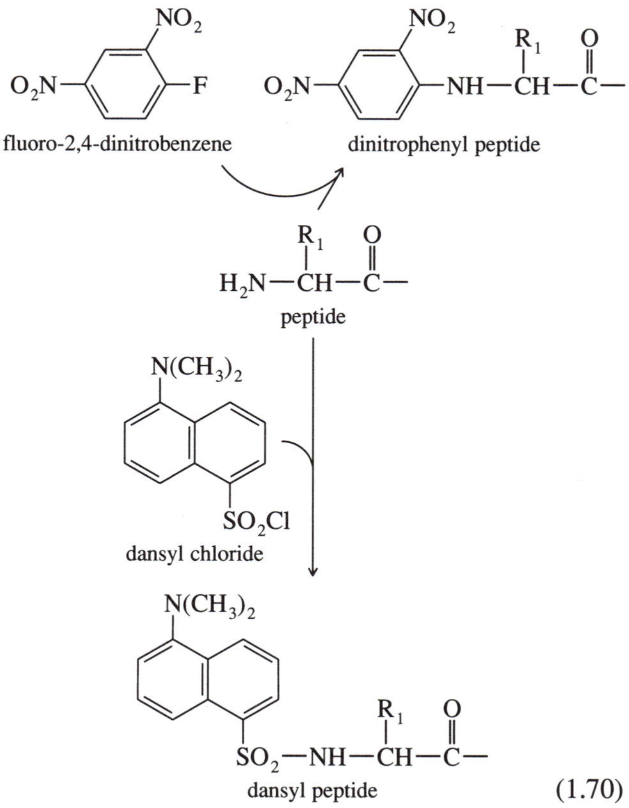

fluoro-2,4-dinitrobenzene   dinitrophenyl peptide

peptide

dansyl chloride

dansyl peptide

(1.70)

The linkages between these reagents and the α-amino group withstand hydrolysis of the peptide to the amino acids. The labeled amino acids are detected by their yellow color in the fluorodinitrobenzene procedure and by fluorescence when the dansyl derivative is generated. They can be identified by chromatographic techniques in which each amino acid derivative has a characteristic mobility.

A related procedure involves reaction of the α-amino group with **cyanate**. Acid hydrolysis converts the modified N-terminal residue to the hydantoin:

$$
\begin{array}{c}
\text{R}^1 \quad \text{O} \qquad \text{R}^2 \quad \text{O} \\
| \qquad \parallel \qquad | \qquad \parallel \\
\text{H}_2\text{N}-\text{CH}-\text{C}-\text{NH}-\text{CH}-\text{C}-
\end{array}
$$

$\downarrow$ HN=C=O

$$
\begin{array}{c}
\text{O} \qquad \text{R}^1 \quad \text{O} \qquad \text{R}^2 \quad \text{O} \\
\parallel \qquad | \qquad \parallel \qquad | \qquad \parallel \\
\text{H}_2\text{N}-\text{C}-\text{NH}-\text{CH}-\text{C}-\text{NH}-\text{CH}-\text{C}-
\end{array}
$$

$\downarrow$ acid hydrolysis

hydantoin

$$+ \; \text{H}_2\text{N}-\overset{\text{R}^2}{\underset{|}{\text{CH}}}-\text{CO}_2\text{H} + \cdots \qquad (1.71)$$

The only amino acid present as the hydantoin is the one that was at the N-terminus.

Procedures for identifying the C-terminal amino acid are not as satisfactory nor as widely used. The most frequently used is **hydrazinolysis**. All the α-carboxyl groups involved in peptide bonds are converted into hydrazides:

$$
\begin{array}{c}
\text{R}^1 \quad \text{O} \\
| \qquad \parallel \\
\text{H}_2\text{N}-\text{CH}-\text{C}- \cdots
\end{array}
$$

$$
\begin{array}{c}
\text{R}^n \\
| \\
-\text{NH}-\text{CH}-\text{CO}_2\text{H}
\end{array}
$$

$\downarrow$ H$_2$N—NH$_2$

$$
\begin{array}{c}
\text{R}^1 \quad \text{O} \\
| \qquad \parallel \\
\text{H}_2\text{N}-\text{CH}-\text{C}-\text{NHNH}_2 + \cdots
\end{array}
$$

$$+ \; \text{H}_2\text{N}-\overset{\text{R}^n}{\underset{|}{\text{CH}}}-\text{CO}_2\text{H} \qquad (1.72)$$

Only the C-terminal residue is released as the unmodified amino acid with a free α-carboxyl group, which permits its identification. The procedure has a number of shortcomings, however: It is technically difficult to free the amino acid from the multitude of hydrazides generated; C-terminal Asn and Gln residues are not recovered as the free amino acids because their side-chain amides are converted to hydrazides; and Arg residues are converted to ornithine (see Eq. 1.30).

The C-terminal residue, which has a unique ability (unless it is Pro) to form an oxazolone on treatment with acetic anhydride (Ac$_2$O), can be labeled selectively with either deuterium or tritium by hydrolyzing it in the presence of ^{2}H$_2$O or ^{3}H$_2$O to incorporate the label:

$$
\begin{array}{c}
\text{R}^1 \quad \text{O} \\
| \qquad \parallel \\
\text{H}_2\text{N}-\text{CH}-\text{C}-\text{NH} \cdots
\end{array}
$$

$$
\begin{array}{c}
\text{O} \qquad \text{R}^{n-1} \quad \text{O} \qquad \text{R}^n \\
\parallel \qquad | \qquad \parallel \qquad | \\
\text{C}-\text{NH}-\text{CH}-\text{C}-\text{NH}-\text{CH}-\text{CO}_2\text{H}
\end{array}
$$

$\downarrow$ Ac$_2$O

$$
\begin{array}{c}
\text{R}^1 \quad \text{O} \\
| \qquad \parallel \\
\text{AcNH}-\text{CH}-\text{C}-\text{NH} \cdots
\end{array}
$$

$\downarrow$ base

$\downarrow$ ^{3}**H**$_2$O (or ^{2}H$_2$O)

$$
\begin{array}{c}
\text{R}^1 \quad \text{O} \\
| \qquad \parallel \\
\text{AcNH}-\text{CH}-\text{C}-\text{NH} \cdots
\end{array}
$$

$$
\begin{array}{c}
\text{O} \qquad \text{R}^{n-1} \quad \text{O} \qquad \text{R}^n \\
\parallel \qquad | \qquad \parallel \qquad | \\
\text{C}-\text{NH}-\text{CH}-\text{C}-\text{NH}-\text{C}-\text{CO}_2\text{H} \\
| \\
^3\text{H}(^2\text{H})
\end{array}
$$

$\downarrow$ hydrolysis

$$
\begin{array}{c}
\text{R}^1 \\
| \\
\text{H}_2\text{N}-\text{CH}-\text{CO}_2\text{H} + \cdots
\end{array}
$$

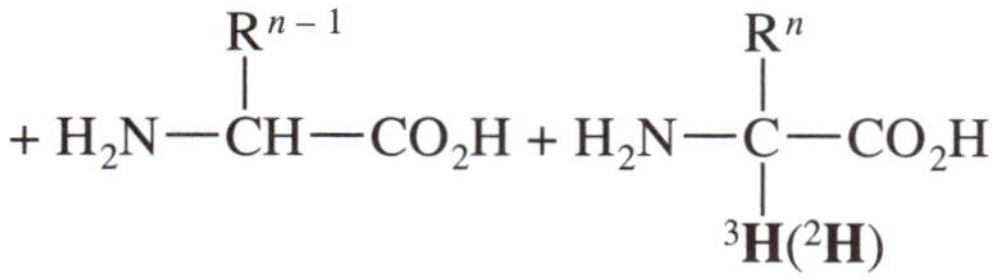

$$+ H_2N - \overset{\overset{\displaystyle R^{n-1}}{|}}{CH} - CO_2H + H_2N - \overset{\overset{\displaystyle R^n}{|}}{\underset{\underset{\displaystyle {}^3H({}^2H)}{|}}{C}} - CO_2H \qquad (1.73)$$

The labeled residue is identified by amino acid analysis after acid hydrolysis.

In an alternative procedure, the protein is hydrolyzed in $H_2^{18}O$; the C-terminal residue is identified by mass spectrometry as the only residue that does not incorporate ^{18}O into its carboxyl group. Another method, involving the release of the C-terminal residue as the thiohydantoin, is described in Section 1.6.2.c.

Short sequences of residues at either the amino or carboxyl ends of peptides can often be determined by using exopeptidases such as **leucine aminopeptidase** and **carboxypeptidase,** which are proteolytic enzymes that sequentially liberate single amino acids from the ends of chains. The amino acids liberated are determined by conventional amino acid analysis. The procedure requires that the kinetics of amino acid release be measured because the second residue can be released only after the first has been removed, and so on. It relies on the residues being removed sequentially with different time courses. Luck is required because the nature of the terminal residue determines the rate at which it is removed. For example, if the second residue is cleaved by the exopeptidase much more rapidly than the first, the two will be observed to be released simultaneously at the rate of the first. In this case, it is impossible to determine the order of release of the two amino acids.

References

A novel method for determination of C-terminal amino acid in polypeptides by selective tritium labeling. H. Matsuo et al. *Biochem. Biophys. Res. Commun.* 22:69–74 (1966).

End group analysis using dansyl chloride. W. R. Gray. *Methods Enzymol.* 25:121–138 (1972).

Use of cyanate for determining the NH_2-terminal residues in proteins. G. R. Stark. *Methods Enzymol.* 25:103–120 (1972).

Hydrazinolysis. W. A. Schroeder. *Methods Enzymol.* 25:138–143 (1972).

Enzymatic hydrolysis with carboxypeptidases. R. P. Ambler. *Methods Enzymol.* 25:143–154 (1972).

A micromethod for the determination of carboxyl-terminal amino acids or peptides and proteins: phenylisothiocyanate reaction with hydrazinolysates. D. J. Strydom. *Anal. Biochem.* 174:679–686 (1988).

Sequencing of peptides and proteins with blocked N-terminal amino acids: N-acetylserine or N-acetylthreonine. D.

Wellner et al. *Proc. Natl. Acad. Sci. USA* 87:1947–1949 (1990).

b. Sequencing from the N-Terminus: The Edman Degradation

One of the most successful methods in protein chemistry is the **Edman degradation** procedure for determining the sequence of amino acids from the N-terminus. It removes and identifies one amino acid at a time from the amino end and can be repeated many times to identify a sequence of amino acids (Fig. 1.10).

This procedure requires that the terminal α-amino group be free so that it can be reacted with phenyliso-

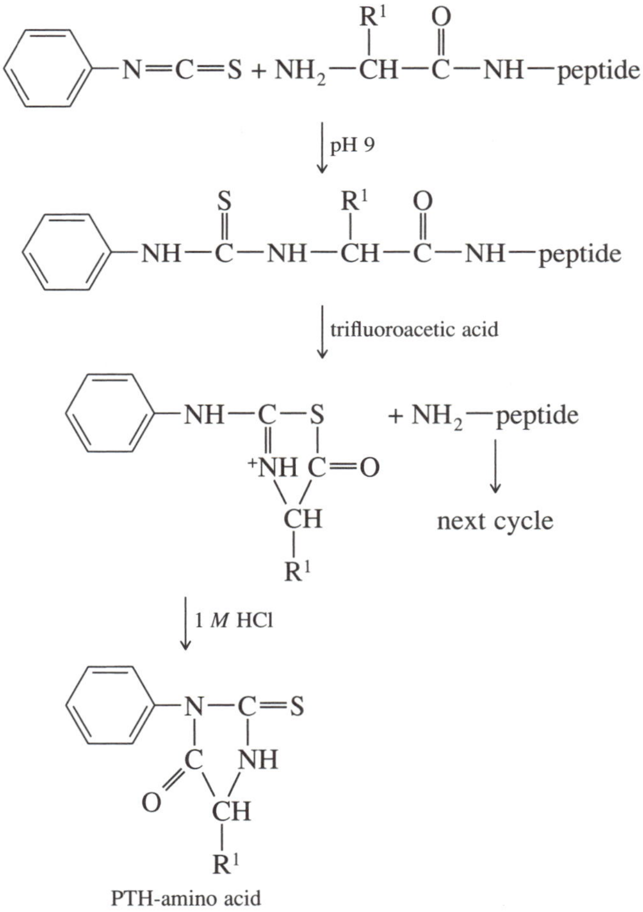

FIGURE 1.10

One cycle of the Edman degradation procedure for sequencing peptides from the N-terminus. The individual reactions are explained in the text.

thiocyanate in alkaline medium to give the phenylthio-carbamyl (PTC) derivative:

$$(1.74)$$

The PTC-peptide is then treated with a strong acid, such as trifluoroacetic acid. The first peptide bond, made relatively unstable by the PTC group, is hydrolyzed under conditions in which the other peptide bonds are stable:

$$(1.75)$$

The liberated cyclic derivative of the N-terminal residue rearranges in aqueous solution to the phenylthiohydantoin (PTH) derivative:

$$(1.76)$$

The original N-terminal residue is identified as the amino acid PTH derivative by chromatographic techniques.

The beauty and great virtue of the method lie in the regeneration of the polypeptide chain, altered only by the loss of the N-terminal residue. This shortened peptide is ready for a second cycle of the same procedure, to identify the second residue in the original sequence. This procedure is repeated sequentially to determine the amino acid sequence.

The primary difficulty with the Edman procedure, as with all repetitive procedures, is that errors are cumulative. If a significant amount of unreacted N-terminal residue remains after any cycle, it will appear in the next cycle, and this error will be propagated in all subsequent cycles. If the same error is repeated in later cycles, the amount and diversity of the background material rise until the residue at the cycle being processed cannot be identified unambiguously. Random breakage of the polypeptide chain under the harsh conditions used also contributes to the background. Side reactions increasingly prevent polypeptide molecules from participating in the reaction, and some polypeptide is dissolved and lost in the organic solvents used for extraction. The latter problem is alleviated in automated instruments by attaching the polypeptide chain covalently to a solid support. There is no ideal method of making the attachment, but proteins are often isolated by blotting them electrophoretically onto membranes, (Sec. 1.5.3); firmly attached to such membranes, they can then be directly subjected to sequence analysis. In the latest gas-phase sequenators, the sample is embedded in a film on the surface of a thin disk. Some of the reagents are delivered as vapors, which minimizes loss of the polypeptide and contamination of the sample. As little as $5-10$ picomoles (10^{-12} mol) of protein are required for analysis in this way. Efficiencies as great as 98% can be obtained for an average cycle, permitting the sequencing of up to 70 residues.

References

A protein sequenator. P. Edman and G. Begg. *Eur. J. Biochem.* 1:80–91 (1967).

Automated Edman degradation: the protein sequenator. H. D. Niall. *Methods Enzymol.* 27:942–1010 (1973).

c. Sequencing from the C-Terminus

Although no method comparable to the Edman degradation has yet been as successful for sequencing from the C-terminus, the **thiocyanate** procedure offers the same possibilities, at least in principle. Acetic anhydride is used to activate the carboxyl group by forming a mixed anhydride. This is then displaced by thiocyanate, to form the corresponding mixed anhydride, which cyclizes to the peptide hydantoin:

$$\underset{O}{\overset{O}{\parallel}} - \overset{}{C} - NH - \underset{|}{\overset{R^n}{\underset{|}{CH}}} - \overset{O}{\overset{\parallel}{C}}OH$$

↓ acetic anhydride

$$-\overset{O}{\overset{\parallel}{C}} - NH - \underset{|}{\overset{R^n}{CH}} - \overset{O}{\overset{\parallel}{C}} - O - \overset{O}{\overset{\parallel}{C}} - CH_3$$

↓ thiocyanate

$$-\overset{O}{\overset{\parallel}{C}} - N \underset{S}{\overset{R^n}{\diagdown}} \overset{O}{\underset{NH}{}} \tag{1.77}$$

Acid or base hydrolysis releases the C-terminal amino acid as the thiohydantoin. The chemistry is similar to that in the cyanate method of determining the N-terminal residue (Eq. 1.71).

Most importantly, the procedure leaves the polypeptide chain shortened by one residue, and the chain can then be subjected to another round of reactions. In spite of its attractions, the procedure suffers from the severity of the conditions required for the complete derivatization of the C-terminal amino acid and for the subsequent release of the C-terminal thiohydantoin. Also, some C-terminal residues are resistant to the procedure. Consequently, the method has not yet been refined to be as sensitive and routine as the Edman degradation.

References

Stepwise sequence determination from the carboxyl terminus of peptides. J. L. Meuth et al. *Biochemistry* 21:3750–3757 (1982).

Carboxy-terminal sequencing: formation and hydrolysis of C-terminal peptidylthiohydantoins. J. M. Bailey and J. E. Shively. *Biochemistry* 29:3145–3156 (1990).

d. Mass Spectrometry

In recent years, mass spectrometry (MS) has become a powerful technique for determining the primary structures of proteins. It was not originally obvious, however, that mass spectrometry could be useful for this purpose. First, the method involves evaporation of the sample into a vacuum, and proteins are not detectably volatile. Second, the sample is fragmented nonspecifi-cally with an electron beam into an assortment of pieces, and a vast number of assorted protein fragments are possible. Third, only the fragments with a net positive charge are detected. Fourth, the fragments are separated solely on the basis of their mass-to-charge ratio, so it is impossible to distinguish Leu and Ile residues and Lys and Gln residues, which have very similar masses (Table 1.1).

Nevertheless, the technique has the great advantages of sensitivity and accuracy. Only picomoles of sample are required, and samples need not be homogeneous. Mass-to-charge ratios can be measured to an accuracy of 1 part in 10,000, up to approximately 15,000 daltons/charge (the presence of various natural isotopes of H, C, N, and O atoms becomes apparent with this technique). The procedure is not hindered by the presence of blocking groups on the terminal amino and carboxyl groups, as are the sequencing techniques just described, and mass spectrometry can identify even the blocking groups. It also can identify atypical amino acid residues and the various posttranslational modifications that often occur in proteins (Sec. 2.4). Consequently, it has been developed into a very useful technique for directly sequencing peptides of up to 15 residues, for identifying the residues released by the sequencing techniques described earlier, and for checking the covalent structures deduced by other methods.

To make peptides sufficiently volatile for standard mass spectrometry, the polar groups of both the backbone and the side chains are modified chemically to abolish their polarity. In the most widely used technique, Arg residues must have their guanido groups removed by hydrazinolysis (Eq. 1.30), and all amino groups are acetylated with acetic anhydride (Eq. 1.22). Then all carboxyl, hydroxyl, and NH groups are permethylated by treating with a strong base and with methyl iodide to abstract the hydrogen atoms as protons and to replace them with methyl groups.

The fragmentation patterns of a peptide induced by an electron beam are potentially very complex, but the modifications introduced tend to cause the peptide backbone to fragment primarily at peptide bonds (Fig. 1.11). Other fragmentations also occur, and the side chains may be cleaved by the electron beam. Fortunately, the typical fragmentation patterns of each type of residue are known, so all the ion fragments produced from a peptide can be used to determine its sequence.

Peptides need not be pure for mass spectrometry. Mixtures of up to five peptides can be effectively subjected to fractional distillation in the spectrometer by gradually increasing the temperature of the sample. The signals in the mass spectrum that are due to any one peptide rise and fall together, so the complex spectra can be resolved into the spectra of the individual com-

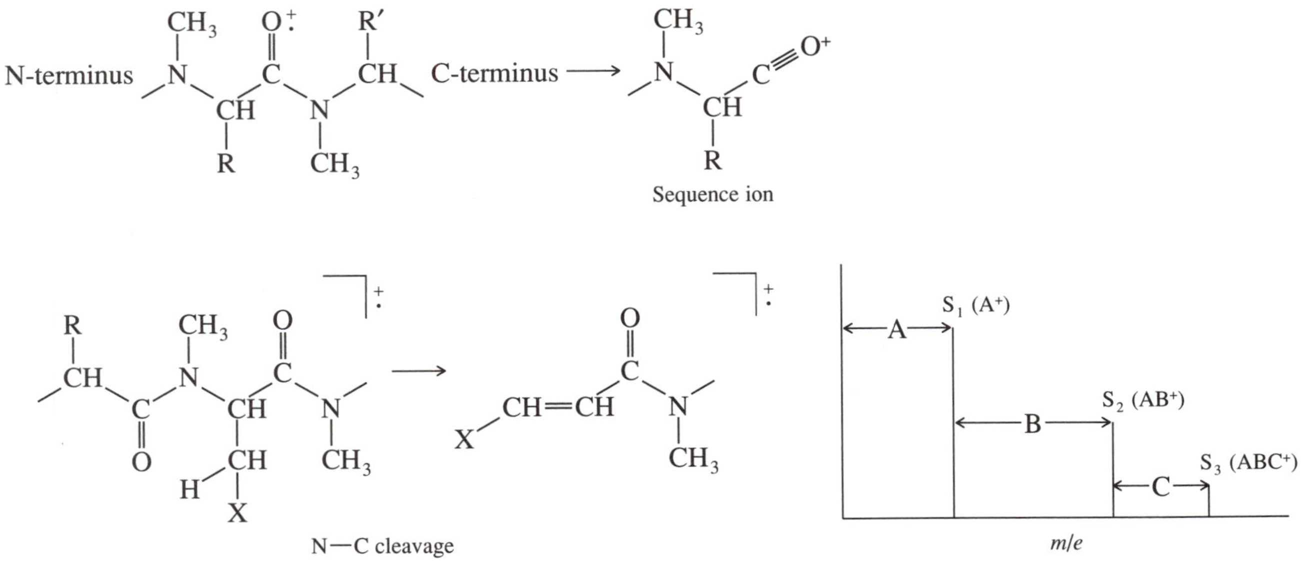

FIGURE 1.11

Fragmentation of a volatile modified peptide in mass spectrometry. At *top* is shown the usual fragmentation of the peptide bond to generate the positively charged "sequence ion" that is detected. However, when the side chain R is that of Phe, Tyr, His, Trp, Asn, or Asp, N—C cleavage occurs instead, as on *lower left*. On *lower right* are shown the three singly charged sequence ions with increasing ratio of mass to charge, *m/e*, that would be expected in the mass spectrum with a peptide A—B—C. The three peaks originate from cleavage at each of the peptide bonds.

ponents. Peptide purification is usually the most difficult step in protein sequencing, so this property of mass spectrometry makes it extremely useful.

The latest techniques of mass spectrometry do not require chemical modification of peptides because they need not be volatile. A sample is placed in the spectrometer in a glycerol matrix or adsorbed to a solid support. It is then ionized by a high-energy beam of atoms or ions. Depending on the method of ionization, such techniques are known as *fast atom bombardment (FAB-MS), plasma desorption (PDMS),* and *field desorption (FDMS) mass spectrometry.* Fragmentation of the sample does not usually take place with these techniques. The original peptide or protein is detected as the ion that is produced, essentially by adding a proton; such ions are known as the $(M + H)^+$ species. If sodium or potassium salts are present in the sample, the ions $(M + Na)^+$ and $(M + K)^+$ are also observed at 22 and 38 mass units greater, respectively, than the $(M + H)^+$ ion; these help to confirm the assignment of the $(M + H)^+$ peak. Molecular weights of peptides and small proteins up to 10,000 can be determined accurately in this way.

The mass of a peptide is often sufficient to identify its composition, but not its sequence, so the techniques of mass spectrometry are used in conjunction with the techniques for sequencing from the ends of the polypeptide chain or with fragmentation of the protein into various smaller peptides.

The problem of involatility of a protein or peptide is overcome by the recent introduction of **electrospray ionization mass spectrometry,** in which a solution of protein molecules in a volatile solvent is sprayed into the mass spectrometer. The solvent in the droplets evaporates rapidly in the vacuum, leaving the protein molecules suspended. The initial solution is usually at acidic pH, in which the protein molecules have a net positive charge. The various intact protein molecules retain varying numbers of positive charges, which can be sufficient to cause them to be separated in a mass spectrometer (Fig. 1.12A). Their mass-to-charge ratios can be within a practical range, in spite of the large mass of the protein. The observed spectrum is interpreted by assuming that adjacent major peaks differ by a unit of charge and by a proton in mass. The accuracy of the measurements makes it possible to assign integral net charges to the various peaks. It is then possible to deconvolute the spectrum to give the distribution of singly charged molecules of different molecular weights (Fig. 1.12B). This procedure works only if all the major peaks arise from different protonation states of the same protein or only a few proteins, so the protein preparation must be relatively homogeneous.

The molecular weights of the constituent polypeptide chains of proteins as large as 100,000 daltons can be measured in this way to an accuracy of about 0.01%. The mass of a protein of 10,000 daltons can therefore be measured to within 1 dalton, which is sufficient to detect most differences in the covalent structures of most proteins. Multimeric proteins tend to dissociate into monomers in the spectrometer, and cofactors do not remain bound to the protein, so the protein molecules probably unfold during the mass measurements.

In **tandem mass spectrometry,** a second spectrometer bombards a selected ion with neutral atoms that

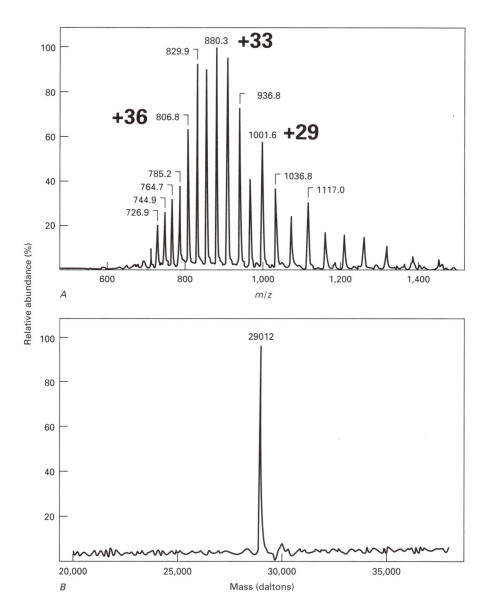

FIGURE 1.12

Electrospray mass spectrometry of carbonic anhydrase. The spectrum of the intact protein is given in *A;* the various peaks correspond to protein molecules with varying net charges *z.* The mass-to-charge ratio *m/z* of each of the major peaks is given by the smaller number. The net charges of three of the peaks deduced from the spectrum are given by the larger numbers. With the assumption that adjacent peaks differ by a single unit charge, the spectrum of *A* was deconvoluted to give the plot in *B* of the relative abundance of protein molecules with different molecular weights. (From I. Jardine, *Nature* 345:747–748, 1990.)

cause further fragmentation. The resultant spectrum of fragments can provide sequence information, as in Figure 1.11. The fragmentations that take place are more complex, although this procedure can distinguish between Leu and Ile residues.

References

Characterization by tandem mass spectrometry of structural modifications in proteins. K. Biemann and H. A. Scoble. *Science* 237:992–998 (1987).

Mass spectrometry of natural and recombinant proteins and glycoproteins. H. R. Morris and F. M. Greer. *Trends Biotech.* 6:140–147 (1988).

Rapid protein sequencing by the enzyme-thermospray LC/MS method. K. Stachowiak et al. *J. Amer. Chem. Soc.* 110:1758–1765 (1988).

Plasma desorption mass spectrometry of peptides and proteins. P. Roepstorff. *Acc. Chem. Res.* 22:421–427 (1989).

New developments in biochemical mass spectrometry: electrospray ionization. R. D. Smith et al. *Anal. Chem.* 62:882–899 (1990).

Elucidation of the primary structures of proteins by mass spectrometry. J. B. Smith et al. *Anal. Biochem.* 193:118–124 (1991).

e. Fragmentation of a Protein into Peptides

None of the techniques that have been described is capable by itself of determining the amino acid sequence of proteins with more than about 50 amino acid residues. It is necessary to cleave larger proteins into smaller peptides and to determine the amino acid sequences of the peptides individually. The basic approach used today is virtually the same as that originally developed by Sanger for the first determination of an amino acid sequence, that of insulin:

1. Fragmentation of the polypeptide chain into peptides

2. Purification of the peptides

3. Determination of the amino acid sequence of the individual peptides

4. Repetition of steps 1–3, using a second type of fragmentation process to produce peptides that overlap those of the first fragmentation

The order of the peptides in a protein can eventually be deduced from these four steps. Although the technology has advanced considerably, it is a sobering thought that the original sequence of insulin is one of the very few protein sequences determined directly that has been

Table 1.5 Methods for Cleaving Polypeptide Chains

Sequence cleaved[a]	Procedure or enzyme
Ala-Yaa	Elastase, bromelain
Arg-Yaa	Trypsin, endoproteinase Arg-C, clostripain
Asn-Gly	Hydroxylamine
Asp-Yaa	V-8 protease
Asp-Pro	Mild acid
Xaa-**Asp**	Asp-N protease
Xaa-**Cys**	Cyanylation
Glu-Yaa	V-8 protease
Gly-Yaa	Elastase
Leu-Yaa	Pepsin
Xaa-**Leu**	Thermolysin
Lys-Yaa	Trypsin, endoproteinase Lys-C, bromelain
Met-Yaa	CNBr
Phe-Yaa	Chymotrypsin, pepsin
Xaa-**Phe**	Thermolysin
Pro-Yaa	Prolylendopeptidase
Trp-Yaa	Iodosobenzoic acid, N-chlorosuccinimide, chymotrypsin
Tyr-Yaa	Chymotrypsin, bromelain

[a] The specific residues are indicated by bold type; Xaa and Yaa can be almost any amino acid, except for Pro in many instances. Cleavage is C-terminal to residue Xaa, N-terminal to Yaa.

substantiated in all respects and that has not required revision.

A variety of techniques for fragmenting proteins have been developed (Table 1.5). Sanger primarily used partial acid hydrolysis because it produces a large number of small overlapping peptides. It is still used frequently with mass spectrometry, which can identify many of the small peptides (Sec. 1.6.2.d). The Edman degradation can be used to sequence much longer peptides, however, so methods for cleaving proteins at more specific locations are generally used today.

Proteolytic enzymes are most commonly used, especially those that cleave proteins at specific locations. One of the most specific enzymes is **trypsin,** which cleaves the peptide bond following Lys and Arg residues. Cleavage by trypsin can be restricted to Arg residues by reversibly blocking the ϵ-amino group of Lys residues by maleylation or citraconylation (Eq. 1.22). Cleavage at Cys residues can be achieved by reacting the thiol group with ethyleneimine to produce a basic side chain (Eq. 1.49). Alternatively, **clostripain** or **submaxillary protease** can be used to cleave only at Arg residues. **Endoproteinase Lys-C** and **lysylendopeptidase** are specific for Lys residues. The **V8 protease** from *Staphylococcus aureus* is complementary to trypsin in that it cleaves only after the acidic residues Asp and Glu. **Chymotrypsin** is fairly specific to the peptide bonds following the aromatic residues Tyr, Phe, and Trp, but it also cleaves after other hydrophobic residues, especially Leu. **Thermolysin** cleaves preferentially before Leu, Val, Ile, and Met residues; **prolylendopeptidase** cleaves after Pro residues; and **Asp-N** protease cleaves before Asp residues (Table 1.5).

Various other chemical procedures cleave polypeptide chains at specific residues. The most widely used is **cyanogen bromide** (CNBr), which cleaves at Met residues. The nucleophilicity of the sulfur atom causes it to react with CNBr:

$$-CH_2-CH_2-S-CH_3 + CNBr$$

$$-CH_2-CH_2-\overset{+}{\underset{|}{S}}-C{\equiv}N + Br^- \quad (1.78)$$

The particular stereochemistry of the Met side chain favors the intramolecular rearrangement of the sulfonium salt:

$$\quad (1.79)$$

The iminolactone generated is readily hydrolyzed by water to cleave the polypeptide chain:

$$\quad (1.80)$$

The N-terminal fragment has a homoserine (Hse) lactone residue at its C-terminus in place of the original Met. The lactone is hydrolyzed reversibly to the free acid:

$$\quad (1.81)$$

This reversible equilibrium can complicate separation of CNBr-generated peptides because the two forms differ in net charge.

Cleavage at Met residues preceding Thr, Ser, and Cys residues is often incomplete. The sulfoxide form of Met residues does not react with CNBr, so this reversible modification can be used to restrict cleavage to nonoxidized Met residues.

Peptide bonds preceding Cys residues can be cleaved by **cyanylation** with 2-nitro-5-thiocyanobenzoic acid:

$$\quad (1.82)$$

Peptide bond cleavage occurs upon incubation at alkaline pH:

$$\quad (1.83)$$

This reaction is specific for Cys residues but has the disadvantage that the new N-terminus generated is blocked and is not amenable to sequence analysis by the Edman procedure.

Peptide bonds following Trp residues can be cleaved with varying efficiencies by several reagents, especially iodosobenzoic acid and BNPA-skatole. Most of these procedures have the disadvantages of side reactions and of oxidizing Cys and Met residues.

Peptide bonds adjacent to Asp, Ser, and Thr residues are especially susceptible to acid hydrolysis because the side chains of these residues can interact with the peptide backbone. In particular, Asp-Pro bonds are cleaved readily by dilute acid. The specificity of this reaction is believed to be due to the ability of the Asp side-chain carboxyl group to interact chemically with the unique tertiary N atom of the Pro residue.

Hydroxylamine cleaves the polypeptide chain at Asn-Gly sequences. This results from the tendency of such Asn residues to deamidate and transiently form succinimide derivatives (Fig. 1.3). If the succinimide ring is opened by hydroxylamine rather than water, peptide cleavage results. The original Asn residue is now the C-terminal Asp of one of the peptides but has a hydroxamate ($-CO-NHOH$) in place of either the α- or the β-carboxyl group.

References

Nonenzymatic methods for the preferential and selective cleavage and modification of proteins. B. Witkop. *Adv. Protein Chem.* 16:221–321 (1961).

The cyanogen bromide reaction. E. Gross. *Methods Enzymol.* 11:238–255 (1967).

Selective cleavage and modification of peptides and proteins. T. F. Spande et al. *Adv. Protein Chem.* 24:97–260 (1970).

Cleavage at cysteine after cyanylation. G. R. Stark. *Methods Enzymol.* 47:129–132 (1977).

Cleavage at Asn-Gly bonds with hydroxylamine. P. Bornstein and G. Balian. *Methods Enzymol.* 47:132–144 (1977).

Cleavage at aspartyl-prolyl bonds. M. Landon. *Methods Enzymol.* 47:145–149 (1977).

Current developments in chemical cleavage of proteins. K. K. Han et al. *Intl. J. Biochem.* 15:875–884 (1983).

f. Mapping the Positions of Residues

The approximate positions in a polypeptide chain of amino acids for which specific chain-cleavage reactions are available (Table 1.5) can be determined. The polypeptide chain is labeled in some way at one of its two ends. The cleavage reaction is then carried out partially so that only a fraction of the molecules are cleaved at each target position. The resulting mixture of cleaved and uncleaved molecules is sorted according to length

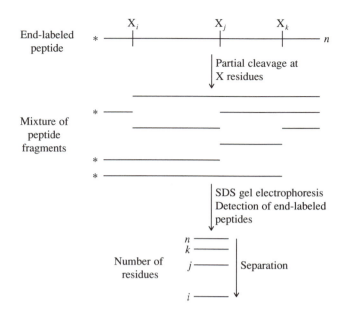

FIGURE 1.13

Determining the positions of specific residues along the polypeptide chain by end-labeling and partial specific cleavage. The polypeptide chain is labeled at one end, indicated by the asterisk; the label can be radioactive or fluorescent, or can bind specifically to some antibody. Partial cleavage at the residue of interest is carried out to generate a wide variety of peptides. The peptides are separated by SDS gel electrophoresis, and the lengths of the labeled peptides are determined. This gives the positions of the cleavage points relative to the labeled end.

by SDS electrophoresis, but only the labeled peptides are detected. The molecular weights of these labeled fragments indicate the positions of the cleavage sites throughout the polypeptide chain (Fig. 1.13).

Such procedures that label the C-terminus can be used to select peptides for sequencing from their N-termini by the Edman degradation. A suitable collection of peptides with N-termini spaced appropriately along the chain requires only a minimum of sequencing reactions to elucidate the entire sequence.

References

A general procedure for the end labeling of proteins and positioning of amino acids in the sequence. D. G. Jay. *J. Biol. Chem.* 259:15572–15578 (1984).

Determination of the relative positions of amino acids by partial specific cleavages of end-labeled proteins. R. A. Jue and R. F. Doolittle. *Biochemistry* 24:162–170 (1985).

A new strategy for primary structure determination of proteins: application to bovine β-casein. C. Carles et al. *FEBS Letters* 229:265–272 (1988).

Landmark mapping: a general method for localizing cysteine residues within a protein. B. Nefsky and A. Bretscher. *Proc. Natl. Acad. Sci. USA* 86:3549–3553 (1989).

g. Purification of Peptides: Diagonal Techniques

Various electrophoretic and chromatographic techniques, especially high-pressure liquid chromatography (HPLC), are useful in purifying peptides. In many systems, the observed behavior of peptides can be rationalized and predicted on the basis of their amino acid compositions.

It is often desirable to purify selectively those peptides that contain a particular amino acid. This is most easily accomplished in many instances by "diagonal" techniques, which rely on a change in properties of peptides caused by selective modification of the amino acid of interest. If a mixture of peptides is separated in one dimension and then subjected to the same procedure a second time, but at right angles to the first, all the peptides will lie on a diagonal because they had the same mobilities in both directions. The mobilities of peptides that have been modified after the first separation can be different in the second dimension, and they will lie off the diagonal of unmodified peptides. If the intervening modification is specific for certain amino acid residues, only peptides containing that amino acid will lie off the diagonal. Consequently, such procedures identify and selectively purify particular peptides.

For example, peptides containing Cys residues can be purified by blocking the thiol groups initially with iodoacetic acid (Eq. 1.47). A two-dimensional diagonal map is then prepared by electrophoresis at pH 3.5, exposing the peptides to performic acid between the two separations. This oxidizes the carboxymethyl-Cys residues to the sulfones:

$$-CH_2-S-CH_2-CO_2H$$
$$\downarrow HCO_3H$$
$$\underset{O}{\overset{O}{\underset{\|}{\overset{\|}{-CH_2-S-CH_2-CO_2H}}}} \qquad (1.84)$$

The oxidation lowers the pK_a value of the carboxyl group so that the peptides containing Cys residues are somewhat more acidic in the second dimension; consequently, they lie off the diagonal (Fig. 1.14A). In the case illustrated, the six peptides off the diagonal contain the six Cys residues of the protein. The procedure can also be used with Cys residues, blocked in other ways, that are suitably modified by performic acid or other reagents.

Similar techniques have been developed for peptides containing Lys, Met, His, Arg, and Trp residues, using the various chemical modifications described in Section 1.3. The amino groups of Lys-containing peptides are initially blocked with trifluoroacetyl or maleyl groups (Eq. 1.22), which alter the charge and are easily removed after the first electrophoresis. Met residues are first alkylated with iodoacetamide to produce the charged sulfonium derivative (Eq. 1.45); after the first electrophoresis, heating causes this derivative to cleave the polypeptide chain by a reaction analogous to that occurring with CNBr (Eqs. 1.79 and 1.80). His residues are first blocked by dinitrophenylation (Eq. 1.70) of the side chain and then restored by exposure to thiol after the first electrophoresis. Arg residues are first blocked by reaction with cyclohexanedione (Eq. 1.29) and then regenerated at alkaline pH. Trp residues are modified by *o*-nitrophenylsulfenyl chloride after the first separation.

Diagonal techniques are ideally suited to two-dimensional separations but can also be used with the more common HPLC analysis; in place of a second dimension, many fractions must be modified individually and reanalyzed.

References

Peptides containing methionine and carboxymethyl cysteine. J. I. Harris. *Methods Enzymol.* 47:390–398 (1967).

Factors affecting retention and resolution of peptides in high-performance liquid chromatography. J. L. Meek and Z. L. Rossetti. *J. Chromatog.* 211:15–28 (1981).

Selective isolation of tryptophan-containing peptides by hydrophobicity modulation. T. Sasagawa et al. *Anal. Biochem.* 134:224–229 (1983).

h. Disulfide Bonds

Although disulfide bonds between Cys residues are not, strictly speaking, part of the amino acid sequence or primary structure, the Cys residues that are linked in this manner can be determined using the same techniques. Disulfide bonds are covalent and are stable under most conditions as long as reductants, oxidants, or agents that catalyze disulfide interchange (especially thiols) are absent. Under appropriate conditions, a protein that contains disulfides can be cleaved into peptides, and the peptides that contain disulfides or are linked by them can be identified. Generally, they are detected as peptides whose separation behavior is modified by breakage of disulfide bonds.

An elegant procedure for determining which Cys residues are disulfide bonded was designed by Brown

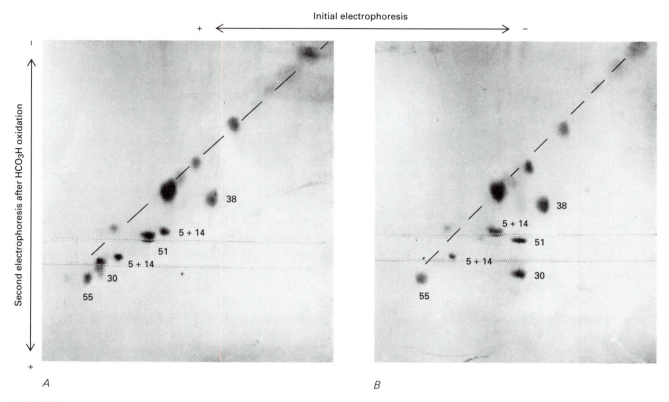

FIGURE 1.14

Diagonal maps demonstrating the selective purification of Cys-containing peptides and identification of any disulfide bonds between them. The proteins were digested with trypsin, then chymotrypsin. The resulting peptides were separated by electrophoresis at pH 3.5 in the horizontal direction (anode at the left). After exposure to performic acid vapor, the electrophoresis was repeated in the vertical direction (anode at bottom); the peptides were visualized by staining with ninhydrin. The diagonal indicated by the dashed line is defined by peptides that were not altered by the performic acid and consequently had the same mobility in both dimensions; many of these peptides have migrated off the paper.

A: Reduced, carboxymethylated bovine pancreatic trypsin inhibitor (BPTI). The six peptides containing the six CM-Cys residues define a CM-Cys diagonal and lie below the diagonal of unmodified peptides because they had slightly lower mobilities in the second dimension, due to the somewhat greater acidity of the CM-Cys residues after oxidation to the sulfones. These peptides are indicated by the Cys residues they contain. Their identities are, starting from *lower left* to *upper right*: residues 54–58 (Cys 55), 27–33 (Cys 30), 5–15 (Cys 5 and 14), 47–53 (Cys 51), 1–15 (Cys 5 and 14), 36–39 (Cys 38). B: The major one-disulfide intermediate in the BPTI folding pathway (see Fig. 7.33). The two peptides containing Cys 30 and Cys 51 were absent from the CM-Cys diagonal, but had the same mobility in the first dimension, indicating that they were originally linked by an intramolecular disulfide bond between these two Cys residues. (From T. E. Creighton, *J. Mol. Biol.* 87:603–624, 1974.)

and Hartley as the original use of the diagonal method described earlier. A protein that contains disulfides is cleaved into peptides, and the peptides are separated by electrophoresis under conditions in which disulfides are stable. The separated peptides are then exposed to performic acid, which cleaves the disulfide bonds and converts the Cys residues to those of cysteic acid (Cya, Eq. 1.52). The peptides originally linked by the disulfides are now independent and more acidic. During electrophoresis in the second dimension, they usually migrate differently than in the first dimension and lie off the diagonal of unmodified peptides (Fig. 1.14B). Which peptides were originally linked is indicated by their common mobility in the first dimension.

References

Location of disulphide bridges by diagonal paper electrophoresis. J. R. Brown and B. S. Hartley. *Biochem. J.* 101:214–228 (1966).

Disulfide bond formation in proteins. T. E. Creighton. *Methods Enzymol.* 107:305–329 (1984).

Analysis for disulfide bonds in peptides and proteins. T. W. Thannhauser et al. *Methods Enzymol.* 143:115–119 (1987).

1.7 Nature of Amino Acid Sequences

Proteins vary enormously in both the lengths and the amino acid sequences of their polypeptide chains. They may be as long as 25,000 amino acid residues or as short as 50. There is no clearly defined lower limit at which a polypeptide becomes a peptide rather than a protein, but it is in the region of 50 residues.

Most protein polypeptide chains contain between 200 and 500 residues. The average molecular weights of polypeptide chains from prokaryotic and eukaryotic cells have been measured by SDS gel electrophoresis to be 24,000 and 31,700, respectively, which correspond to about 212 and 280 residues. These averages are weighted according to the abundance of the various proteins. Estimates of individual proteins have given an average polypeptide molecular weight of about 50,000, with a median of about 40,000.

With 20 different amino acid residues possible at each of 200–500 positions, the number of possible sequences is astronomical. It is not surprising that virtually every protein has a unique amino acid sequence. The number of such sequences known is very large and is growing constantly, and compilations of protein sequences are in the form of computer data bases (see Appendix 1).

Even though each protein sequence is unique, the 20 amino acids are used with similar frequencies in virtually all proteins (Table 1.1); only small proteins or those with unusual structures differ significantly from the common composition. Membrane proteins (see Sec. 7.2) contain somewhat higher levels of hydrophobic amino acids (Leu, Ile, Val, Met, Phe, Tyr) and slightly lower levels of those amino acids that are usually ionized (Lys, Arg, Glu, Asp, His). The most atypical compositions are those of fibrous proteins, such as collagen, which have regular conformations and repetitive amino acid sequences (see Sec. 5.5).

Considering all the known amino acid sequences, one finds the distributions of pairs of adjacent amino acids within typical sequences to be nearly random, which is a hint that the basis of protein structure and function is not simple. In longer stretches of residues, however, nonrandom distributions can be detected.

The primary structure of a protein in isolation is not very meaningful because little can be inferred from just reading the sequence of amino acid residues. Of the many possible short sequences of residues, only a few give clues to the structure or function of a protein. Knowledge of the amino acid sequence of a protein is most useful when the sequence is found to be similar to that of another protein of known structure and function. Such **homologous proteins** always have similar structures and frequently have related functions (Secs. 2.2, 6.4, 8.3).

Although the full implications of the amino acid sequence of a protein are not understood at the present time, it dictates all the structural and functional properties of the protein. How it does so is the main topic of the remainder of this volume.

References

Size distribution of polypeptide chains. E. D. Kiehn and J. J. Holland. *Nature* 226:544–545 (1970).

The independent distribution of amino acid near neighbor pairs into polypeptides. M. H. Klapper. *Biochem. Biophys. Res. Commun.* 78:1018–1024 (1977).

The distribution of subunit sizes of soluble proteins in human tissues. Y. H. Edwards et al. *Ann. Human Genet.* 40D:267 (1977).

The size distribution of proteins, mRNA, and nuclear RNA. S. S. Sommer and J. E. Cohen. *J. Mol. Evol.* 15:37 (1980).

Proteins of *Escherichia coli* come in sizes that are multiples of 14 kDa: domain concepts and evolutionary implications. M. A. Savageau. *Proc. Natl. Acad. Sci. USA* 83:1198–1202 (1986).

Oligopeptide biases in protein sequences and their use in predicting protein coding regions in nucleotide sequences. P. McCaldon and P. Argos. *Proteins* 4:99–122 (1988).

Regularities in the primary structure of proteins. M. Cserző and I. Simon. *Intl. J. Peptide Protein Res.* 34:184–195 (1989).

1.8 Peptide Synthesis

The formation of peptide bonds between amino acids is not in itself a difficult chemical problem (e.g., Eq. 1.13), but forming the bonds among the 20 different amino acids a sufficient number of times to synthesize a protein still taxes the ingenuity of synthetic chemists. The main problem is the diversity of functional groups on the amino acid side chains. To prevent the participation of these groups in undesirable reactions during the formation of a desired peptide bond, all such reactive groups must be blocked during the synthesis, and the blocking groups must be removed completely after the synthesis. A large number of blocking groups have been developed, each with advantages and disadvantages. No perfect, or even universally accepted, set has been developed, so this aspect of peptide synthesis is not described here. Although the synthesis of many peptides is now routine and is performed by automatic instruments, the synthesis of many other peptides requires careful consideration of tactics.

Peptide bond formation between a carboxyl and an amino group does not occur spontaneously under most conditions, so one of these groups must be converted into a more reactive form; almost invariably the carboxyl group is the choice. The usual driving force in this carboxyl activation is the augmentation of the electrophilic character of the carbon atom. Its intrinsically low electron density is decreased in the activated derivatives by a negative inductive effect of the activating substituent X:

$$R-\underset{\delta^+}{C}\overset{\displaystyle O}{\underset{\delta^-}{\parallel}}-X \tag{1.85}$$

The resulting electrophilic center permits attack by a nucleophilic nonionized amino group:

$$R^1-\underset{\delta^+}{C}-X \quad \longrightarrow \quad R^1-\underset{+}{C}-X \quad \longrightarrow \quad R^1-C-NH-R^2 + HX \tag{1.86}$$

Examples of activated carboxyl groups are

- Azides: $-CO-N_3$

- Acid chlorides: $-CO-Cl$

- Mixed anhydrides: $-CO-O-CO-R$

- Activated esters: $-CO-OR'$, where R' is often p-nitrophenyl

Coupling reagents can also be used to cause peptide bond formation between free amino and carboxyl groups. These reagents usually generate transient intermediates similar to those just described, such as acid chlorides and activated esters. The most successful coupling reagents have been the carbodiimides, especially dicyclohexylcarbodiimide, which react with free carboxyl groups to make them reactive with amines (Eq. 1.13).

One of the most troublesome side reactions of peptide synthesis is racemization at the C^α atom of the amino acid. It results primarily from activation of the α-carboxyl group, which assists transient loss of the hydrogen atom on the C^α atom. Consequently, initially pure L isomer of the amino acid gradually reverts to the racemic DL mixture. Racemization has a dramatically deleterious effect on protein conformation (Chap. 5) and biological activity, so much effort has gone into minimizing it. Every known method of forming peptide bonds, however, causes some racemization.

The amino and carboxyl groups not intended to be coupled must be blocked, or uncontrolled polymerization occurs. After the desired amino and carboxyl groups are coupled, the addition of other residues to the chain requires removal of the blocking group from one of its ends. Considerable ingenuity is required to design a useful blocking group because it must be stable under coupling conditions and yet be removable quantitatively when necessary, under conditions that do not affect the blocking groups on the other end and on the side chains.

Virtually all peptide syntheses start with the C-terminal amino acid and add residues to its amino group. This is opposite to the approach that nature takes in protein biosynthesis (Chap. 2).

Synthesis of peptides is often limited by the low solubility of intermediates during coupling reactions and by difficulties in separating peptide products from all the other reagents. The longer the peptide chain becomes, the more severe these problems tend to be. A remarkably successful solution to many of these problems is synthesis in the solid phase, devised initially by Merrifield. The basis of the procedure is to attach the amino acid corresponding to the C-terminus of the peptide to be synthesized to an insoluble resin support, using the α-carboxyl group. The peptide is then extended toward the N-terminus by stepwise coupling of activated amino acid derivatives. The resin serves both to protect the C-terminus of the polypeptide chain during synthesis and to keep the polypeptide chain insoluble. Filtration and thorough washing of the solid phase removes soluble by-products and excess reagents, but the peptide chain remains bound to the resin.

One cycle of the synthesis consists of deprotecting the terminal α-amino group of the peptide and coupling this group to the carboxyl group of the next amino acid residue. The functional groups that are not to be coupled by peptide bonds must still be protected in the usual way. Successive residues are added to the peptide by repeating the cycle until the desired sequence is assembled. Finally, the complete polypeptide is released into solution by cleavage of the link to the solid support, and the protecting groups on the side chains are removed.

Considerable progress has been made in the technology and chemistry of peptide synthesis on solid supports; for example, α-amino protecting groups have been designed that can be removed repeatedly without damaging the nascent peptide, and solid supports have been designed that remain solvated by the various solvents used. The chemistry is now sufficiently reproducible that the procedures have been automated.

The chemistry involved in the various reactions used in peptide synthesis would fill an entire volume and is therefore not described here. The great number of methods proposed is adequate proof that all of them have deficiencies. Current methods fall into two categories, depending on the nature of the group used to block reversibly the amino groups of the amino acids that are being added. The classical method uses the *tertiary-butyloxycarbonyl* [$(CH_3)_3C$—O—CO—, *Boc*] blocking group, which can be removed by acid. Recent methods use the *Fmoc* group (fluoromethoxy carbonyl), which is removed quantitatively by treatment with a mild organic base, R_2NH:

$$(1.87)$$

The other blocking groups on the side chains and the linkage to the resin are stable under basic conditions but are removed by acids. The most recently developed approach is illustrated in Figure 1.15.

The solid-phase approach has the advantages of simplicity, speed, efficiency, and convenience, but it also has certain limitations and disadvantages. The intermediates in the synthesis cannot be purified because they remain bound to the resin. Any incomplete or wrong sequences generated by less-than-perfect procedures must be removed at the end of the synthesis. This can be difficult with long peptides because variant peptides, differing at only one or a few residues, are not substantially different from the genuine product and consequently are difficult to remove.

The efficiency of each coupling cycle is crucial for the success of a solid-phase synthesis. Incomplete deprotection or coupling in any cycle causes a fraction of each of the nascent peptides to be irretrievably lost to the correct sequence. The cumulative effects of such incomplete coupling are easily illustrated with simple calculations. If the overall efficiency of adding each successive amino acid is 90% (which in other chemical syntheses would be considered a very successful synthesis), the proportion of peptide chains of 10 residues with correct sequences would be 35%, of 20-residue chains 12%, and of 100-residue chains only 0.003%. Even with an extremely proficient synthesis of 99% efficiency at each step, the yields of correct polypeptide chains of 100 residues would be only 37%.

The problem of the inefficiency of stepwise synthesis is best circumvented by the preparation of several moderate-sized peptides by the stepwise procedure, which are then purified and joined. Small peptides are much easier to purify than long peptides. Also, the condensation products of fragments are easily purified; for example, the 20-residue condensation product of two decapeptides is very different chemically from unreacted decapeptides or those involved in side reactions. Separating peptides of very different sizes is easier than separating peptides of similar sizes. Condensation reactions also make possible semisynthesis techniques, in which one of the peptides is isolated from a natural protein. Coupling large peptides often presents technical problems, especially the insolubility of one or more of the peptides and reactions between groups other than the amino and carboxyl groups meant to be coupled. The latter problem can usually be circumvented by using enzymes that couple fragments with relative specificity. This recent development usually uses proteolytic enzymes that act in reverse of their normal hydrolytic mechanism (Sec. 9.3.2).

Virtually all known biologically active peptides up to 20 or 30 residues in length have been synthesized, as

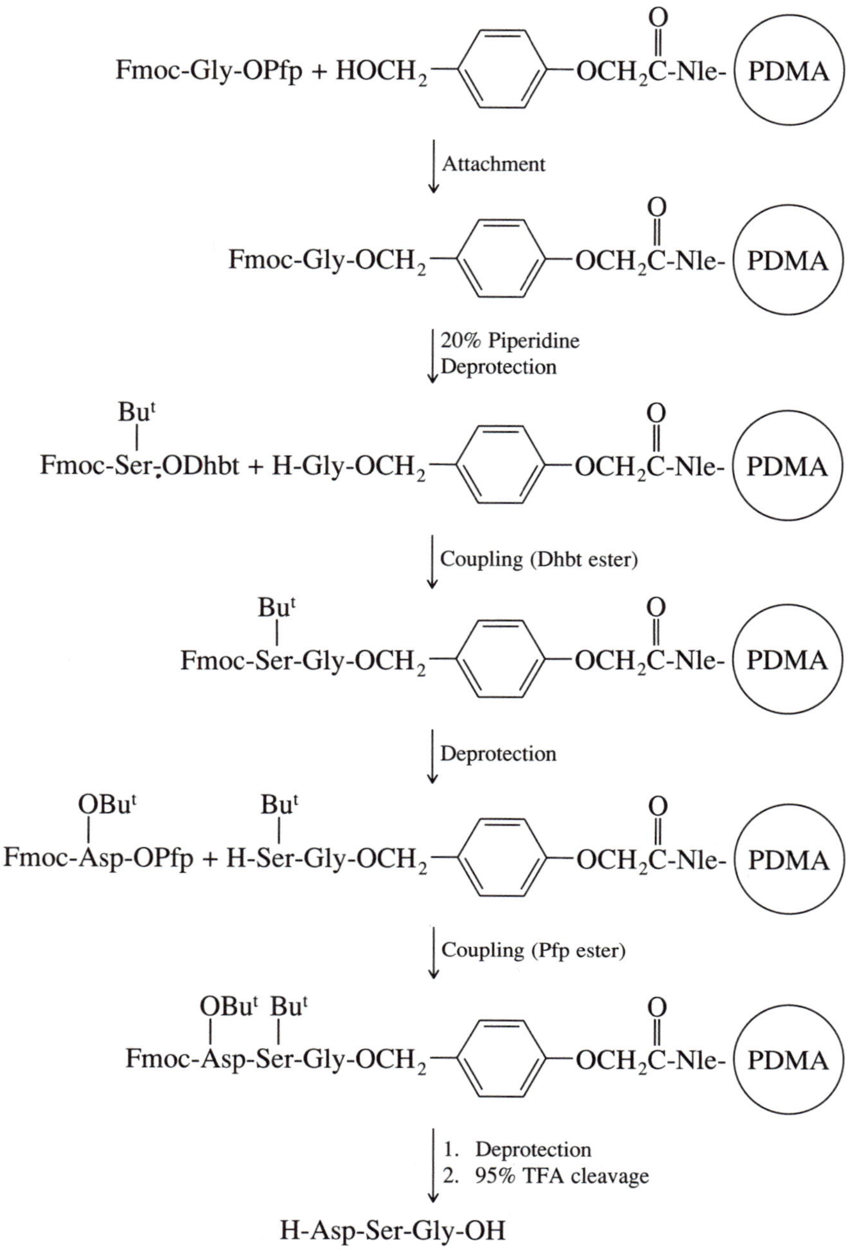

FIGURE 1.15

Illustration of the synthesis of the tripeptide Asp-Ser-Gly. The solid-phase resin is cross-linked polydimethylacrylamide (PDMA). To this is coupled one residue of an unnatural amino acid, norleucine (Nle), which serves as an internal standard when analyzing the synthesis by hydrolyzing a portion of the resin, followed by amino acid analysis. The aromatic moiety attached to the amino group of the Nle residue is an acid-labile linkage agent, from which the assembled chain will be cleaved. The first amino acid is attached to the linkage agent through an ester bond. The amino acids are added sequentially to the resin by first activating their carboxyl groups, in this case either as the pentafluorophenyl (Pfp) ester or as that derived from 3,4-dihydro-3-hydroxy-4-oxo-benzotriazine (Dhbt). Their side chains are protected with acid-labile groups such as *tertiary*-butyl (Bu^t), its ester (OBu^t), or its urethane (Boc). The amino groups of the amino acids are protected by the base-labile Fmoc group, which is removed by piperidine. When the last amino acid has been added, the Fmoc group is removed, then the peptide is released from the resin and from the side-chain blocking groups by treatment with trifluoroacetic acid (TFA). (Figure kindly provided by R. C. Sheppard.)

have scores of their analogues. The frontier that has had the most appeal for synthetic chemists has naturally been the synthesis of larger and larger polypeptides, but it is difficult to demonstrate the purity of the final product. For example, the synthesis of the enzyme ribonuclease A, consisting of 124 amino acid residues, was reported simultaneously in 1969 by two groups, one using stepwise synthesis of fragments in solution, followed by condensation, and the other using stepwise assembly of the entire polypeptide chain by the solid-phase method. The final products were not homogeneous and had lower enzymatic activity than the natural protein; it was impossible to determine chemically the proportion of molecules having the correct sequence.

Chemical synthesis of peptides remains of vital importance in studying many aspects of protein structure and function. It has the advantage that nonbiological amino acids and isotopic labels (^{2}H, ^{3}H, ^{13}C, ^{14}C, and ^{15}N) can be incorporated into specific locations. The synthesis of proteins, however, is done mostly by genetic engineering methods (Chap. 2).

References

In search of new methods in peptide synthesis. M. Bodanszky. *Intl. J. Peptide Protein Res.* 25:449–474 (1985).

Solid phase synthesis. B. Merrifield. *Science* 232:241–247 (1986).

Modern methods of solid phase peptide synthesis. R. C. Sheppard. *Science Tools* 33:9–16 (1986).

Protein engineering by chemical means? R. E. Offord. *Protein Eng.* 1:151–157 (1987).

Using proteases in peptide synthesis. K. Morihara. *Trends Biotech.* 5:164–170 (1987).

Solid-phase peptide synthesis: a silver anniversary report. G. Barany et al. *Intl. J. Peptide Protein Res.* 30:705–739 (1987).

Chemical synthesis of peptides and proteins. S. B. H. Kent. *Ann. Rev. Biochem.* 57:957–989 (1988).

Synthetic approaches to biologically active peptides and proteins including enzymes. E. T. Kaiser. *Acc. Chem. Res.* 22:47–54 (1989).

Peptide and protein synthesis by segment synthesis-condensation. E. T. Kaiser et al. *Science* 243:187–192 (1989).

Exercises

1. Can SDS electrophoresis demonstrate the presence of disulfide bonds in a protein? Of the native disulfides present in the natural protein?

ANSWERS
J. Reynolds and C. Tanford, *J. Biol. Chem.* 245:5161–5165 (1970); W. Fish et al. *J. Biol. Chem.* 245:5166 (1970).

G. Scheele and R. Jacoby, *J. Biol. Chem.* 257:12277–12282 (1982); D. Goldenberg and J. King, *Proc. Natl. Acad. Sci. USA* 79:3403–3407 (1982).

2. After cyanogen bromide cleavage of a protein, two peptides were found to incorporate ^{14}C from iodo[^{14}C]acetic acid but only after exposure to dithiothreitol (P. N. Kao and A. Karlin, *J. Biol. Chem.* 261:8085–8088, 1986). On this basis, it was concluded that the native protein contained intramolecular disulfides between Cys residues within each of these peptides. Is this conclusion valid?

ANSWER
S. Smith and J. Kyte, *Biochemistry* 28:3481–3482 (1989).

3. The thiol groups of proteins are often protected and kept reduced by the addition of reagents such as mercaptoethanol, cysteine, glutathione, and dithiothreitol. Such reagents are gradually oxidized by air to the disulfide form (rapidly at alkaline pH), and their usefulness depends on the thiol form being present in sufficient excess. In view of the relative stabilities of the thiol and disulfide forms of dithiothreitol relative to the other monothiols (Eq. 1.57; W. W. Cleland, *Biochemistry* 3:480–482, 1964), consider the usefulness of 1 mM solutions of the reagents that have been oxidized to the extents of 1%, 10%, and 90%.

If the equilibrium constant for forming a disulfide bond between two Cys residues of a protein,

$$HS-protein-SH + RSSR \Longleftrightarrow$$

$$protein\ SS + 2RSH$$

has the value 100 M (T. E. Creighton, *BioEssays* 8:57–63, 1988), what is the relative usefulness of 1 mM monothiols and dithiothreitol in keeping the Cys residues reduced?

4. During study of the unfolding and refolding of ribonuclease A, which is a very stable protein, an unexpected modification was observed that caused the net charge of the protein to decrease by one unit (T. E. Creighton, *J. Mol. Biol.* 129:411–431, 1979). What modification reactions were most likely to be taking place?

ANSWER
T. W. Thannhauser and H. A. Scheraga, *Biochemistry* 24:7681–7688 (1985).

Why was this modification apparent in this study but not in the many other studies carried out using the same protein?

ANSWER
S. J. Wearne and T. E. Creighton, *Proteins: Struct. Funct. Genet.* 5:8–12 (1989).

5. A useful trick when destaining polyacrylamide gels that have been stained with Coomassie Blue R250 is to include a small piece of polyurethane foam. The dye binds tena-

ciously to the foam, which increases the rate of destaining and makes stain removal from the gel essentially irreversible. What are the drawbacks of this procedure?

ANSWER
Section 1.4.

6. SDS has a variety of effects on the conformational properties of various poly-α-amino acids (e.g., D. K. Igou et al. *Biochem. Biophys. Res. Commun.* 60:140–145, 1974; R. W. McCord et al. *Biopolymers* 16:1319–1329, 1977; T. Overgaard et al. *Biopolymers* 23:1595–1603, 1984; T. H. Lin et al. *Macromolecules* 21:131–136, 1988). What are the implications for the mechanism of action of this detergent?

ANSWER
P. Lundahl et al. *Biochim. Biophys. Acta* 873:20–26 (1986).

7. The sequence of the α subunit of the F_1 ATPase from *Escherichia coli*, determined by two groups from the complementary DNA sequence, contained three Cys residues. Yet the protein was found to contain four thiol groups (H. Stan-Lotter and P. D. Bragg, *Arch. Biochem. Biophys.* 229:320–328, 1984). What possible reasons are there for this discrepancy?

ANSWER
H. Stan-Lotter et al. *FEBS Letters* 119:121–124 (1986).

8. The 188-residue human growth hormone was synthesized chemically and reported to possess 10% of the growth-promoting potency of the natural hormone (C. H. Li and D. Yamashiro, *J. Amer. Chem. Soc.* 92:7608–7609, 1970). Consider the general implications for the use of biological activity as a demonstration of the fidelity of chemical synthesis in view of the subsequent revision of the amino acid sequence of the natural hormone (H. D. Niall, *Nature New Biol.* 230:90–91, 1971).

9. A polypeptide chain corresponding to a drastically shortened form of ribonuclease A was synthesized and was claimed to have significant levels of enzyme activity (B. Gutte, *J. Biol. Chem.* 252:663–670, 1977). Does this affect the claims to have synthesized the complete enzyme (B. Gutte and R. B. Merrifield, *J. Amer. Chem. Soc.* 91:501–502, 1969; R. Hirschmann et al., *J. Amer. Chem. Soc.* 91:507–508, 1969)?

10. The proteins that have had their amino acid sequences determined directly are not a representative selection for a number of reasons. With the sequences of many more proteins being determined from their gene sequences (Chap. 2), how would you expect the known data base of protein sequences to change?

ANSWER
I. Simon and M. Cserzö, *Trends Biochem. Sci.* 15:135–136 (1990).

Biosynthesis
of Proteins

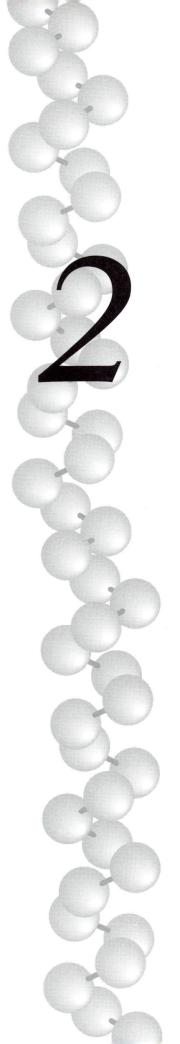

Although the covalent structures of proteins are immense by most chemical standards, they still can be defined very precisely because virtually all molecules in a natural population of a given protein have precisely the same primary structure. This is a result of an elaborate biosynthetic mechanism that specifies and checks many of the details of the large structure as it is being assembled; consequently, biosynthesis takes place with almost no errors or ambiguities. Coded information for the structures of the various proteins is contained in the genetic material of the chromosomes, usually DNA, in the form of linear sequences of nucleotide bases. Three sequential nucleotides contain the code for a single amino acid residue. The genetic information of the DNA is constantly checked and edited by the cell to ensure that it is altered as little as possible. The expression of the coded information also involves checking at various stages to ensure that only the correct amino acid is inserted into each position of the polypeptide chain. A description of protein biosynthesis requires some understanding of nucleic acid structure. This understanding, which can be obtained from current biochemistry textbooks, is assumed for the first half of this chapter.

The primary structure of a polypeptide chain, prescribed completely by the genetic information, is often altered after translation by a variety of covalent modifications. Part of the original primary structure may not be needed or may interfere with the function of a protein, so various segments of a polypeptide chain may be removed; functional groups other than those provided by the 20 biological amino acids may be added; and some covalent modifications may serve as molecular switches to control the biological activity of the protein.

After a protein is synthesized, it may remain in the cytoplasm or be directed to the nucleus, mitochondria, or other cell compartments, to various membranes in the cell or on its surface, or to the cell exterior. The targeting of proteins after biosynthesis is known as **topogenesis.**

Although all known proteins are made in the same way, some unusual natural peptides are made by a different mechanism, which is necessary because they contain amino acids that are very different from the 20 ordinary ones.

References

Genes, 3rd ed. B. Lewin. John Wiley, New York, 1987.

Molecular Biology of the Gene, 4th ed. J. D. Watson et al. Benjamin/Cummings, Menlo Park, Calif. 1987.

Molecular Cell Biology, 2nd ed. J. Darnell et al. Scientific American Books, New York, 1990.

2.1 Assembly of the Primary Structure

2.1.1 Gene Structure

All the information for synthesizing the primary structure of a protein is encoded in the genetic material of the chromosomes, which is double-stranded DNA in most organisms, though some viruses use single-stranded DNA or RNA. In all cases, the information is coded as sequences of the four kinds of nucleotides on one strand of DNA — A (adenine), C (cytosine), G (guanine), and T (thymine) — or of RNA, which has the same nucleotides except that uracil (U) replaces thymine (T). The sequence of the other strand of DNA is chemically complementary to the first; that is, A always pairs with T of the other strand and G with C.

A specific region of DNA or RNA codes for the primary structure of every protein that is synthesized by any organism, and the DNA or RNA segments on both sides of such regions are involved in the regulation and expression of that genetic information (Fig. 2.1). These regions of DNA or RNA are called **genes.** The parts of a gene that code for the primary structure of a protein are well defined, but the regulating parts are not, especially in higher organisms. Consequently, the beginning and end of a gene are often not certain. The information in the nucleic acid sequence of a gene is invariably read in one direction, from the 5′ end of the coding sequence to the 3′ end; so it is possible to describe gene structure in

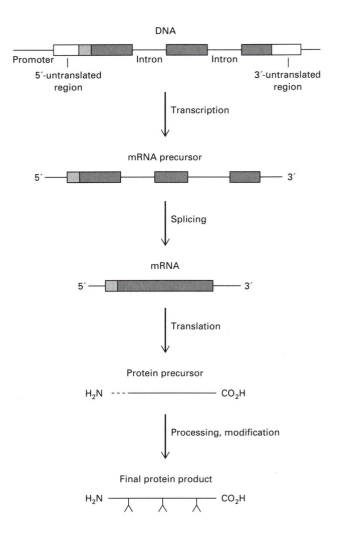

FIGURE 2.1

Steps in the expression of genetic information in biosynthesis of a protein. At the top is a simplified typical eukaryotic gene, with the promoter and other regulatory regions of the gene on the left (upstream). The segments to be transcribed into mature messenger RNA (mRNA) are shown as boxes; those segments coding for protein are shaded, whereas those unshaded correspond to the 5′ and 3′ untranslated regions of the mRNA. After transcription of the gene into the precursor mRNA (second line), the introns are removed by splicing. Prokaryotic genes differ in not having introns and not undergoing splicing. The mature mRNA (line three) is translated into polypeptide chain (line four), which may be the same as the final form of the protein or a precursor form that is modified. Parts of the polypeptide chain can be removed proteolytically from a precursor form, and a variety of other covalent modifications of the chain may also occur; the branched structures here indicate glycosyl groups added to specific Asn residues. In an additional process not indicated here, the polypeptide chain must fold into the appropriate three-dimensional conformation (Chapter 7), to yield the final biologically active gene product.

terms of "upstream" and "downstream" in the coding sequence.

Some genes code only for stable RNA molecules such as ribosomal and transfer RNA (Sec. 2.1.3). The initial steps in the expression of such genes are the same as in the transcription into RNA of the genes that code for proteins, but the resulting RNA is not translated into protein.

It is necessary to distinguish between the eukaryotic cell, which has a nucleus, and the prokaryotic cell, which does not. In a eukaryotic cell, a membrane separates the chromosomes from the cytoplasm; that is, the sites where the genetic information is stored are separated by a membrane from the sites where it is translated. In a prokaryotic cell, there is no such separation.

The DNA that codes for a polypeptide chain in a prokaryote is one continuous segment (the **cistron**), whereas in a eukaryotic cell it is separated into coding segments **(exons)**, between which are noncoding sequences **(introns)** that have no known function. Introns range in size from 50 to 10,000 base pairs, and as many as 41 introns may interrupt a single gene, so genes in eukaryotic cells can be very large. For example, the gene for thyroglobulin is spread over a segment of DNA that contains approximately 200,000 base pairs.

A popular hypothesis is that exons code for structural and functional units of proteins and that the modern eukaryotic genes for proteins have arisen by a genetic process of exon shuffling. This appears to be the case with some complex, multidomain proteins of recent evolution in which the exons clearly correspond to individual protein domains (see Sec. 6.2.1); in other cases there is little evidence for correspondence between exons and protein structure. The possible roles of introns in protein evolution are discussed in Chapter 3 (Sec. 3.4).

The protein biosynthetic systems of the mitochondrial and chloroplast organelles in eukaryotic cells are like those of prokaryotes; it is likely that these organelles have evolved from symbiotic prokaryotes that invaded eukaryotic cells.

For every protein, there is a gene that codes for its synthesis, and most genes are unique; only rarely do two or more genes produce the same protein. All prokaryotic and most eukaryotic genes specify a single protein; however, the phenomenon of alternative splicing of introns causes some genes to produce two or more related polypeptide chains (Sec. 2.1.2).

References

Why genes in pieces? W. Gilbert. *Nature* 271:501 (1978).
Organization and expression of eucaryotic split genes coding for proteins. R. Breathnach and P. Chambon. *Ann. Rev. Biochem.* 50:349–383 (1981).
Genes-in-pieces revisited. W. Gilbert. *Science* 228:823–824 (1985).
The discovery of "split" genes: a scientific revolution. J. A. Witkowski. *Trends Biochem. Sci.* 13:110–113 (1988).

2.1.2 Transcription

The first step in expressing the information in a gene is transcription of the nucleotide sequence into a complementary **messenger RNA (mRNA)** molecule. This occurs, however, only if the appropriate regulatory proteins are present in the correct state. Many regulatory proteins bind to the upstream parts of the gene, known as the **promoter** region, but others act at **enhancer** regions, which can be upstream, downstream, or in the middle of the gene. Some regulatory proteins bind to the promoter region only if certain small regulatory molecules are bound to them, others only if such molecules are absent. Similarly, the binding of regulatory proteins to the DNA molecule can have either a positive or a negative effect on transcription of a gene into mRNA. A variety of mechanisms are used for various genes, often involving the binding of several different regulatory proteins to adjacent parts of a gene. Each mechanism has the simple objective of ensuring that the gene is expressed only under the appropriate conditions, which are determined by the presence or absence of the relevant proteins and regulatory molecules. The subject of gene regulation is central to cell metabolism, to cell differentiation, and to the growth and development of organisms, but it is outside the scope of this volume.

When conditions are right for the expression of a gene, that gene is transcribed by the appropriate **RNA polymerase.** This enzyme copies one strand of the DNA into its complementary RNA sequence. Prokaryotic cells have a single RNA polymerase that transcribes all genes, but it is made specific for different genes by various sigma-factor proteins. Eukaryotic cells have three enzymes, known as RNA polymerases I, II, and III, which transcribe different types of genes. Enzymes I and III transcribe the genes that code for stable RNA molecules such as ribosomal and transfer RNA; RNA polymerase II transcribes genes that code for proteins.

The two strands of the DNA double helix are unwound near the site where transcription begins. Transcription is initiated on one strand, the template strand, at a site dictated by specific sequences in the promoter region of the gene. The template strand of DNA is copied in a 5′ to 3′ direction by assembling nucleoside 5′-triphosphates in a complementary fashion. The first nucleotide of the RNA being assembled is unique in that it retains the 5′-triphosphate group, whereas the others are converted to monophosphates in the process of linking the nucleotides. In eukaryotic cells, this 5′-triphosphate group is rapidly modified to produce a **5′-cap** structure by coupling it with a phosphodiester bond to the 5′ position of a methylated guanine base. Consequently, the capped mRNA has no free 5′ end; this is important for the stability and subsequent translation of the mRNA.

The remainder of the gene is transcribed by adding nucleotides sequentially to the 3′ end of the growing mRNA chain at the rate of approximately 50 nucleotides per second. The mRNA sequence is dictated by complementarity to the nucleotide sequence of the DNA strand being transcribed; that is, during assembly of the mRNA, U, G, C, and A bases attach to the A, C, G, and T bases of the template DNA strand, respectively. Note that the mRNA molecule has the same nucleotide sequence as the unused strand of the DNA molecule, with the usual exception that the T nucleotides in DNA are replaced in RNA by U nucleotides.

The growing mRNA molecule is only transiently associated with the DNA template strand; only some 12 RNA–DNA base pairs are present behind the transcription site. As the mRNA is detached from the template strand after transcription, it is quickly replaced by the other DNA strand as the DNA double helix recouples.

The mRNA molecule is extended until the process reaches signals in the gene sequence that terminate transcription. In eukaryotic cells, the 3′ end of the initial mRNA transcript is trimmed, and a **poly(A)** tail of about 250 adenine nucleotides is added. The role of this poly(A) tail is not known for certain, but it serves in the laboratory as a useful handle for isolating mRNA.

The mRNA in prokaryotic cells can contain the coding regions (cistrons) for more than one protein molecule, but each mRNA molecule in eukaryotic cells codes for a single polypeptide chain. In prokaryotic cells, a single promoter region directs transcription of one or more cistrons (which together are known collectively as an **operon**) into a single mRNA molecule, whereas each eukaryotic gene has its own promoter.

The initial mRNA transcript in eukaryotic cells is cut and spliced to remove the introns. This complicated splicing reaction takes place on a *spliceosome,* which is a large complex containing several proteins and small RNA molecules in addition to the mRNA transcript. Not all the sequences that define the splicing sites are known, but introns invariably start with the sequence GU and end with AG.

Most known genes signal the removal of all introns from all mRNA molecules; this *constitutive splicing*

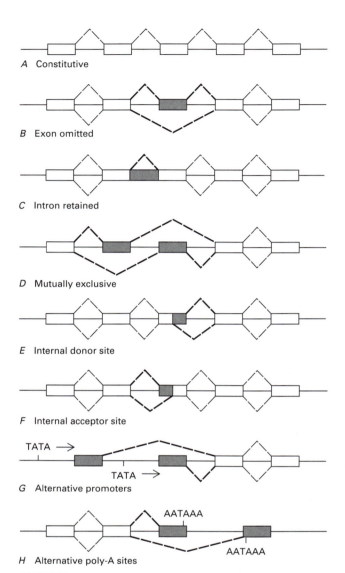

FIGURE 2.2

Observed patterns of precursor mRNA splicing. The DNA of the gene is represented by the horizontal line, with the exons present in any final mRNA boxed. The dashed Vs indicate splicing events to remove the corresponding introns and join the linked exons. Alternative splicing events are indicated by bold Vs. The shaded exons are those that are not always present in the final mRNA as a consequence of alternative splicing. Where the introns interrupt protein-coding sequences, different polypeptide chains are produced as a result of alternative splicing.

A illustrates normal, constitutive splicing, in which every intron is removed from every RNA molecule. B shows the case in which one exon is omitted from some of the mRNA molecules. In C, the segment is removed in some instances but retained in others as an exon. Several such cases can occur in a single gene, and all combinations might result. In D, two exons may be mutually exclusive, in that one is always removed but not both. In cases E and F, alternative donor and acceptor sites are used in some molecules. Alternative promoters G, indicated by the

produces a homogeneous population of mature mRNA molecules. Some introns of certain genes, however, are not spliced out of some of the RNA molecules, leaving new combinations of exons (Fig. 2.2); this *alternative splicing* results in mRNA molecules that have variant sequences. When the introns occur in the protein-coding regions, as is usually the case, variant polypeptide sequences are produced. The mechanism of alternative splicing (in particular, how the alternative splice sites are chosen) is not understood. It appears to be a means of generating a diversity of proteins from a single gene. For example, the gene for fast skeletal troponin T has five consecutive exons that are spliced in all 32 possible combinations, plus two mutually exclusive exons near the region coding for the carboxyl-terminus of the protein. As a result, 64 different troponin T proteins can be coded by a single gene.

Functional differences have been noticed for some of the proteins coded by alternative mRNA splicing, and alternative splicing appears to be biologically regulated. Splicing often generates proteins with different destinations by coding for alternative forms with and without signal peptides or membrane-spanning segments (Sec. 2.3). Why such related proteins are generated by the complexity of alternative splicing of the mRNA of a single gene rather than by different related genes is not known.

The phenomenon of alternative splicing makes the distinction between exon and intron uncertain in some genes (Fig. 2.2). It also constrains the positions of the splice sites so that the reading frame for translation of the mRNA is not altered.

References

The pathway of eukaryotic mRNA formation. J. R. Nevins. *Ann. Rev. Biochem.* 52:441–466 (1983).

Transcription of the mammalian mitochondrial genome. D. A. Clayton. *Ann. Rev. Biochem.* 53:573–594 (1984).

Mechanism and control of transcription initiation in prokaryotes. W. R. McClure. *Ann. Rev. Biochem.* 54:171–204 (1985).

Splicing of messenger RNA precursors. R. A. Padgett et al. *Ann. Rev. Biochem.* 55:1119–1150 (1986).

Transcription termination and the regulation of gene expression. T. Platt. *Ann. Rev. Biochem.* 55:339–372 (1986).

Precision and orderliness in splicing. M. Aebi and C. Weissmann. *Trends Genet.* 3:102–107 (1987).

TATA sequence that is normally present upstream of the transcription initiation site, can produce mRNA molecules with alternative 5′ ends. Likewise, in *H*, alternative sites for transcription termination and polyadenylation, indicated by the usual sequence AATAAA, can result from alternative splicing on the downstream side of the gene.

Alternative splicing: a ubiquitous mechanism for the generation of multiple protein isoforms from single genes. R. E. Breitbart et al. *Ann. Rev. Biochem.* 56:467–495 (1987).

Generation of protein isoform diversity by alternative splicing: mechanistic and biological implications. A. Andreadis et al. *Ann. Rev. Cell Biol.* 3:207–242 (1987).

Structure and function of bacterial sigma factors. J. D. Helmann and M. J. Chamberlin. *Ann. Rev. Biochem.* 57:839–872 (1988).

How the messenger got its tail: addition of poly(A) in the nucleus. M. Wickens. *Trends Biochem. Sci.* 15:277–281 (1990).

Biochemical mechanisms of constitutive and regulated pre-mRNA splicing. M. R. Green. *Ann. Rev. Cell Biol.* 7:559–599 (1991).

2.1.3 Translation

Mature mRNA molecules are translated into polypeptide chains in the cytoplasm by a complex apparatus of ribosomes, **transfer RNA (tRNA),** and various factors. The completed mRNA in eukaryotes migrates from the nucleus into the cytoplasm, whereas in prokaryotic cells the nascent mRNA is not segregated from the cytoplasm and does not undergo splicing or other modifications. Translation in prokaryotes is initiated while the mRNA is being synthesized.

Translation of the mRNA molecule is initiated by the formation of complexes of mRNA, the two subunits of ribosomes, and three initiation-factor proteins. Ribosomes themselves are large particles consisting of 3 RNA molecules and 55 different proteins in prokaryotes (and are even somewhat larger in eukaryotic cells). When they are not bound to mRNA, ribosomes dissociate reversibly into a large and a small subunit. The ribosome binds near the 5′ end of the mRNA molecule, and the sequence there indicates at which point to initiate translation. In prokaryotes this sequence includes a short stretch of nucleotides that is complementary to one of the ribosomal mRNA molecules, whereas the nature of the corresponding sequence in eukaryotic cells is not known. In both cases, translation starts at the triplet sequence AUG.

The life span of mRNA is usually short. In bacteria, the range of half-lives is 2–10 min, which probably reflects the need of bacteria to alter their metabolism quickly in response to drastic environmental changes and their dependence on regulated transcription to accomplish this. In more complex organisms, mRNA is more stable, with half-lives of 0.5–24 h.

a. Activation of Amino Acids

Amino acids couple to appropriate tRNA molecules, which adapt them and supply them in proper sequence to the ribosome – mRNA complex, where they are attached to the polypeptide being synthesized. All tRNA molecules are RNA strands of 73 – 93 nucleotides of various sequences, specified by particular genes, that adopt similar L-shaped folded conformations. Many of the nucleotides of the precursor tRNA molecules are modified covalently after their biosynthesis. Each tRNA molecule is specific for one of the 20 biological amino acids, and there is at least one class of tRNA molecule in every cell for each type of amino acid. All tRNA molecules have the same basic architecture; their specificity is in the sequences of their nucleotides.

The coupling of an amino acid to tRNA occurs through the carboxyl group of the amino acid; this activates the carboxyl group for peptide bond formation, similar to the way peptide chemists synthesize peptides chemically (Sec. 1.8). The chemical energy for activation comes from ATP hydrolysis, and the reaction is catalyzed by an enzyme, the appropriate **aminoacyl tRNA synthetase**. There is one such enzyme for each class of tRNA molecule. The tRNA synthetase first catalyzes the reaction between the appropriate amino acid and ATP, to form an amino acyl adenylate and pyrophosphate (PP_i):

$$
\begin{array}{c}
R \\
| \\
{}^+H_3N - CH - CO_2^- \\
\end{array}
$$

$$\nearrow \text{ATP}$$
$$\searrow \text{PP}_i$$

$$
\begin{array}{c}
R \quad\quad O \quad\quad\quad O \\
| \quad\quad\ \parallel \quad\quad\ \parallel \\
{}^+H_3N - CH - C - O - P - O - \text{ribose} - \text{adenine} \\
| \\
{}_-O
\end{array}
$$

(2.1)

The amino acid moiety is then transferred by the tRNA synthetase to the 2′ or 3′ hydroxyl group at the 3′-terminal A nucleotide of the appropriate tRNA:

(2.2)

Attachment of an amino acid only to its appropriate tRNA molecule is crucial for the fidelity of protein biosynthesis because all subsequent steps are determined only by the interaction between tRNA and mRNA. For example, if the correct amino acid is attached to a tRNA but is subsequently altered chemically, the altered amino acid is incorporated into the nascent polypeptide chain as readily as, and in place of, the correct (unaltered) amino acid. Consequently, aminoacyl tRNA synthetases are very specific for both the amino acid and the tRNA that they join together (see Sec. 9.3.1.a).

Certain organisms use a different mechanism for activating glutamine. Though the amino acids glutamine and asparagine are usually attached directly to their appropriate tRNA molecules, in these atypical organisms the tRNA molecules that are specific for glutamine are first activated with glutamic acid, which is then converted to glutamine by an enzyme. It is not known why this mechanism is used in these species (Chap. 3).

The basis of the selectivity of each tRNA synthetase for its specific tRNA molecule is being elucidated at the present time. All tRNA molecules have similar structures, but they are so large that there are many ways in which they can be distinguished. In the one example of such selectivity that is known in detail, glutaminyl tRNA synthetase, the synthetase makes extensive contacts with its cognate tRNA.

Each tRNA molecule serves as an adaptor between one amino acid and the mRNA: On the side of each tRNA molecule that is opposite to the amino acid bonding site is an **anticodon**, which consists of three nucleotides that base-pair with the complementary three nucleotides of the corresponding **codon** on the mRNA. This base-pairing is always specific for at least two of the three sequential nucleotides, but there is often "wobble" in the interaction with the third. Consequently, some tRNA molecules can recognize two or three codons of the mRNA that differ in the third nucleotide.

References

Enzymic editing mechanisms and the genetic code. A. R. Fersht. *Proc. Roy. Soc., Lond.* 212:351 – 379 (1981).

Recent excitement in understanding transfer RNA identity. L. H. Schulman and J. Abelson. *Science* 240:1591 – 1592 (1988).

Protein biosynthesis in organelles requires misaminoacylation of tRNA. A. Schön et al. *Nature* 331:187 – 188 (1988).

Parameters for the molecular recognition of transfer RNAs. P. Schimmel. *Biochemistry* 28:2747 – 2759 (1989).

Structural basis of anticodon loop recognition by glutaminyl-tRNA synthetase. M. A. Rould et al. *Nature* 352:213 – 218 (1991).

b. Initiation of Translation

The beginning of synthesis of a polypeptide chain on a ribosome–mRNA complex requires the presence of a methionyl tRNA, designated $tRNA_f^{Met}$, which interacts with the initiating AUG codon on the mRNA. In prokaryotic cells, the methionine attached to this tRNA is formylated on its amino group:

$$\mathbf{HC}{-}HN{-}HC{-}\overset{O}{\overset{\|}{C}}{-}O{-}tRNA_f^{Met}$$

$$\overset{O}{\overset{\|}{}}$$

fMet-tRNA$_f^{Met}$ (2.3)

The methionine is formylated enzymatically after its attachment to this tRNA. This formylation does not occur naturally in eukaryotic cells, though it does occur with methionyl-tRNA in vitro in the presence of the appropriate prokaryotic enzymes. The significance of the formylation is not clear, but the initiating tRNA is unique, presumably because it must undergo unique steps in polypeptide initiation. Accordingly, methionine to be incorporated at other positions in the polypeptide chain is attached to a different $tRNA^{Met}$ and is not formylated. Both $tRNA^{Met}$ molecules, however, recognize the same codon, AUG, and the same methionyl tRNA synthetase attaches methionine to both of them.

In prokaryotes, the initiating AUG codon on the mRNA is selected by the ribosome on the basis of a short upstream sequence in the mRNA that is complementary to one of the ribosomal RNA molecules. In some genes the initiating codon may be GUG, UUG, AUU, or AUA, but it is still recognized by Met-tRNA$_f^{Met}$. The signals that define the initiating codon in eukaryotic cells are not so well known; but in 90–95% of genes, it is the first AUG from the 5′ end of the mRNA. It frequently occurs in a sequence like GCCGCC^A/$_G$CCAUGG, in which the boldfaced bases are the initiating codon. A few examples of translation initiation at ACG and CUG codons, but still incorporating a terminal Met residue, have also been observed in eukaryotic cells. The binding of the Met-tRNA$_f^{Met}$ to the ribosome–mRNA complex is a complicated reaction that, in prokaryotes, requires three initiation factors and chemical energy from GTP hydrolysis.

References

Comparison of initiation of protein synthesis in procaryotes, eucaryotes, and organelles. M. Kozak. *Microbiol. Rev.* 47:1–45 (1983).

Initiation of protein synthesis in mammalian cells. V. M. Pain. *Biochem. J.* 235:625–637 (1986).

The scanning model for translation: an update. M. Kozak. *J. Cell Biol.* 108:229–241 (1989).

Translation initiation at non-AUG triplets in mammalian cells. D. S. Peabody. *J. Biol. Chem.* 264:5031–5035 (1989).

Alternative for the initiation of translation. R. C. Herman. *Trends Biochem. Sci.* 14:219–222 (1989).

c. Polypeptide Elongation

The initiating Met-tRNA$_f^{Met}$ binds to the initiating AUG codon of the mRNA and the site on the ribosome (designated P for peptidyl-tRNA) where the nascent peptide chain will be held. An amino acid–charged tRNA with an anticodon that is complementary to the next codon of the mRNA, on the downstream side of the initiating codon, binds to these nucleotides of the mRNA at an adjacent site on the ribosome (A, for aminoacyl-tRNA), again by a complex process. The initiating Met residue now transfers from its tRNA$_f^{Met}$ to the amino group of the second amino acid, which is still attached to its tRNA, to generate the first peptide bond (Fig. 2.3) between the first and second amino acids. The uncharged tRNA$_f^{Met}$ is released from the P site, the dipeptide-tRNA in the A site is translocated to the P site, and the ribosome moves three nucleotides along the mRNA to the next codon.

The ribosome–mRNA complex is now ready to add the third amino acid by inserting the appropriately charged tRNA into the A site of the ribosome, followed by transfer of the dipeptide to this third amino acid. The third tRNA now has a tripeptide attached to it, and it moves from the A site to the P site, displacing the previous deacylated tRNA. There may be a third site on the ribosome (E, for exit) to which the deacylated tRNA moves before dissociation from the ribosome. In the course of adding its amino acid to the polypeptide, a tRNA occupies successively the sites A → P → E.

The polypeptide chain is elongated from its N-terminus by adding one amino acid residue at a time at a rate of 1–10 residues/s. The rate of attachment is somewhat variable for different codons, but there is disagreement about whether the rate-limiting factor is the availability of the appropriately charged tRNA or the need to unravel the three-dimensional structure of the mRNA.

In summary, the nucleotide sequence of the mRNA molecule is read in a 5′ to 3′ direction by the ribosome, three nucleotides (one codon) at a time. Each codon codes for one specific amino acid, which is attached to a tRNA that bears the complementary anticodon. There are three possible reading frames by which trinucleotide codons can be read sequentially on an mRNA molecule, but the reading frame used by the ribosome is determined solely by the position of the initiating codon.

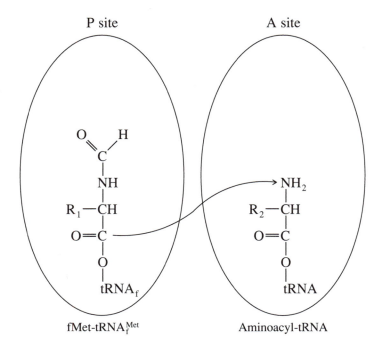

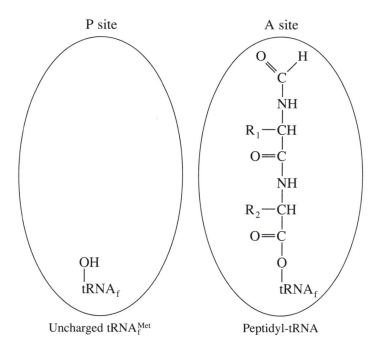

FIGURE 2.3

Formation on the ribosome of the first peptide bond between the initiating N-formyl Met residue and the second residue. After the step shown, the unchanged tRNA$_f^{Met}$ dissociates and the dipeptidyl-tRNA translocates into the P site. Before dissociating, the uncharged tRNA may move to a third site, E. The cycle can then be repeated to add the third amino acid residue, and so forth.

The growing polypeptide chain is thought to pass through a channel in the interior of the ribosome because some 30–40 residues at the growing C-terminus are protected from digestion by added proteases. Other added enzymes and antibodies, however, can act on nascent chains of 15–20 residues. The details of polypeptide growth on the ribosome await further elucidation of the structure of the ribosome.

The polypeptide chain grows until a termination codon (UAA, UGA, or UAG) on the mRNA is reached. These three codons are recognized by release-factor proteins, which cause the polypeptide chain to be hydrolyzed from the last tRNA in the P site. The polypeptide then dissociates from the ribosome, followed by the tRNA and mRNA.

This description of biosynthesis of a polypeptide chain is greatly simplified, explaining only those aspects pertinent to understanding how the final polypeptide chain results. The actual process is remarkably more complex, involving many proteins to catalyze the various steps and to ensure the accuracy of the process.

References

Partial resistance of nascent polypeptide chains to proteolytic digestion due to ribosomal shielding. L. I. Malkin and A. Rich. *J. Mol. Biol.* 26:329–346 (1967).

Stereochemical analysis of ribosomal transpeptidation. V. I. Lim and A. S. Spirin. *J. Mol. Biol.* 188:565–577 (1986).

A new non-crystallographic image-processing technique reveals the architecture of the ribosomes. J. Frank et al. *Trends Biochem. Sci.* 13:123–127 (1988).

Does the channel for nascent peptide exist inside the ribosome? Immune electron microscopy study. L. A. Ryabova et al. *FEBS Letters* 226:255–260 (1988).

Ribosome gymnastics. J. F. Atkins et al. *Cell* 62:413–423 (1990).

The allosteric three-site model for the ribosomal elongation cycle: features and future. K. H. Nierhaus. *Biochemistry* 29:4997–5008 (1990).

d. The Genetic Code

The code relating the 64 possible triplet codons of the mRNA to the amino acids incorporated into the polypeptide chain is shown in Figure 2.4. The code is highly degenerate in that most amino acid residues are designated by more than one triplet. Only Trp and Met residues are designated by single codons. At the other extreme, Leu, Arg, and Ser are each encoded by six codons. In most cases, the degeneracy involves closely related codons, usually differing in the third position. This is the position where the base-pairing phenomenon known as wobble permits the same tRNA molecule to

First position	Second position U	C	A	G	Third position
U	Phe	Ser	Tyr	Cys	U
	Phe	Ser	Tyr	Cys	C
	Leu	Ser	Terminate	Terminate	A
	Leu	Ser	Terminate	Trp	G
C	Leu	Pro	His	Arg	U
	Leu	Pro	His	Arg	C
	Leu	Pro	Gln	Arg	A
	Leu	Pro	Gln	Arg	G
A	Ile	Thr	Asn	Ser	U
	Ile	Thr	Asn	Ser	C
	Ile	Thr	Lys	Arg	A
	Met	Thr	Lys	Arg	G
G	Val	Ala	Asp	Gly	U
	Val	Ala	Asp	Gly	C
	Val	Ala	Glu	Gly	A
	Val	Ala	Glu	Gly	G

FIGURE 2.4

The genetic code of protein biosynthesis. The nucleotides of the 64 possible triplet codons are those on the messenger RNA (example: UUU and UUC both code for Phe).

recognize different codons (Sec. 2.1.3.a). One important exception is Ser, for which two of its codons, AGU and AGC, are very different from its other four codons, UCN where N can be any of the four nucleotides. The two sets cannot be interconverted by single base changes.

Sixty-one of the codons code for the 20 amino acids, and each of the other three designates termination of translation and the carboxyl end of the polypeptide chain. It is surprising in some respects that there are three different termination codons. One would seem to be sufficient if termination were absolute, but this appears not to be the case. In many cells there are minor tRNA species, known as suppressor tRNA, that have anticodons corresponding to the termination codons. They insert amino acid residues where termination normally occurs, though usually with low efficiency, and thus permit the polypeptide chain to be extended until the next termination codon of the mRNA. Some genes avoid this read-through by having two or more termination codons in tandem at the end of the coding region. For most other genes, in which termination read-through occurs at low frequency, there appear to be no physiological consequences of producing a small fraction of polypeptide chains with extended C-termini. In certain viruses, however, the presence of this low level of extended read-through product is important for virus assembly and growth; the viruses appear to use this phenomenon to take over the protein synthetic machinery of their hosts.

The phenomenon of read-through by suppressor tRNA species has long been of interest primarily for genetic studies, in which it is used to suppress the physiological effects of mutations to chain termination codons within a gene, which cause protein fragments to be made (Sec. 3.2.2). In the presence of the suppressor tRNA molecules, various amino acid residues are inserted with varying frequencies into the polypeptide chain at such positions. It is now clear, however, that the phenomenon is of much greater biological significance, in that certain termination codons are used to insert an unusual amino acid in those positions with high efficiency, as in the case of selenocysteine.

References

Suppression of termination codons in higher eukaryotes. D. Hatfield. *Trends Biochem. Sci.* 10:201–204 (1985).

Codon usage and gene expression. J. F. Ernst. *Trends Biotech.* 6:196–199 (1988).

Selenocysteine, the Twenty-First Amino Acid?

The amino acid **selenocysteine** differs from cysteine in having a selenium atom in place of sulfur. Selenium is an essential component of a few important enzymes in both prokaryotes and eukaryotes, such as glutathione peroxidase and formate dehydrogenase. This residue occurs at a specific position in every polypeptide chain of these enzymes. The gene sequences for these enzymes indicate that the codon specifying the selenocysteine residue is UGA, one of the normal termination codons, but this particular codon is translated by a special tRNA. In bacteria this tRNA is charged directly with selenocysteine, but in eukaryotes it is charged initially with serine. The serine residue attached to this tRNA is converted to selenocysteine by exchange of its hydroxyl group for a selenol (—SeH) group, possibly going through a phosphoryl intermediate. Only very few specific UGA codons are translated as selenocysteine, apparently depending on the nucleotides on both sides of the UGA codon, which probably affect the mRNA structure. In addition, a protein is involved in the interaction between the selenocysteinyl tRNA and the mRNA.

References

A new puzzle in selenoprotein biosynthesis: selenoprotein biosynthesis: selenocysteine seems to be encoded by the "stop" codon, UGA. I. Chambers and P. R. Harrison. *Trends Biochem. Sci.* 12:255–256 (1987).

A flexible genetic code, or why does selenocysteine have no unique codon? H. Engelberg-Kulka and R. Schoulaker-Schwarz. *Trends Biochem. Sci.* 13:419–421 (1988).

Selenoprotein synthesis: an expansion of the genetic code. A. Böck et al. *Trends Biochem. Sci.* 16:463–467 (1991).

Variations in the Genetic Code

The genetic code shown in Figure 2.4 is nearly universal in that it is used by most organisms, spanning all the various forms of life. In a number of variations on the theme, however, a slightly different code is used in certain cells and organelles.

The most conspicuous example is the mitochondrion, in which slightly altered codes are used by many species. Mitochondria contain their own DNA genetic material and protein biosynthetic apparatus, with which they produce some 10–20 proteins that are specific to the mitochondrion (many other proteins used by mitochondria are synthesized in the cytoplasm and imported into the mitochondrion, Sec. 2.3.2). The termination codon UGA, which is the codon for selenocysteine in some cases, usually codes for Trp in mitochondria. The AUA codon often codes for Met in mitochondria rather than Ile. Both of these alterations make the usual code more symmetrical in that otherwise identical codons ending in U and C specify the same amino acid, as do those ending in A and G (Fig. 2.4). Other variations have been observed in the mitochondria of certain species.

Some free-living prokaryotes (mycoplasma) and eukaryotes (ciliated protozoa and paramecia) also have variant genetic codes. These variations are still being elucidated, for this requires knowledge of the sequences of both the gene and the corresponding protein.

Variations in the genetic code are surprising because the general universality of the code is part of the evidence for the consensus that all living creatures have descended from a common ancestor (Chap. 3). After the genetic code was fixed, it would seem impossible for it to have been altered without disastrous consequences for gene expression. The mitochondria can be imagined to have been able to do this because they synthesize so few proteins. The other examples cannot be explained so easily; but because they occur in very distantly related species, it is possible that these species diverged from the majority of other species at a very early stage of evolution when the genetic code had not yet been fixed.

References

tRNA, suppression, and the code. E. J. Murgola. *Ann. Rev. Genet.* 19:57–80 (1985).

Natural variation in the genetic code. T. D. Fox. *Ann. Rev. Genet.* 21:67–91 (1987).

Differential use of termination codons in ciliated protozoa. D. S. Harper and C. L. John. *Proc. Natl. Acad. Sci. USA* 86:3252–3256 (1989).

e. Posttranslational Events

The nucleic acid sequence determines directly only the primary structure of a polypeptide chain during biosynthesis, but a polypeptide chain can be altered in many ways during and after biosynthesis, so the structure of the biologically active protein can differ substantially from that coded by its gene. The most common posttranslational modifications are described later in this chapter.

Formation of a biologically active protein requires that its polypeptide chains be folded into appropriate three-dimensional conformations. Very little is known about how this process occurs naturally after biosynthesis, so this topic is discussed in Chapter 7 with regard to refolding in vitro of the complete polypeptide chain. Proteins must also be directed to their appropriate locations inside or outside cells. Much has been learned recently about this phenomenon, which is discussed in a later section of this chapter.

2.2 Protein Engineering

Isolating the gene for a protein or a **complementary DNA (cDNA)** copy of its messenger RNA is now fairly routine. After a gene or cDNA is isolated, the determination of its sequence of nucleotides and, by use of the genetic code (Fig. 2.4), the amino acid sequence of the protein for which it codes is straightforward. Most protein sequences are now determined in this way. Alternatively, if the amino acid sequence of a protein is known, a synthetic gene that codes for it can be constructed. An isolated or synthesized gene can be translated into protein in large quantities by inserting the gene into an appropriate expression system. The nucleotide sequence of a gene can then be modified to make any conceivable variant of the original protein.

These techniques have launched the new technology of protein engineering, which has revolutionized the study of proteins.

References

DNA sequencing and gene structure. W. Gilbert. *Science* 214:1305–1312 (1981).

Determination of nucleotide sequences in DNA. F. Sanger. *Science* 214:1205–1210 (1981).

The human interferons—from protein purification and sequence to cloning and expression in bacteria: before, between, and beyond. S. Pestka. *Arch. Biochem. Biophys.* 221:1–37 (1983).

Protein engineering. K. M. Ulmer. *Science* 219:666–671 (1983).

2.2.1 *Isolating the Gene for a Protein*

It is usually necessary to know part of the sequence of a protein to identify its gene or cDNA, or to have antibodies or some other very sensitive method to detect the protein. Given the amino acid sequence, the genetic code can be used to design an oligonucleotide probe of DNA or RNA that base-pairs with the gene or cDNA coding for that sequence. The degeneracy of the genetic code means, however, that there is uncertainty about which nucleotides should be used at the degenerate positions. This problem is minimized if Trp or Met residues (which have a single codon each) are present. Otherwise, this problem can be overcome by using knowledge about which codons are used most frequently in the species from which the gene is to be isolated and by designing a sufficiently long oligonucleotide that a few wrong nucleotides do not interfere with base-pairing between the oligonucleotide and the gene or cDNA. Alternatively, a mixture of probes can be synthesized by inserting mixtures of nucleotides at the degenerate positions. The synthesized oligonucleotide probes are then used to identify, by specific hybridization, the desired clones in a suitable "library" of cloned genes or cDNA copies.

An alternative procedure to isolating the gene for a particular protein is to clone the genes or cDNA directly into an expression system that produces protein from the cloned fragments. Then, if a suitably sensitive detection procedure is available, the clones that produce the desired protein can be identified. Usually this technique involves antibodies directed against the protein (Sec. 8.3.1.b).

After the gene or cDNA fragment is cloned, it is straightforward to obtain its nucleotide sequence. The methods for sequencing DNA are much more straightforward and universally applicable than are those for sequencing proteins (Sec. 1.6).

Reference

Molecular Cloning: A Laboratory Manual, 2nd ed. J. Sambrook et al. Cold Spring Harbor Lab., New York, 1989.

2.2.2 *Protein Sequences from Gene Sequences*

The gene sequence can be converted into an amino acid sequence using the genetic code, but the coding region and the correct reading frame must first be identified. This identification is most straightforward if segments of the protein sequence are known and if the cDNA has been sequenced, because the introns have been removed from cDNA (Fig. 2.2). Currently, no adequate method exists for identifying introns or the start and stop sites for transcription and translation in gene sequences. In addition, genomes often contain **pseudogenes,** which are not transcribed or translated because of mutations that inactivate them. In some cases, these mutations are not obvious, so it can be difficult to distinguish pseudogenes from real genes. The stop and start sites for translation are more easily identified in cDNA sequences, though it is necessary to be certain that the entire mRNA is represented by a cDNA clone. The segment of the cDNA sequence coding for the protein can usually be identified as the longest open reading frame (lacking termination codons) and as that segment using the particular degenerate codons that are appropriate for the species from which the gene was isolated. Species often differ in the average overall nucleotide content of their DNA and in their use of various codons for each amino acid.

Even under ideal circumstances the amino acid sequence identified from a cDNA sequence must be considered tentative until it is confirmed by at least the amino acid sequence at both ends of the protein. A number of complications can occur during translation of an mRNA. For example, a variant genetic code might be used in the organelle or species from which the gene was isolated; a specific codon might be translated atypically as in the case of UGA coding for selenocysteine (Sec. 2.1.3.d); or an unusual initiation codon might be used. There are also instances in which the mRNA appears to be edited by insertion of nucleotides not present in the gene. In a few other cases, ribosomes translate some mRNA molecules atypically, skipping a nucleotide so that the codon reading frame is shifted. In the mRNA for bacterial release factor 2, for example, the 26th codon is UGA, a normal termination codon, but it is read by ribosomes as the four bases UGAC and is acted upon as a codon for Asp (usually coded by GAC); consequently, the U base is skipped and the reading frame is altered for the remainder of the cistron.

Fortunately, such occurrences are rare, and protein sequences determined from gene sequences — once the open reading frame used is confirmed — are usually more correct than those determined directly by protein sequencing. Gene sequencing is most necessary when proteins are present naturally in only minute quantities, as is usually the case for proteins with potent biological activities, which are often those of greatest interest. The sequences of many hormones and receptors have been determined only from their gene sequences. One exam-

ple (Fig. 2.5) is the hormone precursor pro-opiomelanocortin, which is the precursor to a number of important peptide hormones (Sec. 2.4.1.b).

References

Codon distribution in vertebrate genes may be used to predict gene length. W. Bains. *J. Mol. Biol.* 197:379–388 (1987).

Oligopeptide biases in protein sequences and their use in predicting protein coding regions in nucleotide sequences. P. McCaldon and P. Argos. *Proteins* 4:99–122 (1988).

How "hidden" reading frames are expressed. R. Cattaneo. *Trends Biochem. Sci.* 14:165–167 (1989).

2.2.3 *Identifying a Protein Specified by a Gene of Known Sequence*

Many genes are identified and isolated on the basis of genetic phenomena alone; the product of the gene may not be known. Examples are the genes that produce unusual physiological effects when they are defective, such as oncogenes, developmental genes, and genes causing Huntington's chorea, muscular dystrophy, and other genetic disabilities. The genetic defect can be localized to a particular region of a chromosome, which can then be cloned and its sequence determined. To understand the biochemical basis of the physiological effect, the products of the normal and mutated gene must be identified and characterized.

The entire amino acid sequence of the protein as it is initially produced is known from the gene sequence, but many modifications can take place subsequently (Sec. 2.4). Some properties of a protein that may be useful in identifying it can be predicted from its sequence, such as its molecular weight and isoelectric point, but these properties can be altered drastically by posttranslational modifications. Fortunately, a polypeptide chain in a protein often can be recognized specifically by antibodies prepared against a synthetic peptide that is identical in sequence to a segment of only 8–15 residues of the protein.

For an antibody to bind to a protein, the sequence the antibody identifies must be accessible to the antibody. Antibodies raised against a peptide segment that is buried in the interior of a folded protein (Chap. 6) cannot bind to it. One solution is to unfold the protein before antibody binding. A more satisfactory solution is to select a peptide segment that is likely to be on the surface of the protein and accessible to antibodies, whatever the conformation of the protein; criteria for selecting such peptide segments are being devised. Another consideration in selecting a peptide segment is its

5′----AAGAGAACGAAGGGAAGAAAAGUGACCGAGAGGCCGCUGAACAUCCUCGCCCCGGCGCAGCGGGAGCCGCCCGAGGCAGCUUCCCCGUGACAGAGCCUCAGCCUGCCUGGAAG
−500 −450 −400

−120 ↓
Met Pro Arg Leu Cys Ser Ser Arg Ser Gly Ala Leu Leu Leu Ala Leu Leu Leu Gln Ala Ser Met Glu Val Arg Gly Trp Cys Leu Glu
AUG CCG AGA CUG UGC AGC AGU CGU UCG GGC GCC CUG CUG CUG GCC UUG CUG CUU CAG GCC UCC AUG GAA GUG CGU GGU UGG UGC CUG GAG
−350

−100 −80
Ser Ser Gln Cys Gln Asp Leu Thr Thr Glu Ser Asn Leu Leu Ala Cys Ile Arg Ala Cys Lys Pro Asp Leu Ser Ala Glu Thr Pro Val
AGC AGC CAG UGU CAG GAC CUC ACC ACG GAA AGU AAC CUG CUG GCG UGC AUC CGG GCC UGC AAG CCC GAC CUC UCC GCC GAG ACG CCG GUG
−300 −250

−60
Phe Pro Gly Asn Gly Asp Glu Gln Pro Leu Thr Glu Asn Pro Arg Lys Tyr Val Met Gly His Phe Arg Trp Asp Arg Phe Gly Arg Arg
UUC CCC GGC AAC GGC GAU GAG CAG CCG CUG ACU GAG AAC CCC CGG AAG UAC GUC AUG GGC CAU UUC CGC UGG GAC CGC UUC GGC CGU CGG
−200 −150

−40 −20
Asn Gly Ser Ser Ser Ser Gly Val Gly Gly Ala Ala Gln Lys Arg Glu Glu Glu Val Ala Val Gly Glu Gly Pro Gly Pro Arg Gly Asp
AAU GGU AGC AGC AGC AGC GGA GUU GGG GGC GCG GCC CAG AAG CGC GAG GAG GAA GUG GCG GUG GGC GAA GGC CCC GGG CCC CGC GGC GAU
−100 −50

−1 ↓ 1 ↓ ↓
Asp Ala Glu Thr Gly Pro Arg Glu Asp Lys Arg Ser Tyr Ser Met Glu His Phe Arg Trp Gly Lys Pro Val Gly Lys Lys Arg Arg Pro
GAC GCC GAG ACG GGU CCG CGC GAG GAC AAG CGU UCU UAC UCC AUG GAA CAC UUC CGC UGG GGC AAG CCG GUG GGC AAG AAG CGG CGC CCG
−1 1 50

20 ↓ 40 ↓
Val Lys Val Tyr Pro Asn Gly Ala Glu Asp Glu Ser Ala Gln Ala Phe Pro Leu Glu Phe Lys Arg Glu Leu Thr Gly Glu Arg Leu Glu
GUG AAG GUG UAC CCC AAC GGC GCC GAG GAC GAG UCG GCC CAG GCC UUU CCC CUC GAA UUC AAG AGG GAG CUG ACC GGG GAG AGG CUC GAG
100

60
Gln Ala Arg Gly Pro Glu Ala Gln Ala Glu Ser Ala Ala Ala Arg Ala Glu Leu Glu Tyr Gly Leu Val Ala Glu Ala Glu Ala Glu Ala
CAG GCG CGC GGC CCC GAG GCC CAG GCU GAG AGU GCG GCC GCC CGG GCU GAG CUG GAG UAU GGC CUG GUG GCG GAG GCG GAG GCU GAG GCG
150 200

80 ↓ 100 ↓ ↓
Ala Glu Lys Lys Asp Ser Gly Pro Tyr Lys Met Glu His Phe Arg Trp Gly Ser Pro Pro Lys Asp Lys Arg Tyr Gly Gly Phe Met Thr
GCC GAG AAG AAG GAC UCG GGG CCC UAU AAG AUG GAA CAC UUC CGC UGG GGC AGC CCG CCC AAG GAC AAG CGC UAC GGC GGG UUC AUG ACC
250 300

120
Ser Glu Lys Ser Gln Thr Pro Leu Val Thr Leu Phe Lys Asn Ala Ile Ile Lys Asn Ala His Lys Lys Gly Gln
UCC GAG AAG AGC CAA ACG CCC CUU GUC ACG CUG UUC AAA AAC GCC AUC AUC AAG AAC GCC CAC AAG AAG GGC CAG UGA GGGCGCAGCGGGCAG
350 400

GGGCCUCUCUCCGCGGAAAGUUGACCCUGAAGGCCUCUCUUCUGCCCUCCUACCGCCUCGCAGCCUGGGUGAGGAUUCGCCCAGGCAGUGAUGGCGCCAGGUAUCCCGACUCUUAAAGC
450 500

UGUCUGUAGUUAAGAAAUAAAACCUUUCAAGUUUCACGAAAAAAAA----3′
550

FIGURE 2.5
Nucleotide sequence of the mRNA (single lines and the lower line of each pair of lines) and the inferred amino acid sequence (upper line of each pair of lines, beginning with Met) of bovine preproopiomelanocortin. The mRNA sequence was determined from the sequence of cloned cDNA. The positive numbers of both the nucleotides and the amino acid residues begin at the N-terminus of corticotropin; the negative numbers are for preceding nucleotides and residues. At the 3′ end of the mRNA is a portion of the poly(A) tail. This hormone precursor is processed into smaller fragments by proteolytic cleavage at the arrows. The first 26 residues comprise the signal peptide; subsequent processing occurs as illustrated in Figure 2.16. (From S. Nakanishi et al., *Nature* 278:423–427, 1979.)

immunogenicity; that is, its ability to stimulate the generation of antibodies against it.

Once the antibodies have been generated, the protein to which they bind can be identified after electrophoretic separation ("Western blotting"), by *immunoprecipitation* of soluble protein, *immunofluorescence* of protein fixed in situ, or *immunoaffinity chromatography,* in which antibodies are bound to a resin and protein samples are passed through the resin to determine which proteins bind (see Sec. 8.3.1).

In addition to detecting a natural protein by using antibodies, a protein specified by a cloned gene can be produced directly by expressing the gene in an appropriate expression system.

References

An assessment of prediction methods for locating continuous epitopes in proteins. M. H. V. van Regenmortel and G. Daney de Marcillac. *Immunol. Letters* 17:95–107 (1988).

Immunological detection of proteins of known sequence. K. H. Scheidtmann. In *Protein Function: A Practical Approach,* T. E. Creighton, ed., IRL Press, Oxford, pp. 93–115 (1989).

2.2.4 Expressing a Cloned or Synthetic Gene

After a gene is isolated or synthesized, it is relatively straightforward to express it and to make substantial quantities of the protein for which it codes. The segment of DNA corresponding to the coding region need only be inserted, in the correct reading frame, into a suitable DNA expression vector. Such vectors are usually small, circular DNA molecules that include all the nucleic acid sequences needed to ensure replication of the vector and efficient transcription and translation of the gene in an appropriate host cell. A vast number of expression vectors, too numerous to describe here, are available. They vary in the host cells in which they are designed to be functional, in the way in which expression of the gene is controlled, and in the type of protein they produce.

Expression systems are available for eukaryotic cell cultures, yeast, and various bacteria. They differ primarily in their efficiency of translation and in the way that they process the protein synthesized. For large-scale production of proteins, bacteria such as *Escherichia coli* are favored because of the wealth of knowledge about the genetics of this system and the consequent availability of very efficient expression vectors. It is not uncommon for the protein encoded by the cloned gene to make up 10–50% of the total protein produced by the *E. coli* cells. The main problems with *E. coli* and other bacterial expression systems for producing eukaryotic proteins are that many of the proteins synthesized are degraded and that prokaryotes do not possess systems for posttranslationally modifying the proteins in the way that eukaryotic cells do (Sec. 2.4). This includes, especially, glycosylation and proteolytic cleavage.

Another common property of bacterial expression systems is that proteins being produced in large quantities often accumulate in an insoluble form in **inclusion bodies,** which are amorphous aggregates of the synthesized protein plus other cellular components. Why proteins accumulate in this way is not known, but it is thought to be due to the high concentrations of the protein being made and the relative insolubility of unfolded polypeptides, which aggregate before they can fold to a more soluble form (Sec. 7.5), or to interactions with other components of the cell. Inclusion bodies are not peculiar to eukaryotic proteins but also occur with prokaryotic proteins and even with normal proteins of the host when these are made in abnormally large quantities. Although synthesis of a protein as insoluble inclusion bodies requires that the protein be solubilized and correctly folded in order to convert it to a biologically active form, inclusion bodies have the advantages of aiding purification of the protein and of protecting it from degradation in the cell.

The synthesized proteins may be secreted from the host cells or targeted to locations in the cell if the appropriate signal peptides are included (Sec. 2.3). Secreted proteins are often made in small quantities and can be difficult to recover, but they are not so susceptible to degradation or precipitation as those released into the cytosol.

Synthesis of a desired protein by an expression vector should require only that the protein have an initiating methionine residue because all other known signals for transcription and translation lie outside the protein-coding region of a gene and are part of the expression vector. Many eukaryotic proteins, however, are not made in large quantities in prokaryotic cells. The nucleotide sequence of at least the initial part of the coding sequence determines to some extent the level of expression, which may be due to secondary structure in the mRNA. Also, eukaryotic proteins, especially small ones, are often degraded in bacteria. For these reasons, eukaryotic proteins made in *E. coli* are often fused to the carboxyl end of a host protein that is made in large quantities. The nucleotide sequence of the coding region of a gene that is distant from the initiation point usually does not affect its efficiency of expression.

When the fusion technique is used, the desired protein must be cleaved from the fused protein in a very specific manner. This usually requires the insertion of a suitable sequence that can be cleaved specifically at the junction between the two sections of protein. The favored methods are the most specific of those used in peptide mapping (see Table 1.5), such as CNBr cleavage at Met residues, acid hydrolysis at the sequence Asp-Pro, or cleavage by trypsin or clostripain at basic residues. More specific proteases found to be useful are Factor X_a, which cleaves polypeptides primarily after the sequence -Ile-Glu-Gly-Arg-, and collagenase, which cleaves collagenlike sequences -Pro-Xaa-↓Gly-Pro-Xaa-↓ (Sec. 5.5.3). Which procedure is used depends on what related sequences are present in the protein being made. Some procedures may leave undesirable amino acids at the N-terminus of the final protein product.

One advantage of a fusion protein is that it can be designed to aid purification. The N-terminus of a fusion protein may have specific binding properties that enable the fusion protein to be isolated easily. In a different approach, the gene for a protein can be extended to produce a "tail" of Arg residues at its C-terminus. This can make the protein much more basic, and purification by cation exchange chromatography can be effective. Subsequently, the poly (Arg) tail can be removed enzymatically by carboxypeptidase B to yield the desired protein. The protein's charge properties should be sufficiently altered that it can be purified easily from any contaminating proteins by a repeat chromatographic step.

References

The purification of eukaryotic polypeptides synthesized in *Escherichia coli*. F. A. O. Marston. *Biochem. J.* 240:1–12 (1986).

Formation of recombinant protein inclusion bodies in *Escherichia coli*. J. F. Kane and D. L. Hartley. *Trends Biotechnol.* 6:95–101 (1988).

Engineering proteins for purification. H. M. Sassenfeld. *Trends Biotechnol.* 8:88–93 (1990).

2.2.5 Site-Directed Mutagenesis

The amino acid sequence of a protein being produced in the laboratory by expression of its gene can be changed by **site-directed mutagenesis** of its gene. By this very important technique, the nucleotide sequence of the gene is altered to code for any desired primary structure.

The experimental procedure consists of only four steps (Fig. 2.6), although many variations are possible.

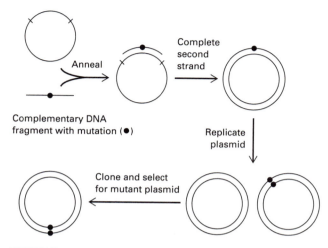

FIGURE 2.6

Site-directed mutagenesis of a cloned gene. A suitable plasmid with a gene to be mutated, in the form of single-stranded DNA, is annealed by base-pairing to a complementary DNA fragment containing the mutation to be incorporated (●). This mutation can be replacement, deletion, or insertion of one or more nucleotides, the only requirements being that the flanking complementary segments of the oligonucleotide be sufficient to bind the fragment specifically to the plasmid. A second strand of DNA is synthesized enzymatically from the first strand, using the oligonucleotide as a primer, to incorporate the mutation. The resulting hybrid plasmid is used to transform cells and to generate homogeneous double-stranded DNA molecules by replication. The DNA molecules containing the mutation on both strands can be selected or simply identified by DNA sequencing.

1. Synthesize an oligonucleotide complementary to the part of the gene to be modified but including the nucleotide changes to be made. These changes can replace, delete, or insert nucleotides. The flanking complementary sequences must be sufficiently long that the mutagenic oligonucleotide still base-pairs to the correct segment of the gene, in spite of the destabilizing effect of the mutation. For single nucleotide changes, flanking sequences of 9–10 nucleotides are usually adequate. For multiple changes, insertions, and deletions, correspondingly longer flanking regions are necessary.

2. Hybridize the mutagenic oligonucleotide specifically to the appropriate segment of the complementary single strand of the gene.

3. Convert the heteroduplex DNA molecule produced by step 2 into a complete, full-length, circular double-stranded DNA molecule by enzymatic DNA synthesis and ligation. One DNA strand will contain the original sequence, the other the desired mutant sequence.

4. Use the double-stranded DNA from step 3 to transform competent host cells; in the process, the heteroduplex DNA molecules will be replicated to form homoduplexes containing either the original or the mutant form of the DNA. Clones containing only the mutant DNA molecules are identified by sequencing various clones or by their preferential hybridization with the mutagenic oligonucleotide.

Ideally, half the resulting double-stranded DNA progeny should be totally mutant, the other half unaltered. In practice, however, the efficiency of mutagenesis is considerably less than might be expected: The mutant oligonucleotide may not have hybridized sufficiently to the original template strand, it may have been displaced during synthesis of the second strand, or it may have been repaired by the host cell. Various methods have been devised to increase the efficiency of mutagenesis. Most of them introduce a difference between the original and mutant DNA strands that can be exploited to select for the strand with the mutation. By any of these methods, the efficiency of cloning can be increased so that any mutation can be easily made.

The alteration of protein sequences by site-directed mutagenesis is generally limited to use of one of the 20 natural amino acids (Fig. 2.4). Yet there are instances in which an unnatural amino acid, such as α-amino butyric acid with a $—CH_2—CH_3$ side chain, would be desirable. Substitutions of unnatural amino acids can be accomplished by carrying out the translation in vitro. The unnatural amino acid is chemically coupled to a suitable suppressor tRNA that recognizes a termination codon such as UAG. This unnatural amino acid is incorporated into the polypeptide chain made in the in vitro expression system at a specific UAG codon incorporated into the coding sequence by site-directed mutagenesis. At present, however, the procedure is limited by the relatively poor yields of in vitro protein translation.

References

In vitro mutagenesis. M. Smith. Ann. Rev. Genet. 19:423–462 (1985).

Strategies and applications of in vitro mutagenesis. D. Botstein and D. Shortle. Science 229:1193–2101 (1985).

A general method for saturation mutagenesis of cloned DNA fragments. R. M. Myers et al. Science 229:242–247 (1985).

Site-directed mutagenesis. P. Carter. Biochem. J. 237:1–7 (1986).

The efficiency of oligonucleotide-directed mutagenesis. T. A. Kunkel. Nucleic Acids and Mol. Biol. 2:124–135 (1988).

A general method for site-specific incorporation of unnatural amino acids into proteins. C. J. Noren et al. Science 244:182–188 (1989).

2.3   Topogenesis

Newly synthesized proteins must be targeted to their appropriate locations. Some proteins function in the cytoplasm of the cell in which they are made, and others are secreted from that cell. In eukaryotic cells, some proteins function in organelles — such as mitochondria, lysosomes, the nucleus, and chloroplasts — though they are synthesized in the cytoplasm; others must be incorporated into an appropriate membrane. The targeting is simplest for proteins that remain in the cytoplasm; they appear to be synthesized on free ribosomes in the cytoplasm and then simply to be released to diffuse in the cell. Reaching other locations usually requires crossing at least one membrane boundary, which is a considerable accomplishment for a long, polar polypeptide chain. One role of membranes is to serve as barriers between compartments that are impermeable to most molecules, and they are usually impermeable to proteins. Yet certain proteins cross such membranes routinely. How and why they do so is the phenomenon of **topogenesis.**

References

Intracellular aspects of the process of protein synthesis. G. E. Palade. Science 189:347–358 (1975).

Intracellular protein topogenesis. G. Blobel. Proc. Natl. Acad. Sci. USA 77:1496–1500 (1980).

Using recombinant DNA techniques to study protein targeting in the eucaryotic cell. H. Garoff. Ann. Rev. Cell Biol. 1:403–405 (1985).

Multiple mechanisms of protein insertion into and across membranes. W. T. Wickner and H. F. Lodish. Science 230:400–407 (1985).

Protein translocation across and integration into membranes. T. A. Rapoport. Crit. Rev. Biochem. 20:73–137 (1986).

Preprotein conformation: the year's major theme in translocation studies. D. I. Meyer. Trends Biochem. Sci. 13:471–474 (1988).

Protein unfolding and the energetics of protein translocation across biological membranes. M. Eilers and G. Schatz. Cell 52:481–483 (1988).

Unity in function in the absence of consensus in sequence: role of leader peptides in export. L. L. Randall and S. J. S. Hardy. *Science* 243:1156–1159 (1989).

2.3.1 Secreted, Lysosomal, and Membrane Proteins, via the Endoplasmic Reticulum

Eukaryotic proteins that are destined for secretion from the cell, for lysosomes and some other organelles, and for the various cellular membranes all begin their synthesis on free ribosomes in the cytoplasm and finish it attached to the **endoplasmic reticulum (ER)** (Fig. 2.7). An N-terminal **signal peptide** is synthesized first; as soon as it appears as the nascent chain on the ribosome, it is recognized by the **signal recognition particle (SRP),** which binds to it and stops peptide elongation. Continued synthesis of the polypeptide chain occurs only after the ribosome–mRNA–nascent-chain–SRP complex interacts with a receptor for SRP on the membrane of the ER (known as the *docking protein*). The numerous ribosomes that attach in this way to the ER membrane give the membrane a microscopic appearance that has caused it to be called the *rough* endoplasmic reticulum.

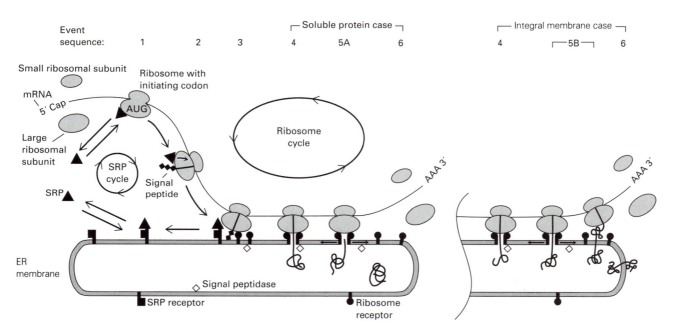

FIGURE 2.7

The mechanism of protein translocation across the endoplasmic reticulum (ER) membrane. The ER is shown at the bottom as a compartment bounded by a bilayer membrane. Above it is the mRNA with ribosomes at various stages of translation. At 1, large and small ribosomal subunits attach near the 5′ end of the mRNA to initiate its translation at the AUG initiation codon. The signal recognition particle (SRP, ▲), having some affinity for the translating ribosome, attaches to it. Upon emergence of the newly synthesized signal peptide from the ribosome at 2, the SRP binds more tightly and inhibits further translation. At 3, the SRP interacts with the SRP-receptor on the ER membrane (▉) and delivers the ribosomal complex to the receptors (●) for ribosomes and signal peptides, also on the ER membrane. The SRP and its receptor dissociate from the ribosome and from each other and recycle via the SRP cycle. At 4, the signal peptide has been cleaved from the nascent chain by the signal peptidase, ◇, and translation proceeds.

  In the case of *soluble proteins*, the completed polypeptide chain is released, at 5A, into the lumen of the ER, possibly through a pore in the membrane, where it folds into its three-dimensional conformation. At 6, having completed translation, the ribosome dissociates from the mRNA and into large and small subunits, which recycle via the ribosome cycle.

  In the case of *integral membrane proteins*, translocation is aborted, at 5B, prior to completion of the polypeptide chain by a hydrophobic stop-transfer segment of the polypeptide chain, and the protein remains embedded in the membrane at one or more segments of the polypeptide chain. (From P. Walter et al. *Cell* 38:5–8, 1984.)

When the ribosome complex binds to the ER membrane, the SRP is displaced, the block on translation is released, and the growing polypeptide chain is extruded directly through the membrane into the lumen of the ER. In the process, the signal peptide is usually cleaved off by a specific *signal peptidase*. The new N-terminus is usually given the residue number $+1$; the residues of the signal peptide have negative numbers. The protein as initially synthesized with the signal peptide is known as the **preprotein.**

After completion of polypeptide synthesis, these proteins are either released into the lumen of the ER or remain in its membrane. They fold to three-dimensional conformations (Sec. 7.5), often assemble into complexes of polypeptide chains, undergo a variety of covalent modifications (Sec. 2.4), and are subsequently directed to their destinations (Sec. 2.3.1.f).

References

Protein translocation across the endoplasmic reticulum. P. Walter et al. *Cell* 38:5–8 (1984).

Protein localization and membrane traffic in yeast. R. Schekman. *Ann. Rev. Cell Biol.* 1:115–143 (1985).

Mechanism of protein translocation across the endoplasmic reticulum membrane. P. Walter and V. R. Lingappa. *Ann. Rev. Cell Biol.* 2:499–516 (1986).

Molecular mechanisms of protein secretion: the role of the signal sequence. M. S. Briggs and L. M. Gierasch. *Adv. Protein Chem.* 38:109–180 (1986).

Protein transport across the ER membrane. T. A. Rapoport. *Trends Biochem. Sci.* 15:355–358 (1990).

a. Signal Peptides

The presence of a signal peptide at the N-terminus of any protein that is being synthesized is usually sufficient to cause the protein to be inserted into the ER. Adding signal peptides to the N-termini of proteins that are normally cytoplasmic diverts many but not all of them to the ER. There is one cellular apparatus (Fig. 2.7) that functions on all proteins with signal peptides. In spite of this unity of function, signal peptides of different proteins have diverse sequences and are not closely related. Indeed, about 20% of random protein sequences have been shown to be able to function as signal peptides when placed at the N-terminus. Although they are diverse, natural signal peptides have a number of characteristics in common:

1. There is a charged region at the N-terminus (often designated as region *n*) of 2–5 residues after the initiating Met residue; this region generally has a net positive charge of $+2$ if the terminal α-amino group is included.

2. The *n* region is followed by a stretch of 7–15 primarily hydrophobic residues, predominantly Leu, Ile, Val, Ala, and Phe. Although Ser and Thr are often present, charged residues are absent; inserting charged residues by mutagenesis at these sites usually inactivates the signal peptide. The hydrophobic segment is often designated as region *h*.

3. The *h* region is followed by approximately five predominantly polar residues. This region, often designated *c*, defines the end of the signal peptide.

4. The total length of the signal peptide is usually 15–26 residues. Much of this length variation occurs in the *n* and *h* regions; the *c* region is less variable in length.

A typical signal peptide is that of human chorionic gonadotropin α subunit:

$$+ \text{Met-Asp-Tyr-}\overset{-}{\text{Arg}}\text{-}\overset{+}{\text{Lys}}\text{-Tyr-}\textbf{Ala-Ala-Ile-Phe-}$$

$$\textbf{Leu-Val-Thr-Leu-Ser-Val-Phe-Leu-}\text{His-Val-}$$

$$\text{Leu-His-Ser-}\downarrow\text{Ala-}\cdots \quad (2.4)$$

The hydrophobic region is in boldface type, the charged groups in the *n* region are indicated, and the cleavage site is indicated by the arrow.

Although the sequences of signal peptides normally are not closely related, the common characteristics just listed are usually sufficient to identify them unambiguously, because the same export apparatus seems to recognize and process all proteins that enter the ER via the diverse signal peptides. Most explanations of how the various signal peptides are recognized by the translocation machinery have involved common physical or conformational properties of signal peptides, rather than specific side-chain recognition.

The positive charge at the N-terminus of the signal peptide is believed to assist insertion into the ER membrane by interacting electrostatically with the negatively charged head groups of the membrane lipids. The hydrophobic region is of greatest importance for the signal peptide and is believed to function primarily by interacting favorably with the nonpolar interior of the membrane, entering and spanning it. This segment is considerably more hydrophobic than the N-terminal regions of proteins that are synthesized and remain in the cytoplasm, yet considerably less hydrophobic than the transmembrane segments of proteins that remain in the membrane (Fig. 2.8).

In exceptional cases, a signal peptide appears not to be at the N-terminus of the polypeptide chain, but to be internal. In the case of hen ovalbumin, the signal peptide is between residues 22 and 41.

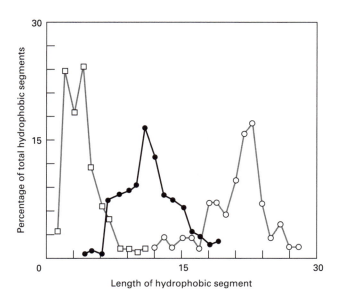

FIGURE 2.8

Distribution of the lengths of the most hydrophobic segments in samples of cytosolic proteins (open squares), signal peptides (solid circles), and C-terminal membrane-spanning segments (open circles). (Kindly provided by G. von Heijne.)

References

Compilation of published signal sequences. M. E. E. Watson. *Nucleic Acids Res.* 12:5145–5164 (1984).

Signal sequences: the limits of variation. G. von Heijne. *J. Mol. Biol.* 184:99–105 (1985).

A new method for predicting signal sequence cleavage sites. G. von Heijne. *Nucleic Acids Res.* 14:4683–4690 (1986).

Many random sequences functionally replace the secretion signal sequence of yeast invertase. C. A. Kaiser et al. *Science* 235:312–317 (1987).

Species-specific variation in signal peptide design. G. von Heijne and L. Abrahmsén. *FEBS Letters* 244:439–446 (1989).

Signal sequences. L. M. Gierasch. *Biochemistry* 28:923–930 (1989).

The functional efficiency of a mammalian signal peptide is directly related to its hydrophobicity. P. Bird et al. *J. Biol. Chem.* 265:8420–8425 (1990).

b. Signal Recognition Particle

The signal recognition particle (SRP) acts as an adaptor between the cytoplasmic protein synthesis machinery of the ribosome and the membrane-bound protein translocation machinery. It binds to ribosomes that are synthesizing polypeptides with signal peptides and then delivers them to the ER. The SRP has an intrinsic affinity for all ribosomes, with a dissociation constant of about $10^{-6}\ M$, but this affinity is increased by three to four

orders of magnitude upon the emergence of a nascent signal sequence. Binding of the SRP to the signal-peptide–ribosome–mRNA complex blocks or slows elongation of the nascent chain, arresting assembly of the nascent chain until the ribosome is in position on the ER membrane through which the protein is to be translocated. This arresting function of SRP can be important, because many nascent polypeptides become incapable of being translocated if they are elongated further. Assembly of the chain is resumed after the ribosome is delivered to the ER membrane by the SRP.

The SRP is a complex of six polypeptides and one RNA molecule of 300 nucleotides, with a total molecular weight of 3×10^5. It is visible in the electron microscope as a rod-shaped particle 240 Å long by 60 Å wide.

References

Mathematical modeling of the effects of the signal recognition particle on translation and translocation of proteins across the endoplasmic reticulum membrane. T. A. Rapoport et al. *J. Mol. Biol.* 195:621–636 (1987).

The affinity of signal recognition particle for presecretory proteins is dependent on nascent chain length. V. Siegel and R. Walter. *EMBO J.* 7:1769–1775 (1988).

Functional dissection of the signal recognition particle. V. Siegel and P. Walter. *Trends Biochem. Sci.* 13:314–316 (1988).

c. Binding to the Endoplasmic Reticulum Membrane and Translocation

The SRP that is bound to a complex of ribosome, mRNA, and nascent signal peptide causes this complex to bind to the membrane of the ER. The SRP is recognized by a receptor, the docking protein, which is an integral membrane protein composed of at least two polypeptide chains. This interaction is believed to be transitory, however, and the ribosome–mRNA–nascent-peptide complex is transferred to other proteins in the ER membrane, such as those that recognize ribosomes (ribophorins) and the signal peptide (signal-sequence receptor). The SRP and its receptor are released from the membrane-bound ribosome and presumably are free to recycle.

The translation arrest is relieved by the dissociation of the SRP. Further elongation of the nascent chain occurs, and it is guided through the membrane and into the lumen of the ER. How this process occurs is largely unknown, but it requires chemical energy in the form of ATP hydrolysis. There may be a proteinaceous pore through the membrane, or the polypeptide may be in contact with the membrane lipids. In any case, the polypeptide chain appears to be in an extended conformation while traversing the membrane.

References

A signal sequence receptor in the endoplasmic reticulum membrane. M. Weidmann et al. *Nature* 328:830–832 (1987).

Signal recognition: two receptors act sequentially. P. Walter. *Nature* 328:763–764 (1987).

d. Cleavage of the Signal Peptide

During or after translocation of the nascent chain across the membrane into the lumen of the ER, the signal peptide is usually cleaved from the preprotein, between residues designated −1 and +1, by a large specific membrane-bound **signal peptidase** on the luminal side of the membrane. This protein is a complex of several different polypeptide chains, though it is believed that only one contains the proteolytic activity. The cleaved signal peptide is probably degraded by a **signal peptide peptidase.**

The site of cleavage at the end of the signal peptide is believed to be determined primarily by the adjacent residues—in particular, residues −1 and −3, which are usually small. Residue −1 is usually Gly, Ala, Ser, Cys, or Thr. The residue at position −3 must not be aromatic (Phe, His, Tyr, Trp), charged (Asp, Glu, Lys, Arg), or large and polar (Asn, Gln). In addition, at least residues +1 and +2 of the protein have some role in determining the cleavage site. Pro residues are not present at position +1 nor usually at positions −1 to −3, but are common at position +2, −4, and −5.

The signal peptides are not cleaved from some membrane proteins, nor from proteins in which the signal peptide is not at the N-terminus. Cleavage of signal peptides is not a requirement for translocation.

References

A putative signal peptidase recognition site and sequence in eukaryotic and prokaryotic signal peptides. D. Perlman and H. O. Halvorson. *J. Mol. Biol.* 167:391–409 (1983).

Patterns of amino acids near signal-sequence cleavage sites. G. von Heijne. *Eur. J. Biochem.* 133:17–21 (1983).

How signal sequences maintain cleavage specificity. G. von Heijne. *J. Mol. Biol.* 173:243–251 (1984).

Evolutionary shift in the site of cleavage of prelysozyme. L. S. Weisman et al. *J. Biol. Chem.* 261:2309–2313 (1986).

Isolation and properties of the signal region from ovalbumin. A. Robinson et al. *FEBS Letters* 203:243–246 (1986).

e. Release, Folding, and Assembly of Proteins in the Endoplasmic Reticulum

In addition to cleavage of the signal peptide, a number of modifications can occur to the nascent chain in the ER (Table 2.1). Especially important is the addition of glycosyl groups to certain Asn residues (Sec. 2.4.3.a). Disulfide bond formation (Sec. 2.4.10) occurs in the ER and is the best diagnostic of folding during biosynthesis; it is clear that the nascent chain undergoes at least partial folding in the case of large polypeptide chains that eventually fold into multiple, independent structural domains (Sec. 7.5.5). The individual domains appear to fold sequentially as they emerge from the ribosome and the ER membrane. There is no evidence for folding of smaller polypeptide chains during biosynthesis of incomplete structural domains, but little direct evidence on the subject is available. The folding of completed polypeptide chains is described in Chapter 7 (Sec. 7.5).

Some proteins in the ER are transiently associated with an intrinsic, soluble protein of the ER known as **heavy-chain binding protein (BiP** or **Hsp 70).** It was originally found to be associated with the heavy chain of immunoglobulins (Sec. 8.3.1), presumably until the heavy chain associates with the appropriate immunoglobulin light chain. BiP is now believed to interact in a similar way with a number of polypeptide chains that must be assembled into multimers of more than one polypeptide chain, either identical or different. This transient association probably assists the assembly process by stabilizing the metastable individual polypeptide chains until they encounter their partner polypeptides. BiP is believed to recognize and interact weakly with the hydrophobic "patches" on the individual, folded chains that need to interact with the other subunits. Similar proteins have been identified in bacteria, mitochondria, chloroplasts, and the cytoplasm and have been given the collective name **molecular chaperones** (Sec. 7.5.6).

References

Formation of an intrachain disulfide bond on nascent immunoglobulin light chains. L. Bergmann and W. M. Kuehl. *J. Biol. Chem.* 254:8869–8876 (1979).

Biosynthesis of dog fibrinogen. Characterization of nascent fibrinogen in the rough endoplasmic reticulum. B. Kudryk et al. *Eur. J. Biochem.* 125:673–682 (1982).

The biosynthesis of rat transferrin. Evidence for rapid glycosylation, disulfide bond formation, and tertiary folding. E. H. Morgan and T. Peters, Jr. *J. Biol. Chem.* 260:14793–14801 (1985).

Protein oligomerization in the endoplasmic reticulum. S. M.

Table 2.1 *Cellular Sites of Major Posttranslational Modifications*

Site	Modifications
Cytoplasm	Removal of initiating Met residue
	Acetylation of α-NH$_2$
	Myristoylation of α-NH$_2$
	O-Glycosylation with GlcNAc
	Addition of palmitoyl groups
	Virus polyprotein processing
Mitochondria, chloroplasts	Cleavage of signal peptide
Endoplasmic reticulum	Cleavage of signal peptide
	Core glycosylation of Asn residues
	Addition of palmitoyl and glycosyl-phosphatidylinositol groups
	Carboxylation of Glu residues
	Hydroxylation of procollagen Pro and Lys residues
	Disulfide bond formation
Golgi apparatus	Modification of N-glycosyl groups
	O-Glycosylation with GalNAc
	Sulfation of Tyr residues
Secretory vesicles and granules	Amidation of α-CO$_2$H
	Proteolytic processing of some precursors

Hurtley and A. Helenius. *Ann. Rev. Cell Biol.* 5:277–307 (1989).

Molecular chaperones: proteins essential for the biogenesis of some macromolecular structures. R. J. Ellis and S. M. Hemmingsen. *Trends Biochem. Sci.* 14:339–342 (1989).

f. Sorting in the Endoplasmic Reticulum and Golgi Apparatus

Proteins that are initially directed to the ER must be further directed to their final destinations (Fig. 2.9). Mechanisms must exist to distinguish among the various proteins destined for various locations: Many of these proteins are secreted by means of one of at least two secretory pathways, one constitutive and occurring continuously, the other regulated and occurring only in response to periodic stimuli; others are bound for lysosomes, the Golgi, and other cellular compartments; and still others remain within the ER, where they are in-

volved in processing newly synthesized proteins or in synthesizing lipids and oligosaccharides. Signals for these destinations are believed to be present in the amino acid sequences of the proteins, but only a few of these signals have been detected. Nevertheless, this is a very active area of research, and the remaining signals are likely to be found in the near future.

References

Progress in unraveling pathways of Golgi traffic. M. G. Farquhar. *Ann. Rev. Cell Biol.* 1:447–488 (1985).

Biosynthetic protein transport and sorting by the endoplasmic reticulum and Golgi. S. R. Pfeffer and J. E. Rothman, *Ann. Rev. Biochem.* 56:829–852 (1987).

Transport of secretory and membrane glycoproteins from the rough endoplasmic reticulum to the Golgi. H. F. Lodish. *J. Biol. Chem.* 263:2107–2110 (1988).

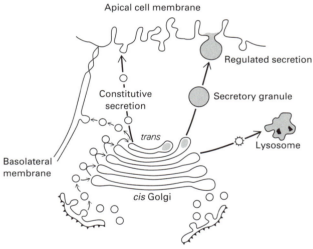

FIGURE 2.9

Possible fates of proteins synthesized on the rough endoplasmic reticulum. Elements of the endoplasmic reticulum are shown at the bottom. Proteins synthesized with signal peptides are extruded into the lumen of the ER. They are then transferred by budding, migration, and fusing of vesicles (the small open circles) to the *cis* side of the Golgi apparatus. By subsequent vesicle movement, the proteins are transferred sequentially from one compartment of the Golgi to the next. On the *trans* side of the Golgi, proteins are directed to granules for either constitutive or regulated secretion or to lysosomes and certain other cell compartments. The proteins in regulated secretory granules are packed more densely than in constitutive secretory vesicles. Other vesicles are believed to recover resident ER proteins from the *cis* Golgi, or an intermediate stage, to the ER. (From M. Farquhar, *Ann. Rev. Cell Biol.* 1:447–488, 1985.)

Regulation of protein export from the endoplasmic reticulum. J. K. Rose and R. W. Doms. *Ann. Rev. Cell Biol.* 4:257–288 (1988).

Heat shock and sorting of luminal ER proteins. H. R. B. Pelham. *EMBO J.* 8:3171–3176 (1989).

Molecular dissection of the secretory pathway. J. E. Rothman and L. Orci. *Nature* 355:409–415 (1992).

Default Pathway   In the absence of any sorting signal, the default pathway for targeting proteins in the ER is believed to be **constitutive secretion.** Proteins following this pathway leave the ER by being incorporated into membrane-bound vesicles that bud off from one side of the ER in an energy-dependent process. The vesicles then migrate through the cytoplasm and fuse with one side (the *cis* side) of the **Golgi apparatus,** a complex cellular organelle composed of a characteristic stack of 3–8 flattened membrane compartments (*cisternae*) that differ among themselves in their compositions. The polarity of the Golgi is usually described by calling the side that is closer to the cell nucleus *cis* and the opposite side, which is usually facing the secretion granules, *trans.* Proteins are thought to be transferred from the ER to the *cis* side of the Golgi and then sequentially from one compartment of the Golgi to the next, by means of vesicles that bud from one compartment and fuse with the next. Upon reaching the *trans* compartment of the Golgi, vesicles containing the proteins that are being constitutively secreted bud off and move to the cell surface, where the secretory vesicles fuse with the cell membrane and release the protein into the extracellular medium. In this way, the proteins remain topologically separate from the cytoplasm after their translocation into the ER and do not cross any other membranes.

This is the "bulk flow" route that all soluble proteins initially inserted into the ER are believed to follow unless they are diverted by a sorting signal. A major question is why, if they are simply following the bulk flow, various proteins move from the ER to the Golgi at rates that differ by as much as an order of magnitude.

A variety of posttranslational modifications occur to proteins as they traverse various parts of the ER and Golgi (Table 2.1). Some of these modifications, described later (Sec. 2.4), have been extremely useful for marking proteins according to which compartments they have traversed.

Regulated Secretion   In regulated secretion, certain proteins follow a similar pathway from the ER to the Golgi but are packaged in a different type of Golgi vesicle that does not migrate directly to the cell membrane but remains in the cell as a storage vesicle. Only after receiving an appropriate signal from the environment, which varies for different cell types, do the vesicles move to the cell membrane and secrete their contents.

Sorting the proteins that are destined for either the regulated or the constitutive pathway takes place on the *trans* side of the Golgi when the different vesicles are formed. Proteins destined for the regulated pathway condense into aggregates, and the secretory granules formed are much more densely packed than are the secretory vesicles of the constitutive pathway. Proteins thought to be involved in the sorting and condensation of the proteins designated for the regulated pathway have been detected in vesicles, but the nature of the signal on the proteins thus designated is not known.

References

Pathways of protein secretion in eukaryotes. R. B. Kelly. *Science* 220:25–32 (1985).

Constitutive and regulated secretion of proteins. T. L. Burgess and R. B. Kelly. *Ann. Rev. Cell Biol.* 3:243–293 (1987).

Molecular sorting in the secretory pathway. K. N. Chung et al. *Science* 243:192–197 (1989).

Lysosomes   Proteins destined for lysosomes also move through the Golgi but are sequestered into vesicles at the *trans* side of the Golgi apparatus. These particular vesicles migrate to the lysosomes and fuse with them. One of the signals that distinguishes lysosomal proteins from all others is the mannose phosphate (mannose-P) residues that are attached to them in the Golgi (Sec. 2.4.3). Receptors for these sugars direct the proteins bearing them to the lysosome. How the lysosomal proteins are recognized for labeling by mannose-P sugars is not known, but the mannose-P groups are not obligatory for targeting proteins to the lysosomes because some lysosomal proteins are targeted to lysosomes in the liver in the absence of this glycosylation.

Targeting to the lysosome is somewhat different from targeting to other organelles in that a fraction of the lysosomal proteins is often secreted, and lysosomal proteins that exist outside of cells can be taken up by a cell by *endocytosis* and transferred to its lysosomes. The mannose-P groups are clearly involved in endocytosis. Whether the sorting that occurs immediately after biosynthesis and after endocytosis occur by the same mechanism is an open question.

Some lysosomal proteins are processed by proteolytic cleavage after reaching the lysosome, but the role of this processing is unknown.

References

Lysosomal enzymes and their receptors. K. von Figura and A. Hasilik. *Ann. Rev. Biochem.* 55:167–193 (1986).

Mannose 6-phosphate receptors and lysosomal enzyme targeting. N. M. Dahms et al. *J. Biol. Chem.* 264:12115–12118 (1989).

Endoplasmic Reticulum Soluble proteins that reside in the ER have similar sequences at their C-termini: -Lys-Asp-Glu-Leu (popularly known by the one-letter abbreviations KDEL). In yeast, the sequence is -His-Asp-Glu-Leu, or HDEL. Deleting this sequence from the C-terminus causes the proteins to be secreted rather than retained in the ER. When added by site-directed mutagenesis to the C-terminus of hen egg-white lysozyme, which is normally a secreted protein, the KDEL sequence caused the extended lysozyme to be retained by the ER.

The KDEL sequence on ER proteins is thought to be recognized transiently by a receptor in a compartment or vesicle encountered after the proteins leave the ER but probably before they reach the Golgi, which returns them to the ER. The ER proteins appear to remain soluble and to migrate from the ER toward the Golgi along with other proteins, but are retrieved and returned by the receptor. This mechanism would have the advantage of requiring relatively small amounts of receptor. An alternative mechanism for retaining proteins in the ER would require stoichiometric amounts of receptor, which would therefore be the main component of the ER; also, the ER proteins would not be soluble but would be bound by the receptor. Not surprisingly, this mechanism does not appear to be used.

References

A C-terminal signal prevents secretion of luminal ER proteins. S. Munro and H. R. B. Pelham. *Cell* 48:899–907 (1987).

Control of protein exit from the endoplasmic reticulum. H. R. B. Pelham. *Ann. Rev. Cell Biol.* 5:1–23 (1989).

Golgi Apparatus A mechanism for keeping soluble proteins in the Golgi apparatus has not yet been identified.

The difficulties in detecting sorting signals on proteins that direct them to the various secretion pathways, to the Golgi apparatus, and to the lysosomes may indicate that the sorting signals are not simple contiguous sequences, like the signal peptides and the KDEL sequence, but patches on the surfaces of the folded proteins. Sorting in the ER differs from the sorting in the cytoplasm that targets proteins to the ER and other organelles, in that proteins in the ER are folded, are likely to retain their folded conformations during subsequent sorting, and need not pass through any other membranes. Consequently, the sorting signals are perhaps composed of residues that are widely separated in the primary structure but are brought together on the surface of the folded protein.

2.3.2 Import of Proteins into Other Organelles

Mitochondria and chloroplasts have their own genetic material and protein synthesis apparatus, but they synthesize only 10–20% of their proteins; the rest are synthesized in the cytoplasm of the cell, coded by genes in the cell nucleus, and then taken up by the mitochondria or chloroplasts. Even the DNA and RNA polymerases, and all but one of the ribosomal proteins, of mitochondria are synthesized in the cytoplasm. Other organelles such as peroxisomes and glyoxysomes have no genetic apparatus and take up all of their proteins from the cytoplasm.

The importation of cytoplasmic proteins by mitochondria and chloroplasts, but not by peroxisomes and glyoxysomes, is similar to that by the ER. The N-terminal signal sequences direct appropriate cytoplasmic proteins to the organelle, and they are usually cleaved by proteolysis (Sec. 2.4.1) after translocation across the membrane. A fundamental difference between translocation of proteins into organelles and into the ER is that translocation into organelles is not co-translational. Instead, the organelle proteins are synthesized on free ribosomes in the cytoplasm, released into the cytoplasm, and then taken up by the organelles.

a. Mitochondria

Translocation requires that a protein be unfolded so that it can traverse the membrane as a linear polypeptide chain. Stabilizing the folded conformation of a protein inhibits uptake by mitochondria. Mitochondrial proteins appear to be prevented from folding into stable three-dimensional conformations after release from ribosomes by their interaction with certain chaperone proteins. Some of these chaperones are of the class that is induced by stress or heat-shock of cells. Part of the energy requirement for translocation (provided by hydrolysis of ATP) is used to unfold the protein transiently or to dissociate it from the chaperones.

Many but not all mitochondrial proteins are synthesized with N-terminal presequences that are cleaved after translocation. In other proteins, the sequences that direct translocation are not known, but they obviously are not removed proteolytically. The presequences of mitochondrial proteins (Fig. 2.10) differ substantially from the signal peptides for translocation into the ER. Consequently, the mitochondrial sequences escape rec-

A Mitochondrial presequences

Cleavage site
↓

Alcohol dehydrogenase III (matrix)

```
         + 0 0 0      0 + +     0      0 +    +    0 0
       + MLRTSSL    FTRRVQPSLF      SRNILRLQST
                 Matrix targeting
```

Subunit IV of cytochrome *c* oxidase (inner membrane)

```
        0  +       0 +   +   0      + 0   0 0+
      + MLSLR    QSIRFFKPAT      RTLCSSRYLL
              Matrix targeting              ↑
```

Cytochrome *c*₁ (intermembrane space)

```
       0    0 + +     + 0   0+0   0 0  0    0+0  +  0    +     0         0  0 0     − 0   0   −     0
     + MFSN  LSKRWAQRTL  SKSFYSTATG  AASKSGKLTQ  KLVTAGVAAA   GITASTLLYA  DSLTAEAMTA
            Matrix targeting      |  Cleavage  |         Stop transport (inner membrane)    ↑
```

70 kDa Outer-membrane protein (outer membrane)

```
           + 0  0 +  + 0        0     0 0                              +   + +
       + M  KSFITRNKTA    ILATVAATGT  AIGAYYYYNQ  LQQQQQRGKK
              Matrix targeting  |       Stop transport (outer membrane)
```

B Chloroplast presequences

Ribulose 1,5-bisphosphate carboxylase/oxygenase small subunit (stroma)

Lemna	+ MASSMMA	STAAVARVGP	AQTNMVGPFN	GLRSSVPFPA	TRKANNDLST	LPSSGGRVSC
Tobacco	+ MASSVLS	SAAVATRSNV	AQANMVAPFT	GLKSAASFPV	SRKQNLDITS	IASNGGRVQC
Pea	+ MASMISS	SAVTTVSRAS	RGQSAAVAPF	GGLKSMTGFP	VKKVNTDITS	ITSNGGRVKC
Soybean	+ MASSM	ISSPAVTTVN	RAGAGMVAPF	TGLKSMAGFP	TRKTNNDITS	IASNGGRVQC
Wheat	+ MAPAVMA	SSATTVAPFQ	GLKSTAGLPI	SCRSGSTGLS	SVSNGGRIRC	
Chlamydomonas	+ MAVI	AKSSVSAAVA	RPARSSVRPM	AALKPAVKAA	PVVAPAEAND	

Ferredoxin (stroma)

Silene	+ MASTLSTLS	VASASLLPKQQ	PMVASSLPTN	MGQALFGLKA	GSRGRVTAM

Light-harvesting chlorophyll binding protein II (thylakoid membrane)

Pea	+ MAASSSS	SMALSSPTLA	GKQLKLNPSS	QELFAARFTM
Lemna	+ MAASSAI	QSSAFAGQTA	LKQRDELVRK	VGVSDGRFSM
Petunia	+ MAAAT	MALSSSSFAG	KAVKLSSSSS	EITGNGKVTM
Wheat	+ MAAT	TMSLSSSSFA	GKAVKNLPSS	ALIGDARVNM

Plastocyanin (thylakoid lumen)

Silene	+ MATVTS	SAAVAIPSFA	GLKASSTTRA	ATVKVAVATP	RMSIKAGLKD	VGVVVAATAA	AGILAGNAMA
Spinach	+ MATVASSA	AVAVPSFTGL	KASGSIKPTT	ALIIPTTTAV	PQLSVKASLK	NVGAAVVATA	AGLLAGNAMA
		Stroma targeting			Processing region		Thylakoid targeting

FIGURE 2.10

Comparisons of the presequences of protein imported into mitochondria *(A)* and chloroplasts *(B)*. The amino acid sequences of the N-terminal segments are given by the one-letter code. The hydroxyl-containing residues Thr (T) and Ser (S) are indicated by Os above them; residues that are usually ionized are indicated by their charges. Cleavage sites are indicated by arrows. The domain structures of complex presequences are indicated by their functions, when known. (Sequences from E. C. Hurt and A. G. M. van Loon, *Trends Biochem Sci.* 11:204–207, 1986; G. W. Schmid and M. L. Mishkind, *Ann. Rev. Biochem.* 55:879–912, 1986; K. Keegstra and C. Bauerle, *BioEssays* 9:15–19, 1988.)

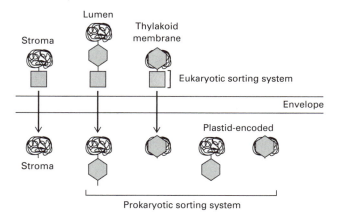

C Sorting systems

FIGURE 2.10 *(Continued)*

C: The two sorting systems proposed to be involved in the import of proteins and their targeting in the chloroplast. A eukaryote-like system in the cytosol uses N-terminal targeting sequences (squares) to direct the proteins into the stroma of the chloroplast. Inside the chloroplast, a prokaryote-like system uses a second targeting sequence (hexagons), which is present either in the mature protein or in a cleavable presequence, to target these proteins to the lumen or the membrane of the thylakoid of the chloroplast. Plastid-encoded proteins are synthesized in the chloroplast. (From S. Smeekens et al., *Trends Biochem. Sci.* 15:73–76, 1990.)

ognition by SRP and diversion to the ER. These sequences are remarkably variable, even more so than signal peptides, and they vary greatly in length, from 12 to 70 residues. The greatest similarities among them are the predominance, all along their length, of basic residues, hydroxyl-containing Ser and Thr residues, and small hydrophobic residues. There is a strong bias against acidic residues but no apparent segregation of the different types of residues into distinct zones.

The N-terminal presequences of mitochondrial proteins interact with one receptor on the surface of the mitochondrion, whereas mitochondrial proteins that lack an N-terminal signal sequence interact with a different receptor. These two receptors are part of a large complex, and both types of protein follow the same process after translocation. Translocation requires an electrochemical potential across the mitochondrial membrane, with the inside of the mitochondrion negative with respect to the outside. The positive charge on the presequence may cause it to be transferred "electrophoretically" across the inner membrane. In most cases the presequence is cleaved after translocation by one or two proteases located inside the mitochondrion, but this is not a requirement.

Mitochondria have two bilayer membranes and an intermembrane space. Protein importation appears to occur at sites where the two membranes are in close contact, and imported proteins probably cross both membranes simultaneously into the interior matrix. Some proteins, however, are destined for the intermembrane space. These proteins appear to have a double presequence (Fig. 2.10A), which is cleaved in two steps. The initial preprotein is transferred to the matrix interior of the mitochondrion, where the N-terminal part of the presequence is cleaved. The second part of the presequence is similar to signal peptides for export from cells and causes the protein to be exported across the inner membrane into the intermembrane space; there the second part of the presequence is removed. This series of events also occurs in chloroplasts (Fig. 2.10C) and is thought to have originated in the prokaryotic organisms that invaded eukaryotic cells. Originally, they synthesized all their own proteins and needed to export some of them. The transfer of the genes for these proteins to the nucleus of the host cell and the subsequent appearance of the apparatus for importing the proteins into mitochondria and chloroplasts were later developments. What is not clear is why the genes for these proteins were transferred to the host nucleus.

Proteins made for the outer membrane of the organelles also have double N-terminal signal sequences. In one example, the N-terminal portion directs the protein to the mitochondrion, but the second segment, a stretch of 28 uncharged residues, causes the protein to be retained within the membrane. Neither portion of the presequence gets cleaved from the protein, perhaps because it does not enter the matrix of the mitochondrion, where the peptidases are located.

Not all imported proteins are taken up by mitochondria in this way. One protein, apocytochrome *c*, destined for the intermembrane space, is taken up by a very different process. The precursor protein seems to insert spontaneously into mitochondrial membranes, where a heme group is attached to it covalently (Sec. 8.3.4.c); then it is pulled into the intermembrane space.

Once in the mitochondrion, the imported proteins must be folded and, in many cases, brought together into functional assemblies of multiple polypeptide chains. Chaperones are involved in this process, but the details are still being elucidated.

References

Targeting proteins into mitochondria. M. G. Douglas et al. *Microbiol. Rev.* 50:166–178 (1986).

Signals guiding proteins to specific intramitochondrial locations. G. Schatz. *Eur. J. Biochem.* 165:1–6 (1987).

Biogenesis of mitochondria. G. Attardi and G. Schatz. *Ann. Rev. Cell Biol.* 4:289–333 (1988)

Import of proteins into mitochondria: a multi-step process. N. Pfanner et al. *Eur. J. Biochem.* 175:205–212 (1988).

Mitochondrial presequences. D. Roise and G. Schatz. *J. Biol. Chem.* 263:4509–4511 (1988).

Protein sorting to mitochondria: evolutionary conservations of folding and assembly. F. U. Hartl and W. Neupert. *Science* 247:930–938 (1990).

Amino acid sequence requirements for the association of apocytochrome *c* with mitochondria. J. R. Sprinkle et al. *Proc. Natl. Acad. Sci. USA* 87:5729–5733 (1990).

The mitochondrial protein import apparatus. N. Pfanner and W. Neupert. *Ann. Rev. Biochem.* 59:331–353 (1990).

b. Chloroplasts

The import of proteins into chloroplasts is remarkably similar to their import into mitochondria, but there are crucial differences. All proteins that are imported into chloroplasts are made with the expected amino-terminal presequences, generally called **transit peptides** in this case, which are invariably cleaved during or after uptake (Fig. 2.10B). The compositions of transit peptides are variable but are similar to those involved in mitochondrial import, containing primarily basic, small hydrophobic, and hydroxyl-containing amino acids. These similarities must be superficial, however, because plant cells can distinguish between mitochondrial and chloroplast presequences. Plant cells have both mitochondria and chloroplasts, and the various presequences have been shown to be specific and to deliver proteins correctly to one or the other organelle. The basis for this specificity is not known, though it has been claimed that chloroplast transit peptides are more similar to each other than to those of plant mitochondria.

A receptor for chloroplast transit peptides has been identified in the outer membrane. After the transit peptides bind to this receptor, chemical energy in the form of ATP is required for further uptake into chloroplasts, but its role is not known. An electrochemical potential across the membranes is not necessary, in contrast to the case of mitochondria.

The proteases that cleave transit peptides are soluble proteins in the stroma of the chloroplasts. Many transit peptides have similar sequences at positions −3 to −5 (Fig. 2.10B), which may be the cleavage signals.

The chloroplast is even more complicated than the mitochondrion in that it has an additional internal compartment, the *thylakoid,* where photosynthetic electron transfer occurs. Proteins are made in the cytoplasm and yet are localized in either the single membrane or the lumen of the thylakoid. During translocation, soluble proteins for the thylakoid interior must cross three membranes, including the inner and outer membranes of the chloroplast. These proteins are synthesized with

double transit peptides. The N-terminal part is like other transit peptides, whereas the following part is more like the signal peptides that target proteins to the ER. The first part presumably directs the proteins across the inner and outer membranes of the chloroplast and is then cleaved. The second part causes the intermediate form of the protein to be translocated across the thylakoid membrane and is then also removed proteolytically. The second part, however, when added to the N-terminus of other proteins, is not sufficient to translocate them across the thylakoid membrane. More targeting signals may be required, or the process may be very sensitive to the nature of the protein being translocated.

References

The transport of proteins into chloroplasts. G. W. Schmidt and M. L. Mishkind. *Ann. Rev. Biochem.* 55:879–912 (1986).

Targeting of proteins into chloroplasts. K. Keegstra and C. Bauerle. *BioEssays* 9:15–19 (1988).

Recent developments in chloroplast protein transport. M. L. Mishkind and S. E. Scioli. *Photosynth. Res.* 19:153–184 (1988).

Transport of proteins into chloroplasts. T. H. Lubben et al. *Photosynth. Res.* 17:173–194 (1988).

Protein transport into and within chloroplasts. S. Smeekens et al. *Trends Biochem. Sci.* 15:73–76 (1990).

c. Peroxisomes, Glyoxysomes, and Glycosomes

These organelles, which carry out a variety of functions in cells, are frequently classed together as *microbodies*. They have no genetic systems and must import all their proteins from the cytoplasm. In each case, this importing occurs posttranslationally, as with mitochondria and chloroplasts, but in many cases no presequences are present and no proteolytic processing is detectable. The signals that direct proteins to these organelles are largely unknown. The targeting signal of peroxisomal proteins, however, has been identified as the sequences Ser/Ala-Lys/Arg/His-Leu, where the slashes separate alternative residues at those positions. These sequences have been shown to function when at the C-terminus of the polypeptide chain, but they are present only at internal positions in some peroxisomal enzymes.

Other peroxisomal and glyoxosomal proteins are synthesized with N-terminal transit peptides that are cleaved after translocation. The transit peptides contain no hydrophobic stretches but are composed primarily of acidic and Ser residues. They are similar in many respects to mitochondrial and chloroplast transit peptides but with the important difference of being acidic rather than basic.

References

Biogenesis of peroxisomes. P. B. Lazarow and Y. Fujiki. *Ann. Rev. Cell Biol.* 1:489–530 (1985).

How proteins get into microbodies (peroxisomes, glyoxysomes, glycosomes). P. Borst. *Biochim. Biophys. Acta* 866:179–203 (1986).

Glycosomes may provide clues to the import of peroxisomal proteins. F. R. Opperdoes. *Trends Biochem. Soc.* 13:255–260 (1988).

A conserved tripeptide sorts proteins to peroxisomes. S. J. Gould et al. *J. Cell Biol.* 108:1657–1664 (1989).

Glyoxysomal malate dehydrogenase from watermelon is synthesized with an amino-terminal transit peptide. C. Gietl. *Proc. Natl. Acad. Sci. USA* 87:5773–5777 (1990).

2.3.3 Nuclear Proteins

Proteins that are localized in the nucleus have very important roles in replicating, maintaining, and expressing the genetic material. There are no ribosomes in the nucleus, so all proteins that function in the nucleus are synthesized in the cytoplasm and then directed to the nucleus. The nucleus is surrounded by a nuclear envelope that is composed of a double membrane with a perinuclear space between. This nuclear envelope is penetrated by nuclear pore complexes that have channels through the middle with functional diameters of about 90 Å. Consequently, proteins need not pass through a membrane to enter the nucleus but can traverse the nuclear pores. The nuclear pore complexes are 1200–1500 Å in diameter and are composed of a number of polypeptide chains with a total molecular weight of about 10^8.

Small proteins with molecular weights of less than about 20,000 can enter the nucleus readily through the pores, though many small cytoplasmic proteins seem to be excluded from the nucleus by an unknown mechanism. Larger proteins are excluded from the nucleus unless they carry specific signals; such signals also accelerate the entry of small proteins. A protein molecule containing the correct sequence binds to certain receptor proteins and causes the pore to open, becoming a larger channel with a diameter of at least 250 Å in a complex process requiring energy from ATP hydrolysis. The nuclear functions of some proteins seem to be regulated by controlling their access to the nucleus, probably by masking or exposing their nuclear location sequences.

The sequences responsible for the nuclear location of several proteins have been localized by site-directed mutagenesis. When these sequences are covalently attached to cytoplasmic proteins or colloidal gold particles, either chemically through side chains or as part of the primary structure, they are taken up by the nucleus. The surprising findings are that the sequences from different proteins are not closely related (Table 2.2) and that they occur in a variety of locations in the primary structures: at the amino end, at the carboxyl end, or in the interior. The most distinguishing aspects of some, but not all, of such sequences are clusters of basic residues separated by nonpolar residues. The identification of several different, apparently unrelated sequences probably indicates that several different mechanisms operate for entry of proteins into the nucleus.

No proteolytic processing associated with nuclear uptake has been observed. This is probably related to the fact that nuclei differ from other organelles in being disrupted during each cell cycle. The nuclear envelope disappears during cell division and reassembles afterward. As a consequence, nuclear proteins must be taken up once again by the reassembled nucleus.

References

Protein import into the cell nucleus. C. Dingwall and R. A. Laskey. *Ann. Rev. Cell Biol.* 2:367–390 (1986).

The nucleus: structure, function and dynamics. J. W. Newport and D. J. Forbes. *Ann. Rev. Biochem.* 56:535–565 (1987).

Identification of specific binding proteins for a nuclear location sequence. S. A. Adam et al. *Nature* 337:276–279 (1989).

Nuclear protein transport. P. Silver and H. Goodson. *Crit. Rev. Biochem.* 24:419–435 (1989).

How proteins enter the nucleus. P. A. Silver. *Cell* 64:489–497 (1991).

2.3.4 Membrane Proteins

Proteins that are integral parts of the membrane and have at least one segment of polypeptide chain traversing the lipid bilayer often undergo translocation across the membrane of the ER using the same type of presequence as water-soluble proteins (Sec. 2.3.1). Translocation of the polypeptide chain across the membrane is interrupted, however, and part of the completed protein is left on the cytoplasmic side of the membrane, one or more segments are within the membrane, and the remainder of the protein is on the extracytoplasmic side (lumen). Translocation is thought to be halted by a strongly hydrophobic *stop-transfer* sequence in the primary structure of the polypeptide chain being synthesized. Although the hydrophobic nature of the stop-transfer sequence may be sufficient to keep the polypeptide within the membrane, the presence of

Table 2.2 *Nuclear Localization Sequences*

Protein	Sequence[a]
	Viral
SV40 T-antigen	—[126]PKKKRKV[132]
SV40 VP1	[1]APTKRKGS[8]
Polyoma large T	—[189]VSRKRPRPA[197]
	—[280]PKKARED[286]
Influenza NP	—[327]QLVWMACNSAAFEDLRVLSFIR[348]
Influenza NS1 NLS1	—[28]DAPFL*DRLRRD*QKSLRG[44]
NLS2	—[203]WGSSNENGGPPLT*PKOKRKM*ARTARSKVRRDKMAD[237]
Adenovirus E1A	—[285]KRPRP[289]
	Saccharomyces cerevisiae
Mat α2	[1]MN*KIPIK*DLLNPQ[13]
Ribosomal L3	[1]MSHRKYEAPRHGHLGFL*PRKR*[22]
GAL4	[1]MKLLSSIEQACDIC*RLKKLK*CSKEKPKCA[29]
Histone H2B	[1]MSAKAEKKPASKAOAEKKPAAKKTSTSTDPG*KKRSKA*[37]
	Other
Xenopus nucleoplasmin	—[153]AV*RPAATKK*AGQAKKK[169]
Rat glucocorticoid receptor	—[497]YRKCLQAGMNLEARK*TKKKIK*CIQQATA[524]
Human lamin A	—[407]SQTQGGG*SVTKKRKLE*STESRSSFSQHARTSGRVAVEE[444]

[a] Italicized sequences are proposed consensus nuclear localization sequences. Numbers indicate the numbers of the residues in the protein primary structure.

From P. Silver and H. Goodson, *Crit. Rev. Biochem.* 24:419–435 (1989).

charged residues flanking the hydrophobic segment is believed to be important for determining the orientation of the segment across the membrane.

Membrane proteins have varying topologies in the membrane. As shown in Figure 2.11, some integral membrane proteins are held by a single polypeptide segment that spans the membrane. The amino and carboxyl ends of the chains are on opposite sides of the membrane, and both of the possible orientations (with the amino or carboxyl on either side) are found in different proteins. Complex membrane proteins have more than one membrane-spanning segment and can have their N- and C-termini on either the same or opposite sides of the membrane; both orientations are found. Membrane-spanning segments can be identified by their amino acid sequence because they are composed of amino acids that are more hydrophobic than are those of other polypeptide segments (see Fig. 2.8).

The assembly into membranes of proteins with single transmembrane segments is best understood. Those proteins made with an N-terminal signal peptide that is cleaved have the new N-terminus on the extracytoplasmic side of the membrane. The original N-terminus probably stays on the cytoplasmic side, and cleavage of the signal peptide occurs on the other side of the membrane. The internal stop-transfer sequence halts trans-

location, and the following part of the polypeptide chain remains on the cytoplasmic side of the membrane.

Proteins with the opposite orientation are synthesized with N-terminal or internal signal peptides that are not cleaved. The signal peptide in this case, which is considerably more hydrophobic than other signal peptides, serves as the membrane anchor.

Members of the third class of single-span membrane proteins have an N-terminal signal sequence that is also a stop-transfer sequence, but in this class the amino end has been translocated across the membrane by an unknown mechanism.

These membrane-spanning segments do not differ markedly in their amino acid sequences or compositions. What determines the polarity of each such segment—that is, which end is translocated across the membrane—appears to depend on the flanking polar residues. In particular, positively charged residues at one end of a transmembrane segment tend to keep that end on the cytoplasmic side of the membrane.

Membrane proteins with multiple transmembrane segments are thought to contain a signal sequence before each extracellular loop and a stop-transfer sequence after it.

Many membrane proteins are not embedded in the membrane but are merely tethered to it by a covalently

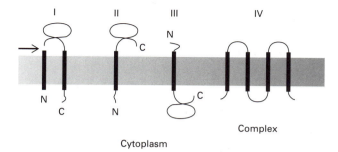

FIGURE 2.11

Topologies of integral membrane proteins. Hydrophobic segments that tend to remain in the membrane are indicated by thick lines. Class I proteins have an N-terminal signal peptide that is cleaved (arrow), generating a new N-terminus on the noncytoplasmic side of the membrane. Classes II and III have only a single hydrophobic signal peptide near the N-terminus that is not cleaved, but the two classes have opposite polarities. Class IV includes all proteins with more than one transmembrane segment; the polypeptide ends can be on the same or opposite sides of the membrane, and either orientation can occur. (From G. von Heijne and Y. Gavel, *Eur. J. Biochem.* 174:671–678, 1988.)

attached fatty acid that becomes part of the lipid bilayer (Sec. 2.4.4).

References

Topogenic signals in integral membrane proteins. G. von Heijne and Y. Gavel. *Eur. J. Biochem.* 174:671–678 (1988).

Predicting the orientation of eukaryotic membrane-spanning proteins. E. Hartmann et al. *Proc. Natl. Acad. Sci. USA* 86:5786–5790 (1989).

Embedded or not? Hydrophobic sequences and membranes. S. J. Singer and M. P. Yaffe. *Trends Biochem. Sci.* 15:369–373 (1990).

Positively charged residues are important determinants of membrane protein topology. R. E. Dalbey. *Trends Biochem. Sci.* 15:253–257 (1990).

2.3.5 Bacterial Secreted and Membrane Proteins

Prokaryotes do not have cellular compartments or organelles, so the only targeting of proteins that occurs is that of secretion out of the cell and insertion into the cell membrane. Bacteria generally have two membranes, inner and outer, with a periplasmic space between them. Proteins are normally secreted across only the inner, plasma membrane; the outer membrane is not permeable to them, so they are generally confined to the periplasmic space.

The export of proteins from the cytoplasm to the periplasm and the outer membrane of gram-negative bacteria such as *E. coli* is similar to insertion of proteins into the ER of eukaryotic cells, in that both systems involve similar signal peptides and there is a prokaryotic equivalent of the SRP and the docking protein. The systems are so similar that each is able to recognize and transfer proteins of the other. For example, rat pre-proinsulin (see Fig. 2.14) cloned into *E. coli* is processed and secreted as proinsulin by the bacterial system. The signal peptides have indistinguishable specifications. Consequently, the prokaryotic and eukaryotic systems are often considered equivalent. Nevertheless, there are small but apparently significant differences between prokaryotic and eukaryotic signal peptides. Prokaryotic signal peptides tend to have an additional positively charged residue near the N-terminus, which may be needed to compensate for the lack of an ionized terminal amino group, due to formylation of the initiator Met residue (Eq. 2.3).

Despite the similarities of their signal peptides, protein translocation in prokaryotes differs in several aspects from that in eukaryotes. The energy source for translocation in bacteria appears to be the electrochemical potential across the membrane, in contrast to the chemical energy of ATP (or GTP) in eukaryotes. There are translocation mechanisms in prokaryotes that are independent of SRP. In these cases, some aspects of the preprotein and a few constituents of the cell seem to keep the preprotein in a conformational state that can be extruded through the cell membrane. In particular, the protein SecB interacts with the completed polypeptide chain to keep it from folding into a stable conformation. SecB transfers the protein to SecA, a peripheral protein of the inner membrane. The protein being synthesized is then translocated through the membrane. Completed polypeptide chains can be translocated in prokaryotic systems with greater efficiency than in eukaryotic systems. Also, release of the final protein from the prokaryotic membrane appears to coincide with folding of the protein.

A remarkable exception to the general prokaryotic mechanism is the export of the hemolysin protein to the cell exterior, across both membranes, by enterobacteria. The hemolysin protein has no N-terminal signal and does not enter the periplasm even transiently. The sequence of this 1024-residue protein that specifies this process has been localized to the last 53 residues. At least two integral membrane proteins are required to carry out this export process.

References

Genetic analysis of protein export in *Escherichia coli* K12. S. A. Benson et al. *Ann. Rev. Biochem.* 54:101–134 (1985).

Cotranslational and posttranslational protein translocation in prokaryotic systems. C. Lee and J. Beckwith. *Ann. Rev. Cell Biol.* 2:315–336 (1986).

Net N-C charge imbalance may be important for signal sequence function in bacteria. G. von Heijne. *J. Mol. Biol.* 192:287–290 (1986).

Trigger factor: a soluble protein that folds pro-OmpA into a membrane-assembly-competent form. E. Crooke and W. Wickner. *Proc. Natl. Acad. Sci. USA* 84:5216–5220 (1987).

Physiological role during export for the retardation of folding by the leader peptide of maltose-binding protein. G. Liu et al. *Proc. Natl. Acad. Sci. USA* 86:9213–9217 (1989).

Isolation and analysis of the C-terminal signal directing export of *Escherichia coli* hemolysin protein across both bacterial membranes. V. Koronakis et al. *EMBO J.* 8:595–605 (1989).

2.4 Posttranslational Covalent Modifications of Polypeptide Chains

Release of a completed polypeptide chain from a ribosome is often not the last chemical step in the formation of a protein. Various covalent modifications often take place, either during or after assembly of the polypeptide chain. One or more residues may be removed proteolytically, and some of the side chains may be altered. Consequently, the amino acid sequence of the final protein need not be the same as that dictated by the gene sequence, and it is usually necessary to characterize the final covalent structure of a protein using the techniques described in Chapter 1.

Proteolytic cleavage of a polypeptide chain has already been encountered in the discussion of topogenesis. Signal peptides that target proteins to various locations are usually removed proteolytically when they are no longer needed. Proteolysis is usually irreversible, so it can be used to sequester proteins in special locations or in particular forms. Many protein precursors are biologically inactive, and the active protein is generated by proteolytic cleavage; this often occurs with hormones and certain enzymes, whose biological activity must be temporarily suppressed for some reason.

The side chains of many amino acids are covalently modified by the addition of various chemical groups. Some of these modifications extend the chemical capa-bilities of the 20 amino acid residues. Others are used as molecular switches, to activate, inactivate, or modify the biological activity of a protein. Others occur for reasons that are not yet known.

Some posttranslational modifications are common, others are rare. Some occur in all the molecules of a protein population, others in just a small fraction. Some are introduced while the polypeptide chain is being assembled, others are manifestations of old age. Some are produced by enzymes, whereas others arise from nonenzymatic chemical processes. Some are reversible, others not. Some are of obvious physiological importance, whereas others probably arise simply by mistreatment during isolation of the protein. This discussion is primarily of the posttranslational modifications that occur to specific residues in particular proteins and are catalyzed by specific enzymes. Nonenzymatic chemical modifications of proteins are discussed in Section 2.4.11 and in Chapter 10.

Although posttranslational modification is variable, it is important to realize that most of the biologically significant, enzymatically catalyzed modifications occur only if a protein is exposed to the relevant enzyme. These enzymes are usually located at specific sites in the cell, so the modifications that take place on any particular protein depend primarily on its location. The cellular sites of the major covalent modifications are listed in Table 2.1.

Well over 200 variant amino acid residues have been detected in proteins. Many of these are easy to overlook with most techniques that are used to characterize proteins (Chap. 1), and only recently has the identification of variant residues become commonplace. Because the list of modifications is constantly growing, only those that occur most frequently or have the most important biological effects will be described here.

References

Posttranslational covalent modification of proteins. R. Uy and F. Wold. *Science* 198:890–896 (1977).

In vivo chemical modification of proteins (posttranslational modification). F. Wold. *Ann. Rev. Biochem.* 50:783–814 (1981).

2.4.1 Proteolytic Processing

Proteolytic cleavage of polypeptide chains after synthesis is a common occurrence with certain classes of proteins, primarily those destined for cellular organelles or for secretion, in addition to removal of the signal peptide (Sec. 2.3.1). For example, proteolytic enzymes in the digestive tract are produced in inactive forms that are generally designated as **zymogens.** Many peptide hormones are synthesized as much larger precursors, often containing several different hormones. Many viruses make their proteins initially as a single *polyprotein* that is then cleaved into individual polypeptide chains.

In all these cases, the precursor is known as the **pro** protein. Most proteins processed in this way are secreted from the cells in which they are made, so they are initially synthesized as **prepro** proteins, with amino-terminal signal peptides (Sec. 2.3.1). For example, serum albumin is synthesized initially as preproalbumin. After removal of the signal peptide, the resulting proalbumin is further processed by removal of the six-residue propeptide from the new N-terminus (Fig. 2.12).

The role of the propeptide in serum albumin is not known, but most proproteins are biologically inactive, and proteolysis is usually a means of regulating the appearance of biological activity. Proteolysis has the advantage that the proteases responsible for the cleavage can be very specific, so that only the desired proteins are activated at the appropriate time and in the correct location. Proteolysis is usually irreversible (unless the new termini are kept in proximity, Sec. 9.3.2.a) and is an excellent "trigger" for activation. It is used extensively in the blood coagulation and immunological complement fixation cascades, in which a small event triggers an activating proteolytic cleavage of a very specific protease that is amplified immensely by sequential activations of other specific proteases. The property of irreversibility is also important in the assembly of large complex structures such as viruses and collagen (Sec. 5.5.3).

Very often, only the processed form of a protein is present in substantial quantities, so it may not be apparent that an activating process has occurred. Discrepancies between the amino acid sequence of the isolated protein and that indicated by its gene sequence are often the first indication of the existence of precursors.

Major questions to be answered about the phenomenon of posttranslational proteolysis are the identity of the proteolytic enzymes responsible, how they are regulated, and what signals define the sites of cleavage.

Only proteolysis that precedes the biological function of a protein is considered here. Inactivation and total proteolysis are the ultimate fates of virtually all proteins; such degradation is described in Chapter 10.

a. Precursors of Proteolytic Enzymes

The mammalian digestive enzymes, such as trypsin, chymotrypsin, and pepsin, are made in the pancreas as the respective inactive zymogens trypsinogen, chymotrypsinogen, and pepsinogen. Being secreted, they are made as the prezymogens; after cleavage of the signal peptide, they are stored and then secreted via secretory vesicles as the folded but inactive zymogens. Activation occurs only after secretion from the cell. Trypsinogen is cleaved by the very specific enzyme enterokinase, and the active trypsin produced is responsible for activation of some of the other zymogens. Enterokinase is not made in the pancreas but is secreted from the brush border of the small intestine. Active trypsin consequently is generated only when the two secretory streams converge. This prevents premature formation of active proteases, which would be very destructive.

The activation cleavage that produces trypsin occurs at the peptide bond after residue 15 of trypsinogen. A slight rearrangement occurs in the folded protein to complete the active site of trypsin (Sec. 9.3.2.e). Only this cleavage occurs to trypsinogen, and the N-terminal

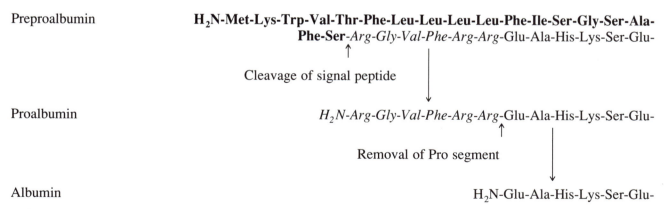

| Preproalbumin | H₂N-Met-Lys-Trp-Val-Thr-Phe-Leu-Leu-Leu-Leu-Phe-Ile-Ser-Gly-Ser-Ala- |
| | **Phe-Ser**-*Arg-Gly-Val-Phe-Arg-Arg*-Glu-Ala-His-Lys-Ser-Glu- |

Cleavage of signal peptide

| Proalbumin | H₂N-*Arg-Gly-Val-Phe-Arg-Arg*-Glu-Ala-His-Lys-Ser-Glu- |

Removal of Pro segment

| Albumin | H₂N-Glu-Ala-His-Lys-Ser-Glu- |

FIGURE 2.12
Proteolytic processing of preproalbumin in the rat liver. The sequences of the prepro and pro segments are indicated, along with the first six residues of the mature protein. (Sequences from A. W. Strauss et al., *J. Biol. Chem.* 252:6846–6855, 1977.)

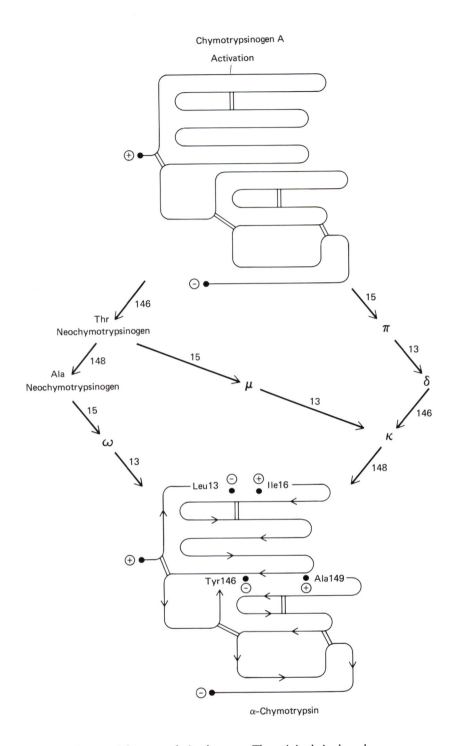

FIGURE 2.13

Activation of bovine chymotrypsinogen A by proteolytic cleavage. The original single poly-peptide chain is cross-linked by five disulfide bonds (double lines, *top*). Four sequential cleavages at the peptide bonds following residues 13, 15, 146, and 148 occur in various orders to generate the three-chain α-chymotrypsin *(bottom)*. The number of the residue at which each cleavage occurs is shown for each step. The process can occur by any of the paths depicted, through the various indicated forms of chymotrypsin. The two dipeptides of residues 14–15 and 147–148 are released, but the three large fragments are held together by the disulfides and by the folded structure of the protein (see Fig. 9.23). Only the cleavage after residue 15 is required to generate enzyme activity, so all of the forms designated by Greek letters are active. (From S. K. Sharma and T. R. Hopkins, *Bioorg. Chem.* 10:357–374, 1981.)

peptide is discarded. Other cleavages occur in related proteases, such as chymotrypsinogen, in which the peptide bonds after residues 13, 146, and 148 are cleaved (Fig. 2.13). The two resulting dipeptides are released, but the three larger polypeptides are held together by the folded conformation of the protein (Sec. 9.3.2.a) and by the disulfide bonds linking them. In this trypsin family of proteases, cleavage sites are determined by the identity of the amino acid residues, by the specificities of the proteases, and especially by the folded conformations of the zymogens. The proteases involved in blood coagulation and complement fixation are also related to this family, and their activation processes are similar.

In the case of pepsin, an unrelated protease, the zymogen pepsinogen is activated by removal of 44 residues from the amino end of the chain. This activation occurs only at acidic pH and presumably ensures that the pepsinogen is activated only when it enters the acidic environment of the stomach (Sec. 9.3.2.c).

Bacterial proteases that are secreted from the cell employ similar processes. For example, subtilisin is produced by various *Bacillus* strains as preprosubtilisin, which has a propeptide of 77 residues between the signal peptide and the mature protein. In this case, the polypeptide chain is released from the bacterial membrane through which it is translocated only if cleavage of the signal peptide and the autocatalytic proteolytic activation also occur. Consequently, the prosubtilisin does not accumulate detectably, and the propeptide is required for folding of the protein (Sec. 7.5).

References

Role of proteolytic enzymes in biological regulation. H. Neurath and K. A. Walsh. *Proc. Natl. Acad. Sci. USA* 73:3825–3832 (1976).

Evolution of proteolytic enzymes. H. Neurath. *Science* 224:350–357 (1984).

Protease pro region required for folding is a potent inhibitor of the mature enzyme. D. Baker et al. *Proteins: Struct. Funct. Genet.* 12:339–344 (1992).

In vitro processing of pro-subtilisin produced in *Escherichia coli.* H. Ikemura and M. Inouye. *J. Biol. Chem.* 263:12959–12963 (1988).

b. Peptide Hormone Precursors

A variety of proteins and peptides serve as hormones, and the list of peptides known to act in the central nervous system is growing rapidly. Virtually all such peptide hormones are made as larger precursors, from which they are released by proteolysis, and are secreted from the cells in which they are made. Synthesis as a large precursor may be a necessity simply due to the small size of some of the peptides; there undoubtedly are lower limits to the size of peptide that can be synthesized on a ribosome and translocated into the ER. Release from a precursor, however, is also a regulatory mechanism. Many precursors contain several active peptide hormones, some with closely related sequences and activities, others with diverse sequences and different hormone actions. The presence of multiple activities on a single polypeptide chain may help to coordinate the separate actions of the linked peptides. Also, different patterns of proteolysis can produce different spectra of active peptides because many of them overlap in the sequence.

The best known peptide hormone is **insulin,** which is synthesized initially as **preproinsulin.** During storage of proinsulin in secretory granules (Fig. 2.14), it is cleaved at two positions to release the internal C-peptide. The mature insulin consists of only 51 of the original 110 residues of preproinsulin, in two chains, A and B, linked by two disulfide bonds. The C-peptide persists in the granules and is eventually released along with the active insulin, but it has no known function. The primary reason for synthesis as proinsulin appears to be to enable folding of the polypeptide chain. After removal of the C-peptide, the separated A and B chains do not reassemble efficiently, whereas proinsulin can refold readily.

Cleavage of proinsulin occurs after pairs of basic residues -Lys-Arg- and -Arg-Arg-, which are then removed; a variety of proteolytic enzymes are likely to be involved. This is the type of proteolytic cleavage most frequently encountered with peptide prohormones (Fig. 2.15). Other cleavage sites can also occur at single, triple, and quadruple basic residues and at a variety of other amino acid residues. Not only the amino acid sequence determines the proteolytic cleavage sites; the three-dimensional conformation of the prohormone must also be involved, though little is known about this in most instances.

Processing generally begins some 10–20 min after completion of synthesis of the prohormone polypeptide chains, after folding has been completed and the prohormones have reached the Golgi apparatus or the secretory granules (see Fig. 2.9). The precursor protein and its complement of processing enzymes are copackaged into the secretory granule, and processing often continues for hours. The timing of processing is crucial because peptides separated in the Golgi could be packaged into separate secretory granules and released separately.

Some of the precursors to peptide hormones are described in Figure 2.15, which also illustrates the basic residues that help define the processing sites. The presence of two or more active peptides on the same precur-

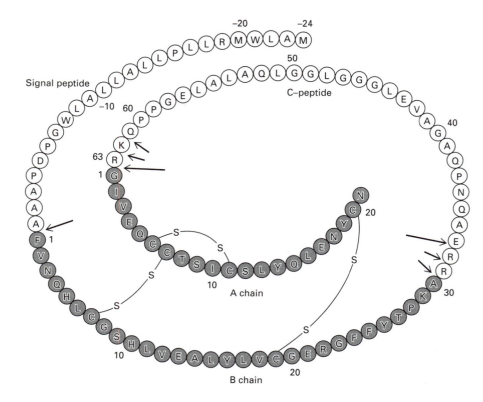

FIGURE 2.14

The primary structure of porcine proinsulin and the proteolytic cleavages that produce active insulin. The proteolytic cleavages are indicated by arrows. The initial cleavages (long arrows) occur after the pairs of basic residues Arg-Arg and Lys-Arg, which are then removed by a carboxypeptidase. The original proinsulin is folded and contains three disulfide bonds.

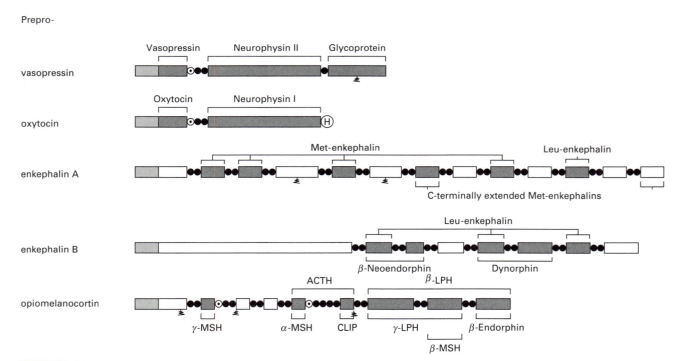

FIGURE 2.15

Precursors to neuropeptides. The lightly shaded boxes indicate signal peptides, dark shaded boxes are active neuropeptides, and the open boxes indicate regions of no known function. Closed circles are basic amino acids. Open circles with dots are Gly residues; those with H are His residues. The processing of proopiomelanocortin is described further in Figure 2.16. (From D. Richter, *Trends Biochem. Sci.* 8:278–281, 1983.)

sor can ensure that they are produced in stoichiometric amounts, as in the cases of provasopressin and prooxytocin. Another segment of each precursor contains a neurophysin, which transports the vasopressin or oxytocin in a 1:1 complex. In both cases, the hormone and neurophysin are separated by the sequence -Gly-Lys-Arg-; the two basic residues are removed, and the Gly residue is converted to a C-terminal amide of the hormone (Eq. 2.7).

A more remarkable case is that of proopiomelanocortin, which contains the sequences of no fewer than eight peptide hormones (Fig. 2.16). The complete amino acid sequence of preproopiomelanocortin is given in Fig. 2.5. Processing occurs primarily at pairs of basic residues, but a unique cleavage pattern occurs at the four contiguous basic residues at positions 15–18. Different patterns of processing occur in different parts of the pituitary to produce different mixtures of peptides (Fig. 2.16). Even greater variation occurs in different

parts of the brain. The basis of this phenomenon of differential processing is not known, so the processing of prohormones cannot be predicted simply by looking for pairs of basic residues in the sequence.

References

Post-translational proteolysis in polypeptide hormone biosynthesis. K. Docherty and D. F. Steiner. *Ann. Rev. Physiol.* 44:625–638 (1982).

Biochemistry of the enkephalins and enkephalin-containing peptides. S. Udenfriend and D. L. Kilpatrick. *Arch. Biochem. Biophys.* 221:309–323 (1983).

Vasopressin and oxytocin are expressed as polyproteins. D. Richter. *Trends Biochem. Sci.* 8:278–281 (1983).

Polyprotein gene expression: generation of diversity of neuroendocrine peptides. J. Douglass et al. *Ann. Rev. Biochem.* 53:665–715 (1984).

Neuropeptides: multiple molecular forms, metabolic pathways, and receptors. D. R. Lynch and S. H. Snyder. *Ann. Rev. Biochem.* 55:773–799 (1986).

Opioid peptide processing and receptor selectivity. V. Höllt. *Ann. Rev. Pharmacol. Toxicol.* 26:59–77 (1986).

Neuropeptides: multiple molecular forms, metabolic pathways, and receptors. D. R. Lynch and S. H. Snyder. *Ann. Rev. Biochem.* 55:773–799 (1986).

Prohormone processing and the secretory pathway. J. M. Fisher and R. H. Scheller. *J. Biol. Chem.* 263:16515–16518 (1988).

Processing of pro-hormone precursor proteins. R. B. Harris. *Arch. Biochem. Biophys.* 275:315–333 (1989).

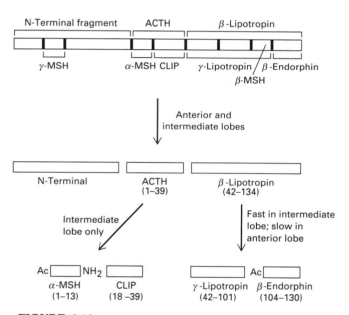

FIGURE 2.16
Different processing of proopiomelanocortin in the anterior and intermediate lobes of the pituitary. The intact precursor is shown schematically *(top)*; dark vertical bars indicate pairs of basic amino acid residues. The abbreviations of the fragments are ACTH, adrenocorticotropic hormone; MSH, melanocyte-stimulating hormone; and CLIP, corticotropin-like intermediate lobe peptide. The complete sequence of the precursor is given in Figure 2.5. The early processing follows the same path in both lobes of the pituitary, but additional cleavages in the intermediate lobe produce smaller final products. In addition, the β-endorphin is acetylated at its amino end; α-MSH is acetylated at its amino end and amidated at its carboxyl end, using Gly 14 (Sec. 2.4.2.f).

c. Virus Assembly

Proteolytic processing of protein precursors is taken to an extreme in viruses that contain RNA as their genetic material in a form that can be used directly as mRNA. Upon infection of a cell by such positive-strand RNA viruses, the viral RNA is used as mRNA to synthesize a single polypeptide chain, known as a **polyprotein,** which is encoded by virtually all of the genetic information of the RNA; in the case of poliovirus, it is 2209 residues long (Fig. 2.17). Hardly any proteases that are capable of making only the required specific cleavages are present in the cytoplasm of the host cells, so these viruses encode their own proteases.

The initial cleavage of the nascent chain usually occurs quickly, before the polyprotein is complete. In the best studied case of poliovirus (Fig. 2.17), this cleavage separates region P1, which codes for structural proteins of the virus, from the proteins that catalyze the many steps in virus replication. This cleavage is between -Tyr-Gly- residues and is catalyzed by a specific protease encoded by protein 2A, which immediately

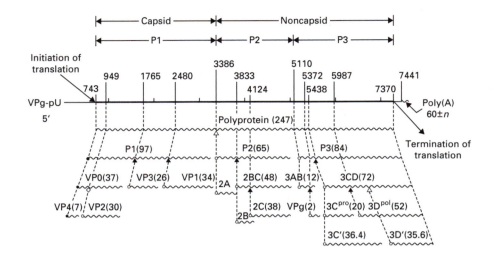

FIGURE 2.17

Translation of the RNA genome of poliovirus and proteolytic processing of the polyprotein. The viral RNA is shown as a solid line, thicker in the translated region; it is terminated at the 5′ end by covalent linkage to the small viral protein VPg, of unknown function, and at the 3′ end by a poly(A) tract, similar to cellular mRNA molecules. The numbers above the viral RNA refer to the first nucleotide of the codon specifying the N-terminal residue of each protein product. Polypeptides are indicated by wavy lines. The coding region is divided into three regions (P1, P2, P3), corresponding to the main cleavage products. The number in parentheses following each polypeptide name is its molecular weight $\times 10^{-3}$. Gln-Gly cleavage sites are indicated by filled triangles, Tyr-Gly by the open triangle, and Asn-Ser by the open diamond. Open circles indicate Gly residues, except for VP2, where it is Ser. Filled circles indicate that the N-terminus is blocked by myristoylation. (Kindly provided by E. Wimmer, after that in N. Kitamura et al., *Nature* 291:547–553, 1981.)

follows the cleavage site. This initial cleavage occurs intramolecularly, presumably as soon as the 2A region has been translated. Most of the subsequent cleavages are produced by another protease, encoded by region 3C^{pro}, either before or after its cleavage from the precursor. These cleavages occur at a number of -Gln-Gly-sequences, and these sequences are cleaved in no particular order in most cases. The final cleavage occurs at an -Asn-Ser- sequence and is catalyzed by an unknown protease.

In another class of viruses, the first proteolytic cleavage of the nascent chain unmasks a signal peptide at the new N-terminus, so the remainder of the polyprotein is translocated into the ER. There it undergoes modifications such as glycosylation (Sec. 2.4.3), which are only possible in the ER and Golgi (Table 2.1).

In some viruses, many of the proteolytic products remain associated and function as part of a large aggregate. In others, the polyprotein is not cleaved but functions as a single polypeptide chain. These extreme examples of synthesis of large polyproteins are probably used by RNA viruses for reasons of genetic economy that are unique to them.

References

Implications of the picornavirus capsid structure for polyprotein processing. E. Arnold et al. *Proc. Natl. Acad. Sci. USA* 84:21–25 (1987).

Viral proteinases. H. G. Kräusslich and E. Wimmer. *Ann. Rev. Biochem.* 57:701–754 (1988).

Proteolytic processing of polyproteins in the replication of RNA viruses. C. U. T. Hellen et al. *Biochemistry* 28:9881–9890 (1989).

d. Polypeptide Splicing of Concanavalin A

Proteolytic processing of concanavalin A, a plant lectin, is followed by a unique event: formation of a peptide bond between the two ends, so that the final protein has an amino acid sequence that is a permuted form of the original sequence (Fig. 2.18).

Concanavalin A is synthesized initially in the ER as preproconcanavalin A with a temporary signal peptide, and the proconcanavalin A traverses the ER and Golgi apparatus in the usual manner. The precursor is then cleaved near the middle of the chain, after an Asn residue, to yield two fragments; and short peptides are cleaved from the carboxyl ends of both fragments. The unusual step follows, in that the two fragments are spliced. A peptide bond is formed between the α-amino group of the N-terminal fragment and the α-carboxyl of the C-terminal fragment. As a consequence, the order of the two fragments is reversed in the mature polypeptide chain.

After the splicing reaction, four residues are removed from the amino end of the spliced chain. All of the proteolytic cleavages occur after Asn residues, so a single enzyme is believed to be involved.

No energy requirement for forming the new peptide bond is apparent, so it is thought to occur at the same time as the cleavage of the short C-terminal fragment, in a transpeptidation reaction. In this reaction, the

α-carboxyl of residue 118 is transferred from the amino group of the C-terminal extension that is being removed to that of the large N-terminal fragment. Presumably, this reaction is also catalyzed by the Asn-specific protease that catalyzes the cleavage reactions.

All of these proteolytic modifications occur after the original precursor has folded to a three-dimensional conformation very similar to that of the final protein (Fig. 2.18B). The new peptide bond can be formed because the various groups are in proximity in this conformation. Whether this unique mechanism of peptide bond formation in a protein is a consequence of this folded conformation or of the catalytic properties of the proteases is not yet clear.

Although this is a unique posttranslational modification, it may have occurred in other cases without yet

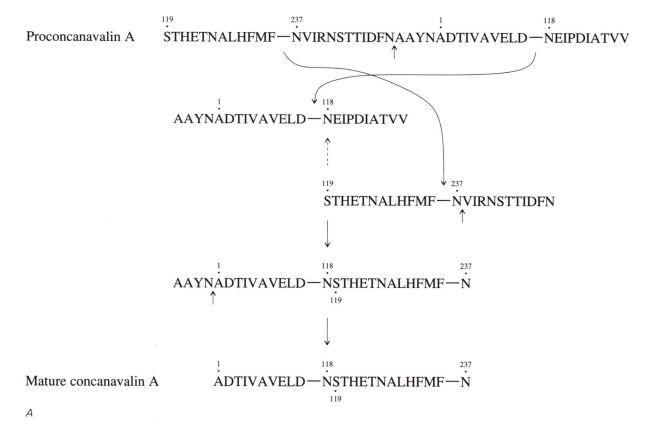

FIGURE 2.18

Proteolytic processing and rearrangement of the primary structure of the polypeptide chain of concanavalin A (ConA). *A:* The polypeptide chain of the precursor form of ConA after removal of the signal peptide *(top).* The numbering system is that of mature ConA *(bottom),* not of the precursor. Proteolytic cleavage after Asn residues is indicated by the short arrows. Formation of the new peptide bond is indicated by the dashed arrow. The rearrangements of the chain depicted after the first cleavage do not occur physically, because the terminal residues of proConA are already in proximity in the three-dimensional structure of the folded protein.

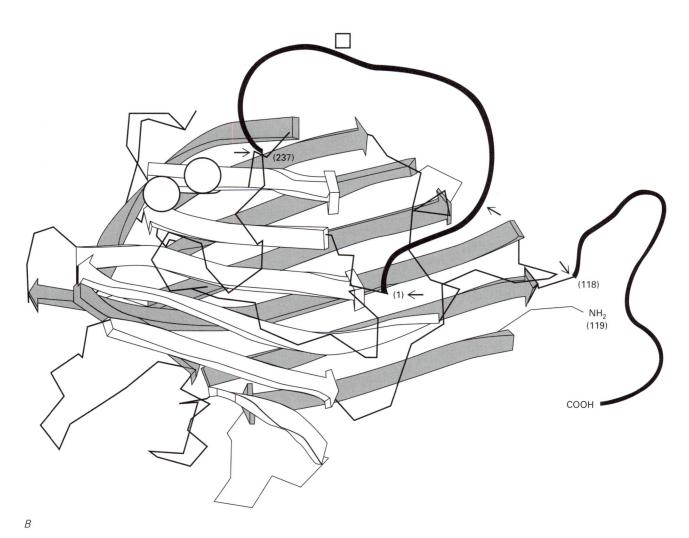

B

FIGURE 2.18 *(Continued)*

B: Postulated three-dimensional structure of proCon A, based on the crystal structure of the mature protein. The polypeptide segments removed from the precursor by proteolysis, and therefore not in the mature protein, are shown by thick solid lines. The new peptide bond is formed between the α-amino group of residue 119 and the α-carboxyl of residue 118, which are in close proximity at the far right. The square indicates the probable position of an oligo-saccharide chain; the circles indicate the positions of the Ca^{2+} and Mg^{2+} ions that are involved in the binding of carbohydrates by the mature protein. (From D. J. Bowles et al., *J. Cell Biol.* 102:1284–1297, 1986.)

having been detected. In any case, it illustrates the range of phenomena that are possible with proteins.

References

Polypeptide ligation occurs during post-translational modification of concanavalin A. D. M. Carrington et al. *Nature* 313:64–67 (1985).

Posttranslational processing of concanavalin A precursors in jackbean cotyledons. D. J. Bowles et al. *J. Cell Biol.* 102:1284–1297 (1986).

Traffic and assembly of concanavalin A. D. J. Bowles and D. J. Pappin. *Trends Biochem. Sci.* 13:60–64 (1988).

2.4.2 Alteration of the Chain Termini

a. N-Terminal Met

The formyl group on the initiating Met residue of polypeptides that are synthesized in prokaryotes (Sec. 2.1.3) is almost always removed by a *deformylase* enzyme.

Only rarely is N-formyl Met found at the N-terminus of a mature protein.

In about half the proteins of both prokaryotic and eukaryotic cells, the initiating Met residue is removed from the nascent chain by a ribosome-associated *Met-aminopeptidase*. Whether it is removed depends primarily on the second amino acid residue. Small residues (Gly, Ala, Ser, Cys, and Thr) favor removal of the Met residue in prokaryotes; large, hydrophobic, and charged residues seem to prevent removal. The enzyme responsible for removal of the Met residue may be saturated in prokaryotes if a protein is being synthesized in large quantities, or the terminus may be inaccessible if the protein is aggregated into inclusion bodies (Sec. 2.2.4). Such proteins are often made as a mixture, with and without the N-terminal Met.

Usually, only the Met residue is removed from the N-terminus, but in some cases an additional residue is also removed. The mechanism for this is not known.

References

Specificity of cotranslational amino-terminal processing of proteins in yeast. S. Huang et al. *Biochemistry* 26:8242–8251 (1987).

Purification and characterization of a methionine-specific amino peptidase from *Salmonella typhimurium*. P. Wingfield et al. *Eur. J. Biochem.* 180:23–32 (1989).

Extent of N-terminal methionine excision from *Escherichia coli* proteins is governed by the side-chain length of the penultimate amino acid. P. H. Hirel et al. *Proc. Natl. Acad. Sci. USA* 86:8247–8251 (1989).

b. Addition of Terminal Residues

The only known instances of posttranslational addition of residues to the ends of polypeptide chains have been shown to occur in vitro by transfer of residues from charged transfer RNA to the α-amino groups of some peptides and proteins. No other components of the protein biosynthetic reaction are required, and the reactions are presumed to occur primarily in the cytoplasm. For example, a single enzyme from *E. coli* transfers Leu and Phe residues from their respective tRNA molecules to the α-amino groups of proteins with Arg or Lys residues at the N-terminus. A similar mammalian enzyme, *arginyl tRNA-protein transferase*, catalyzes the transfer of Arg to peptides with N-terminal Glu or Asp residues.

No functions for these reactions, which have been demonstrated only in vitro, were known for a long time, but they have recently been implicated in tagging proteins for degradation (Sec. 10.3.3).

References

Peptide acceptors in the leucine, phenylalanine transfer reactions. R. L. Soffer. *J. Biol. Chem.* 248:8424–8428 (1973).

Purification and characterization of arginyl-tRNA-protein transferase from rabbit reticulocytes. Its involvement in post-translational modification and degradation of acidic NH_2 termini substrates of the ubiquitin pathway. A. Ciechanover et al. *J. Biol. Chem.* 263:11155–11167 (1988).

c. Acetylation of N-Terminus

Some 59–90% of eukaryotic proteins synthesized in the cytoplasm are isolated with their N-termini acetylated:

$$CH_3-\overset{\overset{\displaystyle O}{\|}}{C}-NH- \qquad (2.5)$$

A variety of N^α-acetyltransferase enzymes are thought to catalyze this reaction, using acetyl-CoA as the acetyl donor. The principal enzyme is loosely associated with ribosomes, consistent with the observation that the nascent chain of only 20–50 residues is usually acted on. Acetylation can occur whether or not the initiating Met residue is still present. Whether acetylation occurs depends to some extent on the nature of the N-terminal residue. In a survey using mutagenesis of the N-terminus of one particular protein, those forms acetylated had N-terminal Gly, Ala, Ser, and Thr residues. The initiating Met was retained and acetylated if the following residue was Asp, Glu, or Asn. Nevertheless, many exceptions to these rules are found in various proteins, and so it must be other properties of the protein that determine whether or not it is acetylated.

One important aspect is whether the acetylating enzyme is present. Acetylation of nascent chains is apparent only in cytoplasmic proteins because the N-terminal targeting peptides of other proteins are subsequently removed. In other cases, acetylation can occur posttranslationally. For example, the melanocyte-stimulating hormone (α-MSH) is acetylated only in some parts of the pituitary. In contrast, another hormone, endorphin, shows just the opposite acetylation properties. In these cases, acetylation occurs only after cleavage of the hormone from a larger precursor (see Figs. 2.5 and 2.16), and the responsible acetyltransferase is present only in secretory granules. Another complication is the presence of enzymes that can remove acetylated residues from the amino end of polypeptides. Acetylation is common only for proteins made in eukaryotic cells, but not in their mitochondria or chloroplasts.

The initiating Met residue is removed from some proteins only after it is acetylated, and then the newly exposed N-terminus is acetylated. This process may take place posttranslationally rather than during synthesis of the polypeptide chain.

Not all the factors that determine whether or not the N-terminus of a protein is acetylated are known, nor is the function of acetylation. For the protein chemist, the primary consequence of acetylation is to make the protein refractory to sequencing by the Edman degradation or by aminopeptidases (Sec. 1.6.2.a,b).

Although acetylation is the main covalent modification made to the amino ends of proteins, a great variety of other modifications have been observed in particular cases, which include addition of formyl, pyruvoyl, fatty acyl, α-keto acyl, glucuronyl, and methyl groups.

Another common modification of the N-terminus is observed when the first residue is Gln. In this case, the side chain reacts with the amino group to generate the pyroglutamyl residue (Eq. 1.19), which has no amino group and is also refractory to sequence analysis. This modification reaction occurs spontaneously, but there are also enzymes that catalyze it in vitro. The enzymes appear to be involved in the posttranslational processing of peptide hormones, which are usually synthesized as large precursors (Sec. 2.4.1.b).

It seems likely, though it is not yet proven, that the state of the amino terminus of a protein has a large effect on the rate of degradation of the protein (Sec. 10.3.3).

References

The mechanism of N-terminal acetylation of proteins. H. P. C. Driessen et al. *Crit. Rev. Biochem.* 18:281–325 (1985).

Structures of N-terminally acetylated proteins. B. Persson et al. *Eur. J. Biochem.* 152:523–527 (1985).

Sequence determinants of cytosolic N-terminal protein processing. C. Flinta et al. *Eur. J. Biochem.* 154:193–196 (1986).

How much sequence information is needed for the regulation of amino-terminal acetylation of eukaryotic proteins? J. Augen and F. Wold. *Trends Biochem. Sci.* 11:494–497 (1986).

Identification of a mammalian glutaminyl cyclase converting glutaminyl into pyroglutaminyl peptides. W. H. Fischer and J. Spiess. *Proc. Natl. Acad. Sci. USA* 84:3628–3632 (1987).

An enzyme(s) that converts glutaminyl-peptides into pyroglutaminyl peptides. W. H. Busby, Jr., et al. *J. Biol. Chem.* 262:8532–8536 (1987).

Cotranslational processing and protein turnover in eukaryotic cells. S. M. Arfin and R. A. Bradshaw. *Biochemistry* 27:7979–7984 (1988).

d. Myristoylation of N-Terminus

A number of cytoplasmic proteins are found with the myristoyl fatty acid linked to their terminal α-amino group:

$$H_3C-(CH_2)_{12}-\overset{\overset{\textstyle O}{\|}}{C}-NH- \qquad (2.6)$$

The myristoyl group is added to the α-amino group during or shortly after biosynthesis of the chain. The myristoyl group is transferred from myristoyl-CoA by the enzyme myristoyl-CoA : protein N-myristoyl transferase.

The information determining whether a protein is myristoylated is thought to reside in the first 7–10 residues of the polypeptide chain, but the rules are not precisely defined. Proteins to which the myristoyl group is added always have Gly as the N-terminal amino acid, and this Gly is the one following the initiating Met residue. The next residue is almost always a small, uncharged residue (Ala, Ser, Asn, Gln, or Val). A broader spectrum of residues occurs at positions 3 and 4, but residue 5 must be small and uncharged (Ala, Ser, Thr, Cys, or Asn).

The long hydrophobic myristoyl group is believed to cause at least some of the attached proteins to be loosely associated with membranes, but these proteins are basically soluble and are not tethered tightly to the membrane. Other myristoylated proteins are not associated with membranes to any significant extent. The consequences of myristoylation are different from those of other lipid attachments, which cause the proteins to be firmly anchored in the membrane (Sec. 2.4.2.e).

Some myristoylated proteins are protein kinases or phosphatases that have important roles in modulating cellular metabolism (Sec. 9.4.3). The myristoyl group may tend to keep its proteins in juxtaposition to particular membranes or other components of cellular regulatory circuits.

References

The biology and enzymology of eukaryotic protein acylation. D. A. Towler et al. *Ann. Rev. Biochem.* 57:69–99 (1988).

Fatty acylation of proteins. A. M. Schultz et al. *Ann. Rev. Cell Biol.* 4:611–647 (1988).

Acylation of viral and eukaryotic proteins. R. J. A. Grand. *Biochem. J.* 258:625–638 (1989).

e. Glycosyl-Phosphatidylinositol and Farnesyl Membrane Anchors at the C-Terminus

Some proteins localized on the cell surface have been found to be firmly anchored to the cell membrane through complex glycosyl-phosphatidylinositols attached to their terminal α-carboxyl groups. The structures of these anchor groups are very complex (Fig.

2.19), involving an ethanolamine group that is attached to the protein α-carboxyl in an amide bond, through a complex glycan and an inositol phosphate to a diacyl glycerol lipid that is embedded in the membrane. The mode of assembly of this complex structure is not known, but it is thought to be preassembled and to be transferred en bloc to the polypeptide chain soon after completion of its translation in the endoplasmic reticulum.

Proteins that are modified in this way are synthesized with signal peptides and are directed into the ER (Sec. 2.3.1). They are intrinsically soluble, hydrophilic proteins, except for a hydrophobic tail at the carboxyl end of the chain. This sequence, plus the 10–12 preceding residues, is cleaved from the protein at about the same time that the glycosyl-phosphatidylinositol group is added to the new terminal carboxyl group. The sequence that is removed appears to be the signal for the modification, analogous to signal peptides for translocation to the ER (Sec. 2.3.1). Adding such a sequence to a protein that is normally secreted causes it to be modified and anchored to the membrane.

The primary purpose of this modification appears to be to anchor otherwise water-soluble proteins to the cell membrane. The hydrophobic C-terminal peptide that is replaced does not seem to be a suitable transmembrane anchor. One advantage of the glycosyl-phosphoinositide type of anchor over that provided by

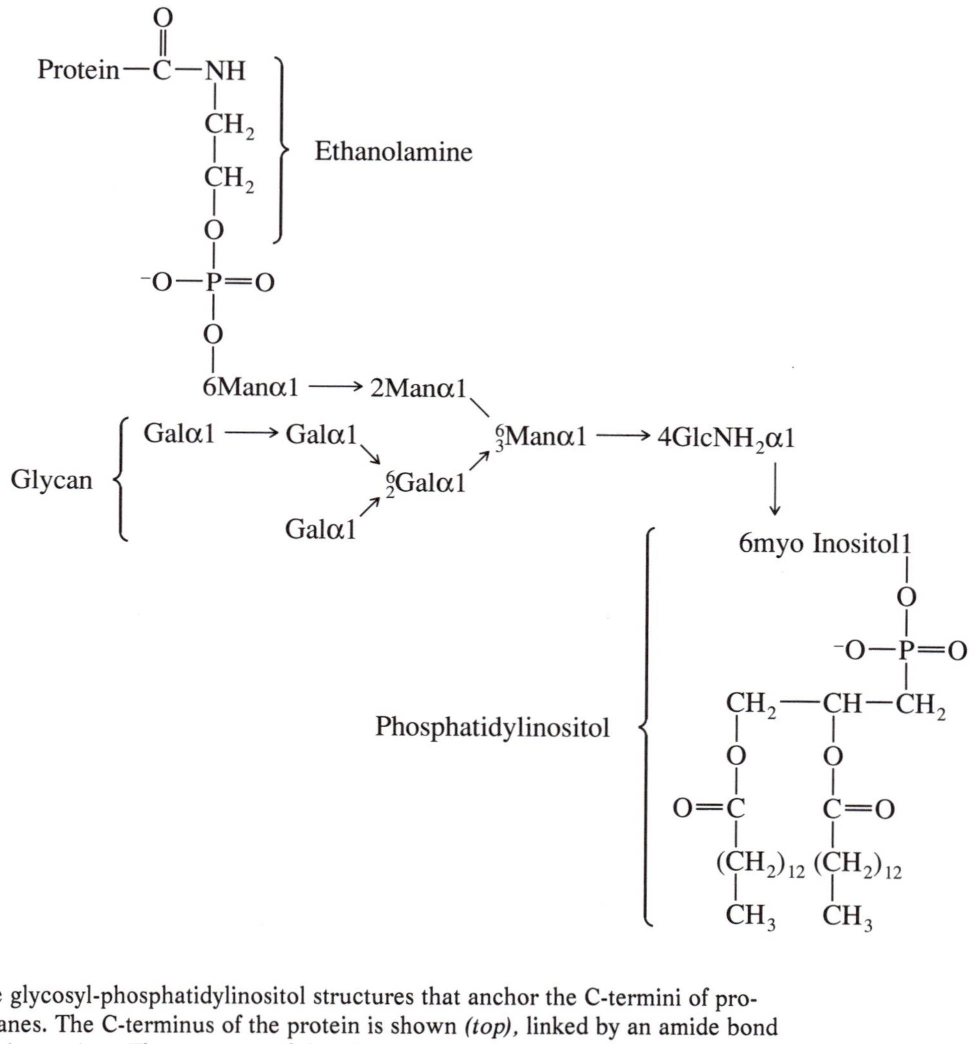

FIGURE 2.19

Structure of the glycosyl-phosphatidylinositol structures that anchor the C-termini of proteins to membranes. The C-terminus of the protein is shown *(top)*, linked by an amide bond to an ethanolamine moiety. The structure of the glycan is indicated schematically (Man, mannose; Gal, galactose; GlcNH$_2$, glucosamine); the numbers refer to the atom numbers through which the covalent bonds occur. The myristyl fatty acids shown can be replaced by others; they are embedded in one layer of the lipid bilayer of the membrane. These structures can vary somewhat, particularly that of the glycan.

transmembrane segments in the polypeptide chain is that it permits the protein to diffuse much more rapidly within the plane of the membrane (Sec. 7.2). Also, the glycosyl-phosphoinositide linkage can be hydrolyzed by specific enzymes, which suggests that anchoring might be regulated. Another intriguing possibility arises from the involvement of phosphoinositides; their breakdown products after cleavage by *phospholipase C* have been identified in recent years as important *second messengers* that are produced in the cell in response to triggers at the cell surface. Cleavage of the glycosylphosphoinositol-anchored proteins produces the same second messengers, but outside the cell. This led to the suggestion that a similar phenomenon might be involved in the interaction of the anchored molecules with ligands or receptors at the cell surface. This possibility remains tantalizing but unproven.

Farnesyl and geranylgeranyl groups are attached to Cys residues at the C-termini of proteins synthesized with C-terminal sequences of the general type -Cys-Axx-Axx-Axx-Xxx, where Axx is a residue of an aliphatic amino acid (most often Leu, Ile, or Val) and Xxx is any C-terminal amino acid residue. The three terminal residues are removed, the farnesyl or geranylgeranyl group is attached to the sulfur atom of the Cys side chain, and the α-carboxyl group is methylated. Farnesyl groups are 15-carbon lipids that are derived from farnesol, which is a branched-chain, polyunsaturated hydrocarbon alcohol intermediate of sterol biosynthesis; geranylgeranyl groups are similar but longer, with 20 carbon atoms. This type of modification is crucial for the biological activities of at least some especially important proteins, such as the products of the *ras* oncogenes, but the chemical basis is not known.

References

Structural and functional roles of glycosyl-phosphatidylinositol in membranes. M. G. Low and A. R. Saltiel. *Science* 239:268–275 (1988).

Cell-surface anchoring of proteins via glycosyl-phosphatidylinositol structures. M. A. J. Ferguson and A. F. Williams. *Ann. Rev. Biochem.* 57:285–320 (1988).

Prenyl proteins in eukaryotic cells: a new type of membrane anchor. J. A. Glomset et al. *Trends Biochem. Sci.* 15:139–142 (1990).

f. Amidation of C-Terminus

A C-terminal amide group in place of the usual carboxyl group is a characteristic feature of many peptide hormones. It is often important for biological activity and contributes to biological stability of the hormone. The amide group is derived from a Gly residue originally present at the C-terminus:

$$-\overset{\overset{\textstyle O}{\|}}{C}-NH-CH_2-CO_2H$$
$$\downarrow$$
$$-\overset{\overset{\textstyle O}{\|}}{C}-NH-CH(OH)-CO_2H$$
$$\downarrow$$
$$-\overset{\overset{\textstyle O}{\|}}{C}-NH_2 + \overset{\overset{\textstyle O}{\|}}{HC}-CO_2H \tag{2.7}$$

The two reactions are catalyzed by two sequential enzymes and involve ascorbic acid (vitamin C), copper, and oxygen. The hydroxylated intermediate is highly unstable.

In many cases, the Gly residue modified in this way occurs at the C-terminus only after proteolytic cleavage of a larger precursor. Proteolytic processing occurs in the Golgi apparatus and maturing secretory granules. The enzymes that catalyze the amidation reaction are in the secretion granules.

Reference

Peptide amidation. A. F. Bradbury and D. G. Smyth. *Trends Biochem. Sci.* 16:112–115 (1991).

g. Reversible Removal of the C-Terminal Tyr of α-Tubulin

Removal of the C-terminal residue in α-tubulin is special because the removal is reversible. Virtually all α-tubulins are synthesized in the cytoplasm with a Tyr residue at the C-terminus. This residue is removed in vivo by the enzyme *tubulinyl tyrosine carboxypeptidase* but is reinstated by another enzyme, *tubulinyl tyrosine ligase*. Addition of the Tyr residue to α-tubulin requires only the free amino acid and ATP.

The details are not certain, but it appears that this reversible modification of α-tubulin is related to the role of this protein in forming microtubules. Tubulin exists as a soluble heterodimer of α- and β-tubulins and reversibly aggregates into filamentous microtubules. The tubulin carboxypeptidase acts preferentially on the polymerized protein, so microtubules gradually become depleted of the C-terminal Tyr residue on α-subunits. The ligase acts primarily on soluble, monomeric tubulin, so monomers with Tyr predominate in vivo and are used primarily in assembling microtubules. Consequently, there is a cycle in which Tyr residues that are added to the soluble tubulin subunits are gradually removed after assembly of the monomers into microtubules. Removal of the Tyr residues affects the dynamic properties of the microtubules, but the precise relationship is not certain.

Microtubules are involved in many cellular functions such as mitosis, morphogenesis, motility, and intracellular organelle transport. To perform so many functions simultaneously, the microtubules may need to be differentiated. The presence or absence of the C-terminal Tyr residue may be one such marker, but it is clear that additional modifications of tubulin may also be significant, such as acetylation of the N-terminus. In addition, a variety of closely related α- and β-tubulins are synthesized by most organisms.

References

Molecular biology and genetics of tubulin. D. W. Cleveland and K. F. Sullivan. *Ann. Rev. Biochem.* 54:331–365 (1985).

Posttranslational modification and microtubule stability. E. Schulze et al. *J. Cell Biol.* 105:2167–2177 (1987).

Differential turnover of tyrosinated and detyrosinated microtubules. D. R. Webster et al. *Proc. Natl. Acad. Sci. USA* 84:9040–9044 (1987).

Microtubules containing detyrosinated tubulin are less dynamic. T. E. Kreis. *EMBO J.* 6:1597–2606 (1987).

2.4.3 Glycosylation

Attachment of carbohydrates is one of the most prevalent posttranslational modifications of eukaryotic proteins, especially of secreted and membrane proteins, yet the process has no well-defined universal purpose. Indeed, the biological activities of many glycoproteins are not detectably different if the carbohydrates are removed, and glycosylation of proteins does not occur at all in prokaryotes. Those functions of the carbohydrates that have been detected thus far appear to be specific to each protein. In a few cases the carbohydrates are involved in biological activity of the protein, but they are more often important for its physical properties, such as solubility. Carbohydrates often lengthen the biological life of a protein by decreasing its rate of clearance from the serum. Being on the surfaces of proteins, the carbohydrates are often involved in interactions with other cells or molecules, such as immunoglobulins, cell-surface receptors, and proteases. The most relevant properties of glycosyl groups attached to proteins are (1) their variable structures, which permit specificity in their interactions with other molecules; (2) their hydro-

philic natures, which keep them in aqueous solution; and (3) their bulk, which markedly affects the surface properties of the protein to which they are attached.

There are two types of glycosylation, called *N*-type or *O*-type depending on the atom of the protein to which the carbohydrate is attached (Fig. 2.20). *N*-type glycosylation occurs exclusively on the nitrogen atom of Asn side chains, whereas *O*-glycosylation occurs on the oxygen atoms of hydroxyls, particularly those of Ser and Thr residues. *N*-glycosylation occurs cotranslationally soon after the Asn residue emerges into the ER, whereas *O*-glycosylation occurs primarily in the Golgi as a posttranslational modification.

Description of glycosylation is made difficult by the complexity of the carbohydrate structures attached to the proteins. Not only are at least eight different sugar monomers used — galactose, glucose, fucose, mannose, *N*-acetylgalactosamine, *N*-acetylglucosamine, sialic acids, and xylose — but they are joined by a variety of glycoside linkages between their various functional groups. Some typical carbohydrate structures that are attached to proteins are described schematically in Figure 2.21. The details of carbohydrate chemistry are omitted in the following discussion, and emphasis is placed on the protein, though this provides only half of the story.

References

Glycoproteins. R. G. Spiro. *Adv. Protein Chem.* 27:349–467 (1973).

Comparative aspects of glycoprotein structure. R. Kornfeld and S. Kornfeld. *Ann. Rev. Biochem.* 45:217–237 (1976).

Sugar residues on proteins. P. V. Wagh and O. P. Bahl. *Crit. Rev. Biochem.* 10:307–377 (1981).

a. *N*-Glycosylation of Asn Residues

Carbohydrate to be attached to Asn residues is preassembled as a core structure (I of Fig. 2.21) attached to a membrane lipid, dolichyl phosphate. The assembly of this core structure in the ER by membrane-bound enzymes is the step that is blocked by the commonly used inhibitor of glycosylation, tunicamycin. This method of attachment of a preformed core structure is probably used because the target Asn residue is encountered only transiently as it emerges through the ER membrane.

The Asn residue that is glycosylated always occurs in a characteristic sequence: -Asn-Xaa-Ser-, -Asn-Xaa-Thr-, or -Asn-Xaa-Cys-. Xaa can be any residue except Pro, which also cannot immediately follow the tripeptide sequence. This characteristic sequence, however, is not the only determinant for glycosylation because not

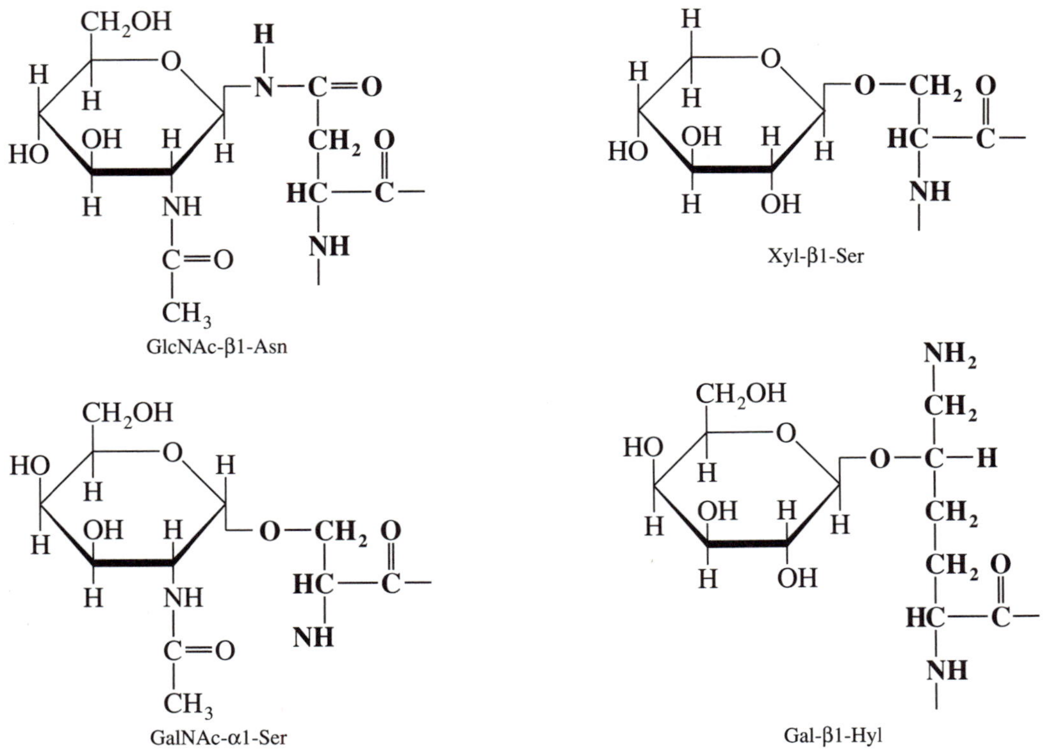

FIGURE 2.20

Structures of carbohydrate–protein linkages commonly found in proteins. GlcNAc-Asn: *N*-acetylglucosaminyl-Asn (*N*-glycosylation); Xyl-Ser: xylosyl-Ser (linkage to glycosamino-glycan in proteoglycans); GalNAc-Ser: *N*-acetylgalactosaminyl-Ser; Gal-Hyl: galactosyl hydroxylysine (*O*-glycosylation).

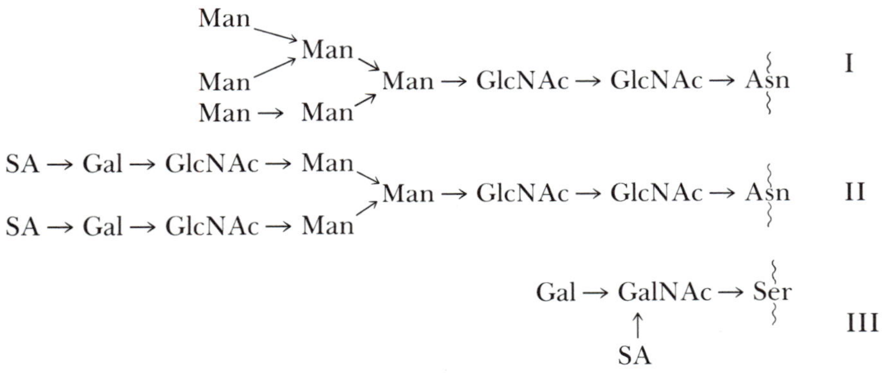

FIGURE 2.21

Structures of typical oligosaccharide chains attached to proteins. I is the polymannose core oligosaccharide transferred to Asn residues in the sequence —Asn-Xaa-Ser/Thr/Cys— from the lipid pyrophosphoryldolichol; the original lipid-linked oligosaccharide contained an additional three glucose residues that are removed after transfer to the protein. II is a typical product of further processing of core structure I, in this case to produce a "complex" type of carbohydrate. In high-mannose types, additional mannose residues are added to the tips of the core structure I. Hybrid types, in which one branch is complex and another is high mannose, are also found. III is a typical *O*-glycosidically linked structure, added to a Ser residue. The abbreviations of the sugar residues are Man, mannose; Gal, galactose; GlcNAc, *N*-acetylglucosamine; GalNAc, *N*-acetylgalactosamine; and SA, sialic acid. The positions and anomeric configurations of the carbohydrate linkages are omitted for simplicity. (From J. A. Hanover and W. J. Lennarz, *Arch. Biochem. Biophys.* 211:1–19, 1981.)

all such Asn residues of proteins that enter the ER are modified. For example, proteins such as ovalbumin and deoxyribonuclease are fully glycosylated at a single site but contain at least one additional sequence that is not glycosylated. Some proteins, such as bovine pancreatic ribonuclease, are glycosylated at an appropriate Asn residue in only a fraction of the molecules (ribonuclease B versus A). Others, such as pancreatic elastase and carboxypeptidase, contain one or more potential glycosylation sites but are not glycosylated at all, even though they are synthesized and secreted by the same pancreatic cells that glycosylate other proteins. Because glycosylation occurs to the nascent chain in the ER, the primary structure would be expected to be the primary determinant, but what distinguishes between the different Asn residues in the same tripeptide sequence is a major unsolved problem.

After attachment of a core glycan to the Asn residue of a protein in the ER, the glycan is extensively modified during passage of the protein through the ER and Golgi. In some cases this modification primarily involves attachment of more mannose groups; in other cases a more complex type of structure is attached (Fig. 2.21). In the case of lysosomal proteins, the core mannose groups are phosphorylated (Sec. 2.3.1.f). The type of processing depends on the identity of the protein, the cell type, and the physiological state of the cell. The processing of the glycosyl groups is typically variable, so an N-glycosylated protein is usually heterogeneous in its carbohydrate groups. The various modifications to the core glycan take place in various parts of the ER and the Golgi (Table 2.1), so the state of its carbohydrate is an excellent marker of where in the cell a protein has traveled.

References

Structural requirements of N-glycosylation of proteins. E. Bause. *Biochem. J.* 209:331–336 (1983).

Glycoprotein synthesis and embryonic development. W. J. Lennarz. *Crit. Rev. Biochem.* 14:257–272 (1983).

Inhibitors of the biosynthesis and processing of N-linked oligosaccharides. A. D. Elbein. *Crit. Rev. Biochem.* 16:21–49 (1984).

Function of glycoprotein glycans. K. Olden et al. *Trends Biochem. Sci.* 10:78–82 (1985).

Assembly of asparagine-linked oligosaccharides. R. Kornfeld and S. Kornfeld. *Ann. Rev. Biochem.* 54:631–664 (1985).

Protein glycosylation in the endoplasmic reticulum. W. J. Lennarz. *Biochemistry* 26:7205–7210 (1987).

Topography of glycosylation in the rough endoplasmic reticulum and Golgi apparatus. C. B. Hirschberg and M. D. Snider. *Ann. Rev. Biochem.* 56:63–87 (1987).

Inhibitors of the biosynthesis and processing of N-linked oligosaccharide chains. A. D. Elbein. *Ann. Rev. Biochem.* 56:497–534 (1987).

Glycobiology. T. W. Rademacher et al. *Ann. Rev. Biochem.* 57:785–838 (1988).

N-glycosylation and the production of recombinant glycoproteins. R. B. Parekh et al. *Trends Biotechnol.* 7:117–122 (1989).

Sequence differences between glycosylated and non-glycosylated Asn-X-Thr/Ser acceptor sites: implications for protein engineering. Y. Gavel and G. von Heijne. *Protein Eng.* 3:433–442 (1990).

The distribution of glycan structures in individual N-glycosylation sites in animal and plant glycoproteins. M. G. Yet and F. Wold. *Arch. Biochem. Biophys.* 278:356–364 (1990).

Environmental effects on protein glycosylation. C. F. Goochee and T. Monica. *Bio/Technology* 8:421–427 (1990).

b. *O*-Glycosylation

Attachment of carbohydrates to the oxygen atoms of amino acid side chains occurs primarily in the Golgi apparatus. N-Acetylgalactosamine (GalNAc) groups are attached to Ser and Thr groups of certain proteins (Fig. 2.20). In the case of collagen, hydroxy-Lys (Hyl) and hydroxy-Pro residues (Sec. 5.5.3) are also modified in this way.

The signals that determine which Ser and Thr residues are glycosylated have not been apparent from just the amino acid residues surrounding them, though the Hyl residues in collagen that are glycosylated occur in a characteristic sequence, -Gly-Xaa-Hyl-Xaa-Arg-, where Xaa is any residue. *O*-Glycosylation occurs in proteins that are already folded and where the three-dimensional structure is probably important. In some proteins, the amino acid residues that are glycosylated are clustered in the primary structure. The carbohydrate content of these proteins can be as high as 65–85% by weight, so the carbohydrate dominates the structure. These regions of the protein tend not to have fixed conformations but to be mostly extended.

O-Glycosylation was thought to be confined to proteins that pass through the ER and Golgi apparatus, but recently it has been found to occur in a surprising number of cytoplasmic and nuclear proteins. In this case, N-acetylglucosamine (GlcNAc) groups are attached to the side chains of Ser and Thr residues, but little is known about the process.

References

Subcellular site of synthesis of the N-acetylgalactosamine (α1-0) serine (or threonine) linkage in rat liver. C. Abeijon and C. B. Hirschberg. *J. Biol. Chem.* 262:4153–4159 (1987).

Why are proteins *O*-glycosylated? N. Jentoft. *Trends Biochem. Sci.* 15:291–294 (1990).

c. Proteoglycans

Proteoglycans are composed of a variety of protein backbones, to which one or many **glycosaminoglycan** chains are covalently attached. The glycosaminoglycans are repeating disaccharide chains of three types: chondroitin sulfate/dermatan sulfate, heparan sulfate/heparin, or keratan sulfate. They are sulfated to various degrees and are usually attached to the protein backbone through a xylose moiety linked to a Ser residue. They can also be attached to Ser residues through *N*-acetylglucosamine residues and through the complex type of carbohydrates *N*-linked to Asn residues. The signal for attachment of chondroitin sulfate chains appears to be the sequence -Ser-Gly-Xaa-Gly- preceded by two or three acidic residues. Other glycosaminoglycan chains are attached to Ser residues that are followed by Gly. No other signals have yet been detected. What chains are attached to each core protein depends on other unknown aspects of the protein structure and on the cell in which it is made.

Proteoglycans are extreme examples of glycosylated proteins. The bulk of their structure is usually the large amount of carbohydrate that is attached to the polypeptide chain at very many sites. Proteoglycans are secreted and in some cases also have *N*-linked oligosaccharides attached to other side chains. Their physical and biological properties consequently are dominated by the carbohydrates, and the protein components have been very difficult to characterize chemically. Information about the protein parts is finally becoming available with the ability to determine their primary structures using recombinant DNA techniques (Sec. 2.2). The greatly varied core proteins in proteoglycans have complex primary structures and are not simply unfolded polypeptide backbones. The core proteins vary in molecular weight from 11,000 to 220,000, and the number of glycosaminoglycan chains can vary from 1 to 100. Besides having segments of the polypeptide chain involved in attachment of glycosaminoglycan chains, the core proteins have other domains that are involved in interactions with membranes, extracellular matrix components, and other molecules.

Proteoglycans are important constituents of the extracellular matrix of multicellular organisms, and they are also associated with most cells, on their surface and inside intracellular storage granules. Most extracellular matrix proteins and many growth factors have binding sites for glycosaminoglycans. The detailed biological

and physical characterization of proteoglycans is only just beginning.

References

Proteoglycan core protein families. J. R. Hassell et al. *Ann. Rev. Biochem.* 55:539–567 (1986).

Proteoglycans in health and disease: structures and functions. A. R. Poole. *Biochem. J.* 236:1–14 (1986).

Structure and function of cell-associated proteoglycans. L. A. Fransson. *Trends Biochem. Sci.* 12:406–411 (1987).

Identification and synthesis of a recognition signal for the attachment of glycosaminoglycans to proteins. M. A. Bourdon et al. *Proc. Natl. Acad. Sci. USA* 84:3194–3198 (1987).

Structure and biology of proteoglycans. E. Ruoslahti. *Ann. Rev. Cell Biol.* 4:229–255 (1988).

Nuclear and cytoplasmic glycosylation: novel saccharide linkages in unexpected places. G. W. Hart et al. *Trends Biochem. Sci.* 13:380–384 (1988).

Immunoglobulin fold and tandem repeat structures in proteoglycan N-terminal domains and link protein. S. J. Perkins et al. *J. Mol. Biol.* 206:737–753 (1989).

Proteoglycans: structures and interactions. L. Kjellén and U. Lindahl. *Ann. Rev. Biochem.* 60:443–475 (1991).

2.4.4 Lipid Attachment

Lipids frequently tether intrinsically soluble proteins to membranes. The polar group of the lipid is attached covalently to the protein, while the hydrophobic portion of the lipid is embedded in the membrane. Examples are the myristoyl groups attached to the N-terminus (Sec. 2.4.2.d) and the glycosyl-phosphatidylinositol and farnesyl groups attached to the C-terminus (Sec. 2.4.2.e) that have been described. In addition, palmitoyl groups can be attached in thioester linkages to the side chains of Cys residues:

$$H_3C-(CH_2)_{14}-\overset{\overset{\displaystyle O}{\|}}{C}-S-CH_2- \qquad (2.8)$$

No similarities among the sequences of proteins modified in this way have yet been noted. Proteins modified with palmitoyl groups are usually firmly anchored in a membrane by the palmitoyl group. In most cases, however, the proteins are intrinsic membrane proteins without the palmitoyl group and are synthesized on the rough endoplasmic reticulum. The palmitoyl groups are added to the completed chain either in the ER or in the *cis* or medial parts of the Golgi apparatus. The Cys residues modified are usually 3–6 residues from the start of transmembrane segments, on the cytoplasmic side. In these cases, the protein would be integrated into the membrane whether or not it were acylated, so the reason for palmitoylation is not known. The palmitoyl group is often labile, being removed and replaced. In other instances, palmitoylation occurs only if the C-terminal Cys residue is farnesylated (Sec. 2.4.2.e).

Other fatty acids can also be esterified to proteins, and other types of linkages are thought to occur. Much remains to be learned about this covalent modification, especially because many proteins involved in cell regulation are modified in this way.

References

The biology and enzymology of eukaryotic protein acylation. D. A. Towler et al. *Ann. Rev. Biochem.* 57:69–99 (1988).

Fatty acylation of proteins. A. M. Schultz et al. *Ann. Rev. Cell Biol.* 4:611–647 (1988).

Acylation of viral and eukaryotic proteins. R. J. A. Grand. *Biochem. J.* 258:625–638 (1989).

The fats of life: the importance and function of protein acylation. R. A. J. McIlhinney. *Trends Biochem. Sci.* 15:387–391 (1990).

2.4.5 Sulfation

Sulfation of Tyr residues is another posttranslational modification that is limited to proteins that pass through the Golgi apparatus. The enzyme responsible, *tyrosyl protein sulfotransferase,* is an integral membrane protein, with its active site in the lumen of the *trans* Golgi network. The donor of the sulfate groups is 3′-phosphoadenosine-5′-phosphosulfate.

Tyr residues that are sulfated do not occur in recognizably similar sequences, though they are usually surrounded by acidic residues, with a paucity of basic residues. Four acidic residues are generally within five residues on either side of the Tyr that is sulfated, and one is usually the preceding residue. It is also important that the Tyr residue be on the surface of the protein conformation and that it be accessible; nearby glycosylation blocks sulfation.

The functional roles of the recently recognized phenomenon of sulfation are just being discovered. There are indications that sulfation affects the biological activities of some neuropeptides, the proteolytic processing of some protein precursors, and intracellular transport of some secretory proteins.

References

Protein tyrosine sulfation. W. B. Huttner. *Trends Biochem. Sci.* 12:361–363 (1987).

Occurrence of tyrosine sulfate in proteins—a balance sheet. A. Hille et al. *Eur. J. Biochem.* 188:577–586 (1990).

2.4.6 γ-Carboxy-Glu Residues

Certain Glu residues, particularly in proteins involved in blood clotting and bone structure, are carboxylated to yield the unusual residue *γ-carboxyglutamic acid,* generally abbreviated as **Gla**:

$$\begin{array}{c}
HO_2C \quad \quad CO_2H \\
\backslash \quad \quad / \\
CH \\
| \\
CH_2 \quad O \\
| \quad \quad \| \\
-NH-CH-C- \\
Gla
\end{array} \qquad (2.9)$$

The enzyme responsible, *vitamin K-dependent carboxylase,* is an integral membrane protein with its active site in the lumen of the endoplasmic reticulum. The carboxyl donor is HCO_3^-, and the enzyme also requires the reduced form of vitamin K and O_2. This is the first biochemical function found for vitamin K; its role appears to be to labilize or remove the γ-hydrogen atom of the Glu side chain that is to be replaced by the carboxyl group.

The Glu residues that are converted to Gla do not occur in any particular amino acid sequence, but they are generally in the first 40 residues of the mature protein. In one case, Factor X, the propeptide has been shown to direct the carboxylation of the 12 Glu residues closest to the amino terminus. There are also some sequence similarities among segments that are modified in various proteins.

The function of Gla residues is almost invariably linked to binding of Ca^{2+} ions (Sec. 8.3.4.a). The second adjacent carboxyl group considerably increases the intrinsic ability of these residues to bind Ca^{2+} ions, and Ca^{2+} binding is invariably involved in the functions of the proteins modified. The modification is essential for the functional properties of the proteins, as can be shown by synthesizing the proteins in the presence of vitamin K antagonists such as warfarin and dicumarol, which inhibit carboxylation and cause the proteins to be biologically inactive.

References

Vitamin K, prothrombin, and γ-carboxyglutamic acid. J. Stenflo. *Adv. Enzymol.* 46:1–31 (1978).

Mechanism of action of vitamin K: synthesis of γ-carboxyglutamic acid. J. W. Suttie. *Crit. Rev. Biochem.* 8:191–223 (1980).

Vitamin K-dependent carboxylase. J. W. Suttie. *Ann. Rev. Biochem.* 54:459–477 (1985).

Recognition site directing vitamin K-dependent γ-carboxylation resides on the propeptide of factor IX. M. J. Jorgensen et al. *Cell* 48:185–191 (1987).

γ-Carboxyglutamate-containing proteins and the vitamin K-dependent carboxylase. C. Vermeer. *Biochem. J.* 266:625–636 (1990).

2.4.7 Hydroxylation

Hydroxylation of certain Pro and Lys residues is an important step in the maturation and secretion of collagen. These modifications occur in the endoplasmic reticulum but only to procollagen. Pro residues that occur in the sequence -Xaa-Pro-Gly- are hydroxylated on the γ-carbon, and Lys residues in the sequence -Xaa-Lys-Gly- are hydroxylated on the δ-carbon:

γ-OH-Pro (4-hydroxy-Pro)   δ-OH-Lys (5-hydroxy-Lys)   (2.10)

Other Pro residues of certain types of collagen, occurring in the sequence -Gly-Pro-, are hydroxylated at the β-carbon:

β-OH-Pro (3-hydroxy-Pro)   (2.11)

Each of these modifications requires O_2, α-ketoglutarate, and ascorbate (vitamin C) and is catalyzed by an enzyme: *prolyl 4-hydroxylase, lysyl 5-hydroxylase,* and *prolyl 3-hydroxylase,* respectively. Each of these enzymes contains ferrous iron, and ascorbate is needed to keep the iron atom reduced. One of the oxygen atoms of O_2 hydroxylates the side chain, and the other oxidizes the α-ketoglutarate to succinate and CO_2.

These modifications are vital for the folding and assembly of mature collagen (Sec. 5.5.3). γ-OH-Pro residues stabilize the collagen triple helix by introducing additional hydrogen bonding, and δ-OH-Lys residues are necessary for the formation of certain cross-links and for the attachment of glycosyl groups (see Fig. 2.20).

A similar hydroxylation has recently been found to occur on certain Asn and Asp residues in a few proteins. In each case, the hydroxyl groups are added posttranslationally in a specific orientation to the β-carbon of the side chains; the hydroxyl group introduces a new center of chirality and is always the *erythro* isomer:

e-β-Hydroxy-Asp (e-βHya)   e-β-Hydroxy-Asn (e-βHyn)   (2.12)

Which Asp and Asn residues are modified in this way seems to depend primarily on the three-dimensional conformation of the protein.

These modifications are thought to participate in Ca^{2+} binding, although little is known.

References

Biochemistry of the hydroxyprolines. R. Kutton and A. N. Radhakrishnan. *Adv. Enzymol.* 37:273–347 (1973).

The biosynthesis of collagen. P. Bornstein. *Ann. Rev. Biochem.* 43:567–603 (1974).

Metabolism of proline and hydroxyproline. E. Adams and L. Frank. *Ann. Rev. Biochem.* 49:1005–1061 (1980).

Hydroxylation of aspartic acid in domains homologous to the epidermal growth factor precursor is catalyzed by a 2-oxoglutarate-dependent dioxygenase. J. Stenflo et al. *Proc. Natl. Acad. Sci. USA* 86:444–447 (1989).

2.4.8 Phosphorylation

An increasing number of proteins are known to be phosphorylated at specific sites, usually reversibly and with important functional consequences (Sec. 9.4.3). The phosphoryl groups are added by specific protein kinases, using ATP as the phosphoryl donor:

$$\text{protein} + \text{ATP} \longrightarrow \text{protein}-PO_3^{2-} + \text{ADP} \quad (2.13)$$

The phosphoryl groups are removed by specific phosphatases:

$$\text{protein}-PO_3^{2-} + H_2O$$
$$\downarrow$$
$$\text{protein} + HPO_3^{2-} \quad (2.14)$$

The sum of these two reactions is simply hydrolysis of ATP, so the two reactions are catalyzed by different enzymes, and their activities are strictly controlled. In fact, it is through control of the kinases and phosphatases that the activities of the phosphorylated proteins are regulated (Sec. 9.4.3). Many hormones act by increasing the intracellular concentration of second messengers — cyclic AMP, diacylglycerol, or Ca^{2+} — which in turn activate protein kinases that phosphorylate Ser and Thr residues of various proteins. The protein products of oncogenes and many growth-factor receptors have protein kinase activities that phosphorylate Tyr residues.

The sites of phosphorylation are usually the hydroxyl groups of specific Ser, Thr, or Tyr residues:

$$-CH_2-O-\overset{\overset{O}{\|}}{\underset{\underset{OH}{|}}{P}}-O^-$$

Ser P

$$-\overset{\overset{CH_3}{|}}{CH}-O-\overset{\overset{O}{\|}}{\underset{\underset{OH}{|}}{P}}-O^-$$

Thr P

$$-CH_2-\langle\!\!\langle\;\rangle\!\!\rangle-O-\overset{\overset{O}{\|}}{\underset{\underset{OH}{|}}{P}}-O^-$$

Tyr P   (2.15)

but Asp, His, and Lys residues may also be phosphorylated:

$$\begin{array}{c}
HO-\overset{\overset{O}{\|}}{P}-O^- \\
| \\
O \\
| \\
-CH_2-CH\begin{array}{c}C-N\\ \diagdown\;\;\diagup\\ N\end{array}CH
\end{array}$$

$$\begin{array}{c}
-CH_2-CH\begin{array}{c}C-N\\\diagup\;\;\diagdown\\N\\|\\O\\|\\HO-\overset{\overset{}{}}{P}-O^-\\\|\\O\end{array}CH
\end{array}$$

His P

$$-(CH_2)_4-NH-O-\overset{\overset{O}{\|}}{\underset{\underset{OH}{|}}{P}}-O^-$$

Lys P

$$-CH_2-\overset{\overset{O}{\|}}{C}-O-\overset{\overset{O}{\|}}{\underset{\underset{OH}{|}}{P}}-O^-$$

Asp P   (2.16)

From the point of view of protein structure and function, the most important aspect of the phosphoryl group appears to be its negative charge (Sec. 9.4.3).

The various kinases have different specificities for different proteins. In any one protein, which residues are phosphorylated depends on the primary structure around them, on their accessibility to the kinase, and on the specificity of the kinase enzyme. The important cyclic AMP-dependent kinase has a strong preference for Ser residues that occur in the sequence -Arg-Arg-$(Xaa)_n$-Ser-, where n is usually 1 but can be 0 or 2. Other Ser/Thr kinases similarly recognize Ser residues following one or two basic residues. In contrast, Tyr phosphorylation usually involves Tyr residues that occur in the sequence -Lys/Arg-$(Xaa)_3$-Asp/Glu-$(Xaa)_3$-Tyr-. The folded conformation of the protein is also important for determining which residues are phosphorylated by any kinase, because short peptides with these sequences are usually relatively poor substrates of the kinases. Phosphorylation almost invariably occurs to folded proteins well after their synthesis has been completed, and in contrast to many other posttranslational modifications, it occurs primarily in the cytoplasm.

Not all phosphorylation is functionally important; that of the milk protein casein is probably primarily of nutritional importance. In this case, the Ser residues phosphorylated are usually to the amino side of a number of acidic residues.

References

Synthesis and properties of N-, O-, and S-phospho derivatives of amino acids, peptides and proteins. A. W. Frank. *Crit. Rev. Biophys.* 16:51–101 (1984).

Protein kinases in the brain. A. C. Nairn et al. *Ann. Rev. Biochem.* 54:931–976 (1985).

Protein-tyrosine kinases. T. Hunter and J. A. Cooper. *Ann. Rev. Biochem.* 54:897–930 (1985).

Protein kinase activity of the insulin receptor. S. Gammeltoft and E. van Obberghen. *Biochem. J.* 235:1–11 (1986).

Protein serine/threonine kinases. A. M. Edelman et al. *Ann. Rev. Biochem.* 56:567–613 (1987).

Protein kinases: a diverse family of related proteins. S. S. Taylor. *BioEssays* 7:24–29 (1987).

Protein phosphorylation and hormone action. P. Cohen. *Proc. Roy. Soc. Lond. B* 234:115–144 (1988).

Histidine phosphorylation and phosphoryl group transfer in bacterial chemotaxis. J. F. Hess et al. *Nature* 336:139–143 (1988).

Calcium/calmodulin-dependent protein kinase II. R. J. Colbran et al. *Biochem. J.* 258:313–325 (1989).

Signal transduction in bacteria. J. B. Stock et al. *Nature* 344:395–400 (1990).

Protein kinase recognition sequence motifs. B. E. Kemp and R. B. Pearson. *Trends Biochem. Sci.* 15:342–346 (1990).

2.4.9 ADP-Ribosylation

Another common modification, ADP-ribosylation, is similar to phosphorylation in that it acts reversibly in the cytoplasm and nuclei of cells to regulate various proteins. All eukaryotic cells seem to have enzymes called *ADP-ribosyl transferases*, which cleave the cofactor NAD^+ and transfer the ADP-ribosyl moiety to various side chains in a number of proteins:

ADP-ribosyl protein   (2.17)

The modification can occur at the nitrogen atoms of His, Arg, Asn, and Lys residues, at the carboxyl group of Glu, and at the α-carboxyl group of terminal Lys residues.

In modifications to carboxyl groups, addition of a first ADP-ribosyl group is followed by addition of others to the 2′ hydroxyls of either of the ribose groups. In this way, linear and branched poly(ADP-ribose) structures containing up to 65 ADP-ribose groups can be generated. This modification occurs primarily in the nucleus.

Multiple ADP-ribosyl groups can be removed by the enzyme poly(ADP-ribose) glycohydrolase, and the group attached directly to the protein can be removed by another enzyme, ADP-ribosyl protein lyase. A multitude of proteins are modified in this way, with a variety of effects on their activities, and no simple, coherent description of the normal physiological effects of ADP-ribosylation can be given. The importance of this modification is illustrated, however, by the toxic effects caused by the ADP-ribosylation of certain proteins by diphtheria, cholera, and pertussis toxins. These toxic modifications mimic and subvert the regulated physiological modifications that normally occur.

References

ADP-ribosylation. K. Ueda and O. Hayaishi. *Ann. Rev. Biochem.* 54:73–100 (1985).
Covalent modification of proteins by ADP-ribosylation. J. C. Gaal and C. K. Pearson. *Trends Biochem. Sci.* 11:171–175 (1986).
ADP-ribosylation of guanyl nucleotide-binding regulatory proteins by bacterial toxins. J. Moss and M. Vaughan. *Adv. Enzymol.* 61:303–379 (1988).

2.4.10 Disulfide Bond Formation

Disulfide bond formation between Cys residues is a common occurrence in proteins synthesized in the ER. The Cys residues linked by disulfide bonds are usually far apart in the primary structure, so disulfide formation between them is intimately linked with three-dimensional folding of the polypeptide chain (Sec. 7.5.4).

The mechanism of disulfide bond formation in vivo is uncertain, but it probably involves thiol-disulfide exchange between the protein synthesized with free SH groups on its Cys residues and small-molecule disulfide compounds. The predominant thiol compound in most cells is **glutathione** (Sec. 2.5, see Eq. 2.22), which exists in both the thiol (GSH) and disulfide (GSSG) forms. Formation of one disulfide bond in a protein requires two sequential thiol-disulfide exchanges involving the mixed-disulfide intermediate:

(2.18)

The protein becomes oxidized and the glutathione is reduced, so it is convenient to define a *disulfide oxidation–reduction potential*, which in this case would be given by the ratio of the concentrations of GSSG and GSH: $[GSSG]/[GSH]^2$.

The chemical reaction of Equation (2.18) can occur rapidly under physiological conditions and is reversible. In a protein with more than two Cys residues, formation of disulfide bonds is often followed by intramolecular rearrangements ("shuffling") of the disulfides among the various Cys residues in the protein. This is frequently the rate-limiting step in the entire process of forming multiple disulfides in a protein (Sec. 7.5.4). Not surprisingly, an enzyme, **protein-disulfide isomerase,** is present in the ER to catalyze disulfide rearrangements in proteins and consequently to assist in their folding.

Which disulfides, if any, are formed in a protein depends on both the conformational properties of the protein, which determines whether and which Cys residues come into appropriate proximity, and the disulfide oxidation–reduction potential, which determines the intrinsic stability of protein disulfide bonds. The observed stabilities of individual protein disulfides are given by the equilibrium constant for Equation (2.18):

$$K_{SS} = \frac{[\text{protein}_S^S]\,[GSH]^2}{[\text{protein}_{SH}^{SH}]\,[GSSG]} \qquad (2.19)$$

The stabilities of protein disulfide bonds vary enormously. For unfolded proteins in which Cys residues are not separated by more than 50 residues, the values of K_{SS} for different pairs of Cys residues are in the region of only 10^{-2} M. For folded proteins in which the folded conformation keeps the Cys residues in proximity for forming disulfides, the values of K_{SS} may be as high as 10^5 M, whereas Cys residues kept apart by the conformation have values of zero.

In the cytoplasm of most cells, glutathione is present at a total concentration of $1-10$ mM, and only about 1% of it is present as GSSG; consequently, the ratio $[GSSG]/[GSH]^2$ has a value between 1 and 10 M^{-1}. Protein disulfides are present under these conditions only if their value of K_{SS} is greater than 1 M. Such large values of K_{SS} occur only if the protein conformation brings pairs of Cys residues into suitable proximity. In this case, disulfides can be stable in folded proteins under such disulfide oxidation–reduction potential conditions, even though the cytoplasm is frequently said to be too reducing because the majority of the glutathione is in the thiol form. Intramolecular protein disulfides can be considerably more stable than the "intermolecular" disulfide of GSSG (Sec. 4.4). In contrast, Cys residues kept apart by the protein conformation largely remain in the thiol form under intracellular redox conditions.

The value of the disulfide oxidation–reduction potential of the lumen of the ER is not known, but it is unlikely to be much more oxidizing than that of the cytoplasm. Otherwise, disulfide bonds would be intrinsically too stable, and polypeptide chains synthesized in the ER would tend to form disulfide bonds between most of their Cys residues, not just those favored by the protein conformation.

References

Native disulphide bond formation in protein biosynthesis: evidence for the role of protein disulphide isomerase. R. B. Freedman. *Trends Biochem. Sci.*9:438–441 (1984).

Disulfide bonds and protein stability. T. E. Creighton. *BioEssays* 8:57–64 (1988).

Thioredoxin and glutaredoxin systems. A. Holmgren. *J. Biol. Chem.* 264:13963–13966 (1989).

Role of protein disulphide-isomerase in the expression of native proteins. R. B. Freedman et al. *Biochem. Soc. Symp.* 55:167–192 (1989).

Protein disulfide isomerase: multiple roles in the modification of nascent secretory proteins. R. B. Freedman. *Cell* 57:1069–1072 (1989).

2.4.11 Common Nonenzymatic, Chemical Modifications

The covalent modifications described in the preceding sections occur to specific residues in certain proteins, a specificity that is possible only because the modifications are catalyzed by enzymes and depend on the detailed structure of the protein modified. A great many chemical modifications occur chemically and spontaneously, in the absence of enzymes. These modifications tend to occur to all appropriate residues in all proteins, depending only on the immediate chemical environment of the residues. Although many such modifications inactivate the modified proteins, most modifications occur at insignificant rates in vivo. Others occur more frequently; but at least in some cases, repair systems are available to minimize their consequences.

One inevitable modification is oxidation by the O_2 that is necessary for most life, and by other oxidants such as peroxides, superoxide and hydroxyl radicals, and hypochlorite ions that are generated during metabolism. Enzymes such as superoxide dismutase, peroxidase, and catalase scavenge many such oxidants, but oxidation of proteins still occurs. Most susceptible are the sulfur atoms of Cys and Met residues (Sec. 1.3.10). The Cys thiol groups of intracellular proteins are generally protected by the high concentrations of glutathione (or other similar thiol compounds in some organisms) that are present in cells and maintained in the reduced thiol form by specific enzyme systems. Met residues are readily oxidized chemically to the sulfoxide (Sec. 1.3.10), which can have drastic functional consequences in some proteins. For example, α_1-proteinase inhibitor (also known as α_1-antitrypsin) has a crucial Met residue in its active site; oxidation of this residue inactivates the inhibitor. The inhibitor normally inhibits serum elastase, a protease that when not controlled destroys the cell walls of the lung by digesting connective tissue proteins. Absence of α_1-proteinase inhibitor from the serum causes pulmonary emphysema. Oxidation of the inhibitor and the severity of emphysema are increased by smoking. Oxidation of Met residues in cells, however, is reversed by the enzyme Met sulfoxide-peptide reductase.

Another common chemical modification of proteins is deamidation of Asn and Gln residues (Sec. 1.3.6). Deamidation of Asn residues usually converts a fraction of them to *iso*-Asp residues, in which the peptide bond occurs through the side chain (see Fig. 1.3). This can have severe effects on protein structure. Many chemical modifications of proteins lead to the degradation of the protein (Chap. 10).

References

Biochemistry and physiological role of methionine sulfoxide residues in proteins. N. Brot and H. Weissbach. *Arch. Biochem. Biophys.* 223:271–281 (1983).

Nonenzymatic covalent posttranslational modification of proteins *in vivo.* J. J. Harding. *Adv. Protein Chem.* 37:247–334 (1985).

Reactions of the amide side-chains of glutamine and asparagine *in vivo.* F. Wold. *Trends Biochem. Sci.* 10:4–6 (1985).

Proteins released from stimulated neutrophils contain very high levels of oxidized methionine. I. Beck-Speier et al. *FEBS Letters* 227:1–4 (1988).

2.5 Nonribosomal Biosynthesis of Unusual Peptides

Some natural peptides have very atypical structures, including D-amino acids, unusual amino acids, covalent linkages other than through peptide bonds, and cyclic polypeptide chains. Most such peptides are antibiotics, which function to inhibit competing or predatory species. In this case, peptides somewhat different from ordinary proteins—and thus resistant to proteases and other defense mechanisms—have probably been very useful in the biological warfare that takes place between species.

For example, the **tyrocidine** antibiotics are cyclic peptides with ordinary peptide bonds, but with residues of both D- and L-phenylalanine and of L-ornithine (Orn):

$$
\begin{array}{ccc}
\text{LPro} & \longrightarrow & \text{LPhe} \\
\nearrow & & \downarrow \\
\text{DPhe} & & \text{DPhe} \\
\uparrow & & \downarrow \\
\text{LLeu} & & \text{LAsn} \\
\uparrow & & \downarrow \\
\text{LOrn} & & \text{LGln} \\
\nwarrow & & \swarrow \\
\text{LVal} & \longleftarrow & \text{LTyr}
\end{array}
\qquad (2.20)
$$

The arrows denote the direction of the polypeptide chain, from the amino to the carboxyl group of each residue. Ornithine is an amino acid that occurs naturally as an intermediate in arginine biosynthesis, but it is not normally used in protein biosynthesis.

The linear **gramicidins** have residues of D-Leu and D-Val, the amino end is formylated, and the C-terminal residue is linked in a peptide bond to ethanolamine:

$$
\begin{array}{l}
\overset{\displaystyle O}{\overset{\displaystyle \|}{HC}}-\text{LVal-Gly-LAla-DLeu-LAla-DVal-LVal-} \\[4pt]
\text{DVal-LTrp-DLeu-LTrp-DLeu-LTrp-DLeu-} \\[4pt]
\text{LTrp}-NH-CH_2-CH_2OH \qquad (2.21)
\end{array}
$$

An example that is not an antibiotic is the tripeptide **glutathione**, γ-Glu-Cys-Gly, in which the N-terminal Glu is linked by a peptide bond through the side-chain carboxyl rather than the usual α-carboxyl:

$$
\begin{array}{c}
\underset{\displaystyle H_2N-CH-CH_2-CH_2-C-NH-CH-C-}{\overset{\displaystyle CO_2H \qquad\qquad\qquad O \qquad\qquad CH_2 \; O}{\overset{\displaystyle |\qquad\qquad\qquad\qquad\quad \| \qquad\qquad |\quad \|}{}}} \\[10pt]
\text{SH above} \\
-NH-CH_2-CO_2H \qquad (2.22)
\end{array}
$$

This small peptide is present in high concentrations in most cells, where it serves a number of purposes. Its main purpose seems to be to protect the cell contents from oxidation, using its thiol group (Sec. 2.4.11). The reason for the atypical peptide bond presumably is to protect glutathione from degradation by distinguishing it from an ordinary peptide such as a degradation product of some protein (Chap. 10).

The question arises as to the mechanism of synthesis of these unusual peptides. Glutathione, for example, is assembled by two specific enzymes. The first, γ-glutamylcysteine synthetase, couples glutamic acid and cysteine in an amide bond through the γ-carboxyl of the glutamic acid. This carboxyl group is first activated using ATP, perhaps by forming γ-glutamyl phosphate transiently. The second enzyme, glutathione synthetase, adds the glycine residue in an ordinary peptide bond.

Activation of the carboxyl group by ATP is also thought to involve the phosphoryl intermediate.

Most peptide antibiotics are synthesized in a similar fashion but usually on large, multifunctional enzymes that carry out virtually all of the steps. The carboxyl groups of the amino acids used are generally activated by ATP to form amino acyl adenylates, as used in ribosomal synthesis (Eq. 2.1), but the amino acids are subsequently bound to the enzyme through a thiol ester to an —SH group rather than to a tRNA. Either D- or L-amino acids can generally be used because the enzymes can also catalyze their isomerization. The activated amino acids are then added sequentially to the growing polypeptide chain, which is attached as a thiol ester to the —SH group of the cofactor 4′-phosphopantetheine. The cofactor is thought to function as a swinging arm to deliver the growing peptide to different sites on the enzyme where the various amino acids are added sequentially. The sequence of the peptide generated is determined by the enzyme, presumably by the amino acid specificities of the sites that add the residues. This type of peptide synthesis has been designated the **thio-template mechanism.**

The enzymes required to produce peptides of 6–15 residues in this way are very large, consisting of multiple polypeptide chains with molecular weights from 100,000 to 450,000. To form the two peptide bonds of glutathione requires two proteins, one of molecular weight 100,000 and the other of molecular weight 118,000. This type of mechanism of peptide synthesis would clearly be impractical for proteins in general; each protein would require a larger protein to make it. Nevertheless, such a mechanism is apparently practical for the biosynthesis of a few atypical peptides.

We would expect there to be a practical limit to the size of protein that could be made in this way. Unusual peptides larger than about 20 residues are made on ribosomes in the usual way, using the normal 20 L-

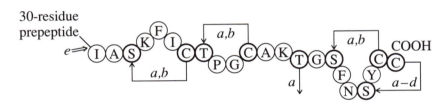

a Water elimination
b Sulfide-ring formation
c Decarboxylation
d Double bond formation
e Proteolytic cleavage

FIGURE 2.22

Structures of the biosynthetic precursor *(top)* and of the mature *(bottom)* antibiotic, staphylococcal epidermin. Abu, aminobutyric acid; Dhb, dehydrobutyrine. The presumed covalent modifications that take place are indicated. They may occur by first eliminating H_2O from all the Ser and Thr side chains *(a)* to form dehydro-Ala and Dhb residues, respectively. The thiol groups of the Cys residues might then add across the indicated double bonds *(b)* to produce the sulfide bridges. The C-terminal residue must also be decarboxylated *(c)* and a double bond introduced *(d)*; this oxidative decarboxylation can be catalyzed by a single enzyme. The precursor also has an unusually hydrophilic, α-helical N-terminal extension of 30 residues that is removed proteolytically *(e)* after the modification of the antibiotic part. (From N. Schnell et al. *Nature* 333:276–278, 1988.)

amino acids, and are then modified posttranslationally. Such modifications can be very dramatic. For example, the bacterial antibiotic *epidermin* is synthesized as a 52-residue prepeptide that is then processed enzymatically to a very unusual 21-residue peptide containing four cyclic structures and the unusual amino acids α-aminobutyric acid (Abu) and dehydrobutyrine (Dhb):

$$
\begin{array}{cc}
\text{CH}_3 & \text{CH}_2 \\
| & \| \\
\text{CH}_2 \quad \text{O} & \text{CH} \quad \text{O} \\
| \quad\quad \| & | \quad\quad \| \\
-\text{HN}-\text{CH}-\text{C}- & -\text{HN}-\text{CH}-\text{C}- \\
\text{Abu} & \text{Dhb}
\end{array}
\qquad (2.23)
$$

The unusual sulfide rings and amino acids are made from ordinary Cys, Ser, and Thr residues of the prepeptide (Fig. 2.22).

References

Polypeptide synthesis on protein templates: the enzymatic synthesis of gramicidin S and tyrocidine. F. Lipmann et al. *Adv. Enzymol.* 35:1–34 (1971).

Nonribosomal polypeptide synthesis on polyenzyme templates. F. Lipmann. *Acc. Chem. Res.* 6:361–367 (1973).

The protein thiotemplate mechanism of synthesis for the peptide antibiotics produced by *Bacillus brevis.* S. G. Laland and T. L. Zimmer. *Essays Biochem.* 9:31–57 (1973).

Biosynthesis of small peptides. K. Kurahashi. *Ann. Rev. Biochem.* 43:445–459 (1974).

Nonribosomal peptide formation on multifunctional proteins. H. Kleinkauf and H. von Döhren. *Trends Biochem. Sci.* 8:281–283 (1983).

Glutathione. A. Meister and M. E. Anderson. *Ann. Rev. Biochem.* 52:711–760 (1983).

Biosynthesis of peptide antibiotics. H. Kleinkauf and H. von Döhren. *Ann. Rev. Microbiol.* 41:259–189 (1987).

Nonribosomal biosynthesis of peptide antibiotics. H. Kleinkauf and H. von Döhren. *Eur. J. Biochem.* 192:1–15 (1990).

Cyclosporin synthetase. The most complex peptide synthesizing multienzyme polypeptide so far described. A. Lawen and R. Zocher. *J. Biol. Chem.* 265:11355–11360 (1990).

Antibiotics—ribosomally synthesized biologically active polypeptides containing sulfide bridges and α,β-didehydroamino acids. G. Jung. *Angew. Chem. Intl. Ed. Engl.* 30:1051–1068 (1991).

Exercises

1. The imminent demise of protein chemistry was predicted in 1978 due to the revolution in recombinant DNA technology (A. D. B. Malcolm, *Nature* 275:90–91, 1978). In what ways has this prediction been borne out and in what ways not?

ANSWERS
Section 1.6 and Sections 2.2 and 2.4.

2. The genetic defect responsible for X-linked chronic granulomatous disease was localized to a particular gene locus solely by genetic methods because the biochemical basis for the disease was unknown (B. Royer-Pokora et al., *Nature* 322:32–38, 1986; S. H. Orkin, *Trends Genetics* 3:149–151, 1987). Complementary DNA clones of the mRNA produced by this gene were sequenced to identify the gene product. A dispute arose subsequently about which AUG codon was used for polypeptide initiation (T. E. Creighton, *Nature* 324:21, 1986; M. Kozak, *Cell* 47:481–483, 1986; D. R. Cavener, *Nature* 325:21, 1987; C. Teahan et al., *Nature* 327:720–721, 1987). What are the lessons of this episode for identifying proteins from gene sequences?

ANSWER
S. H. Orkin, *Trends Genetics* 3:207 (1987).

3. Signal peptides are usually at the N-terminus of the polypeptide chain and are cleaved during or after translocation (Fig. 2.7). Addition of a substantial number of residues to the N-terminus of the signal peptide tends to destroy its activity as a signal peptide (H. Garoff, *Ann. Rev. Cell Biol.* 1:403–445, 1985). Consequently, most models for translocation show the signal peptide traversing the ER membrane once, with the N-terminus at the cytoplasmic surface and the site to be cleaved on the luminal side of the membrane (Fig. 2.7; M. S. Briggs and L. M. Gierasch, *Adv. Protein Chem.* 38:109–180, 1986). Is it possible to reconcile the observations that the signal peptide of one normally secreted protein, fused to the C-terminus of another normally cytoplasmic protein, caused both proteins to be translocated and that cleavage at the carboxyl end of the signal peptide occurred normally (E. Perara and V. R. Lingappa, *J. Cell Biol.* 101:2292–2301, 1985)? What are the implications of this observation for the source of the energy needed to translocate proteins across membranes?

4. Lipoproteins are very hydrophobic extracellular polypeptides that bind noncovalently to lipids and help to transport them through aqueous environments in which neither would be soluble alone. The polypeptides are synthesized in the same way as other secreted proteins (J. I. Gordon et al., *Crit. Rev. Biochem.* 20:37–71, 1986). In view of the hydrophobicity of the lipoproteins, how might they manage to be secreted from the cells in which they are made instead of remaining in the cells as membrane proteins? Does their secretion have implications for the nature of the pore through the ER membrane?

5. It was proposed that the polypeptide chains of proteins that are destined for intracellular organelles such as lysosomes, the ER, peroxisomes, and glycosomes have related sequences at their C-termini when initially synthe-

sized, at least as related as are signal peptides (T. E. Creighton and I. G. Charles, *Cold Spring Harbor Symp. Quant. Biol.* 52:511–519, 1987). In view of the similarities and of the observation that the C-terminal residues are often subsequently removed proteolytically, it seemed plausible that these sequences might be responsible for targeting the proteins to these organelles. How could this proposal be tested?

ANSWER

S. Munro and H. R. B. Pelham, *Cell* 48:899–907 (1987); S. J. Gould et al., *J. Cell Biol.* 108:1657–1664 (1989).

What are the difficulties in envisaging a common mechanism for targeting proteins to peroxisomes, glycosomes, and the other organelles?

ANSWER

Section 2.3.2.

6. The search for signal sequences for targeting proteins to specific locations often makes use of chimeric proteins, in which a segment of amino acid sequence is deleted from or added to a protein (often via its gene). If the chimeric protein follows a path different from that of the original protein, we usually conclude that the altered sequence is involved in directing the proteins to either the old or the new location. What other factors, besides the amino acid sequence, need to be considered? What other phenomena could cause a modified protein to remain within the ER?

ANSWERS

R. W. Doms et al., *J. Cell Biol.* 107:89–99 (1988); C. E. Machamer and J. K. Rose, *J. Biol. Chem.* 263:5955–5960 (1988).

7. A mutation usually causes a single amino acid residue to be altered in the protein product of a gene. Nevertheless, a single nucleotide mutation in the hemoglobin β-chain gene was found to cause two changes to the hemoglobin Long Island β-chain that was produced from the mutant gene. One was an extra Met residue at the N-terminus, and the other was replacement of the normal His2 residue by Pro (R. C. Barwick et al., *Proc. Natl. Acad. Sci. USA* 82:4602–4605, 1985; J. T. Prchal et al., *Proc. Natl. Acad. Sci. USA* 83:24–27, 1986). In mutant hemoglobin South Florida, a single mutation produced three alterations: replacement of the usual N-terminal Val residue by Met, addition of an N-terminal Met residue, and acetylation of the N-terminus (J. P. Boissel et al., *Proc. Natl. Acad. Sci. USA* 82:8448–8452, 1985). What explanation is there for the multiple effect of these single mutations? What do these observations imply for processing the amino ends of proteins?

8. The yeast mating factor hormone called *a*-factor was found from its gene sequence to have the sequence -Cys-Val-Ile-Ala at its C-terminus. This sequence is similar to the sequence at the C-terminus of the product of the *ras*

oncogene, which is palmitoylated at the Cys residue near the C-terminus. Both proteins must be anchored to the cellular membrane to be active biologically (S. Powers et al., *Cell* 47:413–422, 1986; T. Magee and M. Hanley, *Nature* 335:114–115, 1988). A single gene (designated RAM) has been shown to be necessary for the activity of the *a*-factor and for palmitoylation of *ras* proteins (F. Tamamol et al., *J. Cell. Biochem.* 36:261–273, 1988). Is it reasonable to conclude that the *a*-factor is also palmitoylated? Is the RAM gene likely to encode the enzyme responsible for palmitoylation?

ANSWERS

R. J. Anderegg et al., *J. Biol. Chem.* 263:18236–18240 (1988); A. M. Treston and J. L. Mulshine, *Nature* 337:406 (1989).

9. Concanavalin A was observed to be synthesized as a glycosylated precursor (E. M. Herman et al., *Planta* 165:23–29, 1985), yet the mature protein is not a glycoprotein nor does it contain any Asn-Xaa-Thr/Ser/Cys sequences. What possible explanations are there for this observation?

ANSWER

Figure 2.18B and D. J. Bowles et al., *J. Cell Biol.* 102:1284–1297 (1986).

10. Peptide bond hydrolysis is well established to be reversible in the case of protease inhibitors (M. Laskowski, Jr., and I. Kato, *Ann. Rev. Biochem.* 49:593–626, 1980). Does this have any relevance for the mechanism of forming the new peptide bond in concanavalin A (Fig. 2.18)? Is it necessary to postulate a transpeptidation reaction (D. J. Bowles et al., *J. Cell Biol.* 102:1284–1297, 1986)?

11. Biological activities and the correct disulfide bonds of pancreatic secretory proteins were generated in a cell-free translation system only when the preproteins were cotranslationally translocated into microsome preparations of the endoplasmic reticulum, with removal of the signal peptide. When synthesized in the same system lacking microsomes, these proteins formed predominantly incorrect disulfide bonds and acquired no biological activities (G. Scheele and R. Jacoby, *J. Biol. Chem.* 257:12277–12282, 1982; 258:2005–2009, 1983). Can it be concluded from these observations that the signal peptides interfere with the folding of the preproteins? What pertinent factors other than signal peptidase were missing from the experiments without microsomes?

ANSWER

Section 2.4.10.

12. The single polypeptide chain of protein disulfide isomerase is the same as (1) the β subunit of prolyl-4-hydroxylase, an enzyme involved in the posttranslational processing of collagen (J. Koive et al., *J. Biol. Chem.*

262:6447–6449, 1987; K. Tasanen et al., *J. Biol. Chem.* 263:16218–16224, 1988); (2) bovine thyroid binding protein (K. Yamauchi et al., *Biochem. Biophys. Res. Commun.* 146:1485–1492, 1987); (3) a component of the *oligosaccharyl transferase* of the ER (M. Geetha-Habib et al., *Cell* 54:1053–1060, 1988); and (4) a component of the complex that catalyzes the transfer of triglycerides between membranes (J. R. Wetterau et al., *J. Biol. Chem.* 265:9800–9807, 1990). What might be the reasons for this multitude of roles of the disulfide isomerase?

SUGGESTIONS
R. Myllylä et al., *Biochem. J.* 263:609–611 (1989); N. J. Bulleid and R. B. Freedman, *EMBO J.* 9:3527–3532 (1990).

Are there reasons to think that the protein disulfide isomerase might also catalyze prolyl peptide bond isomerization (R. Pain, *Nature* 328:298, 1987)?

ANSWER
R. B. Freedman, *Nature* 329:294–295 (1987); K. Lang and F. X. Schmid, *Nature* 331:453–455 (1988).

13. Most proteins that have a positive net charge are secreted from the cells in which they are synthesized. Virtually all the proteins and other macromolecules present in the cytosol of cells are negatively charged. What would be the likely consequences if a basic protein were synthesized in the cytoplasm of a cell, using protein engineering techniques?

POSSIBLE ANSWER
N. J. Darby and T. E. Creighton, *Nature* 344:715–716 (1990).

14. Histones are basic proteins that organize DNA in the nucleus of eukaryotic cells. They are made on ribosomes in the cytosol in the usual way. Does this invalidate the conclusion of the preceding exercise?

ANSWER
R. A. Laskey et al., *Nature* 275:416–420 (1978).

Evolutionary and Genetic Origins of Protein Sequences

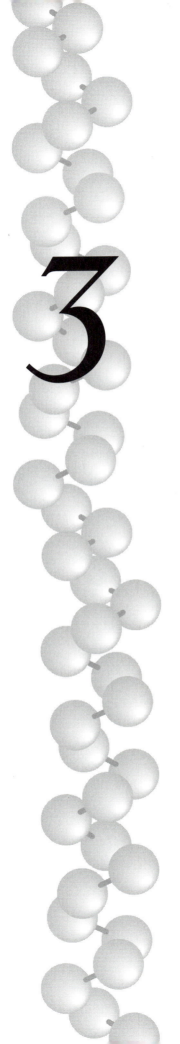

3

*T*he remarkable molecular unity in the living world belies the amazing diversity that we see at the macroscopic level. All living organisms — from the smallest viruses, to unicellular organisms such as bacteria and algae, to plants, and to whales — are similar at the molecular level. In particular, they all use the same 20 amino acids in their proteins and the same nucleotides in their DNA and RNA. Moreover, the amino acids are always the L isomer, and the ribose moieties of DNA and RNA are exclusively the D isomer. The basic properties of proteins, described in Chapter 1 and in the remainder of this volume, pertain equally well to proteins in *Escherichia coli* and in *Homo sapiens*. The functions of DNA and RNA in protein synthesis are also very similar in all organisms, as described in Chapter 2, but at this level the distinction between prokaryote and eukaryote becomes significant. The biochemical similarities extend to even higher levels of biochemical organization, such as biochemical pathways and metabolism, but the differences among organisms become ever more apparent. In general, the higher the level of biochemical organization, the greater the molecular differences among species.

This general pattern can be satisfactorily rationalized with our present knowledge only by postulating that all extant organisms arose from a common ancestor that had already acquired all the basic common biochemical features. It is assumed that the diversity at higher levels of organization has resulted from the Darwinian process of evolutionary divergence (Fig. 3.1) and that, though functional differences among organisms have arisen during evolution, the most basic biochemical processes have been conserved because any alteration to them would have been lethal.

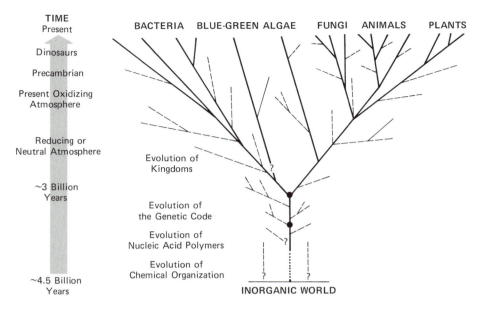

FIGURE 3.1

General phylogenetic tree of how life on earth might have evolved. The dashed lines represent hypothetical lineages that have not survived to the present day. (Adapted from M. O. Dayhoff, *Atlas of Protein Sequence and Structure,* National Biomedical Research Foundation, Washington, D.C., 1972.)

There is considerable evidence for divergent evolution from macroscopic comparisons of organisms and from the fossil record, but evidence from the sequences of proteins and nucleic acids is even more compelling and gives new insights into the process of evolution. Indeed, comparing DNA and protein sequences is becoming accepted as the best means of reconstructing the process of evolution because the genetic information contained in such sequences is so vast. For example, a DNA chain of only 1000 nucleotides, barely enough to code for an average-size protein molecule, can exist with any one of 4^{1000} (or 10^{600}) different nucleotide sequences. For comparison, there are thought to be only 10^{79} atoms in the universe. A very small protein of 100 amino acid residues could have any of 20^{100} (or 10^{130}) different sequences. Just one molecule of each of these proteins, if they were packed together in the most efficient manner, would fill the universe 10^{27} times! The number of possible sequences is vastly greater when different lengths of chains are also considered. Clearly, not all of the possible proteins exist today, nor are they likely to have ever existed on earth, which is only about 10^{17} s old.

As will become apparent, a particular extant protein is merely one of the evolutionary products of many millenia. Its particular amino acid sequence is determined primarily by its ancestry. It probably represents only one of the many related sequences that could serve a given function equally well. Within the constraint that it be functional, it is the product of the stochastic process of mutation and of the haphazard accidents of history rather than a finely tuned, unique entity. It has changed in the past and must be expected to change in the future. Its existence now is a fortuitous and transient occurrence in a constantly changing world. In the contemporary sequence of each protein is stamped the history of that protein and, consequently, part of the history of the species in which it has existed. Collectively, proteins provide a wealth of information about the history of evolution.

Likewise, the evolutionary variations of a protein provide much information about the protein itself. Evolutionary divergence into different species has resulted in many variants of the same protein, all with essentially the same biological function but different amino acid sequences. The differences and similarities of the amino acid sequences of these variants reflect the constraints of structure and function for that protein.

Comparisons of protein structures are so important for understanding their functions that it is vital that their evolution be understood.

References

Molecules as documents of evolutionary history. E. Zuckerkandl and L. Pauling. *J. Theor. Biol.* 8:357–366 (1965).

Evolution of enzymes. E. L. Smith. In *The Enzymes,* 3rd ed., P. D. Boyer, ed., vol. 1, pp. 267–339. Academic Press, New York, 1970.

The information content of protein amino acid sequences. T. T. Wu et al. *Ann. Rev. Biochem.* 43:539–566 (1974).

Biochemical evolution. A. C. Wilson et al. *Ann. Rev. Biochem.* 46:573–639 (1977).

Protein evolution. R. F. Doolittle. In *The Proteins,* 3rd ed., H. Neurath and R. L. Hill, eds., vol. 4, pp. 1–118. Academic Press, New York, 1979.

3.1 Primordial Origins of Proteins

How life arose on this planet will always be something of a mystery, but it is thought that all organic compounds were initially synthesized chemically from the original atmospheric constituents: CO_2, CO, N_2, and H_2O, and perhaps NH_3, CH_4, and H_2. Catalyzed by ultraviolet light from the sun and the electrical discharges of lightning, these compounds could have combined into larger molecules, including the amino acids, nucleotides, and other building blocks of biochemistry. Many such processes have been demonstrated in laboratories under simulated conditions. The surface of the earth could have become a moderately concentrated organic "primordial soup."

The transition from a soup of small molecules, which perhaps even included some polypeptides and nucleic acids, to the genetic systems of today in which "DNA makes RNA, which makes protein," had been difficult to imagine until recently. Life as we know it requires replicating molecules that contain information, such as DNA and RNA, plus catalytic molecules, such as some proteins. Because proteins are necessary for the replication of nucleic acid molecules, and nucleic acid molecules are necessary for the synthesis of proteins, it had seemed that neither could have come into existence without the other. This was the "chicken-and-egg" problem of molecular biology: which came first, nucleic acids or proteins?

A solution seems possible now that RNA molecules have been shown to have catalytic capabilities (in the form of ribozymes) as well as the ability to store genetic information. One can imagine a "living RNA world" populated by RNA molecules that could replicate themselves and also catalyze various chemical reactions. Eventually, however, DNA would have replaced RNA as the storer of genetic information because DNA is chemically more stable than RNA. Likewise, RNA would have been replaced as catalysts by proteins because proteins are intrinsically more suited for that chemical role (Chap. 9). These replacements might have happened gradually, with the early proteins serving only to stabilize the ribozymes and, later, with proteins supplanting the RNA almost entirely as catalysts. Vestiges of this RNA world are seen in (1) messenger RNA; (2) transfer RNA, which in its three-dimensional conformation has the appearance of being nature's attempt to make a protein out of a nucleic acid; (3) hybrid structures such as ribosomes, signal recognition particles, and spliceosomes (Chap. 2), which contain RNA moeties; and (4) the various ubiquitous coenzymes, such as ATP, NAD^+, and FAD, which are composed of RNA substructures.

The synthesis of proteins is imagined to have developed during the RNA world, using the amino acids present in the primordial soup or synthesized by catalytic RNA molecules. A complex process like protein biosynthesis may have evolved slowly, so it is likely that the amino acids used and the genetic code used to specify them would have been selected and refined over a very long time. This long period of evolution would explain the orderliness of the genetic code and how 20 amino acids that are so well suited for their purposes were selected out of some 300 amino acids that occur naturally. Certainly, the 20 biological amino acids are excellent choices, relative to the possible alternatives, in terms of their chemical and physical properties and their stabilities. Only asparagine, glutamine, and histidine would probably not have been in the primordial soup, because they are too unstable; these amino acids might have been synthesized by catalytic RNA molecules and incorporated into the genetic code at a late stage. Most attempts to rationalize the use of the 20 biological amino acids and the use of only the L isomer have concentrated on their occurrence in the primordial soup as a result of solely chemical processes. The true reasons may be very obscure today.

The primary difficulty with the scenario of the RNA world is that it is difficult to explain how RNA molecules could have been synthesized chemically in the primordial soup; attempts to simulate the primordial production of RNA molecules in the laboratory have not been entirely successful. That, fortunately, is a topic beyond the scope of this volume.

By whatever process, an organism that used the biological amino acids, nucleotides, genetic apparatus, protein biosynthetic system, and genetic code must have arisen by about 3×10^9 years ago. It would be the ancestor of all living organisms, and the modern living world would have arisen by a process of genetic divergence since then (see Fig. 3.1). Although most of this divergence occurred millions to billions of years ago, evidence for its occurrence is still detectable in the sequences of the nucleic acids and proteins of contemporary organisms.

References

Reasons for the occurrence of the twenty coded protein amino acids. A. L. Weber and S. L. Miller. *J. Mol. Evol.* 17:273–284 (1981).

What was the first living cell? D. Penny, *Nature* 331:111–112 (1988).

How old is the genetic code? Statistical geometry of rRNA provides an answer. M. Eigen et al. *Science* 244:673–679 (1989).

RNA evolution and the origins of life. G. F. Joyce. *Nature* 338:217–224 (1989).

Catalytic RNA and the origin of genetic systems. A. I. Lamond and T. J. Gibson. *Trends Genetics* 6:145–149 (1990).

Evolutionary changes in the genetic code. S. Osawa et al. *Proc. Roy. Soc. Lond.* B 241:19–28 (1990).

Stereoselective, nonenzymatic, intramolecular transfer of amino acids. N. S. M. D. Nalinie et al. *Biochemistry* 30:2768–2772 (1991).

3.2 Evolutionary Divergence of Proteins

The key to appreciating the usefulness of nucleic acid and protein sequences in reconstructing evolution is to realize that similar sequences, wherever found, imply descent from a common ancestor. The number of possible RNA, DNA, or protein sequences is so great that it is implausible that similar long sequences could have arisen by any mechanism other than evolutionary **divergence** from the same ancestor.

3.2.1 Homologous Genes and Proteins

Nucleic acids and proteins that have evolved from a common ancestor are said to be **homologous**. The sequences of homologous genes and proteins were identical at the time they originated by replication of a single gene. Subsequently the two genes accumulate mutational changes, and the sequences of homologous genes and proteins can be identical, similar to varying degrees, or unrecognizably dissimilar because of extensive mutation.

Proteins and genes are either homologous or not, because they either did or did not descend from a common ancestor. It is not strictly correct to describe two

proteins that have 50% identical sequences as being "50% homologous," for example. Instead, a 50% identity spread throughout a sequence implies total homology. Only when different parts of a single protein or gene sequence have different origins (which happens frequently in large, mosaic proteins composed of multiple domains, Sec. 6.2) is it correct to speak of partial homology. For example, the gene for the low-density lipoprotein (LDL) receptor appears to have obtained 34% of its nucleotides from an ancestral gene for complement component C9, 48% of its nucleotides from an ancestral gene for epidermal growth factor (EGF), and 18% from elsewhere. Consequently, the LDL receptor is truly 34% homologous to C9 and 48% homologous to EGF (see Fig. 3.16).

The only explanation other than divergence for similarities among sequences is **convergence,** in which two or more unrelated sequences have become similar under the pressure of selection for similar functions. Convergence is a fairly common evolutionary phenomenon at the macroscopic level and is even encountered at the level of protein three-dimensional structure (Sec. 9.3.2.a). There are, however, no instances in which nucleic acid or protein sequences have been shown conclusively to have become substantially similar by convergence. On the other hand, proving evolutionary convergence only from extant sequences is inherently difficult, so the possibility cannot be dismissed completely.

References

"Homology" in proteins and nucleic acids: a terminology muddle and a way out of it. G. R. Reeck et al. *Cell* 50:667 (1987).

Homology in classical and molecular biology. C. Patterson. *Mol. Biol. Evol.* 5:603–625 (1988).

a. Detecting Sequence Homology

Aligning two or more sequences for comparison is an essential step in determining if they are homologous, and numerous procedures and computer programs are available for this purpose. Most such programs are designed to search for homologies by comparing a new sequence with all the sequences in the data bases. To make so many comparisons in a reasonable time is not a trivial accomplishment, and the most popular programs are exceedingly sophisticated.

If most of the residues of two sequences are identical, they are almost certainly homologous, but interpretation of a slight similarity is not straightforward. The number of amino acid identities between distantly related sequences, in which many changes have occurred, may be almost insignificant. It is difficult to give simple rules for the degree of similarity necessary to demonstrate unambiguously that two protein sequences are homologous because this depends on the lengths of the sequences and their amino acid compositions. Unrelated sequences, chosen at random, would be expected to be identical in 5% of their residues if all 20 amino acids occurred with the same frequency, and in about 6% if the amino acids occur with their observed normal frequencies (see Table 1.1). In addition, there are mutations in which residues are inserted or deleted, so it is necessary to allow for some insertion and deletion when comparing proteins; this will increase the percent identity, and overzealous allowance can make even unrelated sequences seem similar. Nevertheless, for long sequences with typical amino acid compositions and with only limited introduction of insertions and deletions, an identity of greater than 20% probably indicates homology.

It is helpful in comparing distantly related proteins to consider the chemical natures of the amino acid residues because analyses of closely related proteins demonstrate that chemically similar amino acids most often replace each other in substitution mutations. For example, the basic residues Lys and Arg often replace each other, as do the acidic residues Asp and Glu, the hydroxyl residues Ser and Thr, the aromatic residues Tyr, Phe, and Trp, and the nonpolar residues Ala, Val, Leu, Ile, and Met. A *similarity index* matrix can be used to weight each pairing of different amino acid residues in an alignment of sequences. A variety of such matrices have been devised, but the one most frequently used is semiempirical and based on which replacements have actually occurred during the evolution of closely related proteins (Fig. 3.2). These replacements are found to be quite different from what would be expected from the probability of random mutations of the genetic code (see Fig. 2.4); instead they reflect a bias that is due to the chemical properties of the amino acids and their frequencies of occurrence in proteins (see Table 1.1). From all these considerations, a *mutation data (MD)* or *point accepted mutation (PAM)* matrix can be prepared for evaluating the significance of each alignment of two sequences.

To estimate, by a given method, the significance of any similarity found between two protein sequences, the result should be compared with the values obtained by applying the same method to randomly permuted versions of one of the sequences. This procedure preserves the exact length and amino acid composition of the protein, and the statistical variation of the random comparisons provides a measure of the significance of the observed similarity.

An effective way of searching for significant similarities between two sequences is to generate a **diagonal dot-plot** comparison of the two (Fig. 3.3). A short segment of one sequence (perhaps up to 30 residues) is compared with all possible segments of the same length in the second sequence. All possible segments of the first sequence are compared in this way to generate a matrix of all segments of the first sequence versus all those of the second. The similarities between each pair of segments are scored by the number of identities between the aligned residues, by the number of mutations necessary to convert one to the other, or by weighting the differences between them. If the similarity between any two segments is greater than some threshold value, a positive score is registered in the comparison matrix. This score is exhibited as a dot when the matrix is displayed with the sequence of one protein aligned vertically and the other horizontally. Alternatively, the magnitude of the score can be entered and only the highest scores displayed (Fig. 3.3).

Homology is apparent as an obvious diagonal run of dots or high scores, which is especially evident to the eye. One of the advantages of this method is that insertions and deletions of amino acid residues do not interfere with the detection of homologies but merely cause the diagonal to break and shift. Also, all possible comparisons of the sequences are made, so no preconceptions about the alignment of the sequences are necessary. The significance of any similarities can be judged by comparison with other regions of the dot-plot, and the scoring threshold can be raised or lowered to search for weak signals of homology.

References

A general method applicable to the search for similarities in the amino acid sequences of two proteins. S. B. Needleman and C. D. Wunsch. *J. Mol. Biol.* 48:443–453 (1970).

Random sequences. W. M. Fitch. *J. Mol. Biol.* 163:171–176 (1983).

Similar amino acid sequences: chance or common ancestry? R. F. Doolittle. *Science* 214:149–159 (1981).

A sensitive procedure to compare amino acid sequences. P. Argos. *J. Mol. Biol.* 193:385–396 (1987).

Detecting homology of distantly related proteins with consensus sequences. L. Patthy. *J. Mol. Biol.* 198:567–577 (1987).

Profile analysis: detection of distantly related proteins. M. Gribskov et al. *Proc. Natl. Acad. Sci. USA* 84:4355–4358 (1987).

Improved tools for biological sequence comparison. W. R. Pearson and D. J. Lipman. *Proc. Natl. Acad. Sci. USA* 85:2444–2448 (1988).

Basic local alignment search tool. S. F. Altschul et al. *J. Mol. Biol.* 215:403–410 (1990).

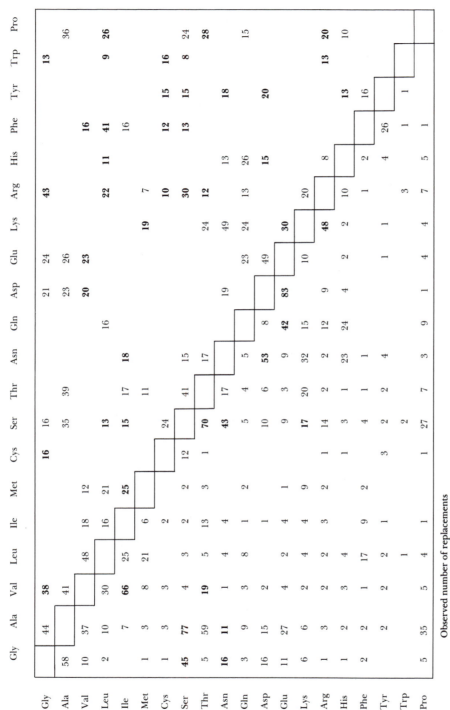

FIGURE 3.2

Relative frequences of amino acid replacements in a total of 1572 examples between closely related proteins that are observed (*bottom, left*) and those expected for random single-nucleotide mutations (*top, right*). The greatest discrepancies between the observed and random replacements are shown in boldface type. Replacements involving chemically similar amino acids are generally observed to be much more frequent than expected for random mutations. (Observed replacements compiled by M. O. Dayhoff, *Atlas of Protein Sequence and Structure*, vol. 5, suppl. 3, National Biomedical Research Foundation, Washington, D.C., 1978.)

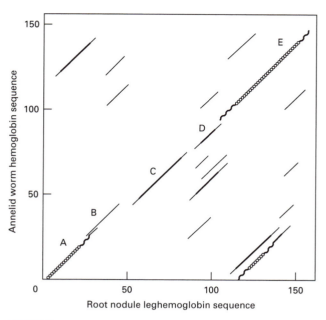

FIGURE 3.3
Diagonal dot-plot comparison of the amino acid sequences of the distantly related annelid worm hemoglobin and the root nodule hemoglobin, which have no more than 20% of their residues identical. Special comparison methods were used to detect this very marginal similarity. The comparison of each alignment calculated the probability with which the various pairs of aligned residues replace each other in other homologous proteins (e.g., Fig. 3.2), and the similarity in five physical parameters of the amino acid residues. The results from window lengths of between 7 and 25 residues of each protein were overlaid. The most significant similarities found are indicated by the diagonal lines, in the following order of increasing significance: *thin lines* < *thick lines* < *wavy lines* < *circles*. The diagonal segments labeled A–E correspond to the five structurally homologous segments that are known from the crystal structures of the two proteins (see Fig. 6.30). (From P. Argos, *J. Mol. Biol.* 193:385–396, 1987.)

b. Aligning Homologous Sequences

Aligning two or more sequences is straightforward only if they are very similar. The general criterion used in aligning sequences is to use the alignment that minimizes the number of genetic mutations needed to account for the differences between the sequences. This is the principle of **maximum parsimony,** or **minimal mutations.** The assumption that this alignment is the correct one is most valid when there have been the fewest mutations. As the number of assumed mutations increases, the possibility of there having been multiple mutations at some sites becomes much greater. In this case, reverse mutations may have occurred, which will not be apparent from the contemporary sequences.

Consequently, the greater the number of differences between two sequences, the greater the probability that any alignment of their sequences is incorrect.

The greatest complication in aligning sequences arises from allowances for insertions and deletions of amino acid residues to make an alignment work. The introduction of gaps in a sequence diminishes the significance of any alignment, so a penalty is usually assigned for each gap when assessing the significance of an alignment.

The sequences of distantly related proteins can be aligned most accurately if the three-dimensional structures of one or more of them are known. The three-dimensional structure of a protein is better conserved during divergence than its primary structure (Sec. 6.4).

References

Alignment of the amino acid sequences of distantly related proteins using variable gap penalties. A. M. Lesk et al. *Protein Eng.* 1:77–78 (1986).

A method for multiple sequence alignment with gaps. S. Subbiah and S. C. Harrison. *J. Mol. Biol.* 209:539–548 (1989).

3.2.2 Mutations and Protein Structure

Mutations are both the raw material and the driving force of evolution, whereas natural selection modulates the rate of divergence. Although the spontaneous mutation rate is generally low (only 10^{-8} per nucleotide per generation in humans), the large size of the human genome (3×10^9 nucleotide pairs) implies that some 60 mutations occur on average in each individual. A variety of mutations occur to DNA and RNA, by numerous and complex mechanisms. Simplest and most frequent is the replacement of one nucleotide by another; insertions and deletions of one or more nucleotides are also common. More complex rearrangements of DNA are less common but of greater consequence. Our knowledge of mutation and its consequences in higher organisms is greatest for humans, due to extensive genetic screening. The most thoroughly studied genes are those of human hemoglobin because this protein is both easy to study and susceptible to substantial functional variation. Because the function of hemoglobin is to carry oxygen from the air in the lungs to the tissues of the body (Sec. 8.4.3), it interacts closely with the environment. Perhaps for this reason, it exhibits a variety of mutational and physiological phenomena.

The very diverse consequences of mutations on protein structure depend on the location of the affected nucleotides in the gene that codes for the polypeptide

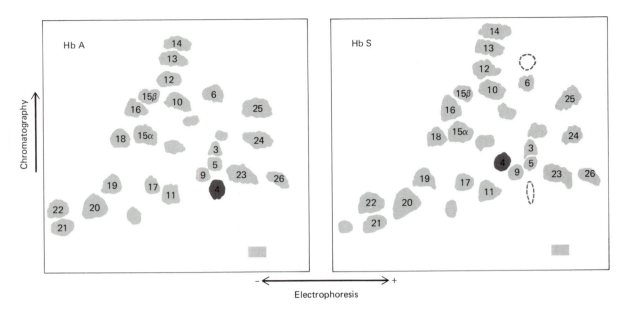

FIGURE 3.4

Two-dimensional peptide maps, or fingerprints, of normal hemoglobin (Hb A, *left*) and sickle-cell hemoglobin (Hb S, *right*). The two proteins were digested into peptides by trypsin, and the resulting mixture of peptides was applied at the rectangular origin of the chromatography paper. The peptides were separated in the horizontal dimension by electrophoresis in pyridine acetate buffer, pH 4.6. After the paper was dried, it was subjected to chromatography in the second dimension, using a 35:35:30 mixture of isoamyl alcohol, pyridine, and water. The peptides were detected by staining with nihydrin. The two fingerprints differ significantly only in the position of peptide 4; the other peptides are identical in the two proteins. Peptide 4 comprises residues 1–8 of the β chain, with the sequence Val-His-Leu-Thr-Pro-**Glu**-Glu-Lys in Hb A, but Val-His-Leu-Thr-Pro-**Val**-Glu-Lys in Hb S. The Val residue replacing Glu makes the Hb S peptide more basic at pH 6.4, so it migrates electrophoretically faster toward the cathode. It is also less polar, so it migrates more rapidly during the chromatography. Consequently, peptide 4 from Hb S lies further to the left and above that from Hb A. (Adapted from C. Baglioni, *Biochim. Biophys. Acta* 48:392–396, 1961.)

chain. A single nucleotide replacement might have no effect on a protein sequence because of the degeneracy of the genetic code (see Fig. 2.4). On the other hand, mutation to a termination codon causes the polypeptide chain to end at that position. More frequently, mutation replaces one amino acid with another. The classic example of replacement, the first genetic disease to be characterized at the molecular level, is **sickle-cell hemoglobin.** The demonstration that the normal Glu at residue 6 of the hemoglobin β chain is replaced by Val as a result of this mutation was accomplished by comparing two-dimensional peptide maps, or *fingerprints,* of the normal and mutant proteins (Fig. 3.4). In this method of comparing closely related protein sequences, each protein is cleaved into small peptides by using a specific protease, such as trypsin (Sec. 9.3.2). The peptides are then separated and compared; classically, two different separation procedures have been used, such as chromatography and electrophoresis, sequentially and at right

angles to each other. Peptides that differ in covalent structure should have different mobilities in at least one of the two dimensions and, consequently, lie in different positions in the final fingerprint. The variant peptide is usually apparent because of its altered position relative to the nearby unaltered peptides. The nature of the difference between the peptides from the two proteins can be determined from their amino acid compositions or, if necessary, by amino acid sequencing (Sec. 1.6.2). The mutation responsible for the change of Glu to Val at residue 6 of sickle-cell hemoglobin was subsequently found to be the result of replacement of an A nucleotide by T in the codon specifying Glu 6 of the β chain.

Even more drastic consequences follow if the single amino acid replacement affects a posttranslational modification (Sec. 2.4) such as proteolytic cleavage. Mutation to a termination codon results in a shortened polypeptide chain, which is usually nonfunctional unless the mutation is very near the C-terminus of the chain. Alteration of an initiation or termination codon abolishes protein synthesis or results in an elongated polypeptide chain. For example, at least four different mutations have occurred in the termination codon of the gene for the α chain of some human hemoglobins. The normal α chain is 141 residues long, but the mutants are elongated to 172 residues; another termination codon is encountered at this point in the normally untranslated 3′ region of the mRNA. The added sequences of the four mutant hemoglobins are identical except for the new residue 142, which corresponds to the termination codon that was mutated in four different ways.

Insertion or deletion of one or two nucleotides alters the reading frame of the coding sequence and, consequently, changes all the residues of the polypeptide chain on the carboxyl side of the mutation, including the end of the chain. Usually, this type of mutation produces a polypeptide chain abnormal enough to be unrecognizable by most criteria. Insertions and deletions near the C-terminus, however, result in only slightly altered, functional proteins. Three examples are the abnormal human hemoglobins *Wayne, Cranston,* and *Tak,* in which insertions occurred near the carboxyl end of the hemoglobin β chain (Fig. 3.5). The normal termination codon UAA is missed, and translation of the incorrect sequence continues until the next termination codon is encountered. These mutant β chains consequently contain 157 residues rather than 147.

Mutations of parts of a gene other than the protein-coding region can affect production of and even the structure of its protein. Mutations in promoter and enhancer regions (Sec. 2.1.2) affect the transcription of the gene, and other mutations can affect mRNA stability; both affect the amount of protein made. Mutations can also alter the sites used in RNA splicing (Sec. 2.1.2), either abolishing the normal sites or introducing new ones; in both cases, the sequence of the resulting protein can be dramatically altered.

More substantial and complex alterations of genes have correspondingly greater effects on protein structure. Large deletions and insertions remove or insert numbers of residues but also can introduce frameshifts or termination codons. Gene rearrangements can link different polypeptide segments (Sec. 3.4), but in most cases, gross alterations will result in the gene not being expressed or the gene product being unrecognizable.

References

Molecular genetics of human hemoglobin. D. J. Weatherall and J. B. Clegg. *Ann. Rev. Genet.* 10:157–178 (1976).

The sensitivity of gel electrophoresis as a detector of genetic variation. J. A. M. Ramshaw et al. *Genetics* 93:1019–1037 (1979).

The molecular genetics of human hemoglobins. T. Maniatis et al. *Ann. Rev. Genet.* 14:145–178 (1980).

The mutation and polymorphism of the human β-globin gene and its surrounding DNA. S. H. Orkin and H. H. Kazazian, Jr. *Ann. Rev. Genet.* 18:131–171 (1984).

Molecular charge and electrophoretic mobility in Cetacean myoglobins of known sequence. T. McLellan *Biochem. Genetics* 22:181–200 (1984).

The rate with which spontaneous mutation alters the electrophoretic mobility of polypeptides. J. V. Neel et al. *Proc. Natl. Acad. Sci. USA* 83:389–393 (1986).

3.2.3 Genetic Divergence during Evolution

Although gene replication is usually precise, genes are continually undergoing the inevitable stochastic pro-

Cranston
-Lys -Ser -Ile -Thr -Lys -Leu -Ala -Phe -Leu -Leu -Ser -Asn -Phe -Tyr -terminate
AAG **AG**U AUC ACU AAG CUC GCU UUC UUG CUG UCC AAU UUC UAU UAA

AAG UAU CAC UAA GCU CGC UUU CUU GCU GUC CAA UUU CUA UUA A

Normal
-Lys -Tyr -His -terminate
AAG UAU CAC UAA GCU CGC UUU CUU GCU GUC CAA UUU CUA UUA A

Tak
AAG UAU CAC **AC**U AAG CUC GCU UUC UUG CUG UCC AAU UUC UAU UAA
-Lys -Tyr -His -Thr -Lys -Leu -Ala -Phe -Leu -Leu -Ser -Asn -Phe -Tyr -terminate

FIGURE 3.5
Insertion mutations that alter the reading frame of the gene for the β chain of hemoglobin. The normal sequences of mRNA and of the protein product are shown, as well as the sequences for two mutants, *Cranston* and *Tak.* The nucleotides inserted in the mutants are in boldface type; note that both insertions apparently occurred by duplication of the two preceding nucleotides. (Sequences from N. Proudfoot and G. Brownlee, *Brit. Med. Bull.* 32:251–256, 1976.)

cess of mutation, and every gene is subject to selective pressures. Any gene that suffers a mutation that diminishes its chance, relative to other copies of the gene in the population, of being passed on to the next generation is subject to negative selection and has a greater chance of disappearing from the population. Conversely, any gene with a mutation that gives it a greater than average chance of being passed on is subject to positive selection and has an increased chance of remaining in the population. Natural selection generally operates by means of the somewhat different probabilities that various forms of a gene have of being passed from one generation to the next. It is important to understand that selection usually affects the probability of survival of a mutant gene only slightly, unless the mutation is lethal to the individual in which it occurs. Except in such drastic cases, the fate of a gene is determined primarily by probabilities and, consequently, by chance.

The probability that a particular copy of a gene will be passed on to the next generation varies between zero and nearly one, depending on the average number of progeny produced by the individual bearing the copy. The greater the number of progeny produced, the greater the probability that all the genes of the parents will be retained. When a population of diploid eukaryotes is stable in size, the probability that any particular gene copy will be passed on is approximately 0.75 in the absence of selection; therefore, most genes tend to disappear after a very few generations from any population that is not expanding greatly. The genes of contemporary individuals are mostly those that just happened by chance to survive, having been selected primarily for not having suffered any lethal mutations.

Mutations that affect a gene's ability to survive can be either deleterious or beneficial, but many mutations appear to have insignificant effects; these are known as **neutral mutations.** It is inevitable that neutral mutations will have occurred to every gene during its descent, so the rate of change in the sequence of a gene should be at least that of its neutral mutation rate. The overall mutation rate is probably similar for most nucleotides in most organisms and is of the order of 10^{-8} per nucleotide per year. The neutral mutation rate is some fraction of the overall rate: the fraction of mutations that have no significant effect on the function of the gene. It differs for each nucleotide in each gene.

References

Evolutionary rate at the molecular level. M. Kimura. *Nature* 217:624–626 (1968).

Non-Darwinian evolution. J. L. King and T. H. Jukes. *Science* 164:788–798 (1969).

The Neutral Theory of Molecular Evolution. M. Kimura. Cambridge University Press, Cambridge, 1983.

3.3 Reconstructing Evolution from Contemporary Sequences

If a sufficient number of related sequences are available, the origin of any gene and protein can be determined, and the evolutionary history of its divergence can be inferred. Reconstructing an evolutionary pathway by comparing contemporary sequences is to some extent a matter of educated guesswork because the sequences of the ancestral genes or proteins are usually not available. Nevertheless, gene and protein sequences contain so much information that evolutionary reconstructions are possible. Obviously, the more closely related the available sequences are, the more plausible the reconstruction of their evolution.

3.3.1 Variation among Species

Because of the biochemical unity of life, most organisms are similar at the molecular level. The more closely related they are evolutionarily, the more similar they are biochemically and genetically, and the more genes and proteins of essentially the same functions they share. For example, it is possible to find comparable hemoglobins, cytochromes *c*, glyceraldehyde phosphate dehydrogenases, DNA polymerases, and histones in a wide variety of organisms and to compare their gene and protein sequences.

The more closely related the organisms, the more similar the sequences of their genes and proteins are found to be. Individuals of the same interbreeding population, or species, tend to have identical genes and proteins because rare variants of genes tend to be lost, and it is possible to speak of a human hemoglobin and a horse cytochrome *c*. There are, however, exceptions in which two or more slightly different versions are present in substantial numbers of individuals in a species (Sec. 3.3.2). Closely related species also have closely related proteins. For example, chimpanzees and humans have identical cytochromes *c*, hemoglobin-α, -β, and -γ chains, and the fibrinopeptides, whereas their myoglobins and hemoglobin-δ chains each differ by one amino acid residue. As less closely related species are compared, the differences among their genes and proteins increase. A number of sequences of cytochromes *c* from a few selected eukaryotic species are given in Figure 3.6, and a *difference matrix* that gives the number of

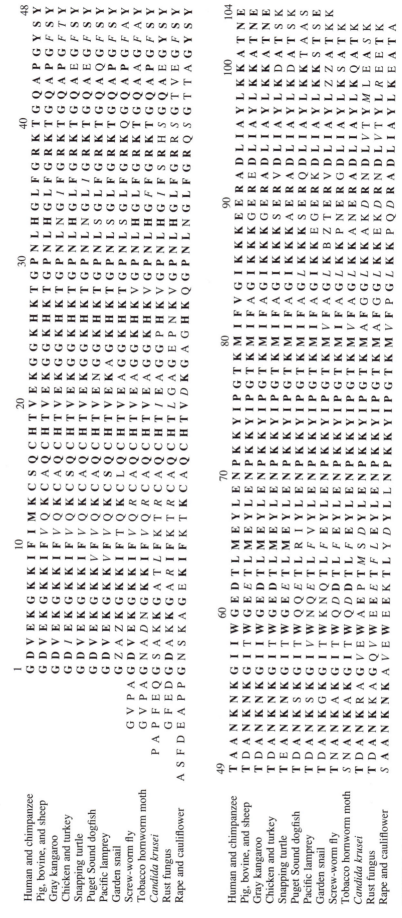

FIGURE 3.6

Amino acid sequences of some cytochromes c from various eukaryotes. The one-letter code (see Table 1.1) is used; Z is either E or Q (Glu or Gln), B is either D or N (Asp or Asn). Residues that are identical to those in the human protein are boldface; chemically similar residues are in italics.

	Chimpanzee	Sheep	Rattlesnake	Carp	Snail	Moth	Yeast	Cauliflower	Parsnip
Human	0	10	14	18	29	31	44	44	43
Chimpanzee		10	14	18	29	31	44	44	43
Sheep			20	11	24	27	44	46	46
Rattlesnake				26	28	33	47	45	43
Carp					26	26	44	47	46
Garden snail						28	48	51	50
Tobacco hornworm moth							44	44	41
Baker's yeast (iso-1)								47	47
Cauliflower									13

FIGURE 3.7

Amino acid difference matrix for cytochromes c of a few representative species. These proteins all consist of at least 104 amino acid residues, so at least half the residues of each pair are identical.

amino acid differences among a few species is shown in Figure 3.7.

References

Comparative aspects of primary structures of proteins. C. Nolan and E. Margoliash. *Ann. Rev. Biochem.* 37:727–790 (1968).

Evolution at two levels in humans and chimpanzees. M. C. King and A. C. Wilson. *Science* 188:107–116 (1975).

a. Nature of Sequence Differences among Related Proteins

Closely related proteins generally differ only by replacements of one amino acid by another at a few positions in the polypeptide chain. Less frequent are differences in the total number of residues, which are due to the deletion or insertion of residues within or at either end of the polypeptide chain. In more distant relationships, the numbers and natures of sequence differences can increase greatly. A composite of the amino acid sequences for cytochromes c from widely divergent eukaryotic sequences is shown in Figure 3.8, in which every amino acid that has been observed to occur at each position is listed. Great variation occurs at some positions, such as at residues 60, 89, and 92, whereas only amino acids with chemically similar side chains are found at other positions, such as the large nonpolar residues at positions 35 and 36. No variation is found at other positions, such as the Cys residue at position 17 and the nearly invariable stretch from residue 67 to 80.

For this particular protein, a surplus of basic over acidic residues is also conserved within rather narrow limits. There obviously have been constraints on the evolutionary changes that have occurred in cytochrome c, but the amount of variation that has occurred is perhaps the more amazing; all the proteins listed in Figure 3.6 function as cytochromes c, many with indistinguishable properties.

As less closely related species are compared, the constraints on variation are less apparent. For example, cytochromes of various bacteria have been given designations such as c_2, c_5, c_6, c_{550}, c_{551}, and c_{555} because they are detectably different from cytochromes c in their physical and functional properties. Nevertheless, their three-dimensional structures (see Fig. 6.31) demonstrate clearly that all are homologous to the cytochromes c, and the functional and physical differences among them are probably due to the substantial alteration of amino acid sequences that they have undergone. When 29 of these sequences are compared with those of cytochromes c, only 3 residues are found to be invariant in all the sequences: Cys 17, His 18, and Met 80. Some pairs of these cytochromes differ at 88% of their residues. Additions and deletions are also common, and 162 residues are required to account for all the cytochrome c-like sequences, whereas only 116 are required in Figure 3.8 to account for those in the cytochromes c.

The greater differences observed among proteins from distantly related species probably reflect primarily the greater time since the species diverged, time during which mutations have accumulated (Fig. 3.1). Different functional constraints may also have existed

```
                                                                                Met
                        Asn                        Lys                           Phe
                        Asp                        Ile                           Lys
                        Gln                        Val Phe Val      Gly          Val Gly        Leu
            Ala Gly Glu Gln Phe Ser Asp            Ala Thr Arg      Ala Thr      Ile Ala        Glu
            Ser Ser Gly Thr Tyr Lys Lys            Ser Ser Lys      Arg Ser      Ser Asn        Gln Thr
        Ala Ser Glu Glu Val Ser Val Asn Gln        Glu Ala Asn      Lys Asn Thr  Thr Gln        Asn Ala
        Pro Pro Glu Pro Pro Ala Gly Ile Glu Glu    Ser Asp Asp Thr  Asp Asp Val  Glu Glu Gln    Ser Gln
    Gly Lys Lys Lys Lys Arg Thr Pro Leu Ala Ala    Asn Asn Glu Ser  Gln Arg Leu  Asp Ser Arg Ala Gly Glu
   Acetyl-Ala-Thr-Phe-Ser-Glu-Ala-Pro-Pro-Gly-Asp-Pro-Lys-Ala-Gly-Glu-Lys-Ile-Phe-Lys-Thr-Lys-Cys-Ala-Glx-
      -10                              -1  +1                       10                           16

                                                        Trp
        Glu      Δ           Cys                        Tyr
        Cys Gly Ala          Pro                        Phe
        Gly Ala Val          Lys                        Gln              His               Glu
        Ala Asn Asn  Δ    Δ  Glu          Ile           Asp      Ala Ile Arg Cys           Val
        Gly Tyr Asp Gln Lys Leu Ile Ser Gly Thr     Val Asn      Phe Leu Asn Lys Gln   Lys Ile
        Ala Ile Gln Glu Asn Glu Leu Gln Thr Ala     Ser Ser      Val Val Ser Thr Thr   Gln Ala
        Ser Leu Glu Arg Ala Gly Ala Asn Ser Glu Ala Ala Gly      Ile Tyr Val Ala Ala   Ser Ser
       -Cys-His-Thr-Val-Glx-Lys-Gly-Ala-Gly-His-Lys-Gln-Gly-Pro-Asn-Leu-His-Gly-Leu-Phe-Gly-Arg-Gln-Ser-Gly-Thr-Thr-
          20                          30                           40

                                            Gly
                                            Ala               Ser
    Pro                      Ser            Leu   Lys Pro Pro
    Val                      Ala            Val   Gln Ala Ala
    Lys          Pro         Gly            Ile   Glu Tyr Gly        Asp
    Ser          Lys         Arg Met        Thr   Asp Lys Lys His    His Ile
    Gln      Lys Glu         Lys Ala        Lys   Asn Asn Asn Val    Arg Val            Lys
    Glu      Val Asp         Ile Gln Ser Thr Gln Gln Thr Asp Asp Asn Leu Ala            Glu
    Asp  Pro Ala Asn         Ala Glu Asn Asn Ala Glu Ser Gln Gly Asp Phe Met Lys        Thr
    Asn Ser Phe Thr     Thr Ala Ala Met Arg Asp Arg Gly Ile Asp Ala Glu Glu Ser Met Phe Asp Phe
   -Ala-Gly-Tyr-Ser-Tyr-Ser-Thr-Gly-Asn-Lys-Asn-Lys-Ala-Val-Asn-Trp-Gly-Glx-Glx-Thr-Leu-Tyr-Glu-Tyr-Leu-Leu-Asn-
      50                          60                           70

                                                            Thr
                                                            Ser               Val
                                                    Lys Ala                   Lys
                                                    Glu Gly                   Gln
                                            Val     Asp Asn                   Glu
                            Ser     Thr     Phe     Asn Thr Asp Asp   Thr         Lys
            His Met         Ala     Ala     Met Pro Ser Ser Gln Gln   Asn His Val Val Ala
    Ser     Phe Val     Lys Asn     Ile     Gly     Ile Ser Ala Ala Glu Glu Gly Asn Ile Leu Thr Phe
   -Pro-Lys-Lys-Tyr-Ile-Pro-Gly-Thr-Lys-Met-Val-Phe-Pro-Gly-Leu-Lys-Lys-Pro-Glx-Glx-Arg-Ala-Asp-Leu-Ile-Ser-Tyr-
                          80                           90

        Δ
        Ala
        Ile
        Val
    Δ   Ser             Lys
    Glu Thr Leu Leu Gln Lys
    Ala Lys Val Lys Glu Leu
    Thr His Lys Asn Asn Ala
    Leu Asp Glu Cys Asp Ser
    Val Asn Thr Ala Ala Asp
    Met Arg Glu Ser Ser Gly Glu
   -Leu-Lys-Gln-Ala-Thr-Ser-Gln-Glu
        100
```

FIGURE 3.8

Composite amino acid sequence of cytochromes c from 92 eukaryotic species. The continuous sequence of 113 residues is the longest cytochrome c, that of *Gingko bilboa*. The numbering of the residues is that of the mammalian proteins, which start at position 1 and end at 104. The proteins included here start at positions −11 to +1 and end at positions 101 to 105. Deletions of residues are indicated by Δ. (The sequences were obtained from M. O. Dayhoff, *Atlas of Protein Sequences and Structure*, National Biomedical Research Foundation, Washington, D.C., 1972; and from D. M. Hampsey et al., *J. Biol. Chem.* 261:3259–3271, 1986.)

Table 3.1   *Relative Variabilities of Amino Acid Residues during Divergence of Homologous Proteins*[a]

Residue	Variability	Residue	Variability	Residue	Variability
Asn	100	Met	70	Gly	37
Ser	90	Gln	69	Tyr	31
Asp	79	Val	55	Phe	31
Glu	76	His	49	Leu	30
Ala	75	Arg	49	Cys	15
Thr	72	Lys	42	Trp	13
Ile	72	Pro	42		

[a] The number of times that a given amino acid residue has changed during the evolution of various proteins has been divided by the number of times the residue occurred. These values have been normalized by setting the highest to 100.

Adapted from M. O. Dayhoff, *Atlas of Protein Sequence and Structure*, vol. 5, suppl. 3, National Biomedical Research Foundation, Washington, D.C., 1978.

among different evolutionary branches, varying somewhat with the particular sequences that were present and with the environment in which each gene and protein had to function.

The constraints on change in the amino acid sequences of cytochromes result from the necessity that they function as cytochromes of the *c*-type: if a cytochrome fails to function, then its gene does not survive for long. The more important an amino acid residue is to function, the lower its probability of changing during evolution. The three invariant residues (Cys 17, His 18, and Met 80) of the *c*-type cytochromes interact directly with the heme group, which is involved in their electron transport function (Sec. 8.3.4.c). Other residues are involved in interacting with the other proteins with which cytochrome c transfers electrons (Sec. 8.3.4.c). Others have roles in maintaining the folded conformation necessary for function. The folded conformations of proteins are observed to be more highly conserved during evolution than are their primary structures (Sec. 6.4).

Constraints on divergence are also apparent in the gene sequences coding for closely related proteins. Nucleotide changes at the third position of a codon that do not alter the specified amino acid occur more frequently than those that do. The sequences of introns, which have no known function, change much more rapidly than do those of exons. The upstream and downstream untranslated regions of genes also change more rapidly than the regions translated into polypeptide chains. Flanking sequences that are conserved are usually those involved in regulation of gene expression.

In general, the amino acid replacements that occur during protein divergence are nonrandom, both in the extent to which various residues change and in the number of amino acids that replace each other. As noted earlier in the discussion of homology, the most prevalent replacements occur between amino acids with similar side chains: Gly/Ala, Ala/Ser, Ser/Thr, Ile/Val/Leu, Asp/Glu, Lys/Arg, Tyr/Phe, and so forth (Fig. 3.2). The numbers of observed replacements also reflect the frequency with which each amino acid occurs in proteins. Correcting for the different frequencies gives the normalized values for the relative mutabilities of the different amino acids (Table 3.1), which vary by a factor of 7. The acidic and hydrophilic residues are most frequently changed, whereas the large, hydrophobic residues are replaced least often. The other conserved residues are Gly, Pro, and Cys.

This chemical bias, by which certain amino acids change to varying extents and are replaced nonrandomly by certain other amino acids, does not arise from the nature of the genetic code or from the types of mutations that occur. The amino acid replacements that would be expected from random single nucleotide replacements are shown in the upper half of Figure 3.2; this distribution is remarkably different from that actually observed among homologous proteins, shown in the lower half of the figure.

This bias in amino acid replacements presumably reflects the role of selection; only those mutations that do not disrupt the function of the protein survive. Consequently, chemically similar replacements are found most frequently. The roles of the most conserved residues, the hydrophobic residues, plus Cys, Gly, and Pro, will become evident from the three-dimensional structures of proteins described in Section 6.2.

References

The molecular evolution of cytochrome *c* in eukaryotes. W. M. Fitch. *J. Mol. Evol.* 8:13–40 (1976).

On the evolution of myoglobin. A. E. Romero-Herrera et al. *Phil. Trans. Roy. Soc. B* 283:61–163 (1978).

Molecular evolution, intracellular organization, and the quinary structure of proteins. E. H. McConkey. *Proc. Natl. Acad. Sci. USA* 79:3236–3240 (1982).

Conservation and change in the DNA sequences coding for alcohol dehydrogenase in sibling species of *Drosophila*. M. Bodmer and M. Ashburner. *Nature* 309:425–430 (1984).

Structure and evolution of the insulin gene. D. F. Steiner et al. *Ann. Rev. Genet.* 19:463–484 (1985).

Molecular evolution of the ribonuclease superfamily. J. J. Beintema et al. *Prog. Biophys. Mol. Biol.* 51:165–192 (1988).

b. Phylogenetic Trees

The evolutionary process of divergence, as illustrated generally in Figure 3.1, can be traced by reconstructing the phylogenetic tree from the sequences of nucleic acids or proteins. A phylogenetic tree constructed from the sequences of cytochromes *c* is illustrated in Figure 3.9. Constructing a phylogenetic tree is not a simple exercise because a vast number of trees are possible for even small numbers of sequences. More than 2 million trees are possible for only 10 sequences, and this number increases exponentially with the number of sequences.

The simplest and most widely used method of constructing phylogenies employs a difference matrix of all the sequences, like that in Figure 3.7. The total length of the branches connecting two sequences is assumed to be proportional to the number of differences between them. By considering all possible pairs of sequences,

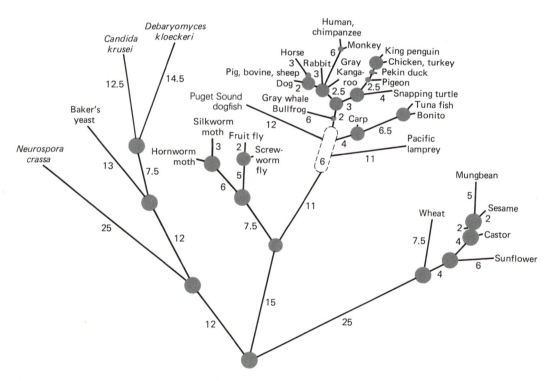

FIGURE 3.9

Phylogenetic tree constructed from the sequences of cytochromes *c* by minimizing the total number of amino acid replacements. The sequences of the ancestors at each branch point were inferred from the present-day sequences. The length of each branch is proportional to the indicated number of amino acid changes that are believed to have occurred. The branch points within the dashed oval were not adequately defined by the sequences. (Adapted from M. O. Dayhoff, *Atlas of Protein Sequence and Structure,* National Biomedical Research Foundation, Washington, D.C., 1972.)

one can construct a complete tree. This method generates the correct tree if the rates of change along all the branches of the tree were nearly equal during divergence, although statistical fluctuations with small numbers of differences along the various branches can produce mistakes. The nature of the difference is not considered, so multiple, independent occurrences of the same mutation along different branches, as might have occurred in positive selection for such a mutation, can also produce incorrect trees.

A more rigorous method of reconstructing phylogenies compares sequences to infer the ancestral sequences that occurred at each of the branch points of the tree, and chooses the tree that minimizes the total number of mutations required throughout the tree. This produces the most accurate tree if changes in the sequences have not occurred at nearly equal rates or if the same mutation has tended to be fixed independently in different branches. Just such an occurrence is shown by the different numbers of mutations along the different branches of the tree of Figure 3.9. This type of phylogenetic tree also makes it possible to examine the mutations that have been fixed along each branch in order to look for nonrandom events suggestive of selection and to consider the primary structures of the ancestral proteins. For example, the inferred sequence of the common ancestor of the cytochromes c in Figure 3.9 is compared with that of contemporary human cytochrome c

in Figure 3.10. The entire sequence of the ancestral protein cannot be determined unambiguously in this way, but only about a dozen amino acid replacements need to have occurred during the 10^9 years of evolution of human cytochrome c from the common ancestor of most eukaryotic cytochromes c (Fig. 3.1).

A phylogeny based on the sequences of cytochromes c should be the same as that of the species involved, so long as the genes for cytochrome c and the species diverged at the same time. Such phylogenies will be incorrect, however, if later gene duplications in a species (Sec. 3.4) produced some of the sequences compared. A phylogeny based on just one gene or protein can also be misleading if a small number of replacements have occurred or if the changes that took place were not random but were positively selected. It is necessary, therefore, to use a number of genes or proteins to construct accurate phylogenies of species. Usually, however, for members of the same species there is reasonable agreement among the phylogenies that are based on different genes or proteins.

Phylogenies based on nucleic acid or protein sequences are generally similar to those constructed by classical techniques of taxonomy. Those based on molecular sequences are coming to be accepted as the more reliable due to the vast amount of information in nucleic acid and protein sequences. Consequently, molecular techniques are increasingly being used to deter-

	1		5			10			15			20			25
Ancestral	Pro-Ala-**Gly-Asp**- ? -**Lys-Lys-Gly-Ala-Lys-Ile-Phe**-Lys-Thr - ? -**Cys-Ala-Gln-Cys-His-Thr-Val-Glu**- ? -**Gly-Gly**- ? -
Human	**Gly-Asp**-Val-**Glu**-**Lys-Gly-Lys-Lys-Ile-Phe**-Ile -Met-Lys-Cys-Ser-**Gln-Cys-His-Thr-Val-Glu**-Lys-**Gly-Gly**-Lys-

	30		35		40		45		50
	His-Lys-Val-**Gly-Pro-Asn-Leu-His-Gly-Leu-Phe-Gly-Arg-Lys**- ? -**Gly-Gln-Ala**- ? -**Gly-Tyr-Ser-Tyr-Thr**-Asp-
	His-Lys-Thr-**Gly-Pro-Asn-Leu-His-Gly-Leu-Phe-Gly-Arg-Lys**-Thr-**Gly-Gln-Ala**-Pro-**Gly-Tyr-Ser-Tyr-Thr**-Ala -

	55		60		65		70		75
	Ala-Asn-Lys-Asn-Lys-Gly- ? - ? -**Trp**- ? -**Glu-Asn-Thr-Leu-Phe** -**Glu-Tyr-Leu-Glu-Asn-Pro-Lys-Lys-Tyr**-Ile -
	Ala-Asn-Lys-Asn-Lys-Gly-Ile-Ile-**Trp**-Gly-**Glu**-Asp-**Thr-Leu**-Met-**Glu-Tyr-Leu-Glu-Asn-Pro-Lys-Lys-Tyr**-Pro-

	80		85		90		95		100
	Pro-Gly-Thr-Lys-Met- ? -**Phe**- ? -**Gly-Leu-Lys-Lys**- ? - ? -**Asp-Arg-Ala-Asp-Leu-Ile-Ala-Tyr-Leu-Lys**- ? -
	Pro-Gly-Thr-Lys-Met-Ile-**Phe**-Val-**Gly-Ile** -**Lys-Lys**-Lys-Glu-Glu-**Arg-Ala-Asp-Leu-Ile-Ala-Tyr-Leu-Lys**-Lys-

Ala-Thr-Ala -
-**Ala-Thr**-Asn-Glu

FIGURE 3.10

Comparison of the sequences of contemporary human cytochrome c and of the common ancestor of the species of Figure 3.9, which is that at the lowest node of the phylogenetic tree. Its sequence was inferred from the present-day sequences and the minimum number of mutational events required to account for the phylogenetic tree. Positions for which it is impossible to decide between two or more alternatives are indicated by a question mark. (Sequences from M. O. Dayhoff, *Atlas of Protein Sequence and Structure*, National Biomedical Research Foundation, Washington, D.C., 1972.)

Table 3.2 *Relative Extents of Divergence for Several Proteins from Different Species*

Protein	Residues Differing from Human Protein (%)								
	Rhesus monkey	Cow	Pig	Rabbit	Chicken	Frog	Fish	Fruit fly	Yeast
Cytochrome *c*	1	10	10	9	13	17	20	27	41
Hemoglobin α chain	3	12	13	18	25		50		
Hemoglobin β chain	5	17	16	10	26	46			
Fibrinopeptides A and B	30	70	67	70					

mine evolutionary relationships of closely related species, such as human, chimpanzee, and gorilla, and among those of the major groupings of all living organisms. More detailed phylogenies should become available soon because it is possible to clone genes or gene fragments from fossil materials, which should make it possible to determine directly the amino acid sequences of at least some ancestral proteins.

References

Construction of phylogenetic trees. W. M. Fitch and E. Margoliash. *Science* 155:279–284 (1967).

Congruency of phylogenies derived from different proteins. E. M. Prager and A. C. Wilson. *J. Mol. Evol.* 9:45–57 (1976).

Testing the theory of evolution by comparing phylogenetic trees constructed from five different protein sequences. D. Penny et al. *Nature* 297:197–200 (1982).

Phylogenies from molecular sequences: inference and reliability. J. Felsenstein. *Ann. Rev. Genet.* 22:521–565 (1988).

Ancient DNA and the polymerase chain reaction. The emerging field of molecular archaeology. S. Pääbo et al. *J. Biol. Chem.* 264:9709–9712 (1989).

c. Rates of Divergence

Different proteins have evolved at different rates (Table 3.2). For example, humans and rhesus monkeys differ in 1% of the residues of their cytochromes *c* but in 3–5% of the residues of their hemoglobin-α and -β polypeptide chains and in 30% of the residues of their fibrinopeptides. A similar ordering of the relative degrees of variation of these proteins is observed when other species are compared. Therefore, if the times of divergence of the various species are known from other data, such as the fossil record, the rate at which each protein or gene has changed can be calculated. Some examples are given in Table 3.3. The rate of change of a gene is expressed as the number of single base-pair mutations per nucleotide position per unit of time. It is analogous to a mutation rate.

It has been suggested that the rate of occurrence of mutations in a gene is constant in all species, and that, therefore, the rate of evolutionary change of any protein or gene can serve as an **evolutionary clock,** with each gene or protein clock ticking at a particular rate in all species in which it occurs. Many data indicate that this is very often the case (Fig. 3.11), at least to a first approximation. If this is true, it has great implications for studying the mechanism of molecular evolution and the evolutionary relationships among species, especially the great majority of species for which the fossil record is poor or nonexistent. There has been much controversy about the validity of the molecular clock, with claims that the rate of divergence of a particular gene or protein has been different along various lineages. In most cases, however, deficiencies of the fossil record have led to considerable uncertainty about the dates at which various species diverged, which may account for the apparently different rates of divergence.

An alternative approach to comparing the rates of evolutionary divergence that does not require knowledge of the times of divergence is known as the **relative rate test.** Two homologous sequences of a protein or a gene are compared with a third, reference sequence that diverged from the common ancestor of the three sequences at an earlier time. If the two test sequences have diverged at a constant rate, they should have the same number of differences from the reference sequence, because both have evolved for the same length of time since their divergence from the reference sequence. This is usually found to be the case, within the limits of expected statistical variation, and is evident in Figure 3.7.

The greatest exceptions to constant rates of divergence are those instances in which the function of the protein has changed. For example, the protein α-crystallin has evolved at an anomalously high rate

Table 3.3   *Rates of Evolution of Some Proteins*

Protein	Accepted point mutations [(number)/(100 residues) · (10⁸ years)]	Protein	Accepted point mutations [(number)/(100 residues) · (10⁸ years)]
Histones		Hormones *cont.*	
H4	0.25	Proparathyrin	14
H3	0.30	Prolactin	20
H2A	1.7	Growth hormone	25
H2B	1.7	Lutropin β chain	33
H1	12	Insulin C peptide	53
Fibrous proteins		Oxygen-binding proteins	
Collagen (α-1)	2.8	Myoglobin	17
Crystallin (αA)	4.5	Hemoglobin α chain	27
Intracellular enzymes		Hemoglobin β chain	30
Glutamate dehydrogenase	1.8	Secreted enzymes	
Triosephosphate		Trypsinogen	17
dehydrogenase	5.0	Lysozyme	40
Triosephosphate isomerase	5.3	Ribonuclease A	43
Lactate dehydrogenase H	5.3	Immunoglobulins	
Lactate dehydrogenase M	7.7	κ chains (V region)	100
Carbonic anhydrase B	25	κ chains (C region)	111
Carbonic anhydrase C	48	λ chains (V region)	125
Electron carriers		γ chains (V region)	143
Cytochrome *c*	6.7	Snake venom toxins	
Cytochrome *b₅*	9.1	Long neurotoxins	111
Plastocyanin	14	Cytotoxins	111
Ferredoxin	17	Short neurotoxins	125
Hormones		Other proteins	
Glucagon	2.3	Parvalbumin	20
Corticotropin	4.2	Albumin	33
Insulin	7.1	α-Lactalbumin	43
Thyrotropin β chain	11	Fibrinopeptide A	59
Lipotropin β chain	13	Casein κ chain	71
Lutropin α chain	14	Fibrinopeptide B	91

From A. C. Wilson et al., *Ann. Rev. Biochem.* 46:573–639 (1977).

in the blind mole rat. This protein is a major structural component of the eye lens, but this function was lost when the blind mole rat lost its sight during its evolution. The greatly increased rate of divergence since then presumably reflects the drastic easing of functional constraints on the changes in this protein that are permitted by selection in the blind mole rat.

What is most surprising about the constant rate of evolutionary change of each protein or gene is that it remains constant per unit of time in species with greatly different generation times. For example, mutations appear to have taken place at the same rate in bacteria, in which a generation may be as short as 20 min, and in higher vertebrates, in which a generation is one or more years. The genetic basis for this constant rate of change is not yet established.

The importance of a constant genetic clock is that once the rate of change is established for a gene or a protein, that rate can be used with the appropriate sequences to estimate the time of divergence of any pair of species. Rapidly changing proteins such as the fibrinopeptides are useful for studying closely related species, and slowly changing proteins and genes divulge more distant evolutionary relationships. No longer does the study of evolution depend only on finding fossils.

References

Nucleotide sequence divergence and functional constraint in mRNA evolution. T. Miyata et al. *Proc. Natl. Acad. Sci. USA* 77:7324–7332 (1980).

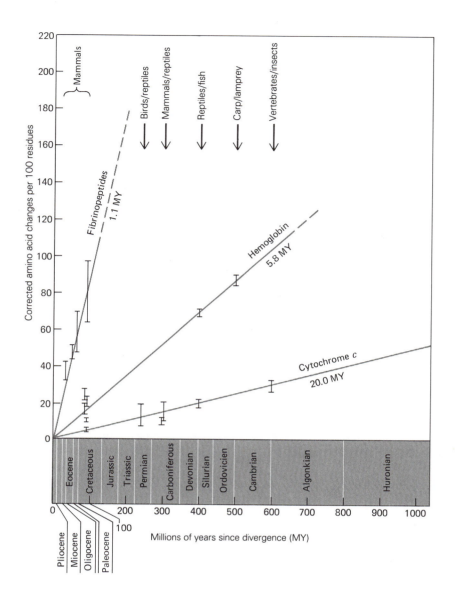

FIGURE 3.11

Rates of evolutionary divergence of fibrinopeptides A and B, hemoglobin-α and -β chains, and cytochromes *c*. The number of amino acid changes per 100 residues, corrected for multiple changes, is plotted versus the estimated time since separation of the genes for the proteins compared. Below each line is the *unit evolutionary period* in millions of years (MY), the time required for a change of 1% of the residues. The times of divergence of some evolutionary lineages are indicated at the top. (Adapted from R. E. Dickerson, *J. Mol. Evol.* 1:26–45, 1971.)

Evidence for higher rates of nucleotide substitution in rodents than in man. C. I. Wu and W. H. Li. *Proc. Natl. Acad. Sci. USA* 82:1741–1745 (1985).

Evolution of cytochrome *c* genes and pseudogenes. C. I. Wu et al. *J. Mol. Evol.* 23:61–75 (1986).

Molecular time scale for evolution. A. C. Wilson et al. *Trends Genet.* 3:241–247 (1987).

Rate constancy of globin gene evolution in placental mammals. S. Easteal. *Proc. Natl. Acad. Sci. USA* 85:7622–7626 (1988).

d. Roles of Selection

If mutations occur at a constant rate in all genes, how can we explain the wide range of evolutionary rates of change among different proteins (Table 3.3) and the nearly constant rate for each protein? The most plausible explanation is that the observed differences among proteins are largely due to neutral mutations that do not significantly affect protein function and so have not been selected for or against. This is not to say that natural selection has not been important, because it certainly must have selected against adverse mutations.

According to the neutral mutation hypothesis, the constant rate of divergence of a protein is the same as its particular neutral mutation rate per gene copy, which is the total mutation rate times the fraction of mutations that are effectively neutral. Even if the total mutation rate is the same for all genes, the neutral mutation rate would differ for each gene because of the different fractions of mutations in the various proteins that are effectively neutral; every gene and protein would differ from every other in how much its amino acid sequence can vary without affecting its function. If the exact amino acid sequence is not critical for the function of a protein, a large fraction of its total mutations would be neutral, and the sequence of the protein would evolve rapidly. Fibrinopeptides are examples of such proteins. They appear to function primarily to block the aggregation of the precursor protein, fibrinogen. They are cleaved proteolytically from the amino ends of two of the three fibrinogen polypeptides in the first step of blood clotting and play no further known role. As a consequence of their removal, the fibrinogen is converted to fibrin, which aggregates and forms the framework of the blood clot. The only known functional constraints on the amino acid sequences of the fibrinopeptides are a carboxyl-terminal Arg residue, which is required for proteolytic cleavage by thrombin, and a somewhat acidic net charge, which probably inhibits aggregation of the precursor, fibrinogen. Within these minor limitations, many amino acid sequences are functional, which explains why these protein segments have evolved at relatively rapid rates (Table 3.3).

At the other extreme, proteins for which very few amino acid replacements are acceptable evolve at very slow rates. An example is cytochrome c, which must interact with a number of other proteins in its function of transferring electrons (Sec. 8.3.4.b, Table 3.3). Variation has occurred at only a few sites (Fig. 3.6), which are presumed not to play crucial roles in this protein's function.

Generally, the degree of change in a protein's primary structure is found to be inversely proportional to the biological importance of each residue. The most variable residues are those that occur on the surface of a protein but are not involved in functional interactions with other molecules (Chaps. 6 and 7). The most conserved amino acid residues are those that are most directly involved in the biological function of the protein, for example, the residues in cytochrome c that interact directly with the heme group (Figs. 3.6 and 3.8) and the active sites of enzymes. The same considerations apply to gene sequences. The untranslated regions of genes,

particularly the introns, vary much more than the regions coding for proteins. Within the regions coding for protein, the most frequent nucleotide changes are those that do not alter the amino acid sequence.

The neutral mutation rate differs for each nucleotide in a gene and is usually a good indicator of the functional importance of each part of the amino acid sequence. Proinsulin is a good example (Fig. 2.14). The C peptide has evolved at a rate that is seven times more rapid than that of the A and B chains, which make up the functional hormone (Table 3.3). The C peptide is removed proteolytically from the middle of the proinsulin polypeptide chain after it has folded to its correct conformation. The primary role of the C peptide appears to be to ensure correct folding of the protein; it has no other known role, and other cross-links are able to function in the refolding of insulin in vitro. The greater rate of divergence of the C peptide than of the A and B chains, therefore, reflects the fewer constraints on its precise amino acid sequence, relative to the functional parts of the hormone.

All the preceding observations indicate that the type of changes at the molecular level that have occurred during evolution are those that are least likely to have functional consequences (and least likely to have been selected). Thus, the occurrence of primarily nonfunctional changes is most readily explained as being the result of the accumulation of neutral mutations. Natural selection at the molecular level seems primarily to be negative, weeding out the deleterious mutations that affect function.

Of course, functional changes have occurred during evolution, as evidenced by the diversity of organisms. This diversity is often not evident at the molecular level, in that proteins with the same function that are from different species usually have very similar properties. There are exceptions, however; the hemoglobins of vertebrates vary widely in the ways that their oxygen-binding properties are regulated (Sec. 8.4.3). For example, fish hemoglobins are used for respiration in the usual way, but they also secrete oxygen into the swim bladder and the eye in order to regulate buoyancy. This release of oxygen, which occurs in response to a decrease in the pH of the swim bladder, is known as the *Root effect* and does not occur in the hemoglobins of nonfish species. In another example, crocodiles are able to stay underwater for as long as an hour because their hemoglobins have evolved to liberate oxygen to cells only when absolutely required. Also, some birds are able to fly at a very high altitude because their hemoglobins have very high oxygen affinities. These are just a few of the ways that hemoglobins have evolved to permit species to occupy extreme environments, and such evolutionary changes would be expected to have been hastened by natural selection. All of these functional differences can be attributed to mutational alterations of just a few residues (Sec. 8.4.3), however, and most of the evolutionary divergence that has occurred in the hemoglobins is believed to be neutral.

There are remarkably few other instances at the molecular level in which natural selection has had a positive effect in selecting *for* favorable mutations. One of the best candidates is the insulin of the guinea pig, which has evolved at a much greater rate than in other species. The guinea pig hormone has an unusually low biological potency but is present at relatively high levels. Positive selection may be enhancing a novel biological property of the insulin at the expense, perhaps, of its potency as an insulin. The two related hormones glucagon and pancreatic polypeptide have also evolved in the guinea pig at greater than normal rates, giving the guinea pig a number of biochemical peculiarities. Many of these apparent anomalies, however, can be explained by an alternative evolutionary origin for the guinea pig.

In another possible instance, two groups of mammals (ruminants and colobine monkeys) have independently evolved a fermentative foregut in which the enzyme lysozyme apparently digests bacteria. The lysozymes from these two groups share certain similarities in their functional properties and in their amino acid sequences that are not present in other lysozymes. These unusual lysozymes have evolved at twice the normal rate, suggesting that at least several of the amino acid changes are functional and were selected for.

The most dramatic evidence for positive selection for functional differences is found in the protein inhibitors of proteolytic enzymes. These proteinase inhibitors act by binding at the active site of the proteolytic enzyme and blocking its access to substrate (Sec. 9.3.2). The evolutionary variation observed in closely related proteinase inhibitors and their genes is just the opposite of that usually observed, in that the functionally important regions have changed the most. The most variable parts of the genes are those coding for the protein, in which most of the nucleotide replacements change the amino acid coded for. Those residues known to interact directly with the proteolytic enzymes have changed the most. Some of the inhibitors have been shown to be specific for different proteases. A corresponding hypervariability of the active site regions of certain proteolytic enzymes has also been observed, so some proteases and their inhibitors may be coevolving by positive selection.

It is probable that other examples of the role of positive selection pressure for functional differences will be discovered, but most evolutionary divergence of proteins is probably of the neutral variety and of no functional significance.

References

Evolutionary rate at the molecular level. M. Kimura. *Nature* 217:624–626 (1968).

Non-Darwinian evolution. J. L. King and T. H. Jukes. *Science* 164:788–798 (1969).

Biochemical peculiarities of the guinea pig and some examples of convergent evolution. J. C. Wriston. *J. Mol. Evol.* 17:1–9 (1981).

Guinea pig preproinsulin gene: an evolutionary compromise? S. J. Chan et al. *Proc. Natl. Acad. Sci. USA* 81:5046–5050 (1984).

Species adaptation in a protein molecule. M. F. Perutz. *Adv. Protein Chem.* 36:213–244 (1984).

Adaptive evolution in the stomach lysozymes of foregut fermentors. C. B. Stewart et al. *Nature* 330:401–404 (1987).

Ovomucoid third domains from 100 avian species: isolation, sequences, and hypervariability of enzyme-inhibitor contact residues. M. Laskowski, Jr., et al. *Biochemistry* 26:202–221 (1987).

Functional evolutionary divergence of proteolytic enzymes and their inhibitors. T. E. Creighton and N. J. Darby. *Trends Biochem. Sci.* 14:319–324 (1989).

Concerted evolution of ruminant stomach lysozymes. Characterization of lysozyme cDNA clones from sheep and deer. D. M. Irwin and A. C. Wilson. *J. Biol. Chem.* 265:4944–4952 (1990).

Is the guinea pig a rodent? A. D. Graur et al. *Nature* 351:649–652 (1991).

3.3.2   Variation within Species

It is usually possible to describe a human insulin, a bovine ribonuclease, or a horse cytochrome *c* because members of a species tend to have the same genes and proteins. The reason for this is genetic, due to the finite number of individuals in any species. As mentioned earlier, each gene in an individual has only a moderate probability of being passed on to the next generation. When the population is stable in size, this probability is 0.75 on average, in which case it is improbable that any particular copy of a gene will be passed on for very many generations, even if there are any typically moderate selective pressures. As a consequence, all copies of a particular gene that are present at any instant in a population are likely to have descended from a single copy that was present a limited number of generations previously; in this case, the genetic variation among the copies of a gene in the individuals of a population is limited to that which has arisen by mutation in the meantime.

The variant genes generated by mutation are known as **alleles.** Over 300 human hemoglobin variants have been found by genetic screening. Most are exceedingly rare and produce little or no clinical effect, but some are responsible for genetic diseases, such as sickle cell anemia and thalassemia. The hemoglobin alleles generally differ from the normal gene by only a single mutational event. It is likely that most of these rare alleles arose by recent spontaneous mutation and have not yet been subjected fully to natural selection. Only those that are lethal are missing from the population. For this reason, the variation among allelic forms of genes in a population is often different from the variation observed between closely related species, for which only neutral or beneficial mutations are likely to be observed.

Two or more different alleles of some proteins are present in high frequency in populations, depending on the species. This genetic heterogeneity within a population is known as **polymorphism.** It conceivably could have arisen by random processes and involve functionally indistinguishable alleles, or it might be maintained by selection. The classic example of maintenance by selection is sickle-cell hemoglobin in humans, a variant present in high frequency in African populations. The change of Glu to Val at residue 6 of the β chain of sickle-cell hemoglobin (Fig. 3.4) causes its deoxygenated form to polymerize within the red blood cell, thereby producing the characteristic "sickling" of the cell. Sickling causes blockage of the capillaries carrying blood to vital regions, producing a clinical crisis for the affected person. Individuals with sickle-cell anemia are unlikely to survive beyond the age of 30 without intensive treatment, yet the sickle-cell allele of the hemoglobin-β gene is present in African populations with a frequency of about 0.05.

The reason for the maintainance of sickle-cell hemoglobin is that **heterozygotes,** those individuals with one copy of the sickle-cell and one copy of the normal hemoglobin-β gene allele, do not suffer from sickle-cell anemia and are resistant to malaria. The parasite responsible for malaria, *Plasmodium falciparum,* spends part of its reproductive cycle in human erythrocytes, which, if they contain significant amounts of sickle-cell hemoglobin, can resist infection for reasons that are not clear. Heterozygotes do not suffer from sickle-cell anemia because about half of their hemoglobin is the normal type, which inhibits polymerization of the sickle-cell type. **Homozygotes** with two copies of the sickle-cell gene also are resistant to malaria but suffer from sickle-cell anemia, whereas homozygotes with two copies of the normal gene tend to suffer from malaria but not from sickle-cell anemia. Consequently, the heterozygotes have a selective advantage over both homozygotes in malarial regions, and as a result, the sickle-cell gene is maintained within populations that are exposed to malaria.

It is tempting to invoke similar arguments to account for the other instances of human polymorphisms, but it is unlikely that there could be similar selection for many such genes in one species. If many other human genes had polymorphisms subject to selection like that of sickle-cell hemoglobin, all humans would tend to be homozygous for some fraction of these genes and would be very ill indeed. Thus, the number of genes that can be simultaneously subjected to such selective pressures is limited by a population's ability to survive.

Polymorphisms have been studied extensively in other species, but usually only by searching for variant alleles on the basis of the electrophoretic properties of the proteins; such techniques cannot detect all allelic forms. Only a few polymorphisms have been studied at the sequence level. In the alcohol dehydrogenase gene of the fruit fly, *Drosophila melanogaster,* the two alleles detected by electrophoresis differ by only a single amino acid, Thr versus Lys at residue 192. Various copies of each allelic gene, however, also differ in having one or more silent mutations in the coding region that do not alter the protein sequence and other mutations in the noncoding regions of the gene. The esterase-6 gene of this species also demonstrates a number of alleles; the proteins differ at an average of 4.1 residues, along with numerous silent mutations. The occurrence of multiple mutations in these genes suggests that they have all been present in the population for a considerable length of time, making it unlikely that they are neutral variants. Indeed, both the alcohol dehydrogenase and the esterase-6 polymorphisms are thought to be maintained by selection, the heterozygotes having a selective advantage. The physiological basis for this presumed selection is not known. Other explanations for polymorphisms are possible, such as the founder effect in initiating populations and species. In this case, the presence of neutral alleles in a population is caused by the random genetic heterozygosity of a few individuals that founded the population. The complexity of natural populations makes it difficult to prove either the absence or the presence of selective pressures.

References

The complete amino acid sequence of the three alcohol dehydrogenase alleloenzymes (Adh[N-11], Adh[S], and Adh[UF]) from the fruitfly *Drosophila melanogaster.* D. R. Thatcher. *Biochem. J.* 187:875–886 (1980).

The interaction of malaria parasites with red blood cells. G. Pasvol and R. J. M. Wilson. *Brit. Med. Bull.* 38:133–140 (1982).

Nucleotide polymorphism at the alcohol dehydrogenase locus of *Drosophila melanogaster*. M. Kreitman. *Nature* 304:412–417 (1983).

Two-dimensional electrophoresis of plasma polypeptides reveals "high" heterozygosity indices. B. B. Rosenblum et al. *Proc. Natl. Acad. Sci. USA* 80:5002–5006 (1983).

The mutation and polymorphism of the human β-globin gene and its surrounding DNA. S. H. Orkin and H. H. Kazazian, Jr. *Ann. Rev. Genet.* 18:131–171 (1984).

Population genetics. R. C. Lewontin. *Ann. Rev. Genet.* 19:81–102 (1985).

Observing the founder effect in human evolution. J. M. Diamond and J. I. Rotter. *Nature* 329:105–106 (1987).

The Drosophila alcohol dehydrogenase gene-enzyme system. G. K. Chambers. *Adv. Genetics* 25:39–107 (1988).

Amino acid polymorphisms for esterase-6 in *Drosphila melanogaster*. P. H. Cooke and J. G. Oakeshott. *Proc. Natl. Acad. Sci. USA* 86:1426–1430 (1989).

3.4   Gene Rearrangements and the Evolution of Protein Complexity

Both genetic and biochemical complexity are likely to have increased during evolution, and present-day organisms are assumed to be considerably more complex anatomically than ancestral organisms and also to vary more in their biology and DNA content. It is difficult to imagine how useful genes could arise de novo, but it probably happened to generate the original living organisms, and it probably was a rate-limiting step in the appearance of life.

After some functional genetic material existed, its complexity would increase more easily by duplication of existing genes than by the generation of new genetic material. DNA is now known to be a very dynamic molecule that undergoes a wide variety of alterations and modifications, and gene duplications occur naturally and frequently. With two copies of a gene available in a genome, one copy could provide the necessary original function while the other accumulated mutations that altered its function. If this altered copy evolved eventually to serve a new function, it would tend to be retained in the genome and passed on to later generations.

It is plausible that most modern genes originated from one or a few ancestral genes, but this is difficult to prove or disprove because the amount of change that would have occurred since then will have obscured any similarities among their modern descendants.

References

Origins of repeated DNA. J. Rogers. *Nature* 317:765–766 (1985).

Molecular drive in multigene families: how biological novelties arise, spread and are assimilated. G. A. Dover. *Trends Genet.* 2:159–165 (1986).

Birth of a unique enzyme from an alternative reading frame of the preexisted, internally repetitious coding sequence. S. Ohno. *Proc. Natl. Acad. Sci. USA* 81:2421–2425 (1984).

How big is the universe of exons? R. L. Dorit et al. *Science* 250:1377–1382 (1990).

3.4.1   Products of Gene Duplications

Many genes and proteins of an organism are homologous and are obviously the products of gene duplication. The genes of such homologous proteins in a genome are said to comprise a **gene family**. As the number of known protein and gene sequences increases, more and more such gene families are being found.

The most studied gene family, and the most striking for the similarities among its polypeptide chains, is that of the globins of higher organisms; this family includes single-chain myoglobin and the various polypeptides of tetrameric hemoglobins. The globins have related functions in that they store oxygen; myoglobin stores oxygen in muscle and hemoglobins in erythrocytes, while transporting it. The various polypeptide chains of hemoglobin function at different stages of life, and they associate to form comparable tetrameric molecules. The ϵ and ζ chains function in the very young embryo as an $\epsilon_2\zeta_2$ complex, and the α and γ chains as an $\alpha_2\gamma_2$ complex in the fetus at a later stage. The β and δ chains replace the γ chain in the adult, which has $\alpha_2\beta_2$ and $\alpha_2\delta_2$ tetramers. The many similarities among the amino acid sequences of these polypeptide chains (Fig. 3.12) leave little doubt that they must have arisen by gene duplications of a common ancestor.

The probable sequence of events can be inferred from the relative similarities of the amino acid sequences of these proteins (Table 3.4). The myoglobin sequence is least like the others, so it probably diverged first by duplication of the ancestral globin gene, to give

Table 3.4 *Number of Amino Acid Identities between Various Members of the Human Globin Family*

| | Hemoglobin | | | | |
	α	β	γ	δ	ϵ
Myoglobin	38	36	36	37	34
Hemoglobin α		64	59	63	55
Hemoglobin β			107	136	110
Hemoglobin γ				105	116
Hemoglobin δ					106

rise to the ancestral myoglobin and hemoglobin genes (Fig. 3.13). The hemoglobin gene probably duplicated again at a later time and gave rise to the ancestral α-globin gene and that of the β-type genes (β, γ, δ, and ϵ). The next duplication, of the β-type gene (at least 200 million years ago), would have produced the line of descent to the γ and ϵ genes plus that to the β and δ genes; both gene lines were duplicated again (about 100 million and 40 million years ago, respectively) to produce each pair of genes (γ and ϵ; β and δ).

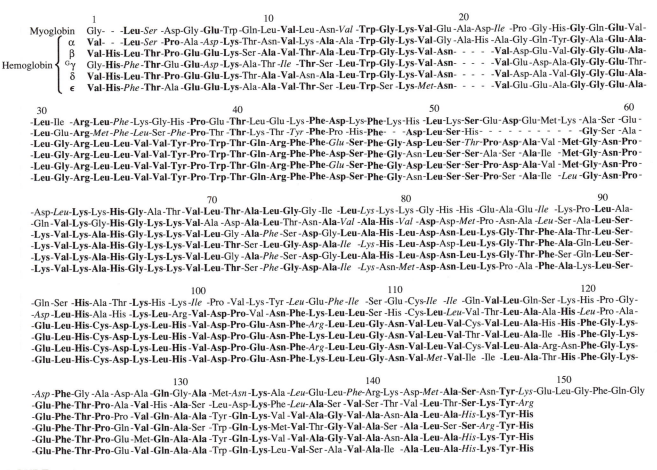

FIGURE 3.12

Amino acid sequences of members of the human globin family, myoglobin plus the hemoglobin polypeptide chains. Positions where three or more of the amino acid residues are identical are in boldface; chemically similar residues are in italics.

The sequences of the genes that code for these proteins confirm and extend these inferences. Two human genes have been found for γ chains, $^G\gamma$ and $^A\gamma$, which produce γ chains that differ only in having either Gly or Ala at position 136. There are also two α genes, but they produce identical α polypeptide chains. Duplication of the α genes is common in primates and probably occurred before these species diverged from their common ancestor.

The scheme in Figure 3.13 is for the evolutionary divergence of duplicated genes *within* a species, whereas that of Figure 3.9 is for divergence *between* species. Gene duplication and consequent divergence must be distinguished from species divergence, although many phylogenetic trees do not do so. It seems appropriate to use different dimensions for the two processes, as illustrated in Figure 3.14. When two species arise from one species by mechanisms that are not known, they are presumed to retain most, if not all, of the original genome. Therefore, all of the globin genes present at the time of species divergence are present in both potential species. The *presence* of homologous duplicated genes in various extant species is a good indication that the gene duplication occurred before the species separated. The *absence* of homologous duplicated genes in one or more species, however, could be

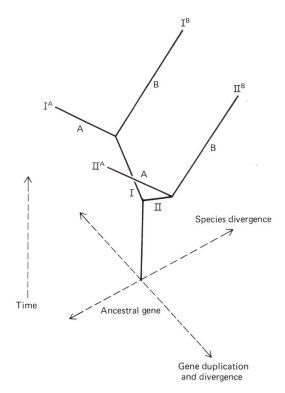

FIGURE 3.14
Simple example of a three-dimensional phylogenetic tree showing both gene duplication and species separation. The ancestral gene in the ancestral species is depicted as first being duplicated in the ancestral genome to give two homologous genes, I and II. The ancestral species then separated into two different species, A and B. Both genes I and II would also separate at this stage and consequently would be present in both species A and B. Either gene could be lost at any subsequent stage, however, so both need not be present in contemporary species. Also, further gene duplications and species separations could occur along any branch of the tree. This example, with just one gene duplication followed by one species separation, and no loss of genes, would yield at the present time four genes: I^A and II^A (in species A) and I^B and II^B (in species B). Each gene will independently accumulate mutational alterations along its branch; the differences between any two contemporary genes would be expected to be proportional to the length of the branches joining them.

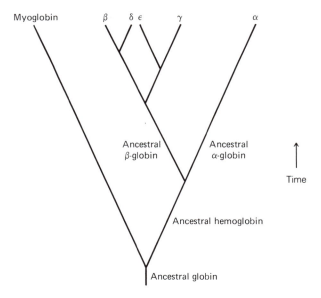

FIGURE 3.13
Evolutionary scheme for the divergence of the globin chains, inferred from the relative similarities of their gene and protein sequences. Time increases from the bottom to the top of the figure. The separation of the $^A\gamma$ and $^G\gamma$ genes is not depicted here. (Adapted from A. Efstratiadis et al., *Cell* 21:653–668, 1980.)

due either to the occurrence of duplication after speciation or to the loss of one or more of the duplicated genes during subsequent evolution of the species, as has occurred for the δ-globin gene. In at least some cases, the inactive gene is still present and is detectable by genome sequencing.

The presence of several related genes and proteins in an organism presents potential pitfalls in evolutionary studies. For example, the α and β chains of hemoglobin are the products of a gene duplication that took place long before mammals diverged. Both sequences have been evolving independently in all mammalian species since the time of this gene duplication. Unwitting comparison of the α chain of one mammal with the β chain of another would be comparing two sequences whose divergence took place much earlier than that of the species containing them.

Genes that occupy the same gene locus in different species, and protein products of such genes, are said to be **orthologous,** whereas genes at different loci that are related by gene duplication are designated **paralogous.** The phylogenies of species can be reconstructed only by comparing orthologous genes. The differences between paralogous genes or proteins from different species are not related to the time since divergence of the species, but to the time since gene duplication (Table 3.5). Comparisons of paralogous genes confirm that the number of differences among homologous genes depends primarily on the length of time since duplication, irrespective of whether they have existed in the same or different species. For example, the human and dog α-hemoglobin chains are nearly equally similar to the β, γ, and myoglobin chains of both species (Table 3.5). Other such comparisons give the same result, indicating that paralogous genes have accumulated mutations at the same constant rate since the time of gene duplication.

Many other gene families are known. In most, the general function of the protein has been preserved; for example, all the globin chains function in binding oxygen (Sec. 8.4.3), but with slight variations on the theme. It is generally assumed that a newly sequenced gene or protein will have the same type of function as any other protein to which it is found to be homologous. Such preservation of function is not necessarily the case, however, and varying degrees of functional divergence are known. For example, the proteases of the trypsin family catalyze the same basic reaction but with markedly different substrate specificities (Sec. 9.3.2.a). Haptoglobin, homologous with the trypsin proteases, undergoes similar proteolytic processing, but it has no proteolytic activity; its physiological function is to bind α-β complexes of the hemoglobin chains. Lysozyme from hen eggs is homologous with α-lactalbumin from bovine milk: of 123 residues, 47 are identical and many others are physically similar. Yet lysozyme functions to degrade the polysaccharides of bacterial cell walls, whereas α-lactalbumin has no such activity but instead regulates lactose synthesis by binding to galactosyl transferase, a membrane-bound enzyme. The latter enzyme ordinarily glycosylates proteins in the Golgi apparatus, but it synthesizes lactose when complexed with α-lactalbumin. The only apparent similarity in the func-

Table 3.5 *Paralogous Comparisons of Globin Chains*[a]

Comparison protein	Number of Differences in Amino Acid Sequence	
	Human α-globin	Dog α-globin
Human β-globin	84	84
Human γ-globin	89	83
Human myoglobin	115	119
Dog β-globin	88	87
Dog myoglobin	115	118

[a] The globin genes duplicated and diverged long before the human and dog species diverged, so comparisons of the different globins are relevant to the time since the genes duplicated. Consequently, very similar results are obtained irrespective of which species is used.

tions of lysozyme and α-lactalbumin is that they both act in reactions involving saccharides, although there is no evidence that α-lactalbumin actually interacts directly with the saccharide. Ovalbumin, a seemingly functionless food storage protein of egg white, is homologous to the *serpin* class of protease inhibitors, as is angiotensinogen, the precursor to the hormone angiotensin; neither have protease inhibitor activities, nor do the serpins act as hormones.

In other cases, homologous proteins have developed auxiliary functions. The most striking examples are several enzymes (e.g., lactate dehydrogenase, argininosuccinate lyase, aldose reductase, enolase, and glutathione transferase) that have been duplicated and adapted to serve as structural proteins of the eye lens or have simply added this role as an additional function.

All of these examples indicate that some caution should be exercised in suggesting or accepting that primary structure homology is indicative of similarity of function.

References

The structure and evolution of the human β-globin gene family. A. Efstratiadis et al. *Cell* 21:653–668 (1980).

A history of the human fetal globin gene duplication. S. H. Shen et al. *Cell* 26:191–203 (1981).

Rise and fall of the delta globin gene. S. L. Martin et al. *J. Mol. Biol.* 164:513–528 (1983).

Evolution of the mammalian β-globin gene cluster. S. C. Hardies et al. *J. Biol. Chem.* 259:3748–3756 (1984).

The genealogy of some recently evolved vertebrate proteins.

R. F. Doolittle. *Trends Biochem. Sci.* 10:233–237 (1985).

Relationships between gene trees and species trees. P. Pamilo and M. Nei. *Mol. Biol. Evol.* 5:568–583 (1988).

More molecular opportunism. R. F. Doolittle. *Nature* 336:18 (1988).

Complete nucleotide sequence of the rabbit β-like globin gene cluster. J. B. Margot et al. *J. Mol. Biol.* 205:15–40 (1989).

Kinetic analysis of duck ε-crystallin, a lens structural protein with lactate dehydrogenase activity. S. H. Chiou et al. *Biochem. J.* 267:51–58 (1990).

Lysozyme and α-lactalbumin: structure, function, and interrelationships. H. A. McKenzie and F. H. White, Jr. *Adv. Protein Chem.* 41:173–315 (1991).

3.4.2 Evolution of Metabolic Pathways

If homologous proteins arose from a common ancestor, how many different ancestral proteins were there originally? The answer is not known, but it is conceivable that there was a single ancestral protein. In which case, the homologies that predate the last common ancestral organism would not be apparent today; the absence of sequence similarity does not disprove homology. Distant homologies are most apparent in the three-dimensional structures of the proteins, which have generally been more conserved during evolution than have primary structures (Chap. 6). Indications of homology among proteins that must have diverged very long ago suggest that the many proteins that catalyze modern metabolic pathways might have been produced by duplication of just a few genes.

How might the biochemical complexity of metabolic pathways have evolved? In the case of the biosynthetic pathways that produce the building blocks of amino acids, nucleotides, sugars, and so forth, it is likely that these building blocks were originally present in the primordial soup and were used directly. As organisms increased in number, however, these constituents would have become scarce. Any organism that could produce one of them from some unused component of the primordial soup, using a newly evolved enzyme, would have had a selective advantage. Once the availability of that component became limiting, there would have been selection for any organism that could produce it from some other component of the primordial soup. According to this scenario, the enzymes of metabolic pathways would have evolved in a sequence opposite to the one they have in the modern pathway.

In 1945, N. H. Horowitz proposed that the enzymes of pathways might have arisen in this manner by a process of gene duplication. In this hypothesis, the protein that would become the terminal enzyme of a modern metabolic pathway had a binding site for the substrate that it used from the soup (Chap. 9). If the gene for this protein were duplicated, the protein from the second gene could still use this binding site for the same substrate molecule but evolve the capability of producing it as a product from another component of the primordial soup. Thus, the substrate for the terminal enzyme has become an intermediate in a developing metabolic pathway. In this way, the various enzymes of a pathway would have evolved from each other in a reverse order to that in which they appear in the modern pathway. There are a few remarkable cases in which such homologies have been detected and a few other cases in which homology has been detected among proteins from different pathways that have related catalytic functions but different substrate specificities. In these latter cases, evolution of the duplicated proteins has altered the substrate specificity while retaining the similar catalytic mechanism.

In no case is the mechanism of such functional divergence of duplicated genes known in detail. The ancestral protein might have had just one of the present-day functions, and the others evolved later in the duplicated versions. Alternatively, the ancestral protein might have had a relatively nonspecific range of functions that encompassed those of the present-day proteins. After gene duplication, this nonspecific function might have been refined in each case to the present-day specificity.

References

The evolution of biochemical systems. N. H. Horowitz. *Proc. Natl. Acad. Sci. USA* 31:153–157 (1945).

Protein differentiation: a comparison of aspartate transcarbamoylase and ornithine transcarbamoylase from *Escherichia coli* K-12. J. E. Houghton et al. *Proc. Natl. Acad. Sci. USA* 81:4864–4868 (1984).

Biochemical pathways in prokaryotes can be traced backward through evolutionary time. R. A. Jensen. *Mol. Biol. Evol.* 2:92–108 (1985).

Evolution in biosynthetic pathways: two enzymes catalyzing consecutive steps in methionine biosynthesis originate from a common ancestor and possess a similiar regulatory region. J. Belfaiza et al. *Proc. Natl. Acad. Sci. USA* 83:867–871 (1986).

Evolution of biosynthetic pathways: a common ancestor for threonine synthase, threonine dehydratase and D-serine dehydratase. C. Parsot. *EMBO J.* 5:3013–3019 (1986).

The evolution of the glycolytic pathway. L. A. Fothergill-Gilmore. *Trends Biochem. Sci.* 11:47–51 (1986).

Complete nucleotide and deduced amino acid sequence of bovine phenylethanolamine *N*-methyl transferase: partial amino acid homology with rat tyrosine hydroxylase. E. E. Bartge et al. *Proc. Natl. Acad. Sci. USA* 83:5454–5458 (1986).

A common origin for enzymes involved in the terminal step of the threonine and tryptophan biosynthetic pathways. C. Parsot. *Proc. Natl. Acad. Sci. USA* 84:5207–5210 (1987).

3.4.3 Protein Elongation by Intragene Duplication

Duplication of sequences within a gene elongates the gene. Part or all of the amino acid sequence in the protein product is also duplicated if the reading frame of the gene sequence is maintained. The duplicated portions can then accumulate mutational changes and diverge. Such gene elongation by duplication is often most evident in diagonal dot-plots (Fig. 3.3), in which two or more diagonals of homology are observed for a single sequence.

A striking example of gene elongation by duplication is shown by the ferredoxins, in which the homology between the two halves of the amino acid sequence is very apparent (Fig. 3.15); the three-dimensional structure of the protein shows it to be a remarkable "Siamese twin" (see Fig. 6.36 and Sec. 6.4.3). In transferrin, the two 340-residue halves are 50% identical, and each half binds a single iron atom. Serum albumin has a triplicated structure derived from a simpler 190-residue form. Some other proteins have multiple repeats of smaller segments. For example, the α chain of human fibrinogen has a succession of 13-residue repeats with similar sequences; the enzyme protein phosphatase 2A has 15 imperfect repeating units of 39 residues each. The keratin structural proteins have multiple repeats of 10-residue segments. Collagens have sequences that tend to be tandem repeats of the sequence -Gly-X-Y-,

where X and Y are often Pro; this may have arisen by duplication of the three codons for this tripeptide, although there is also a structural basis for this repeating sequence (Sec. 5.5.3).

Partial gene duplications in which the reading frame has been altered will not be detectable in the protein primary structure. They might be detected from the gene sequence if divergence has not proceeded too far, but any functional role of the duplicated segment would be accidental, and there are no well-documented examples.

References

Proteins of *Escherichia coli* come in sizes that are multiples of 14 kDa: domain concepts and evolutionary implications. M. A. Savageau. *Proc. Natl. Acad. Sci. USA* 83:1198–2102 (1986).

Early genes that were oligomeric repeats generated a number of divergent domains on their own. S. Ohno. *Proc. Natl. Acad. Sci. USA* 84:6486–6490 (1987).

3.4.4 Gene Fusion and Division

Many proteins, especially those unique to vertebrates, have mosaic structures in which various segments appear to have had different origins. One segment might be homologous to one protein family, another to a different family; this is the one case in which it is correct to describe fractional homologies (Sec. 3.2.1). Such a protein gives the impression of having been assembled by stringing modules together, presumably by fusing their genes. Each module usually corresponds to an entire structural and functional domain of a protein. For example, a calcium-sensitive protease was apparently constructed by fusing the genes for a calcium-binding protein and a thiol protease. Some other examples, pri-

1   10

Ala-*Tyr*-Lys-**Ile-**   -**Ala-Asp**-*Ser*-**Cys**-*Val*-Ser -**Cys-Gly**-Ala

-Ile -*Phe*-Val-**Ile**-Asp-**Ala-Asp**-*Thr*-**Cys**-*Ile* -Asp-**Cys-Gly**-Asn

30   40

20

-**Cys-Ala**-Ser -Glu-**Cys-Pro-Val**-Asn-**Ala**-Ile -Ser -**Gln**-Gly-Asp-Ser

-**Cys-Ala**-Asp-Val-**Cys-Pro-Val**-Gly -**Ala**-Pro-Val-**Gln**-Glu

50   55

FIGURE 3.15
Homology between the two halves of the primary structure of *Clostridium pasteurianum* ferredoxin. Identical residues are in boldface, similar residues in italics.

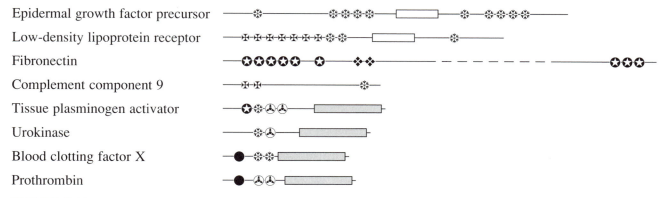

FIGURE 3.16

Mosaic structures of some vertebrate proteins. Each symbol represents members of a set of homologous segments of polypeptide chain, which in these examples are also independent structural units, or domains, of the protein. Those with distinct structural or functional properties are the serine protease domains (shaded rectangles), Ca^{2+}-binding domains containing γ-carboxy-Glu residues (●), so-called kringle domains (⊕), and epidermal growth factor (EGF) domains (✳).

marily of proteins that are active in the plasma of vertebrates, are illustrated schematically in Figure 3.16.

Gene fusion occurs more readily in eukaryotes than in prokaryotes, probably because the two DNA coding regions need not be joined exactly in the correct reading frame if the fusion occurs in introns. Two genetic regions coding for protein can be effectively fused by genetic recombination in an intron between them that can be excised correctly. The existence of introns in the genome would be expected to facilitate fusion, duplication, and rearrangement of gene segments. There are many indications that vertebrate mosaic proteins have arisen by shuffling introns and exons. The more general importance of introns and exons in the evolution of smaller proteins without mosaic structures has important implications for protein structure but is, however, a contentious question for which there is no clear answer. In some cases, there are indications that exons of the gene may code for distinct elements of protein structure that plausibly could have arisen by exon shuffling in the genome, whereas in other cases there is no such obvious correlation.

Gene fusion is not restricted to eukaryotic cells, and it has also occurred in prokaryotes, where there are no introns and where gene fusion consequently must be precise. For example, biosynthesis of the amino acid tryptophan requires five unique chemical reactions that are catalyzed by five enzymatic activities, which in some bacteria occur on five different polypeptide chains encoded by different genes. In other bacteria, such as *Escherichia coli*, two pairs of these genes are fused to produce two bifunctional polypeptides. In each bifunctional polypeptide, the two enzymatic activities are carried out by separate regions of the polypeptide chain, which are homologous to the corresponding proteins of other bacteria. The two halves of the fused polypeptide chain can often be separated by proteolytic treatment. Each fragment appears to maintain its functional and structural integrity and carries out the pertinent enzymatic reaction. In the original fused protein, the independent functional and structural units are simply linked by the connecting polypeptide chain. Physiological reasons for these gene fusions have not been established.

References

Why genes in pieces? W. Gilbert. *Nature* 271:501 (1978).

Gene fusion during the evolution of the tryptophan operon in Enterobacteriaceae. G. F. Miozzari and C. Yanofsky. *Nature* 277:486–489 (1979).

Evolution of the proteases of blood coagulation and fibrinolysis by assembly from modules. L. Patthy. *Cell* 41:657–663 (1985).

The genealogy of some recently evolved vertebrate proteins. R. F. Doolittle. *Trends Biochem. Sci.* 10:233–237 (1985).

Exons and the evolution of proteins. C. C. F. Blake. *Int. Rev. Cytology* 93:149–185 (1985).

Intron-dependent evolution: preferred types of exons and introns. L. Patthy. *FEBS Letters* 214:1–7 (1987).

How were introns inserted into nuclear genes? J. H. Rogers. *Trends Genetics* 5:213–216 (1989).

Evolution of a biosynthetic pathway: the tryptophan paradigm. I. P. Crawford. *Ann. Rev. Microbiol.* 43:567–600 (1989).

Evolution of the tryptophan synthetase of fungi. D. M. Burns et al. *J. Biol. Chem.* 265:2060–2069 (1990).

3.4.5 *Genetic Consequences of Duplicated Genes*

Any gene duplication results in the presence of duplicated segments of DNA in a genome, which can have important genetic consequences. Genetic phenomena such as mitosis and meiosis involve the pairing of homologous sequences within the two genomes (or sets of chromosomes) present in diploid individuals, and recombination frequently occurs between them. If two duplicated regions are present in one genome, the possibility exists that the duplicated segments will be paired incorrectly between the two genomes and that recombination will take place between them. In such recombination, known as *unequal crossing-over,* one genome ends up with three copies of the duplicated region and the other with one. In this way, the numbers of duplicated genes within a genome can rapidly increase or decrease.

Color blindness is common in humans probably because of the presence of two homologous genes close together on the same chromosome. Human color vision is due to three different light-sensitive proteins (called *opsins*) that respond to the three colors red, green, and blue. These three proteins are clearly homologous. The genes for the red- and green-sensitive opsins occur in a tandem array on the X chromosome. They occur in variable numbers in different individuals, presumably because of unequal crossing-over between them. Some chromosomes lack a red-sensitive gene or a green-sensitive gene; male individuals with such a chromosome are red–green color-blind.

Hybrid proteins can also result from unequal crossing-over. Two well-known examples occur in hemoglobins Lepore and Kenya. In the first, crossing-over between the genes for β- and δ-hemoglobin produced a single hybrid gene and a polypeptide chain that is β-like at one end and δ-like at the other. In the case of hemoglobin Kenya, the gene and protein are hybrids of β and $^A\gamma$ (Fig. 3.17). Similarly in the opsins of the eye, hybrid proteins resulting from crossing-over between the adjacent, homologous red- and green-sensitive genes have altered color sensitivities.

Another consequence of the mismatching of duplicated genes is that one gene can be "corrected" by replacing its DNA sequence with that of the other, an ill-defined genetic phenomenon known as **gene conversion.** As a result of gene conversion, two duplicate genes that have diverged since their duplication are suddenly made identical. This phenomenon can happen to all or just a part of a gene. Consequently, the protein products of duplicated genes may be much more similar than would be expected for the time since they began to diverge. Fortunately for evolutionary studies, gene conversion is a relatively rare event: During divergence of the hominids, seven gene conversion events are thought to have occurred in the γ-hemoglobin genes. Nevertheless, this possibility must be considered when estimating the times of gene duplications from contemporary sequences, especially those that are very similar.

References

The evolution of multigene families: human haptoglobin genes. N. Maeda and O. Smithies. *Ann. Rev. Genet.* 20:81–108 (1986).

Molecular genetics of human color vision: the genes encoding blue, green, and red pigments. J. Nathans et al. *Science* 232:193–202 (1986).

Molecular genetics of inherited variation in human color vision. J. Nathans et al. *Science* 232:203–210 (1986).

Gene conversions in the horse α-globin gene complex. J. B. Clegg. *Mol. Biol. Evol.* 4:492–503 (1987).

3.5 *Using Genetics to Probe Protein Structure*

In genetic approaches to the study of proteins, the gene that codes for a protein is altered and the effect of the alteration on the organism is determined, so the function of the protein product of the gene is being studied in its natural, in vivo environment. Genetic alterations of protein structures have the advantage of being extremely specific and can be made to order (Sec. 2.2). Specific site-directed mutagenesis (see Fig. 2.6) is ideally suited for testing hypotheses about well-characterized genes and proteins. Alternatively, alterations of the gene and protein that produce a desired physiological effect can often be selected after random mutagenesis. Such an approach is ideally suited for initially characterizing the functional properties of a newly discovered gene and protein. No prior knowledge of the gene or protein is required, and there is little possibility of overlooking important residues of the protein so long as the mutagenesis and selection procedures are sufficiently rigorous. Many genetic studies uncover genes without discovering any clues about their functions or products. In this case, the gene can often be disrupted at specific sites by inserting appropriate DNA sequences using the principles of site-directed mutagenesis. If such a di-

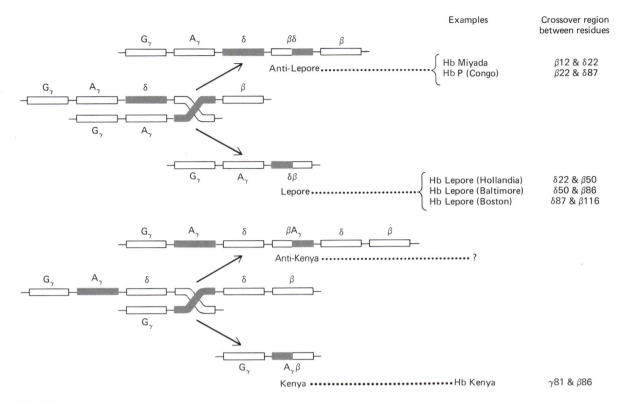

FIGURE 3.17
Genetic mechanism for production of hybrid proteins by mismatched recombination between duplicated genes. The chromosomal arrangements of the β-like globin genes are depicted, with recombination between the δ and β genes at the top and between the ᴬγ and β genes at the bottom. These events are inferred from the amino acid sequences of the hybrid proteins produced. (From D. J. Weatherall and J. B. Clegg, *Cell* 16:467–479, 1979.)

rected mutation inactivates the gene and its product, the effect on the organism can give clues to the normal gene's function.

The most critical aspect of genetic studies is the method of mutagenesis. Single, specific mutations are usually used for site-directed mutagenesis, whereas genetic screens are most informative if all possible mutations are generated. Many methods with varying selectivities for mutating genes in vivo exist, but mutagenesis is more satisfactorily carried out in vitro on the cloned gene. Chemical agents can be used to modify the structure of the DNA, or mutations can be generated by replicating or synthesizing the gene under conditions in which the "wrong," mutant nucleotides are incorporated at a certain frequency. Such procedures can be carried out on the entire gene, on short fragments of it, or at selected sites. In the latter cases, all possible single-nucleotide replacements can be easily generated by incorporating the three wrong nucleotides with a significant frequency at each of the desired sites during synthesis of the mutagenic DNA fragment (see Fig. 2.6).

References

Strategies and applications of *in vitro* mutagenesis. D. Botstein and D. Shortle. *Science* 229:1193–1201 (1985).

In vitro mutagenesis. M. Smith. *Ann. Rev. Biochem.* 19:423–462 (1985).

Functional inactivation of genes by dominant negative mutations. I. Herskowitz. *Nature* 329:219–222 (1987).

The new mouse genetics: altering the genome by gene targeting. M. R. Capecchi. *Trends Genetics* 5:70–76 (1989).

3.5.1 Selecting for Functional Mutations

Having generated a large number of mutated forms of a gene, one selects for those mutations that disrupt the structure or some function of the protein. No general procedures for this selection can be given because the procedure used is necessarily specific for each gene and protein and also depends on the organism in which they occur. However, mutations that disrupt some function of the normal protein usually are selected initially. If the mutation is lethal to the organism, "conditional lethal" mutations, which act only under certain conditions, can be sought; the most commonly used conditional lethal mutations are temperature sensitive and act only at high or low temperatures. After the first selection, genes with mutations that reverse the effect of the first mutation can be selected; a few of these will be simply reversals of the first mutation, but others that are mutations at second sites in the gene and the protein also may be found.

Mutations that disrupt a particular function of a protein may do so by altering specific residues that affect that function, but they also can act indirectly, for example by abolishing synthesis of the protein or by destroying its three-dimensional structure. Mutations that abolish the enzyme activity of some proteins, such as the α subunit of tryptophan synthase, occur at only a limited number of amino acid residues that are part of the enzyme's active site region. Many of the mutations that disrupt the function of other, presumably less stable proteins do so indirectly by affecting the stability of the folded protein structure; these mutations occur throughout the primary structure. Most mutations, however, have no substantial effect on the structure or activity of a protein. The nature of the mutations that are selected by the procedure, as well as those that produce proteins sufficiently normal to avoid the selection procedure, is pertinent for studying protein structure (Chaps. 6 and 7) and function (Chaps. 8 and 9).

References

Genetic analysis of protein folding pathways. J. King. *Bio/Technology* 4:297–303 (1986).

Genetic and structural analysis of the protein stability problem. B. W. Matthews. *Biochemistry* 26:6885–6888 (1987).

Genetic studies of protein stability and mechanisms of folding. D. P. Goldenberg. *Ann. Rev. Biophys. Biophys. Chem.* 17:481–507 (1988).

Genetic analysis of protein stability and function. A. A. Pakula and R. T. Sauer. *Ann. Rev. Genetics* 23:289–310 (1989).

Genetic studies of the *lac* repressor. XIII Extensive amino acid replacements generated by the use of natural and synthetic nonsense suppressors. L. G. Kleina and J. H. Miller. *J. Mol. Biol.* 212:295–318 (1990).

A genetic screen to identify variants of bovine pancreatic trypsin inhibitor with altered folding energetics. L. J. Coplen et al. *Proteins: Struct. Funct. Genet.* 7:16–31 (1990).

3.5.2 Simulation of Evolution

Many questions about protein evolution need to be answered. How have the functional and physical properties of a given protein changed during evolution? Why have certain residues been replaced and others have not? What effect did a certain amino acid replacement have on the function of the protein? Which mutations are responsible for a known change in function? By what sequence of events do the proteins of duplicated genes acquire new functions? How many amino acid replacements does it take to alter the function of a protein in a particular way?

All of these questions are open to experimental investigation. Any ancestral protein can be made with the technology of protein engineering (Sec. 2.2) if its amino acid sequence can be inferred from phylogenetic trees derived from contemporary sequences. Its functional properties can then be studied in vitro, although not in vivo in the ancestor. Mutations that are thought to have occurred during evolution can be introduced into an appropriate gene and their effects on the protein studied. Hybrid proteins from different species can be constructed and compared with the parental proteins.

Questions about the likely mechanism of evolution can also be answered by simulating it in the laboratory. Microorganisms, with their rapid growth rates and short generation times, can be made to evolve on a relatively short time scale. Mutagens will increase the mutation rate, and selection can be applied intensively. Population studies of microorganisms are usually made in a chemostat, where a growing culture is constantly diluted with fresh medium so that a steady state of growth is maintained. If a culture is not growing under optimal conditions, better-adapted mutant strains with faster growth rates will take over the population.

Several such studies have examined the alterations that occur to a protein when a nonoptimal condition, such as using a substrate that is inadequate for its normal enzyme activity, is made to limit growth of the microorganism. For example, the enzyme ribitol dehydrogenase normally catalyzes the conversion of ribitol to D-ribulose, but it can also utilize, though less efficiently, the similar sugar xylitol. The product, xylulose, can be metabolized further, so xylitol can be used as a sole carbon source by bacteria that possess ribitol dehydrogenase. The bacteria grow very slowly under these conditions because xylitol is such a poor substrate for ribitol dehydrogenase, so strains that are able to grow much faster on xylitol can be selected.

The most common mutations in such studies have simply caused an increased production of the normal enzyme, frequently by multiple duplications of its gene.

Mutated proteins with improved catalytic capabilities in the inefficient reaction have also been isolated. In some cases, a single mutation was sufficient to alter the activity whereas in others multiple mutations were necessary. In still other instances, a cryptic gene is expressed whose gene product has the required catalytic activity. Selection can also be applied under conditions in which some other property of a protein is rate-determining for growth of the organism. For example, a protein can be made more thermostable by introducing its gene into a thermophilic bacterium that requires the protein's function for growth, and then selecting for strains able to grow at elevated temperatures. In no case, however, has the evolution simulated in these situations been thoroughly characterized.

References

Evolution of a regulated operon in the laboratory. B. G. Hall. *Genetics* 101:335–344 (1982).

Isolation of a thermostable enzyme variant by cloning and selection in a thermophile. H. Liao et al. *Proc. Natl. Acad. Sci. USA* 83:576–580 (1986).

Fusion of *trpB* and *trpA* of *Escherichia coli* yields a partially active tryptophan synthetase polypeptide. C. Yanofsky et al. *J. Biol. Chem.* 262:11584–11590 (1987).

Recruitment of substrate-specificity properties from one enzyme into a related one by protein engineering. J. A. Wells et al. *Proc. Natl. Acad. Sci. USA* 84:5167–5171 (1987).

An experiment in enzyme evolution. Studies with *Pseudomonas aeruginosa* amidase. P. H. Clarke and R. Drew. *Biosci. Reports* 8:103–120 (1988).

Evolving theories of enzyme evolution. D. L. Hartl. *Genetics* 122:1–6 (1989).

The ribonuclease from an extinct bovid ruminant. J. Stackhouse et al. *FEBS Letters* 262:104–106 (1990).

Ancestral lysozymes reconstructed, neutrality tested, and thermostability linked to hydrocarbon packing. B. A. Malcolm et al. *Nature* 345:86–89 (1990).

Exercises

1. Of all the phylogenetic studies based on the sequences of genes and proteins, none have been more controversial than those concerning the origin of *Homo sapiens*. Anatomical and morphological criteria place all the great apes in a closely related group, separate from humans. However, molecular data suggest that humans and chimpanzees are more closely related than either is to the gorillas (R. L. Cann et al., *Nature* 325:31–36, 1987; M. M. Miya-

moto et al., *Science* 238:369–373, 1987; C. G. Sibley and J. E. Ahlquist, *J. Mol. Evol.* 26:99–122, 1987). What aspects of the molecular approach and what aspects of the phylogenetic relationships have contributed to this controversy?

ANSWERS

N. Barton and J. S. Jones, *Nature* 306:317–318 (1983); R. Lewin, *Science* 226:1179–1182 (1984), 229:743–745 (1985); J. Diamond, *Nature* 310:544 (1984); J. S. Jones and S. Rouhani, *Nature* 319:449–450 (1986); R. Lewin, *Science* 238:24–26 (1987); J. M. Diamond, *Nature* 332:685–686 (1988); J. Felsenstein, *Nature* 335:118 (1988); R. Holmquist et al., *Mol. Biol. Evol.* 5:201–216 (1988).

2. Ser residues are unique in that it is impossible to interconvert by single nucleotide substitutions all of the six Ser codons without having to pass through an intermediate codon that does not encode Ser (Sec. 2.1.3.d). The two groups of codons that specify Ser are UCN, where N is any of the four nucleotides, and AGY, where Y is U or C (see Fig. 2.4). It might then be expected that homologous genes for proteins with an essential Ser residue would retain only one of the two groups of codons (S. Brenner, *Nature* 334:528–530, 1988). How can it be explained that this apparently is not the case?

ANSWERS

P. S. Goldfarb, *Nature* 336:429 (1988); D. M. Irwin, *Nature* 336:429–430 (1988); E. V. Koonin and A. E. Gorbalenya, *Nature* 338:467–468 (1989).

3. Evaluate the primary structures of elephant, pig, and human growth hormones (J. D. Hulmes et al., *Intl. J. Peptide Protein Res.* 33:368–372, 1989) in view of the sizes of these three mammals.

4. Deriving phylogenetic trees from contemporary protein and gene sequences assumes that they have arisen only by the "vertical" process of gene divergence. What would be the consequences if genes were transferred "horizontally" between species?

ANSWERS

R. Lewin, *Science* 217:42–43 (1982); J. J. Hyldig-Nielsen et al., *Nucleic Acids Res.* 10:689–701 (1982); J. Landsmann et al., *Nature* 324:166–168 (1986); R. F. Doolittle et al., *J. Mol. Evol.* 31:383–388 (1990).

5. A plausible gene for an unsuspected hemoglobin polypeptide chain, designated θ1, was discovered in the region of the genome coding for the β-globin family (J. Marks et al., *Nature* 321:785–788, 1986). How is it possible to decide whether this gene is expressed and the putative polypeptide chain is functional?

ANSWER

J. B. Clegg, *Nature* 329:465–466 (1987); N. Fischel-Ghodsian et al., *Nature* 329:397 (1987); S. L. Hsu et al., *Nature* 331:94–96 (1988).

6. A useful procedure for identifying distantly related members of a gene family is to look for the conservation of just a few crucial residues (see, for example, S. K. Hanks et al., *Science* 241:42–52, 1988). What are the potential pitfalls of such an approach?

ANSWERS

S. Brenner, *Nature* 329:21 (1987); A. Bairoch and J. M. Claverie, *Nature* 331:22 (1988); D. P. Leader, *Nature* 333:308 (1988).

7. The intensely sweet protein thaumatin has been claimed to be homologous both to proteases (T. Skern et al., *Nature* 344:26, 1990) and to proteins that inhibit proteases (M. Richardson et al., *Nature* 327:432–434, 1987; B. T. C. Cornelissen et al., *Nature* 321:531–532, 1986). Evaluate the evidence and judge whether either or both claims are likely to be correct.

HINTS

M. Richardson, *Nature* 345:299 (1990).

8. On the basis of homology, a protein encoded by a gene from *Plasmodium falciparum* was concluded to be a protease of the thiol type (D. G. Higgins et al., *Nature* 340:604, 1989; see Sec. 9.3.2.b). Evaluate the evidence for this conclusion by aligning the sequences of the various proteins in the light of the known three-dimensional structures of such proteases (I. G. Kamphuis et al., *J. Mol. Biol.* 182:317–329, 1985). Into what category of protease is this protein likely to fall? What does this categorization imply about the evolution of thiol proteases and the serine codons discussed in Exercise 2?

ANSWERS

A. E. Eakin et al. and J. C. Mottram et al., *Nature* 342:132 (1989).

9. Alcohol dehydrogenase in humans has a multiplicity of forms (isozymes). Three classes of molecules can be distinguished. In one of these classes, three polypeptide chains are synthesized from three distinct genes; these polypeptide chains associate to form homo- and heterodimers. More variability is introduced because some allelic variants of each of these genes occur with high frequencies in certain populations. In particular, an allelic form of the β chain that has altered functional properties predominates in Oriental populations. Amino acid sequence analysis of the aminoethylated polypeptide chain indicated that two separate but adjacent mutations occurred at positions 46 and 47, converting the sequence -Cys-Arg- to -His-Cys- (A. Yoshida et al., *J. Biol. Chem.*

256:12430–12436, 1981). Evaluate the evidence that this sequence is correct.

VERDICT

H. Jörnvall et al., *Proc. Natl. Acad. Sci. USA* 81:3024–3228 (1984); Y. Matsuo et al., *Eur. J. Biochem.* 183:317–320 (1989).

10. Genes whose protein products are no longer used, due to changed circumstances of a species, are usually lost. The gene may no longer be expressed, the protein may become inactive, and the gene may eventually disappear from the genome. What mechanisms are likely to be involved in such changes? How could the various possibilities be tested experimentally?

ANSWER

J. M. Diamond, *Nature* 321:565–566 (1986).

11. Amino acid residues that are invariant in widely divergent homologous proteins are assumed to have been retained because of their functional importance. Yet a number of such residues have been experimentally altered in contemporary proteins by site-directed mutagenesis, and the altered proteins are found to be functional (G. J. Pielak et al., *Nature* 313:152–154, 1985; D. Valenzuela et al., *Nature* 313:698–700, 1985; D. Holzschu et al., *J. Biol. Chem.* 262:7125–7131, 1987). What is the likely explanation for this apparent paradox?

ANSWER

S. A. Sawyer et al., *Proc. Natl. Acad. Sci. USA* 84:6225–6228 (1987).

12. Within the past few decades the chemical industry has produced many useful new products and thereby has introduced novel chemicals into the environment. Yet some natural microorganisms can degrade these unnatural chemicals. What are the most likely ways that they could have acquired this capability? Is it conceivable that such metabolic activities could have arisen in the short time since the chemicals have been produced?

ANSWER

S. Ohno, *Proc. Natl. Acad. Sci. USA* 81:2421–2425 (1984); K. Tsuchiya et al., *J. Bacteriol.* 171:3187–3191 (1989).

13. Three enzyme families—trypsin proteases, subtilisins, and lipases—have catalytically important Ser residues at their active sites that have the respective sequences Gly-Asp-Ser-Gly-Gly, Gly-X-Ser-X-Ala, and Gly-X-Ser-X-Gly, where residues X are variable in various members of the family. What is the probability that these three enzyme families are evolutionarily related?

ANSWER

D. Blow, *Nature* 343:694–695 (1990).

Physical Interactions That Determine the Properties of Proteins

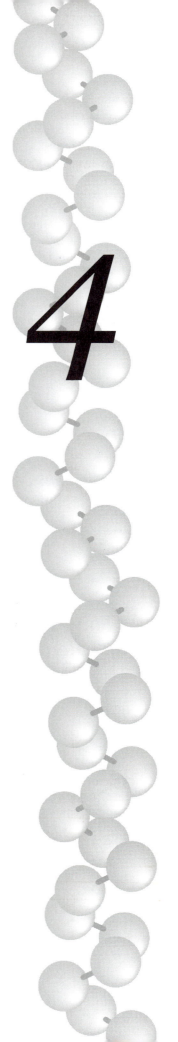

4

The first three chapters of this volume have discussed proteins only in terms of their covalent structures. Knowledge of this structure is usually adequate for characterizing the chemistry of small molecules, but not for proteins. The large size of polypeptide chains enables them to fold back on themselves so that many simultaneous interactions take place among different parts of the molecule. A complex, three-dimensional structure results, which provides the unique environments and orientations of the functional groups that give proteins their many special properties. The biological activities of proteins are also mediated by their interactions with their environment: with water, salts, membranes, other proteins, nucleic acids, and the numerous other large and small molecules in living systems. All of these interactions arise from a limited set of fundamental noncovalent forces, but with many variations on the theme, so it is important to understand the physical basis of the interactions.

The physical natures of the noncovalent interactions between atoms are understood fairly well for individual molecules in a vacuum and in a regular solid, but not in liquids. This is a consequence of the complexity of the liquid state, with its constantly changing interactions among many molecules in transient ensembles. The complexities of liquids are especially relevant to proteins because their folded conformations usually occur only in a liquid-water environment or in membranes; the latter occur as a result of the relatively poor interactions of lipids with water. In spite of water's biological importance and the many studies of it, water is not one of the best understood liquids, and our limited understanding of water limits our understanding of proteins. The most important characteristic of all forces between molecules dissolved in water is that these forces are often due more to the properties of this extraordinary solvent than to the intermolecular interactions themselves. The interactions of water with ions, dipoles, and hydrogen bond acceptors and donors are so strong as to diminish greatly most of the forces that occur among such groups in a vacuum or in a nonpolar solvent. Water produces what is commonly considered to be a unique force between nonpolar atoms, the hydrophobic interaction, which is often thought of as primarily a consequence of the strong interaction of the water molecules with each other, rather than of any direct interaction between the nonpolar, hydrophobic molecules themselves.

This chapter briefly and simply describes the types of noncovalent interactions that take place among atoms, reviews the structure and properties of liquid water, and examines the various physical interactions in a polypeptide chain in an aqueous environment.

References

Noncovalent interactions. P. A. Kollman. *Acc. Chem. Res.* 10:365–371 (1977).

Intermolecular and Surface Forces; With Applications to Colloidal and Biological Systems. J. N. Israelachvili. Academic Press, London, 1985.

Weakly polar interactions in proteins. S. K. Burley and G. A. Petsko. *Adv. Protein Chem.* 39:125–189 (1988).

Computer modeling of the interactions of complex molecules. P. A. Kollman and K. M. Merz, Jr. *Acc. Chem. Res.* 23:246–252 (1990).

4.1   The Physical Nature of Noncovalent Interactions

4.1.1   Short-Range Repulsions

The most important interaction, energetically and structurally, between atoms and between molecules is the repulsion that eventually takes place between them as they approach each other. As they come near enough for their electron orbitals to begin to overlap, the repulsion increases enormously because the electrons on the different molecules cannot be in the same part of space at the same time, as stated by the Pauli exclusion principle. The repulsive energy is often said to increase with the inverse of the 12th power of the distance between the centers of the two atoms. A more realistic description has the energy varying exponentially with the inverse of the distance, but there is little practical difference between the two descriptions.

Table 4.1   *Van der Waals Radii of Atoms Found in Proteins*

Atom	Observed range (Å)	Radius when singly bonded (Å)
Hydrogen	1.0–1.54	1.17
Oxygen	1.4–1.7	1.40
Nitrogen	1.55–1.60	1.55
Carbon	1.70–1.78	1.75
Sulfur	1.75–1.80	1.80

Values from A. Bondi, *J. Phys. Chem.* 68:441–451 (1964) and A. Gavezzotti, *J. Amer. Chem. Soc.* 105:5220–5225 (1983).

Table 4.2   *Van der Waals Surface Areas and Volumes of Chemical Groups When Bonded to Carbon Atoms*

Chemical group	Area (Å²)	Volume (Å³)
—C— (with two additional bonds)	1.0	5.5
—CH (with two additional bonds)	10.9	11.5
—CH₂—	20.9	16.8
—CH₃	33.4	22.3
Phenyl	94.9	76.1
—OH	19.3	12.6
—C(=O)—	22.3	18.2
—C(=O)—OH	43.4	
—SH		24.6
—NH₂	26.5	17.5
—NH—	16.4	13.4

Values from A. Bondi, *J. Phys. Chem.* 68:441–451 (1964) and A. Gavezzotti, *J. Amer. Chem. Soc.* 107:962–967 (1985).

Because the repulsive energy rises so steeply, it is possible to consider atoms and molecules as having definite dimensions and occupying volumes that are impenetrable to other atoms and molecules at ordinary temperatures. Individual atoms are usually modeled as

Table 4.3   *Volume Properties of Individual Amino Acid Residues*

Residue	Van der Waals volume[a] (Å^3)	Partial volume in solution[b] (Å^3)	Partial specific volume[b] (cm^3/g)
Ala (A)	67	86.4	0.732
Arg (R)	148	197.4	0.756
Asn (N)	96	115.6	0.610
Asp (D)	91	108.6	0.573
Cys (C)	86	107.9	0.630
Gln (Q)	114	142.0	0.667
Glu (E)	109	128.7	0.605
Gly (G)	48	57.8	0.610
His (H)	118	150.1	0.659
Ile (I)	124	164.6	0.876
Leu (L)	124	164.6	0.876
Lys (K)	135	166.2	0.775
Met (M)	124	160.9	0.739
Phe (F)	135	187.3	0.766
Pro (P)	90	120.6	0.748
Ser (S)	73	86.2	0.596
Thr (T)	93	113.6	0.676
Trp (W)	163	225.0	0.728
Tyr (Y)	141	190.5	0.703
Val (V)	105	136.8	0.831
Weighted average[c]			0.703

[a] Volume enclosed by van der Waals radius.

[b] Increase in volume of water after adding either one molecule or one gram of residue (A. A. Zamyatnin, *Ann. Rev. Biophys. Bioeng.* 13:145–165, 1984.)

[c] Weighted by frequency of occurrence in proteins, to give the value for an average residue in globular proteins.

spheres, and their impenetrable volumes are usually defined by the **van der Waals radius.** Values of van der Waals radii are usually measured from the smallest distances that can exist between neighboring, but not covalently bonded, atoms in the crystalline state; this distance is the sum of their respective van der Waals radii. Some typical values of van der Waals radii are given in Table 4.1. A range of values is given because the observed radius depends on the way in which the atom is covalently bonded. For example, the van der Waals radius of a hydrogen atom varies from 1.0 Å when bonded to an aromatic carbon atom to 1.54 Å when bonded to a negative ion. Fortunately, these are extreme variations, and two atoms are generally in close van der Waals contact when the distance between them is approximately 0.8 Å greater than when they are covalently bonded. The van der Waals radius is a minimal estimate of the size of an atom or molecule. Optimal van der Waals interactions (Sec. 4.1.3) generally occur at a distance that is about 1.2 Å greater than the covalent bond length.

The van der Waals radius also defines the van der Waals surface area and volume of an atom or molecule; some pertinent values for various atoms and chemical groups are given in Table 4.2. The volumes and surface areas of entire molecules can be estimated by summing the parameters for the constituent parts, so long as the molecule is not structurally strained. The van der Waals volumes of the amino acid residues are given in Table 4.3.

The van der Waals surface of a molecule is not especially relevant chemically, because it inevitably includes many nooks and crannies between atoms that are not accessible to other atoms or molecules that may be present. A more practical concept of surface is that which is defined by the solvent molecules in contact with the molecule, the **accessible surface area.** It is generally described by the center of a solvent molecule of

Table 4.4 *Accessible Surface Areas of Amino Acid Residues in a Gly-X-Gly Tripeptide in an Extended Conformation*

Residue	Total ($Å^2$)	Main-chain atoms ($Å^2$)	Side-Chain Atoms ($Å^3$)		
			Total	Nonpolar	Polar
Ala	113	46	67	67	
Arg	241	45	196	89	107
Asn	158	45	113	44	69
Asp	151	45	106	48	58
Cys	140	36	104	35	69
Gln	189	45	144	53	91
Glu	183	45	138	61	77
Gly	85	85			
His	194	43	151	102	49
Ile	182	42	140	140	
Leu	180	43	137	137	
Lys	211	44	167	119	48
Met	204	44	160	117	43
Phe	218	43	175	175	
Pro	143	38	105	105	
Ser	122	42	80	44	36
Thr	146	44	102	74	28
Trp	259	42	217	190	27
Tyr	229	42	187	144	43
Val	160	43	117	117	

From S. Miller et al, *J. Mol. Biol.* 196:641–656 (1987).

radius 1.4 Å, representative of a water molecule, in van der Waals contact with the molecule (see Sec. 6.2.5). Estimated values for the accessible surface areas of the amino acid residues are given in Table 4.4.

References

Van der Waals volumes and radii. A. Bondi. *J. Phys. Chem.* 68:441–451 (1964).

Crystal environments and geometries of leucine, isoleucine, valine and phenylalanine provide estimates of minimum nonbonded contact and preferred van der Waals interaction distances. R. O. Gould et al. *J. Amer. Chem. Soc.* 107:5921–5927 (1985).

The size of molecules. A. Y. Meyer. *Chem. Soc. Rev.* 15:449–474 (1986).

Properties of atoms in molecules: atomic volumes. R. F. W. Bader et al. *J. Amer. Chem. Soc.* 109:7968–7979 (1987).

Calibration of effective van der Waals atomic contact radii for proteins and peptides. H. Iijima et al. *Proteins: Struct. Funct. Genet.* 2:330–339 (1987).

4.1.2 Electrostatic Forces

a. Point Charges

All intermolecular forces are thought to be essentially electrostatic in origin, and the most fundamental noncovalent attraction is that between electrostatic charges. Coulomb's law states that the energy of the electrostatic interaction between two atoms A and B in a vacuum is simply the product of their two charges divided by the distance between them r_{AB}:

$$\Delta E = \frac{Z_A Z_B \epsilon^2}{r_{AB}} \tag{4.1}$$

where ϵ is the charge of an electron and Z the number of such charges on each atom. The energy ΔE is expressed relative to the energy when the two charges are very far apart.

If the two charges are of opposite sign, the energy decreases as they approach each other, and the interaction is favorable; if the charges are of the same sign, there is repulsion between them. Because the interaction varies inversely with only the first power of the distance, it is effective over relatively large distances; at twice the distance, the interaction energy is only halved. It is also a very strong interaction. The attractive interaction between an isolated Na^+ ion in optimum contact with a Cl^- ion, 2.76 Å apart, has the value 2.0×10^{-19} calorie, which is equivalent to 120 kcal/mol (502 kJ/mol).

Coulomb's law (Eq. 4.1) is valid only for two point charges in a vacuum. For other environments, the electrostatic interaction is modulated by other interactions. In homogeneous environments, the electrostatic interaction is decreased by the **dielectric constant D**:

$$\Delta E = \frac{Z_A Z_B \epsilon^2}{D r_{AB}} \qquad (4.2)$$

The values of dielectric constants are invariably greater than unity, so the electrostatic interaction in media other than a vacuum is always less than that stated by Coulomb's law. This is especially important to consider when dealing with liquids, which usually have dielectric constants in the range of 2–110 (that of water is about 80). The concept of a dielectric constant is valid only for homogeneous liquids; less homogeneous environments must be treated explicitly. Special problems arise at interfaces between regions with very different bulk dielectric constants, such as between an aqueous solvent and a folded protein. Two charges on opposite sides of a sphere of low dielectric constant immersed in a medium of high dielectric constant interact not through the shortest distance of low dielectric, but around the outside of the sphere. The very long path length through the high dielectric medium results in the energy of this electrostatic interaction being much smaller than might otherwise be expected.

Because Coulomb's law ignores the finite sizes of ions, it is valid only at distances that are significantly greater than atomic dimensions. The charge of an atom is separated between the nucleus and the diffuse electron cloud, so it cannot be treated as a point charge at short distances. This problem of charge separation is even more severe with proteins, in which the charges of the ionized groups of the side chains of Lys, Arg, His, Asp, and Glu residues, plus the α-amino and α-carboxyl groups, are distributed over two or more hydrogen or oxygen atoms. Consequently, interactions between very close, oppositely charged groups in a protein, known as **salt bridges,** usually consist not only of electrostatic interactions but also of some degree of hydrogen bonding (Fig. 4.1).

The charged groups in polypeptide chains titrate in accessible pH regions, so electrostatic effects in proteins invariably involve changes in their ionization

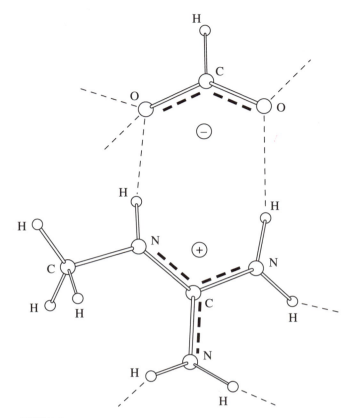

FIGURE 4.1

Hydrogen bonds involved in the interaction between ionized carboxyl and guanidinium groups, illustrated by the crystal structure of methylguanidinium formate. The thin dashed lines indicate hydrogen bonds assumed to be present from the short distances between the adjacent atoms. The thick dashed lines indicate partial double bonds arising from resonance. The distances between the nitrogen and oxygen atoms of the two complete hydrogen bonds shown are 2.81 Å *(left)* and 2.89 Å *(right)*. (Adapted from D. D. Bray et al., *Int. J. Peptide Protein Res.* 24:414–418, 1984.)

tendencies, or pK_a values (see Table 1.2). Favorable electrostatic interactions increase the tendency of any group to ionize. Other factors, however, such as the accessibility to the solvent and the solvent's polarity, also affect the pK_a value. For example, adding dioxane to decrease the polarity of the solvent inhibits ionization of accessible amino and carboxyl groups (Table 4.5). Ionization also becomes less favorable with increasing bulkiness of the surrounding aliphatic groups (Table 4.6).

b. Dipoles

A molecule need not have a net charge to participate in electrostatic interactions because electron density can

Table 4.5 *Effect of Nonaqueous Environment on the pK_a Values of Amino and Carboxyl Groups*

	pK_a Values for Various Wt % Dioxane in Water				
Acid or base	0	20	45	50	70
Acetic acid	4.76	5.29	6.31		8.34
$(HOCH_2)_3C-NH_2$	8.0	8.0	8.0	8.0	8.0
Benzoylarginine	3.34			4.59	4.60
Glycine $\{$ $-CO_2H$	2.35	2.63	3.11		3.96
$-NH_2$	9.78	9.29	8.49		7.42

From A. Fersht, *Enzyme Structure and Mechanism.* W. H. Freeman, Reading, England, 1977.

Table 4.6 *Steric Effects on the Ionization of Carboxyl Groups*

Model compound	pK_a[a]
H_3C-CO_2H	5.55
CH₃—C(CH₃)(CH₃)—C(H)—CO₂H	6.25
$CH_3-CH_2-C(CH_2CH_3)(CH_2CH_3)-CO_2H$	6.44
$CH_3-C(CH_3)(CH_3)-C(CH_3)(CH_3)-CO_2H$	6.71
complex branched compound	6.97

[a] The pK_a values were measured in equal volumes of methanol and water at 40°C by G. S. Hammond and D. H. Hogle, *J. Amer. Chem. Soc.* 77:338–340 (1955).

be localized if the atoms have different electronegativities. Atoms with the greatest electronegativities have an excess of negative charge, $\delta-$, others an excess of positive charge, $\delta+$. The electronegativities of the atoms in proteins are O, 3.45; N, 2.98; C, 2.55; S, 2.53; and H, 2.13. The importance of these values is illustrated by the peptide bond, which has partially double-bonded character due to resonance with a form in which the more electronegative oxygen atom acquires a net partial negative charge, whereas the $-NH-$ group is somewhat positively charged.

$$\underset{C^\alpha}{\overset{O}{\|}}C-N\underset{H}{\overset{C^\alpha}{}} \rightleftharpoons \underset{C^\alpha}{\overset{O^-}{}}C=N\underset{H^+}{\overset{C^\alpha}{}} \quad (4.3)$$

The double-bonded charged species is populated about 40% of the time, so the peptide group can be represented as having partial charges of up to $\pm 0.4\epsilon$. Polar oxygen and nitrogen atoms in other molecules have partial charges as great as $\pm 0.35\epsilon$, but those in aliphatic amino acid side chains are probably no greater than $\pm 0.1\epsilon$.

The π electrons in the aromatic rings of Phe, Tyr, and Trp residues are localized above and below the face of the ring. This excess of electrons gives the face of the ring a small net negative charge of approximately -0.15ϵ, whereas the hydrogen atoms on its edge have a corresponding positive net charge. The electrostatic interactions between these partial charges dominate the interactions between aromatic rings. Such rings preferentially interact with the positively charged edge of one ring pointing at the negatively charged face of another, or with their rings parallel but offset so that the edge of each ring is interacting with the face of the other. The commonly held impression that aromatic rings interact favorably by stacking their rings one above the other is incorrect. Electronegative oxygen and sulfur atoms

Electrostatic interaction b/w aromatic rings

tend to interact favorably with the edges of aromatic rings, —NH groups with their faces.

The separation of charge in a molecule determines its **dipole moment** μ_D, which is given by the product of the magnitude of the separated excess charge Z and the distance d by which it is separated:

$$\mu_D = Zd \qquad (4.4)$$

One electron unit of positive and negative charge separated by 1 Å has a dipole moment of 4.8 Debye units (D). The dipole moment of a peptide bond is 3.5 D and that of a water molecule 1.85 D. The dipole moment has direction as well as magnitude and is usually depicted as a vector along a straight line from the negative to the positive charge. The dipole moment of the peptide bond can be represented as

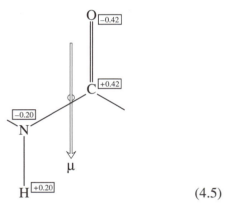

$$(4.5)$$

Dipoles interact with point charges, with other dipoles, and with more complex charge separations known as quadrupoles, octupoles, and so on, in a complex manner that is determined by the relative orientations of the two interacting groups. The interactions can be computed by considering the individual charges of the groups, including those of the dipole, quadrupole, and so on, and calculating the coulombic interactions among all of them. The interactions among the four partial charges of two dipoles are analogous to those between two bar magnets. Two side-by-side dipoles repel each other when parallel, whereas there is an equivalent attraction between them when they are antiparallel. Maximum interactions, repulsive or attractive, occur in a head-to-tail orientation. Dipolar interactions are, however, weaker than those between ions because both attractions and repulsions occur between the two separated charges of each of the dipoles. This combination of attraction and repulsion also has the effect of making the energy of the interactions depend inversely on the second to third power of the distance between interacting molecules when they are in fixed orientations, and on the sixth power when they are free to rotate in response to the interaction. Consequently, electrostatic interactions between dipoles fall off much more abruptly as the distance between them increases than do interactions between point charges; the interactions are, however, modulated in the same way by the dielectric properties of the medium.

Interactions involving dipoles also modify the nature of the dipole charge distribution in the interacting molecules. Being simply an unequal distribution of electrons, that distribution is easily perturbed. For example, a nearby charged group induces a dipole moment even in a spherical molecule. The tendency of the charge distribution of a molecule to be altered by an electric field is called the electronic **polarizability** of the molecule, the value of which depends primarily on how tightly the electrons are held by the nuclei; in general, the larger an atom, the greater its polarizability. The importance of polarizability is that the induced dipole always interacts favorably with the field that induced it, so there is an attraction between them. The energy of this interaction is only half that which would have occurred if the dipole preexisted because some of the energy of interaction must be used in inducing the dipole.

The multiple interactions that take place among the point charges and dipoles on a number of atoms are mutually dependent and turn the very simple relationship of Coulomb's law into a very complex phenomenon. The electrostatic interactions among molecules in a homogeneous liquid can be averaged and expressed as a simple dielectric constant of the liquid. This concept is not valid, however, when the environment is not homogeneous at the molecular level, as is always the case for proteins, whose electrostatic interactions invariably involve interactions among the multiple charges and dipoles of the protein, and between these and the bulk solvent and any ions in it. In this case, interactions between individual charges and dipoles of the protein must be calculated directly. This is impractical with the many molecules of the solvent, and there is still considerable controversy about how to analyze and calculate electrostatic interactions in proteins.

References

Calculations of electrostatic interactions in biological systems and in solutions. A. Warshel and S. T. Russell. *Quart. Rev. Biophys.* 17:283–422 (1984).

Electrostatic effects in proteins. J. B. Matthew. *Ann. Rev. Biophys. Bioeng.* 14:387–417 (1985).

Aromatic–aromatic interactions: a mechanism of protein structure stabilization. S. K. Burley and G. A. Petsko. *Science* 229:23–28 (1985).

The modelling of electrostatic interactions in the function of globular proteins. N. K. Rogers. *Prog. Biophys. Mol. Biol.* 48:37–66 (1986).

Treatment of electrostatic effects in macromolecular modeling. S. C. Harvey. *Proteins: Struct. Funct. Genet.* 5:78–92 (1989).

The nature of $\pi-\pi$ interactions. C. A. Hunter and J. K. M. Sanders. *J. Amer. Chem. Soc.* 112:5525–5534 (1990).

Electrostatic interactions in macromolecules: theory and applications. K. A. Sharp and B. Honig. *Ann. Rev. Biophys. Biophys. Chem.* 19:301–332 (1990).

4.1.3 Van der Waals Interactions

All atoms and molecules attract each other, even in the absence of charged groups, as a result of mutual interactions related to the induced polarization effects described in the preceding section. These ubiquitous attractions, known as **van der Waals interactions,** are weak and close-range, varying as the sixth power of the distance between them, d^{-6}. They arise from three types of interactions: those between two permanent dipoles, those between a permanent and an induced dipole, and those between two mutually induced dipoles, known as **London** or **dispersion forces.** The first two types have been described, but the third is the most important because it occurs among all atoms and molecules.

The dispersion force is complex because it is quantum mechanical in nature. A greatly simplified description can be derived from the classical picture of an atom with electrons orbiting the nucleus. Such a spherical atom has no net dipole moment, but it can have a transient dipole moment resulting from a temporarily asymmetric orientation of the electrons and nucleus. This transient dipole polarizes any neutral atom nearby, creating an attraction between them. Although the transient dipole of the first atom constantly and rapidly changes, that of the other atom tends to follow it, and the two are correlated. This dispersion force is basically electrostatic in nature and varies as d^{-6}, as do the other two components of van der Waals interactions. This distance dependence breaks down at distances greater than about 50 Å, however, because the correlation between the two electron distributions diminishes as the time it takes the field from one atom to reach the other increases.

Van der Waals interactions are often represented by an energy potential as a function of distance d that includes both the attractive force and the repulsion at close range. The most well known of these is the **Lennard–Jones potential** of the form:

$$E(d) = \frac{C_n}{d^n} - \frac{C_6}{d^6} \qquad (n > 6) \qquad (4.6)$$

where C_n and C_6 are constants. The first term gives the repulsions, the second the van der Waals attractions. The most common potential has $n = 12$, which is

known as the Lennard–Jones 6,12 potential and is computationally efficient because $12 = 2 \times 6$ (Fig. 4.2).

The optimal distance for the interaction of two atoms, given by the minimum in Figure 4.2, is usually 0.3–0.5 Å greater than the sum of their van der Waals radii as measured from the closest contact distance in crystals (Table 4.1). The van der Waals radius is given by the steeply ascending repulsive interaction at closer distances. The van der Waals interaction is generally considered to be independent of the orientation of the interacting molecules, but this is only approximately true, when the interacting molecules are independent and tumbling rapidly in a gas or a liquid. Otherwise, the magnitude of the interaction of even a nonpolar group

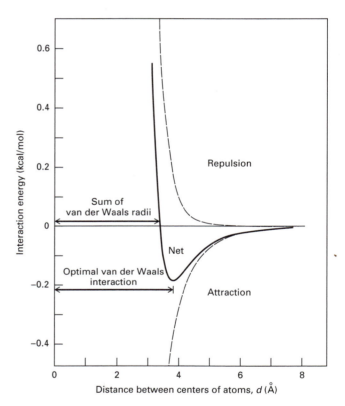

FIGURE 4.2

Representative profile of the energy of the van der Waals interaction as a function of the distance d between the centers of the two atoms. The individual attractive and repulsive components are indicated by the dashed lines, the net interaction by the solid line. The optimal interaction between the two atoms occurs where the energy is at a minimum. The sum of the van der Waals radii of the two atoms is given by the distance at which the energy increases sharply. The interaction energy was calculated using the Lennard–Jones 6,12 potential (Eq. 4.6) with $C_{12} = 2.75 \times 10^6$ Å^{12} kcal/mol and $C_6 = 1425$ Å^6 kcal/mol for the interaction between two carbons atoms (M. Levitt, *J. Mol. Biol.* 82:393–420, 1974).

like —CH₃ can vary with orientation because the polarizability of a C—H bond is nearly twice as great along the bond as perpendicular to it.

References

Van der Waals forces in biological systems. J. N. Israelachvili. *Quart. Rev. Biophys.* 6:341–387 (1973).

Van der Waals picture of liquids, solids and phase transformations. D. Chandler et al. *Science* 220:787–794 (1983).

Theoretical studies of van der Waals molecules and intermolecular forces. A. D. Buckingham et al. *Chem. Rev.* 88:963–988 (1988).

4.1.4 Hydrogen Bonds

A hydrogen bond occurs when two electronegative atoms compete for the same hydrogen atom:

$$—D—H\cdots A— \qquad (4.7)$$

The hydrogen atom is formally bonded covalently to one of the atoms, the *donor* D, but it also interacts favorably

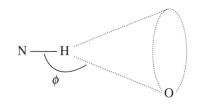

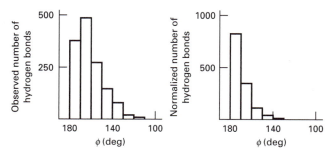

FIGURE 4.3

Linearity of N—H $\cdots$ O hydrogen bonds observed in crystal structures of small molecules. The degree of linearity is measured by the angle ϕ *(top)*, which would have a value of 180° for a perfectly linear hydrogen bond. The histogram of observed values of intermolecular hydrogen bonds *(left)* is affected by a geometric factor in which the various ranges of the value of ϕ include different volumes of three-dimensional space. Correcting for this gives the histogram on the *right*, which illustrates the marked tendency of hydrogen bonds to be linear. (Adapted from R. Taylor and O. Kennard, *Acc. Chem. Res.* 17:320–326, 1984.)

with the other, the *acceptor* A. In a few strong, short hydrogen bonds, the hydrogen atom is symmetrically placed between the two electronegative atoms, but usually it is covalently bonded to one of the atoms, with a normal covalent bond length.

The main component of the hydrogen bond is an electrostatic interaction between the dipole of the covalent bond to the hydrogen atom, in which the hydrogen atom has a partial positive charge, and a partial negative charge on the other electronegative atom:

$$—D^{\delta-}—H^{\delta+}\cdots^{\delta-}A— \qquad (4.8)$$

The hydrogen atom is special in being able to interact strongly with one electronegative atom while being covalently attached to another. It can do this because of its small size and its substantial charge, which results from its tendency to be positively polarized. In strong hydrogen bonds, an additional covalent aspect arises from a transfer of electrons. The electrostatic and covalent aspects of the hydrogen bond cause the most common and, presumably, most energetically favorable hydrogen bonds to keep the three bonded atoms collinear (Fig. 4.3). There is considerable uncertainty, however, about how the strength of the hydrogen-bond interaction varies with departures from linearity.

Oxygen atoms are frequently observed to participate simultaneously as acceptors in two hydrogen bonds:

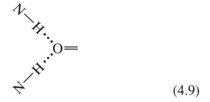

$$(4.9)$$

The partial negative charge at the electronegative acceptor atom such as an oxygen atom is localized on the two lone-pair electron orbitals (Sec. 4.2.2), and an intrinsic preference for one hydrogen atom to point toward each of these orbitals might be expected. Hydrogen bonds between two molecules in the gas phase show this geometry, and it is also frequently observed in crystals with carbonyl oxygen acceptors (Fig. 4.4). It is not observed so frequently in other circumstances, however, and the energetic preference for hydrogen atoms to be directed at the lone-pair electrons is probably so small as to be easily overwhelmed by other considerations. Much less frequently, single hydrogen atoms are shared between two acceptors.

The lengths and strengths of hydrogen bonds depend on the electronegativities of the acceptor and donor; the greater their electronegativities, the shorter the distance between them and the stronger the hydrogen bond. Charged groups also give shorter and stronger hydrogen bonds (Table 4.7). Hydrogen bonds in proteins most frequently involve the C=O and N—H groups of the polypeptide backbone. In this type of hydrogen bond, the H ··· O distance is most often 1.9–2.0 Å. The hydrogen atom is generally not observed directly in protein crystal structures, however (see Chap. 6), so hydrogen-bond distances in proteins are usually expressed as the distance between the donor and acceptor atoms. The covalent N—H distance is

1.03 ± 0.02 Å, so a typical N—H ··· O=C hydrogen bond has the nitrogen and oxygen atoms 3.0 Å apart.

The strengths of hydrogen bonds are generally said to be within the rather broad range of 2–10 kcal/mol at room temperature. Part of this variation is due to the variety of hydrogen bonds, but much is also a result of uncertainty. The chemical groups in proteins that most commonly serve as hydrogen-bond donors are N—H, O—H, and, much less frequently, S—H and C—H groups. The most common acceptors are O=, —O—, —N=, and, much less frequently, —S⁻, —S—, and the π electrons of aromatic groups.

References

Hydrogen bonding. G. C. Pimentel and A. L. McLellan. *Ann. Rev. Phys. Chem.* 22:347–385 (1971).

The origin of hydrogen bonding. An energy decomposition study. H. Umeyama and K. Morokuma. *J. Amer. Chem. Soc.* 99:1316–1332 (1977).

Hydrogen-bond geometry in organic crystals. R. Taylor and O. Kennard. *Acc. Chem. Res.* 17:320–326 (1984).

Directional character, strength, and nature of the hydrogen bond in gas-phase dimers. A. C. Legon and D. J. Millen. *Acc. Chem. Res.* 20:39046 (1987).

Hydrogen bond stereochemistry in protein structure and function. J. A. Ippolito et al. *J. Mol. Biol.* 215:457–471 (1990).

4.2   The Properties of Liquid Water and the Characteristics of Noncovalent Interactions in This Solvent

The preceding discussion of intermolecular interactions concentrated on interactions between pairs of molecules in a vacuum, where the nature of the interaction is relatively straightforward. In condensed media, liquids, solids, and macromolecules, numerous atoms and molecules are interacting simultaneously and usually inducing alterations in each other, so the exact mathematical treatment of interactions becomes much more problematic. For example, involvement of an O—H group as donor in a hydrogen bond increases the negative charge on the donor oxygen atom, so it becomes a better hydrogen-bond acceptor in a second hydrogen bond. Van der Waals interactions among atoms affect their polarizabilities, and the magnitude of the van der Waals interaction between two molecules is about 30% greater when they are part of a liquid than when isolated in a vacuum.

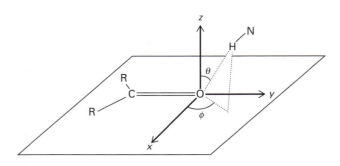

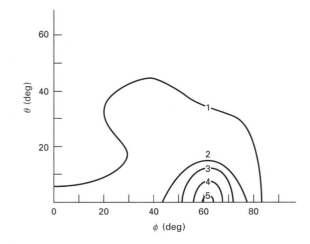

FIGURE 4.4

The geometries of C=O ··· H—N hydrogen bonds observed in crystal structures of small molecules. The definitions of the angles ϕ and θ are illustrated at the *top*, and the relative frequencies of their observed values in intermolecular hydrogen bonds (R. Taylor et al., *J. Amer. Chem. Soc.* 105:5761–5766, 1983) are given by the contours. The angle ϕ measures departures from linearity of the C=O bond and the H atom; the most frequently observed values are in the region of 50°–60°. The angle θ measures the extent to which the H atom lies out of the plane defined by the R, C, and O atoms; the most commonly observed values are in the region of 0°–7°. The lone-pair electrons of the oxygen atom are believed to project at angles of $\phi = 60°$, $\theta = 0°$. The spherical polar coordinate system used here gives a bias toward small values of θ that could be corrected by plotting sin θ.

Table 4.7   *Lengths of* H—N ··· O=C *hydrogen bonds[a]*

Donor	Mean H ··· O Distance for Different Acceptors (Å)		
	Carboxyl[b]	Carboxylate[c]	Amide
$\diagdown$ N—H[d] $\diagup$	2.002 ± 0.012	1.928 ± 0.012	1.934 ± 0.005
$\diagdown$ N$^+$—H[e] $\diagup$	1.983 ± 0.055	1.869 ± 0.028	1.858 ± 0.043
NH_4^+	1.916 ± 0.041	1.886 ± 0.018	1.988 ± 0.075
$R—NH_3^+$	1.936 ± 0.014	1.841 ± 0.008	1.891 ± 0.034
$R_2—NH_2^+$	1.887 ± 0.047	1.796 ± 0.014	1.793 ± 0.070
$R_3—NH^+$		1.722 ± 0.025	1.845 ± 0.014

[a] The N—H distance is generally 1.03 Å; adding this value to the tabulated distances gives the distance between the N and O atoms.

[b] C=O oxygen atom of unionized carboxylic acids and esters.

[c] Oxygen atom of carboxyl anions ($—CO_2^-$).

[d] Uncharged donor.

[e] Charged donor with trigonal geometry.

From R. Taylor and O. Kennard, *Acc. Chem. Res.* 17:320–326 (1984).

4.2.1   Liquids

Liquids have no clearly defined structure and are further complicated by the many interactions occurring simultaneously among neighboring molecules. A liquid is usually lighter than the corresponding solid by 5–15%, indicating increased distances and decreased interactions between molecules in a liquid. Liquids usually have similar volume densities (the ratio of the van der Waals volume of a molecule to the average total volume occupied by it in the liquid) between 0.48 and 0.61. These densities barely change with pressure, indicating that the molecules are generally in van der Waals contact with their neighbors; for example, doubling atmospheric pressure generally decreases the volume of a liquid by only 0.01%. On the other hand, the liquid volume generally increases by about 0.1% for each 1°C rise in temperature, so the molecules are also fluctuating substantially.

The best experimental description of liquid structure comes from the scattering of X rays or neutrons, which yields a **radial distribution function g(r)** that gives the symmetrically averaged density of atoms as a function of radial distance from the center of one molecule, *r*, relative to the bulk density of the liquid (Fig. 4.5). The value of $g(r)$ is zero at $r = 0$, and its value becomes substantial when *r* approaches twice the van der Waals radius of the molecule. At about this distance, $g(r)$ gen-

erally reaches a maximum, with a value indicating that, in a regular liquid consisting of nearly spherical molecules, 9–11 nearest-neighbor molecules are packed in nearly van der Waals contact around the central molecule. At somewhat greater values of *r*, the value of $g(r)$ drops to a minimum, indicating that there are few spaces for molecules to penetrate the first shell of nearest neighbors. The density again increases to another maximum when *r* is just under two molecular diameters, corresponding to the second shell of neighboring atoms. The second maximum is markedly lower than the first, indicating that the structural order is diminishing with increasing distance. This trend continues, and a third shell may be apparent, but the atom density rapidly approaches that of the bulk liquid. The lower the temperature, the greater the degree of order in liquids unless the volume is kept constant by decreasing the pressure. More detailed descriptions of liquids come from numerical simulations, but their validity depends on that of the model used for the calculations.

Current models depict liquids as consisting of close-packed, hard-sphere molecules in which the packing is both irregular and constantly changing. The shapes of molecules and the harsh repulsive forces between them largely determine the properties of a liquid. The structure of liquid argon is well represented by a box of marbles, that of liquid benzene by a box of small O-rings. Although the attractive interactions between

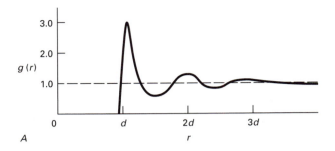

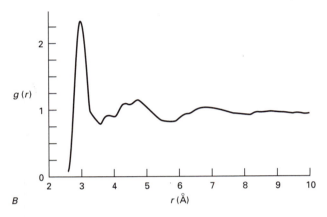

FIGURE 4.5
Radial distribution function of a normal liquid *(A)* and of water at 4°C *(B)*. The probability of finding other atoms at radial distance *r* from a central molecule is given by *g(r)* times the density of the liquid. The distance *r* is expressed in *A* in terms of the van der Waals diameter of the molecule, *d*. The experimental curve for water measured using X rays is given in *B*; X rays are scattered primarily by the oxygen atom of the water molecule. (From A. H. Narten and H. A. Levy, *J. Chem. Phys.* 55:2263–2269, 1971.)

molecules stabilize the liquid phase, they play a minor role in determining its structure unless they include hydrogen bonds or strong ionic interactions, which is the case with the biologically most important liquid, water.

References

The structure of liquids. J. D. Bernal. *Proc. Roy. Soc. London* A280:299–322 (1964).

Structures of molecular liquids. D. Chandler. *Ann. Rev. Phys. Chem.* 29:441–471 (1971).

4.2.2 Water

The H_2O molecule is unique in having the same number of hydrogen atoms and lone-pair electron acceptors. As a result, hydrogen bonding is preeminent. The H_2O molecule has a bent geometry, with an O—H bond length of 0.957 Å and a bond angle of 104.5°:

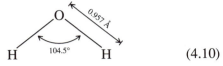

(4.10)

With van der Waals radii of 1.2 Å and 1.45 Å for the hydrogen and oxygen atoms, respectively, the molecule has a volume of 17.7 Å³. The molecule in this stick representation appears very asymmetric, but oxygen has eight electrons and a share of the single electron of both hydrogens, so the electronic structure of the molecule is nearly spherical (Fig. 4.6); it is often represented as a sphere of radius 1.4 Å. The net charge is distributed asymmetrically, however, with excess electrons on the more electronegative oxygen atom (Fig. 4.6). The molecule has a dipole moment of 1.85 D units, and each O—H bond can be considered to have 33% ionic character.

The predominance of hydrogen bonding in determining the properties of water is amply illustrated by the structure of ice (Fig. 4.7). This structure gives the impression of being determined exclusively by the four hydrogen bonds in which each H_2O molecule participates, two as hydrogen donor and two as acceptor. The angle between the hydrogen atoms in the water molecule (104.5°) is very close to the ideal for tetrahedral packing, 109.5°. Due to the hydrogen bonding, the crystal structure of ice is much more open than other crystals because each molecule of ice has only four immediate neighbors, instead of the usual 12 in crystalline closest-packing of spheres. Only 42% of the volume of ice is filled by the van der Waals volume of the molecules, rather than the 74% observed in spherical close-packing. (The hydrogen bonds of ice decrease the van der Waals volume of each water molecule by 4.0 Å³. If this were not taken into account, the density of ice would suggest a packing volume of 54%.) The four hydrogen-bonded water molecules have their oxygen atoms 2.76 Å from the central oxygen atom, with the next nearest neighbors 4.5 Å away. The hydrogen bonds are shorter in ice than between isolated H_2O dimers (2.98 Å) because of cooperativity; participation of the hydrogen atom in a hydrogen bond as a donor causes the oxygen atom to be much more effective as an acceptor in a second hydrogen bond, and vice versa. There may also be a contribution from the order of the crystal, with so many hydrogen bonds being present simultaneously (see Sec. 4.4).

The partial negative charge on the oxygen atom is frequently described as being localized primarily on two lone-pair electrons that effectively project above and below the plane of the molecule, giving water a tetrahedral structure. Hydrogen bonding in ice and in H_2O

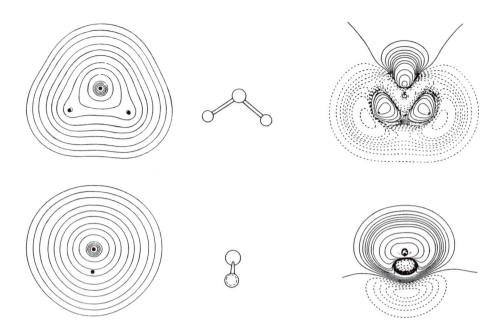

FIGURE 4.6

The electronic structure of the water molecule, shown as contour maps of electron density through the center of the molecule, viewed from two perpendicular angles. At *left* is shown the total electron density, illustrating the nearly spherical shape of the molecule. At *right* is shown the difference between this total electron density of the molecule and the density that would result from the superposition of individual spherical atoms. This gives the effect of covalent bonding on the electron density. The shift of electrons to the oxygen atom is indicated by the positive electron density (solid curves) on the O, the negative electron density (dashed lines) on the hydrogen atoms. (Adapted from I. Olovsson, *Croatica Chem. Acta* 55:171–190, 1982.

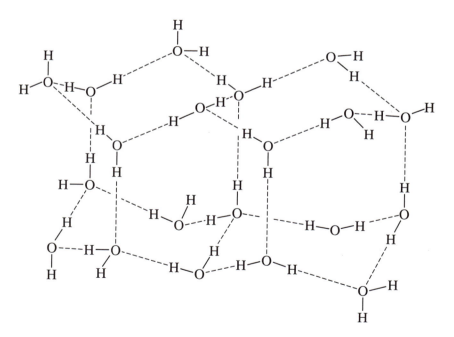

FIGURE 4.7

Structure of normal ice. Each H_2O molecule is involved in four hydrogen bonds (thin, dashed lines), each 2.76 Å between oxygen atoms. The water molecule is donor in two hydrogen bonds, acceptor in the other two. Substantial empty channels run between the molecules.

dimers in the gas phase directs the hydrogen atoms toward the positions of the lone-pair electrons, but this geometry is now thought to result primarily from repulsions between the hydrogen atoms. The lone-pair electrons on the oxygen atom are smeared out between the tetrahedral positions, lying in the plane bisecting the molecule (Fig. 4.6). There is probably only a slight energetic preference for a tetrahedral arrangement of hydrogen bonds, which in ice probably results primarily from packing considerations.

Although our understanding of the structure of liquid water is still somewhat uncertain, the unusual properties of water are well known, particularly its anomalous density change with temperature. The radial distribution function $g(r)$ of water shows substantial differences from those of other liquids (see Fig. 4.5). Measured with X rays, $g(r)$ reveals primarily the relative positions of the oxygen atoms. The value of $g(r)$ is zero for r less than 2.5 Å but rises to a maximum at $r = 2.82$ Å at low temperatures; r at this maximum is just slightly greater than the distance between hydrogen-bonded neighbors in normal ice, and this distance increases somewhat at higher temperatures, up to 2.94 Å at 200°C. The number of nearest neighbors is approximately 4.4, greater than the 4.0 in ice but less than in most liquids. A second maximum of $g(r)$ occurs at 3.5 Å, but this peak is now thought to be an artifact of the diffraction analysis. The next maximum is at about 4.5 Å, which corresponds to the distance between pairs of oxygen atoms that are hydrogen bonded in ice to the same water molecule. Another maximum occurs at about 7 Å, which would be the next nearest neighbor in ice, but there is little evidence for more order. The maxima at 4.5 and 7 Å largely disappear at temperatures greater than 50°C, indicating a breakdown of even the local structure, and the number of nearest neighbors increases to about 5 at 100°C. Neutron diffraction analysis reveals the relative positions of the hydrogen atoms. They are not fixed beyond about 5 Å, suggesting that the water molecules rotate freely about one hydrogen bond—in contrast to ice, in which the tetrahedral crystal lattice keeps them fixed.

The X-ray and neutron-scattering data confirm the importance of hydrogen bonding in the structure of water, but how a disordered liquid state can exist is not clear. Part of water's structural disorder probably arises from flexibility in the water molecule itself, both in variations of O—H bond lengths and in an average variation of about ±15° in the normal H—O—H bond angle of 104.5°. The local tetrahedral arrangement of four hydrogen-bonded near neighbors of ice appears to persist in the liquid but with a fifth neighbor frequently present. Although in ice the rotation about the hydrogen bond is limited to one of three angles by the crystal

lattice, many orientations of neighboring molecules probably exist in the liquid, so the relative positions of hydrogen atoms on neighboring molecules are not well defined.

Many models of liquid water are built on the assumption that all the molecules are hydrogen bonded all of the time, but with a great variety of hydrogen-bond geometries and energies. In other models, each group is hydrogen bonded only a fraction of the time. Some models incorporate the experimental evidence that water is a mixture of two states in equilibrium: one state with relatively low enthalpy, low entropy, and large volume, similar to hydrogen-bonded ice; the other state with relatively high enthalpy, high entropy, and small volume, analogous to a normal liquid with much less hydrogen bonding. At present, no particular model is obviously more realistic than the others.

The strong hydrogen bonding among water molecules is clearly the basic explanation for many of the peculiar properties of this solvent. Hydrogen bonding also causes thermodynamic studies of phenomena in water to be particularly complex because changes in entropy and enthalpy tend to be mutually compensating, with relatively little change in free energy. For example, formation of hydrogen bonds in water should produce a favorable decrease in enthalpy H, but an unfavorable decrease in entropy S, because the molecules participating in the hydrogen bonds must be relatively fixed in orientation and proximity. These two contributions to the Gibbs free energy G tend to cancel out, because

$$\Delta G = \Delta H - T\,\Delta S \qquad (4.11)$$

where T is the temperature. Even with large changes in enthalpy and entropy, there may be relatively little or no change in free energy. Other, relatively small effects may thus predominate in determining the free energy of any such transition in water. Rationalization of thermodynamic data in water is very tricky indeed.

References

Observed diffraction pattern and proposed models of liquid water. A. H. Narten and H. A. Levy. *Science* 165:447–454 (1969).

The structure of ordinary water. H. S. Frank. *Science* 169:635–641 (1970).

Water—A Comprehensive Treatise. F. Franks, ed. Plenum Press, New York, 1972.

Water revisited. F. H. Stillinger. *Science* 209:451–457 (1980).

Atom pair distribution functions of liquid water at 25°C from neutron diffraction. A. H. Narten et al. *Science* 217:1033–1034 (1982).

Supercooled water. C. A. Angell. *Ann. Rev. Phys. Chem.* 34:593–630 (1983).

The structure of polar molecular liquids. P. J. Rossky. *Ann. Rev. Phys. Chem.* 36:321–346 (1985).

Model for the structure of the liquid water network. E. Grunwald. *J. Amer. Chem. Soc.* 108:5719–5726 (1986).

Equation of state of a random network, continuum model of liquid water. A. R. Henn and W. Kauzmann. *J. Phys. Chem.* 93:3770–3783 (1989).

4.2.3 *Aqueous Solutions*

a. Solubilities

The solvent properties of liquid water are dominated by the polar character of the water molecule and by its relatively ordered structure. To be soluble in water, a molecule must occupy a certain volume, thereby disrupting the water structure at least within that volume. But the volume that is occupied by such a molecule in solution (the **partial molecular volume**) reflects not only the atomic volume of the molecule but also any changes it produces by rearranging the liquid around it. The measured partial volumes occupied by the various amino acid residues are given in Table 4.3 and are compared with their van der Waals volumes. The partial volumes occupied by the residues are generally 28–38% greater than their van der Waals volumes, which presumably reflects the open packing of water molecules around the molecules. The values are somewhat less when hydrogen bonding or charged groups are present on the residue side chain, presumably because they interact more strongly with water.

The solubility of a molecule in water depends on how much of the unfavorable aspects of creating a cavity in water are compensated by favorable interactions with the surrounding water molecules. A measure of the favorable interactions of a molecule with water, its **hydrophilicity,** is its relative tendency to equilibrate between being dissolved in the aqueous phase and being free in vapor (Fig. 4.8). The relative concentrations $[X]$ of the molecule in the vapor phase and in the aqueous phase at equilibrium gives the **partition coefficient K_D:**

$$K_D = \frac{[X]_{water}}{[X]_{vapor}} \qquad (4.12)$$

The free energy of transfer from vapor to water is given by $-RT \ln K_D$, which is a measure of the hydrophilicity of a molecule; the most hydrophilic molecules have the most negative values. Only the relative values of the partition coefficients and hydrophilicities are relevant.

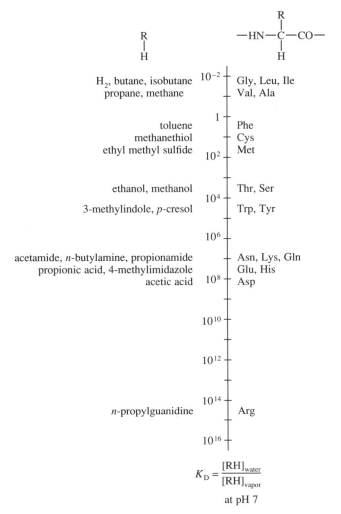

$$K_D = \frac{[RH]_{water}}{[RH]_{vapor}}$$

at pH 7

FIGURE 4.8

Relative hydrophilicities of amino acid side chains, measured by the partition coefficient between vapor of the appropriate small molecule and water at pH 7. The scale gives the equilibrium constant between the vapor and aqueous phase. The model compounds used for the side chains of the amino acid residues *(right)* are given on the *left.* (From R. Wolfenden et al., *Biochemistry* 20:849–855, 1981.)

The hydrophilicities of the amino acid side chains have been measured using model compounds in which the main chain was replaced by a hydrogen atom. For example, CH_4 was the model for the Ala side chain, toluene for Phe. A model for the peptide backbone is *N*-methylacetamide:

$$CH_3-\overset{\overset{\displaystyle O}{\|}}{C}-NH-CH_3 \qquad (4.13)$$

Ionized molecules have a negligible tendency to vaporize, so the observed partition coefficients for the corre-

Table 4.8 *Relative Hydrophilicities and Hydrophobicities of Amino Acid Side Chains*

| Residue | Hydrophilicity[a] (kcal/mol) | Hydrophobicity (kcal/mol) | | | |
		Side-chain analogues[a]	Amino acids[b]	N-acetyl amides[c]	Calculated[d]
Arg	−22.31	15.86	3.0	1.01	3.95
Asp	−13.34	9.66	2.5	0.77	3.81
Glu	−12.63	7.75	2.5	0.64	2.91
Asn	−12.07	7.58	0.2	0.60	1.91
Lys	−11.91	6.49	3.0	0.99	2.77
Gln	−11.77	6.48	0.2	0.22	1.30
His	−12.66	5.60	−0.5	−0.13	0.64
Ser	−7.45	4.34	0.3	0.04	1.24
Thr	−7.27	3.51	−0.4	−0.26	1.00
Tyr	−8.50	1.08	−2.3	−0.96	−1.47
Gly	0	0	0	0	0
Pro			−1.4	−0.72	−0.99
Cys	−3.63	−0.34	−1.0	−1.54	−0.25
Ala	−0.45	−0.87	−0.5	−0.31	−0.39
Trp	−8.27	−1.39	−3.4	−2.25	−2.13
Met	−3.87	−1.41	−1.3	−1.23	−0.96
Phe	−3.15	−2.04	−2.5	−1.79	−2.27
Val	−0.40	−3.10	−1.5	−1.22	−1.30
Ile	−0.24	−3.98	−1.8	−1.80	−1.82
Leu	−0.11	−3.98	−1.8	−1.70	−1.82

[a] Hydrophilicity was measured by the partition coefficient K_D of the model for each side chain (backbone replaced by hydrogen atom, Fig. 4.8) from vapor → water; hydrophobicity from water → cyclohexane. For ionizing side chains, the values were corrected for the fraction of each side chain that is ionized at pH 7. Both scales were normalized to zero for the value of Gly (A. Radzicka and R. Wolfenden, *Biochemistry* 27:1664–1670, 1988).

[b] Some values were measured from the relative solubilities of the amino acids in water and ethanol or dioxane (Y. Nozaki and C. Tanford, *J. Biol. Chem.* 246:2211–2217, 1971); others were extrapolated from these data (M. Levitt, *J. Mol. Biol.* 104:59–107, 1976).

[c] Measured from the partition coefficient between water and octanol of the N-acetyl amino acid amides (J. Fauchère and V. Pliska, *Eur. J. Med. Chem.* 18:369–375, 1983).

[d] Calculated from the hydrophobicities of the individual groups that make up each side chain, using data for the partition coefficient between water and octanol of many model compounds (M. A. Roseman, *J. Mol. Biol.* 200:513–522, 1988).

sponding nonionized molecules are corrected to the fraction of nonionized form present at pH 7. The measured hydrophilicity values of the amino acid side chains, normalized so that of Gly is zero, are given in Table 4.8.

Molecules with polar hydrogen-bond donors or acceptors strongly prefer the aqueous environment because they form hydrogen bonds to water approximately as well as other water molecules do. Maximum hydrophilicity is observed with ionized molecules and with those that can act as both donor and acceptor in hydrogen bonds with water. The peptide bond is as hydrophilic as the side chains of Asn and Gln residues.

Nonpolar molecules do not interact as favorably with water. Some favorable van der Waals interactions probably take place between water and nonpolar solutes, but they are probably relatively weak because of the relatively open structure of water and because they are less favorable than the interactions between water molecules. Nonpolar molecules have low solubilities in

Table 4.9 *Association in Water of Small Molecules Typical of Noncovalent Interactions in Proteins*

Type of interaction	Example	Association constant (M^{-1})
Salt bridge	$CH_3-CO_2^- \cdot H_2^+N=\overset{\overset{NH_2}{\mid}}{C}-NH_2$	0.5^a 0.37^b
	$CH_3-CO_2^- \cdot H_3^+N-(CH_2)_3-CH_3$	0.31^b
	phenyl$-O^- \cdot H_3^+N-CH_2-CH_2-OH$	0.20^c
Hydrogen bond^i	Formic acid dimers	0.04^d
	Urea dimers	0.04^e
	N-Methylacetamide dimers	0.005^f
	δ-Valerolactam dimers	0.013^g
Van der Waals	Benzene dimers	0.4^h
	Cyclohexane $\cdot$ cyclohexanol	0.9^h
	Benzene $\cdot$ phenol	0.6^h

[a] C. Tanford, *J. Amer. Chem. Soc.* 76:945–946 (1954).

[b] B. Spriggs and P. Haake, *Bioorg. Chem.* 6:181–190 (1977).

[c] N. Stahl and W. P. Jencks, *J. Amer. Chem. Soc.* 108:4196–4205 (1986).

[d] A. Katchalsky et al., *J. Amer. Chem. Soc.* 73:5889–5890 (1951).

[e] J. A. Schellman, *Compt. Rendu Trav. Lab. Carlsberg Ser. Chim.* 29:223–229 (1955).

[f] I. M. Klotz and J. S. Franzen, *J. Amer. Chem. Soc.* 84:3461–3466 (1962).

[g] H. Susi et al., *J. Biol. Chem.* 239:3051–3054 (1964).

[h] S. D. Christian and E. E. Tucker, *J. Solution Chem.* 11:749–754 (1982).

[i] Interactions other than hydrogen bonding may contribute to the dimerization of these molecules, so the association constants are maximum values for hydrogen bonding.

water (Fig. 4.8), which can be ascribed to the hydrophobic effect, described later (see Sec. 4.3).

b. Interactions between Molecules in Water

Interactions between model compounds in aqueous solution provide information for analyzing the corresponding interactions in peptides and proteins. They are usually measured by the equilibrium constant K_{AB} for the association of two appropriate molecules A and B:

$$A + B \underset{\phantom{K_{AB}}}{\overset{K_{AB}}{\rightleftharpoons}} A \cdot B$$

$$K_{AB} = \frac{[A \cdot B]}{[A][B]} \qquad (4.14)$$

For two molecules to interact favorably in solution, they must overcome a loss of entropy and must interact with each other more strongly than they do individually with

water. For both of these reasons, interactions between individual molecules in solution are usually very weak; some values are given in Table 4.9. For comparison, small molecules that interact with each other as well as they do with water would be expected to have $K_{AB} = \frac{1}{55} M = 0.02\ M^{-1}$, where $55\ M$ is the concentration of water molecules in liquid water.

By altering the water structure between them, molecules in water can interact to varying extents over a distance, without coming into contact and forming a complex. For example, large nonpolar surfaces have been shown to interact in water over distances as great as 25 Å. The interaction energy does not vary gradually with distance between the surfaces but exhibits oscillations with an average periodicity of 2.5 Å, approximately the diameter of a water molecule. The atomic structure of the water between such surfaces largely determines the apparent interaction between them; the most favorable interactions occur when the distance between the surfaces is compatible with integral numbers of layers of water molecules. In these cases, the apparent interaction between the two surfaces is largely indirect and occurs because of the unfavorable situation of the water molecules between them.

Electrostatic interactions in water are less than those in other solutes because of water's high dielectric constant, which results from the tendency of the large dipoles of water molecules to align with any electric field. The dielectric constant of pure water at 25°C is 78.5, and it decreases at higher temperatures because thermal motion overcomes the orienting effects of the water dipoles. When small diffusible ions such as Na^+ and Cl^- are present in water, the apparent dielectric constant of the solution increases because the ions tend to concentrate in the vicinity of charges of the opposite sign. This **Debye–Hückel screening** is often described by an effective dielectric constant that increases with increasing distance d between the charges:

$$D_{eff} = D_{H_2O} \exp(+\kappa d) \qquad (4.15)$$

where κ is a parameter that is proportional to the square root of the ionic strength. The parameter $1/\kappa$ is known as the *Debye screening distance,* a measure of the distance over which electrostatic effects are damped out by the mobile ions. At physiological ionic strength, about 150 mM, the Debye distance is about 8 Å, but it ranges from about 300 Å in 10^{-4} M salt to 3 Å at 1 M.

The relatively ordered structure of liquid water is susceptible to alteration by high concentrations of other molecules. Changes in the physical properties of water, such as its surface tension, are reflected in the interactions of aqueous solutions with other molecules; both reflect the structure of water at a physical interface, but a detailed explanation is not currently available. Salts in the **Hofmeister series** are known to have such effects when present at concentrations in the range of 0.01–1 M. This series was first observed over 100 years ago by Hofmeister in his work on the effectiveness of salts in precipitating serum globulins and has been encountered over and over again in a variety of phenomena. The Hofmeister series for cations and anions are

Cations: $NH_4^+ > K^+ > Na^+ > Li^+ > Mg^{2+}$
$> Ca^{2+} >$ guanidinium   (4.16)

Anions: $SO_4^{2-} > HPO_4^{2-} >$ acetate $>$ citrate
$>$ tartrate $> Cl^- > NO_3^- > ClO_3^-$
$> I^- > ClO_4^- > SCN^-$ (4.17)

The effects of cations and anions are usually independent and additive, with the anion having the larger effect. The first ions of each series disrupt the structure of

water, markedly increase its surface tension, and decrease the solubility of nonpolar molecules (i.e., "salt out"). The last ions of each series generally increase the structure of water, have less effect on its surface tension, and increase the solubility of nonpolar molecules (i.e., "salt in"). The dividing points between the two effects are usually taken as Na^+ and Cl^-; NaCl is approximately neutral in this respect.

Nonionic molecules can also affect the physical properties of water. The best example is urea,

$$\begin{array}{c} O \\ \parallel \\ H_2N-C-NH_2 \end{array} \qquad (4.18)$$

which hydrogen-bonds to water and at high concentrations disrupts the usual aqueous hydrogen-bond network.

Although all these additives seem to affect the properties of bulk water, a major part of their effects on the solubilities of nonpolar compounds arises from the fact that they are excluded from the solvent–nonpolar-solute interface. This interface is crucial for determining aqueous solubility. Exclusion of an additive from the air–water interface is the reason for the increase in surface tension of the bulk solvent.

References

Water and proteins. I. The significance and structure of water; its interaction with electrolytes and non-electrolytes. II. The location and dynamics of water in protein systems and its relation to their stability and properties. J. T. Edsall and H. A. McKenzie. *Adv. Biophys.* 10:137–207 (1978); 16:53–183 (1983).

Waterlogged molecules. R. Wolfenden. *Science* 222:1087–1093 (1983).

Amino acid, peptide, and protein volume in solution. A. A. Zamyatnin. *Ann. Rev. Biophys. Bioeng.* 13:145–165 (1984).

Theory of protein solubility. T. Arakawa and S. N. Timasheff. *Methods Enzymol.* 114:49–77 (1985).

The Hofmeister effect and the behaviour of water at interfaces. K. D. Collins and M. W. Washabaugh. *Quart. Rev. Biophys.* 18:323–422 (1985).

Solvation forces and liquid structure, as probed by direct force measurements. J. Israelachvili. *Acc. Chem. Res.* 20:415–421 (1987).

Forces between surfaces in liquids. J. N. Israelachvili and P. M. McGuiggan. *Science* 241:795–800 (1988).

Characteristic thermodynamic properties of hydrated water for 20 amino acid residues in globular proteins. M. Oobatake and T. Ooi. *J. Biochem.* 104:433–439 (1988).

4.3   The Hydrophobic Interaction

Electrostatic, hydrogen-bond, and van der Waals interactions between two molecules in an aqueous environment are not particularly favorable energetically (Table 4.9) because there are comparable competing interactions between the molecules and the water surrounding them. The interactions with water are not so favorable, however, with nonpolar surfaces. Water is a very poor solvent for nonpolar molecules compared with most organic liquids. Nonpolar molecules cannot participate in the hydrogen bonding that appears to be so important in liquid water, and aqueous solutions of such molecules have many anomalous physical properties. This relative absence of interactions between nonpolar molecules and water causes interactions among the nonpolar groups themselves to be much more favorable than would be the case in other solvents, so nonpolar molecules greatly prefer nonpolar environments. This preference of nonpolar atoms for nonaqueous environments has come to be known as the **hydrophobic interaction**. It is a major factor in the stabilities of proteins, nucleic acids, and membranes, and it has some unusual characteristics.

4.3.1   The Hydrophobic Interaction in Model Systems

The magnitude of the hydrophobic interaction is usually measured by the free energy of transfer ΔG_{tr} of a nonpolar molecule in the gas, liquid, or solid state to water. The free energy of transfer is positive, indicating that the nonpolar molecule prefers a nonaqueous environment. The thermodynamics of transfer between these phases of a molecule the size of cyclohexane are illustrated in Figure 4.9 at two temperatures; the thermodynamics of transfer to water are anomalous in being markedly temperature dependent. In considering the thermodynamics of this transfer, keep in mind that the enthalpy change ΔH reflects the difference in the magnitude of the noncovalent interactions between molecules that occur in the two phases, whereas the entropy change ΔS

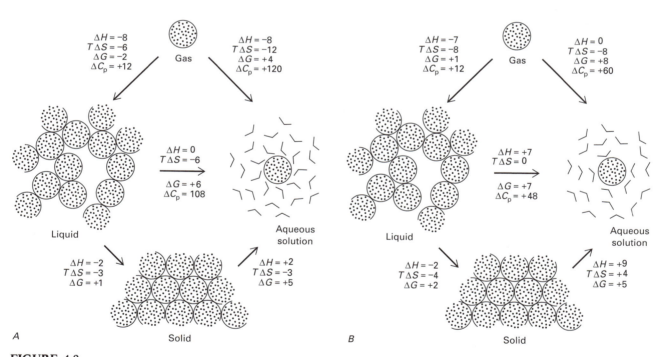

FIGURE 4.9

Typical thermodynamics of transfer of a nonpolar molecule the size of cyclohexane between the gas, liquid, and solid phases and aqueous solution at two temperatures: (A) T_H, approximately 20°C, where $\Delta H_{tr} = 0$ for transfer between liquid and water; and (B) T_S, approximately 140°C, where $\Delta S_{tr} = 0$. The values of ΔH, $T\Delta S$, and ΔG are in units of kcal/mol, that of ΔC_p in units of cal/(K·mol).

reflects the difference in disorder. Transfering a solute molecule to a liquid involves (1) creating a suitable cavity in the liquid, (2) introducing the solute molecule into the cavity, and (3) rearranging the solute and the surrounding liquid molecules to maximize the interactions between them. The observed thermodynamics of transfer between the two phases are the net effect of all three factors, so physical interpretation of thermodynamic parameters for transfer is not always straightforward. This is especially the case with water, in which the high degree of hydrogen bonding results in a more negative enthalpy and also a more positive entropy due to the necessity of fixing the positions of the interacting molecules; the enthalpy and entropy have compensating effects on the free energy (Eq. 4.11). Nevertheless, analysis of the temperature dependence of the hydrophobic interaction is crucial for understanding its physical basis.

The thermodynamics of transfer data indicate that it is the aqueous solution of nonpolar molecules that shows the anomalous temperature-dependent physical properties. The differences in the thermodynamic parameters for transfer of nonpolar molecules to water from their gaseous, liquid, or solid states simply reflect the normal thermodynamic differences between these three states. For example, a nonpolar liquid has favorable van der Waals interactions among its molecules that are essentially absent in the gas phase, but the liquid also has less disorder than the gas; such differences are apparent in the negative changes of both ΔH and ΔS upon transfer from the nonpolar gas to the nonpolar liquid. Changes of the same type, but of smaller magnitude, occur upon solidification of the liquid.

At room temperature, the unfavorable transfer of a nonpolar molecule from a nonpolar liquid to water is primarily a result of the unfavorable change in entropy. The enthalpy change is approximately zero at this temperature, so there are similar enthalpic interactions in the aqueous solution and in the nonpolar liquid. The precise temperature at which $\Delta H_{tr} = 0$ is known as T_H (Fig. 4.9). The unfavorable entropy change is thought to result from increased ordering of water molecules around the nonpolar molecule. These water molecules appear to be more tightly packed than those of normal bulk water because the measured partial volumes of nonpolar molecules are smaller in water than in other liquids. Water molecules cannot make hydrogen bonds to nonpolar solutes so they are imagined to satisfy their hydrogen-bond potential by forming a hydrogen-bonded "iceberg" network among themselves around the nonpolar surface in a "water-ordering effect." Extreme examples of such ordered water cages are observed in **clathrates,** ordered water structures that in-

corporate apolar gases at low temperatures and high gas pressures. The water molecules are fully hydrogen-bonded in the clathrates, as they are in ice, although with nonoptimal geometries. A similar ordering of water molecules is thought to take place around a nonpolar solute molecule in aqueous solution in normal conditions, although to a lesser degree; the water molecules become more ordered and lose entropy, but their increased hydrogen bonding compensates by decreasing their enthalpy. Because the entropic factor dominates the unfavorable ΔG_{tr} to water, it was previously thought that the water-ordering effect is responsible for the low solubility of nonpolar molecules in water, but later analysis demonstrated just the opposite, that the water ordering occurs because it makes the interaction between water and a nonpolar molecule more favorable.

As the temperature is increased, the ordered water shell around the nonpolar solute tends to melt out and to become more like bulk water. This melting of the ordered water produces the anomalously large heat capacity C_p of this type of aqueous solution. The large C_p is the thermodynamic hallmark of aqueous solutions of nonpolar molecules. It causes the thermodynamics of such solutions to be markedly temperature dependent (Fig. 4.10) because the heat capacity defines the temperature dependence of both the enthalpy and entropy:

$$C_p = \frac{\partial H}{\partial T} = \frac{T \partial S}{\partial T} \tag{4.19}$$

Its value is generally found to be proportional to the nonpolar surface area of the solute molecule exposed to water, as are the other thermodynamic parameters (Fig. 4.11).

The temperature dependence of the hydrophobic interaction provides important clues to its physical nature. At temperatures above T_H, the entropy of transfer decreases and becomes less unfavorable for transfer to water, but the enthalpy change becomes more unfavorable. The entropy of transfer from the nonpolar liquid to water becomes zero at temperature T_S (Fig. 4.9). The value of T_S was originally thought to be about 110°C, when ΔC_p was thought to be independent of temperature. The value of ΔC_p is now known to decrease at higher temperatures, and T_S is thought to be about 140°C (Fig. 4.10). The temperature dependence of ΔC_p primarily affects extrapolations to nonphysiological temperatures, however, so it is often convenient to approximate ΔC_p as a constant.

The large changes with temperature of ΔH_{tr} and ΔS_{tr} mostly compensate, and the value of ΔG_{tr} changes much less than they do (Fig. 4.10). Nevertheless, the magnitude of the hydrophobic interaction has a maxi-

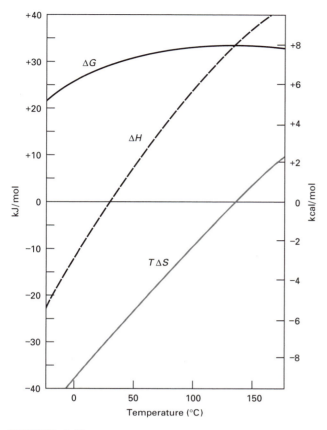

FIGURE 4.10

Typical thermodynamics of the free energy of transfer of a hydrocarbon from the liquid to aqueous solution, using pentane as an example. The strong temperature dependence of both the enthalpy and entropy difference between the two phases is a result of the different heat capacities of the two phases. The free-energy difference is the net difference between the enthalpic (ΔH) and entropic ($T \Delta S$) contributions. It reaches a maximum where $\Delta S = 0$, whereas the equilibrium constant (which is proportional to $-\Delta G/T$) reaches a maximum where $\Delta H = 0$. (Adapted from P. L. Privalov and S. J. Gill, *Adv. Protein Chem.* 39:191–234, 1988.)

mum value at one temperature, above and below which it decreases. Hydrophobicity measured as ΔG_{tr} is at a maximum at temperature T_S, whereas hydrophobicity as measured by the equilibrium constant for transfer is at a maximum at temperature T_H. It is important, therefore, to specify which measure of hydrophobicity is being used.

At the higher temperature T_S, the value of zero for the ΔS_{tr} from nonpolar liquid to water suggests that the net water-ordering effect has disappeared and that the water has now become a normal solvent. The difficulty with this interpretation is that the ΔC_p, which is thought to arise from the water-ordering effect, is still substan-

tial, although diminished (Fig. 4.9). Nevertheless, there appear to be no net interactions between a nonpolar solute molecule and water at this temperature because the ΔH_{tr} from the gas phase to water is approximately zero (Fig. 4.9). A possible explanation for this apparent discrepancy is that totally disordered water would have net repulsions with nonpolar solutes and that sufficient ordering persists at T_S to balance such repulsions. Not all of the details are clear, but the important point is that water becomes a much more normal solvent at high temperatures.

At the high temperature T_S, in the apparent absence of the net water-ordering effect, the low solubility of nonpolar molecules in water is observed to be due to the much weaker enthalpic interactions between the nonpolar molecules and water than between the nonpolar molecules in the nonpolar liquid and between the water molecules in bulk liquid water. The van der Waals interactions in a nonpolar liquid or solid are much greater than those between water and the nonpolar solvent. Likewise, the hydrogen bonds existing in water cannot be transferred to the nonpolar molecule but are disrupted by it when it is dissolved in the water. This is the cause of the hydrophobic interaction.

The hydrophobic interaction does not result from repulsions between water and nonpolar molecules as

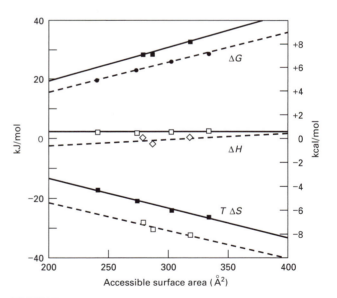

FIGURE 4.11

Thermodynamics of dissolution of hydrocarbon liquids into water at 25°C as a function of the accessible surface area of the hydrocarbon. The enthalpy change is virtually zero at this temperature. The dashed lines are for aliphatic molecules, the solid lines for aromatics. (Kindly provided by S. J. Gill.)

implied by the term. There are, in fact, favorable interactions between water and a nonpolar molecule at all accessible temperatures, those less than T_s (Fig. 4.9). The magnitudes of these interactions, however, are less than those of the van der Waals interactions in nonpolar liquids and those of the hydrogen bonding in liquid water.

The hydrophobic interaction results in a tendency of nonpolar atoms to interact with each other rather than with water. This interaction has the unusual property of decreasing in magnitude at lower temperatures, which results from the increasing tendency of water at lower temperatures to form the hydrogen-bonded clathrate-like structures around the nonpolar molecule. The water ordering effect increases the solubility of nonpolar molecules in water because it seems to be water's attempt to improve its interactions with the nonpolar molecule. The water-ordering effect is not responsible for the low solubility of nonpolar molecules, as has often been assumed from the way the entropy change dominates the thermodynamics of transfer at room temperature. The water-ordering effect is responsible for the decrease in ΔG_{tr}, that is, the decrease in magnitude of the hydrophobic interaction, at low temperatures. It is important, therefore, not to use terms such as *hydrophobicity, hydrophobic interaction,* or *hydrophobic effect* to refer to the water-ordering effect or to the resultant anomalous thermodynamic parameters. The water-ordering effect increases the solubility of nonpolar molecules in water and has opposite implications to the usual meaning of the term *hydrophobic interaction.*

Although the exact nature of the water-ordering that occurs in the solvation of nonpolar surfaces by water is uncertain, it is the primary cause of the complex thermodynamics of the hydrophobic interaction.

References

Some factors in the interpretation of protein denaturation. W. Kauzmann. *Adv. Protein Chem.* 14:1–63 (1959).

"Iceberg" formation and solubility. K. Shinoda. *J. Phys. Chem.* 81:1300–1302 (1977).

Temperature dependence of the hydrophobic interaction in protein folding. R. L. Baldwin. *Proc. Natl. Acad. Sci. USA* 83:8069–8072 (1986).

Stability of protein structure and hydrophobic interaction. P. L. Privalov and S. J. Gill. *Adv. Protein Chem.* 39:191–234 (1988).

The hydrophobic effect: a reappraisal. P. L. Privalov and S. J. Gill. *Pure Appl. Chem.* 61:1097–1104 (1989).

Search for a realistic view of hydrophobic effects. N. Muller. *Acc. Chem. Res.* 23:23–28 (1990).

4.3.2 Hydrophobicities of Amino Acid Residues

The hydrophobicities of the individual amino acid side chains have been measured experimentally in a variety of ways, using the free amino acids, amino acids with the amino and carboxyl groups blocked, and side-chain analogues with the backbone replaced by a hydrogen atom (Fig. 4.8), and using a variety of nonpolar solvents including ethanol, octanol, dioxane, and cyclohexane. Because the hydrophobic interaction is an important component in stabilizing protein folded conformations (see Sec. 7.4), it will hereafter be defined as the free energy of transfer from water to a nonpolar liquid. Consequently, the more hydrophobic molecules have the more negative hydrophobicities. The relative values of ΔG_{tr} are relevant, not their absolute values, so the side-chain hydrophobicities are obtained by subtracting the value measured for Gly. Unfortunately, the hydrophobicity values measured in various ways differ substantially, so several representative scales are given in Table 4.8.

The apparent hydrophobicities of the amino acid side chains vary enormously, depending primarily on whether or not polar groups are present. Ionized and polar side chains interact strongly with water and have lower solubilities in nonpolar solvents because of the unfavorable energetics of placing a polar group in a nonpolar environment. The magnitude of this effect varies enormously, depending on the solvent and the molecule, and is probably the main source of the variation in the hydrophobicity scales. In a nonaqueous solvent, the polar groups of the side chains can have varying interactions with other polar groups of the peptide backbone and any of the solvent, including any water that is present. In this case, molecules that have polar groups can appear to be more hydrophobic than they really are. Perhaps for that reason, the most extreme values of hydrophobicity have been measured using models for the side chain alone, without the polypeptide backbone, and with the apolar solvent cyclohexane, which has the least polar nature (Table 4.8). Because of the widely varying hydrophobicities of amino acid side chains, the more neutral term **hydropathy** is often used to describe their relative preferences for aqueous and nonpolar environments.

The diversity of polar and nonpolar groups in amino acid side chains and in the polypeptide backbone makes it advisable to consider the individual groups rather than the side chain as a whole (Table 4.10). The free energy of transfer from water to nonpolar solvents of the nonpolar side chains is correlated with their surface areas, and there are remarkably large differences

Table 4.10 *Thermodynamic Parameters for the Transfer at 25°C from Nonpolar Solvent to Water of 1 Å² of Accessible Surface Area of Various Chemical Groups*

Chemical group	ΔG_{tr} (cal·mol⁻¹·Å⁻²)	ΔH_{tr} (cal·mol⁻¹·Å⁻²)	ΔC_p (cal·K⁻¹·mol⁻¹·Å⁻²)
Aliphatic: —CH₃, —CH₂—, CH	+8	−26	0.370
Aromatic	−8	−38	0.296
Hydroxyl	−172	−238	0.008
Amide & amino: —NH—, NH₂	−132	−192	−0.012
Carbonyl C: C=	+427	+413	0.613
Carbonyl O: =O	−38	−32	−0.228
Thiol and sulfur: —SH, —S—	−21	−31	−0.001

From T. Ooi and M. Oobataka, *J. Biochem.* 103:114–120 (1988).

among the hydrophobicities as measured in various ways (Fig. 4.12A). The values measured for the transfer of the other amino acids are dominated by their polar groups. That solvation by water is the predominant factor in hydrophobicity is illustrated by the excellent correlation of the surface areas of all the amino acid side chains with their free energy of transfer from vapor to cyclohexane (Fig. 4.12B); the polar groups are presumably equally uncomfortable in both phases. The heat capacities of aqueous solutions of the side-chain analogues are directly proportional to their nonpolar accessible surface areas (Fig. 4.13), except that the side chains with ring structures, and methionine with its sulfur atom, give slightly lower values. There can be little doubt that the unusually large heat capacities of aqueous solutions of nonpolar molecules and the anomalous thermodynamics of the hydrophobic effect arise from the interactions of water with nonpolar atoms.

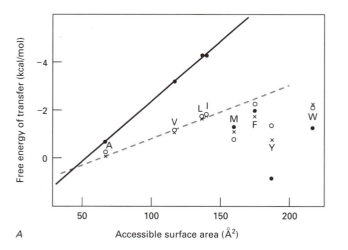

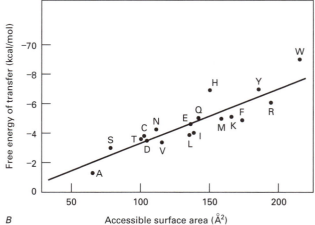

FIGURE 4.12
Relationship between the accessible surface areas of the nonpolar amino acid side chains and their free energies of transfer from water to nonaqueous solvent *(A)* and from vapor to cyclohexane *(B)*. The amino acid residues are designated by the one-letter code (see Table 4.3). The free energies of transfer in *A* are from Table 4.8: the ● points represent measurements with the side-chain analogues, using cyclohexane as the nonpolar solvent; the x points were obtained with the free amino acids and ethanol and dioxane as the nonpolar solvent; the ○ points were calculated from the hydrophobicities of the parts of each side chain. In *A* the slope of the solid line is 43 cal/Å², the dashed line 20 cal/Å². The free energies of transfer in *B* were from A. Radzicka and R. Wolfenden, *Biochemistry* 27:1664–1670 (1988). The slope of the line in *B* is 41 cal/Å².

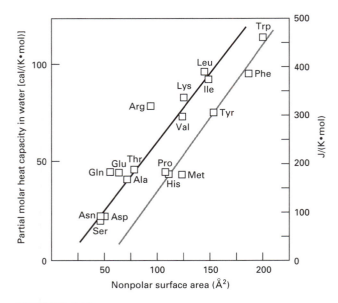

FIGURE 4.13
Correlation between the heat capacities in aqueous solution at 25°C with the accessible surface area of the nonpolar atoms of analogues of the amino acid side chains. The upper straight line fits all the side chains except those with ring structures and the sulfur-containing Met (lower line). The slope of the upper line is 0.72 cal/K·mol Å² (300 J/ K·mol nm²). (Adapted from G. I. Makhatadze and P. L. Privalov, *J. Mol. Biol.* 213:375–384, 1990.)

References

Extension of the fragment method to calculate amino acid zwitterion and side chain partition coefficients. D. J. Abraham and A. J. Leo. *Proteins: Struct. Funct. Genet.* 2:130–152 (1987).

Characteristic thermodynamic properties of hydrated water for 20 amino acid residues in globular proteins. M. Oobatake and T. Ooi. *J. Biochem.* 104:433–439 (1988).

Heat capacity of proteins. I. Partial molar heat capacity of individual amino acid residues in aqueous solution: hydration effect. G. I. Makhatadze and P. L. Privalov. *J. Mol. Biol.* 213:375–384 (1990).

4.4   Intramolecular Interactions

The analysis presented in this chapter of the electrostatic, van der Waals, and hydrogen-bond interactions among the atoms of proteins indicates that all such interactions are weak in the presence of water. Ionized or polar groups interact with water almost as favorably as they interact with other suitable ionized or polar groups (Table 4.9), and it is energetically unfavorable to remove them from aqueous solution (Fig. 4.8). Nonpolar groups prefer to interact with each other rather than with water (Table 4.8), but even the resulting hydrophobic interaction is not very strong (Table 4.9). Yet such interactions will be shown in Chapter 6 to produce stable folded conformations of proteins. How can they do this?

For molecules to interact, they must lose entropy, which is energetically unfavorable. Were it not for entropy, all matter would be solid. This contribution to the free energy, which arises from molecules having freedom, makes the liquid and gaseous states possible. How much entropy is lost in an interaction depends on the number of degrees of freedom that must be fixed. For example, van der Waals interactions require the least entropy loss because only the distance between two atoms is fixed (Fig. 4.2), whereas hydrogen bonding requires that both proximity and orientation be fixed to some extent (Figs. 4.3 and 4.4).

Entropic considerations are especially important when two or more interactions are possible simultaneously in a single molecule because, in favorable cases, much less entropy need be lost in the second and subsequent interactions than in the first. Two interactions that can occur simultaneously can be much more favorable energetically than might be expected from their individual strengths.

The amino acid side chains with polar groups are **amphiphilic,** having both polar and nonpolar segments. Other amphiphiles include lipids and detergents. Amphiphilic molecules tend to interact in aqueous solution in such a way that their nonpolar segments interact with other nonpolar groups and their polar groups are in contact with water. This is the basic principle of the formation of membranes, lipid bilayers, and micelles. A similar phenomenon can be seen to produce the folded conformations of proteins (see Chap. 6).

A useful concept in considering amphiphilic molecules is the **hydrophobic moment,** which is exactly analogous to the dipole moment of electrical charge (Sec. 4.1.2.b) but represents a vector from the hydrophilic to the hydrophobic parts of a molecule. The hydrophobic moment of a polypeptide chain in a particular three-dimensional structure is calculated from the vector sum of the contributions of each of the amino acid residues. This contribution is given by a vector that points from the C^α atom to the center of the side chain and whose length is proportional to the hydrophobicity of the side chain. This parameter can often account for the architecture and interactions of large molecules such as proteins.

4.4.1 Effective Concentrations

The magnitude of entropic cooperativity can be illustrated by intramolecular interactions. Two parts of the same molecule can interact without losing as much entropy as must be lost to bring two independent molecules together. Because the two parts of a molecule are already fixed to some degree in proximity and orientation, only some fraction (which depends on the molecule and the interaction) of the internal flexibility of the molecule has to be lost in the intramolecular interaction.

Intramolecular and bimolecular examples of the same interaction can be compared by means of the ratio of their equilibrium constants, which for the intramolecular interaction is dimensionless and for the intermolecular interaction has dimensions of (concentration)$^{-1}$. Therefore, the ratio of the two has the dimensions of concentration, which can be thought of as the **effective concentration** of the two groups when they are part of the same molecule in the intramolecular interaction:

$$A - B \underset{\longleftarrow}{\overset{K_{intra}}{\rightleftharpoons}} A \cdot B \qquad (4.20)$$

$$A + B \underset{\longleftarrow}{\overset{K_{inter}}{\rightleftharpoons}} A \cdot B \qquad (4.21)$$

$$\frac{K_{intra}}{K_{inter}} = \text{effective concentration of } A - B \qquad (4.22)$$

It was thought for a long time that the maximum effective concentration of two groups in aqueous solution was about 55 M, the concentration of pure water, when one group could be considered to be immersed in a liquid environment of the second component. Consequently, one often finds instances of the magnitudes of intramolecular and intermolecular interactions being interconverted by using the factor of 55 M.

Many experimental measurements have been made for various chemical reactions, however, and much greater values of effective concentrations are generally found; representative examples are given in Table 4.11. These examples represent chemical reactions involving reversible covalent bond formation that can be considered analogous to noncovalent interactions. The covalent nature of these interactions, however, probably increases the magnitude of the entropic effect, due to the more stringent geometrical requirements of covalent bond formation. In any case, the first three examples in the table (A–C) involve flexible molecules with relatively free rotations about three single bonds, which must be restricted to form the product. In spite of this considerable entropic loss, the effective concentrations measured are in the range of 10^2 to 10^5 M. Therefore, merely keeping two groups in reasonable proximity by linking them covalently through several bonds causes their concentration relative to each other to be much higher than would be possible if the groups were on separate molecules, even in the most concentrated liquid state. The last example of Table 4.11 (D) has an enormous effective concentration of 5×10^9 M, which is undoubtedly due primarily to the small entropy difference between the molecule with and without the anhydride interaction. In this case, the planar aromatic structure of the molecule keeps the carboxyl groups in close proximity whether or not the anhydride is present. The very small increase in flexibility and entropy that occurs when the anhydride interaction is broken results in an enormous effective concentration that is close to the maximum considered possible theoretically (approximately 10^{10} M). Of course, other factors can cause large apparent effective concentrations, such as strain in the molecule that is relieved upon forming the interaction, but numerous examples of large values are known that illustrate the entropic contribution.

When there is no entropic difference between the molecules with and without the interaction, the effective concentration is at its maximum value. This value depends on the type of interaction. Those in which the proximity and orientation of the interacting groups are very important, as in a hydrogen bond and especially when a covalent bond is formed, have very high maximum effective concentrations, up to 10^{10} M. When these factors are not so important, as in van der Waals interactions, the interacting groups have significant degrees of freedom and have less entropy to gain upon dissociating, so lower values of maximum effective concentrations apply. Even in this last case, though, the maximum values are substantially greater than 55 M. The reason for this high maximum value is that the molecules of a liquid have a high degree of translational and rotational freedom, so they are far from being in the optimal situation for interacting.

Unfortunately, the magnitudes of the effective concentrations expected for interactions of the type observed in proteins are not known. Only in the case of the disulfide interaction between thiol groups have values been measured in proteins (Sec. 7.5.4). The maximum value observed is somewhat greater than 10^5 M, but the disulfide bond is a covalent interaction, which tends to enhance the effective concentration. Hydrogen bonds are moderately sensitive to orientation and probably have a partial covalent character, so substantial maximum values would be expected, but probably much less than 10^{10} M and less than those involving disulfide bonds. Ionic and hydrophobic interactions are not stereochemically very stringent, so maximum values of 10^2–10^3 M may apply in these instances.

Table 4.11 *Selected Examples of Measured Values of Effective Concentrations of Two Reactive Groups in Small Molecules*

Example	Equilibrium reaction	Effective concentration (M)
A		3.7×10^3 [a]
B		2×10^2 [b]
C		1.9×10^5 [a]
D		5.4×10^9 [a]

[a] From A. J. Kirby, *Adv. Phys. Org. Chem.* 17:183–278, (1980).

[b] From M. H. Chau and J. W. Nelson, *FEBS Letters* 291:296–298 (1991).

In contrast to the very high effective concentrations that are possible when interacting groups are held in the appropriate proximity and orientation, constituent groups that are kept apart by the structure of their molecules have very low, or zero, effective concentrations. Intramolecular interactions are much more sensitive to their environment than are interactions between independent molecules in the liquid state.

Detailed explanations of most values of the effective concentrations measured are complicated by the presence of unfavorable steric or physical interactions in the molecules with or without the interaction. Consequently, there is no ideal example with which to illustrate the solely entropic contribution to the effective concentration, but the many experimental examples available indicate that the effect is very substantial.

References

Entropic contributions to rate accelerations in enzymic and intramolecular reactions and the chelate effect. M. I. Page and W. P. Jencks. *Proc. Natl. Acad. Sci. USA* 68:1678–1683 (1971).

Effective molarities for intramolecular reactions. A. J. Kirby. *Adv. Phys. Org. Chem.* 17:183–278 (1980).

effects on water activity. M. T. Record et al. *Quart. Rev. Biophys.* 11:183–278 (1980).

4.4.2 Multiple Interactions

Multiple groups on a molecule can behave very differently from the same groups in isolation. For example, individual ions do not associate very strongly in aqueous solution (Table 4.9) because of their favorable entropies and strong interactions with water; but a **polyelectrolyte** molecule that has a number of such charged groups binds ions of the opposite charge very tightly due to interactions among the charged groups. Being part of the same molecule, the charged groups are constrained to be close to each other by the covalent bonds. The electrostatic repulsion between groups with the same charge is compensated by the counterions that they attract from the solution and bind very tightly; just how tightly depends on the charge density of the polyelectrolyte and the valence of the counterions. The degree of binding is almost independent of the concentration of counterions in the bulk solvent because the diminution of the electrostatic repulsions in the polyelectrolyte is energetically much more important than is the equilibration of the ions with the bulk solvent. The counterions are not necessarily bound at specific sites, and they can retain their water of hydration and move in an unrestricted and random manner along the polyelectrolyte chain. This phenomenon can be very important for binding other ligands that have the same charge as the counterions; displacement of the counterions into a very dilute bulk solution by the binding of the ligand can provide a strong driving force for its binding.

Another means of compensating unfavorable electrostatic repulsions in a polyelectrolyte is suppression of the ionization of a fraction of its groups. Consequently, groups on a polyelectrolyte can have pK_a values that are very different from those found when they are isolated. These electrostatic effects among multiple groups on a polyelectrolyte are important for the function of proteins, but especially for the function of nucleic acids, with their numerous phosphate groups, and for the interactions of nucleic acids with proteins (Sec. 8.3.2).

References

The molecular theory of polyelectrolyte solutions with applications to the electrostatic properties of polynucleotides. G. S. Manning. *Quart. Rev. Biophys.* 11:199–246 (1978).

Thermodynamic analysis of ion effects on the binding and conformational equilibria of proteins and nucleic acids: the roles of ion association or release, screening and ion

4.4.3 Cooperativity of Multiple Interactions

The simultaneous presence of multiple interactions in a single molecule produces cooperativity between them, and together they can be much stronger than might be expected from the sum of their individual strengths. Cooperativity is essential for proteins, in which the noncovalent interactions are intrinsically very weak (Table 4.9). Only when such interactions cooperate is a stable single conformation produced.

Consider an unfolded polypeptide chain in which two groups A and B are capable of interacting favorably as in a hydrogen bond, a salt bridge, or a nonpolar hydrophobic interaction:

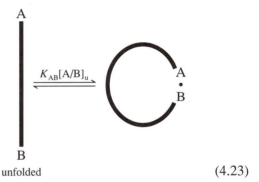

$$\text{unfolded} \qquad\qquad (4.23)$$

The observed equilibrium constant for interaction of the two groups, $K_{obs,U}$, can be expressed as

$$K_{obs,U} = K_{AB}[A/B]_U \qquad (4.24)$$

where K_{AB} is the association constant measured with groups A and B on individual molecules (Table 4.9) and $[A/B]_U$ is the effective concentration of the two groups relative to each other on the unfolded polypeptide U. Groups attached to moderate sized random polypeptides have effective concentrations in the range 10^{-2}–10^{-5} M, depending on their relative positions in the polypeptide chain (see Sec. 5.2). With typical values of K_{AB} (Table 4.9), values for the observed equilibrium constant $K_{obs,U}$ of between 4×10^{-3} and 10^{-7} are expected for individual hydrogen bonds, salt bridges, and so on. Consequently, a single interaction between two groups on a polypeptide chain is not expected to be stable unless the groups are close in the covalent structure in such a way that they have an especially high effective concentration.

Multiple interactions among two or more pairs of groups on the same molecule are often not independent but assist or interfere with each other. The following equilibria are possible with two pairs of groups on a polypeptide:

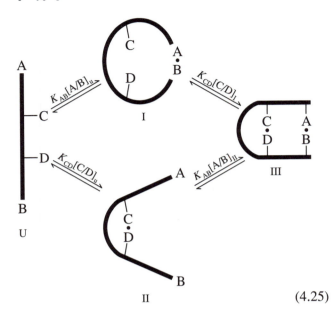

$$(4.25)$$

If both interactions $A \cdot B$ and $C \cdot D$ are possible simultaneously, the interaction between one pair of groups will frequently increase the effective concentration of the other pair. This will occur in a mutual manner, with both interactions having the same effect on each other because the thermodynamics of cyclic equilibria (see Sec. 8.4.1.d) require that the two be linked:

$$\frac{[A/B]_{II}}{[A/B]_U} = \frac{[C/D]_I}{[C/D]_U} = Coop \qquad (4.26)$$

The factor *Coop* is the degree of cooperativity between the two interactions. Consequently, each interaction can be more stable in the presence of the other interaction than when it takes place alone.

If additional groups that may also interact simultaneously are present on the polypeptide chain, these equilibria are extended in a similar way. The overall equilibrium constant between the final state (with all the interactions present) and the unfolded state (with none) is the product of the individual equilibrium constants along any of the conceivable reaction paths; for example,

$$K_{net} = (K_{AB}[A/B]_U)(K_{CD}[C/D]_I)(K_{EF}[E/F]_{III}) \cdots$$
$$(4.27)$$

The value of K_{net} is independent of the reaction path, so we need not know or propose a specific "folding pathway."

The final conformation is stable—that is, populated by most of the molecules—only if the value of K_{net} is greater than unity. Consider a series of weak interactions. The first will be very weak, with an equilibrium

constant of 10^{-3}–10^{-7}. But the presence of the first interaction can increase the effective concentration of the second pair of groups, so the equilibrium constant for the second interaction may be somewhat larger than that of the first, by the factor Coop. If the second equilibrium constant is also less than unity, however, the product of the two equilibrium constants is even smaller than the first (Fig. 4.14). Similarly, the net stabilities of conformations with additional weak interactions are even lower than that of the conformation with a single interaction. This process continues until the effective concentrations of additional interacting groups are suf-

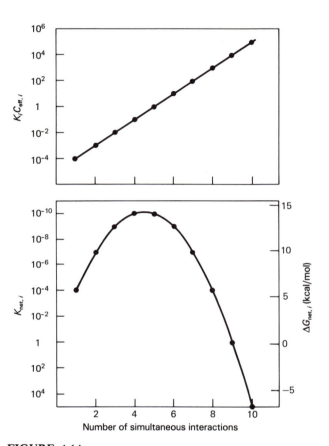

FIGURE 4.14

Hypothetical illustration of the cooperativity produced by multiple weak interactions. Up to 10 interactions are possible simultaneously, and the contribution of the ith interaction is given at the *top*. The initial interaction has an equilibrium constant of 10^{-4}, and each additional interaction is 10 times stronger than the previous one as a result of an increase in the effective concentration. The overall equilibrium constant $K_{net,i}$ *(bottom)* is the product of the contributions of the i interactions present (see Eq. 4.27). Only with 10 such interactions is $K_{net,i} > 1$, implying stability of the folded structure. The free energy of each state relative to U is given by $\Delta G°_{net,i} = -RT \ln K_{net,i}$, with the scale on the right pertaining to 25°C.

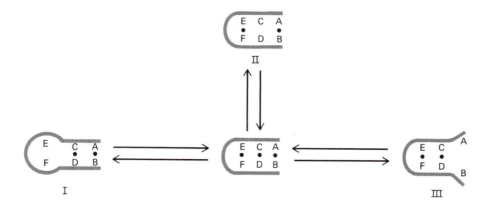

FIGURE 4.15

Simple schematic diagram of cooperativity among three simultaneous interactions occurring between groups A and B, C and D, and E and F. The strength of each interaction is determined by the effective concentration of the two groups when they are not interacting, as in conformations I, II, and III. Assuming there are no other considerations, the value of the effective concentration will be inversely proportional to the degree of flexibility permitted. Therefore, the most stable interactions should be those between groups that are held most rigidly by the other interactions, in this case C and D, and the stability of each interaction should depend on the stabilities of all the others.

ficiently increased to make the equilibrium constant for each additional interaction greater than unity. The value of K_{net} then increases in magnitude with each additional interaction. A sufficient number of simultaneous weak interactions can make the value of K_{net} greater than unity and the folded conformation stable.

An example with Coop = 10 for each additional interaction is given in Figure 4.14. Partially folded structures, those with incomplete stabilizing interactions, are unstable relative to the initial and final states, which means that the transition is cooperative. The degree of cooperativity will be even greater if the intermediate structures have nonbonded groups in unfavorable environments, such as polar groups present in nonpolar environments without being hydrogen bonded. Such situations have been ignored here.

In summary, weak interactions are expected to stabilize a particular folded conformation only when they cooperate so that the interacting groups have very high effective concentrations in that structure. The effective concentration of two groups in a folded structure depends on the extent to which the groups are held in proximity when not interacting (Fig. 4.15), which in turn depends on the stability of all the surrounding interactions. All parts of such a structure, therefore, are expected to be mutually dependent to varying degrees.

As just described, the contribution of each interaction to net stability of the folded structure should depend on the effective concentration of the interacting groups in that folded structure. If the groups are on the surface or in a flexible part of the folded structure, their effective concentration will be low and the interaction will provide little, if any, net stability. Breaking that interaction will have little effect on the folded state. On the other hand, groups in relatively rigid parts of the folded structure will have high effective concentrations, and their interaction will provide a substantial contribution to the net stability of the conformation; removing or altering such an interaction would have a large effect on the stability of the folded conformation.

References

An empirical approach to protein conformation stability and flexibility. T. E. Creighton. *Biopolymers* 22:49–58 (1983).

Dissecting the roles of individual interactions in protein stability: lessons from a circularized protein. D. P. Goldenberg. *J. Cell. Biochem.* 29:321–335 (1985).

Strategy for analysing the co-operativity of intramolecular interactions in peptides and proteins. A. Horovitz and A. R. Fersht. *J. Mol. Biol.* 214:613–617 (1990).

Exercises

1. Ion pairs in proteins involving Arg residues have been observed to be energetically stronger than those involving Lys residues. If this were the case, in what ways might

Arg and Lys residues be used differently in proteins? How could this hypothesis be tested?

ANSWER
D. B. Wigley et al., *Biochem. Biophys. Res. Commun.* 149:927–929 (1987).

2. The hydrogen bond between two water molecules in isolation is about ten times stronger energetically than a van der Waals contact between two xenon atoms, yet water molecules dimerize in the gas phase only about 30% more frequently than xenon atoms. Why?

ANSWER
Section 4.4.

3. The ammonia molecule (NH_3) might be considered analogous to H_2O in its hydrogen-bonding properties. In what way is it different?

ANSWER
D. D. Nelson, Jr., et al., *Science* 238:1670–1674 (1987).

4. Early models of liquid water envisaged mixtures of icelike and disordered collections of H_2O molecules. What are the implications for such models of the observation that liquid water can be supercooled to $-41°C$?

ANSWER
C. A. Angell, *Ann. Rev. Phys. Chem.* 36:321–346 (1985).

5. Many measurements of hydrophobicity measure the partition of a molecule between aqueous and octanol phases (e.g., C. Hansch and A. Leo, *Substituent Constants for Correlation Analysis in Chemistry and Biology*, John Wiley, New York, 1979). What is the likely significance of the presence of 2.3 M water in the octanol phase? Water-saturated cyclohexane contains only 2.5 mM water, just slightly more than the vapor phase of water. Why is the partitioning of hydrophobic molecules between the aqueous and cyclohexane phases similar to that between the aqueous and vapor phases (Fig. 4.12)?

ANSWER
A. Radzicka and R. Wolfenden, *Biochemistry* 27:1664–1670 (1988).

6. Some measurements of hydrophobicity indicate that the Trp side chain is less hydrophobic than the Phe side chain; others indicate just the opposite (J. L. Fauchere, *Trends Biochem. Sci.* 10:268, 1985). What is the most likely explanation?

ANSWER
R. Wolfenden and A. Radzicka, *Trends Biochem. Sci.* 11:69–70 (1986); *Biochemistry* 27:1664–1670 (1988).

7. Many properties of larger molecules can be calculated from the properties—such as volumes (Table 4.2) and hydrophilicities and hydrophobicities (Table 4.10)—of the smaller groups from which they are constructed. Under what conditions can this be valid, and when is it not?

ANSWER
R. Wolfenden et al., *J. Amer. Chem. Soc.* 109:463–466 (1987); *Science* 222:1087–1093 (1983). M. A. Roseman, *J. Mol. Biol.* 200:513–522 (1988).

8. Some small proteins are observed to act as antifreeze agents, depressing the freezing point of water in polar fish that live in sea water at $-1.9°C$. Other proteins promote the crystallization of water. How might proteins have these effects?

ANSWER
R. E. Feeney et al., *Ann. Rev. Biophys. Biophys. Chem.* 15:59–78 (1986); D. S. C. Yang et al., *Nature* 333:232–237 (1988); P. Wolber and G. Warren, *Trends Biochem. Sci.* 14:179–182 (1989).

9. The hydrophobicity of a molecule measured by its ΔG_{tr} between aqueous and nonpolar solvents is increased if it has multiple polar groups that can interact in the nonpolar solvent ("self-solvation"), for example by hydrogen bonding (M. A. Roseman, *J. Mol. Biol.* 200:513–522, 1988). In this case, is it better to measure the hydrophobicities of amino acid side chains using the free amino acids (Y. Nozaki and C. Tanford, *J. Biol. Chem.* 246:2211–2217, 1971), N-acetyl amino acid amides (J. L. Fauchére and V. Pliska, *Eur. J. Med. Chem.* 18:369–375, 1983), analogues of the side chain only (A. Radzicka and R. Wolfenden, *Biochemistry* 27:1664–1670, 1988), or some other model compound?

10. The carbonyl and amide —NH— groups of the protein backbone are very hydrophilic and readily form hydrogen bonds with water. However, it has been suggested that when they are hydrogen-bonded to each other, they become essentially hydrophobic and their surface area can be considered nonpolar (C. Chothia, *Nature* 248:338–339, 1974). How could this idea be tested?

ANSWER
M. Roseman, *J. Mol. Biol.* 201:621–623 (1988).

11. Individual model molecules (say, acceptor A and donor B) do not often form hydrogen bonds in aqueous solution (Table 4.9), presumably because of the loss of entropy involved in bringing two molecules together and because the molecules form comparable hydrogen bonds with water. The hydrogen bonds in a folded protein often occur in the nonpolar interior, where hydrogen bonds might be expected to be much more stable than in aqueous solution. The question of the energetics of hydrogen bonding in folded relative to unfolded proteins then becomes that of the energetics of hydrogen bonding in a nonpolar envi-

ronment, designated by the superscript np, relative to the individual polar groups in water, designated by superscript w. This is often analyzed using the free energies of transfer of the bonded and nonbonded molecules from water to nonpolar solvent, and by the extent of their hydrogen bonding in the two solvents. The data for dimerization of N-methylacetamide designated as $A \cdot B$ (Eq. 4.14) in water and in CCl_4 as the nonpolar solvent give the following picture (M. Roseman, *J. Mol. Biol.* 201:621–623, 1988):

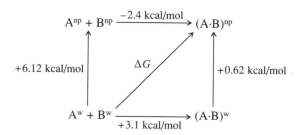

From the requirement that free-energy changes around any cycle be zero, the free energy of forming a hydrogen bond in a nonaqueous environment from donors and acceptors in water, ΔG, has the value $+3.72$ kcal/mol; that is, it appears to be unfavorable energetically. Does this indicate that "hydrogen bonding opposes folding" of proteins (K. Dill, *Biochemistry* 29:7123–7155, 1990)?

ANSWER

T. E. Creighton, *Curr. Opinion Struct. Biol.* 1:5–16 (1991).

12. The nature of the genetic code (Fig. 2.4) is such that an amino acid specified by a gene sequence in the normal way has opposite hydropathic characteristics to the amino acid that would be specified by the complementary DNA strand when read in either the 3′ to 5′ or 5′ to 3′ directions (J. E. Blalock and E. M. Smith, *Biochem. Biophys. Res. Commun.* 121:203–207, 1984; J. E. Blalock and K. L. Bost, *Biochem J.* 234:679–683, 1986; J. E. Blalock, *Trends Biotechnol.* 8:140–144, 1990). Such "antisense" peptides are not normally synthesized in biological systems (Chap. 2), but when they are made in the laboratory, they have been claimed to form specific complexes with the normal "sense" peptide (e.g., K. L. Bost et al., *Proc. Natl. Acad. Sci. USA* 82:1372–1375, 1985; R. R. Brentani et al., *Proc. Natl. Acad. Sci. USA* 85:1364–1367, 1988). What are the implications of this observation?

ANSWER

Y. Shai et al., *Biochemistry* 28:8804–8811 (1989); G. Markus et al., *Arch. Biochem. Biophys.* 272:433–439 (1989); E. S. Najem et al., *FEBS Letters* 250:405–410 (1989); J. E. Zull and S. K. Smith, *TIBS* 15:257–261 (1990); E. Roubos, *Trends Biotechnol.* 8:279–281 (1990); R. R. Brentani, *Trends Biotechnol.* 8:281 (1990).

Conformational Properties of Polypeptide Chains

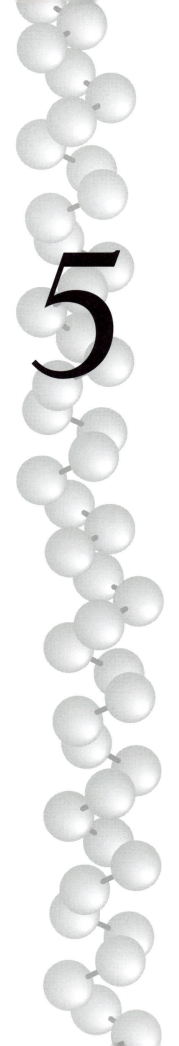

5

The structures of all molecules extend to three dimensions. The ordinary chemical representations used in Chapters 1–4 usually suffice for small molecules because their three-dimensional structures are implied reasonably well by the well-defined bond lengths and bond angles of their covalent structures. These representations are not sufficient for larger molecules, however, because rotations about bonds occur that alter the relative positions of the constituent atoms. Various nonsuperimposable three-dimensional arrangements of atoms that are interconvertible without breaking covalent bonds are generally described as **conformations**. Different conformations of a protein molecule must have the same chirality of atoms, that is, the same **configuration;** for example, the D and L isomers of an amino acid residue are not different conformations. The two terms *conformation* and *configuration* are not interchangeable and should not be confused.

Three-dimensional aspects of structure are especially important for macromolecules, in which many bonds can rotate and can make many conformations. Macromolecules tend in general to be very flexible, for no one conformation usually is sufficiently more stable than all others to predominate. Proteins, on the other hand, have used this flexibility to adopt relatively fixed conformations that are determined by noncovalent interactions among atoms that are distant in the covalent structure (Chap. 6). There is a hierarchy of levels of protein structure. The **primary structure** is the covalent structure, which is defined by the amino acid sequence and any posttranslational covalent modifications, as discussed earlier. The **secondary structure** is the local conformation of the polypeptide backbone, which is discussed in this chapter. The **tertiary** and **quaternary** structures will be described in the next chapter.

Reference

Design of peptides and proteins. W. F. DeGrado. *Adv. Protein Chem.* 39:51–124 (1988).

5.1 Three-Dimensional Conformations

How to define a conformation is not obvious. Even a simple molecule might be considered to exist in an infinite number of conformations if the positions of the atoms are defined with sufficient accuracy, because bond lengths vary by ± 0.05 Å and bond angles by about $\pm 5°$ at room temperature. For this reason, only the energetically most stable arrangements (i.e., energy minima that are separated by distinct energy barriers) are usually classified as individual conformations. For example, small molecules such as CH_4 are considered to have a single conformation, as is ethane, H_3C-CH_3; although rotations occur about the C—C bond of ethane, the most stable structure is that in which the six hydrogen atoms are staggered, because then the repulsions between them are minimized. The six hydrogen atoms are equivalent, so the three stable rotations are indistinguishable and there is said to be a single conformation. This can be illustrated by a Newman projection, in which the spatial arrangements of the groups attached to two adjacent atoms are viewed in projection down the bond joining them:

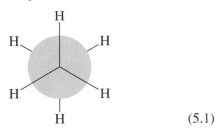

$$(5.1)$$

A molecule such as 1,2-dichloroethane is similar, but has nonequivalent chlorine and hydrogen atoms and can be considered to have three conformations, in which the two chlorine atoms are *trans* (*t*), *gauche*$^+$ (*g*$^+$), and *gauche*$^-$ (*g*$^-$), respectively.

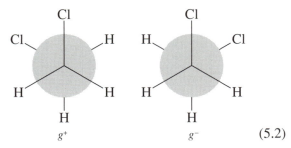

$$(5.2)$$

In the case of proteins, each amino acid residue in a polypeptide chain contains three bonds of the polypeptide backbone (Fig. 5.1), plus those of the side chain. The peptide bond of the backbone has partial double-bond character (Sec. 5.2.1) and consequently is limited to planar *cis* or *trans* rotations (Eq. 5.5). The other two bonds of the backbone have single-bond character, as do many of the bonds of the various side chains. Rotation about any of these bonds produces a different conformation.

Each amino acid residue in a chain can exist in a number of conformations, perhaps eight on average. A small polypeptide chain of 100 residues might, therefore, be able to adopt up to 8^{100} (or 2×10^{90}) conformations. Some of these theoretical conformations are not possible because they would have atoms overlapping in space (the **excluded volume effect**). It is not yet possible, unfortunately, to calculate the number of conformations that are actually possible, and only rough estimates can be given. There is, however, much scope to be conservative and still predict an astronomical number of conformations. For example, even an average of only two conformations per residue would add up to over 10^{30} conformations for a 100-residue polypeptide chain. If all these conformations have similar energies, each would have only a very small probability of ever being adopted by any molecule, simply because there are so many possible conformations. Consequently, every molecule of a finite population of unfolded polypeptide chains is likely to be in a unique conformation at any instant. For example, a 1-mg sample of a certain protein with a molecular weight of 10,000 contains only 6×10^{16} molecules. A sample of such a protein containing 10^{30} molecules would weigh more than 10^7 kg (or 10,000 tonnes).

This potential for such huge numbers of distinct conformations makes the adoption by any molecule of any particular conformation unlikely owing to the **conformational entropy**, ΔS_{conf}, that would have to be lost by the polypeptide chain. It has the value

$$\Delta S_{conf} = R \ln N \qquad (5.3)$$

where R is the gas constant and N the number of possible conformations. The contribution of conformational entropy to the free energy of the disordered polypeptide chain is given by $T \Delta S_{conf}$, which has a value of -123

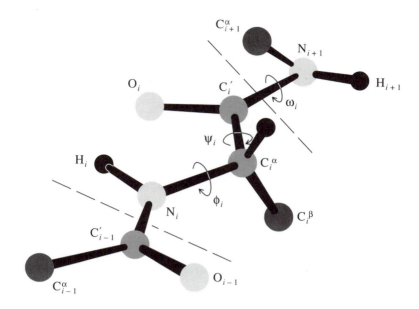

FIGURE 5.1

Perspective drawing of a segment of polypeptide chain comprising two peptide units. Only the C^β atom of each side chain is shown. The limits of a single residue (number i of the chain) are indicated by the dashed lines. The recommended notations for atoms and torsion angles are indicated. The polypeptide chain is shown in the fully extended conformation, where $\phi = \psi = \omega = 180°$.

kcal/mol at 25°C for 100 residues and an average of eight conformations per residue. For any one conformation to be stable (i.e., assumed by all the molecules in a sample), it would need to have stabilizing interactions greater than $T\Delta S_{conf}$ to overcome this loss of conformational entropy. In the absence of such stabilizing interactions, a polymer such as a polypeptide chain tends to exist in many different conformations. Yet proteins and some regular polypeptides assume a particular conformation that is stabilized by the weak interactions described in Chapter 4.

These complexities of protein structure have caused many of the most fundamental studies to be carried out using synthetic **polyamino acids,** in which all the residues are identical, or simple repeating polymers of more than one amino acid, such as poly(Pro-Ala), which has Pro and Ala residues alternating along the sequence. The regularity of the covalent structures of these molecules causes the conformations they assume to be either very regular or very random.

5.2 Polypeptides as Random Polymers

Because so many conformations are possible, the conformational properties of random polypeptides are best

calculated statistically using the mathematical procedures developed for synthetic polymers. Such calculations require detailed knowledge of the conformational properties of the monomeric unit of the polymer; that is, the relative energies of all its possible conformations.

References

Principles of Polymer Chemistry. P. J. Flory. Cornell University Press, Ithaca, New York, 1953.

Conformations of Macromolecules. T. M. Birshstein and O. B. Ptitsyn. Wiley-Interscience, New York, 1964.

The configuration of random polypeptide chains. II. Theory. D. A. Brant and P. J. Flory. *J. Amer. Chem. Soc.* 81:2791–2800 (1965).

Statistical Mechanics of Chain Molecules. P. J. Flory. John Wiley, New York, 1969.

5.2.1 Local Restrictions on Flexibility: The Ramachandran Plot

A portion of the backbone of a polypeptide chain is shown in Figure 5.1, illustrating the conventions used in describing polypeptide conformation. The peptide bond is usually planar because of its partial double bond, and the group of atoms

$$
\begin{array}{c}
\text{H} \qquad \text{C}^{\alpha}_{i+1}\!\!-\!\! \\
\diagdown \quad \diagup \\
\text{N—C}' \\
\diagup \qquad \diagdown \\
-\text{C}^{\alpha}_i \qquad \text{O}
\end{array}
\qquad (5.4)
$$

usually acts as a rigid unit. For this reason, this group is often designated a **peptide unit.** The unit more commonly used is the **residue,** in which all the atoms originate from the same amino acid.

Bond angles are usually denoted by the symbol τ, with a subscript i that gives the number of the residue in the chain, followed by symbols, in brackets, of the atoms that define the bond angle. For example, $\tau_i[\text{NC}^{\alpha}\text{C}']$ denotes the angle formed by the NC^{α} and $\text{C}^{\alpha}\text{C}'$ bonds at the C^{α} atom of the ith residue.

Rotations about bonds are described as **torsion** or **dihedral** angles, which are usually taken to lie in the range $-180°$ to $+180°$. Rotation about the N—C^{α} bond of the peptide backbone is denoted by the torsion angle ϕ, rotation about the C^{α}—C' bond by ψ, and that about the peptide bond (C'—N) by ω. The maximum value of $180°$ (which is the same as $-180°$) is given to each of the torsion angles in the maximally extended chain, as shown in Figure 5.1, when the N, C^{α}, and C' atoms are all *trans* to each other. At the other extreme, the *cis* conformation, the rotation angles have the value of zero. Rotation about a bond from this position so that the atoms behind the rotated bond when viewed in a Newman projection move clockwise are given positive values; counterclockwise rotations are given negative values.

Torsion angles of the side chain are designated by χ_j, where j is the number of the bond counting outward from the C^{α} atom of the main chain. The commonly used designations of the side-chain atoms are given in Figure 1.1.

Rotations about single bonds are intrinsically equivalent, and the relative preference for each particular torsion angle is determined by the energetics of the noncovalent interactions among the atoms and of the atoms with their environment. In the case of the peptide bond, the *trans* form ($\omega = 180°$) is favored by a ratio of approximately 1000:1 over the *cis* form ($\omega = 0°$) because in the *cis* form the C^{α} atoms and the side chains of neighboring residues are in too close proximity:

trans

cis   (5.5)

When residue $i + 1$ is Pro, however, there is very little difference between the two forms:

trans

cis   (5.6)

and the *trans* form is only slightly favored, generally by a ratio of about 4:1. The peptide bond preceding a Pro residue does not have the double-bond character that specifies the planar form (Sec. 1.2). Small deviations from planarity of either the *cis* or *trans* form, with $\Delta\omega = -20°$ to $+10°$, are thought to be only slightly unfavorable energetically in most peptide bonds.

The values of ϕ and ψ that are possible are constrained geometrically due to steric clashes between nonneighboring atoms. The permitted values of ϕ and ψ were first determined by Ramachandran and colleagues, using hard-sphere models of the atoms and fixed geometries of the bonds. The permitted values of ϕ and ψ are usually indicated on a two-dimensional map of the ϕ–ψ plane, in what has come to be known as a **Ramachandran plot.** An example for an Ala residue is illustrated in Figure 5.2A. The normally allowed values, for which there is no steric overlap, are shaded; the extreme limits for some unfavorable contacts are indicated by the other solid lines. The connecting region, indicated by the dashed lines, is permitted if slight alterations of bond angles are allowed. The part of the total area that is fully allowed is about 7.5% and the part that is partially allowed is 22.5%, which gives a quantitative

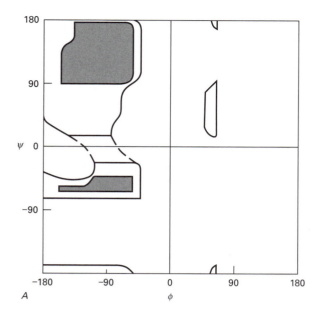

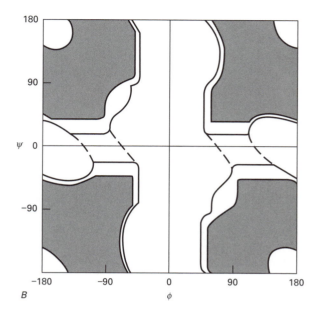

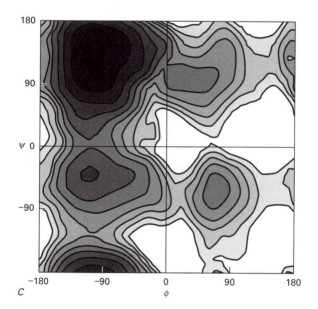

FIGURE 5.2

Ramachandran plots of the permitted values of ϕ and ψ for different residues. Each two-dimensional plot is continuous at the edges, because a rotation of $-180°$ is the same as one of $+180°$. The original plots that considered only repulsions between hard-sphere atoms are shown in A and B for Ala and Gly residues, respectively. The fully allowed regions are shaded; the partially allowed regions are enclosed by a solid line. The connecting regions enclosed by the dashed lines are permissible with slight flexibility of bond angles. The much greater flexibility of the Gly residue compared with Ala is apparent, as is the symmetry of the plot for Gly residues resulting from the absence of a chiral side chain. (From G. N. Ramachandran & V. Saisekharan, *Adv. Protein Chem.* 23:283–437, 1968.) C: Ramachandran plot computed for the dipeptide N-acetyl-Ala-Ala-amide using molecular dynamics simulations and including water as the solvent. The apparent free energies for the various values of ϕ and ψ are given as contours of 2 kJ/mol (0.5 kcal/mol) relative to the lowest free energy in the upper left-hand corner that is shaded black. Note that the differences between allowed and disallowed regions are much less distinct than in A and B. (Figure kindly provided by J. Hermans.)

measure of the limitations on flexibility of the polypeptide chain.

Gly residues have no C^β atom and so the restrictions on allowed conformations are much less severe (Fig. 5.2B), and 45% of the total area is fully allowed, 61% within the extreme limits. In this case, the Ramachandran plot is symmetric as a result of the absence of chirality of Gly residues.

Additional restrictions arise for the longer, larger side chains of the other amino acids, and the allowed region is somewhat smaller. The Pro residue is a special case because the relatively rigid five-membered ring drastically limits the value of ϕ to approximately $-60°$.

More precise Ramachandran plots result from calculations of the relative energies of each conformation, permitting appropriate flexibility of bond lengths and angles and evaluating all favorable and unfavorable interactions, including those with the solvent (Fig. 5.2C). The more detailed calculations indicate much smaller energy differences between the so-called allowed and disallowed regions than might be expected solely from steric considerations. Flexibility of bond lengths and angles permits conformations that are not possible for hard-sphere atoms and rigid bonds, and the variety of interactions that occur among atoms can cause the energies to vary substantially. Nevertheless, the classical Ramachandran plots illustrating the fully and partially allowed regions are remarkably appropriate for proteins.

References

Conformation of polypeptides and proteins. G. N. Ramachandran and V. Sasisekharan. *Adv. Protein Chem.* 23:283–437 (1968).

The mean geometry of the peptide unit from crystal structure data. G. N. Ramachandran et al. *Biochim. Biophys. Acta* 359:298–302 (1974).

An explanation for the rare occurrence of *cis* peptide units in proteins and polypeptides. G. N. Ramachandran and A. K. Mitra. *J. Mol. Biol.* 107:85–92 (1976).

Microfolding: conformational probability map for the alanine dipeptide in water from molecular dynamics simulations. A. G. Anderson and J. Hermans. *Proteins: Struct. Funct. Genet.* 3:262–265 (1988).

5.2.2 Statistical Properties of Random Polypeptides

Given the intrinsic conformational properties of an individual amino acid residue (Fig. 5.2), it is possible to calculate the conformational properties of a random polypeptide chain, statistically averaged over all its many possible conformations. The most easily calculated state is the **unperturbed random coil,** which takes into account the three-dimensional structure of the amino acid residues and their intrinsic conformational properties but not the interactions among distant parts of the polypeptide chain. It is a "random coil" in that the conformation of each part of the polypeptide chain is assumed to be independent of the conformation of the remainder of the polypeptide. In the "unperturbed" state, no account is taken of the excluded volume effect, so impossible conformations in which nonbonded atoms occupy the same space are included.

The average properties of such polymers are often compared with those of the hypothetical **random-flight chain,** or **freely jointed chain.** This is simply a mathematical string of vectors of fixed length representing the bonds between atoms; the atoms are not included, the chain has no volume, all bond angles have equal probability, and all rotations about the bonds are equally likely. Most calculations of polypeptide conformations simplify the detailed chemical architecture of the polypeptide backbone and use a single **virtual bond** for each residue. This is a vector joining adjacent C^α atoms; with a planar *trans* peptide bond, it has a length of 3.8 Å.

a. End-to-End Distances

Of greatest interest with random polymers are the average and the variation of their physical dimensions. The root-mean-square (rms) value of the distance, r, between two atoms of the hypothetical random-flight chain separated by n bonds of length l is given by the relationship

$$\langle r^2 \rangle_0^{1/2} = n^{1/2} l \tag{5.7}$$

where the angle brackets indicate the statistical mechanical average over all conformations; the subscript zero refers to the unperturbed state.

The other statistical measure often used is the average **radius of gyration (R_G),** which is defined as the rms distance of the collection of atoms from their common center of gravity. For the random-flight chain,

$$\langle R_G \rangle_0^2 = \frac{nl^2}{6} \frac{n+2}{n+1} \tag{5.8}$$

For large values of n this becomes

$$\langle R_G \rangle_0^2 = \frac{nl^2}{6} = \frac{\langle r^2 \rangle_0}{6} \tag{5.9}$$

This relationship exists between the average radius of gyration and the average end-to-end distance in the unperturbed states of all polymers, so the radius of gyration is simply 0.408 ($= 6^{-1/2}$) times the average end-to-end distance.

The limited flexibility of an actual random coil causes its dimensions to be considerably greater than those of the random-flight chain (Eqs. 5.8 and 5.9). This factor is usually expressed as the **characteristic ratio** C_n, which is defined as

$$C_n = \frac{\langle r^2 \rangle_0}{nl^2} \qquad (5.10)$$

where $\langle r^2 \rangle_0$ is the observed average end-to-end distance squared for the actual random coil, and nl^2 is this value for the random-flight chain (Eq. 5.7). The value of C_n increases with increasing n but in very long chains approaches a limit designated as C_∞, which is a measure of the inherent stiffness of the chain.

For long polypeptide chains of n residues of amino acids other than Gly and Pro, the rms end-to-end distance is given approximately by

$$\langle r^2 \rangle_0^{1/2} = (130n)^{1/2} \qquad (5.11)$$

which gives a value of 9.0 for C_∞. The radius of gyration for random-coil polypeptide chains of various lengths is illustrated in Figure 5.3. The relatively large value of C_∞ for polypeptides is due not only to the steric restrictions on rotations about ϕ and ψ (Fig. 5.2) but also to the presence of only L-amino acid residues. The presence of D-amino acids would cause the direction of the chain to be effectively reversed from that favored by L-amino acids, and a polymer of alternating D- and L-Ala residues has a C_∞ of only 0.9. The value of C_∞ for poly(Gly) is only 2.0, owing to the increased flexibility possible with Gly residues and the symmetry of the allowed torsion angles (Fig. 5.2B). A small percentage of randomly distributed Gly residues in a polypeptide chain has a large effect on the chain dimensions of random polypeptides. The presence of Pro residues also lowers the value of C_∞ because the *cis* form causes the chain to change its direction.

The calculated distribution of end-to-end distances is usually expressed as either the **Gaussian distribution function** or the **radial distribution function,** which are illustrated in Figure 5.4 for a hypothetical random-flight polypeptide chain. The Gaussian distribution function, $W(x,y,z)\,dx\,dy\,dz$, gives the probability that the end of the chain is within the volume $dx\,dy\,dz$ at coordinates x, y, and z; the other end of the chain is taken as the origin. This distribution is spherically symmetrical, so it is usually expressed as the radial distribution function $W(r)\,dr$, the probability that the two ends of the chain are within a distance r and $r + dr$ from each other. For unperturbed random-coil chains, the scale for r in random-flight chains is simply increased by the factor $C_n^{1/2}$.

References

Conformational energy estimates for statistically coiling polypeptide chains. D. A. Brant et al. *J. Mol. Biol.* 23:47–65 (1967).

Random coil configurations of polypeptide copolymers. W. G. Miller et al. *J. Mol. Biol.* 23:67–80 (1967).

Dimensions of protein random coils. W. G. Miller and C. V. Goebel. *Biochemistry* 7:3925–3935 (1968).

How alike are the shapes of two random chains? A. D. McLachlan. *Biopolymers* 23:1325–1331 (1984).

b. Fluorescence Energy Transfer

The three-dimensional distance between two residues in a polypeptide chain can be measured if one residue has a fluorescence energy donor and the other a suitable energy acceptor. If the absorption spectrum of the acceptor overlaps the fluorescence emission spectrum of the donor, fluorescent light emitted by the donor is absorbed by the acceptor. The efficiency of this process depends on the sixth power of the distance between the

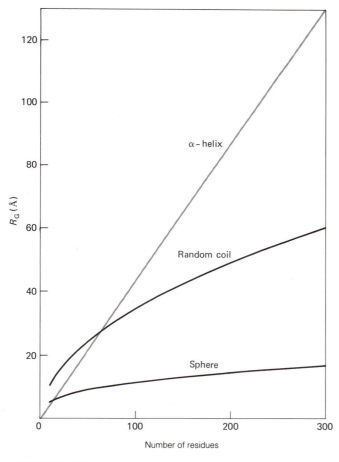

FIGURE 5.3

The radius of gyration R_G for polypeptide chains of various lengths in α-helical, random coil, and compact spherical conformations.

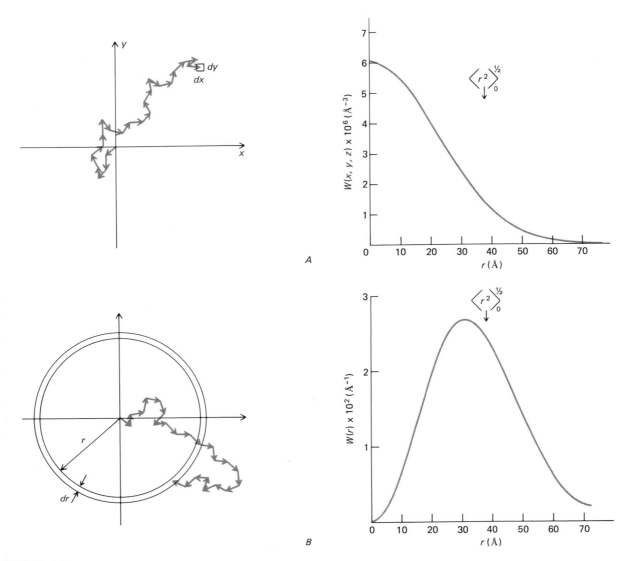

FIGURE 5.4

Illustration of Gaussian (*A*) and radial (*B*) distribution functions for the end-to-end distance of a freely jointed chain. On the *left* of each figure is a two-dimensional representation of how each distribution function is defined, giving the probability that the other end of the chain lies within the enclosed area. The distribution functions are given by

$$W(x, y, z)\ dx\ dy\ dz = (\beta/\sqrt{\pi})^3\ e^{-\beta^2 r^2}\ dx\ dy\ dz$$

$$W(r)\ dr = (\beta/\sqrt{\pi})^3\ e^{-\beta^2 r^2}\ 4\pi r^2\ dr$$

where $r^2 = x^2 + y^2 + z^2$, $\beta = (3/2nl^2)^{1/2}$, and n is the number of freely jointed bonds of length l. On the *right* of each figure is the calculated distribution for a freely jointed polypeptide chain of 100 residues and a virtual bond length of 3.8 Å. The rms distance $\langle r^2 \rangle_0^{1/2} = 38$ Å is indicated.

The probability of the Gaussian distribution function reaches a maximum near the origin whereas that of the radial distribution function approaches zero. The latter is simply a mathematical consequence of the decreasing volume of the spherical shell between r and $r + dr$ as r decreases.

donor and acceptor; for practical reasons, distances of between 10 and 60 Å can be measured most accurately. To measure the distances between various portions of a flexible polypeptide chain, a naphthalene donor and a dansyl acceptor were attached to the ends of peptides of N-(hydroxyethyl)glutamine:

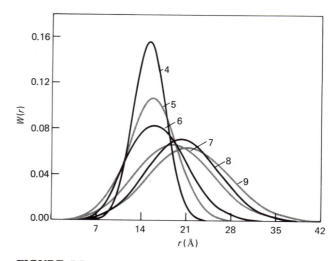

FIGURE 5.5

Radial distribution function of the distances between naphthalene and dansyl groups attached to the ends of peptides of 4–9 residues of N-hydroxyethyl-Gln, measured by fluorescence energy transfer. (From E. Haas et al., *Proc. Natl. Acad. Sci. USA* 72:1807–1811, 1975.)

$$\tag{5.12}$$

where n was 4–9. The observed efficiencies of energy transfer could be used to reconstruct the distribution function of the distances between the groups (Fig. 5.5).

References

Distribution of end-to-end distances of oligopeptides in solution as estimated by energy transfer. E. Haas et al. *Proc. Natl. Acad. Sci. USA* 72:1807–1811 (1975).

Fluorescence energy transfer as a spectroscopic ruler. L. Stryer. *Ann. Rev. Biochem.* 47:819–846 (1978).

Diffusion-enhanced fluorescence energy transfer. L. Stryer et al. *Ann. Rev. Biophys. Bioeng.* 11:203–222 (1982).

Resolution of end-to-end distance distributions of flexible molecules using quenching-induced variations of the Förster distance for fluorescence energy transfer. I. Gryczynski et al. *Biophys. J.* 54:577–586 (1988).

c. Cross-links

The probability that the ends of a chain are spatially near each other in the protein conformation gives the probability that two functional groups on a polypeptide separated by that length of chain will interact, for example, to form a covalent cross-link. This probability can be expressed as the effective concentration of the residues at the two ends with respect to each other (see Sec. 4.4.1). Effective concentration values calculated for cyclization, by forming a peptide bond between the ends of a chain, of unperturbed random polypeptides of 6–20 residues are in the range of 30–60 mM for poly(Gly) and 0.4–0.9 mM for poly(Ala). For longer chains, the effective concentration decreases in proportion to $n^{-3/2}$. Experimental measurements using disulfide-bond formation between Cys residues (Sec. 7.5.4) give values in the range of 1–70 mM, depending on the number of residues between the two Cys residues.

Cross-linking two parts of a polypeptide chain with a covalent bond, such as a disulfide bond between Cys residues, decreases the conformational entropy of the disordered polypeptide. The magnitude of this effect has been estimated by considering the probability that the ends of a random coil corresponding to the points of cross-linkage would lie simultaneously within a small volume element V in the absence of the cross-link (e.g., Figs. 5.4 and 5.5). The smaller this probability, the greater the effect on conformational entropy of constraining the ends to lie within this volume by a covalent cross-link. The farther apart in the primary structure the residues that are cross-linked, the greater the decrease in conformational entropy. Equations for the loss in conformational entropy, ΔS_{conf}, on cross-linking residues that are separated by n residues are generally of the form

$$\Delta S_{conf} = -b - \frac{3}{2} R \ln n \qquad (5.13)$$

Unfortunately, it is not certain what volume element V is appropriate, and equations such as (5.13) use values of b that vary from 2.1 to 7.9 cal/(mol °C). These calculations also ignore any steric constraints on the orientations of the cross-linked segments.

The main interest in cross-links is that they stabilize any ordered conformations with which they are compatible because they destabilize the disordered state by decreasing its conformational entropy. This is often considered to be the sole effect of cross-links, but there is also evidence that cross-links affect the energetics of the ordered conformation to varying extents.

References

Macrocyclization equilibria of polypeptides. M. Mutter. *J. Amer. Chem. Soc.* 99:8307–8314 (1977).

Ring closure reactions of bifunctional chain molecules. G. Illuminati and L. Mandolini. *Acc. Chem. Res.* 14:95–102 (1981).

Dissecting the roles of individual interactions in protein stability: lessons from a circularized protein. D. P. Goldenberg. *J. Cell. Biochem.* 29:321–335 (1985).

Dependence of formation of small disulfide loops in two-cysteine peptides on the number and types of intervening amino acids. R. Zhang and G. H. Snyder. *J. Biol. Chem.* 264:18472–18479 (1989).

5.2.3 Rates of Conformational Change

The description of disordered polypeptides just presented takes into consideration their average equilibrium properties, in which case all the possible conformations should be rapidly interconvertible by bond rotations. The rate at which any reaction occurs, including conformational rotations about bonds, depends on the free energy of its transition state.

a. Transition-State Theory

For any chemical reaction, the step in which the intermediate with the highest free energy occurs determines the overall rate of the reaction. This high-energy intermediate is known as the **transition state**. Because it has the highest free energy, it is also the least populated species along the reaction pathway; it is essentially hypothetical because it cannot be observed directly. In the simplest transition-state theory, the highest energy intermediate is postulated to break down to product at the vibrational frequency of a covalent bond, v:

$$v = \frac{k_B T}{h} \qquad (5.14)$$

where k_B is Boltzmann's constant (1.38046×10^{-16} erg · K^{-1}) and h is Planck's constant (6.6254×10^{-27} erg · s). At 25°C, $v = 6.2 \times 10^{12}$ s^{-1}. This frequency is assumed to be the same for all transition states, so the observed rate constant for a particular reaction is determined by the extent to which the transition state is populated; that is, its free energy relative to the initial reactants, $\Delta G^{\ddagger}$. The superscript $\ddagger$ is used to designate the transition state and the thermodynamic quantities relating to it. The value of $\Delta G^{\ddagger}$ is obtained by considering the transition state to be in equilibrium with the normal ground state of the reactant A:

$$A \underset{}{\overset{K^{\ddagger}}{\rightleftharpoons}} [A^{\ddagger}] \overset{v}{\longrightarrow} P \qquad (5.15)$$

$$\Delta G^{\ddagger} = -RT \ln K^{\ddagger} = -RT \ln \frac{[A^{\ddagger}]}{[A]} \qquad (5.16)$$

The kinetic equation for the reaction $A \rightarrow P$ is then

$$\frac{-d[A]}{dt} = \frac{d[P]}{dt} = k_{obs}[A] = \frac{k_B T}{h} [A^{\ddagger}] = \frac{k_B T K^{\ddagger}}{h} [A]$$

$$= \frac{k_B T}{h} [A] \exp \frac{-\Delta G^{\ddagger}}{RT} \qquad (5.17)$$

The observed rate constant k_{obs} is related to the energy of the transition state by

$$k_{obs} = \frac{k_B T}{h} \exp \frac{-\Delta G^{\ddagger}}{RT} \qquad (5.18)$$

With the measured value of k_{obs}, the relative free energy of the hypothetical transition state can be calculated:

$$\Delta G^{\ddagger} = RT \ln \frac{k_B T}{k_{obs} h} \qquad (5.19)$$

At 25°C with k_{obs} expressed in seconds^{-1}, this equation has the form

$$\Delta G^{\ddagger} = (17.4 - 1.36 \log k_{obs}) \text{ kcal/mol} \quad (5.20)$$

The higher the energy of the transition state, the slower the reaction. A reaction with no free-energy barrier occurs at the rate v, 6.2×10^{12} s^{-1} at 25°C.

The transition state can be imagined to have not only free energy but also all the other thermodynamic quantities: enthalpy, entropy, and heat capacity. The enthalpy of the transition state, $\Delta H^{\ddagger}$, relative to that of the initial state, determines the temperature dependence of the rate of the reaction:

$$\ln k_{obs} = \frac{-\Delta H^{\ddagger}}{R} \left(\frac{1}{T} \right) + \frac{\Delta S^{\ddagger}}{R} \qquad (5.21)$$

Just as the enthalpy change for a reaction can be obtained from the slope of a plot of log K_{eq} versus temperature^{-1} (**van't Hoff plot**), so the value of $\Delta H^{\ddagger}$ can be obtained by plotting log k_{obs} versus temperature^{-1} (**Arrhenius plot**).

b. Intrinsic Rates of Bond Rotation in Polypeptides

The free-energy barriers between the energetically favored conformations of a single residue in a polypeptide chain, by rotations of ϕ and ψ (see Fig. 5.2C), are only of the order of 0.5–1.5 kcal/mol (2–6 kJ/mol), so these rotations would be expected to occur at rates of the order 10^{12} s^{-1}. Movements in polymers, however, are complex and not entirely understood. Each conformational parameter of an ideal random coil can be independent of all others at equilibrium, but this cannot be the case for fluctuations on a short time scale. If only one bond near the middle of a chain were to rotate by 180°, the ends of the chain would have to undergo extremely large movements, and such a process is implausible in a viscous solution. It seems intuitively obvious, therefore, that the rotations of all bonds must be coordinated in such a way as to produce more plausible types of movements, but a complete description of how this might occur is not available.

The average rates at which individual bonds in disordered polypeptides change conformations are known from their **relaxation times** measured by ^{13}C nuclear magnetic resonance (Table 5.1). The relaxation time is the average time it takes a population of molecules to change in some way by e^{-1} of their equilibrium positions. For a bond rotation, this change is 68°, the angle for which the cosine has the value e^{-1}. The relaxation times of the C$^{\alpha}$ atoms of the backbone of a disordered polypeptide chain have been measured to be 1.4–2.6 ns, indicating that the bonds of the polypeptide backbone are rotating by more than 1° every 2 × 10^{-11} s. These rotations of the backbone occur more slowly than they would in a very small molecule; the relaxation times of the side chains are shorter and decrease for atoms that are farther from the backbone.

The rates at which the ends of a polypeptide chain are moving relative to each other by diffusion have been measured by using the efficiency of fluorescence energy transfer between a fluorescent donor and an acceptor (see Fig. 5.5). Their rates of relative motion were found to be an order of magnitude lower than the diffusion of a fluorescent donor and acceptor that are not tethered by the polypeptide chain, indicating that the polypeptide chain possesses appreciable internal friction that resists motion. Nevertheless, parts of a disordered polypeptide chain that are separated by 50–100 residues tend to

Table 5.1 *Rotational Relaxation Times in a Random Polypeptide Chain*

Carbon atom		Relaxation time[a] (10^{-9} s)
Ala	C$^{\beta}$	0.21
Thr	C$^{\beta}$	1.56
	C$^{\gamma}$	0.18
Lys	C$^{\beta}$	0.81
	C$^{\gamma}$	0.54
	C$^{\delta}$	0.60
	C$^{\epsilon}$	0.27
Peptide	C$^{\alpha}$	1.4–2.6

[a] The values were measured at 45°C on performic acid–oxidized ribonuclease A by ^{13}C nuclear magnetic resonance.

From V. Glushko et al., *J. Biol. Chem.* 247:3176–3185 (1972).

move through distances that are comparable to their average separation (Fig. 5.5) in 10^{-5}–10^{-6} s. Therefore, two groups on a disordered polypeptide chain come into proximity about 10^5–10^6 times per second.

The only conformational transition in disordered polypeptide chains known to have an intrinsically high free-energy barrier is rotation about the peptide bond, interconverting *cis* ($\omega = 0°$) and *trans* ($\omega = 180°$) forms. This rotation requires disruption of the normal double-bonded nature of the peptide bond (Eq. 1.7), and the rate of interconversion is not known because the *cis* form is not usually populated significantly with a single bond (although the *cis* population might be significant when there are many peptide bonds in a molecule, Sec. 7.5.3.a). In the case of Pro residues, however, the situation is different because the *cis* form of the preceding peptide bond is of nearly the same energy as the *trans* form (Sec. 5.2.1). This peptide bond might not be expected to have double-bond character, due to the absence of an N—H on the Pro residue, but its rate of *cis*–*trans* interconversion is very slow, with a half-time of 20 min at 0°C. The free-energy barrier therefore is 20.4 kcal/mol. The rate is very temperature dependent, having an activation energy of 20 kcal/mol, so the rate increases by a factor of 3.3 for each 10°C rise in temperature within the normal range.

References

Conformational states of bovine pancreatic ribonuclease A observed by normal and partially relaxed carbon-13 nuclear magnetic resonance. V. Glushko et al. *J. Biol. Chem.* 247:3176–3185 (1972).

Carbon 13 nuclear magnetic resonance of pentapeptides of glycine containing central residues of aliphatic amino acids. P. Keim et al. *J. Biol. Chem.* 248:6104–6113 (1973).

Determination of rotational mobilities of backbone and side-chain carbons of poly(γ-benzyl L-glutamate) in the helical and random-coil states from measurements of carbon-13 relaxation times and nuclear Overhauser enhancements. A. Allerhand and E. Oldfield. *Biochemistry* 12:3428–3433 (1973).

Brownian motion of the ends of oligopeptide chains in solution as estimated by energy transfer between the chain ends. E. Haas et al. *Biopolymers* 17:11–31 (1978).

End-to-end cyclization of polymer chains. M. A. Winnick. *Acc. Chem. Res.* 18:73–79 (1985).

Segmental relaxation in macromolecules. A. Perico. *Acc. Chem. Res.* 22:336–342 (1989).

5.3 Regular Conformations of Polypeptides

The random coil might be considered the natural state of a polymer, favored by its conformational entropy and by any favorable interactions with solvent. Certainly a single favorable hydrogen bond, salt bridge, or van der Waals interaction in a random polypeptide chain is unlikely to be stable in most instances. The values of the intrinsic association constants K_{AB} for these interactions in water (see Table 4.9) range from 0.01 to 0.9 M^{-1}, and the effective concentrations C_{eff} of pairs of groups in random coils are generally no greater than 0.1 M (Sec. 5.2.2.c). The expected value for the equilibrium constant K_{eq} for such interactions in random polypeptide chains is

$$K_{eq} = K_{AB}C_{eff} \qquad (5.22)$$

The maximum expected value of K_{eq} is then about 0.09, and even such a favorable interaction might be present only about 8% of the time.

On the other hand, multiple interactions that occur simultaneously in any particular conformation would be expected to stabilize each other (Sec. 4.4), and that conformation might be sufficiently stable to predominate over all other possible conformations. Many synthetic polyamino acids, in which one or a few amino acids are polymerized in a regular sequence, have been observed to adopt a few such regular conformations. The regularity of the conformation is a result of the regularity of the primary structure in these cases because all the identical residues tend to adopt the same conformation. That conformation can be specified by just a few dihedral angles for a single repetitive unit. If all residues have

the same conformation, a helical conformation results, which can be characterized by the number of residues per turn of the helix and by the distance traversed along the helix axis by each residue. These values for the regular conformations described here are presented in Table 5.2.

The regular conformations observed in polyamino acids, in spite of their special properties, are also found in natural proteins, in which they are known collectively as **secondary structure**.

5.3.1 The α-Helix

The right-handed α-helix is the best known and most easily recognized of the polypeptide regular structures (Fig. 5.6). This helix has 3.6 residues per turn and a translation per residue of 1.50 Å, which gives a translation of 5.41 Å per turn. The torsion angles ϕ and ψ are favorable for most residues (Fig. 5.7), and the atoms of the backbone pack closely, making favorable van der Waals interactions. Most conspicuously, the backbone

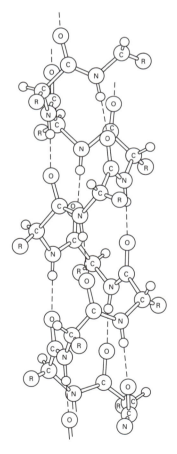

FIGURE 5.6
The classical right-handed α-helix.

Table 5.2   *Parameters for Regular Polypeptide Conformations*

	Bond Angle (deg)			Residues per turn	Translation per residue (Å)
	ϕ	ψ	ω		
Antiparallel β-sheet	−139	+135	−178	2.0	3.4
Parallel β-sheet	−119	+113	180	2.0	3.2
Right-handed α-helix	−57	−47	180	3.6	1.50
3_{10}-helix	−49	−26	180	3.0	2.00
π-helix	−57	−70	180	4.4	1.15
Polyproline I	−83	+158	0	3.33	1.9
Polyproline II	−78	+149	180	3.00	3.12
Polyglycine II	−80	+150	180	3.0	3.1

Adapted from G. N. Ramachandran and V. Sasisekharan, *Adv. Protein Chem.* 23:283–437 (1968); IUPAC–IUB Commission on Biochemical Nomenclature, *Biochemistry* 9:3471–3479 (1970).

carbonyl oxygen of each residue hydrogen-bonds to the backbone —NH of the fourth residue along the chain. These hydrogen bonds are 2.86 Å long from the oxygen atom to the nitrogen atom, are very nearly straight, and are nearly parallel to the helix axis in the classical α-helix. The exact hydrogen-bond geometry may not be optimal, however, being neither linear nor directed toward the lone-pair electrons of the oxygen atom (see Fig. 4.4).

The detailed geometry of the α-helix is found to vary somewhat in folded proteins, depending on its environment, so the classical α-helix just described is only one variant of a family of very similar structures. A slightly different geometry is usually adopted by natural proteins, with the carbonyl groups tending to point outward, away from the helix axis, and with hydrogen bonds that are less straight; the torsion angles usually have values of $\phi = -62°$ and $\psi = -41°$, instead of the classical values of −57° and −47°, respectively. This geometry is thought to be more favorable than that of the classic α-helix because it permits each carbonyl oxygen to hydrogen-bond to both the —NH of residue $i + 4$ and the aqueous solvent, or to other donors.

All the hydrogen bonds and peptide groups point in the same direction in the α-helix; consequently, the dipoles of each peptide bond should be cumulative, and an α-helix of n residues should have a macrodipole moment of $n \times 3.5$ Debye units. Furthermore, the polarization effects of hydrogen bonding may increase the dipole moment of each peptide bond by as much as 50%. A dipole moment of 3.5 Debye units is equivalent to 0.5 unit charge separated by 1.5 Å, the axial length of each residue in the helix; so the macrodipole of an α-helix is expected to correspond to the presence of about 0.5–0.7 unit charge at each end of the helix. The amino end of the helix would be positive, the carboxyl end negative.

The side chains project outward into solution from the α-helix, although they are tilted toward the amino end of the helix, and need not interfere with the helical backbone. There are, however, varying restrictions on

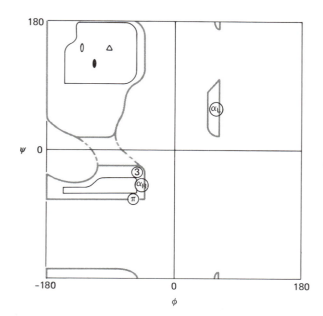

FIGURE 5.7
The positions of the regular conformations of polypeptides on a Ramachandran plot. The regular conformations are α_R, the right-handed α-helix; α_L, the left-handed α-helix; O, the antiparallel β-sheet; ●, the parallel β-sheet; 3, the right-handed 3_{10}-helix; π, the right-handed π-helix; △, polyPro I, polyPro II, and polyGly II. (From G. N. Ramachandran and V. Sasisekharan, *Adv. Protein Chem.* 23:283–437, 1968.)

conformations of the side chains. In particular, the g^+ rotamer of angle χ_1 between the C^α and C^β atoms (Eq. 5.2) is almost forbidden because any side chain would overlap atoms of the previous turn of the helix. Side chains with branched C^β atoms (Val, Ile, Thr) are most restricted in their conformations. Polar groups on some side chains, such as those of Ser, Thr, Asp, and Asn, can hydrogen-bond to the backbone peptide groups and interfere with the hydrogen bonding of the α-helix. Such interactions are often found at the ends of α-helices in folded proteins, where there are usually four carbonyl oxygens or peptide —NH groups that are not involved in the hydrogen bonding of the helix. Only Pro residues are incompatible with the α-helix conformation because the Pro side chain is bonded to the backbone nitrogen atom, which interferes with the backbone packing, and there is no —NH for hydrogen bonding. Nevertheless, Pro residues fit well at the N-terminus of an α-helix, and single Pro residues can be accommodated in long α-helices by distorting the helical geometry locally.

Many α-helices are **amphipathic**, in that they have predominantly nonpolar side chains along one side of the helical cylinder and polar residues along the remainder of its surface. Such helices have substantial hydrophobic moments (Sec. 4.3.2) and often aggregate with each other or with other nonpolar surfaces. Whether a particular amino acid sequence can produce an α-helix with such properties can be tested by plotting the sequence along a **helical wheel**, which is a projection down the helical axis of the positions of the side chains (Fig. 5.8)

A left-handed α-helix is also possible sterically; its ϕ and ψ have the same absolute values as the right-handed helix, but opposite signs (Table 5.2). This backbone conformation is also intrinsically favorable (Fig. 5.7), but the side chains are too close to the backbone, so this conformation is less stable than the right-handed version and is rarely encountered.

Although they appear structurally to be very ordered, isolated α-helices are usually only marginally stable, if at all, in aqueous solution under normal circumstances. A helix is usually formed very rapidly, within 10^{-5}–10^{-7} s, but its unraveling is usually just as rapid. The rate of formation is generally independent of the length of the polypeptide chain, but the rate of unraveling is strongly length dependent.

The explanation for the kinetics of helix formation was given by Zimm and Bragg and by Lifson and Roig. Initiation of the helix in a random coil is the slowest step, and subsequent growth of the helix is rapid. A helix can be initiated anywhere in a random coil, but growth and unraveling occur only at the ends of helices. For a polyamino acid of moderate length with a single

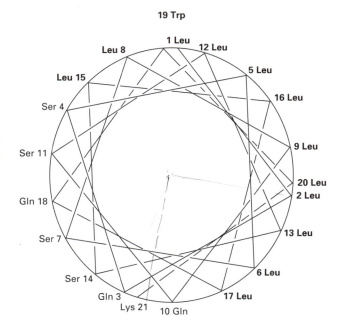

FIGURE 5.8

Helical wheel representation of an α-helix. The positions of the side chains are shown in projection down the helix axis. In an ideal α-helix, there are 3.6 residues per complete turn, or a rotation of $100°$ per residue. The helical wheel consequently repeats after five turns of 18 residues; residues 19–21 are offset slightly here to make them visible. In the amphipathic helix of the peptide shown, the hydrophobic residues are indicated in bold, and they can be seen to lie solely on one side of the helix; the opposite side is composed solely of polar residues. (From W. F. DeGrado et al., *J. Amer. Chem. Soc.* 103:679–681, 1981.)

helical region, this can be expressed simply by interconversion of two conformations, coil C and helical H_i:

$$C \underset{k_b}{\overset{\sigma k_f}{\rightleftarrows}} H_1 \underset{k_b}{\overset{k_f}{\rightleftarrows}} H_2 \underset{k_b}{\overset{k_f}{\rightleftarrows}} \cdots \underset{k_b}{\overset{k_f}{\rightleftarrows}} H_{n-1} \underset{k_b}{\overset{k_f}{\rightleftarrows}} H_n$$

$$\frac{k_f}{k_b} = s \tag{5.23}$$

where i is the number of hydrogen bonds involved in the helical conformation, k_f is the rate constant for adding a residue to the helix, and k_b is the rate constant for removing a residue. The rate constants have very similar values, in the range of 10^8–10^{11} s^{-1}, so the equilibrium constant for each residue added to the helix, s, has a value not far from unity. The nucleation factor σ reflects the difficulty in initiating a helix; the most accurately measured values are between 2×10^{-3} and 3.5×10^{-3}.

Why is initiating a helix so much more unfavorable than adding a residue? An α-helix is defined by the hydrogen bonds between residues that lie four apart in the polypeptide chain. Consequently, the two residues that participate in the first hydrogen bond (residues 1 and 5) and the three intervening residues must be appropriately fixed in space before the first turn of the helix can form. Adding an additional residue requires that only one residue be fixed; this residue is already in reasonable proximity to the end of the helix because it is the next residue in the amino acid sequence. In other words, the entropic cost of forming the first turn and first hydrogen bond of the helix is much greater than that of adding residues; the effective concentrations of two residues lying four apart in a random coil are much lower than when one is part of a helix and the other is the next residue in the polypeptide chain.

Other factors can also play a role in the energetics of initiating an α-helix. For example, the dipoles of residues in one turn of an α-helix are aligned parallel; that is, unfavorably. Once the helix is formed, however, this parallel alignment is compensated by favorable head-to-tail interactions between the peptide dipoles that lie three and four residues apart. The unfavorable dipole interactions predominate in the first turn of helix, but the subsequent favorable dipole interactions assist subsequent helix propagation.

Owing to the difficulty of nucleation, the α-helix–coil transition is cooperative. The equilibrium constant between the two conformations for n residues is given by

$$K_n = \frac{[H_n]}{[C]} = \sigma s^n \tag{5.24}$$

With σ values of about 2×10^{-3} and s values not much greater than unity, large values of n are required to give values of K_n greater than unity (i.e., a stable α-helix); only long polypeptides are expected to be helical under such conditions. The total average helix content of polypeptides, however, is given by the sum of all the possible helices of various lengths. For example, with $\sigma = 2 \times 10^{-3}$ and $s = 1.2$, a 13-residue peptide is totally helical only 2.1% of the time, but the partial helices increase the average helix content of a peptide to nearly 20%.

At the midpoint of the helix–coil transition in very long chains, with 50% of the residues in the helical conformation, the average length of a helix is given by $\sigma^{-1/2}$. Therefore, with $\sigma = 2 \times 10^{-3}$, the average helical segment at the midpoint of the transition is 22 residues long, followed by an average of 22 coil residues. When $s > 1$, the observed rate constants for helix formation (k_{+1}) and unraveling (k_{-1}) are given by

$$k_{+1} = \sigma k_f \frac{s-1}{s}$$

$$k_{-1} = k_b \frac{s-1}{s} \left(\frac{1}{s}\right)^{n-1} \tag{5.25}$$

The term $(s-1)/s$ gives the probability that helix formation will be completed once a nucleus has formed. The value of σ is nearly independent of temperature, but the value of s usually decreases with increasing temperature. Consequently, the α-helix is most stable at low temperatures and can usually be melted out by heating. Although they appear to be rigid structures (Fig. 5.6), isolated α-helices are usually dynamic systems in aqueous solution, being rapidly folded and unfolded some $10^5 - 10^7$ times per second.

The models of Zimm and Bragg and Lifson and Roig just described (Eqs. 5.23 to 5.25) apply only to certain regular polyamino acids because they are based on the assumption that the addition of a residue to either end of the helix is equally probable, which is not the case with mixed sequences or with ionized amino acid residues. Charged side chains interact with the macro-dipole of the α-helix; adding a positively charged residue to the carboxyl, negative end of a helix is more favorable, by a factor of about four, than adding such a residue to the amino end; the opposite holds for negatively charged side chains. Other polar side chains can also interact with the peptide backbone to various extents at the two ends of the helix. Hydrophobic or ionic interactions between side chains can also occur and can have significant effects on helix stability.

The various amino acids are thought to have different tendencies to form α-helices, but it has not been possible to measure their intrinsic values of σ and s by using the various polyamino acids because most of them are insoluble or preferentially form other conformations. Values of s have been estimated by incorporating the amino acids to varying extents as "guests" in a water-soluble "host" polypeptide, but this method is susceptible to interactions of the guest side chains with the peptide dipole, with each other, and with those of the host. The best recent values of the relative intrinsic helix-forming tendencies of various amino acids (Table 5.3), measured using short peptides of defined sequences, indicate that they vary more than was thought previously. For example, Ala has been estimated to have an s value of approximately 1.56 whereas that of Gly is only 0.15. Values of s are not yet available for most of the residues, and they will probably depend on the sequence in which the amino acid occurs. Nevertheless, it is clear that one factor limiting the helix-forming tendencies of residues with branched side chains is that only certain conformations of the side chain are compatible with the α-helical conformation. It is not yet possible, however, to rationalize the helical propensities of all the residues on physical grounds, and it is unlikely that all the factors that determine the stability of the α-helix have been discovered.

Table 5.3 *Relative Helical Tendencies of the Amino Acids Measured in One Peptide*

Amino acid residue	Relative stabilization of α-helical conformation[a] (kcal/mol)
Ala	−0.77
Arg	−0.68
Lys	−0.65
Leu	−0.62
Met	−0.50
Trp	−0.45
Phe	−0.41
Ser	−0.35
Gln	−0.33
Glu	−0.27
Cys	−0.23
Ile	−0.23
Tyr	−0.17
Asp	−0.15
Val	−0.14
Thr	−0.11
Asn	−0.07
His	−0.06
Gly	0
Pro	≈ 3

[a] Measured by substituting each amino acid residue into a solvent-accessible position in a synthetic peptide that forms an α-helix dimer that is in equilibrium with a randomly coiled monomeric state. The equilibrium constants for the monomer–dimer equilibrium were determined with the various peptides to provide a measure of the differences in stabilities of the α-helical conformation. The value for Gly was arbitrarily chosen as being zero.

Data from K. T. O'Neil and W. F. DeGrado, *Science* 250:646–651 (1990).

References

The structure of proteins: two hydrogen-bonded helical configurations of the polypeptide chain. L. Pauling et al. *Proc. Natl. Acad. Sci. USA* 37:205–211 (1951).

Theory of the phase transition between helix and random coil in polypeptide chains. B. H. Zimm and J. R. Bragg. *J. Chem. Phys.* 31:526–535 (1959).

On the theory of the helix–coil transition in polypeptides. S. Lifson and A. Roig. *J. Chem. Phys.* 34:1963–1974 (1961).

On the kinetics of the helix–coil transition of polypeptides in solution. G. Schwarz. *J. Mol. Biol.* 11:64–77 (1965).

Thermodynamic parameters of helix–coil transitions in polypeptide chains. O. B. Ptitsyn. *Pure Appl. Chem.* 31:227–244 (1972).

The α-helix as an electric macro-dipole. A. Wada. *Adv. Biophys.* 9:1–63 (1976).

Kinetics of the helix–coil transition of a polypeptide with nonionic side groups derived from ultrasonic relaxation measurements. B. Gruenewald et al. *Biophys. Chem.* 9:137–147 (1979).

The helical hydrophobic moment: a measure of the amphiphilicity of a helix. D. Eisenberg et al. *Nature* 299:371–374 (1982).

The role of the α-helix dipole in protein function and structure. W. G. J. Hol. *Prog. Biophys. Mol. Biol.* 45:149–195 (1985).

A thermodynamic scale for the helix-forming tendencies of the commonly occurring amino acids. K. T. O'Neill and W. F. DeGrado. *Science* 250:646–651 (1990).

Large differences in the helix propensities of alanine and glycine. A. Chakrabartty et al. *Nature* 351:586–588 (1991).

5.3.2 β-Sheets

After the α-helix, the second most regular and identifiable conformation adopted by homopolypeptides is the β-sheet (Fig. 5.9). The basic unit is the **β-strand**, with the polypeptide almost fully extended; this can be considered a special type of helix with 2.0 residues per turn and a translation of 3.4 Å per residue (Table 5.2). This extended conformation of a single chain is not stable, however, because no interactions occur among atoms that are not close in the covalent structure. The β-strand conformation is stable only when incorporated into a β-sheet, where hydrogen bonds with close to optimal geometry are formed between the peptide groups on adjacent β-strands; the dipole moments of the strands are also aligned favorably. Adjacent strands can be either parallel or antiparallel, and the stereochemistries of the strands in the two cases are slightly different. Antiparallel sheets are thought to be intrinsically more stable than parallel sheets, although this may depend on which amino acid side chains are present. Side chains from adjacent residues of the same strand protrude from opposite sides of the sheet and do not interact with each other, but they do have significant interactions with their backbone and with the side chains of neighboring strands.

The classical β-sheets originally proposed are planar and flat, but most of those observed in natural proteins have a right-handed twist, with slightly more positive values of ϕ and ψ (Table 5.2). This is believed to result from the intrinsic tendencies of the polypeptide

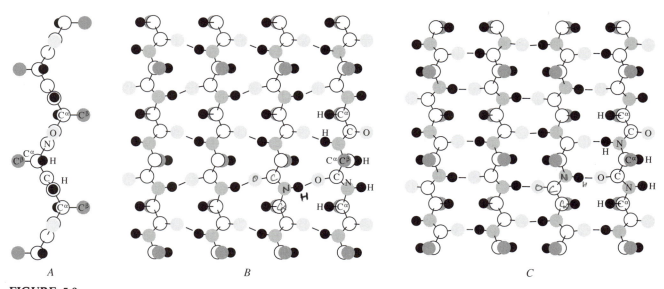

FIGURE 5.9

A single β-strand (*A*) and its incorporation into flat parallel (*B*) and antiparallel (*C*) β-sheets.

backbone and from its interactions with the side chains. Consequently, the tendency to twist may depend on what amino acid residues are present.

β-sheets are considered to be a type of secondary structure, but they differ significantly from α-helices. Helices are virtually one-dimensional structures with interactions taking place solely between residues that are close in the primary structure. In β-sheets, interactions take place among residues that are from different strands and are distant in the primary structure or that are from different molecules, depending on whether the β-sheet is intra- or intermolecular. An intramolecular β-sheet is not a completely regular structure because the segments of polypeptide chain that connect the various strands must adopt very different conformations. For that reason, an intramolecular β-sheet is not an ideal structure for a homopolypeptide unless the amino acid residue used is adept at forming both the extended β-strand conformation and suitable connections between the strands.

Poly(Tyr), poly(Lys), and poly(S-carboxymethyl-Cys) form soluble β-sheets under some conditions, but they also have a tendency to form large intermolecular aggregates because a β-sheet can grow indefinitely. For these reasons, there is no good model system for studying β-sheet formation, and very little is known about the mechanism and rate at which they are formed and about how the various amino acids differ in their β-sheet propensities. Under the most favorable conditions, intramolecular β-sheets have been observed to form on the 10^{-2} s time scale.

References

Configurations of polypeptide chains with favored orientations around single bonds: two new pleated sheets. L. Pauling and R. B. Corey. *Proc. Natl. Acad. Sci. USA* 37:729–740 (1951).

Role of interchain interactions in the stabilization of the right-handed twist of β-sheets. K. C. Chou et al. *J. Mol. Biol.* 168:389–407 (1983).

Kinetics of pH-induced random coil-β structure conversion of poly[S-(carboxymethyl)-L-cysteine]. K. Fukada et al. *Macromolecules* 22:640–645 (1989).

The β-sheet to coil transition. W. L. Mattice. *Ann. Rev. Biophys. Biophys. Chem.* 18:93–111 (1989).

5.3.3 *Other Regular Conformations*

a. The 3_{10} and π-Helices

Variations on the α-helix in which the chain is either more tightly or more loosely coiled, with hydrogen bonds to residues $i + 3$ and $i + 5$, are designated the **3_{10}-helix** and the **π-helix**, respectively. The packing of the backbone atoms is somewhat too tight in the 3_{10}-helix, and the hydrogen bonds are nonlinear; it is observed only in special circumstances (e.g., with the unnatural amino acid α-aminoisobutyric acid) or at the ends of α-helices, where the final turn may have this conformation. The name of this helix refers to the occurrence of 3 residues per turn and the 10 atoms between the hydrogen-bond donor and acceptor. (According to this nomenclature, the α-helix is a 3.6_{13}-helix.) The π-helix is wound less tightly and would have a hole down the middle, so the backbone atoms are not in contact. This helix has never been observed in proteins, nor have the other more extreme helical variants that can be imagined.

b. Poly(Pro) I and II

Pro residues are not ideally suited for either the α-helix or β-sheet conformations, so it is not surprising that poly(Pro) forms other regular conformations known as poly(Pro) I and II. Pro residues are special in permitting both *cis* and *trans* peptide bonds (Sec. 5.2.1), and the two forms of poly(Pro) differ in this respect. Poly(Pro) I contains all *cis* peptide bonds whereas form II has all *trans*. The values of ϕ and ψ are very similar for both (Table 5.2), but form I is a right-handed helix with 3.3 residues per turn, whereas form II is a left-handed helix with 3.0 residues per turn (Fig. 5.10). The values of ϕ ($-83°$ and $-78°$ for forms I and II, respectively) are compatible with that dictated by the cyclic Pro side chain. The values for ψ are constrained by steric repulsions and are similar for both forms.

Whether form I or II is adopted by poly(Pro) depends primarily on the solvent; form I predominates in propanol and butanol; form II in water, acetic acid, and benzyl alcohol. Interconversion of poly(Pro) I and II can be followed kinetically after altering the solvent. The interconversion is very slow, frequently occurring on a time scale of hours. The observed rate is nearly independent of the concentration of the initial form (i.e., the reaction is zero-order), and the half-time for the reaction is greater for longer chains. The interconversion occurs by a "zipper" mechanism in which the intrinsically slow *cis–trans* isomerization of each of the peptide bonds starts at one end and progresses sequentially along the polypeptide chain. The interconversion is so slow that a proline-specific peptidase able to cleave only amino-terminal *trans* Pro-Pro peptide bonds could be used to demonstrate that the I $\rightarrow$ II interconversion starts at the amino end of the chain, whereas the reverse interconversion starts at the carboxyl end.

c. Poly(Gly) I and II

Gly residues have unique conformational flexibility, and poly(Gly) likewise forms two regular conformations in the solid state, designated I and II. The former has a β-sheet conformation; the latter is a helix with three residues per turn like that of poly(Pro) I (Table 5.2).

References

Structure of polyglycine II. F. H. C. Crick and A. Rich. *Nature* 176:780–781 (1955).

Structure of poly-L-proline II. V. Sasisekharan. *Acta Crystallog.* 12:897–903 (1959).

Molecular structure of polyglycine II. G. N. Ramachandran et al. *Biochim. Biophys. Acta* 112:168–170 (1966).

Kinetic mechanism for conformational transitions between poly-L-prolines I and II: a study utilizing the *cis–trans* specificity of a proline-specific protease. L. N. Lin and J. F. Brandts. *Biochemistry* 19:3055–3059 (1980).

α-Helix and mixed $3_{10}/\alpha$-helix in cocrystallized conformers of Boc-Aib-Val-Aib-Aib-Val-Val-Val-Aib-Val-Aib-OMe. I. L. Karle et al. *Proc. Natl. Acad. Sci. USA* 86:765–769 (1989).

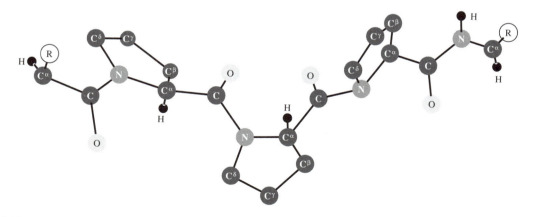

FIGURE 5.10

The poly(Pro) II helix. (From A. G. Walton, *Polypeptides and Protein Structure,* Elsevier-North Holland, New York, 1981.)

5.4 Experimental Characterization of Polypeptides in Solution

Natural polypeptides can only approximate truly random coils because each residue is subject to varying interactions with the solvent and with other parts of the polypeptide chain. Consequently, different disordered conformations do not have exactly the same energy or probability of occurrence as is assumed for the random coil. Strong favorable interactions with the solvent cause a polypeptide chain to adopt more expanded dimensions, on average, whereas relatively poor interactions with the solvent cause the polypeptide chain to collapse or aggregate. A *θ*-solvent is a solvent whose interactions with a polypeptide chain exactly balance the repulsive and attractive forces among the different parts of the chain. The polymer dimensions in a *θ*-solvent are those calculated for the unperturbed state. Natural proteins, however, are much more complex than other polymers because of the diversity of their amino acid side chains. A solvent that is good for polar side chains is likely to be poor for hydrophobic groups, and vice versa, so the existence of a *θ*-solvent for proteins is virtually impossible. The best solvents for disordered polypeptides appear to be very concentrated solutions of urea or guanidinium salts (see Fig. 7.16).

Which, if any, regular conformation is adopted by a peptide is partially determined by the solvent. Homopolypeptides of ionizable residues are expanded random coils at pH values for which they have a significant charge density. For example, poly(Lys) is disordered at neutral and acidic pH values, where its side chains tend to be ionized. At high pH values, however, the nonionized form adopts the *α*-helical conformation; raising the temperature (e.g., to 52°C for 15 min) converts this helix to aggregated antiparallel *β*-sheets. Addition of trifluoroethanol to aqueous solutions of most polypeptides tends to enhance the stability of the *α*-helical conformation. Most polypeptides can be induced into the *α*-helical conformation by dissolving them in dichloroacetic acid, whereas *β*-sheets tend to be favored in solutions with formic acid.

Long, regular polypeptides can adopt regular conformations, but most short peptides remain largely unfolded in aqueous solution. At most, they have been observed to adopt *α*-helical conformations partially at low temperatures or to show indications of other nonrandom conformations in that certain hydrogen bonds tend to be present. The physical basis for these nonrandom conformations is not yet understood but is being studied intensely.

Reference

Conformation of peptide fragments of proteins in aqueous solution: implications for initiation of protein folding. P. E. Wright et al. *Biochemistry* 27:7167–7175 (1988).

5.4.1 Hydrodynamic Properties

The dimensions of polymers are usually measured experimentally by means of their hydrodynamic properties. Such measurements are affected, however, by interactions of the polymer with the solvent; solvent molecules that are bound to the surface or trapped in crevices or holes move with the polymer and increase its effective mass. The relative contributions of the physical dimensions of a polypeptide and its solvation can be determined only by a number of different measurements.

The **intrinsic viscosity [η]** of a molecule is a measure of its effective hydrodynamic volume per unit of mass, determined by its resistance to flow. It is related to the radius of gyration R_G and the mean end-to-end distance $\langle r^2 \rangle^{1/2}$ (Sec. 5.2.2). For an ideal random polymer of molecular weight M,

$$[\eta] = \Phi \frac{\langle r^2 \rangle^{3/2}}{M} = \Phi \frac{6^{3/2} R_G^3}{M} \qquad (5.26)$$

where Φ is a constant that is independent of the nature of the polymer. For very long polymers, Φ has the value 2.1×10^{23}, but somewhat smaller values apply for shorter polypeptides.

A related measure is the **frictional coefficient f,** which is generally evaluated from the sedimentation coefficient or from the diffusion coefficient (Sec. 7.1.2). For a random coil, the value of f should be given by

$$f = 5.1 \eta_0 \langle r^2 \rangle^{1/2} \qquad (5.27)$$

where η_0 is the viscosity of the solvent. The frictional coefficient is usually expressed as the **frictional ratio,** the ratio of the observed value of f to that expected for a molecule of the same mass, but with a solid sphere with radius r, f_0:

$$f_0 = 6\pi\eta_0 r = 6\pi\eta_0 \left(\frac{3M\bar{v}}{4\pi N_A} \right)^{1/3} \qquad (5.28)$$

where $\bar{v}$ is the partial specific volume (Sec. 7.1.2.b) and N_A is Avogadro's number. The frictional ratio f/f_0 gives a measure of the departure of the overall shape from a sphere.

Viscosity and frictional coefficients have been used most frequently to measure molecular weights of synthetic polymers of known conformation or to infer the conformation of the polymer from the dependence on the molecular weight, which is usually different for different overall shapes (Fig. 5.3). Viscosity measurements are useful in following conformational changes because random polypeptides have much greater viscosities than do folded, globular proteins; viscosities of long helices are even greater (see Fig. 5.3).

5.4.2 Spectral Properties

The intrinsic spectral properties of the amino acid side chains (Table 1.3) are barely affected by the conformation of the polypeptide backbone. On the other hand, the polypeptide backbone absorbs light at wavelengths of less than 240 nm and has numerous groups that contribute to vibrational spectra. Both are affected by the polypeptide conformation.

a. Circular Dichroism

The L-amino acids in polypeptides and proteins interact differently with beams of left- and right-circularly polarized light, which causes two such beams to travel at different speeds through these molecules, thereby rotating the polarized light. **Optical rotatory dispersion (ORD)** results from the dependence of this rotation on wavelength λ. In a wavelength region where the molecule does not absorb light, the rotation varies gradually with wavelength. In a wavelength region where the light is absorbed, the absolute magnitude of the rotation ($[m']$ in Fig. 5.11) at first varies rapidly with λ, crosses zero at the absorption maximum, and then again varies rapidly with λ but in the opposite direction; this phenomenon is known as the **Cotton effect**. Left- and right-circularly polarized beams of light are also absorbed to different extents by chiral molecules, which is the basis of **circular dichroism (CD)**. These phenomena all have the same cause and are related as illustrated in Figure 5.11.

Early ORD studies used the gradual variation of rotation $[m']$ with wavelength at wavelengths greater than 240 nm, where the peptide bond does not absorb; the curve was fitted to the **Moffitt equation:**

$$[m']_\lambda = \frac{a_0 \lambda_0^2}{\lambda^2 - \lambda_0^2} + \frac{b_0 \lambda_0^4}{(\lambda^2 - \lambda_0^2)^2} \qquad (5.29)$$

where a_0, b_0, and λ_0 are constants. When λ_0 was taken to be 212 nm, the b_0 value of polypeptides was found to

be proportional to the α-helix content; with no helix, $b_0 = 0$, whereas with total α-helix, $b_0 = -630$. At shorter wavelengths, where the peptide bond absorbs, the ORD was more sensitive to conformation, including the β-sheet (Fig. 5.12). Circular dichroism has come to displace ORD, however, because it has the advantage of giving discrete spectral bands that can be positive or negative; consequently, the usual type of CD spectrum composed of multiple absorption bands is easier to resolve.

Examples of the ORD and CD spectra of poly(Lys) in the random, α-helix, and β-sheet conformations are illustrated in Figure 5.12. Similar spectra are generally

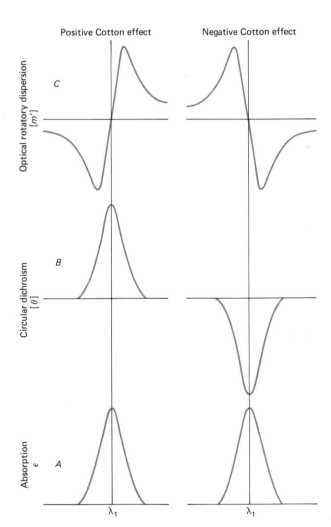

FIGURE 5.11

A typical electronic absorption band (*A*), which can have either a positive or negative Cotton effect. The two types of circular dichroism (*B*) and optical rotatory dispersion (*C*) spectra that result are illustrated. (From A. J. Adler et al., *Methods Enzymol.* 27:675–735, 1973.)

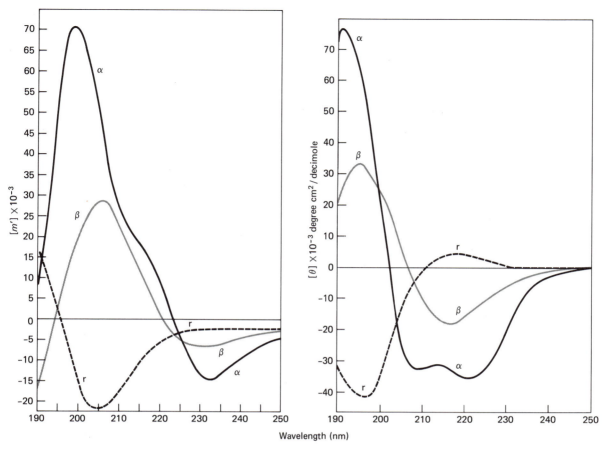

FIGURE 5.12

Optical rotatory dispersion *(left)* and circular dichroism *(right)* spectra of poly(Lys) in the α-helical (α), antiparallel β-sheet (β), and random-coil (r) conformations. (From N. J. Green-field et al., *Biochemistry* 6:1630–1637, 1967; 8:4108–4116, 1969.)

obtained for other polyamino acids in these conformations, so the CD spectra reflect primarily the conformation of the backbone. When more than one conformation is present, the observed spectrum is a combination of the spectra of the individual conformations, so the standard spectra of Figure 5.12 can often be used to deconvolute spectra of other polypeptides or proteins and to determine the relative amounts of the disordered, helical, and β-strand (often designated *extended*) conformations present. CD spectra are most sensitive to the amount of α-helix present (see Sec. 7.1.3). Problems can be encountered, however, if side chains or other conformations also contribute to the measured spectrum.

References

Determination of protein secondary structure in solution by vacuum ultraviolet circular dichroism. S. Brahms and J. Brahms. *J. Mol. Biol.* 138:149–178 (1980).

Optical spectroscopy of proteins. C. R. Cantor and S. N. Tima-sheff. In *The Proteins*, 3rd ed., H. Neurath and R. L. Hill, eds., vol. 5, pp. 145–306. Academic Press, New York, 1982.

Theoretical determination of the CD of proteins containing closely packed antiparallel β-sheets. M. C. Manning and R. W. Woody. *Biopolymers* 26:1731–1752 (1987).

Circular dichroism studies of distorted α-helices, twisted β-sheets, and β-turns. M. C. Manning et al. *Biophys. Chem.* 31:77–86 (1988).

b. Infrared and Raman Spectra

The frequencies with which bonded atoms vibrate relative to each other determine the vibrational spectrum of a molecule. Vibrational spectra are obtained as either **infrared (IR)** spectra or **Raman** spectra. The spectral bands obtained by the two techniques are due to the same transitions and are found at the same frequencies, but their intensities are governed by different factors.

Table 5.4 *Characteristic Infrared Bands of the Peptide Linkage*

Designation	Approximate frequency (cm⁻¹)	Description
A	~3300	NH stretching in resonance with (2 × amide II) overtone
B	~3100	
I	1600–1690	C=O stretching
II	1480–1575	CN stretching, NH bending
III	1229–1301	CN stretching, NH bending
IV	625–767	OCN bending, mixed with other modes
V	640–800	Out-of-plane NH bending
VI	537–606	Out-of-plane C=O bending
VII	~200	Skeletal torsion

From H. Susi, *Methods Enzymol.* 26:455–472 (1972).

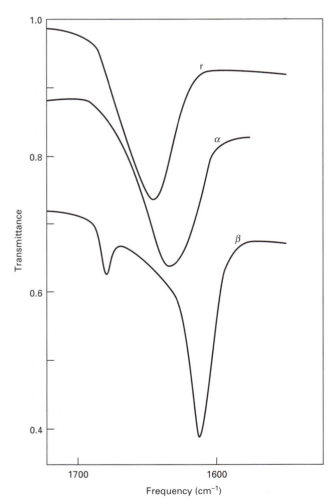

FIGURE 5.13

The amide I band of poly(Lys) in the random coil (r), α-helix (α), and antiparallel β-sheet (β) conformations, as measured by infrared spectroscopy in ²H₂O. The characteristic frequencies for this synthetic polyamino acid are somewhat different from those found in other polypeptides and proteins. (From H. Susi, *Methods Enzymol.* 26:455–472, 1972.)

Both spectra are difficult to measure because of the high background signal due to water, which absorbs strongly. Infrared spectra are often measured in ^2H_2O, which absorbs less but tends to shift the wavelengths of the polypeptide bands.

The vibrational bands of the peptide linkage are listed in Table 5.4, with a description of their origins. The amide I, II, and III bands are the most prominent, are easily measured, and are sensitive to conformation of the backbone. The amide II band is the least sensitive to conformation, and the amide III band is relatively weak and is affected by other vibrations, so most studies have concentrated on the amide I band.

The amide I band represents primarily the C=O stretching vibrations of the amide groups coupled to the in-plane NH bending and C—N stretching modes. The exact frequency of this vibration depends on the nature of hydrogen bonding involving the C=O and NH moieties, which varies for the different secondary structures of the polypeptide chain (Fig. 5.13). The α-helical conformation usually gives a band centered between 1650 and 1658 cm⁻¹, β-sheets between 1625 and 1640 cm⁻¹, and random coils between 1640 and 1648, all in ^2H_2O. These bands are not fully resolved, and detailed analysis of the spectrum is usually necessary to resolve components that are present simultaneously. Nevertheless, the amide I band is one of the spectral probes that is most sensitive to the presence of β structure.

References

Infrared spectroscopy-conformation. H. Susi. *Methods Enzymol.* 26:455–472 (1972).

Raman spectroscopy. M. C. Tobin. *Methods Enzymol.* 26:473–497 (1972).

Vibrational spectroscopy and conformation of peptides, polypeptides, and proteins. S. Krimm and J. Bandekar. *Adv. Protein Chem.* 38:181–364 (1986).

New insight into protein secondary structure from resolution-enhanced infrared spectra. W. K. Surewicz and H. H. Mantsch. *Biochim. Biophys. Acta.* 952:115–130 (1988).

Protein secondary structures in water from second-derivative amide I infrared spectra. A. Dong et al. *Biochemistry* 29:3303–3308 (1990).

5.5 Fibrous Proteins

Most fibrous proteins play structural roles and have regular, extended structures that represent a level of complexity intermediate between pure secondary structure and the tertiary structures of globular proteins. The basis for their regular conformations can be sought in regularities in their amino acid sequences. On the other hand, the large sizes of fibrous proteins and their general insolubility make them more difficult than other proteins to characterize experimentally.

Described here are the classical fibrous proteins that are best understood structurally. There are a great many other proteins with structural roles, often of medical importance, that are only now being identified and characterized. Many of these proteins are unrelated to any of the fibrous proteins described here, and many of them have properties that fit neither the fibrous nor the globular category, undergoing remarkable shape changes in response to minimal environmental perturbations. The following discussion of fibrous proteins is simply a starting point for understanding the behavior of the more complex molecules.

References

Protean proteins perceived. W. Gratzer. *Nature* 328:669–670 (1987).

Structure and hydrodynamic properties of plectin molecules. *J. Mol. Biol.* 198:515–531 (1987).

Spectrin and related molecules. S. R. Goodman et al. *Crit. Rev. Biochem.* 23:171–234 (1988).

Dystrophin: the gene and its product. J. L. Mandel. *Nature* 339:584–586 (1989).

5.5.1 Silk Fibroin

Silk fibroin is the structural protein synthesized by spiders for their webs and by silkworms for their cocoons, and used by humans for their clothes. The silkworm protein is synthesized as polypeptide chains of 350,000–415,000 M_w that are disulfide-bonded to light chains of 25,000 M_w. The silk polypeptide is stored initially as a concentrated aqueous solution, in which it has a predominantly random coil-like conformation.

The final, stable structure is adopted only after the silk is spun by the silkworm, in a process that is not understood.

The final protein is thought to consist of antiparallel β-sheets, interspersed with irregular regions of unknown structure. The segments in the β-sheet conformation consist of about 50 repeated -(Gly-Ala)$_2$-Gly-Ser-Gly-Ala-Ala-Gly-(Ser-Gly-Ala-Gly-Ala-Gly)$_8$-Tyr-sequences. The tendency for Gly residues to alternate with Ala or Ser in these sequences suggests that the silk β-sheets have predominantly Gly residues on one side, Ala and Ser residues on the other. The sheets are stacked on top of each other, with the surfaces with Gly residues tending to be in contact with each other, alternating with faces of Ala and Ser in contact; this arrangement gives alternating distances between sheets of 3.5 Å and 5.7 Å. The larger side chains cannot be accommodated between these close-packed sheets, so the presence of amino acids other than Gly, Ala, and Ser causes the polypeptide chain to adopt other irregular conformations.

References

An investigation of the structure of silk fibroin. R. E. Marsh et al. *Biochim. Biophys. Acta* 16:1–34 (1955).

Comparative studies of fibroins. II. The crystal structures of various fibroins. J. O. Warwicker. *J. Mol. Biol.* 2:350–362 (1960).

The silkworm, a model for molecular and cellular biologists. J. P. Garel. *Trends Biochem. Sci.* 7:105–108 (1982).

The spinning of silk. P. Calvert. *Nature* 315:17–18 (1985).

Primary structure of the silk fibroin light chain determined by cDNA sequencing and peptide analysis. K. Yamaguchi et al. *J. Mol. Biol.* 210:127–139 (1989).

Structure of a protein superfiber: spider dragline silk. M. Xu and R. V. Lewis. *Proc. Natl. Acad. Sci. USA* 87:7120–7124 (1990).

5.5.2 Coiled Coils

A wide variety of the structural proteins involved in maintaining cell shape, in organizing cytoplasm, and in movement are coiled coils of two or three α-helices wound around each other to form a left-handed superhelix (Fig. 5.14). The individual α-helices are distorted from their normal geometry, but only slightly, into the left-handed superhelix conformation. The α-helices interact through hydrophobic residues that form an apolar stripe along one side of each helix. There also are stabilizing electrostatic interactions between the side chains on either side of the apolar surface.

The presence of this structure in a protein is apparent from the amino acid sequence alone because of the

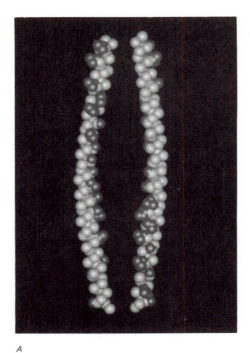

A

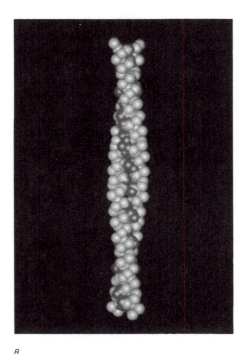

B

regularity of the α-helix structure (Fig. 5.15), which repeats every seven residues, the **heptad repeat**. The apolar stripe between helices is defined by hydrophobic side chains at residues *a* and *d* of the heptad repeat, *a-b-c-d-e-f-g*, and the electrostatic interactions occur primarily between residues *e* and *g*. Position *a* is most frequently Leu, Ile, or Ala, and *d* is primarily Leu or Ala. Residues *e* and *g* are often Glu or Gln, with Arg and Lys prominent at position *g* (Table 5.5). Charged residues also predominate at the other positions, which are in contact with the solvent. There are exceptions to this repeating heptad, however, and Pro residues are occasionally found within heptads; the exceptions are thought to cause functional variations of this regular structure.

The chains in coiled-coil proteins that have two polypeptide chains are aligned in parallel and in exact

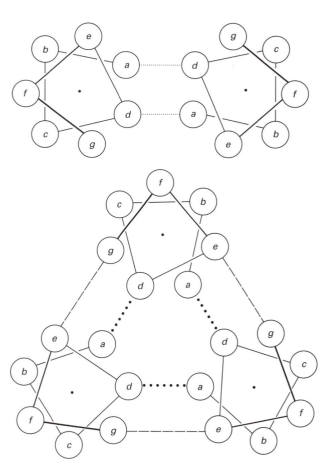

FIGURE 5.14

The structure of α-helical coiled coils. *A* shows two parallel α-helices twisted slightly as they are when interacting. The shaded atoms make up the apolar stripe of residues *a* and *d* of the heptad repeat that pack together in the parallel dimer (*B*). (From C. Cohen and D. A. D. Parry, *Proteins: Struct. Funct. Genet.* 7:1–15, 1990.)

FIGURE 5.15

Helical wheel representations of one heptad repeat of the individual α-helices, illustrating how the side chains of residues *a* and *d* pack together in the two- or three-stranded coiled coils. (From A. C. Steven et al., *J. Mol. Biol.* 200:351–365, 1988.)

Table 5.5 *Frequency of Occurrence of Amino Acids in the Heptad Repeats of Coiled-Coil Proteins*

Amino acid	Average Occurrence at Different Positions (%)						
	a	*b*	*c*	*d*	*e*	*f*	*g*
Ala	10.2	12.3	8.1	22.2	4.4	11.1	9.3
Cys	0.9	0.0	0.4	0.3	0.2	0.5	0.1
Asp	0.1	13.4	13.0	1.0	4.0	9.8	8.0
Glu	0.8	21.2	19.3	5.5	31.5	14.7	20.1
Phe	2.0	0.4	1.4	2.1	0.4	0.4	0.0
Gly	0.6	1.7	3.5	1.0	1.3	4.2	1.2
His	1.2	2.7	1.6	1.1	0.8	2.6	0.7
Ile	13.2	0.9	2.2	6.3	2.4	2.2	2.2
Lys	7.7	15.3	11.5	0.6	9.0	10.5	14.9
Leu	32.2	1.6	3.9	34.7	6.4	3.9	5.6
Met	4.9	0.9	0.9	2.3	0.9	1.0	0.4
Asn	3.6	4.3	4.6	1.1	5.7	4.8	2.7
Pro	0.0	0.2	0.1	0.0	0.0	0.0	0.0
Gln	0.8	8.7	7.5	4.1	14.0	6.1	13.2
Arg	5.5	6.2	8.2	0.9	6.6	13.2	10.7
Ser	2.1	3.6	8.1	2.2	5.2	8.1	4.5
Thr	1.2	3.8	3.7	2.2	5.1	3.6	3.5
Val	8.9	1.9	1.9	6.0	1.9	3.1	2.7
Trp	0.1	0.0	0.0	0.7	0.0	0.1	0.0
Tyr	4.1	0.9	0.3	5.7	0.3	0.2	0.2

The average occurrences of the various amino acid residues at positions *a-b-c-d-e-f-g* were tabulated from the heptad repeats of tropomyosin, myosin, paramyosin, and intermediate filament proteins.

From C. Cohen and D. A. D. Parry, *Proteins: Struct. Funct. Genet.* 7:1–15 (1990).

axial register; this arrangement is expected to be the only way to maximize coiled-coil formation between the two chains. The polypeptide chains can be identical or different. Proteins with three-stranded coiled coils are few in number and all are extracellular. Again, the chains may be identical or not. Four-stranded coils occur in the silks of bees, wasps, and ants.

As mentioned earlier, isolated α-helices are fairly unstable, even if they are very long (Sec. 5.3.1), but they are very stable in coiled coils because of the many interactions between them. The architecture of a right-handed α-helical structure, bonded by three chains of hydrogen bonds with a left-handed twist (Fig. 5.6) and convolved into a left-handed coil, produces a very stable structure in the way that extremely strong ropes can be made by coiling individually weak strands together.

Many of the proteins with coiled-coil structures have segments of polypeptide chain with different conformations at one or both ends. For example, myosin of muscle is a long coiled coil that has globular heads at each of the two carboxyl ends, which carry out the enzymatic and movement functions. Intermediate filaments have central double-helical rods 310 residues long but other segments at both ends. Many of the coiled-coil proteins also participate in higher order structures, aggregating with identical or differing molecules. These higher order structures are often dynamic, assembling and disassembling in response to circumstances.

References

The packing of α-helices: simple coiled coils. F. H. C. Crick. *Acta Cryst.* 6:689–697 (1953).

Compound helical configurations of polypeptide chains: structure of proteins of the α-keratin type. L. Pauling and R. B. Corey. *Nature* 171:59–61 (1953).

A four-strand coiled-coil model for some insect fibrous proteins. E. D. T. Atkins. *J. Mol. Biol.* 24:139–141 (1967).

Tropomyosin crystal structure and muscle regulation. G. N. Phillips et al. *J. Mol. Biol.* 192:111–131 (1986).

Intermediate filament proteins: a multigene family distinguishing major cell lineages. M. Osborne and K. Weber. *Trends Biochem. Sci.* 11:469–472 (1986).

α-Helical coiled coils—a widespread motif in proteins. C. Cohen and D. A. D. Parry. *Trends Biochem. Sci.* 11:245–248 (1986).

Molecular and cellular biology of intermediate filaments. P. M. Steinert and D. R. Roop. *Ann. Rev. Biochem.* 57:593–625 (1988).

α-Helical coiled coil and bundles: how to design an α-helical protein. C. Cohen & D. A. D. Parry. *Proteins: Struct. Funct. Genet.* 7:1–15 (1990).

5.5.3 *Collagen Triple Helix*

Collagen is the main constituent of higher animal frameworks—the bones, tendons, skin, ligaments, blood vessels, and supporting membranous tissues. In spite of this great diversity of roles, there are only about a dozen distinct, but closely related, types of collagen polypeptide chains. Collagen polypeptides are distinctive in their repetitive sequences, in which every third residue is Gly (i.e., (-Gly-Xaa-Yaa-)$_n$), with a preponderance of Pro residues as Xaa and Yaa. Many of the Pro residues (and also Lys residues) at Yaa are hydroxylated as a posttranslational modification (Sec. 2.4.7).

The reason for the repeating sequence of the collagen chain is apparent from its three-dimensional structure (Fig. 5.16). Three polypeptide chains are coiled together, each with a slightly twisted, left-handed, threefold helical conformation like that of poly(Pro) II and poly(Gly) II (Table 5.2 and Fig. 5.10). The Gly residues at every third position come into close proximity with the other two chains in the triple helix, too close to permit a side chain. The chains are linked by hydrogen bonds between the backbone —NH of the Gly residues and the backbone carbonyl group of residue Xaa of another chain. The Pro residues at positions Xaa and Yaa probably impart rigidity and stability to the structure, because this conformation is one of the few accessible to Pro residues. The hydroxyl groups on hydroxy-Pro residues (often abbreviated Hyp) are involved in hydrogen bonding between chains, which consequently stabilize the triple helix.

The three polypeptides in a typical collagen molecule are not identical; the most common collagen, collagen 1, contains two identical polypeptides (α_1) and a third (α_2) that is very similar in any individual animal. Collagen polypeptide chains typically have just over 1000 residues, but considerable variation occurs in certain types. Polypeptide chains of 1000 residues produce a collagen triple helix 14 Å in diameter and 3000 Å in length, sufficiently large to be seen in the electron microscope. Collagen triple-helix molecules up to 28,000 Å (2.8 μm) long have been observed.

Heating the collagen triple helix unfolds it and converts it to gelatin, in which the polypeptide chains are dissociated, unraveled, and disordered. Cooling the gelatin regenerates the triple-helical conformations for short stretches by means of random recombination of the polypeptide chains, but these conformations are not in the proper axial register. Collagens are assembled correctly in vivo because the polypeptide chains are synthesized with nonhelical, globular extensions of just over 100 and 300 residues at the amino and carboxyl ends, respectively. The C-terminal globular extensions associate specifically and serve to align the three polypeptide chains and to nucleate assembly of the triple helix. Folding propagates zipperlike from the C-terminus to the N-terminus. The rate-limiting step is propagation of *cis* → *trans* isomerization of the peptide bonds preceding the many Pro residues. Posttranslational modifications, especially hydroxylation and glycosylation of Pro and Lys residues at position Yaa, occur before the polypeptide chains are assembled into the triple helix. After completion of folding of the procollagen, the propeptide extensions at both ends of the chains are removed proteolytically, and specific aggregation of multiple triple-helical chains occurs, followed gradually by further cross-linking.

A number of mutant collagens have been identified in which various single amino acid replacements have varying effects on the stability of the collagen triple helix. Many of these involve replacement of the essential Gly residues by other amino acids. Besides destabilizing the triple helix, they decrease the rate of folding of the triple helix at the point of the mutation; consequently, residues on the N-terminal side of the mutation are often subject to much more posttranslational modification than normal, with further adverse effects on stability of the final structure.

Collagen functions by aggregating side-by-side into microfibrils, which assemble into larger supramolecular arrays. The dimensions of these arrays vary widely with type of collagen and site of assembly. The surface of the collagen triple helix is defined primarily by the side chains of residues Xaa and Yaa. Nonpolar and salt-bridge interactions among these side chains are believed to control the side-by-side interaction of triple helices in the microfibril. Once assembled, microfibrils are stabilized by a variety of covalent cross-links among the triple helices, involving primarily the hydroxy-Lys

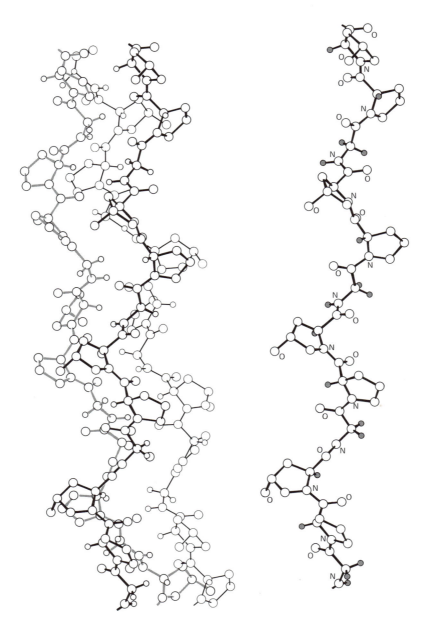

FIGURE 5.16

Model for the three-stranded collagen structure, represented as the repeating sequence -Gly-Pro-γOH Pro-. The three-stranded structure is on the *left,* a single strand on the *right.* (From R. D. B. Fraser et al., *J. Mol. Biol.* 129:463–481, 1979.)

side chains. The fibrils are assembled into many different types of connective tissue, often by combining with other types of molecules. Tendon is almost pure collagen, with the fibrils parallel to give great tensile strength. The fibrils in skin are woven into sheets that can be stretched. The fibrils in cartilage are embedded in a matrix of proteoglycans. Collagen also supplies the matrix of bone, which is cemented into a rigid structure by deposits of inorganic crystals similar to calcium hy-

droxyapatite. The molecular details of these higher levels of collagen architecture remain to be elucidated.

References

The chemistry and biology of collagen. P. Bornstein and W. Traub. In *The Proteins,* 3rd ed., H. Neurath and R. L. Hill, eds., vol. 4, pp. 411–632. Academic Press, New York, 1979.

Structurally distinct collagen types. P. Bornstein and H. Sage. *Ann. Rev. Biochem.* 49:957–1003 (1980).

Collagen: molecular diversity in the body's protein scaffold. D. R. Eyre. *Science* 207:1315–1322 (1980).

Crystal and molecular structure of a collagen-like polypeptide (Pro-Pro-Gly)$_{10}$. K. Okuyama et al. *J. Mol. Biol.* 152:427–443 (1981).

Molecular conformation and packing in collagen fibrils. R. D. B. Fraser et al. *J. Mol. Biol.* 167:497–521 (1983).

Cross-linking in collagen and elastin. D. R. Eyre et al. *Ann. Rev. Biochem.* 53:717–748 (1984).

New collagens, new concepts. R. E. Burgeson. *Ann. Rev. Cell Biol.* 4:551–577 (1988).

The family of collagen genes. E. Vuorio and B. de Crombrugghe. *Ann. Rev. Biochem.* 59:837–872 (1990).

Analysis of structural design features in collagen. E. Y. Jones and A. Miller. *J. Mol. Biol.* 218:209–219 (1991).

The zipper-like folding of collagen triple helices and the effects of mutations that disrupt the zipper. J. Engel and D. J. Prockop. *Ann. Rev. Biophys. Biophys. Chem.* 20:137–152 (1991).

Exercises

1. The first edition of this book contained an illustration of the structure of a β-sheet that contained D-amino acid residues. This illustration was taken from the classical work on this conformation, so how could such a mistake possibly have been made?

ANSWER
A. S. Edison, *Trends Biochem. Sci.* 15:216–217 (1990).

2. To test the role of salt bridges between oppositely charged side chains separated by three and four residues in stabilizing the α-helical conformation, two pairs of peptides of the type

$$\overset{-}{A}E\overset{+}{A}A\overset{-}{A}K\overset{+}{E}A\overset{-}{A}A\overset{+}{K}E\overset{-}{A}A\overset{+}{A}K A$$
$$\overset{+}{A}K\overset{-}{A}A\overset{+}{A}E\overset{-}{K}A\overset{+}{A}A\overset{-}{E}K\overset{+}{A}A\overset{-}{A}E A$$

$$\overset{-}{A}E\overset{+}{A}A\overset{-}{K}A\overset{+}{E}A\overset{-}{A}K\overset{+}{A}E\overset{-}{A}A K A$$
$$\overset{+}{A}K\overset{-}{A}A\overset{+}{E}A\overset{-}{K}A\overset{+}{A}E\overset{-}{A}K\overset{+}{A}A E A$$

were compared (S. Marqusee and R. L. Baldwin, *Proc. Natl. Acad. Sci. USA* 84:8898–8902, 1987). The α-amino and α-carboxyl groups were blocked with acetyl and amide groups, respectively. How would the first and second peptides of each pair be expected to differ in their helicity? What objection could be made to the use of these peptides to measure helix propensity?

ANSWER
M. F. Perutz and G. Fermi, *Proteins: Struct. Funct. Genet.* 4:294–295 (1988).

3. A polypeptide chain containing about 70 amino acid residues demonstrates a circular dichroism spectrum typical of an α-helix under normal conditions, but in the presence of 8 M urea its spectrum is like that of a random coil. Yet no conformational transition can be detected by hydrodynamic methods upon adding urea. What is the most likely explanation?

ANSWER
Figure 5.3.

4. Especially long α-helices that appear to be intrinsically stable are found in the proteins troponin C and calmodulin (see Sec. 8.3.4.a). It was proposed that the seemingly unusual stability of these helices is the result of salt bridges between charged amino acid side chains (M. Sundaralingam et al., *Proc. Natl. Acad. Sci. USA* 82:7944–7947, 1985). How would the charged residues be expected to be spaced in the amino acid sequence?

ANSWER
A. Holtzer and M. E. Holtzer, *Macromolecules* 20:671–675 (1987); M. Sundaralingam et al., *Proteins: Struct. Funct. Genet.* 2:64–71 (1987).

5. Polyamino acids of undefined length gave distinct ^{1}H-NMR signals for the α-helical and random-coil structures when both were present at equilibrium (E. M. Bradbury et al., *Nature* 217:812–816, 1968; F. J. Joubert et al., *Biochemistry* 9:2197–2211, 1970). This is surprising if the molecules are rapidly equilibrating between the two conformations on the 10^{-5} to 10^{-7} s time scale, because then a single, averaged NMR signal should be observed. What is the most plausible explanation?

ANSWER
R. Ullman, *Biopolymers* 9:471–487 (1970).

6. The C-peptide of ribonuclease A, corresponding to residues 1–13, was concluded to be 1000-fold more helical than expected from Zimm–Bragg theory (P. S. Kim et al., *J. Mol. Biol.* 162:187–199, 1982). Other workers consider the helicity of the C-peptide to be compatible with a slightly extended version of the Zimm–Bragg model that takes into account interactions of side chains with the helix dipole and with each other (A. V. Finkelstein et al., *Nature* 345:300, 1990). Can these views be reconciled?

POSSIBLE ANSWER
A. V. Finkelstein et al., *Proteins: Struct. Funct. Genet.*, 10:287–299 (1991); A. Chakrabartty et al., *Nature* 351:586–588 (1991).

7. What is the dipole moment of a β-sheet? Is it plausible that parallel and antiparallel sheets could have substantially different dipole interactions? (W. G. Hol et al., *Nature* 294:532–536, 1981)

ANSWER
P. T. van Duijnen et al., *Biopolymers* 24:735–745 (1985).

8. What is the dipole moment of a random-coil polypeptide?

ANSWER
A. L. Pineiro and E. Saiz, *Intl. J. Biol. Macromol.* 5:37–41 (1983).

9. The peptide bond has been concluded to be formed on the ribosome during protein biosynthesis with the amino acid residues in an α-helical conformation (V. I. Lim and A. S. Spirin, *J. Mol. Biol.* 188:565–577, 1986). What is the likely relevance of this conclusion for the conformation of the nascent polypeptide chain?

10. Lys residues stabilize the α-helical conformation when they are at the carboxyl end more than when they are at the amino end. How might helices be expected to grow in poly(Lys) molecules? With a population of long poly(Lys) molecules at equilibrium with both random-coil and α-helical conformations, what would be the expected distribution of α-helices throughout the individual poly(Lys) molecules? How might this be tested experimentally?

11. Peptides and proteins have been observed to adopt a helical conformation in the presence of sodium dodecyl sulfate (SDS) (E. Bairaktari et al., *J. Amer. Chem. Soc.* 112:5383, 1990). What are the implications of this observation for the use of SDS-polyacrylamide gel electrophoresis in determining the molecular weights of polypeptides? (See Figs. 1.5 and 1.7.)

12. The *leucine zipper* was originally noticed as a characteristic repetition of Leu residues every seventh residue (W. H. Landschulz et al., *Science* 240:1759–1764, 1988). It was proposed to be a new protein structural motif in which "the leucine side chains extending from one α-helix interdigitate with those displayed from a similar α-helix of a second polypeptide, facilitating dimerization." What would have been a more plausible model?

ANSWER
Table 5.5; E. K. O'Shea et al., *Science* 254:539–544 (1991).

13. How might you expect the helical conformation of a two-chain coiled coil to change as the temperature is increased?

ANSWER
J. Skolnick and A. Holtzer, *Macromolecules* 18:1549–1559 (1985); M. E. Holtzer et al., *Biochemistry* 25:1688–1692 (1986); A. Holtzer et al., in *Protein Folding*, L. M. Gierasch and J. King, eds., pp. 177–190. AAAS, Washington, 1990.

The Folded Conformations of Globular Proteins

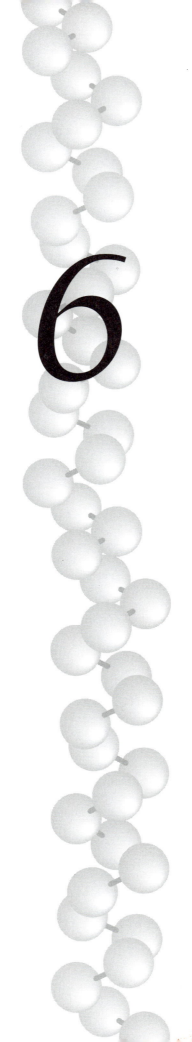

6

Most biological proteins differ dramatically from the polypeptides with random or simple repetitive conformations and from the structural proteins described in Chapter 5. Most natural proteins in solution are much smaller in their dimensions than comparable polypeptides with random or repetitive conformations and have roughly spherical shapes; hence they are generally referred to as **globular.** Their physical properties do not change gradually as the environment is altered (e.g., by changes in temperature, pH, or pressure) as do the properties of random polypeptides. Instead, globular proteins usually exhibit little or no change until a point is reached at which there is a sudden drastic change and, invariably, a loss of biological function. This phenomenon is known as **denaturation.** A denatured polypeptide chain is much more like a random polypeptide than the original protein, which must have possessed a folded three-dimensional conformation that was disrupted by denaturation and on which its biological properties were critically dependent. The nature of these folded conformations is the subject of this chapter.

Few techniques are available for determining the structures of proteins to atomic resolution. A structure can be visualized only if light with a wavelength comparable to its dimensions is used. At the atomic scale of angstroms, visualization requires X rays, electrons, or neutrons. X rays and neutrons are scattered by a protein in solution, but they cannot be focused to reconstruct an image of a single molecule. X-ray or neutron scattering provides a measure of a protein's molecular weight and radius of gyration and some information about its overall shape, but only to a resolution of 20–40 Å. Electrons can be focused in an electron microscope, but only large proteins can be seen, and these only to low resolution. Conventional electron microscopy destroys the sample with its high doses of electrons, so proteins usually must be outlined by an electron-dense material and then only the imprint left by the protein is visualized. This procedure also overcomes the problem of the lack of contrast of the protein molecules, but it limits observation of the detail of the structure. Newer techniques, such as scanning tunneling and atomic force microscopy, show promise of visualizing structures at the atomic level, but only at their surfaces.

At present, only two techniques can elucidate the three-dimensional structure of a protein to high resolution: X-ray or neutron diffraction analysis of single crystals and NMR analysis of small proteins in solution, which are described in this chapter. An understanding of the physical basis of these techniques and their strengths and weaknesses is essential for evaluation of the structures produced using them. This chapter also describes the general properties of globular protein structures. The few known structures of proteins that are integrated in membranes are described in Chapter 7.

References

Three-dimensional structure determination by electron microscopy of two-dimensional crystals. L. A. Amos et al. *Prog. Biophys. Mol. Biol.* 39:183–231 (1982).

A new non-crystallographic image-processing technique reveals the architecture of ribosomes. J. Frank et al. *Trends Biochem. Sci.* 13:123–127 (1988).

Structural studies of proteins by high-flux X-ray and neutron solution scattering. S. J. Perkins. *Biochem. J.* 254:313–327 (1988).

Probing biological structure. R. A. Crowther. *Nature* 339:426–427 (1989).

6.1 Three-Dimensional Structures by X-ray Diffraction

6.1.1 Crystallizing Proteins

Elucidation of a protein structure by X-ray or neutron diffraction analysis requires three-dimensional crystals of the protein, and not just any type of crystal. Crystals suitable for diffraction must be large, individual, and well ordered, with virtually every protein molecule held in a specific position in the three-dimensional lattice. Unfortunately, most proteins have been selected during evolution on the basis of properties other than their ability to crystallize, and little is known about the mechanism of protein crystallization. A strictly empirical approach is generally taken, searching as systematically as possible the many parameters that affect crystal formation.

In general, crystals grow only in supersaturated solutions. The concentration of a protein must be high initially, and its solubility limit must be decreased to below this concentration. The solubility of a protein can be diminished by making its interactions with the solvent less favorable, by adding agents that compete with the protein for water (Sec. 7.1.1). Protein molecules must bind water to remain in solution; when their interactions with water are lessened, they tend to associate with other protein molecules instead. By decreasing the solubility of the protein sufficiently to reach the supersaturated level, crystallization can occur after a suitable nucleation event to produce the first semblance of a crystal lattice. Nucleation must occur spontaneously to produce the first crystals but can be replaced subsequently by "seeding" with microcrystals. A great variety of physical approaches are used to alter the conditions gradually to induce crystallization. During all of these procedures, the protein must be maintained in its folded conformation.

Most protein crystals consist of about 50% solvent by volume, although this may vary between 25% and 90% (Fig. 6.1). Consequently, protein crystals are fragile, with only a few weak intermolecular contacts between the protein molecules holding the crystal together. Protein crystals depend on the solvent being present, so they must be kept wet with the mother liquor solution. Although these are technical disadvantages, the protein molecules are in an aqueous environment not vastly different from the one they normally encounter, so their three-dimensional structures are not usually altered by their incorporation into the crystal lattice. Essentially the same folded structure is obtained irrespective of the method of crystallization. Furthermore, most of the agents that diminish interactions of the protein with the solvent also tend to stabilize folded conformations (Sec. 7.4.3). The presence of large channels of solvent in protein crystals also means that small molecules may be added to the mother liquor and will diffuse throughout the crystal interior. Their interactions with the protein can then be determined crystallographically (Chap. 8).

References

Preparation and Analysis of Protein Crystals. A. McPherson. John Wiley, New York, 1982.

Comparison of the solution and crystal structures of staphylococcal nucleases with ^{13}C and ^{15}N chemical shifts used as structural fingerprints. H. B. R. Cole et al. *Proc. Natl. Acad. Sci. USA* 85:6362–6365 (1988).

Studies of crystal growth mechanisms of proteins by electron

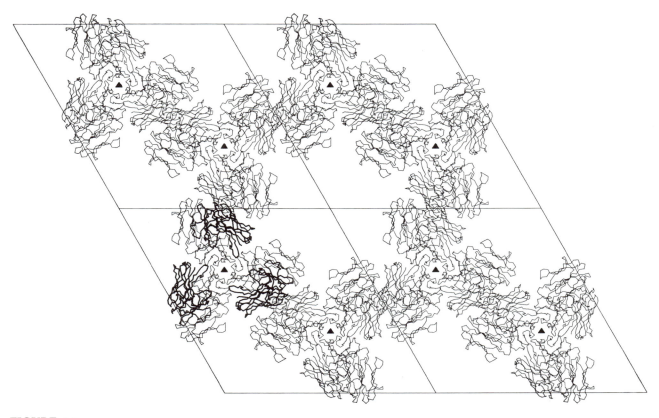

FIGURE 6.1

Crystal structure of the immunoglobulin Fab McPC603 in which 70% of the volume is solvent. Four unit cells are shown, with the Fab molecules indicated by the backbones of the heavy and light polypeptide chains that comprise each molecule. The heavy chains have been drawn bolder in the three molecules that are clustered about a crystallographic three-fold axis *(triangle)* in the lower left-hand cell. The clusters of three Fab molecules are maintained principally by a set of hydrogen bonds and van der Waals contacts between the molecules. The space group is $P6_3$; twofold screw axes are located midway between each adjacent pair of threefold axes, and a 6_3 axis is located at each corner of the unit cell. (From Y. Satow et al., *J. Mol. Biol.* 190:593–604, 1986.)

microscopy. S. D. Durbin and G. Feher. *J. Mol. Biol.* 212:763–774 (1990).

Current approaches to macromolecular crystallization. A. McPherson. *Eur. J. Biochem.* 189:1–23 (1990).

Physical principles of protein crystallization. P. C. Weber. *Adv. Protein Chem.* 41:1–36 (1991).

Comparison of the crystal structure of bacteriophage T4 lysozyme at low, medium, and high ionic strengths. J. A. Bell et al. *Proteins: Struct. Funct. Genet.* 10:10–21 (1991).

6.1.2 Basic Principles of Diffraction

Diffraction analysis determines the structure not just of one molecule but of the approximately 10^{15} molecules that make up a suitable crystal. The molecules in a crystal are arranged in specific positions and orientations on a three-dimensional lattice, so the diffraction patterns of the individual molecules add up; the crystal thus amplifies the diffraction pattern of the molecules. The basic unit of a crystal is the **unit cell,** which is the smallest parallelepiped in a crystal that, when repeated by translations that are parallel to its edges in three directions without rotating the unit, makes the **crystal lattice** (Fig. 6.2). The three directions in which the unit cell is translated in the crystal lattice define the three crystal axes. In some lattices, however, the definitions are relaxed somewhat to allow orthogonal axes (at right angles to each other), to make unit cells that are described as *face centered* or *body centered*.

The unit cell in many crystal lattices has internal symmetry in that two or more identical structures are related by axes or planes of symmetry. For example, two identical molecules may occur in a unit cell and may

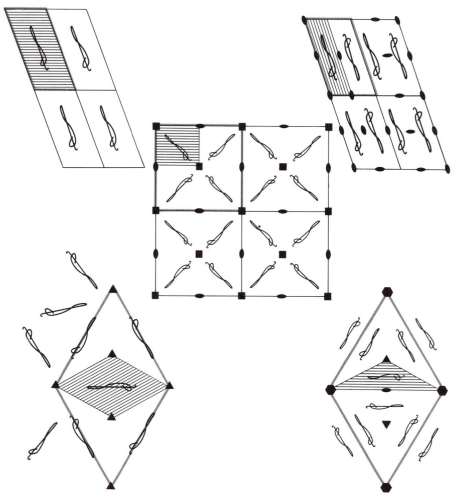

FIGURE 6.2

Some examples of unit cells and asymmetric units. A single unit cell is outlined heavily; one asymmetric unit is shaded. The following symbols represent the rotational symmetry elements: ●, twofold; ▲, threefold; ■, fourfold; ⬢, sixfold.

have different orientations with respect to the crystal axes. The symmetry axes or planes that relate the molecules often coincide with those of the crystal axes; such symmetry within the unit cell is designated *crystallographic*. The individual unit that, when repeated an integral number of times by such crystallographic operations, yields the entire unit cell is known as the **asymmetric unit**. There are 65 possible types of crystal lattices (i.e., **space groups**) in which asymmetric molecules can crystallize. Noncrystallographic symmetry is also possible, in which two or more identical molecules are related by symmetry axes that differ from those of the crystal lattice; in this case the asymmetric unit consists of more than one molecule, but always an integral number of molecules.

Not all of the space groups are possible with proteins, which are made of L-amino acids and have a unique handedness; for example, mirror planes are not possible with proteins. Furthermore, a single polypeptide chain of L-amino acids may have no exact internal symmetry; therefore, the asymmetric unit of a protein crystal must include at least one molecule of a complete polypeptide chain.

In the remainder of this discussion, the radiation is assumed to be X rays because they have been used in most of the work on protein structure. The more limited results from neutron diffraction are described briefly later. For technical reasons, the X rays used have nearly always been those of wavelength 1.542 Å, the so-called Cu K_α X rays, which are emitted upon electron bom-

Table 6.1 *Amplitudes of X-ray and Neutron Waves Scattered by the Atoms Comprising Proteins*

Element or isotope	Number of electrons	Amplitude $(10^{-12}$ cm/atom$)^a$	
		X-ray	Neutron
H	1	0.28	−0.374
^{2}H (deuterium)	1	0.28	0.667
C	6	1.69	0.665
N	7	1.97	0.94
O	8	2.25	0.58
S	16	4.5	0.28

^{a} All amplitudes are of positive sign unless otherwise noted. Only coherent scattering is considered, which predominates with these atoms.

From B. P. Schoenborn and A. C. Nunes, *Ann. Rev. Biophys. Bioeng.* 1:529–552 (1972); B. P. Schoenborn, *Methods Enzymol.* 144:510–529 (1985).

bardment of copper. The increasing use of the X-ray radiation produced by synchrotrons, however, which has a continuous spectrum of wavelengths, is resulting in the use of X rays of other wavelengths. The important feature of X rays is that they are scattered primarily by the electrons of the object; the scattering by each atom is then proportional to its atomic number (Table 6.1), and the structure determined by X-ray diffraction analysis is that of the electron density of molecules in the asymmetric unit.

The diffraction of X rays is analogous to the scattering of visible light on a larger scale, and the general principles of diffraction can be illustrated in two dimensions using **optical transforms**. These are prepared by drilling a mask with holes to represent the atoms of a molecule and illuminating it with visible light, as shown in Figure 6.3. An example of a diffraction pattern of a single molecule of six atoms is illustrated in Figure 6.3A; the important point is that this pattern is a continuous function of varying intensity in two dimensions that is determined by the structure of the molecule. When a molecule is repeated in a crystal lattice, the lattice acts as a diffraction grating, and the scattered waves are confined to a number of discrete directions — those in which the waves scattered by all the molecules of the lattice are in phase. The continuous diffraction pattern of a single molecule is then sampled at only a few points, as in Figure 6.3B,C, even though the scattered radiation is increased by the inclusion of more molecules. The directions of these scattered X rays, generally designated **reflections**, depend only on the

crystal lattice and not on the structure of the molecule (see Fig. 6.3C–F). A three-dimensional crystal lattice gives a three-dimensional lattice of scattered X rays in what is termed **reciprocal space**, because the reciprocal lattice has parameters that are the inverse of the crystal lattice (compare Figs. 6.3D,E). This is a result of the **Bragg law of diffraction**:

$$2d \sin \theta_n = n\lambda \qquad (6.1)$$

where d is the spacing between molecules in the crystal, θ_n is the angle of diffraction for the nth diffraction order, and λ is the wavelength of the radiation. For larger spacings (i.e., greater values of d), the scattering angle is decreased and the spacing is decreased in the reciprocal lattice. The directions of the three axes of the reciprocal lattice are also related reciprocally to the directions of the crystal axes. From measurements of the spacings and the axes of the reciprocal lattice of the diffraction pattern produced by diffraction of a crystal with radiation of a known wavelength, the dimensions and axes of the crystal lattice in real space can be determined. The reciprocal lattice of X-ray crystallography can be visualized directly in **precession** patterns, which record one two-dimensional layer of the three-dimensional reciprocal lattice (Fig. 6.4).

The individual reflections of the reciprocal lattice are generally designated by their **Miller indices h, k, l**, which are integers (either positive or negative) that give the number of spacings from the origin ($h = 0$, $k = 0$, $l = 0$) in the reciprocal lattice at which the reflection occurs. A specific reflection is then referred to as (h, k, l), giving the index values: (1, 2, 1), (4, 3, 2), and so on. Two-dimensional layers of the reciprocal lattice as recorded in precession photographs are designated by the Miller index of the layer; for example, ($h0l$) for the layer of all reflections for which $k = 0$.

The structural information about the contents of a unit cell is contained in the intensities of the individual reflections (compare Figs. 6.3B,C). In the first instance, any crystallographic symmetry in the unit cell produces a corresponding but inverse symmetry in the reciprocal lattice. Any such symmetry is in addition to one element that is always present, even with a nonsymmetric unit cell, because the intensity of reflection (h, k, l) is always the same as that of ($\bar{h}$, $\bar{k}$, $\bar{l}$) (the latter designation is for negative values of h, k, and l), except for the small effects of anomalous scattering (Sec. 6.1.3). Crystallographic symmetry of a unit cell can also result in systematic absence of some reflections; for example, ($0k0$) reflections are present in some space groups only if k is an even integer.

It is usually possible to determine the space group of the lattice and the dimensions of the unit cell from the

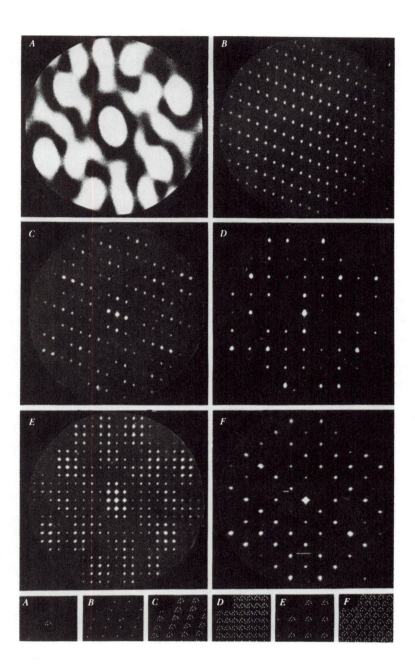

FIGURE 6.3

The effect of the crystal lattice on the diffraction pattern, illustrated with optical transforms. The masks (bottom row) were used to generate the six optical transforms at the top. The holes in the masks represent the atoms of one or more molecules; the optical transform gives the diffraction pattern of the mask. *A* shows the continuous transform of a single six-atom molecule. *B* shows the pattern from a lattice only, which gives just a uniform series of reciprocal lattice points. The consequences of incorporating the "molecule" of *A* into various lattices (*C–F*) is to sample the continuous transform of *A* at the points of the corresponding reciprocal lattice. (From C. A. Taylor and H. Lipson, *Optical Transforms,* Bell & Hyman Ltd., London, 1964.)

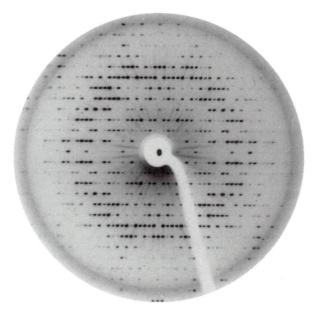

FIGURE 6.4

Example of an X-ray diffraction precession photograph. The *h0l* plane from the diffraction pattern of a crystal of the pore-forming fragment of colicin A is illustrated. The crystal was tetragonal, with space group $P4_32_12$ and unit cell dimensions $a = b = 73.0$ Å and $c = 171.6$ Å. (Photograph kindly provided by Dmitri Tsernoglou.)

diffraction pattern of a crystal. This gives the number of asymmetric units in the unit cell and the volume of each. The molecular weight of the protein in the asymmetric unit can also be estimated if the relative volumes of protein and solvent are known. The volume of solvent can be found by varying the density of the crystal mother liquor and measuring the effect of the variation on the density of an isolated crystal. The density of the crystal can be determined by finding the density at which it floats in a density gradient of liquids not miscible with the crystal solvent. Such measurements need not be very accurate because the mass of protein in the asymmetric unit must consist of an integral number of protein polypeptide chains; the correct value of the integer, therefore, is usually apparent even from approximate measurements. It is often adequate to assume that the solvent content in a protein crystal is in the usual range of 40–60%, even though extremes of 25–90% have been observed.

References

The use of X-ray diffraction in the study of protein and nucleic acid structure. K. C. Holmes and D. M. Blow. *Methods Biochem. Anal.* 13:113–239 (1965).

Protein Crystallography. T. L. Blundell and L. N. Johnson. Academic Press, New York, 1976.

X-ray structure of proteins. B. W. Matthews. In *The Proteins,* 3rd ed., H. Neurath and R. L. Hill, eds., vol. 3, pp. 403–590. Academic Press, New York, 1977.

Diffraction methods for biological macromolecules, parts A and B. H. W. Wyckoff et al., eds. *Methods Enzymol.* 114 and 115 (1985).

6.1.3 Phase Determination

The goal of diffraction analysis is reconstruction of the detailed structure of the asymmetric unit from a diffraction pattern. The diffraction pattern breaks down the structure into discrete sine waves, as in Fourier analysis (Fig. 6.5). Any shape can be represented in three dimensions as the sum of sine waves of varying amplitudes and phases. The individual reflections of a diffraction pattern represent such waves, which have wavelength components in the three dimensions inversely proportional to their values of h, k, and l, respectively. The image of the object can be reconstructed by recombining the individual sine waves, as occurs in the objective lens of a microscope. It is not possible to focus X rays, however, so the diffraction pattern itself must be recorded. Only the intensities of the reflections are recorded, not their relative phases, so it is impossible to recombine them directly. This is the well-known **phase problem** of crystallography.

The phase problem in the case of proteins was overcome initially by the method of **isomorphous replacement.** This technique depends on the preparation of protein crystals into which additional heavy atoms that scatter X rays very strongly (e.g., uranium, platinum, or mercury) have been introduced at a few specific positions, without otherwise affecting the crystal structure. The modified crystals must have the same crystal lattice as (i.e., be isomorphous with) the unmodified crystals, so that their diffraction patterns are sampled at the same points. The heavy atoms must contribute to the diffraction pattern sufficiently to alter the intensities of the reflections but cause minimal alterations of the protein structure. The diffraction pattern is then the crystallographic sum of the protein and the heavy atoms. The heavy atoms must also be in only one or a few positions of each asymmetric unit so that their positions in the unit cell can be deduced from the way they alter the protein diffraction pattern.

The preparation of heavy-atom derivatives generally relies on trial-and-error testing of many compounds rather than on design. The need for specific binding at only one or two sites on a protein makes it impossible to use reagents that react with all the residues of one type.

The affinity of heavy metal ions for thiol groups (Sec. 1.3.10) makes Cys residues logical targets, but they are useful only if there are just one or two that can react with the heavy metal without disturbing the protein structure or the crystal lattice. Individual Cys residues can be introduced by site-directed mutagenesis, but the residues that would be appropriate must be guessed. In most cases, selective binding of the heavy atom must occur at a few sites that are produced by the protein three-dimensional structure; the nature of the structure and its likely binding sites are not known, however, when the heavy-atom derivatives must be prepared. Considerable experience has yielded a collection of compounds of uranium, lead, platinum, mercury, gold, silver, and the lanthanides that have been successful in the past and are worth testing. After the protein structure has been determined, the mode of binding of the heavy atoms can be determined and can usually be rationalized, although surprises still occur.

Difference Patterson maps are usually used initially to determine the positions of the heavy atoms in a protein crystal. Only the intensities of the reflections, not their phases, are needed for the calculation of a Patterson map. Such maps show the vectors between all atoms of the unit cell, superimposed and emanating from the origin of the map (Fig. 6.6). There are $n(n-1)$ such vectors from a crystal containing n atoms in the unit cell; consequently, only Patterson maps of very simple structures can be interpreted. To determine the positions of the heavy atoms, the Patterson map is calculated using the differences in diffraction intensities produced by the heavy atoms; in the ideal case the protein structure is unchanged by the heavy atom, so the difference Patterson map is equivalent to determining the structure of only the heavy atoms.

After the positions of the heavy atoms have been determined, both the intensities and the phases of their diffraction pattern can be calculated. How the heavy-atom contribution affects the intensity of each reflection of the protein crystal depends on the relative phases of the two contributions; their intensities are combined if both have the same phase but are subtracted if they have opposite phases, with intermediate effects for other combinations of phases. If the phase of the heavy-atom contribution is known, the phase of the protein contribution can be determined, except that two solutions are usually possible. It is necessary to have a second, different heavy-atom derivative, or further information, to resolve the ambiguity. Even with two heavy-atom derivatives, the phases are usually not determined very accurately, owing to imperfect isomorphism of the crystals and to experimental errors in the measurements of the small differences in intensities produced by the heavy atoms. The phases determined by isomorphous

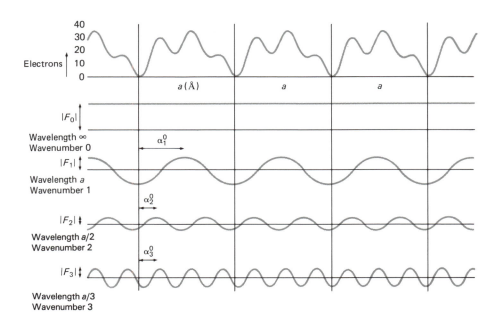

FIGURE 6.5

Fourier synthesis of a one-dimensional electron density profile *(top)* using four terms, F_0 through F_3. The unit cell dimension is a. $|F_i|$ and α_i are, respectively, the amplitude and the phase of the ith term. The sum of these four terms gives the electron density at the top. (Adapted from C. C. F. Blake, *Essays Biochem.* 11:37, 1975.)

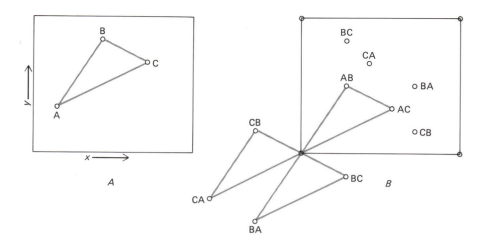

FIGURE 6.6

Two-dimensional Patterson map. *A*: Hypothetical molecule with three atoms (A, B, and C) in its unit cell in the crystal. *B*: Theoretical Patterson map of *A* with the positions of the peaks indicated. Each peak is designated as arising from the vector from the first atom to the second. The map can be generated by placing each of the three atoms, in turn, at the origin with the molecule in the correct orientation and marking the positions of the other two atoms; the resulting three representations of the molecule are indicated by the lines joining the appropriate peaks. The Patterson map is repetitive, just as the crystal is, and one unit cell is indicated by the rectangle. Large peaks at each of the unit cell origins are due to the self-vectors (AA, BB, and CC).

replacement are generally accurate only to within about 40° out of a possible range of 180°, so additional derivatives are usually desirable.

Other methods can be used to complement phase determination by heavy-atom derivatives. **Anomalous scattering** of X rays by the inner electrons of many heavy atoms is caused by resonance between the electronic vibrations excited by the incident X rays and the natural oscillations of the bound electrons. The magnitude of this effect is greatest for an X-ray wavelength close to the atomic absorption edge of an atom. It causes small differences between the otherwise identical intensities of reflections (h, k, l) and $(\bar{h}, \bar{k}, \bar{l})$. If the positions of the anomalous scattering atoms in the unit cell are known, these differences provide information about the phase of each reflection. One isomorphous heavy-atom derivative with anomalous scattering is often sufficient to define the phases. In favorable cases, anomalous scattering at several X-ray wavelengths by the sulfur atoms of Cys residues or by certain metal atoms can be sufficient for initial phase determination. Also, selenium atoms give substantial anomalous scattering and can be incorporated into proteins biosynthetically as seleno-Met residues, replacing the sulfur atom. Anomalous scattering effects are small, and their use requires very accurate measurement. Nevertheless, the approach is becoming more widely used.

When the asymmetric unit of a crystal contains two or more identical molecules that are related by noncrystallographic symmetry, the possible values of the phases are restricted. This restriction can be used to refine the initial phases to improve the electron density map. The higher the degree of symmetry, the greater the restrictions. An extreme example is the tobacco mosaic virus coat protein disk, which has 17-fold noncrystallographic symmetry; this symmetry was important for determining its structure.

Even in the absence of noncrystallographic symmetry, the information present in only the intensities of a complete set of reflections to high resolution is redundant in that there are usually 10 reflections per atom of the protein to be fixed in space. Also, the positions of the atoms are not independent, being linked by the known covalent structure. Furthermore, the electron density can never be negative. The intensities alone, therefore, provide a considerable amount of information, and measurement of the initial phases need not be particularly accurate to determine the protein structure. The initial phases can be refined, by using the measured intensities and the constraints of the covalent structure and nonnegative electron density, to give an improved determination of the structure.

The crystal structure of a protein can often be determined without measuring the phases of the reflections if the three-dimensional structure of a similar molecule is known, by means of the **molecular replacement** technique. The known structure is used as an initial model for the new structure, and the information in the amplitudes of the reflections of the new crystal is used to find the position and orientation within the unit cell of the model structure. This initial model is then refined using the measured amplitudes and the phases calculated from the model (Sec. 6.1.6).

References

Use of site-directed mutagenesis to obtain isomorphous heavy-atom derivatives for protein crystallography: cysteine-containing mutants of phage T4 lysozyme. S. DaoPin et al. *Protein Eng.* 1:115–123 (1987).

Selenomethionyl proteins produced for analysis by multi-wavelength anomalous diffraction (MAD): a vehicle for direct determination of three-dimensional structure. W. A. Hendrickson et al. *EMBO J.* 9:1665–1672 (1990).

The application of the molecular replacement method to the *de novo* determination of protein structure. M. C. Lawrence. *Quart. Rev. Biophys.* 24:399–424 (1991).

6.1.4 Calculation of the Electron Density Map

After the phases of each of the reflections are determined, the structure of the unit cell of the crystal is reconstructed by recombining mathematically the individual reflections of the diffraction pattern, a computation known as a **Fourier synthesis.** The electron density ρ at a point (x, y, z) in the unit cell, where x, y, and z are expressed as fractions of the unit cell dimensions a, b, and c, is given by

$$\rho(x, y, z) = \frac{1}{V} \sum_h \sum_k \sum_l F(h, k, l) e^{i\alpha(h,k,l)} \, e^{-2\pi i(hx+ky+lz)}$$
$$(6.2)$$

where V is the volume of the unit cell; $F(h, k, l)$ is the amplitude (the square root of the intensity) of the reflection with indices h, k, and l; and $\alpha(h, k, l)$ is its phase. The only reason for presenting this equation here is to note that calculation of the electron density at each point (x, y, z) includes the amplitudes and phases of all the reflections of the diffraction pattern. Consequently, every reflection contains information about all parts of the unit cell, just as each atom of the crystal contributes to each reflection. Therefore, a portion of the diffraction pattern does not correspond to a specific part of the crystal. Determination of a crystal structure is therefore an all-or-none process, in that all of the structure is

determined simultaneously, because the data obtained are equally relevant to all parts of the crystal.

The quality of a crystal structure determination depends on the number of reflections included in the summation of the Fourier synthesis of Equation (6.2). The Fourier synthesis consists of adding sinusoidal waves (representing electron density) with varying wavelengths (Fig. 6.5). The waves of decreasing wavelength (i.e., increasing values of h, k, and l) confer increasing detail to the electron density map. It is necessary in practice to terminate a Fourier synthesis at some maximum values of h, k, and l, using only the reflections included in a sphere about the origin of the reciprocal lattice; some errors in the calculated map are inevitable from this series termination, but the primary effect is to limit the apparent detail. This can be illustrated visually, as in Figures 6.7 and 6.8, using optical transforms in which the optical diffraction pattern is converted directly into the image, as in a microscope. The clarity of the reconstructed image depends on the extent of the diffraction pattern included in the reconstruction.

Likewise, the detail visible in electron density maps depends on the extent of the three-dimensional diffraction pattern included in the Fourier synthesis. The nominal resolution of such a structure determination is taken to be the minimum interatomic spacing (d of Eq. 6.1) that gives rise to reflections included in the Fourier synthesis. Thus crystallographic structure determinations often proceed in stages of increasing resolution, each step including higher order reflections in the calculation of the electron density map. Because the crystal lattice and the reciprocal lattice are three-dimensional, doubling the resolution requires the inclusion of eight times as many reflections.

Of course, the quality of the electron density map depends not only on the nominal resolution but also on the accuracy of the data — the amplitudes and phases of the reflections. The degree of accuracy depends on both the skill of the crystallographer and the regularity of the crystals that are used. Ultimately, the latter is the limiting factor because the resulting electron density map is an average over all the unit cells of all the crystals and over the substantial period of time during which the data were collected, a time during which the X rays may be damaging the protein. Disorder of the crystal is usually evident as a smearing of the electron density. Disorder also limits the resolution attainable because the high-order reflections fade out most rapidly as a result of disorder. Protein crystals rarely diffract to a resolution better than 1.5 Å, although for some the resolution is < 1.0 Å.

In spite of the technical difficulties and limitations of the X-ray diffraction technique, it can produce sur-

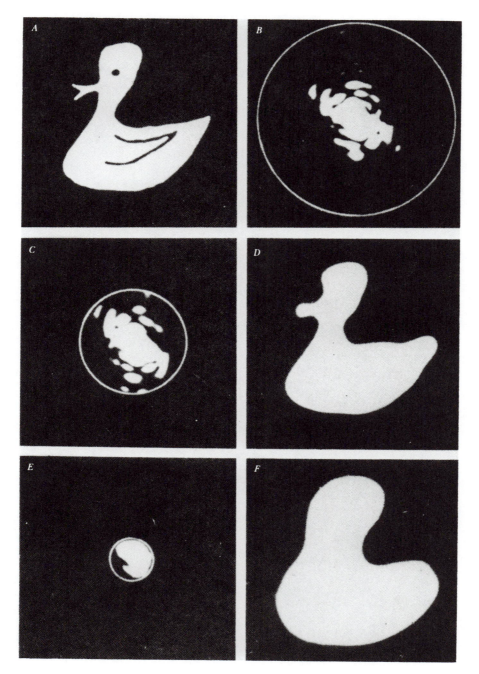

FIGURE 6.7

The diffraction pattern of the irregular object in *A* is shown in *B*. Only portions of the pattern were used in *C* and *E* to regenerate the original image in *D* and *F*, respectively, with correspondingly lower resolutions. (From C. A. Taylor and H. Lipson, *Optical Transforms*, Bell & Hyman Ltd., London, 1964.)

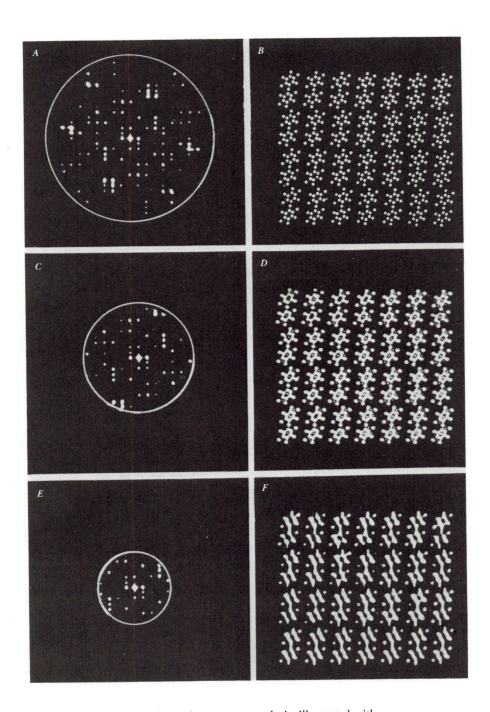

FIGURE 6.8
The effects of using fewer data of lower resolution in structure analysis, illustrated with optical transforms. The diffraction patterns enclosed in circles in *A*, *C*, and *E* were used to regenerate the original crystal lattice in *B*, *D*, and *F*, respectively; the clarity is correspondingly decreased upon using less of the diffraction pattern. (From C. A. Taylor and H. Lipson, *Optical Transforms*, Bell & Hyman Ltd., London, 1964.)

prisingly accurate and beautiful electron density maps, owing primarily to the large number of reflections usually included in the data. As an extreme example, 200,000 independent reflections were included in determining the structure at 2.8 Å resolution of the tobacco mosaic virus coat protein disk, which consists of 34 identical polypeptide chains of only 158 residues each.

6.1.5 *Interpretation of the Electron Density Map*

The all-or-nothing aspect of crystallography requires an investment of a substantial amount of work in protein crystallization, collection of X-ray diffraction data, search for isomorphous heavy-atom derivatives, calculation and deconvolution of difference Patterson maps, and calculation of the phase of each reflection, with virtually no relevant information about the protein obtained in the process. Yet after calculating the electron density map, the crystallographer is presented with detailed structural information about the entire molecule. Interpreting that awesome amount of information is not a simple task.

The protein molecule is apparent in the map as higher electron density than that of the surrounding solvent; its clarity depends on the accuracy and the resolution of the data. Covalent bonds are usually less than 1.5 Å long, so individual atoms are not resolved at resolutions less than this. At a resolution of at least 3 Å, the peptide group and the general shape of the side chains can usually be discerned, and the covalent structure can usually be traced as a continuous ribbon of electron density. At lower resolution, the polypeptide chain may not be traceable throughout the structure, except where there are dense, regular features, such as helices or β-sheets.

The clarity depends not only on the accuracy and resolution of the data but also on the degree of order of the protein in the crystal. The electron densities of portions of the protein that are mobile, or that can adopt alternative conformations in different molecules, are smeared. Groups that attain four or more very different orientations are usually invisible, but two alternative conformations can often be identified.

Owing to the nature of the electron density map, its interpretation requires knowledge of the primary structure of the protein. Some side chains can be tentatively identified by their shapes, but this is seldom unambiguous. No primary structure of any protein has been determined crystallographically, although errors in the chemically determined primary structure have been corrected from crystallographic data.

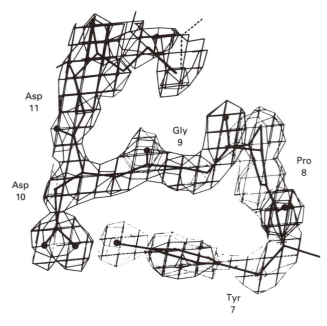

FIGURE 6.9
Fitting a protein model to a refined electron density map at 1.4 Å resolution. Residues 7–11 of avian pancreatic polypeptide are shown fitted to the map displayed as a three-dimensional net at a constant electron density. (From T. L. Blundell et al., *Proc. Natl. Acad. Sci. USA* 78:4175–4179, 1981.)

The model of a protein is usually "assembled" in a computer, using a graphics system to display the model and the electron density map simultaneously (Fig. 6.9). The standard covalent structures of amino acid residues are generally assumed, with standard bond lengths and angles. Interpreting electron density maps is a subjective procedure requiring considerable judgment and skill, and it has not yet been possible to automate this procedure. Consequently, mistakes in interpreting electron density maps do occur, so it is permissible to question details of any protein model determined crystallographically.

References

Interactive computer graphics: FRODO. T. A. Jones. *Methods Enzymol.* 115:157–171 (1985).

Using known substructures in protein model building and crystallography. T. A. Jones and S. Thirup. *EMBO J.* 5:819–822 (1986).

Modelling the polypeptide backbone with "spare parts" from known protein structures. M. Claessens et al. *Protein Eng.* 2:335–345 (1989).

Between objectivity and subjectivity. C. I. Brändén and T. A. Jones. *Nature* 343:687–689 (1990).

6.1.6 Refinement of the Model

After determination of the best fit of the protein's covalent structure to the electron density map, it is necessary to check that the model is consistent with the experimental diffraction data and to refine the model. The amplitudes of the X-ray reflections expected from the protein model can be determined by computing the Fourier transform of the model when in the crystal lattice. The agreement between the amplitudes calculated (F_{calc}) and those observed (F_{obs}) is usually measured by the **R** value:

$$R = \frac{\Sigma_{h,k,l}|F_{obs}(h, k, l) - F_{calc}(h, k, l)|}{\Sigma_{h,k,l}F_{obs}(h, k, l)} \quad (6.3)$$

where the summations are over all reflections. The smaller the R value, the better the agreement; a random assemblage of atoms in a model would be expected to give an R value of 0.59.

The R values computed for the initial protein models generally fall in the range of 0.4–0.5. One reason for this seemingly large discrepancy between the model and the data is the usual omission of the solvent in the model at this stage. The solvent generally has a substantial uniform electron density, indicative of disorder of the individual solvent molecules; increasing order of the electron density, however, is usually apparent closer to the protein molecules. Including solvent molecules at these positions in the model usually decreases the R value significantly, but including too many solvent molecules can produce R values that are artificially low and do not reflect the quality of the protein structure.

The extent of disorder of each atom affects its contribution to the diffraction pattern; this is taken into account by weighting its contribution by the factor

$$\exp\left(-B_i\frac{\sin^2\theta}{\lambda^2}\right) \quad (6.4)$$

where θ is the angle of each reflection and λ is the wavelength of the X rays (Eq. 6.1). B_i is the **temperature factor** (or Debye–Waller factor) for atom i; the greater its value, the less localized the atom in the crystal. In the ideal case of a perfect crystal lattice, the disorder is due solely to harmonic thermal vibrations of the atom, so the B value is related to the root-mean-square (rms) amplitude of vibration, U_i^2:

$$B_i = 8\pi^2 U_i^2 \cong 79 U_i^2 \quad (6.5)$$

A value of 79 Å² for B_i implies an rms vibration of 1 Å. Although termed a temperature factor, B_i includes not only temperature-dependent vibrations of the protein but also all kinds of static disorder within and between various unit cells. Consequently, B values cannot be simply interpreted according to Equation (6.5).

By varying the coordinates and the temperature factor of each atom of the protein, the initial model can be refined by several procedures to minimize the disagreement between F_{calc} and F_{obs}. The phases of each reflection have not been determined directly and are not very accurate when determined by isomorphous replacement, so their agreement with the calculated values is usually not determined. Instead, a new electron density map is calculated using the observed amplitude of each reflection (F_{obs}) and the phase value calculated from the model (α_{calc}). The model is then adjusted to give the best fit to this electron density map. The model can be altered manually or by carrying out molecular dynamics simulations of the molecule; in the latter case, a constraint is introduced to minimize the differences between the observed and calculated X-ray amplitudes in addition to the usual energy parameters that are believed to stabilize protein conformations (Sec. 7.4). The new model is then used to calculate new phases, and the process is repeated. An electron density map is often calculated with amplitudes of $|2F_{obs} - F_{calc}|$, so that differences between the observed and calculated data contribute more to the electron density map.

The agreement between observed and calculated amplitudes can be improved by a combination of refinement procedures to give R values as low as 0.10. This is still large, however, compared with the R values of 0.01 that are obtained routinely with crystallographic analyses of small molecules. If refinement does not decrease an R value to less than 0.25, the model of the protein is probably substantially incorrect.

Refined crystal structures must be interpreted cautiously because the initial model is retained to some extent in the electron density map, even if incorrect, since it determines the phases of the reflections. Fortunately, it is possible to check whether any particular feature of such a map is authentic by omitting it from the initial model. If a map is calculated with F_{obs} and α_{calc} calculated without the suspect feature in the model, then a genuine feature should reappear, although with lower prominence, because every part of the crystal structure contributes to both the amplitude and the phase of each reflection.

By fully using such refinement procedures, fairly accurate models of proteins can be derived. Examples of how an electron density map can be improved by refinement and of how well the map agrees with the model obtained are illustrated in Figure 6.10. Comparison of independent models of the same protein show that the coordinates of refined models are generally accurate to within 0.15 Å, even though data to a nominal resolution

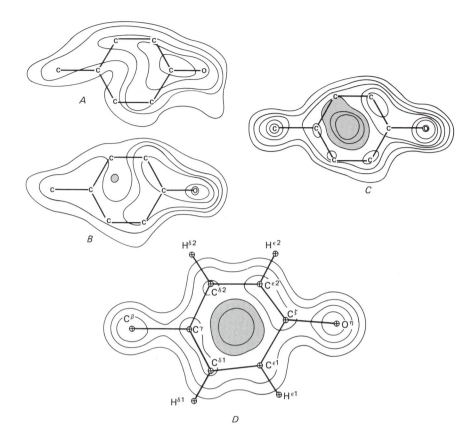

FIGURE 6.10

Effect of increased resolution and phase refinement on the electron density map. One section through the plane of the side chain of Tyr 11 of rubredoxin is shown, with the skeleton structure of the model superimposed. The inner, shaded contours are of decreasing electron density. The resolution was 2.0 Å in both A and B, using phases determined by isomorphous replacement in A and phases calculated from the model in B. In C the phases were calculated, but the resolution was 1.5 Å. The fully refined map at 1.2 Å resolution is shown in D, with the optimum positions of the atoms indicated; the ideal geometry of a Tyr side chain was not imposed during the refinement. (From K. D. Watenpaugh et al., *Acta Cryst.* B29:944–956, 1973; *J. Mol. Biol.* 138:615–633, 1980.)

of only 1.5–2.0 Å were used. The reason that the accuracy exceeds the resolution is that a resolution of 2 Å means only that all reflections (i.e., distances) of less than 2 Å are excluded. Some information about the shorter distances is contained in the lower resolution reflections, however, and the detailed geometries of the amino acid residues are known; hence, the apparent discrepancy.

References

Overview of refinement in macromolecular structure analysis. L. H. Jensen. *Methods Enzymol.* 115:227–234 (1985).

Improving crystallographic macromolecular images: the real-space approach. A. D. Podjarny et al. *Ann. Rev. Biophys. Biophys. Chem.* 16:351–373 (1987).

Inclusion of thermal motion in crystallographic structures by restrained molecular dynamics. P. Gros et al. *Science* 249:1149–1152 (1990).

6.1.7 Rapid Diffraction Measurements

The use of X-ray diffraction methods has been limited by the averaging of the data that occurs over all the unit cells of the crystal and over the time required to collect the data. Little can be done about the first limitation, so efforts have been concentrated on diminishing the second. Whereas it could take weeks to months to measure a complete diffraction data set a few years ago, the development of intense X-ray sources, efficient data collection methods, and sensitive X-ray detectors have reduced the data acquisition time remarkably.

More attention is now being given to **Laue diffraction,** which does not require rotation of the crystal and thus makes possible a further reduction in data collection time. Ordinarily, the wavelength of a monochromatic X-ray beam is kept constant while the crystal is rotated to satisfy the Bragg relationship (Eq. 6.1) by varying θ_n; thus the various reflections appear transiently. The Laue method uses a stationary crystal and a polychromatic beam of X rays with a wide range of wavelengths, usually from a synchrotron. The range of X-ray wavelengths causes many reflections to satisfy the Bragg relationship simultaneously. With a high-symmetry space group, a large fraction of the diffraction data can be obtained from a single diffraction pattern measured in a very short time. There are technical problems with Laue diffraction, primarily as a result of overlap of reflections, but it is possible to acquire data within nanoseconds to seconds, depending on the intensity of the X-ray source and the crystal. X-ray diffraction analysis is becoming a dynamic technique.

References

Time-resolved macromolecular crystallography. K. Moffat. *Ann. Rev. Biophys. Biophys. Chem.* 18:309–332 (1989).

Progress with Laue diffraction studies on protein and virus crystals. J. Hajdu and L. N. Johnson. *Biochemistry* 29:1669–1678 (1990).

6.1.8 Neutron Diffraction

X-rays distinguish among different atoms only by the difference in the number of their electrons. In proteins, sulfur atoms are the most effective scatterers of X rays, having twice as many electrons as the other atoms (Table 6.1). Carbon, nitrogen, and oxygen atoms differ so little they are not distinguishable in practice by X-ray diffraction; consequently, it is difficult to model the side chains of Asn, Gln, and Thr residues. Hydrogen atoms, which have only a single electron, barely scatter X rays at all and are visible only in very high resolution maps; their positions, however, can be extremely important to the functions of proteins.

Diffraction by neutrons is not limited by the preceding problems. Neutrons interact in a complex manner with the nuclei of atoms, and their scattering amplitudes vary widely (Table 6.1). In particular, hydrogen

atoms have a negative scattering factor, whereas the deuterium isotope has a large positive factor, as do most of the other atoms in proteins. Consequently, hydrogen atoms are discernible with neutron diffraction as negative peaks and can be readily identified (Fig. 6.11). The replacement of ^1H atoms in the protein by ^2H atoms from the solvent can also be detected (Sec. 7.3.1). The other advantage of neutron over X-ray diffraction is that the map is considerably sharper because neutrons scatter off the nucleus of an atom, which is virtually a point source, whereas X rays scatter from the more diffuse electron cloud.

Neutron diffraction does have technical difficulties, however, such as the need for a neutron reactor and

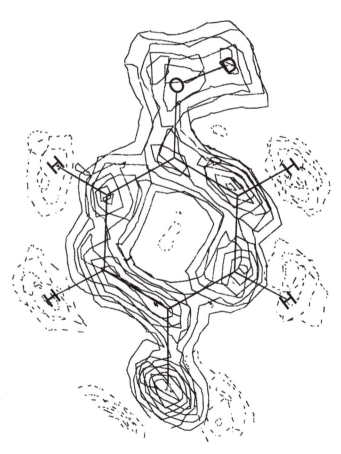

FIGURE 6.11
Neutron scattering of a well-ordered Tyr side chain in the 1.4 Å resolution structure of crambin, with the model superimposed on the map. The ^1H atoms of the aromatic ring and of the $C^\beta H_2$ group are represented by negative scattering density, whereas that of the hydroxyl group (labeled D) has exchanged with the ^2H_2O solvent and has positive density. The carbon, nitrogen, and oxygen atoms also give positive density. (Adapted from A. A. Kossiakoff, *Ann. Rev. Biochem.* 54:1195–1227, 1985.)

the low beam fluxes possible with neutrons; consequently, long exposure times and large crystals are required for obtaining data. Nevertheless, the importance of the information obtained by this technique has prompted many recent studies.

References

Neutron diffraction of crystalline proteins. A. Wlodawer. *Prog. Biophys. Mol. Biol.* 40:115–159 (1982).

The application of neutron crystallography to the study of dynamic and hydration properties of proteins. A. A. Kossiakoff. *Ann. Rev. Biochem.* 54:1195–1227 (1985).

Hydroxyl hydrogen conformations in trypsin determined by the neutron diffraction solvent difference map method: relative importance of steric and electrostatic factors in defining hydrogen-bonding geometries. A. A. Kossiakoff et al. *Proc. Natl. Acad. Sci. USA* 87:4468–4472 (1990).

6.2 The General Properties of Protein Structures

The structures of approximately 200 proteins have been determined to high resolution; references to these structures are given in Appendix 2. Data on the structures of proteins determined to atomic resolution are usually deposited in the Brookhaven Data Bank (F. C. Bernstein et al., *J. Mol. Biol.* 112:535–542, 1977) at Brookhaven National Laboratory, Upton, New York, from which all the parameters needed to describe a protein structure can be obtained.

The most striking feature of the folded conformation of a protein as determined by X-ray diffraction analysis is its complexity (Fig. 6.12), which makes it difficult to comprehend a protein structure and to communicate it to readers by means of two-dimensional media. Simplified representations are often useful in describing the gross or regular features of a structure, but they are no substitute for examining a detailed three-dimensional model, either a graphics display or a physical model. It is better still to construct such a model; especially recommended are the Nicholson model components supplied by Labquip (Labquip, Ashridgewood Place, Forest Road, Wokingham, Reading RG11 5RA, England).

It is customary to discuss proteins in terms of four levels of structure. The **primary structure** is the amino acid sequence described in the first three chapters of this book. **Secondary structure** refers to regular local structures of linear segments of polypeptide chains, such as a helix or an extended strand (described in Chap. 5). **Tertiary structure** is the overall topology of the folded polypeptide chain, and **quaternary structure** is the aggregation of the separate polypeptide chains of a protein. The crystal structure of a protein gives information about all four levels of structure, although independent knowledge of the primary structure is necessary for interpreting the electron density map. With regard to quaternary structure, however, there is often uncertainty about whether interactions between protein molecules present in a crystal lattice are relevant to the protein structure or only to the crystallization process. Fortunately, most protein crystals contain 40–60% solvent, so the crystal lattice interactions are not strong (see Fig. 6.1). Lattice interactions between independent protein molecules are not generally extensive, and many of them involve bridging by solvent molecules. More extensive direct interactions between protein molecules usually represent interactions that are also significant in solution. In general, the more extensive the contacts between molecules, the stronger the interaction. The strength of the interactions between polypeptide chains involved in strong quaternary interactions is comparable to that in the interiors of the individual protein molecules.

The properties of proteins of known structure described here give a generalized picture of protein structure, to which there are many exceptions. The common properties illustrate the general rules of protein architecture, but each protein is unique, and most of them attain their functional properties by means of specific exceptions to these generalities. This chapter briefly describes the general properties, and the remainder of the volume is concerned with the exceptions.

References

Principles of Protein Structure. G. E. Schulz and R. H. Schirmer. Springer-Verlag, New York, 1977.

The anatomy and taxonomy of protein structure. J. S. Richardson. *Adv. Protein Chem.* 34:167–339 (1981).

Principles that determine the structure of proteins. C. Chothia. *Ann. Rev. Biochem.* 53:537–572 (1984).

An Introduction to Protein Structure. C. I. Brändén and J. Tooze. Garland Publishing, New York, 1991.

Protein Architecture: A Practical Approach. A. M. Lesk. Oxford University Press, 1991.

6.2.1 The Tertiary Structure

The folded structures of most small proteins are roughly spherical and remarkably compact, with very irregular surfaces. The structures of most proteins that have more than about 200 residues appear to consist of two, three,

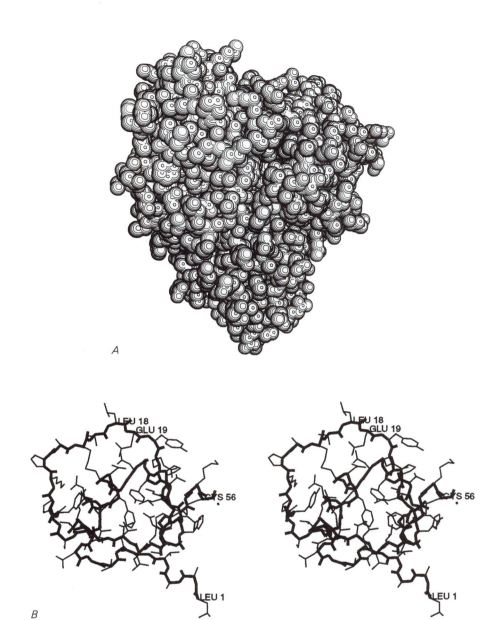

FIGURE 6.12
Various representations of the three-dimensional structures of globular proteins.
A: Space-filling model of cytochrome P450, with 414 amino acid residues, in which each atom (except hydrogen) is represented by a sphere of the appropriate van der Waals radius. Note that only the surface atoms are visible and that it is impossible to trace the polypeptide chain. (T. L. Poulos et al., *J. Mol. Biol.* 195:687–700, 1987.) *B*: Stereo picture of a skeletal model of the isolated 56-residue third domain of silver pheasant ovomucoid. Bonds between all the atoms (except hydrogen) are illustrated; the polypeptide backbone is indicated by a thicker line. The illustration can be viewed in stereo by using a stereo viewer or by diverging or converging the two eyes. With such a small protein, all of the structure can be viewed, but little information about the packing of the atoms is apparent. The terminal residues are labeled, as are residues 18 and 19 that flank the peptide bond that is cleaved reversibly by proteases. (Kindly provided by W. Bode.) *C*: Skeletal model of the two-domain 344-residue leucine/isoleucine/valine–binding protein. Bonds of the polypeptide backbone are shown as thick lines. The dots are fixed water molecules. Note that it is impossible to trace the chain in this larger protein without stereo. *D*: Same view of leucine/isoleucine/ valine–binding protein as in *C*, but a schematic representation of the polypeptide backbone, with arrows for β-strands and coils for α-helices. (Kindly provided by F. Quiocho.)

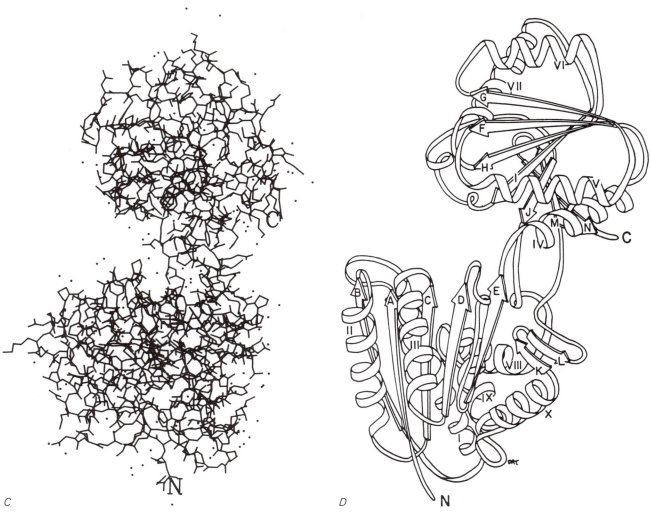

C

D

FIGURE 6.12 *(Continued)*

or more structural units, usually referred to as **domains.** The domains of a protein molecule interact to varying extents, but less extensively than do structural elements within domains. Often a single segment of polypeptide chain links the domains, and each domain consists of a single stretch of polypeptide chain. Domains along a polypeptide chain are not always segregated in this way, however; in phosphofructokinase, pyruvate kinase, and arabinose-binding protein, for example, there are two or three polypeptide connections between domains. In some cases, the end of a polypeptide chain in one domain interacts with another domain, appearing to serve as a strap that holds the domains together. The definition of a domain is not rigorous, and the division of a structure into domains is a subjective process that is done in different ways by different people. Other terms and subdivisions, such as *subdomain* and *folding unit,* are also encountered in the literature. Nevertheless, the

presence of domains in many protein molecules is clear to all observers (Fig. 6.12C, D). Domains are most evident by their compactness, which can be expressed quantitatively as the ratio of the surface area of a domain to the surface area of a sphere with the same volume: observed values are 1.64 ± 0.08 for obvious domains.

The course of a polypeptide backbone through a domain is irregular, but it generally follows a moderately straight course across the entire domain and then makes a U-turn and recrosses the domain in a more or less direct but different path to the other side (Fig. 6.13). The observer receives an impression of segments of a somewhat stiff polypeptide chain interspersed with relatively tight turns or bends, which are almost always on the molecule's surface. This general type of structure has been compared to the behavior of a fire hose when dropped in one spot. It can be contrasted with other possible limiting situations: one more irregular, such as

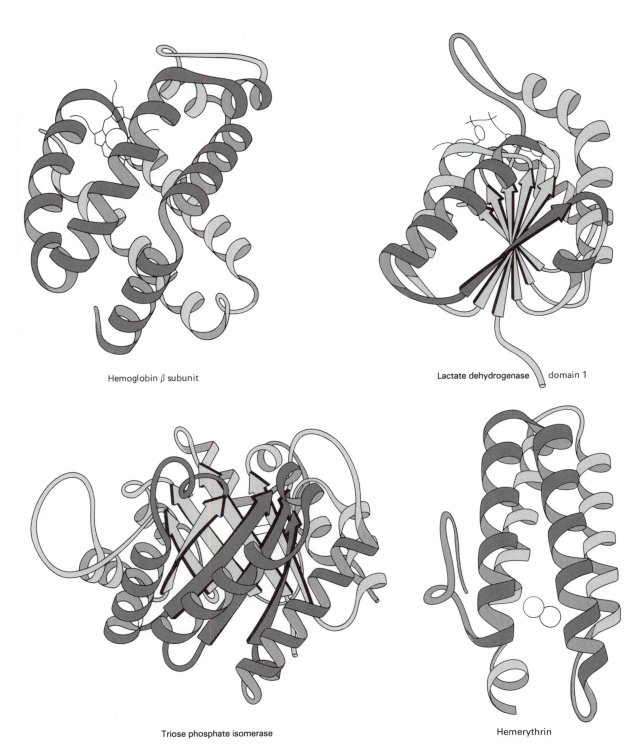

Hemoglobin β subunit

Lactate dehydrogenase   domain 1

Triose phosphate isomerase

Hemerythrin

FIGURE 6.13

Schematic drawings of the polypeptide backbones of four proteins. α-helices are depicted as helical ribbons, β-strands as arrows. The hemoglobin-β subunit has a heme group bound; hemerythrin has two iron atoms. Lactate dehydrogenase domain 1 has a molecule of the coenzyme NAD illustrated schematically. (From J. S. Richardson, *Adv. Protein Chem.* 34:167–339, 1981.)

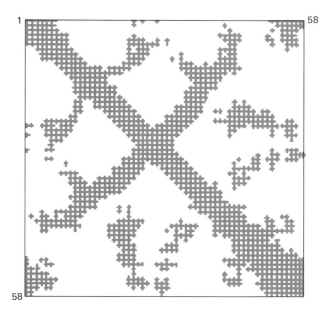

FIGURE 6.14
Contact map for bovine pancreatic trypsin inhibitor, BPTI. A cross is present whenever two residues are within 10 Å of each other. (From M. Levitt, *J. Mol. Biol.* 104:59–107, 1976.)

that obtained by dropping a flexible string, and the other more regularly curved, such as a ball of string.

Two-dimensional plots of the distances between the C^α atoms of all pairs of residues i and j (Fig. 6.14) are useful descriptions of the overall folding of polypeptide chains, especially for comparing different proteins. In such plots, called **contact maps** or **distance maps,** the distances between all pairs of C^α atoms in the protein, in order of the amino acid sequence, are represented as contours (see Fig. 6.32), or only those a certain distance apart, usually 10 Å, may be represented (Fig. 6.14).

The polypeptide backbone has never been observed to form a well-defined knot in a folded protein; that is, if a polypeptide chain were grasped at both ends and pulled straight, a linear chain would always result. Apparent exceptions occur in carbonic anhydrase and catalase where the two ends of the chain are somewhat entwined. These exceptions probably occur because only the very ends of the polypeptide chain are involved.

Rotations about the individual bonds of both the backbone and the side chains are generally close to one of the conformations favored by the isolated structural unit. Consequently, the peptide bonds of the backbone are nearly always planar and are the *trans* isomer ($\omega = 180°$), unless the next residue is Pro, when *cis* peptide

bonds ($\omega = 0$) are not so energetically unfavorable (Sec. 5.2.1). *Cis* peptide bonds occur in folded proteins at about 5% of the bonds that precede Pro residues, primarily at tight bends of the polypeptide backbone. Very few peptide bonds that do not involve Pro residues, no more than 0.05%, have been found to be *cis*. It could be, however, that many *cis* peptide bonds have been missed in earlier protein structure determinations at low and moderate resolutions, and that more *cis* peptide bonds will be observed as more refined, high-resolution structures are made. In carboxypeptidase A, for example, three non-Pro *cis* peptide bonds are present in a polypeptide chain of 307 residues; these were not apparent in the structure determined at 2 Å resolution but were found in refined structures at 1.75 Å and 1.54 Å resolution.

The dihedral angles ϕ and ψ of the polypeptide backbone generally lie within the limits deduced for the isolated peptide unit (Fig. 6.15). Similarly, rotations about the bonds of the side chains are generally close to one of the three conformations in which the attached atoms are staggered, with the conformation that gives the greatest separation of the bulkiest groups being favored (Fig. 6.16). Unfavorable stereochemistry appears to be used in proteins only when it is required for their functions.

Many proteins also contain various ligands as intimate parts of their structures: prosthetic groups, coenzymes, metal ions, and so forth. The roles of these groups are discussed in Chapters 8 and 9.

References

Directional structural features of globular proteins. G. M. Crippen and I. D. Kuntz. *J. Theor. Biol.* 66:47–61 (1977).

Conformation of amino acid side-chains in proteins. J. Janin et al. *J. Mol. Biol.* 125:357–386 (1978).

Continuous compact protein domains, M. H. Zehfus and G. D. Rose. *Proteins: Struct. Funct. Genet.* 2:90–110 (1987).

Occurrence and role of *cis* peptide bonds in protein structures. D. E. Stewart et al. *J. Mol. Biol.* 214:253–260 (1990).

Analysis of the steric strain in the polypeptide backbone of protein molecules. O. Herzberg and J. Moult. *Proteins: Struct. Funct. Genet.* 11:223–229 (1991).

6.2.2 Secondary Structure

The relatively straight segments of the polypeptide chain that traverse folded domains often have regular conformations like those observed in model polypeptides (Sec. 5.3), especially right-handed α-helices and extended β-strands associated into β-sheets. Approxi-

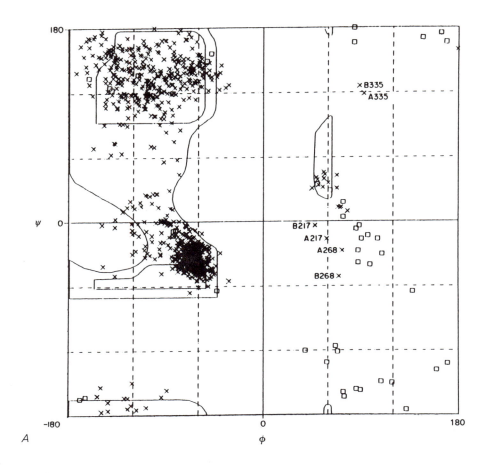

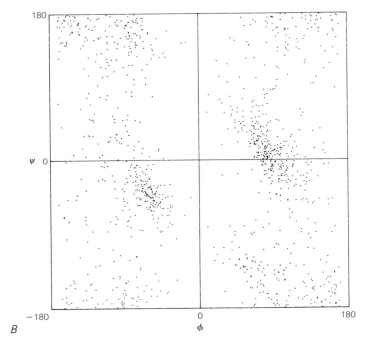

FIGURE 6.15

Ramachandran plots of the observed values of ϕ and ψ for (A) all residues in creatine amidinohydrolase refined at 1.9 Å with an R value of 17.7% and (B) Gly residues in many proteins. Gly residues in A are indicated by squares. The only other residues to lie well outside the areas considered usually allowed are residues Asp 217, Asp 268, and Arg 335, which are labeled. The A and B in A refer to the two crystallographically distinct subunits. (From H. W. Hoeffken et al., *J. Mol. Biol.* 204:417–433, 1988; C. Ramakrishnan et al., *Intl. J. Peptide Protein Res.* 29:629–637, 1987.)

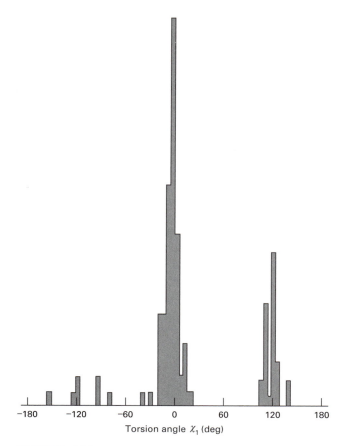

FIGURE 6.16

Histogram of the values of the C^α–C^β torsion angle χ_1 observed for 151 Val residues in highly refined protein structures. Similar sharp histograms of the three preferred rotamers are observed for other residues, but Val is special in having the greatest predominance of just one rotamer. (From J. W. Ponder and F. M. Richards, *J. Mol. Biol.* 193:775–791, 1987.)

mately 31% of the residues in known globular proteins are located in α-helices and 28% in β-strands. Other regular conformations are much less frequent. A poly(Pro) helix is frequently observed when two or more Pro residues are close to each other in a sequence. Short segments of left-handed, collagen-like helices with Pro as every third residue are found occasionally, as are short three-stranded, keratin-like coiled coils. Regular secondary structure is a useful focal point for comprehending a complex protein structure, so many schematic drawings of protein emphasize it, often in a very idealized manner. Helices are often indicated by cylinders or coiled ribbons and extended strands of β-sheets by broad arrows, indicating the amino-to-carboxyl polarity of the polypeptide backbone. Some simplified representations of several proteins are shown in Figure 6.13.

The segments of α-helices and β-sheets in globular proteins are generally short, being limited to the diameter of the protein globule. The length of an α-helix is usually 10–15 residues, and that of a β-strand is 3–10 residues. Exceptions are the 50-residue α-helix of influenza hemagglutinin and the 31-residue α-helices of calmodulin and troponin C, which dominate these structures. In a more typical α-helix of 12 residues, only the central four residues have both of the hydrogen bonds that characterize this structure (Fig. 5.6). The four residues at each end of the α-helix have only one hydrogen bond each, and the conformations of the polypeptides near these residues are often irregular, frequently forming a local 3_{10}-helix. Which residues should be counted as part of the helix, therefore, is often not clear. Various criteria—such as hydrogen bonding, conformational angles ϕ and ψ, and distances between C^α atoms—have been used for defining secondary structure in proteins.

Secondary structures of proteins are generally somewhat distorted, but the hydrogen bond lengths are most constant. In solvent-exposed α-helices, the plane of the peptide bond is often rotated so that the carbonyl group points outward from the helix axis toward the solvent. The helix axis is often curved, with the surface on the outside of the globular structure somewhat extended, possibly because the hydrogen bonds there are exposed to water and consequently are weaker and slightly longer. The values of ϕ and ψ in α-helices in folded proteins average $-62°$ and $-41°$, instead of $-48°$ and $-57°$ in the standard α-helix (Table 5.2).

Most β-sheets in folded proteins are twisted rather than planar, with a right-handed twist of $0°–30°$ between strands (see Fig. 6.13). The conformational parameters can also deviate considerably from ideality. More positive values of both ϕ and ψ than the standard values (Table 5.2) are generally observed in twisted sheets. Further distortions occur in β-sheets that consist of both parallel and antiparallel strands, because the ideal backbone conformations for the two types of sheet differ (Table 5.2). An extra residue is often present in a β-strand at the edge of a sheet, interrupting the hydrogen bond pattern and producing a **β-bulge**.

β-sheets can consist entirely of parallel or antiparallel strands or can have a mixture of the two. Purely parallel sheets are least frequent; purely antiparallel sheets are most common. Antiparallel sheets often consist of just two or three strands, whereas parallel sheets always have at least four. Mixed sheets usually contain 3–15 strands. Adjacent strands in a sheet tend to be the strands that are also adjacent in the primary structure. This correlation is greatest for antiparallel strands and least for parallel strands.

Purely parallel or antiparallel sheets of six or eight strands are often said to curve around to close up the sheet into a continuous **β-barrel**, although in some cases they are probably described more accurately as a "sandwich" of two β-sheets packed against each other. The most spectacular examples of barrels are those of eight parallel β-strands, with an α-helix on the outside of the barrel connecting each pair of β-strands. This type of structure has been found in at least 16 proteins but was first encountered in the enzyme triose phosphate isomerase (TIM) (Fig. 6.13); it is often called the **TIM barrel**, the **$(\beta\alpha)_8$ barrel**, or the **$(\alpha/\beta)_8$ barrel**. In each case, the eight parallel strands slope at an angle of about 36° to the barrel axis, and adjacent strands are offset, or sheared, by the same amount relative to each other (Fig. 6.17).

Secondary structure is most apparent in large proteins, where it comprises most of the interior. One important property of secondary structure is that it provides an efficient means of pairing, by hydrogen bonds, the polar groups of the polypeptide backbone that must be located in the protein interior, which is a prerequisite for stability of the folded conformation (Sec. 7.4).

In contact maps (see Fig. 6.14), α-helices are evident by a greater spread of close contacts along the diagonal (i.e., among nearby residues in the amino acid sequence) because C_i^α is in close proximity to C_{i-4}^α, C_{i-3}^α, C_{i+3}^α, and C_{i+4}^α, where the subscript is the number of the residue in the polypeptide chain. Within a parallel β-sheet, in which the first two residues of two adjacent extended strands are i and j, C_i^α is next to C_j^α, C_{i+1}^α is adjacent to C_{j+1}^α, and so on; this gives rise to a series of close contacts on a diagonal line that is parallel to the main diagonal but offset from it by $(i - j)$ residues. In the case of two strands of an antiparallel β-sheet, where residues i and j are the first and last hydrogen-bonded residues, respectively, C_i^α is next to C_j^α, C_{i+1}^α, is next to C_{j-1}^α, and so forth. This gives rise to a series of contacts that define a diagonal line that is perpendicular to the main diagonal (Fig. 6.14). Structural domains along the polypeptide backbone are often apparent as segregated areas of contacts on the distance plots (see Fig. 6.32 for the two-domain proteins chymotrypsin and elastase).

Protein structures have been divided into four classes on the basis of their secondary structures: (α), containing only α-helices; (β) containing primarily β-sheet structure; (α + β), containing helices and sheets in separate parts of the structure; and (α/β), in which helices and sheets interact and often alternate along the polypeptide chain. In (α) proteins, about 60% of the residues are in α-helices, and the helices are usually in contact with each other. In (β) proteins, there are always two β-sheets, both usually antiparallel, that pack against each other. In the (α + β) proteins, there may be a single β-sheet, usually antiparallel; the helices often cluster together at one or both ends of the β-sheet. The

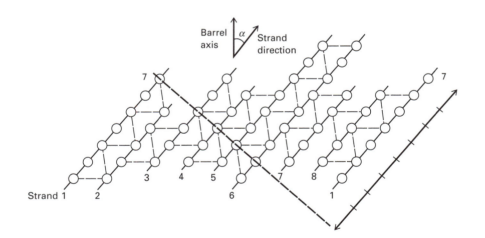

FIGURE 6.17

The β-sheet structure of TIM, $(\beta\alpha)_8$, or (α/β) barrels. The β-sheet barrel of triosephosphate isomerase has been unrolled schematically; the positions of the residues along the backbone are depicted by circles, hydrogen bonds of the β-sheet are indicated by dashed lines. The N-terminal β-strand (number 1) is shown twice, at both sides of the sheet. The thick dashed line connects residues that are opposite each other in the β-sheet, starting on the left with residue number 7 of strand 1. At the other side of the sheet, this line would intersect a residue that would be number −1 of strand 1, eight positions away from residue 7. The shear number is therefore 8. If the strands were vertical, with no shear, the dashed line would connect residue 7 in strand 1 at both sides of the sheet. The angle α gives the tilt of the β-strands from the vertical; its value is 36°. (From C. Chothia, *Nature* 333:598–599, 1988.)

(α/β) proteins have one major β-sheet of primarily parallel strands; a helix usually occurs in each of the segments of polypeptide chain connecting the β-strands, probably owing to the necessarily long lengths of these connections. The helices pack on both sides of the sheet unless the sheet is closed into a barrel, in which case the α-helices pack around the outside of the barrel (see Figs. 6.13 and 6.17).

References

Dictionary of protein secondary structure: pattern recognition of hydrogen-bonded and geometrical features. W. Kabsch and C. Sander. *Biopolymers* 22:2577–2637 (1983).

Structural properties of protein β-sheets. F. R. Salemme. *Prog. Biophys. Mol. Biol.* 42:95–133 (1983).

Hydrogen bonding in globular proteins. E. N. Baker and R. E. Hubbard. *Prog. Biophys. Mol. Biol.* 44:97–179 (1984).

Helix geometry in proteins. D. J. Barlow and J. M. Thornton. *J. Mol. Biol.* 201:601–619 (1988).

6.2.3 Reverse Turns

Nearly one-third of the residues of globular proteins are involved in the tight turns that reverse the direction of polypeptide chains at the surfaces of the molecules and make possible the overall globular structure. Because of their prevalence, these **reverse turns** or **loops** are frequently classified as a third type of secondary structure.

Various types of reverse turns occur, involving different numbers of residues and depending on which types of secondary structure they link. The best characterized are the **β hairpins** that link adjacent strands in an antiparallel β-sheet. If only one residue in a chain is not involved in the hydrogen-bonding pattern of the sheet, there is a **γ turn,** of which two types are possible (Fig. 6.18). This very tight turn requires unfavorable geometry for the adjacent hydrogen bond of the β-sheet and unusual values of ϕ and ψ in the central residue of the turn (Table 6.2). More common are **β turns,** in which two residues are not involved in the hydrogen bonding of the β-sheet; the two residues on either side of the non-hydrogen-bonded residues are included in the β turn, which, therefore, is defined by four residues at positions designated i to $i + 3$ (Fig. 6.18; Table 6.2). The existence of three ideal β turns, designated types I, II, and III, was predicted by Venkatachalam on the basis of allowed polypeptide geometry with planar *trans* peptide bonds. Mirror images of the backbone — but not the side chains, of course — occur in variants I', II', and III'. Type-I β turns occur most frequently, two to three times more frequently than type II. The mirror-image types I'

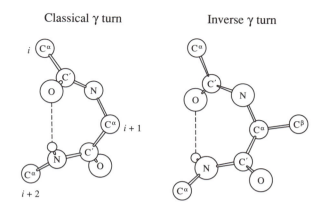

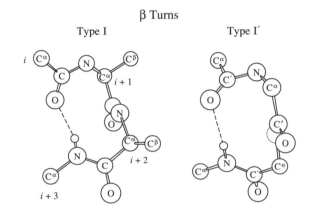

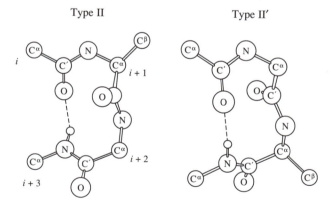

FIGURE 6.18

The most common γ turns and β turns connecting adjacent strands of an antiparallel β-sheet. The three and four residues, respectively, that are considered to define the turns are shown, with the first residue designated i. C^β atoms are shown only in positions where non-Gly residues occur frequently. The last hydrogen bond of the β-sheet is shown as a dashed line. (From G. D. Rose et al., *Adv. Protein Chem.* 37:1–109, 1985.)

Table 6.2   *Structural Features of γ and β Turns*

Bend type	Dihedral Angles of Central Residues (deg)[a]			
	ϕ_{i+1}	ψ_{i+1}	ϕ_{i+2}	ψ_{i+2}
γ				
classical	70 to 85	−60 to −70		
inverse	−70 to −85	60 to 70		
β				
I	−60	−30	−90	0
I′	60	30	90	0
II	−60	120	80	0
II′	60	−120	−80	0
III	−60	−30	−60	−30
III′	60	30	60	30
IV	Any bend with two or more angles differing by >40° from those given here			
V	−80	80	80	−80
V′	80	−80	−80	80
VIa[b]	−60	120	−90	0
VIb[b]	−120	120	−60	0
VII	Kink in chain created by $\psi_2 \approx 180°$, $|\phi_3| < 60°$; or $|\psi_2| < 60°$, $\phi_3 \approx 180°$			
VIII	−60	−30	−120	120

[a] The central residue of a γ turn is numbered $i + 2$; the two central residues of a β turn are $i + 2$ and $i + 3$.

[b] The peptide bond between residues $i + 1$ and $i + 2$ is *cis*, and residue $i + 2$ is Pro.

Data from P. Y. Chou and G. D. Fasman, *J. Mol. Biol.* 115:135–175 (1977); C. M. Wilmot and J. M. Thornton, *J. Mol. Biol.* 203:221–232 (1988); J. S. Richardson, *Adv. Protein Chem.* 34:167–339 (1981).

and II′ are rare, but type I′ is preferred in β hairpins, presumably because it fits the twist of the β-sheet. The type-III β turn is a short portion of a 3_{10} helical conformation.

The conformations of short loops, such as γ and β turns, depend primarily on the positions of certain residues in the loop — usually Gly, Asn, or Pro — that allow the chain to take up an unusual conformation. The type-I β turn is compatible with any amino acid at positions i through $i + 3$, except that Pro cannot occur at position $i + 2$. Gly predominates at position $i + 3$, and Pro predominates at position $i + 1$ of both type-I and type-II turns. Asp, Asn, Ser, and Cys residues frequently occur at position i, where their side chains often hydrogen-bond to the —NH of residue $i + 2$. Gly and Asn occur most frequently at position $i + 2$ of type-II turns because they adopt the required backbone angles most easily. Ideally, type-I′ turns have Gly at positions $i + 1$ and $i + 2$, and type-II′ turns have Gly at position $i + 1$. Type-III turns can have most amino acid residues, but type III′ requires Gly at positions $i + 1$ and $i + 2$.

The conformations of reverse turns observed in proteins are often very unlike those of the ideal turns expected (Table 6.2), probably because the restrictions on the flexibility of the peptide backbone were initially overestimated (see Fig. 5.2A). In particular, the hydrogen bond between residues i and $i + 3$ in the β turn is often missing. Several other classes of β turns have been proposed: IV, V, VIa, VIb, VII, and VIII (Table 6.2). Types VIa and VIb have a *cis* peptide bond and Pro as the next residue. Large loops have less well defined conformations, which often are determined by interactions with the rest of the protein. In all reverse turns, the peptide groups are not paired by regular hydrogen bonds, so they must be accessible to the solvent, which is why reverse turns occur on the protein surface.

References

Turns in peptides and proteins. G. D. Rose et al. *Adv. Protein Chem.* 37:1–109 (1985).

Loops, bulges, turns and hairpins in proteins. E. J. Milner-White and R. Poet. *Trends Biochem. Sci.* 12:189–192 (1987).

Analysis and prediction of the different types of β-turn in proteins. C. M. Wilmot and J. M. Thornton. *J. Mol. Biol.* 203:221–232 (1988).

Conformation of β-hairpins in protein structures. B. L. Sibanda et al. *J. Mol. Biol.* 206:759–777 (1989).

Structural determinants of the conformations of medium-sized loops in proteins. A. Tramontano et al. *Proteins: Struct. Funct. Genet.* 6:382–394 (1989).

6.2.4 Supersecondary Structures

Certain assemblies of a number of secondary-structure elements, including the segments of polypeptide chain that connect the secondary-structure elements, have been observed often enough that they are becoming recognized as another level of structure, termed **supersecondary structures.** These structures are a higher level of structure than secondary structure but do not constitute entire structural domains.

In one common supersecondary structure, often designated β-α-β, two β-strands are parallel in a β-sheet but are not necessarily adjacent. The α-helix occurs in the connecting segment, which must be roughly antiparallel to the β-strands (Fig. 6.19A). The connecting segment lies on the same side of the β-sheet in virtually all structures, even when there is no α-helix in the connecting strand; this topology is defined as right-handed. This particular topology may arise from the tendency of β-strands to twist in a right-handed direction (Sec. 6.2.2); this topology can be visualized by twisting a ribbon and bringing its ends together.

In antiparallel β-sheets, a common occurrence is the **Greek key** topology (Fig. 6.19B), named after a pattern found on Greek pottery. Once again, the topology almost always occurs with only one of the two possible orientations. The adjacent β-strands in antiparallel β-sheets are often those strands that are sequential in the primary structure of the polypeptide chain. The adjacent β-strands are connected by β turns to produce a **β-meander** (Fig. 6.19C). Here there is no asymmetry, and all connections are equivalent.

The particular supersecondary structures that are observed can be rationalized as resulting simply from the general occurrence of right-handed connections between parallel β-strands, from a tendency for elements of secondary structure that are adjacent in the amino acid sequence also to be adjacent in the tertiary structure, and from a tendency of connections between secondary-structure elements not to cross each other or make knots in the polypeptide chain.

The architectures of most proteins appear to be made up of segments of secondary structure packed together, so efforts have been directed at understanding the basis for the interactions between helices, between β-sheets, and between helices and sheets. Their general architectures cause the interactions between elements of secondary structure to be governed primarily by the amino acid side chains on their surfaces. For example, interdigitation of the amino acid side chains of two ideal α-helices would be expected only when the axes of the helices cross at angles of $-82°$, $-60°$, or $+19°$ (a rotation is positive when a "lower" helix is rotated clockwise relative to an "upper" one). Helices pack onto β-sheets with their axes nearly parallel to the β-strands because in that way the twist of the β-sheet matches the surface of the helix. Two normal twisted β-sheets should pack together face-to-face with the upper sheet rotated clockwise between 20° and 50°, whereas two sheets that are folded over onto themselves should be at angles of about 90°. Proteins are found to observe these ideals to a first approximation, but substantial nonideal variation is found in real secondary structures, and the amino acid side chains involved in the contacts are variable. In the great majority of cases, only nonpolar amino acid side chains are involved in these interactions.

References

Comparison of super-secondary structure in protein. S. T. Rao and M. G. Rossmann. *J. Mol. Biol.* 76:241–256 (1973).

Packing of α-helices onto β-pleated sheets and the anatomy of α/β proteins. J. Janin and C. Chothia. *J. Mol. Biol.* 143:95–128 (1980).

The classification and origins of protein folding patterns. C. Chothia and A. V. Finkelstein. *Ann. Rev. Biochem.* 59:1007–1039 (1990).

6.2.5 Interiors and Exteriors

Detailed analyses of the complex surfaces of folded proteins and the internal packing of atoms in proteins generally use the procedure of Lee and Richards (Fig. 6.20A). Every atom of the protein is depicted as a sphere of the appropriate van der Waals radius; overlapping regions of the spheres, where the atoms are covalently bonded, are truncated. The complex surface that results is called the **van der Waals surface;** it has a strictly

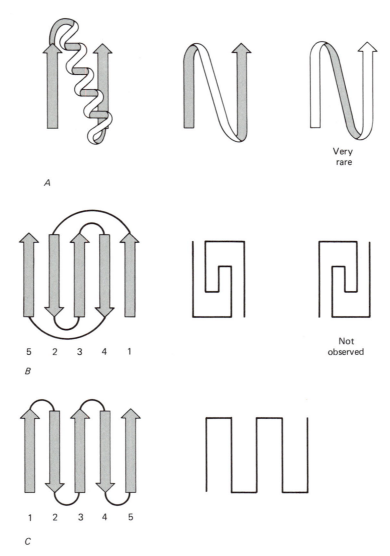

FIGURE 6.19
Supersecondary structures observed in proteins. *A:* A β-α-β unit; the segment joining the
two β-strands is almost always above their plane, not below. *B:* The Greek-key motif of antiparallel β-sheets, which also is always of one "handedness." *C:* A so-called β-meander.

defined surface area and encloses a definite volume
(Table 4.2). No chemical procedure can measure this
area or volume directly, however, because any chemical probe has significant dimensions. The surface that is
accessible to a chemical probe, such as the water molecules of the solvent, is explored by imagining a spherical
probe of appropriate radius R rolling on the outside of
the molecule, maintaining contact with the van der
Waals surface. In the hypothetical protein of Figure
6.20A the probe does not contact atoms 3, 9, or 11; such
atoms are considered to be interior atoms, *not* part of
the surface of the molecule. Those parts of the van der

Waals surface that make contact with the surface of the
probe are designated the **contact surface;** they comprise
a series of disconnected patches. When the probe is
simultaneously in contact with more than one protein
atom, its interior surface defines the **reentrant surface.**
The contact surface and the reentrant surface together
define a continuous surface known as the **molecular
surface.** The surface that is discussed most often is the
accessible surface, which is defined by the center of the
probe as it moves over the surface of the protein. The
probe is frequently taken to be a water molecule, approximated as a sphere with a radius of 1.4 Å (Fig. 4.6).

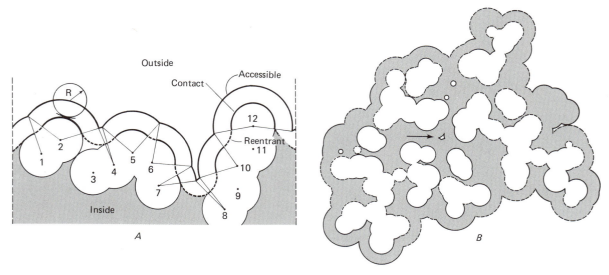

FIGURE 6.20

Analysis of protein surfaces. The various surfaces are defined in *A* for a two-dimensional slice through a hypothetical protein including atoms 1–12, using a solvent probe of radius *R*. The van der Waals and accessible surfaces of a section through ribonuclease S are shown in *B*. The solid lines outline carbon and sulfur atoms; the dashed lines, nitrogen and oxygen. In places, the accessible surface is controlled by atoms above or below the section shown. The arrow indicates a cavity inside the protein large enough to accommodate a water molecule with a radius of 1.4 Å, like that used for the probe. (Adapted from F. M. Richards, *Ann Rev. Biophys. Bioeng.* 6:151–176, 1977; B. Lee and F. M. Richards, *J. Mol. Biol.* 55:379–400, 1971.)

The van der Waals surface and the accessible surface of a section through the crystallographic structure of ribonuclease S are illustrated in Figure 6.20B.

The total accessible surface areas of proteins are approximately proportional to the two-thirds power of their molecular weights, as would be expected for objects of similar shapes. The accessible surface area of a protein is, however, nearly two times greater than that of a sphere of the same volume, which is some indication of the roughness of the protein surface. The accessible surface area A_s (in Å^2) of a typical small monomeric protein is usually related to its molecular weight M_w by the approximate relationship

$$A_s = 6.3(M_w)^{0.73} \qquad (6.6)$$

This is only 23–45% of the surface area of the unfolded polypeptide chain. The volume V of a typical monomeric protein is given by

$$V = 1.27 M_w \text{ Å}^3 \cdot \text{dalton}^{-1} \qquad (6.7)$$

The interiors of proteins are densely packed, with adjacent atoms frequently in van der Waals contact (Fig. 6.21). About 75% of the interior volume is filled with atoms, as defined by their van der Waals radii. This is close to the value of 74% possible with close packing of identical spheres and is within the range of 70–78% found for crystals of small organic molecules. The average volumes occupied by residues in folded proteins (Table 6.3) are virtually the same as those they occupy in crystals of the amino acids (Fig. 6.22). The packing in protein interiors contrasts with the lower values observed for liquids, such as water (58%) and cyclohexane (44%). In the case of proteins, however, many more of the atoms are close because they are covalently bonded, which exaggerates the close packing. Adjacent atoms are not always in van der Waals contact, and the packing density varies somewhat throughout the interior, generally being highest in areas where the polypeptide topology is most regular. In few instances are there unfilled cavities large enough to accommodate other molecules; one is shown in Figure 6.20B. The dense packing of atoms in protein interiors is still impressive in light of the fact that it must be compatible with the covalent connectivity of the polypeptide chain.

Virtually all polar groups in the protein interior are paired in hydrogen bonds; most of these polar groups are on the polypeptide backbone, and they usually are hydrogen bonded in secondary structure. Water mole-

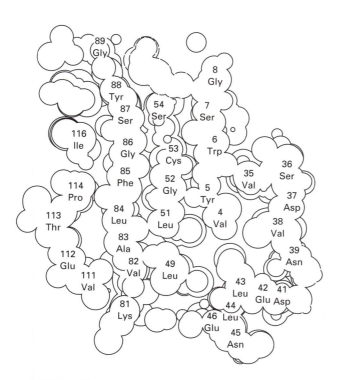

FIGURE 6.21
Serial section through the interior of flavodoxin, demonstrating the dense packing of atoms. Atoms of the protein are drawn with their van der Waals radii as solid lines; the dashed line shows the prosthetic group FMN. Three successive sections cut 1 Å apart are shown. The labels indicate the positions of the C^α atoms. Several parallel β-strands are visible. (Kindly provided by A. M. Lesk.)

cules are generally excluded from protein interiors and, when present, appear to be integral parts of the protein structure (Fig. 6.23). These solvent molecules are fixed in internal cavities of the protein that are isolated from the bulk solvent; they invariably form hydrogen bonds to polar groups of the protein, and they tend to be conserved in homologous structures. Some water molecules occur singly, some in pairs, and others in extended networks. Most make two or three hydrogen bonds to polar groups or other water molecules, although some do make four, involving the oxygen atom as hydrogen acceptor in two hydrogen bonds and the two hydrogen atoms as donors in two other hydrogen bonds. Many water molecules bind to the NH and CO groups of the backbone, although some are attached to polar side chains. Such buried water molecules appear to be important for filling holes and, probably more important, for pairing with internal polar groups of the protein in hydrogen bonds.

Virtually all ionized groups in water-soluble proteins are on the surface of the molecule, exposed to the solvent. On average, Asp, Glu, Lys, and Arg residues comprise 27% of the protein surface residues and only 4% of the interior residues. Most proteins have one to two charged groups per 100 Å² of surface, but the charge density varies between 0.5 and 25 charges per 100 Å². There is a slight tendency for oppositely charged groups to be near each other on the surface, where they could form salt bridges, but they are rarely observed to do so in protein crystal structures unless the side chains are held in position by the rest of the protein. Ionized pairs of acidic and basic groups hardly ever occur in the interiors of proteins, even though such a pair might be expected to have no net charge.

Integral membrane proteins differ from water-soluble proteins primarily in having extremely nonpolar surfaces that are in contact with the nonpolar membrane interior. The interior of the only membrane protein known in detail, the photosynthetic reaction center, is normal in the types of side chains present and in its packing density (Sec. 7.2.2).

Nonpolar side chains predominate in the protein interior; Val, Leu, Ile, Phe, Ala, and Gly residues comprise 63% of the interior residues. Because of the large size and complexity of many amino acid side chains and

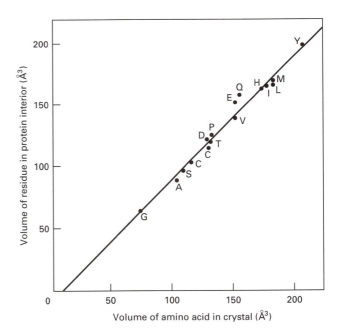

FIGURE 6.22
The mean volumes of the various amino acid residues buried in protein interiors plotted against the volume of the corresponding amino acid in crystals. The line has a slope of 1 and an intercept of 11 Å³ (the volume of the water molecule lost by an amino acid upon becoming a residue). (From C. Chothia, *Ann. Rev. Biochem.* 53:537–572, 1984.)

Table 6.3 *The Packing of Residues in the Interior of Proteins*

Residue	Average volume of buried residues (Å³)[a]	Fraction of residues at least 95% buried[b]	Relative free energy of residue in interior to that on surface (kcal/mol)[c]
Gly	66	0.36	0
Ala	92	0.38	−0.14
Val	142	0.54	−0.55
Leu	168	0.45	−0.59
Ile	169	0.60	−0.68
Ser	99	0.22	0.40
Thr	122	0.23	0.32
Asp	125	0.15	0.78
Asn	125	0.12	0.75
Glu	155	0.18	1.15
Gln	161	0.07	0.80
Lys	171	0.03	2.06
Arg	202	0.01	1.40
His	167	0.17	0.02
Phe	203	0.50	−0.61
Tyr	204	0.15	0.28
Trp	238	0.27	−0.39
Cys	106[d]	0.40[d]	−0.61[d,e]
	118[e]	0.50[e]	—
Met	171	0.40	−0.65
Pro	129	0.18	0.50

[a] From C. Chothia, *Nature* 254:304–308 (1975).

[b] Average for 12 proteins. From C. Chothia, *J. Mol. Biol.* 105:1–14 (1976).

[c] Calculated as $-RT \log_e f$, where f is the ratio of the occurence of this amino acid residue on the interior to that on the surface. The values were normalized with that for Gly set to zero. From S. Miller et al., *J. Mol. Biol.* 105:641–656 (1987).

[d] When in disulfide form.

[e] When in thiol form.

the relatively small size of folded proteins, it is often difficult to classify residues simply as buried or exposed. For example, the ionized terminal groups in the long side chains of Lys and Arg are almost invariably exposed to the solvent, but the other hydrophobic methylene carbons are often buried in the interior; consequently, the C^α atoms of these residues may be far from the surface, and most of the side chain may be buried. Small proteins have only one or two completely buried residues, and even in large proteins only 15% of the residues are totally inaccessible to solvent. A residue is usually considered to be buried if more than 95% of its surface area is inaccessible to solvent. For example, 54% of Val and 60% of Ile residues are at least 95% buried, but only 1% of Arg and 3% of Lys residues

(Table 6.3). The hydrophobic residues are primarily involved in packing together the elements of secondary structure.

The 20 different amino acid side chains show no conspicuous tendencies to be adjacent to each other or to the peptide backbone in protein interiors, other than the general tendencies just described for hydrophobic groups to cluster, polar groups to be paired in hydrogen bonds, and oppositely charged groups to be near each other. The only other tendencies noted are for pairs of Cys residues to occur in positions where they can form disulfide bonds and for aromatic residues to undergo favorable electrostatic interactions with each other and with sulfur, oxygen, and amino groups. No simple structural rules appear to relate conformation to the

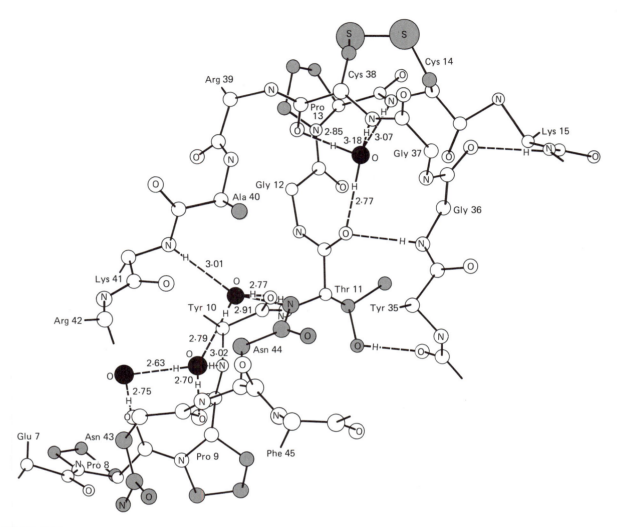

FIGURE 6.23
The four internal water molecules of bovine pancreatic trypsin inhibitor (BPTI). The oxygen atoms observed crystallographically are shown in black, with the presumed positions of the hydrogen atoms indicated. The lengths in angstroms of the hydrogen bonds between nitrogen and oxygen atoms are indicated. (From T. E. Creighton, *Prog. Biophys. Mol. Biol.* 33:231–297, 1978.)

amino acid sequence. This point will become even more obvious in Section 6.4, in the discussion of the variety of amino acid sequences that can produce the same folded conformation.

References

Areas, volumes, packing and protein structure. F. M. Richards. *Ann. Rev. Biophys. Bioeng.* 6:151–176 (1977).

Cavities in proteins: structure of a metmyoglobin–xenon complex solved to 1.9 Å. R. F. Tilton et al. *Biochemistry* 23:2849–2857 (1984).

Internal cavities and buried waters in globular proteins. A. A. Rashin et al. *Biochemistry* 25:3619–3625 (1986).

The distribution of charged groups in proteins. D. J. Barlow and J. M. Thornton. *Biopolymers* 25:1717–1733 (1986).

Interior and surface of monomeric proteins. S. Miller et al. *J. Mol. Biol.* 196:641–656 (1987).

π–π Interactions: the geometry and energetics of phenylalanine–phenylalanine interactions in proteins. C. A. Hunter et al. *J. Mol. Biol.* 218:837–846 (1991).

6.2.6 Quaternary Structure

Many proteins exist naturally as aggregates of two or more polypeptide chains, either identical or different. Different polypeptide chains can be called subunits, monomers, or protomers and are usually designated by letters; for example, normal adult hemoglobin is $\alpha_2\beta_2$ (Sec. 8.4.3), aspartate transcarbamoylase is r_6c_6 (Sec. 9.4.2.b), and many proteins are dimers, trimers, tetramers, or even high-order aggregates of identical polypeptide chains. Such quaternary structure is invariably observed in the crystal structure of the protein, and the detailed nature of the interactions that produce the quaternary structure is also apparent.

Each polypeptide-chain subunit is usually folded into an apparently independent globular conformation, which then interacts with other monomers (Fig. 6.24); one of the greatest exceptions is the dimeric *trp* repressor in which two identical polypeptide chains are intimately entwined (Fig. 6.24C). The centers of the interfaces between monomers are usually similar to the interiors of the individual monomers in that they are closely packed and involve primarily hydrophobic interactions between nonpolar side chains. The periphery of the interface, however, is usually similar to the exteriors of the monomers and involves many hydrogen bonds and salt bridges between ionized side chains. Some interfaces are also like the interiors of protein

molecules in that they involve interactions between elements of secondary structure, whereas others involve interactions between loops on the surfaces of the monomers. The only aspect that is common to the interacting surfaces of monomers in all quaternary structures is that they are highly complementary, both in shape and in pairing of polar groups.

Two fundamental types of interaction between identical monomers are possible, which have been designated as **isologous** and **heterologous** (Fig. 6.25). Isologous association involves the same surfaces on both monomers, which associate to produce a dimer with a twofold rotation axis of symmetry. The two monomers are equivalent, which requires that the two halves of the interface between the monomers be complementary about a "mirror plane" that coincides with the symmetry axis in the dimer. Nonequivalent association of the two surfaces to form a nonsymmetrical dimer is unlikely because some of the complementary sites would not be paired. No further association is possible using the binding sites of isologous interactions. Association to produce tetramers requires the monomers to have another type of binding surface. Most tetramers

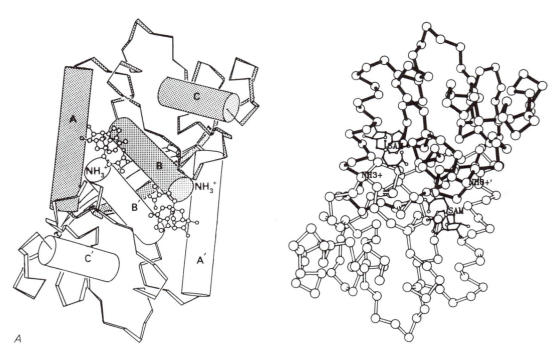

FIGURE 6.24

Examples of protein quaternary structure. *A*: The *met* repressor dimer, showing the course of the polypeptide chain schematically *(left)* and with virtual bonds connecting the C$^\alpha$ atoms, depicted as open circles *(right)*; the bonds of one monomer are white and those of the other are shaded. (J. B. Rafferty et al., *Nature* 341:705–710, 1989.)

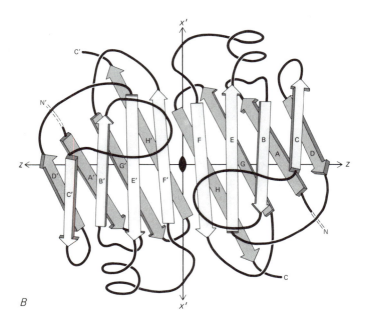

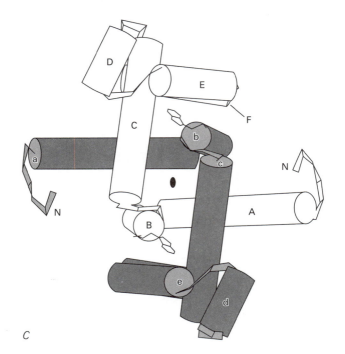

FIGURE 6.24 *(Continued)*

B: The prealbumin dimer, in which arrows depict the β-strands, labeled A–H in one monomer, A′–H′ in the other. The two monomers (*left* and *right*) associate to extend the two β-
sheets, by strands F and H of one monomer hydrogen-bonding to the corresponding strands
of the other molecule. Two dimers further associate at nearly right angles to each other to
form a tetramer by isologous interactions between the side chains protruding outward from
β-sheet D′A′G′H′HGAD. (C. C. F. Blake et al., *J. Mol. Biol.* 88:1–12, 1974.) *C*: Unusual
quaternary structure of the *trp* repressor. Instead of two independently folded polypeptide
chains, the two polypeptide chains of the *trp* repressor are entwined; one polypeptide chain
is white, the other shaded. α-helices are shown as cylinders, the connecting turns depicted
as a folded tape. The twofold axis is shown, as are the two molecules of the amino acid tryptophan that bind to the two identical binding sites. (R. W. Schevitz et al., *Nature* 317:782–
786, 1985.)

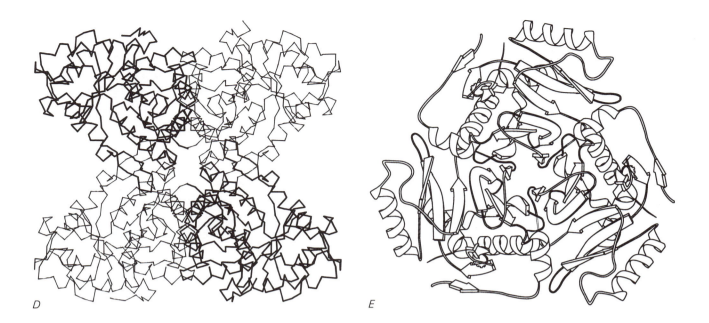

D

E

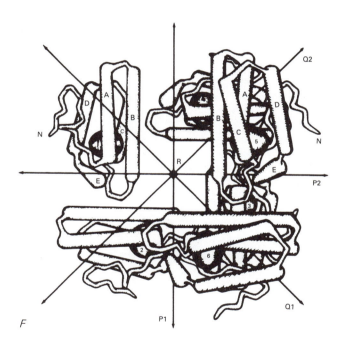

F

FIGURE 6.24 *(Continued)*

D: The isologous tetrameric structure of rabbit skeletal muscle aldolase, with three twofold symmetry axes. The course of the polypeptide backbone of each 363-residue subunit is traced light in two of the subunits, bold in the other two. The primary contacts between subunits involve hydrophobic side chains and water molecules between them. (J. Sygusch et al., *Proc. Natl. Acad. Sci. USA* 84:7846–7850, 1987.) *E*: The heterologous trimeric structure of chloramphenicol acetyltransferase, in which subunits interact by extending by one strand the six-stranded β-sheet of each subunit. (A. G. W. Leslie, *J. Mol. Biol.* 213:167–186, 1990.) *F*: The octameric structure of hemerythrin; two of the subunits of the top layer are omitted to show the lower layer. Each layer is a heterologous tetramer, with fourfold symmetry. The two layers are joined by isologous interactions. Each monomer is a four-helix bundle, with two iron atoms between the helices (see Fig. 6.13). (J. L. Smith et al., *Nature* 303:86–88, 1983.)

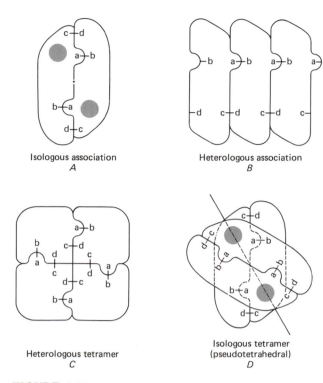

FIGURE 6.25
Schematic illustrations of isologous and heterologous association between protein subunits. *A*: Isologous association to form a dimer with a twofold symmetry axis perpendicular to the plane of the figure. *B*: Heterologous association leading possibly to infinitely long polymers. *C*: Heterologous association to form a closed, finite structure, in this case a tetramer, with a fourfold symmetry axis. *D*: Forming a tetramer by isologous association, using two different bonding interactions: a—b and c—d, plus the shaded circles of *A*. (From J. Monod et al., *J. Mol. Biol.* 12:88–118, 1965.)

are formed by two sets of isologous interactions to give three twofold axes of symmetry. For example, prealbumin, concanavalin A, and a few other proteins associate isologously to dimers by extending the β-sheets of different molecules (Fig. 6.24); then two such dimers pack together in another isologous manner to produce a tetramer.

Heterologous association involves two different sites that are complementary and largely nonoverlapping. In such a dimer, the two monomers are not equivalent, and each has one unpaired binding site, which tends to bind to another monomer. Indefinite polymerization results unless the geometry of the association leads to a closed ring (Figs. 6.25B,C). Any oligomeric protein with a fixed number of identical subunits that is not a power of two is likely to result from heterologous association. Some elongated structural proteins, such as actin, are formed by indefinite heterologous polymeri-

zation of globular monomers. Trimers (bacteriochlorophyll protein and KDPG aldolase), some tetramers (manganese superoxide dismutase and neuraminidase), pentamers (muconolactone isomerase), and a 17-mer (tobacco mosaic virus disk) are produced by closed heterologous association.

More complex interactions often occur in higher order structures, such as the icosahedral viruses. An icosahedron, which has 60 monomers in identical symmetrical positions, is the largest symmetrical closed structure that is possible with identical monomers. But the shells of many icosahedral viruses are composed of more than 60 monomers, so they cannot all have equivalent positions. For example, the tomato bushy stunt and southern bean mosaic viruses have 180 identical monomers, which occur in three nonequivalent positions in the symmetrical structure. In particular, two-, three-, and fivefold symmetry axes are present between subunits. This structural **quasiequivalence** is possible because the virus-coat protein molecule has several domains that can move relative to each other and thus can adopt these three packing arrangements.

Association of two molecules requires only spatial and physical complementarity of the interacting surfaces. Proteins have probably evolved not to interact with the many other proteins that they normally come in contact with but should not bind to.

References

The symmetry of self-complementary surfaces. R. S. Morgan et al. *J. Mol. Biol.* 127:31–39 (1979).

The structure of spherical viruses. L. Liljas. *Prog. Biophys. Mol Biol.* 48:1–36 (1986).

The structure of interfaces between subunits of dimeric and tetrameric proteins. S. Miller. *Protein Eng.* 3:77–83 (1989).

The structure of protein–protein recognition sites. J. Janin and C. Chothia, *J. Biol. Chem.* 265:16027–16030 (1990).

6.2.7 Flexibility Detected Crystallographically

The protein models that result from crystallographic structure determinations are usually static, but this should not be understood to imply that the molecule is also static, even within the crystal lattice. Electron density maps provide direct evidence for flexibility in protein molecules, by the extent to which the electron density is smeared out. This smearing is usually expressed as the temperature factor B (Eq. 6.4), although the name of this factor should not be taken literally. Proteins have considerably larger B values than those determined for crystals of small molecules, whose value of B is usually $3-4$ Å² and is seldom greater than 10 Å². The B values for the interior atoms of well-ordered protein crystals are usually in the range of $10-20$ Å². If they were due solely to thermal vibrations of atoms within molecules, such B values would imply rms amplitudes of vibrations of $0.3-0.5$ Å, and instantaneous deviations of up to $1.0-1.5$ Å would be expected frequently.

The value of B varies throughout a protein structure, with the atoms on the surface of the molecule usually having the largest values. In a few instances, surface side chains or small portions of a polypeptide chain may be invisible, presumably because they adopt a number of conformations. In the cases of trypsinogen and IgG(Kol), entire portions of their structures were not visible, which is taken to indicate that they have extreme flexibility.

The value of the temperature factor is affected not only by the flexibility of the protein molecule but also by errors in the crystallographic phases and by disorder in the crystal lattice, so interpretation of B values is not straightforward. Lattice disorder and internal flexibility can be distinguished by comparing different parts of the electron density map to determine which variation can be explained by a slightly different packing of rigid molecules in the unit cell, by comparing maps of the same molecule in different crystal lattices or in different environments in the same lattice (e.g., when there is noncrystallographic symmetry), or by varying the temperature to see if the flexible portions of the molecule can be frozen in fixed conformations at low temperatures. Identical protein molecules that were crystallized in various ways—often using different solvents or observed in different environments in the same crystal—usually show very similar conformations of their individual domains, which usually differ only in surface side chains and loops. Significant differences in the conformation of a structural domain are observed only when there is intrinsic flexibility, with alternative conformations that have very similar free energies, and only among the smallest proteins.

Increasing the temperature causes both the crystal lattice and the protein structure to expand; increasing the temperature from 80 K to $255-300$ K in the case of myoglobin caused volume expansions of 5% in the crystal and 3% in the protein. This degree of expansion of the protein is intermediate between that of liquid water and that of hydrocarbons. Most of the expansion of the protein occurred as a result of a loosening in the hydrophobic packing between helices; the helices themselves did not change much.

Increasing the hydrostatic pressure on lysozyme crystals to 1000 atmospheres (atm) decreased the volume of the protein molecule by only 0.3%, but the decrease varied substantially throughout the molecule. The elements of secondary structure, the β-sheet and individual α-helices, changed the least, and the compression was accomplished by increasing the packing density of side-chain atoms. Both decreasing the temperature and increasing the pressure caused many of the B values to decrease, indicating that the B values in this case reflect primarily the inherent flexibility of the protein molecule. In the case of myoglobin at 145 atm, N_2 molecules from the pressurizing gas appeared to fill small cavities in the protein interior, causing slight readjustments of the internal packing. Nevertheless, the most impressive conclusion from these and other studies of compressibility is that protein molecules are no more compressible than ice.

Indirect evidence for protein flexibility in the crystal comes from chemical exchange of buried hydrogen atoms of the proteins with those of the solvent, which can be observed by neutron diffraction using the isotopes ^{1}H and ^{2}H. This phenomenon is studied much more thoroughly in solution (Sec. 7.3.1).

References

Transmission of conformational change in insulin. C. Chothia et al. *Nature* 302:500–505 (1983).

Fluctuations in protein structure from X-ray diffraction. G. A. Petsko and D. Ringe. *Ann. Rev. Biophys. Bioeng.* 13:331–371 (1984).

Mapping protein dynamics by X-ray crystallography. D. Ringe and G. A. Petsko. *Prog. Biophys. Mol. Biol.* 45:197–235 (1985).

Structural heterogeneity in protein crystals. J. L. Smith et al. *Biochemistry* 25:5018–5027 (1986).

Thermal expansion of a protein. H. Frauenfelder et al. *Biochemistry* 26:254–261 (1987).

Low temperature X-ray investigation of structural distributions in myoglobin. F. Parak et al. *Eur. Biophys. J.* 15:237–249 (1987).

Crystal structure of hen egg-white lysozyme at a hydrostatic pressure of 1000 atmospheres. C. E. Kundrot and F. M. Richards. *J. Mol. Biol.* 193:157–170 (1987).

A structure of sperm whale myoglobin at a nitrogen gas pressure of 145 atmospheres. R. F. Tilton, Jr., and G. A. Petsko. *Biochemistry* 27:6574–6582 (1988).

Atomic motions in molecular crystals from diffraction measurements. J. D. Dunitz et al. *Angew. Chem. Intl. Ed. Engl.* 27:880–895 (1988).

Crystallography of biological macromolecules at ultra-low temperature. H. Hope. *Ann. Rev. Biophys. Biophys. Chem.* 19:107–126 (1990).

A mutant T4 lysozyme displays five different crystal conformations. H. R. Faber and B. W. Matthews. *Nature* 348:263–266 (1990).

6.2.8 *The Solvent*

The structure of the aqueous solvent surrounding protein molecules is an important aspect of protein structure, but one that is very difficult to determine. The electron density of the unit cell determined crystallographically is averaged over the extensive time required to collect the data and over all the many molecules of the crystal lattice. The molecules of solvent in the crystal are particularly mobile and by themselves would give a uniform, average electron density throughout the nonprotein areas of the crystal, but distinct electron density peaks corresponding to relatively fixed solvent molecules are observed near the surfaces of protein molecules. In well-refined and well-ordered crystal structures, an average of two fixed water molecules are observed for each amino acid residue. Of course, the electron density map does not identify the fixed molecules, and most crystal solvents contain high concentrations of salts or other agents that were added to induce crystallization. Most such fixed molecules appear from their interactions to be water molecules, however, and they are assumed to be water in the absence of indications to the contrary.

Fixed water molecules occur primarily in positions where they can hydrogen-bond to polar groups, and the degree of order of such solvent molecules is generally proportional to their proximity to the protein surface and to the extent of their participation in hydrogen bonding. Water molecules are most highly ordered, with B values as low as 13 $Å^2$, when extensively hydrogen-bonded in crevices on the protein surface or when bridging between molecules in the crystal lattice. The fixed water molecules are almost invariably anchored by hydrogen-bonding to fixed polar groups on the protein surface, and one or more firmly anchored water molecules can apparently fix adjacent water molecules in a hydrogen-bonded lattice. Beyond this, there is a continuum of degree of order of solvent molecules, with B values increasing to the point at which the molecules can be considered part of the bulk solvent and it is no longer deemed worthwhile to include them in the protein structure. All except the most ordered water molecules have only partial occupancies, indicating considerable flexibility in the solvent structure.

Only in the exceptionally hydrophobic protein, crambin, has water ordering of the type that might be expected from the hydrophobic interaction (Sec. 4.3) been observed around nonpolar surfaces. Because they usually are not observed, such ordered networks normally must be transient, or they might "slip" along the nonpolar surface and thus appear smeared out in the electron density map.

References

Structure and dynamics of water surrounding biomolecules. W. Saenger. *Ann. Rev. Biophys. Biophys. Chem.* 16:93–114 (1987).

Distributions of water around amino acid residues in proteins. N. Thanki et al. *J. Mol. Biol.* 202:637–657 (1988).

Crystal structure of low humidity tetragonal lysozyme at 2.1 Å resolution. Variability in hydration shell and its structural consequences. R. Kodandapani et al. *J. Biol. Chem.* 265:16126–16131 (1990).

Water–protein interactions: theory and experiment. M. M. Teeter. *Ann. Rev. Biophys. Biophys. Chem.* 20:577–600 (1991).

6.3 Protein Structure Determination by Nuclear Magnetic Resonance Spectroscopy

The general topology of the polypeptide chain in solution can be determined by nuclear magnetic resonance (NMR). The structure obtained in this way is not as detailed and accurate as that obtained crystallographically, but NMR has the advantage of using a protein in solution rather than in a crystal lattice. Though it is important to understand the basic principles of how protein structures are obtained by this technique, a deep understanding of the physical principles of NMR is not required.

6.3.1 *Nuclear Magnetic Resonance Spectra of Proteins*

NMR spectra are generated by placing a sample in a magnetic field and applying radio-frequency pulses, which perturb the equilibrium nuclear magnetization of those atoms with nuclei of nonzero spin. Transient *time domain* signals are detected as the system returns to equilibrium. Fourier transformation of the transient signal into a *frequency domain* yields a one-dimensional NMR spectrum, which is a series of resonances from the various nuclei at different frequencies, or chemical shifts (relative to some standard, and usually expressed as parts per million, ppm). The chemical shift of an atom depends on the electronic environment of its nucleus.

The ^{1}H atom is the only atom normally present in proteins that can be observed by NMR, although the low natural abundance of the ^{13}C atom can be used to some extent, and ^{13}C and ^{15}N atoms can be incorporated into the protein during its biosynthesis. All the ^{1}H atoms of a protein can be observed, except those labile hydrogen atoms of —NH—, —NH$_2$, —OH, and —SH groups that are exchanging with hydrogen atoms in the aqueous solvent at rapid rates (Sec. 7.3.1). Hydrogen atoms of these groups can be observed by NMR only when the pH is slightly acidic so that hydrogen exchange with the solvent is sufficiently slow. These exchangeable hydrogen atoms can be made invisible to NMR by placing the protein sample in ^{2}H$_2$O; this assists in identifying these atoms and in simplifying the ^{1}H-NMR spectra.

Each hydrogen atom in a protein can be resolved by NMR if it has a unique chemical shift, which depends on which atom it is bonded to and on its environment. In the absence of a unique environment, as in an ideal random polypeptide chain, all hydrogen atoms of a particular type will be very similar. The chemical shifts for the hydrogen atoms of natural proteins in the random coil conformation are listed in Table 6.4. They fall into various classes, as illustrated in Figure 6.26.

The environmental factors that affect the chemical shift are not entirely known, but the main factors are believed to be hydrogen bonding and proximity to aromatic rings. The chemical shifts of the backbone NH resonances are particularly sensitive to their environment. Two equivalent hydrogen atoms, such as the ϵ^1 and ϵ^2 CH atoms of Phe and Tyr rings, have separate chemical shifts if they are in unique environments and if the aromatic rings are not rotating rapidly on the NMR time scale (which depends on the chemical shift but is roughly of the order of a few milliseconds). Each hydrogen atom contributes equally to the spectrum, in the area under its resonance, unless it is exchanging with the solvent. Some resonances may be very broad, however, and not apparent under certain conditions, such as when the atom is tumbling only slowly in solution.

The information about protein conformation that is present in NMR spectra arises from the interactions between hydrogen atoms that occur through the covalent bonds **(through-bond J couplings)** or through space (the **nuclear Overhauser effect, NOE**). J couplings are observed only between hydrogen atoms that are separated by three or fewer covalent bonds; that is, the hydrogen atoms must be covalently bonded to the same or neighboring atoms. The residues in polypeptides are isolated entities because no J coupling occurs through a peptide bond. The value of the three-bond coupling constant ^3J is related to the dihedral angle θ of the bond between the atoms to which the hydrogen atoms are bonded; the relationship is of the form

$$^3J = A \cos^2 \theta + B \cos \theta + C \qquad (6.8)$$

For example, the value of $^3J_{\mathrm{HNC^{\alpha}H}}$ between the NH and C$^\alpha$ hydrogen atoms gives information about the torsion angle ϕ:

$$^3J_{\mathrm{HNC^{\alpha}H}} = 6.4 \cos^2 \theta - 1.4 \cos \theta + 1.9 \qquad (6.9)$$

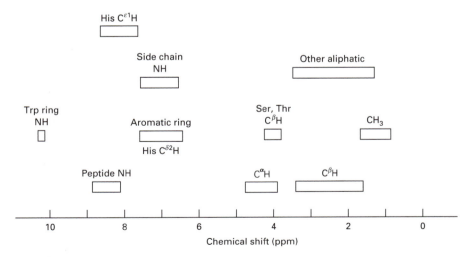

FIGURE 6.26
The range of ^{1}H-NMR chemical shifts observed for different hydrogen atoms of peptides in the random coil conformation.

Table 6.4 ^1H *Chemical Shifts of the Amino Acid Residues in the Random Coil Conformation*[a]

Residue	Chemical Shift (ppm)			
	NH	$C^\alpha H$	$C^\beta H$	Others
Gly	8.39	3.97		
Ala	8.25	4.35	1.39	
Val	8.44	4.18	2.13	$C^\gamma H_3$ 0.97, 0.94
Ile	8.19	4.23	1.90	$C^\gamma H_2$ 1.48, 1.19
				$C^\gamma H_3$ 0.95
				$C^\delta H_3$ 0.89
Leu	8.42	4.38	1.65, 1.65	$C^\gamma H$ 1.64
				$C^\delta H_3$ 0.94, 0.90
Pro (*trans*)		4.44	2.28, 2.02	$C^\gamma H_2$ 2.03, 2.03
				$C^\delta H_2$ 3.68, 3.65
Ser	8.38	4.50	3.88, 3.88	
Thr	8.24	4.35	4.22	$C^\gamma H_3$ 1.23
Cys	8.31	4.69	3.28, 2.96	
Asp	8.41	4.76	2.84, 2.75	
Glu	8.37	4.29	2.09, 1.97	$C^\gamma H_2$ 2.31, 2.28
Asn	8.75	4.75	2.83, 2.75	$N^\gamma H_2$ 7.59, 6.91
Gln	8.41	4.37	2.13, 2.01	$C^\gamma H_2$ 2.38, 2.38
				$N^\delta H_2$ 6.87, 7.59
Met	8.42	4.52	2.15, 2.01	$C^\gamma H_2$ 2.64, 2.64
				$C^\epsilon H_3$ 2.13
Lys	8.41	4.36	1.85, 1.76	$C^\gamma H_2$ 1.45, 1.45
				$C^\delta H_2$ 1.70, 1.70
				$C^\epsilon H_2$ 3.02, 3.02
				$N^\epsilon H_3^+$ 7.52
Arg	8.27	4.38	1.89, 1.79	$C^\gamma H_2$ 1.70, 1.70
				$C^\delta H_2$ 3.32, 3.32
				NH, NH_2^+ 7.17, 6.62
His	8.41	4.63	3.26, 3.20	$C^{\delta 2}H$ 7.14
				$C^{\epsilon 1}H$ 8.12
Phe	8.23	4.66	3.22, 2.99	$C^\delta H$ 7.30
				$C^\epsilon H$ 7.39
				$C^\zeta H$ 7.34
Tyr	8.18	4.60	3.13, 2.92	$C^\delta H$ 7.15
				$C^\epsilon H$ 6.86
Trp	8.09	4.70	3.32, 3.19	$C^{\delta 1}H$ 7.24
				$C^{\epsilon 3}H$ 7.65
				$C^{\zeta 3}H$ 7.17
				$C^\eta H$ 7.24
				$C^{\zeta 2}H$ 7.50
				$N^{\epsilon 1}H$ 10.22

[a] Measured at pH 7.0 and 35°C as peptide Xaa in tetrapeptide Gly-Gly-Xaa-Ala.

From K. Wüthrich, *NMR of Proteins and Nucleic Acids,* Wiley Interscience, New York, 1986.

where $\theta = |\phi - 60°|$. The torsion angle χ_1 between C^α and C^β is given by the coupling constant:

$$^3J_{\alpha\beta} = 9.5 \cos^2 \chi_1 - 1.3 \cos \chi_1 + 1.6 \quad (6.10)$$

These relationships are not unambiguous, however, because some values of J correspond to more than one dihedral angle.

The NOE is a consequence of dipole–dipole coupling between different nuclear spins, which causes spin polarization to be transferred from one nucleus to any nearby nucleus. The atoms do not have to be in the same residue; they simply have to be near each other in space. If everything else is equal, the magnitude of the NOE is proportional to r^{-6}, where r is the distance between the interacting nuclei. Unfortunately, the magnitude of the NOE is also affected by a number of other phenomena, such as the rate of tumbling, which can diminish its magnitude and even make it zero. It is only possible, therefore, to put an upper limit on the distance r indicated by any given NOE. In practice, NOEs are observed in proteins between hydrogen atoms that are no more than 5 Å apart.

The through-bond and through-space interactions are usually measured using two-dimensional spectra, in which the second dimension is the frequency from a second radio-frequency pulse. Different sequences of pulses and delays lead to different types of two-dimensional spectra. Such spectra are square because both axes correspond to the same chemical shift and are symmetrical about the diagonal. The diagonal contains the normal one-dimensional spectrum, with both coordinates corresponding to the chemical shift of each hydrogen atom. Interactions between two hydrogen atoms produce two symmetrical cross-peaks off the diagonal, with coordinates corresponding to the chemical shifts of the two interacting hydrogen atoms.

The commonly used two-dimensional NMR spectra that arise from through-bond interactions are J-correlated spectrocopy (COSY), spin echo correlated spectroscopy (SECSY), relayed coherence transfer spectroscopy (RELAY), double quantum spectroscopy (DQNMR), homonuclear Hartmann–Hahn spectroscopy (HOHAHA), and total correlation spectroscopy (TOCSY). The simplest spectrum is COSY, in which the cross-peaks arise from hydrogen atoms bonded to the same or adjacent atoms in a residue, such as NH/C^αH, C^αH/C^βH, C^βH/C^γH. The commonly used two-dimensional spectra of through-space interactions are nuclear Overhauser effect spectroscopy (NOESY) and the closely related rotating frame NOESY (ROESY). The NOE is independent of the covalent bonding of the hydrogen atoms, so hydrogen atoms close together in the covalent structure almost invariably give substantial

NOE cross-peaks, and most of the COSY cross-peaks are also observed in NOESY spectra.

Assigning the resonances of the spectrum of a protein or polypeptide to particular hydrogen atoms usually occurs in two steps. The first step identifies resonances according to their residue type, and the second uses the amino acid sequence of the protein to identify each resonance with its position in the primary structure. The first step uses through-bond interactions, usually from COSY spectra, and identifies residues according to type; some examples are given in Figure 6.27. Gly, Ala, Ile, Leu, Val, Pro, Lys, Arg, and Thr residues can give unique connectivity patterns between atoms of the side chain and usually can be identified unambiguously. Cys, Asp, Asn, Ser, His, Phe, Tyr, and Trp residues give indistinguishable COSY patterns, however, because hydroxyl, carboxyl, amino, and thiol groups exchange their hydrogens rapidly with the aqueous solvent and because the coupling between aromatic rings and C^βH atoms is too weak to be observed. These residues ex-

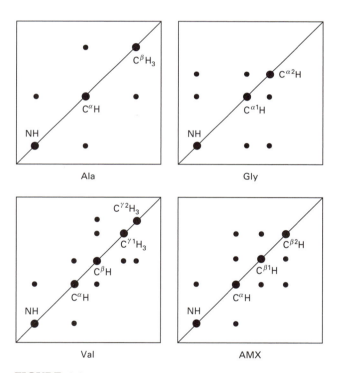

FIGURE 6.27
Examples of COSY patterns observed for Ala, Gly, Val, and AMX-type residues. The strong resonances on the diagonal, indicated by large circles, give the chemical shifts of the indicated hydrogen atoms. The cross-peaks off the diagonal give the through-bond interactions between hydrogen atoms separated by no more than three covalent bonds. The chemical-shift differences between the various types of hydrogen atoms are not to scale (see Fig. 6.26).

hibit only NH, C^αH, and two C^βH resonances and are classed together as AMX spin systems (Fig. 6.27). For similar reasons, Glu, Gln, and Met residues give similar patterns (coupling through the sulfur atom of the Met side chain is too weak), with NH, C^αH, two C^βH, and two C^γH resonances, and are classed together as AM(PT)X systems.

Having identified the various spin systems with the various classes of amino acids, the spin systems are assigned to individual residues of the protein or polypeptide by using the amino acid sequence to indicate which residues are adjacent in the primary structure and by looking for interresidue NOEs between appropriate spin systems. On the basis of the distance between hydrogen atoms, the NH of residue $i + 1$ is certain to be ≤ 3.0 Å from, and to have a strong NOE with, at least one of the NH, C^α, or C^β hydrogens of residue i, irrespective of the backbone conformation (Fig. 6.28), and correspondingly short distances are unlikely between residues that are not adjacent in the sequence. If a search is made for strong NOEs between residues that are adjacent in the sequence, the spin systems can usually be assigned to particular residues. Pro residues, of course, do not have NH groups, so the sequential assignments made in this way are brought to a halt at Pro residues in the sequence.

When all the resonances of the NMR spectra are assigned, the data from J couplings and NOE distances are used to infer the conformation of the polypeptide. Secondary structure is usually apparent from the strong NOEs used to make the assignments. Stretches of residues in an α-helix have strong NOEs between NH_i/ NH_{i+1} and $C^\beta H_i/NH_{i+1}$ but not between $C^\alpha H_i/NH_{i+1}$. In β-strands, adjacent residues give strong NOEs between $C^\alpha H_i/NH_{i+1}$ but not between NH_i/NH_{i+1}. More detailed three-dimensional information comes from NOEs between hydrogen atoms that are distant in the primary structure. For example, in a β-sheet there are strong NOEs between C^α hydrogen atoms that are juxtaposed in adjacent β-strands but are distant in the primary structure. In α-helices, NOEs are observed between residues that are three apart in the sequence.

With sufficient NOE data, a folded conformation can be determined in great detail (Sec. 6.3.2), but only if it is sufficiently stable to be present in a substantial fraction of the molecules. If there is substantial interconversion of conformations on the NMR time scale, all the NMR parameters are averages. In the case of NOEs, the dependence of intensity on r^{-6} means that any distance indicated is not simply the average of the distances present in the interconverting structures. Different conformations can be distinguished by NMR only if they are slowly interconverted.

NMR spectra are interpretable only if the hydrogen atoms have unique chemical shifts. Considerable overlap of resonances occurs if there are too many resonances in any area of a two-dimensional spectrum, either due to the protein being so large that it has a great many hydrogen atoms or due to the various hydrogen atoms being indistinguishable because there is so little nonrandom conformation that the chemical shifts are very similar. The resolution of NMR spectra can be increased by including ^{2}H, ^{13}C, and ^{15}N atoms in the protein and by going to third and fourth dimensions. The ^{2}H atoms are not observed by NMR, so their inclusion at certain positions can simplify otherwise complex two-dimensional spectra. Couplings between ^{1}H, ^{13}C, and ^{15}N atoms provide the third and fourth dimensions. In this way, larger and larger proteins are being characterized in detail by NMR techniques.

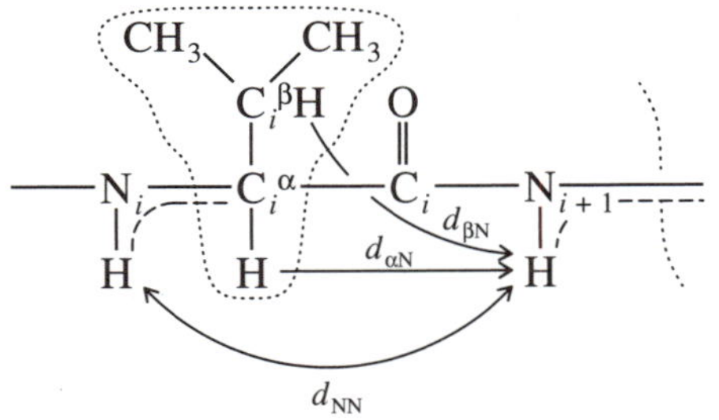

FIGURE 6.28
NMR interactions in and between residues that are used in sequential assignments of individual hydrogen atoms in proteins. The broken lines indicate through-bond connectivities between hydrogen atoms of the same residue. The solid arrows indicate the through-space distances between the NH of residue $i + 1$ and the C^αH ($d_{\alpha N}$), NH (d_{NN}), and C^βH ($d_{\beta N}$) of residue i that give strong NOEs and are used in identifying residues that are adjacent in the primary structure. (From K. Wüthrich, *NMR of Proteins and Nucleic Acids*, Wiley Interscience, New York, 1986.)

References

Sequential resonance assignments in protein ^{1}H nuclear magnetic resonance spectra. Computation of sterically allowed proton–proton distances and statistical analysis of proton–proton distances in single crystal protein conformations. M. Billeter et al. *J. Mol. Biol.* 155:321–346 (1982).

Polypeptide secondary structure determination by nuclear magnetic resonance observation of short proton–proton distances. K. Wüthrich et al. *J. Mol. Biol.* 180:715–740 (1984).

Main-chain-directed strategy for the assignment of ^{1}H NMR spectra of proteins. S. W. Englander and A. J. Wand. *Biochemistry* 26:5953–5958 (1987).

Assignment strategies in homonuclear three-dimensional ^{1}H NMR spectra of proteins. G. W. Vuister et al. *Biochemistry* 29:1829–1839 (1990).

Heteronuclear three-dimensional NMR spectroscopy of isotopically labelled biological macromolecules. S. W. Fesik and E. P. Zuiderweg. *Quart. Rev. Biophys.* 23:97–131 (1990).

Uniform and selective deuteration in two-dimensional NMR of proteins. D. L. LeMaster. *Ann. Rev. Biophys. Biophys. Chem.* 19:243–266 (1990).

Defining solution conformations of small linear peptides. H. J. Dyson and P. E. Wright. *Ann. Rev. Biophys. Biophys. Chem.* 20:519–538 (1991).

6.3.2 Determining a Protein Structure by Nuclear Magnetic Resonance

The three-dimensional structure of a protein can be determined by ^{1}H-NMR primarily because hydrogen atoms within 5 Å of each other usually give an NOE cross-peak in NOESY spectra, with an intensity proportional to the inverse sixth power of the distance between them. Other factors affect the magnitude of the NOE, however, so it is only possible to put an upper limit on the distance. Usually, NOEs are classified as being strong, intermediate, or weak, which implies that the distance is, respectively, <2.5, <3.5, or <5 Å. A lower limit can be imposed from the sum of the van der Waals radii of the atoms.

The presence of NOEs between groups that are close in the covalent structure can define the secondary structure (Sec. 6.2.2), and the tertiary structure is defined primarily by NOEs between hydrogen atoms that are distant in the primary structure but close in space. The presence of NOEs between various residues can be used to construct a contact map like that in Figure 6.14, which gives some impression of the structure.

More importantly, NOEs can be used to generate a three-dimensional structure, although this is not a trivial mathematical exercise. The usual procedure is to start with a randomly generated conformation of the protein, built using the known primary structure and standard geometries of the peptide backbone and side chains, but with random dihedral angles. This starting structure is incompatible with most of the NMR data, and the discrepancy can be quantified. The structure is then modified gradually by changing the dihedral angles to minimize the discrepancies; local constraints between groups that are close in the primary structure are satisfied first, and then those between groups that are increasingly distant. At some stage, the structure is incapable of being improved further, but this might be because of simple geometric problems that cannot be easily corrected, such as initially putting one group on the wrong side of another group; interchanging their positions is not feasible by means of a few small structural rearrangements. For this reason, the procedure is repeated many times, using the same NMR data but different random starting conformations. The correct structures are usually apparent by their greater consistency with the NMR data and by their similar final structures. Such structures are usually superimposed and displayed simultaneously (Fig. 6.29); the closer their agreement, the more well defined the structure.

The accuracy of the determination of protein structures by NMR depends not so much on the accuracy of the NOE distance constraints as on their number. There is no lower limit on the number of NOE distance constraints necessary for a structure calculation, and there is no equivalent of the resolution of an X-ray structure. The quality of structures determined by NMR varies enormously. Often, the initial NMR structures are "improved" by minimizing their "energies," by optimizing their dihedral angles, packing, and hydrogen bonds.

The validity of the NMR method was established conclusively by determining the three-dimensional structure of the protein *tendamistat* independently using NMR and normal X-ray diffraction analysis. The structures were closely similar, with only 4 of the 69 residues having deviations of their backbone atoms of greater than 1.5 Å (Fig. 6.29). Although NMR does not determine structure in such fine detail as X-ray diffraction, its advantages are that the protein is free of constraints of the crystal lattice and that it can unambiguously define the orientations of certain side chains, such as those of Asn, Gln, and Thr, which are barely distinguishable by X rays.

NMR also gives much more detailed information about the flexibility of the protein structure in solution. Alternative conformations or the absence of a fixed conformation are usually apparent from the NOE data. The rates at which groups are rotating rapidly can be determined by ^{13}C-NMR in many instances (Table 5.1). In addition, ^{1}H-NMR can distinguish whether or not Tyr and Phe aromatic rings are rotating rapidly by whether or not the two *ortho* ($C^{\delta 1}$ and $C^{\delta 2}$) and *meta* ($C^{\epsilon 1}$ and $C^{\epsilon 2}$) hydrogen atoms of Tyr and Phe residues give individual resonances (Sec. 6.3.1). Although NMR provides much dynamic data, the inclusion of flexibility in calculations of the protein structure from NOE data is not straightforward.

References

Distance geometry and related methods for protein structure determination from NMR data. W. Braun. *Quart. Rev. Biophys.* 19:115–157 (1987).

FIGURE 6.29

Stereo diagram of nine structures of the protein tendamistat generated by distance geometry analysis of NMR data. Only the positions of the C^α atoms are indicated, with virtual bonds between them. The circles indicate the positions of the C^α atoms indicated by the X-ray crystal structure, each with a radius defined by the crystallographic temperature factor. (From M. Billeter et al., *J. Mol. Biol.* 206:677–687, 1989.)

Protein structures from NMR. R. Kaptein et al. *Biochemistry* 27:5389–5395 (1988).

The development of nuclear magnetic resonance spectroscopy as a technique for protein structure determination. K. Wüthrich. *Acc. Chem. Res.* 22:36–44 (1989).

Comparison of the high-resolution structures of the α-amylase inhibitor tendamistat determined by nuclear magnetic resonance in solution and by X-ray diffraction in single crystals. M. Billeter et al. *J. Mol. Biol.* 206:677–687 (1989).

Two-, three-, and four-dimensional NMR methods for obtaining larger and more precise three-dimensional structures of proteins in solution. G. M. Clore and A. M. Gronenborn. *Ann. Rev. Biophys. Biophys. Chem.* 20:29–63 (1991).

6.4 Proteins with Similar Folded Conformations

From the very wide diversity of amino acid sequences observed in proteins, it might be expected that there would be a corresponding diversity of folded conforma-tions, but similarities among protein structures are observed more frequently than are differences. In particular, the tertiary structures of proteins have been much more conserved during evolution than have their primary structures. The extent of this evolutionary conservation is a matter of intense debate and speculation because similarities in tertiary structures are being found in cases where there is no evolutionary or functional reason to expect them.

6.4.1 Evolutionarily Related Proteins

Homologous proteins that have substantial similarities in their primary structures and that are almost certain to have arisen from a common ancestor (see Chap. 3) have invariably been found to have very similar folded conformations. For example, the three-dimensional structures of horse and human hemoglobin are virtually identical, even though they differ in 43 of the 287 residues of the α and β chains. Similarly, the cytochromes *c* of horse and tuna have very similar structures, even though they differ at 17 of the 104 amino acid residues. Such conservation of conformation is understandable in

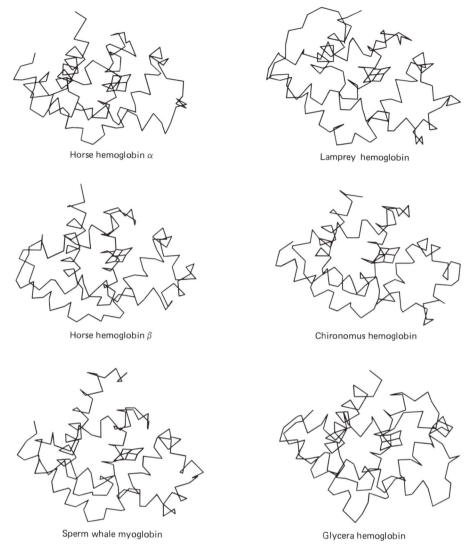

Horse hemoglobin α

Lamprey hemoglobin

Horse hemoglobin β

Chironomus hemoglobin

Sperm whale myoglobin

Glycera hemoglobin

FIGURE 6.30

Comparison of the similar polypeptide backbone foldings in the globins, some of which are related only distantly. (From W. Love, *Cold Spring Harbor Symp. Quant. Biol.* 36:349, 1971.)

that the related proteins probably serve the same function in the different species, with very similar conformational requirements.

That more distantly related proteins also have very similar conformations was shown by the very first two proteins whose structures were determined crystallographically: sperm whale myoglobin and horse hemoglobin. The single polypeptide chain of myoglobin and the α and β chains of hemoglobin are remarkably similar in the general topologies of their polypeptide backbones (Fig. 6.30). The amino acid sequences of these proteins are similar enough to indicate that the proteins are related evolutionarily (Fig. 3.12), and they have similar functions in reversibly binding oxygen at a bound heme

group. Other oxygen-binding, heme-containing proteins from a variety of vertebrates, a marine annelid worm, a larval insect, and lupine root nodules also have similar conformations (Fig. 6.30). These proteins are usually grouped as the globin family. The primary structures of some of these globins are not detectably homologous. Only two amino acid residues are common to all the globins, but the existence of intermediate sequences to link them evolutionarily, and their closely similar conformations, indicate that they have all diverged from a common ancestor.

Another family of proteins with similar conformations is the cytochrome *c* proteins. The conformations of the closely similar cytochromes *c* from the mitochondria

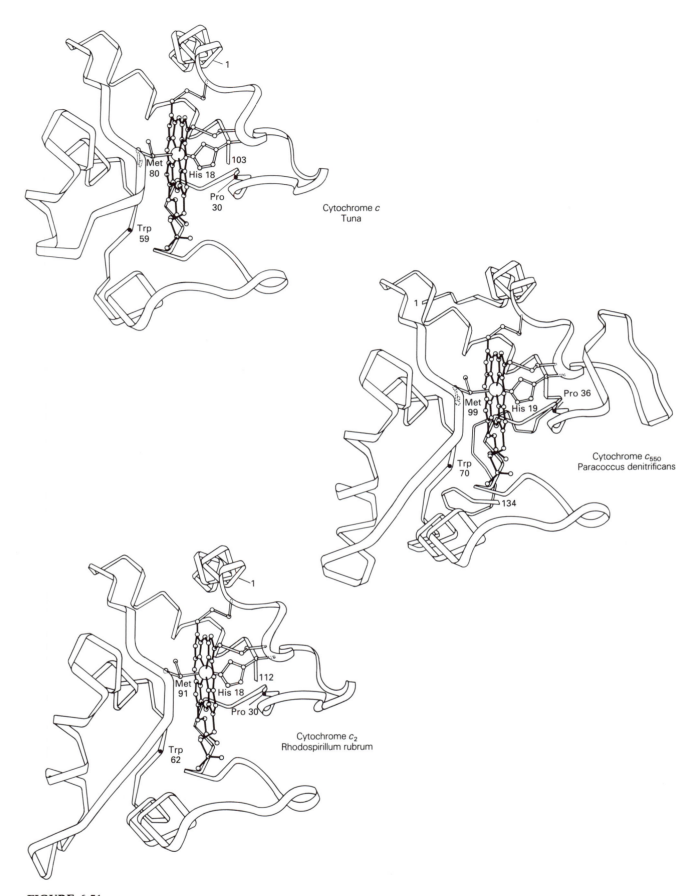

FIGURE 6.31

The similar folded conformations of distantly related cytochromes *c*. (From F. R. Saleme, *Ann. Rev. Biochem.* 46:299–329, 1977.)

of vertebrates (Fig. 3.6) resemble those in a variety of functionally related cytochromes from bacteria (Fig. 6.31). The bacterial cytochromes were recognized by their properties to be similar to cytochrome c but were sufficiently different in primary structure and in physical and functional properties to be given names such as c_2, c_{550}, c_{551}, and c_{555}. The polypeptide chains of these proteins range in length from 82 to 135 residues and have only 5 residues that are identical in all, yet their three-dimensional structures are remarkably similar.

The other well-characterized family of homologous proteins is that of the trypsinlike serine proteases, so named because they all have an important serine residue at the active site and similar catalytic mechanisms (Sec. 9.3.2). This family includes the related proteins of higher organisms — trypsin, chymotrypsin, elastase, and thrombin — as well as similar enzymes in microbes. In the case of bovine α-chymotrypsin and elastase, 39% of their residues are identical, and their conformations can be superimposed so that on average the atoms of their polypeptide backbone differ in relative positions by only 1.80 Å. This similarity in topology is immediately obvious in their distance plots (Fig. 6.32). Trypsin and α-chymotrypsin can be juxtaposed to within 0.75 Å, even though only 44% of their residues are identical. The bacterial serine proteases have diverged more than the mammalian proteins in both primary sequence and overall conformation; fewer than 20% of their residues are the same as those of any of the mammalian proteins, with numerous insertions and deletions. Although 20% is not much more similar than random sequences would be, the three-dimensional conformations are more obviously similar than are their primary structures; 55 – 64% of the residues of the

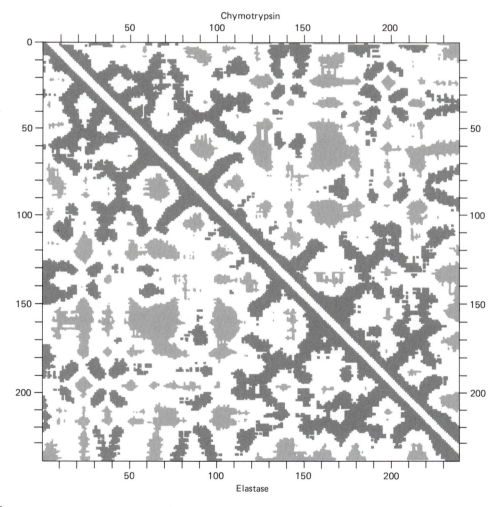

FIGURE 6.32

Similar contact maps for the homologous structures of α-chymotrypsin *(top right)* and elastase *(bottom left)*. Distances between residues greater than 30 Å are light gray; distances less than 15 Å are dark gray. (From L. Sawyer et al., *J. Mol. Biol.* 118:137–208, 1978.)

mammalian and bacterial proteins are topologically equivalent in that their C^α atoms occupy the same relative positions. With the greater divergence of amino acid sequence, there is a somewhat greater change in structure, but the structure is more conserved.

The first question to be asked about these observations is how such wide variation in amino acid sequences can be compatible with such similarity in folded conformations. The greatest dissimilarities in amino acid residue sequences occur at positions in the polypeptide chains where the side chains are on the surface. Here the chemical nature of the side chains has often changed. There is a tendency, however, for the residues at reverse turns to have a short, polar side chain or to be Gly or Pro (the residues found most often at these positions). The conformational flexibility in these regions can lead to surprising differences in local conformations of homologous proteins. For example, bovine and porcine phospholipases A_2 have homologous sequences and very similar three-dimensional structures, except for a segment of 12 residues on their surfaces; the two proteins have very different conformations in this segment even though the sequences differ at only one position. Interior residues differ least frequently, and the nonpolar nature of the side chains is highly conserved. The side-chain torsion angles are also conserved to a remarkable degree, as are interior polar groups involved in hydrogen bonds. There is generally a common core of residues, comprising the center of the molecule and the main elements of secondary structure, that are highly conserved. Only very slight structural changes occur in this common core as the primary structure of the protein diverges, and there is a good correlation between the extent of the sequence divergence and the structural differences between the common cores (Fig. 6.33).

The most highly conserved residues are those involved directly in the functional properties of the protein. Examples are the invariant cytochrome c Cys residues 14 and 17, to which the heme group is attached, and the ligands to the heme iron atoms, His 18 and Met 80 (Figs. 3.8 and 6.31). Likewise, the His side chains of hemoglobin that bind the heme iron (Fig. 8.22) are conserved, as are the catalytically important residues in the active sites of the serine proteases (Fig. 9.19). Site-directed mutagenesis of residues that are conserved in many homologous sequences usually has substantial effects on the function or stability of a protein, although not always the dramatic effects that might be predicted.

Insertions and deletions of residues in homologous polypeptide chains occur most frequently at reverse turns on the surface of the folded protein, usually with little perturbation of the interior. Such changes in sequences are apparent in the various c-type cytochromes

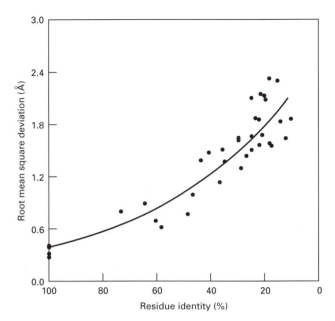

FIGURE 6.33

The relationship between identity of residues (in %) in the primary structure and the rms deviation of the backbone atoms of the common cores of 32 pairs of homologous proteins. The common core is that part of the related proteins that retains the same fold. With only 20% of residues identical, the common core is generally only about half of the residues of each protein. (From C. Chothia and A. M. Lesk, *EMBO J.* 5:823–826, 1986.)

(Fig. 6.31). Disulfide bridges are often added or deleted. For example, in the serine protease family, trypsin has six bridges, chymotrypsin five, elastase four, α-lytic protease three, and proteases A and B of *Streptomyces griseus* two each. The Cys residues have never been observed to change their disulfide pairings, however. If a disulfide bridge has been deleted, the Cys residues on both homologous proteins are generally changed to other amino acid residues, presumably as a result of two mutational events, but their relative three-dimensional positions in the folded structure remain the same. Obviously, disulfide bonds do not determine conformations, nor are they absolutely necessary to them.

Elements of secondary structure can move relative to each other, can change in length, or can even disappear altogether, but a helix is not replaced by a β-sheet, or vice versa. In particular, neither the order nor the orientation (parallel or antiparallel) of strands in a β-sheet has ever been observed to differ between proteins with homologous primary structures. The relative positions of α-helices are retained, but often within wide limits. For example, corresponding pairs of α-helices in some of the widely differing globins may differ by up to 7 Å in proximity and by up to 30° in relative orientation.

Single nonlethal mutations of primary structure seem to produce very small changes in secondary structure, but the cumulative effects of many such mutations can produce substantial differences among homologous conformations.

In summary, the three-dimensional structures of evolutionarily related proteins appear to have been remarkably conserved during their evolution, suggesting that the three-dimensional structure was well defined at an early stage of evolution in a common ancestor and that it is crucial for the function of the protein. Changes in primary structure due to genetic mutations have accumulated during evolutionary divergence, but only those that were compatible with the folded conformation and with its function have been retained. Nevertheless, the great variety of amino acid sequences that produce a particular folded conformation is indicative of the flexibility in the rules of protein folding.

Knowledge of the three-dimensional structures of evolutionarily related proteins is very useful for inferring the genetic basis of the evolutionary divergence, in that it identifies which residues of primary structures of homologous proteins should be aligned for comparison of sequences. Structurally equivalent residues are assumed to be those that are evolutionarily related. Alignment of sequences of distantly related proteins is usually uncertain, especially when there have been numerous deletions and insertions. The usual approach of minimizing the number of genetic mutations required to interconvert two sequence often gives alignments that are inconsistent with structural homology.

References

Polypeptide chains with similar amino acid sequences but a distinctly different conformation. B. W. Dijkstra et al. *FEBS Letters* 164:25–27 (1983).

Evolution and the tertiary structure of proteins. M. Bajaj and T. Blundell. *Ann. Rev. Biophys. Bioeng.* 13:453–492 (1984).

Amino acid replacements in yeast iso-1-cytochrome *c*. Comparison with the phylogenetic series and the tertiary structure of related cytochromes *c*. D. M. Hampsey et al. *J. Biol. Chem.* 261:3259–3271 (1986).

The relation between the divergence of sequence and structure in proteins. C. Chothia and A. M. Lesk. *EMBO J.* 5:823–826 (1986).

The response of protein structures to amino-acid sequence changes. A. M. Lesk and C. H. Chothia. *Phil. Trans. R. Soc. Lond.* A317:345–356 (1986).

Analysis of side-chain orientations in homologous proteins. N. L. Summers et al. *J. Mol. Biol.* 196:175–198 (1987).

Protein structure alignment. W. R. Taylor and C. A. Oregno. *J. Mol. Biol.* 208:1–22 (1989).

6.4.2 Conformational Similarity without Apparent Sequence Homology

Have the folded conformations of proteins been so conserved during evolution that similar conformations in two proteins are indicative of a common ancestor, even when there is no detectable homology between them in primary structure? If so, what degree of conformational similarity is significant? How much of the similarity among three-dimensional structures simply reflects the general principles of protein conformation? Do a common backbone topology and a similar function indicate that two proteins had a common ancestor? Alternatively, can similar conformations arise by convergent evolution as a result of selection for a common function and common structural requirements for such a function? These are just a few of the questions raised by the existence of similarly folded conformations when none was expected on the basis of the primary structures of the proteins. No examples of convergent evolution of protein conformation have been well documented, but this may be because convergence is difficult to prove. The best evidence for convergence is usually taken to be when elements of secondary structure that have the same position in similar three-dimensional structures have different orders in the primary structure. The rearrangements of genes that would be required for this are considered unlikely to be compatible with retention of three-dimensional structures, although it is possible in some instances (see Fig. 2.18). The most plausible candidates for evolutionary convergence are the $(\beta\alpha)_8$ barrels that were first observed in triose phosphate isomerase (Fig. 6.17). This structure has now been found in at least 16 proteins with no detectable amino acid homology among them. Two fundamentally different types of barrel have been identified on the basis of the packing of side chains in the barrel. In one protein structure, the order in the primary structure of the first helix and the second β-strand is interchanged. The regularity of the $(\beta/\alpha)_8$ structure makes it plausible that it might be an example of convergent evolution for structural reasons. Proteins with this structure also have some functional similarities, however, in that they are all enzymes with their active sites in the same position at the C-terminal end of the β-barrel. This might indicate extreme divergence from a common ancestor.

Other examples of conformational similarities without detectable primary structure homology also involve functional similarities. The **mononucleotide-binding domain** consists of three parallel β-strands with two intervening α-helices, and it occurs in many proteins that bind nucleotides (Fig. 6.34). Generally, no amino acid sequence homology is detectable among

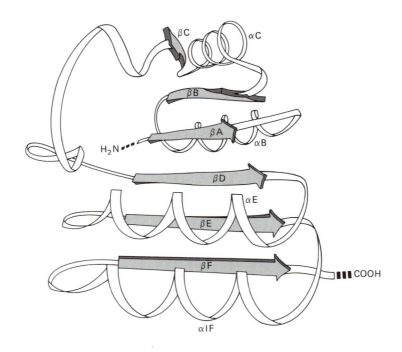

FIGURE 6.34

The NAD-binding domain of the dehydrogenases, which is composed of two β-α-β-α-β nu-cleotide-binding units. Nucleotides are generally bound on the right, near the carboxyl ends of the β-strands (see also Fig. 6.35). (From I. Ohlsson et al., *J. Mol Biol.* 89:339–354, 1974.)

them except that necessary for structural and functional reasons (Fig. 6.35), but almost all of these supersecondary structures bind nucleotides in a similar way, and they are now considered to have arisen by divergence, rather than convergence.

Although the lysozymes of chicken and of bacteriophage T4 have similar functions, they were initially concluded to be structurally unrelated. Later, closer comparison revealed significant structural similarities that probably indicate they diverged long ago from a common ancestor; this was confirmed by the discovery of lysozymes with intermediate structures. This is an example of the conservation of a function even though the structure has diverged substantially.

The best examples of evolutionary convergence in proteins have similar active sites and catalytic mechanisms but have no sequence or conformational homology. The bacterial protease *subtilisin,* members of the mammalian trypsin protease family, and a serine carboxypeptidase have very similar geometries of residues at their active sites but no other structural similarities; remarkably, the serine carboxypeptidase is structurally homologous to a different class of proteases that includes carboxypeptidase A (Sec. 9.3.2). Thermolysin and bovine carboxypeptidase A have similar catalytic mechanisms with similar active sites but no other de-

tectable similarities. The possibility that these similarities are simply vestiges of distant evolutionary origins obscured by extreme divergence is thought to be ruled out by the different positions occupied by the equivalent catalytic site residues in the primary structures.

References

Similarities of protein topologies: evolutionary divergence, functional convergence or principles of folding? O. B. Ptitsyn and A. V. Finkelstein. *Quart. Rev. Biophys.* 13:339–386, (1980).

Relation between structure and function of α/β proteins. C. I. Brändén. *Quart. Rev. Biophys.* 13:317–338, (1980).

Common precursor of lysozymes of hen egg-white and bacteriophage T4. B. W. Matthews, et al. *Nature* 290:334–335, (1981).

Structure of phage P22 gene *19* lysozyme inferred from its homology with phage T4 lysozyme. Implications for lysozyme evolution. L. H. Weaver et al. *J. Mol. Biol.* 184:739–741 (1985).

Structure of wheat serine carboxypeptidase II at 3.5 Å resolution. A new class of serine protease. D. I. Liao and S. J. Remington. *J. Biol. Chem.* 265:6528–6531 (1990).

The evolution of α/β barrel enzymes. G. K. Farber and G. A. Petsko. *Trends Biochem. Sci.* 15:228–234 (1990).

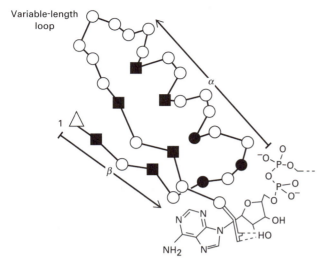

FIGURE 6.35

A typical β-α-β nucleotide-binding domain with an ADP moiety bound at lower right. Residues in the α-helix and β-strands indicated by squares are usually small and hydrophobic (Ala, Ile, Leu, Val, Met, or Cys) and form the hydrophobic core between the β-strands and the α-helix. Filled circles indicate Gly residues that permit a sharp turn between the first β-strand and the helix. The first residue of the polypeptide chain shown (indicated by a triangle and numbered) is usually basic or hydrophilic (Lys, Arg, His, Ser, Thr, Gln, or Asn). The last residue of this segment of chain is either Asp or Glu, with the side chain hydrogen-bonded to the ligand. The loop between the helix and the second β-strand can be variable in length. Proteins containing this structure can be identified from their sequences on the basis of this pattern of residues. (From R. K. Wierenga et al., *J. Mol. Biol.* 187:101–107, 1986.)

6.4.3 Structural Homology within a Polypeptide Chain

Amino acid sequence homology between parts of a polypeptide chain indicates that elongation of the gene, and therefore of the protein, occurred by gene duplication (Sec. 3.4.3). All the structures of such duplicated proteins—for example, ferredoxin (Fig. 3.15), parvalbumin (Fig. 8.21), and the immunoglobulins (Fig. 8.9)—also have demonstrable internal homologies at the level of tertiary structure (Fig. 6.36).

Other instances of structural homology among parts of a protein, with a corresponding symmetry of the total structure, have been observed even between parts in which there is no apparent homology in their primary structures. For example, the two domains of rhodanese,

each composed of about 140 residues, are not detectably homologous, but 117 of their C^α atoms can be superimposed with an average deviation of only 1.95 Å. Other proteins with two homologous halves, but no apparent sequence homology, are tobacco mosaic virus coat protein, cytochrome b_5, arabinose-binding protein, NAD-binding domains (Fig. 6.34), trypsin proteases (Figs. 6.32 and 9.23), acid proteases, glutathione reductase, hexokinase, γ-crystallin, and rubredoxin. A three-fold repeat is observed in soybean trypsin inhibitor.

As with homologies between proteins that have no detectable sequence homology, it is difficult to determine whether these instances arose by gene duplication and subsequent extreme divergence of the amino acid sequence or by convergence. In all these intramolecular cases, however, the similar conformations are related by symmetry axes, very close to those expected if the similar structures were identical subunits in an oligomeric protein (Fig. 6.25). For example, the two parts of twofold repeats are related by rotations close to 180°; in threefold repeats the rotations are very close to 120°. These findings suggest that such proteins were originally oligomers of identical chains and that their genes

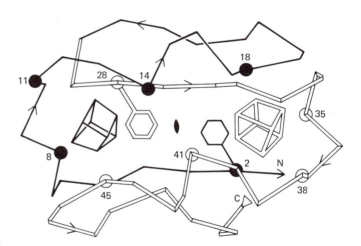

FIGURE 6.36

Internal symmetry in ferredoxin, resulting from internal homology in its primary structure. The sequence homology between the two halves of the polypeptide chain is shown in Figure 3.15; the amino-terminal half of the polypeptide backbone is dark and drawn with solid lines, the other half is drawn with open lines. Note that each of the two symmetry-related halves of the structure involves both halves of the primary structure. The distorted cubes are the iron–sulfur clusters used to bind electrons reversibly (see Fig. 8.19). Circles indicate Cys residues that are attached to the iron–sulfur clusters. (From A. D. McLachlan, *J. Mol. Biol.* 128:49–79, 1979.)

were duplicated to give a single fused polypeptide chain; otherwise, there is no apparent reason for the existence of such symmetry axes. The original amino acid homologies of these chains would have disappeared by divergence, but the three-dimensional structures and the packing of the original monomers have been retained.

The structure of one ferredoxin demonstrates the twofold symmetry (Fig. 6.36) and further strengthens the argument for a dimer precursor during evolution. The two halves of the primary structure do not form separate domains, but both halves of the tertiary structure contain residues from both halves of the polypep-

tide. Therefore, the two original polypeptide chains were probably adapted to fit together before they fused, which implies a dimeric structure. In other ferredoxins, one of the iron–sulfur complexes has disappeared, some of the iron-ligating Cys residues have been changed, and one segment of polypeptide chain has become an α-helix (Fig. 6.37). The protein structure in this case has lost its twofold symmetry and is an extreme example of tertiary structure evolution.

Internal symmetry in proteins thus provides further evidence for the predominant role of evolutionary divergence and the greater preservation of the folded conformations than of the amino acid sequences. These

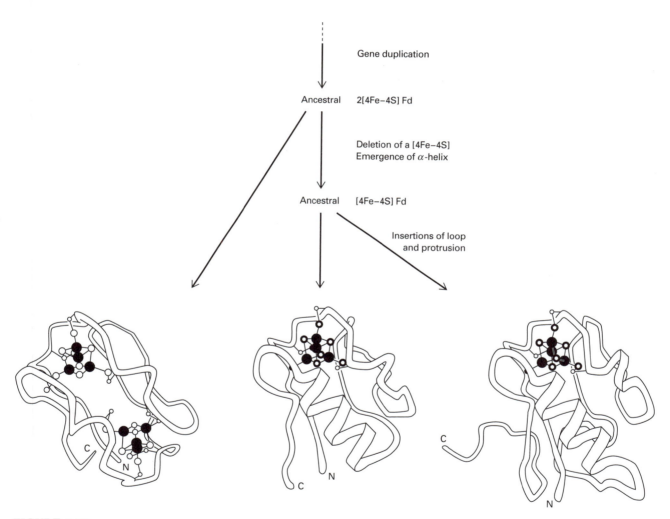

FIGURE 6.37

Change in tertiary structure during evolution of bacterial ferredoxins. The structure at the left has two iron–sulfur clusters and is observed in the ferredoxins from *Peptococcus aerogenes* and *Azotobacter vinelandii*. The structure on the right is the one-cluster ferredoxin from *Bacillus thermoproteolyticus*. The structure in the middle is that of other one-cluster ferredoxins. (From K. Fukuyama et al., *J. Mol. Biol.* 199:183–193, 1988.)

observations strengthen the conclusion that conformational similarity implies homology and a common origin of polypeptides, but the generality of this conclusion is still a matter of debate.

References

Structural evidence for gene duplication in the evolution of the acid proteases. J. Tang et al. *Nature* 271:618–621 (1978).

Gene duplications in the structural evolution of chymotrypsin. A. D. McLachlan. *J. Mol. Biol.* 128:49–79 (1979).

Gene duplication in glutathione reductase. G. E. Schulz. *J. Mol. Biol.* 138:335–347 (1980).

Pseudo 2-fold symmetry in the copper-binding domain of arthropodan haemocyanins. Possible implications for the evolution of oxygen transport proteins. A. Volbeda and W. G. J. Hol. *J. Mol. Biol.* 206:531–546 (1989).

6.5 Rationalization and Prediction of Protein Structure

The folded conformation of a protein is thought to be dictated solely by its primary structure, which opens the tantalizing possibility of being able to predict three-dimensional structures from amino acid sequences. Prediction eludes us so far, however, because of the astronomical number of possible conformations and because of uncertainty about how and why polypeptide chains fold (Secs. 7.4 and 7.5). The great variation in amino acid sequences that can occur in essentially identical folded conformations indicates that there is substantial flexibility, or redundancy, in the rules relating primary, secondary, and tertiary structures. Clearly, the particular amino acid side chain at each residue of a protein is not necessarily important for determining the folded conformation. Some internal residues that are conserved in all related proteins may be important, but only the general type of side chain (e.g., aromatic, nonpolar, hydrogen-bonding, or ionic) is conserved at other residues. Nevertheless, the general rules of protein structure noted in Section 6.2 are always followed: The interior is densely packed, with no large empty spaces; charged and polar groups are on the surface, accessible to the solvent, unless they are paired by hydrogen bonds; and the torsion angles for bond rotations are generally favorable, so there is minimal conformational strain. These principles are not surprising because they clearly contribute to lowering the free energy of the folded conformation (Sec. 7.4). The folded conformation may be the one that has the lowest possible free energy, although it may only be the lowest energy conformation of those that are accessible by the kinetics of folding (Sec. 7.5).

It is not possible at present to search all the possible conformations of a polypeptide chain to find the one with the lowest free energy, because the number of possibilities is too great and it is not yet possible to calculate the free energies of protein molecules in solution. Empirical potential energies can be calculated that consider bond lengths and angles, van der Waals interactions, hydrogen bonds, and electrostatic interactions, using simplified energy parameters derived from model systems. Such calculations generally ignore the solvent and omit the entropy of the systems; they are sensitive primarily to bad geometries and to too-close contacts between atoms. The relevance of such empirical energy calculations is brought into question by their inability to distinguish between correct and incorrect conformations. For example, in one calculation the sequences and three-dimensional structures of two unrelated proteins of the same size were interchanged. The primary structure of hemerythrin, which normally has a 4-helix bundle structure, was built into the 4-stranded β-sheet sandwich structure of an immunoglobulin domain, and vice versa. The initial structures were refined by minimizing their empirical potential energies, and the incorrect structures reached values comparable to those of the correct structures. Yet the refined, incorrect structures clearly violated the general principles of close packing, hydrogen bonding, minimal exposed nonpolar surface area, and solvent accessibility of charged groups, all of which are found in natural folded proteins. Subsequently, attention has focused on packing and accessible-surface criteria as a measure of the stability of protein conformations. For example, the hydrophobicities of the surface areas buried in authentic folded structures are remarkably consistent and depend only on the size of the protein (Fig. 6.38); misfolded structures do not obey this relationship.

At present, it is possible to rationalize the energetic and structural basis of the folded conformations of proteins only semiquantitatively. For example, pairing of internal polar groups in hydrogen bonds is energetically necessary (Sec. 4.2.3), and the prevalent secondary structure in protein interiors is an efficient way of doing so. The nonpolar surfaces between these elements of secondary structure minimize the less favorable interactions with water that they would have in the unfolded state by packing together. Most protein domains have layered structures with two or three layers of either α-helices or β-sheets, but not both in one sheet, packed together by nonpolar side chains. A mixed layer would

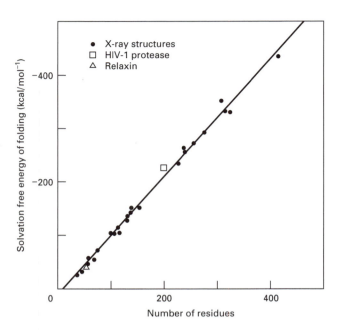

FIGURE 6.38
Hydrophobicity of the surface buried in folded protein structures relative to the number of residues in the polypeptide chain. The hydrophobicity is expressed as the negative solvation free-energy contributions of the various groups, like those of Table 4.10. The closed circles correspond to structures determined by X-ray crystallography, the open symbols to models built from homologous structures. (From L. Chiche et al., *Proc. Natl. Acad. Sci. USA* 87:3240–3243, 1990.)

be unfavorable because the end strands of β-sheets could not be hydrogen-bonded to protein atoms or to water. Similarly, reverse turns occur only at the surfaces because there the backbone groups of residues involved in the turn can hydrogen-bond to water; reverse turns generally connect nearly antiparallel segments in the same or adjacent layers because this minimizes their length. The packing of a β-sheet against an α-helix or another β-sheet is governed by the right-handed twist of the sheet. In α-helical proteins, the helices are packed around a hydrophobic core. Here the helices are amphipathic (Fig. 5.8), with the interior face nonpolar, the outer hydrophilic. The diameter of the cores in these proteins is about the length of two side chains. The right-handed connections between β-strands (Fig. 6.19) are favored because the twist of the β-sheet makes them the shortest connections. For large polypeptide chains, the number of layers would become too great, so a more stable arrangement is to divide the globule into smaller individual domains.

The amino acid sequences of water-soluble globular proteins do not have the periodicities typical of fibrous proteins such as keratin and collagen (Sec. 5.5)

that prevent them from adopting compact conformations. Nor do they have the large segments of polar and nonpolar residues typical of membrane proteins (Sec. 7.2). Instead, the nonpolar residues of globular proteins are more or less uniformly mixed with polar residues throughout the sequence. The method of burying the maximum number of nonpolar side chains, and the way they pack together while keeping the ionized side chains on the surface, appears to determine the chain fold. There is a reasonable correlation between the hydrophobicities of the amino acid side chains and their tendencies to be buried in protein structures (Fig. 6.39).

Not all aspects of protein conformation can be rationalized on the basis of maximizing stability. The tendency for elements of secondary structure that are adjacent in the sequence also to be adjacent in the tertiary structure, the absence of knots, and the tendency for domains to be formed by a single continuous segment of polypeptide chain are more likely to be produced by the kinetic process of folding (Sec. 7.5).

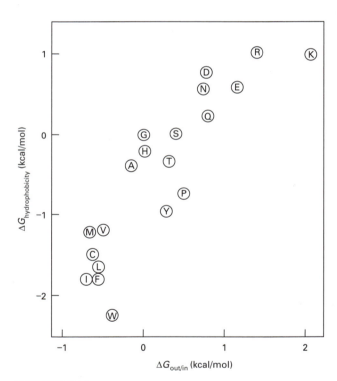

FIGURE 6.39
Correlation between the hydrophobicities of the amino acid residues ($\Delta G_{hydrophobicity}$) as measured by the free energy of transfer from water to nonpolar solvent and the corresponding value for occurrence at protein surfaces rather than interiors ($\Delta G_{out/in}$). The values of $\Delta G_{hydrophobicity}$ are from J. Flauchére and V. Pliska, *Eur. J. Med. Chem.* 18:369–375 (1983) given in Table 4.8; values of $\Delta G_{out/in}$ are from S. Miller et al., *J. Mol. Biol.* 105:641–656 (1987) given in Table 6.3.

References

An analysis of incorrectly folded protein models. Implications for structure predictions. J. Novotny et al. *J. Mol. Biol.* 177:787–818 (1984).

Hydrophobicity of amino acid residues in globular proteins. G. D. Rose et al. *Science* 229:834–838 (1985).

An empirical examination of potential energy minimization using the well-determined structure of the protein crambin. M. Whitlow and M. M. Teeter. *J. Amer. Chem. Soc.* 108:7163–7172 (1986).

Why do globular proteins fit the limited set of folding patterns? A. V. Finkelstein and O. B. Ptitisyn. *Prog. Biophys. Mol. Biol.* 50:171–190 (1987).

Criteria that discriminate between native proteins and incorrectly folded models. J. Novotny et al. *Proteins: Struct. Funct. Genet.* 4:19–30 (1988).

The classification and origins of protein folding patterns. C. Chothia and A. V. Finkelstein. *Ann. Rev. Biochem.* 59:1007–1039 (1990).

6.5.1 Predicting Secondary Structure

Most attempts to predict protein structure have concentrated on predicting the elements of secondary structure, because 90% of the residues in most proteins are involved either in α-helices (38%), β-strands (20%), or reverse turns (32%). If the secondary-structure elements could be predicted accurately, it might be feasible to pack them together to generate the correct folded conformation.

The various amino acid residues do demonstrate conformational preferences. The relative tendencies of the various residues to be involved in α-helices, β-strands, and reverse turns are given by the conformational preferences P_α, P_β, and P_t, respectively (Table 6.5). These preferences, however, are only marginal. For example, the most helix-preferring amino acid, Glu, occurs in α-helices only 59% more frequently than random, and even Gly and Pro residues are found in helices about 40% as often as random, even though they are not stereochemically compatible with the helical conformation.

Fortunately, helices, β-sheets, and reverse turns are not determined by a single residue but by a number of them adjacent in the sequence. A segment of a particular secondary structure is much more probable when several adjacent residues prefer that structure. A number of prediction schemes based on such empirical observations have been proposed. The easiest to use, and the best known, is that of P. Y. Chou and G. D. Fasman, which classifies the amino acids as favoring, breaking, or being indifferent to each type of conformation. An α-helix is predicted if four out of six adjacent residues are helix-favoring and if the average value of P_α is greater than 1.0 and greater than P_β; the helix is extended along the sequence until either Pro or a run of four sequential residues with average value of $P_\alpha < 1.0$ is reached. A β-strand is predicted if three out of five residues are sheet-favoring and if the average value of P_β is greater than 1.04 and greater than P_α; the strand is extended along the sequence until a run of four residues with an average value of $P_\beta < 1.0$ is reached. A reverse turn is the likely conformation when sequences of four residues characteristic of reverse turns are found (Table 6.5).

These procedures can be improved somewhat by also considering the various positions of the residues in the elements of secondary structure. For example, there are preferences of particular amino acid residues for specific positions in α-helices. Acidic Asp and Glu residues predominate at the N-terminus, and basic Lys, Arg, and His residues predominate at the C-terminus, presumably as a result of favorable interactions with the partial charges of the helix dipole (Sec. 5.3.1). Pro residues rarely occur in the interiors of helices, where their unusual backbones interrupt the helix and cause it to kink, but they occur frequently at the N-terminal first turn of the helix, where their particular geometry fits well. Asn, Asp, Ser, and Thr residues often occur in the first turn, where their side chains tend to hydrogen-bond to the backbone of the third residue farther along; the side chain provides hydrogen bonds that would have been provided by the backbone if the helix were extended farther at the amino end. Gly residues occur at the carboxyl end of about a third of all α-helices, where the more flexible backbone of this residue tends to disrupt the helix by tending toward a 3_{10} type of conformation (Table 5.2). In reverse turns, there are even greater preferences, but they depend on the type of turn (Table 6.5); it is necessary, therefore, to calculate predictions for each basic type of turn.

A more a priori theory, based on stereochemical considerations of the hydrophobic, hydrophilic, and electrostatic properties of the side chains in terms of the structural rules of folded proteins, was proposed by V. I. Lim. Interactions between side chains separated by up to three residues in the sequence are considered in terms of their packing in either the helical or β-sheet conformations. For example, a sequence with alternating hydrophobic and hydrophilic side chains is likely to be a strand in a β-sheet, in which the orientations of the side chains alternate, with the hydrophilic side exposed to the solvent and the hydrophobic side buried in the interior of the protein. A helical segment must have hydrophobic residues every three or four residues to give at least one side of the helix a hydrophobic surface

Table 6.5   *Conformational Preferences of the Amino Acids*

Amino acid residue	Preference[a]			α-Helix Preference[b]			Turn Preference		
	α-helix (P_α)	β-strand (P_β)	Reverse turn (P_t)	N-term	Middle	C-term	Type I	Type II	Other
Glu	**1.59**	0.52	1.01	2.12	1.18	1.21	1.12	0.84	1.06
Ala	**1.41**	0.72	0.82	1.33	1.60	1.46	0.74	0.94	0.58
Leu	**1.34**	1.22	0.57	1.03	1.50	1.46	0.61	0.53	0.75
Met	**1.30**	1.14	0.52	0.75	1.44	1.92	0.66	0.73	0.96
Gln	**1.27**	0.98	0.84	1.39	1.22	1.24	0.79	1.45	1.02
Lys	**1.23**	0.69	1.07	0.98	1.05	1.68	0.70	0.73	1.04
Arg	**1.21**	0.84	0.90	1.26	1.25	1.23	0.88	1.22	0.84
His	**1.05**	0.80	0.81	0.68	0.97	1.57	0.78	0.64	1.00
Val	0.90	**1.87**	0.41	1.00	1.09	1.08	0.39	0.61	0.48
Ile	1.09	**1.67**	0.47	0.96	1.31	0.99	0.39	0.43	0.93
Tyr	0.74	**1.45**	0.76	0.63	0.61	1.00	0.71	0.91	0.97
Cys	0.66	**1.40**	0.54	0.78	0.66	0.56	1.38	0.99	0.78
Trp	1.02	**1.35**	0.65	1.20	1.34	0.78	1.35	0.15	0.52
Phe	1.16	**1.33**	0.59	0.94	1.45	1.20	0.77	0.76	0.53
Thr	0.76	**1.17**	0.90	0.75	0.87	0.80	1.25	0.67	0.93
Gly	0.43	0.58	**1.77**	0.60	0.47	0.31	1.14	2.61	1.38
Asn	0.76	0.48	**1.34**	0.80	0.80	0.75	1.79	0.99	1.37
Pro	0.34	0.31	**1.32**	0.90	0.19	0.06	0.95	1.80	1.51
Ser	0.57	0.96	**1.22**	0.67	0.44	0.73	1.47	0.76	1.49
Asp	0.99	0.39	**1.24**	1.35	1.03	0.67	1.98	0.71	1.28

[a] The normalized frequences for each conformation (e.g., P_α, P_β, P_t) were calculated from the fraction of residues of each amino acid that occurred in that conformation, divided by this fraction for all residues. Random occurrence of a partricular amino acid in a conformation would give a value of unity.

[b] *N-term* and *C-term* include the four helical residues at the ends of a helical segment eight or more residues long, and three residues at the ends of segments six or seven residues long. *Middle* includes all helical residues between N-term and C-term.

From R. W. Williams et al., *Biochim Biophys. Acta.* 916:200–204 (1987); C. M. Wilmot and J. M. Thornton, *J. Mol. Biol.* 203:221–232 (1988).

with which to interact with the rest of the protein structure.

Many variations of these two approaches to predicting secondary structure have been devised, with varying degrees of improvement. At present, secondary structure can be predicted correctly to an extent greater than expected for a random process, but not markedly so. Predictions that sequences will have helical, extended, or nonregular conformations are usually about 60% correct. Reverse turns have been predicted exactly in about 47% of the instances, but the occurrence of a turn within ±2 residues in the sequence has been pre-

dicted correctly 72% of the time, and the correct type of turn has been predicted in about 60% of the cases. Some improvement can be obtained by aligning a number of homologous sequences and averaging predictions.

All of these prediction procedures assume that local sequences determine secondary structures, that it is the intrinsic tendency of each residue to adopt a certain secondary structure that is important. That this is not strictly correct, at least for segments of five or fewer residues, was shown by the observation that identical pentapeptides adopted the same secondary structure in unrelated proteins of known structure in only 20% of the cases. The tendency for amino acids to occur preferentially in one type of secondary structure (Table 6.5) need not imply simply that they "prefer" that conformation intrinsically; it could also reflect more general rules of protein structure. For example, there is only a moderate correlation between the occurrence of residues in α-helices in proteins (Table 6.5) and their helical propensities measured in small peptides (Table 5.3). Val, Ile, Tyr, Phe, Trp, and Leu, which have bulky nonpolar side chains, tend to occur most frequently in β-sheets; β-sheets are generally internal structures, occurring just where nonpolar side chains are most expected from the general principles of protein structure. Similarly, the amino acids with short hydrophilic side chains, such as Asn, Asp and Ser, occur most often in reverse turns, which are usually at the protein surface, where such side chains are most likely to be. The frequent occurrences of Gly and Pro in reverse turns are clearly due to their unique conformational properties (Sec. 5.2.1).

It is clear that it is not only the local amino acid sequence but also the tertiary structure that determines the secondary structure. Structure predictions are being attempted that incorporate at least some elements of the tertiary structure, as in supersecondary structures (Fig. 6.19).

References

Algorithms for prediction of α-helical and β-structural regions in globular proteins. V. I. Lim. *J. Mol. Biol.* 88:873–894 (1974).

Empirical predictions of protein conformation. P. Y. Chou and G. D. Fasman. *Ann. Rev. Biochem.* 47:251–276 (1978).

Theory of protein secondary structure and algorithm of its prediction. O. B. Ptitsyn and A. V. Finkelstein. *Biopolymers* 22:15–25 (1983).

The hydrophobic moment detects periodicity in protein hydrophobicity. D. Eisenberg et al. *Proc. Natl. Acad. Sci. USA* 81:140–144 (1984).

On the use of sequence homologies to predict protein structure: identical pentapeptides can have completely different conformations. W. Kabsch and C. Sander. *Proc. Natl. Acad. Sci. USA* 81:1075–1078 (1984).

Analysis of sequence-similar pentapeptides in unrelated protein tertiary structures. P. Argos. *J. Mol. Biol.* 197:331–348 (1987).

Prediction of protein secondary structure and active sites using the alignment of homologous sequences. M. J. Zvelebil et al. *J. Mol. Biol.* 195:957–961 (1987).

Analysis and prediction of the different types of β-turn in proteins. C. M. Wilmot and J. M. Thornton. *J. Mol. Biol.* 203:221–232 (1988).

A critical evaluation of methods for prediction of protein secondary structures. G. E. Schulz. *Ann. Rev. Biophys. Biophys. Chem.* 17:1–21 (1988).

6.5.2 Modeling Homologous Protein Structures

The one instance in which a tertiary structure can be predicted with reasonable certainty is when the amino acid sequence is homologous to that of a protein of known tertiary structure. Homologous proteins almost certainly have very similar conformations; thus far, there are no exceptions to this rule. A model of the unknown protein can then be constructed using its amino acid sequence and the backbone conformation of the known structure. The accuracy of such a model depends on the degree of similarity between the two primary structures, especially the number of insertions and deletions of residues, but the ability to construct such models correctly is improving as the general rules of protein structure become understood. Prediction of protein conformation in this way should become increasingly feasible as more three-dimensional structures are determined. Furthermore, enough similar conformations are being discovered to suggest that only a limited number of folded conformations are used by all proteins, perhaps 200–500, over 100 of which are already known (Appendix 2).

For many cases in which sequences have diverged greatly, the difficulty of predicting conformation lies in recognizing the homology. In this case, multiple sequences and secondary-structure predictions can help. For example, 10 homologous sequences of tryptophan synthase α subunit from various microorganisms were aligned, and the secondary structure was predicted for each and averaged. This suggested the presence of eight β-strands, with α-helices interspersed. The structure was correctly predicted to be a $(\beta\alpha)_8$ barrel of the type found so often (Fig. 6.17), even though the sequence of the α subunit is not homologous with any of the other known $(\beta\alpha)_8$ barrel proteins.

A more general approach to recognizing whether any given amino acid sequence is compatible with a known tertiary structure is to set up a so-called **tertiary template,** in which the sequence requirements for that structure are specified. Residues thought to be crucial for that structure are specified in the template as being a particular amino acid or an amino acid of a certain type (e.g., polar or nonpolar, small or large). Other residues are not specified, but the minimum and maximum number of residues in the sequence between the crucial

residues might be. The extent to which any given sequence fits the template pattern is then calculated. Such
procedures have been most useful in detecting supersecondary structures, such as nucleotide-binding folds
(Fig. 6.35).

Another promising method aligns in all possible
ways the sequence of the unknown protein on the three-
dimensional structure of a protein that might be structurally homologous. For each alignment, the probability
that each amino acid residue would occur in such an
environment in a folded conformation is calculated,
based on the preferences of the 20 amino acid residues
observed in all known protein structures. The value
calculated gives the probability that a polypeptide chain
with this sequence would adopt that particular three-
dimensional structure. This method was able to detect
the structural similarity of proteins known to have similar structures, even though they have no detectable sequence similarity.

References

Protein differentiation: emergence of novel proteins during
    evolution. G. E. Schulz. *Angew. Chem. Int. Ed.* 20:143–
    151 (1981).

Prediction of secondary structure by evolutionary compari-
    son: application to the α subunit of tryptophan synthase.
    I. P. Crawford et al. *Proteins: Struct. Funct. Genet.*
    2:118–129 (1987).

Determinants of a protein fold—unique features of the globin
    amino acid sequences. D. Bashford et al. *J. Mol. Biol.*
    196:199–216 (1987).

Tertiary templates for proteins. Use of packing criteria in the
    enumeration of allowed sequences for different structural
    classes. J. W. Ponder and F. M. Richards. *J. Mol. Biol.*
    193:775–791 (1987).

A template based method of pattern matching in protein se-
    quences. W. R. Taylor. *Prog. Biophys. Mol. Biol.*
    54:159–252 (1989).

Flexible protein sequence patterns. A sensitive method to de-
    tect weak structural similarities. G. J. Barton and M. J. E.
    Sternberg. *J. Mol. Biol.* 212:389–402 (1990).

A method to identify protein sequences that fold into a known
    three-dimensional structure. J. U. Bowie et al. *Science*
    253:164–170 (1991).

6.5.3 De Novo Protein Design

Although it is not yet possible to predict the three-
dimensional structure of a protein from its amino acid
sequence alone, the rules of protein structure are now
sufficiently understood to make it feasible to design protein sequences that adopt one of a few folded conforma-

tions. The amino acid sequence must be designed not
only to be capable of adopting a desired conformation
but also to avoid forming incorrect structures.

The procedure is illustrated (Fig. 6.40) by the initial
design of a four-α-helix bundle conformation found in
several natural proteins (Fig. 6.13). In the first step, a
16-residue peptide was designed to adopt a tetrameric
α-helical bundle. Only Leu residues were used to form
the hydrophobic interface between the desired helices
because this is also a helix-favoring residue (Tables 5.3
and 6.5). Gly residues were incorporated at both ends of
the helix to terminate it and to anticipate reverse turns
to be incorporated there. The other residues, on the
solvent-facing side of each helix, were either Lys or Glu,
in positions to interact favorably with each other and
with the helix dipole. After the peptide sequence had
been optimized for forming the tetrameric helix bundle,
a loop was designed to connect two helices in an antiparallel orientation with the sequence -Gly-Pro-Arg-
Arg-Gly-. This two-helix peptide formed stable dimers.
Finally, a similar loop was inserted to link two such
peptides. The final 74-residue peptide adopts a stable
monomeric conformation that is at least approximately
like that anticipated.

A similar approach has produced a synthetic $(\beta\alpha)_8$
barrel, designated *octarellin*. Because of the intense ac-

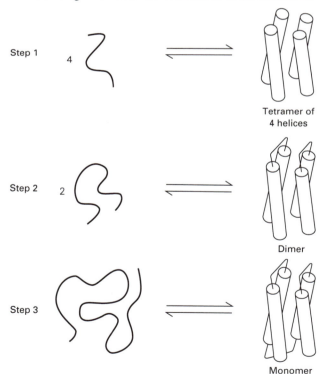

FIGURE 6.40
Schematic illustration of the incremental approach to the
design of a four-helix bundle protein. (Kindly provided by
W. F. DeGrado.)

tivity in this field, this description of the state of progress is certain to be out of date very quickly.

References

Design of peptides and proteins. W. F. DeGrado. *Adv. Protein Chem.* 39:51–124 (1988).

Characterization of a helical protein designed from first principles. L. Regan and W. F. DeGrado. *Science* 241:976–978 (1988).

Crystal structure of α_1: implications for protein design. C. P. Hill et al. *Science* 249:543–546 (1990).

Synthesis, purification and initial structural characterization of octarellin, a *de novo* polypeptide modelled on the α/β barrel proteins. K. Goraj et al. *Protein Eng.* 3:259–266 (1990).

Exercises

1. The regular growth of protein crystals is thought to be perturbed by convection currents in the crystallization medium and by sedimentation of the crystal nuclei. How might these disturbing influences be minimized?

   ANSWER
   V. A. Erdmann et al., *FEBS Letters* 259:194–198 (1989); L. J. DeLucas et al., *Science* 246:651–654 (1989).

2. The termini of a polypeptide chain are most distant in the primary structure and tend to be the most distant atoms in the random coil state, yet they are frequently found to be close together in folded proteins (J. M. Thornton and B. L. Sibanda, *J. Mol. Biol.* 167:443–460, 1983). What are the implications of this observation?

3. Aromatic side chains were concluded to interact in a specific manner in the interiors of proteins by S. K. Burley and G. A. Petsko (*Science* 229:23–28, 1985), but not by J. Singh and J. M. Thornton (*FEBS Letters* 191:1–6, 1985). What factor had been overlooked?

   ANSWER
   T. Blundell et al., *Science* 234:1005 (1986).

4. The β-sheet structures of immunoglobulin domains and of the copper, zinc superoxide dismutase subunit were compared as β-barrels and were proposed to be homologous (J. S. Richardson et al., *J. Mol. Biol.* 102:221–235, 1976). Is this proposal still valid if the β-structures are treated as β-sheet sandwiches rather than barrels?

5. Hydrophobic residues predominate in the interiors of folded proteins, and charged groups are almost entirely on the surface. Yet the surface that is buried upon folding comprises a greater area of polar atoms than is on the surface, whereas the proportion of area from nonpolar atoms is virtually the same in the interior and on the surface (B. K. Lee and F. M. Richards, *J. Mol. Biol.* 55:379–400, 1971; A. Shrake and J. A. Rupley, *J. Mol. Biol.* 79:351–371, 1979). What is the basis for this apparent paradox?

   ANSWER
   S. Miller et al., *J. Mol. Biol.* 196:641–656 (1987).

6. The side-chain nitrogen atoms of Asn, Gln, His, Lys, and Arg residues were concluded to show preferred packing arrangements with aromatic rings in proteins (S. K. Burley and G. A. Petsko, *FEBS Letters* 203:139–143, 1986), and energy calculations were used to propose that the interaction between the NH group and the aromatic ring is a type of hydrogen bond (M. Levitt and M. F. Perutz, *J. Mol. Biol.* 201:751–754, 1988). In this case, what would be the expected geometry of the two groups in folded proteins?

   ANSWER
   J. Singh and J. M. Thornton, *J. Mol. Biol.* 211:595–615 (1990).

7. The TIM-like, $(\beta\alpha)_8$ barrels have been encountered at least 16 times, and all have very similar structures, always with eight β-strands in the barrel and with similar twists and shear numbers (Fig. 6.17; I. Lasters et al., *Proteins: Struct. Funct. Genet.* 7:249–256, 1990). How might you experimentally investigate the basis for the stability of this particular structure?

   ANSWER
   K. Luger et al., *Protein Eng.* 3:249–258 (1990); K. Goraj et al., *Protein Eng.* 3:259–266 (1990).

   The TIM-like, $(\beta\alpha)_8$ barrel is the best candidate for evolutionary convergence to the same structure. How might evidence of this still be available in contemporary structures?

   ANSWER
   A. M. Lesk et al., *Proteins: Struct. Funct. Genet.* 5:139–148 (1989).

   How might such evolutionary convergence have occurred?

   ANSWER
   N. B. Tweedy et al., *Biochemistry* 29:1539–1545 (1990).

   What are the counterarguments against convergence and in favor of divergence?

   ANSWER
   G. K. Farber and G. A. Petsko, *Trends Biochem. Sci.* 15:228–234 (1990).

8. Various proteins with $(\beta\alpha)_8$ structures were concluded to have arisen by a process of evolutionary divergence be-

cause the structures were most similar when the strands were superimposed in register (strand 1 with 1, strand 2 with 2, etc.) rather than with the other superpositions that are possible with 45° rotations about the eightfold axis (L. Lebioda and B. Stec, *Nature* 333:683–686, 1988). What was the greatest weakness in this argument?

ANSWER
L. Lebioda et al., *J. Biol. Chem.* 264:3685–3693 (1989).

What are the implications of the yeast enolase barrel structure for the evolutionary history of TIM-like barrels? What would be the expected tertiary structure of an enzyme with a primary structure that is homologous to yeast enolase?

9. Non-Gly residues in proteins are rarely observed to have the "left-handed helical" conformation, although this conformation is common for Gly residues (Figs. 5.2B and 6.15). What would be the expected consequences of replacing Gly residues with this conformation in folded proteins with other residues?

ANSWER
H. Nicholson et al., *J. Mol. Biol.* 210:181–193 (1989).

10. NMR analysis shows the methyl groups of protein side chains to be rotating rapidly (G. Wagner, *Quart. Rev. Biophys.* 16:1–57, 1983), yet neutron diffraction analysis shows the three hydrogen atoms of methyl groups to be in well-defined positions in protein crystal structures (A. A. Kossiakoff and S. Shteyn, *Nature* 311:582–583, 1984). Are these observations inconsistent? What are the implications for protein structure?

11. One crystal structure of the protease precursor trypsinogen showed the "activation domain" to be disordered and invisible (H. Fehlhammer et al., *J. Mol. Biol.* 111:415–438, 1977), whereas another showed the domain to be fixed in position (A. A. Kossiakoff et al., *Biochemistry* 16:654–664, 1977). What are the possible reasons for this discrepancy?

ANSWER
W. S. Bennett and R. Huber, *Crit. Rev. Biochem.* 15:291–384 (1984).

12. The protease chymotrypsinogen would be expected to be similar in structure to homologous trypsinogen but has its activation domain well defined in crystal structures (S. T. Freer et al., *Biochemistry* 9:1997–2009, 1970), in contrast to the disordered activation domain of trypsinogen (see Exercise 11). Does this indicate that the activation domain of chymotrypsinogen is ordered in solution?

ANSWER
W. Bode and R. Huber, *FEBS Letters* 90:265–269 (1978); D. Wang et al., *J. Mol. Biol.* 185:595–624 (1985).

13. Ferredoxins from *Peptococcus aerogenes* (E. T. Adman et al., *J. Biol. Chem.* 248: 3987–3996, 1973) and *Azotobacter vinelandii* (D. Ghosh et al., *J. Biol. Chem.* 256:4185–4192, 1981) were found to have different crystal structures, even though their amino acid sequences are homologous. What is the most likely explanation for this paradox?

ANSWER
C. D. Stout, *J. Biol. Chem.* 263:9256–9260 (1988); G. H. Stout et al., *Proc. Natl. Acad. Sci. USA* 85:1020–1022 (1988).

14. The tryptophan synthase α subunit was known from its circular dichroism spectra to contain substantial amounts of both α-helical and β-sheet secondary structure. More experimental data and the primary sequences of five homologous sequences were used to predict that it had a structure of a parallel β-sheet flanked on both sides by α-helices (M. R. Hurle et al., *Proteins: Struct. Funct. Genet.* 2:210–224, 1987). Why was the correct $(\beta\alpha)_8$ barrel structure overlooked?

ANSWER
Appendix, M. R. Hurle et al., *Proteins: Struct. Funct. Genet.* 2:210–224 (1987); I. P. Crawford et al., *Proteins: Struct. Funct. Genet.* 2:118–129 (1987).

15. Examine the correlation between the intrinsic propensity of the various amino acid residues to adopt the helical conformation in model peptides (Table 5.3) and their tendency to occur in that conformation in folded proteins (Table 6.5). Can you account for any correlation and for the exceptions?

ANSWER
K. T. O'Neil and W. F. DeGrado, *Science* 250:646–651 (1990).

Proteins in Solution
and in Membranes

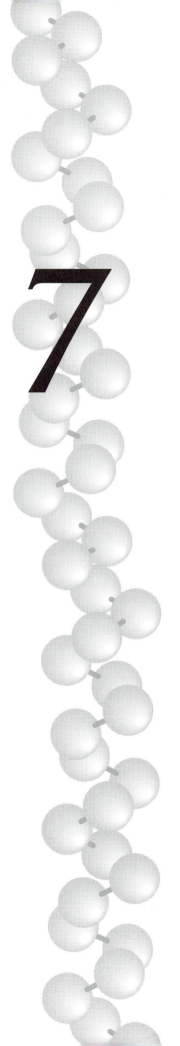

7

Proteins normally exist in solution or embedded in membranes, where they interact in a variety of ways with the environment. These interactions can have numerous effects on the physical and biological properties of proteins.

7.1 Physical and Chemical Properties of Soluble Proteins

The folded conformations of native proteins give them physical, chemical, and biological properties that are quite unlike those of unfolded polypeptides and are not the sum of the properties of their constituent amino acids. Owing to the compactness of their folded conformations, native proteins can diffuse and rotate rapidly. The individual domains of proteins are relatively resistant to proteases, which is frequently used as a criterion for whether a protein is folded. Multi-domain proteins often can be cleaved between domains. Some domains are cleaved by proteases at peptide bonds in mobile surface loops, but the folded structures generally remain intact. If dissociated, the fragments often recombine spontaneously under the appropriate conditions to regenerate the folded structure. The folded conformation places the atoms of a protein in unique environments that often markedly affect their physical and chemical properties. Two or more functional groups are often held in proximity by the folded conformation, making their effective concentrations relative to each other so high that reactions occur between them that would be negligible if the functional groups were on separate molecules.

Many of these properties are not evident when proteins are crystallized, but appear in solution or in membranes where the proteins are more flexible. Nevertheless, knowing the crystal structure of a protein is necessary to understand its properties under other conditions. Much to the collective relief of protein crystallographers, it is now clear that the conformation of a protein is not substantially altered by its inclusion in a crystal lattice, except possibly for the positions of intrinsically flexible side chains and loops on the surface of the molecule. A protein that is crystallized in a number of different crystal lattices, often by very different crystallization procedures, invariably has been found to have essentially the same structure in all of them. The same is true of related proteins; indeed, many instances have been found of surprising similarities in protein crystal structures (Sec. 6.4.2), whereas essentially no instances have been found of differing structures that were reasonably expected to be similar. The interactions between adjacent molecules that determine the form of a crystal lattice appear to be too weak to perturb the overall structure of a stable folded conformation. The intermolecular interactions in a crystal lattice are often similar to the intramolecular forces that specify the folded conformation; crystallization conditions that favor the crystal lattice, therefore, often tend also to stabilize the folded structure.

Very small proteins are the most likely exceptions to the generality that a protein has the same conformation in a crystal lattice and in solution, because small proteins appear to have the most mobile conformations. Some small proteins do not maintain a single conformation in solution and are probably designated more appropriately as peptides. The 29-residue hormone glucagon is a notable example; it approximates a random coil in dilute solution and tends to acquire a trimeric helical structure like that observed in the crystal lattice only in concentrated solution. There appears to be a lower limit to the number of amino acid residues that are necessary for a protein to maintain a single conformation in solution. In the case of very small proteins (which may turn out to be large peptides), it is necessary to determine the biological relevance of their crystal structures by examining their conformational properties in solution.

With the exception of very small proteins and local areas of protein surfaces, the available evidence indicates that there is only one compact folded structure for each protein domain. There is only one case, plasminogen activator inhibitor 1 of the serpin family of protease inhibitors, in which a folded domain has been observed to adopt two substantially different folded conformations without a change in covalent structure. Nevertheless, the literature is full of references to conformational changes in proteins; protein chemists tend to suggest changes in conformation to rationalize observations of any unexpected protein behavior. Many of these so-called conformational changes may involve strictly localized alterations in conformation or changes in the degree of flexibility. In all cases, when substantial conformational changes in folded proteins have been characterized, they have been found to involve primarily quaternary rearrangements of subunits or of structural domains relative to each other. The subunits and domains themselves maintain their overall singular conformations, and it is these individual structural units for which the architecture is apparently unique.

Reference

Structural basis of latency in plasminogen activator inhibitor-1. J. Mottonen et al. *Nature* 355:270–273 (1992).

7.1.1 Aqueous Solubility

The solubilities of proteins in aqueous solutions vary enormously. Some proteins are so soluble in water that they can compose up to 35% of the volume of a saturated solution. Others, especially structural proteins, are essentially insoluble under physiological conditions and exist normally as solids, aggregated into complexes of varying sizes and specificities. Many proteins that are relatively insoluble in water are sequestered into membranes. The solubility of a protein in water is determined by its free energy when surrounded by aqueous solvent relative to its free energy when interacting in an amorphous or ordered solid state with any other molecules that might be present, or when immersed in membranes. This is a very complex situation, for which no quantitative explanations of protein solubility are available.

The interactions of a protein molecule with solvent or with other molecules are determined primarily by its surface. The most favorable interactions with aqueous solvent are provided by charged and polar groups of the hydrophilic side chains (Table 4.8). The surfaces of most water-soluble globular proteins are covered uniformly by charged and polar groups, and their solubilities are governed primarily by the interactions of the polar groups with water. Structural proteins also have polar surfaces, but they interact with other protein molecules more avidly than they do with water. Membrane proteins are more complex; their interactions with membranes are described later.

The solubility of a globular protein in water generally increases at pH values farther away from its isoelectric point, which is the pH at which the protein has zero net charge. Thus, the greater the net charge on the protein molecule, the greater the electrostatic repulsions between molecules, which tends to keep them in solution. Most proteins unfold at some pH value, however, often with drastic consequences for their solubility, because unfolding exposes many nonpolar surface areas to the solvent. Most proteins can be solubilized in aqueous solutions by adding detergents or denaturants such as urea or guanidinium salts, but the proteins are then usually unfolded. The remainder of this discussion of solubility is limited to conditions that do not cause conformational changes.

Most proteins in aqueous solution are surrounded by a tightly bound hydration layer that has properties that differ from those of bulk water. The bound hydration layer is more ordered and less mobile than bulk water and has a 10% greater density and a 15% greater heat capacity. This hydration layer contains 0.3 g of water per gram of protein, which is equivalent to nearly two H_2O molecules per amino acid residue. The interactions of water with protein surfaces have been investigated by gradually adding water to a dry protein, usually hen egg-white lysozyme, which contains 129 residues. The first 60 molecules of water interact primarily with the charged groups on the lysozyme surface, which can then ionize normally. The next 160 water molecules bind primarily to the polar groups on the lysozyme surface. In both types of interaction, approximately one water molecule is involved per charged or polar atom. The next 80 water molecules complete a

monolayer around the protein molecule, which now has virtually all the properties of a lysozyme molecule in dilute aqueous solution. These observations are generally consistent with the relative hydrophilicities of the protein groups (Table 4.8) and with the crystallographic observations of the ordered solvent around crystalline proteins (Sec. 6.2.8). Although the water molecules that are interacting strongly with the charged and polar groups on the protein surface are held tightly, they are kinetically labile and exchange on the 10^{-9} s time scale with the bulk solvent. Even those water molecules that are buried and are integral parts of the protein structure exchange with the solvent, although somewhat more slowly.

The solubilities of globular protein are affected by the addition of cosolvents, especially salts (Fig. 7.1). A protein molecule in a low-ionic-strength aqueous solution is surrounded by an ionic atmosphere described by Debye–Hückel theory (Sec. 4.2.3.b), with an excess of ions of charge opposite to the net charge of the protein molecule. This ion screening decreases the electrostatic free energy of the protein and increases its solubility. Consequently, increasing the ionic strength at low values tends to increase the solubility of a protein; this salting-in effect is independent of the nature of the salt (Fig. 7.1). The solubility of proteins tends to decrease at higher salt concentrations. The magnitude of this salting-out effect depends on the nature of the salt and generally follows the Hofmeister series (Sec. 4.2.3.b).

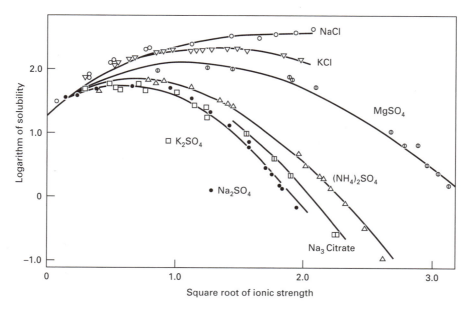

FIGURE 7.1

The solubility of hemoglobin (with carbon monoxide bound) in various electrolytes at different concentrations and 25°C. Solubility is expressed as grams per 1000 grams H_2O. (From A. A. Green, *J. Biol. Chem.* 95:47–66, 1932.)

This phenomenon is thought to arise from a number of effects. There is thought to be a general electrostatic repulsion between the surrounding salt ions and the less polar protein interior. Salts also interact preferentially with the bulk water, which affects its structure in a manner that depends on the nature of the salt. Cations and anions usually act independently, and their effects are additive. As mentioned earlier, one factor in determining the solubility of any substance is the free energy that is required to produce the cavity in the solvent necessary to accommodate the solute molecule, which is related to the surface tension of the solvent. The surface tension of water is increased by high concentrations of most inorganic salts to varying extents that parallel the Hofmeister series. The degree to which a salt increases the surface tension is proportional to its tendency to salt-out proteins. Such effects of salts on the bulk solvent can be overcome, however, by direct interactions between the salt and the protein. For example, divalent cation salts such as $CaCl_2$, $MgCl_2$, and $BaCl_2$ greatly increase the aqueous surface tension of water but do not decrease protein solubility, because they bind directly to proteins.

Many inorganic salts are very useful for reversibly precipitating proteins because they do not unfold or inactivate them. Those salts that are most effective at salting out also tend to increase the stabilities of folded conformations, because precipitation and folding both tend to minimize the protein surface area in contact with the solvent. Ammonium sulfate is the classical reagent for precipitating proteins, owing to its effectiveness and its own great solubility in water.

Organic solvents also tend to decrease the solubilities of proteins, primarily by lowering the dielectric constant of the solvent. Polar interactions between the solvent and the protein surface are consequently less favorable. The stability of the folded state is also lowered, however, so organic solvents tend to denature proteins (Sec. 7.4).

Other polymers also tend to decrease the solubilities of proteins. Any two polymers interact unfavorably in solution because of **volume exclusion,** which is the impossibility of any parts of the two molecules occupying the same space at the same time (Fig. 7.2). Consequently, the second polymer is sterically excluded from the solvent near the protein's surface. Two liquid phases can be produced in mixtures of certain polymers; the polymers and any other molecules present may be concentrated in just one of the phases. The simple steric phenomenon of volume exclusion appears to be the primary basis for the ability of water-soluble polymers, such as poly(ethylene glycol), to precipitate proteins. In addition, unfavorable interactions between such a polymer and charged groups on the protein surface also

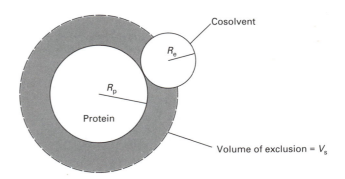

FIGURE 7.2

Schematic illustration of preferential hydration due to exclusion of large cosolvent molecules for steric reasons. The cosolvent is effectively excluded from the shaded volume around the protein, producing effective preferential hydration. Most large cosolvents are not spherical, however, but are disordered polymers; it is more realistic to model them as long rods or as flexible segmented chains. (From T. Arakawa and S. N. Timasheff, *Biochemistry* 24:6756–6762, 1985.)

contribute to forces driving the protein from solution. The excluded volume effect might be important in vivo; for example, muscle cells contain approximately 23% protein by weight, red blood cells 35%. Such high concentrations tend to stabilize the more compact structures of proteins that result from their folding and association with other molecules.

References

How crowded is the cytoplasm? A. B. Fulton. *Cell* 30:345–347 (1982).

Theory of protein solubility. T. Arakawa and S. N. Timasheff. *Methods Enzymol.* 114:49–77 (1985).

Partitioning in aqueous two-phase systems: an overview. H. Walter and G. Johansson. *Anal. Biochem.* 155:215–242 (1986).

Binding of dimethyl sulfoxide to lysozyme in crystals, studied with neutron diffraction. M. S. Lehmann and R. F. D. Stansfield. *Biochemistry* 28:7028–7033 (1989).

Protein hydration in aqueous solution. G. Otting et al. *Science* 254:974–980 (1991).

Protein hydration and function. J. A. Rupley and G. Careri. *Adv. Protein Chem.* 41:37–172 (1991).

7.1.2 Hydrodynamic Properties in Aqueous Solution

Protein molecules in solution at normal temperatures are not static but undergo a variety of movements, ranging from rotations about single bonds, bond stretching, and bond-angle vibrations to translation and rotation of the entire molecule.

References

Hydrodynamics in biophysical chemistry. V. A. Bloomfield. *Ann. Rev. Phys. Chem.* 28:233–259 (1977).

Hydrodynamics and protein hydration. P. G. Squire and M. E. Himmel. *Arch. Biochem. Biophys.* 196:165–177 (1979).

a. Diffusion

Molecules undergo random rotation and translation because of Brownian motion, which subjects them to repeated collisions with the atoms of their environment. **Translational movement** is characterized by the diffusion coefficient D, which is classically defined as, and measured by, the decrease with time t of a concentration gradient (dc/dx) of the protein:

$$\frac{\delta c}{\delta t} = D \frac{\delta^2 c}{\delta x^2} \qquad (7.1)$$

The diffusion coefficient is the average of the square distance traveled per molecule, $\overline{x^2}$, per unit of time:

$$D = \frac{\overline{x^2}}{2t} \qquad (7.2)$$

Owing to the random-walk nature of diffusion, the average distance traveled is proportional to the square root of time. Theoretical values of the average distances traveled as a function of time are given in Figure 7.3 for molecules with diffusion coefficients of 10^{-6}, 10^{-7}, and 10^{-8} cm²/s, the range observed for proteins and related molecules (Table 7.1).

The rate of translational movement depends on the size of the molecule, its shape, and its interactions with the solvent. The observed rate of translational movement is often expressed as the **frictional coefficient f,** which is inversely related to the diffusion coefficient by the Einstein–Sutherland equation:

$$f = \frac{k_B T}{D} \qquad (7.3)$$

where k_B is Boltzmann's constant and T is the absolute temperature. Although many proteins are nearly spherical, their rates of movement are not as great as those expected for a sphere of the same size, f_0 (Eq. 5.28). The **frictional ratio f/f_0** is greater than unity; values of 1.05–1.38 are generally observed for globular proteins (Table 7.1). Two factors other than departure from spherical shape are responsible for the large frictional ratios: the roughness of the surface and the bound sol-

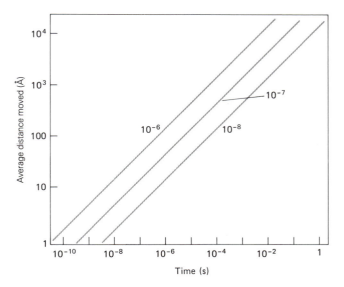

FIGURE 7.3

Average distance moved as a function of time by molecules with typical translational diffusion coefficients of 10^{-6}, 10^{-7}, and 10^{-8} cm²/s. Values were calculated with Equation (7.2).

vent. The surface roughness increases the resistance to movement through the solvent, whereas the tightly bound solvent in the hydration layer effectively increases the mass of the particle. Until recently, attempts to distinguish the contributions of shape and solvent to protein diffusion treated the protein molecule as a smooth ellipsoid and neglected the surface roughness. Consequently, the effects of hydration were overestimated. Only recently has it become possible to calculate the shape factor from the known crystal structures of proteins. Taking shape into account, and including water molecules bound tightly to surface polar groups of the type seen in crystal structures, gives reasonable agreement between calculated and measured frictional coefficients.

References

Measurement and interpretation of diffusion coefficients of proteins. L. J. Gosting. *Adv. Protein Chem.* 11:429–554 (1956).

The application of intensity fluctuation spectroscopy to molecular biology. F. D. Carlson. *Ann. Rev. Biophys. Bioeng.* 4:243–264 (1975).

The translational friction coefficient of proteins. D. C. Teller et al. *Methods Enzymol.* 61:103–124 (1979).

Tracer diffusion of globular proteins in concentrated protein solutions. N. Muramatsu and A. P. Minton. *Proc. Natl. Acad. Sci. USA* 85:2984–2988 (1988).

Table 7.1 *Hydrodynamic Properties of Proteins of Known Structure*

Protein (source)	$s^0_{20,w}$ [a] (S)	$D^0_{20,w}$ [b] (10^{-7} cm²/sec)	$\bar{v}$ [c] (ml/g)	Molecular Weight Structure [d]	Molecular Weight Measured [e]	f/f_0 [f]	Dimensions [g] (Å)
Pancreatic trypsin inhibitor (bovine)	1.0	12.9	0.718	6,520	6,670	1.321	29 × 19 × 19
Cytochrome c (equine)	1.83	13.0	0.715	12,310	11,990	1.116	25 × 25 × 37
Ribonuclease A (bovine)	1.78	10.7	0.703	13,690	13,600	1.290	38 × 28 × 22
Lysozyme (hen)	1.91	11.3	0.703	14,320	13,800	1.240	45 × 30 × 30
Myoglobin (sperm whale)	1.97	11.3	0.745	17,800	16,600	1.170	44 × 44 × 25
Adenylate kinase (porcine)	2.30	10.2	0.74	21,640	21,030	1.167	40 × 40 × 30
Trypsin (bovine)	2.50	9.3	0.727	23,200	23,890	1.187	50 × 40 × 40
Bence Jones REI (human) [h]	2.6	10.0	0.726	23,500	23,020	1.156	40 × 43 × 28
Chymotrypsinogen (bovine)	2.58	9.48	0.721	25,670	23,660	1.262	50 × 40 × 40
Elastase (porcine)	2.6	9.5	0.73	25,900	24,600	1.214	55 × 40 × 38
Subtilisin novo (B. amyloliq.)	2.77	9.04	0.731	27,530	27,630	1.181	48 × 44 × 40
Carbonic anhydrase (human)	3.23	10.7	0.729	28,800	27,020	1.053	47 × 41 × 41
Superoxide dismutase (bovine)	3.35	8.92	0.729	33,900	33,600	1.132	72 × 40 × 38
Carboxypeptidase A (bovine)	3.55	9.2	0.733	34,500	35,040	1.063	50 × 42 × 38
Phosphoglycerate kinase (yeast)	3.09	6.38	0.749	45,800	46,800	1.377	70 × 45 × 35
Concanavalin A	3.8	6.34	0.732	51,260	54,240	1.299	80 × 45 × 30
Hemoglobin, oxy (equine) [i]	4.22	6.02	0.750	64,610	67,980	1.263	70 × 55 × 55
Malate dehydrogenase (porcine) [h]	4.53	5.76	0.742	74,900	73,900	1.344	64 × 64 × 45
Alcohol dehydrogenase (equine) [h]	5.08	6.23	0.750	79,870	79,070	1.208	45 × 55 × 110
Lactate dehydrogenase (dogfish) [j]	7.54	4.99	0.74	146,200	141,000	1.273	74 × 74 × 84

[a] Sedimentation coefficient at 20°C in water, extrapolated to zero protein concentration, in Svedberg units.

[b] Translational diffusion coefficient at 20°C in water, extrapolated to zero protein concentration.

[c] Partial specific volume.

[d] Calculated from covalent structure.

[e] Calculated from hydrodynamic data (see Eq. 7.5).

[f] Frictional ratio from experimental hydrodynamic data.

[g] From crystal structure.

[h] Dimer of identical polypeptide chains.

[i] $\alpha_2\beta_2$ tetramer.

[j] Tetramer of identical polypeptide chains.

From P. G. Squire and M. E. Himmel, *Arch. Biochem. Biophys.* 196:165–177 (1979).

b. Sedimentation Analysis

The hydrodynamic properties of protein molecules are often measured by their **sedimentation coefficient,** the rate at which they sediment in a gravitational field. The rate dr/dt of sedimentation in a centrifugal field, where r is the radius at which the protein is situated and t is time, is given by

$$\frac{dr}{dt} = \frac{M_W(1 - \bar{v}\rho)}{N_A f} \omega^2 r \qquad (7.4)$$

where the pertinent parameters of the protein are M_W, its molecular weight; $\bar{v}$, its partial specific volume; and f, its translational frictional coefficient. Alternatively, the frictional coefficient can be replaced by the diffusion constant D using Equation (7.3). The density of the solvent is ρ, ω is the radial velocity of the rotor in radians per second, and N_A is Avogadro's number.

Equation (7.4) simply states that the rate of sedimentation of any molecule is proportional to its molecular weight M_W, to the centrifugal force $\omega_2 r$, and to the density difference between the molecule and the solvent, $1 - \bar{v}\rho$. On the other hand, the greater the frictional coefficient, the lower the rate of sedimentation. The sedimentation coefficient, s, of a molecule is defined as its rate of sedimentation in a given centrifugal force, $(dr/dt)(1/\omega^2 r)$, to yield the well-known **Svedberg equation:**

$$s = \frac{M_W(1 - \bar{v}\rho)}{N_A f} = \frac{M_W(1 - \bar{v}\rho)}{DRT} \qquad (7.5)$$

The value of s is usually expressed in the Svedberg unit S ($\equiv 10^{-13}$ s).

The s value of a protein depends on three parameters: M_W, f (or D), and $\bar{v}$. The first two are related to the size and shape of the protein molecule, and the third is its partial specific volume. The partial specific volume is given in units of cm^3/g and is essentially the inverse of the protein density. The rate of sedimentation depends on the density difference between the protein and the solvent. If the protein molecule has the same density as the solvent, it does not sediment, irrespective of centrifugal force; if it is lighter (i.e., $\bar{v} > 1/\rho$), it moves opposite to the centrifugal force. The partial volume of a protein is given by the net increase in volume of a solution caused by dissolving the protein in it. The value of $\bar{v}$ is determined by three factors: (1) the intrinsic volume of the protein given by its van der Waals volume; (2) changes in the volume of the surrounding solvent when the protein is added; and (3) interactions of the protein with all other molecules of the solution. Factor (3) depends on the particular situation, but (2) is generally important. For example, water surrounding a nonpolar surface has a lower density than bulk water (Sec. 4.3.1),

whereas charged groups tightly bind water and cause electrostriction and a diminished volume. The value of $\bar{v}$ can be measured from the densities of aqueous solutions of the protein of known concentrations, but this requires large quantities of the protein. More frequently, $\bar{v}$ is estimated from the amino acid composition of the protein by calculating the weighted average of the $\bar{v}$ values of the constituent amino acid residues (Table 4.3); in this respect a protein appears to be the sum of its parts. The volumes of amino acid residues in folded proteins determined crystallographically are virtually the same as the volumes of the constituent residues in solution (Fig. 7.4). The volumes in proteins of polar residues appear to be slightly greater than expected; this may be due to differences in electrostriction or because polar residues are at the surfaces of folded proteins; there are difficulties in defining the boundary between a protein and the solvent in a protein crystal structure. Values of $\bar{v}$ calculated from the amino acid compositions of proteins are useful, but they are approximations. Accurate sedimentation studies require more accurate measured values because uncertainty in the value of $\bar{v}$ is magnified in Equations (7.4) and (7.5).

Determination of the molecular weight of a protein from hydrodynamic analysis requires measurements of

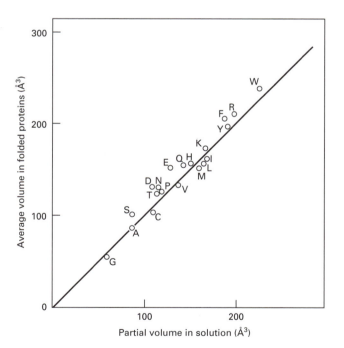

FIGURE 7.4

Correspondence between the average volume occupied by each amino acid residue in solution and in folded proteins. The line has a slope of unity. The values for the partial molar volumes in solution are from Table 4.3, those for folded proteins from Table 6.3.

both s and f (or D), or the use of sedimentation equilibrium (Sec. 1.5.1). The value of s is easier to measure, so it is often tempting to estimate the size of a protein from just its sedimentation coefficient. Empirical equations for this purpose have been proposed. They assume, however, that the protein is typical in its amino acid composition, overall shape, hydration, and surface roughness. There are sufficient examples of exceptions to make such calculations unwise. At best, they can give a minimal estimate of a protein's molecular weight because the most significant variations cause the s value to be decreased.

References

Ultracentrifugation of Macromolecules. J. W. Williams. Academic Press, New York, 1972.

Partial specific volume changes of proteins: ultracentrifugal and viscometric studies. H. Durchschlag and R. Jaenicke. *Intl. J. Biol. Macromol.* 5:143–148 (1983).

Implications for protein folding. Additivity schemes for volumes and compressibilities. M. Iqbal and R. E. Verrall. *J. Biol. Chem.* 263:4159–4165 (1988).

Analytical centrifugation with preparative ultracentrifuges. A. P. Minton. *Anal. Biochem.* 176:209–216 (1989).

c. Gel Filtration

The average overall dimensions of a protein can be determined by molecular sieving, or gel filtration, as described in Section 1.5 and Figures 1.5 and 1.6. There is no satisfactory method of directly calculating the size of a protein from just its elution position with a given column material, so a column is usually calibrated by passing a set of proteins of known dimensions through it. Consequently, the accuracy of the dimensions of a protein determined in this way depends on how closely the protein of interest resembles those used to calibrate the column.

References

Interpretation of the Stokes radius of macromolecules determined by gel filtration chromatography. K. Horiike et al. *J. Biochem.* 93:99–106 (1983).

Size exclusion chromatography and universal calibration of gel columns. M. LeMaire et al. *Anal. Biochem.* 177:50–56 (1989).

d. Rotation

The rate at which a molecule rotates is very sensitive to its shape. The rates of such dynamic processes of molecules are frequently expressed as a **relaxation time,** which is the average time required for a population of molecules to change from their original positions to within $1/e$ ($=0.37$) of their equilibrium positions. The average change for a rotation would be 68°, that angle for which the cosine has the value $1/e$. Many resonance techniques use **correlation times,** which are one-third of the corresponding relaxation times, or **rotational diffusion constants,** which are the reciprocals of twice the relaxation time.

Smooth spherical molecules of a certain size rotate most rapidly, with a rotational relaxation time τ_R predicted by the equation

$$\tau_R = \frac{3V\eta_0}{k_B T} \tag{7.6}$$

where V is the volume of the sphere, η_0 is the viscosity of the solvent, k_B is Boltzmann's constant, and T is the temperature. For spheres with radii of up to 50 Å in water at room temperature, this time is expected to be up to 10^{-7} s. Folded, globular proteins are observed to rotate at rates close to those expected. Rodlike molecules such as α-helices rotate about their helical axis at comparable rates, but they rotate at right angles to the helix axis much more slowly. For example, the rod-shaped tobacco mosaic virus, 180 Å in diameter and 3000 Å long, rotates in this way with a relaxation time of about 10^{-3} s (Table 7.2). Which rotation is measured depends on the technique used.

Rotation of proteins has been measured primarily by **fluorescence depolarization.** A molecule absorbs light with maximum efficiency when the electric vector of the impinging light wave vibrates along certain directions in the molecule, depending on its electronic structure; the fluorescent light emitted is also oriented in certain directions. If the light used for excitation is plane polarized, with the vectors of all waves the same, it excites only those molecules in a solution that happen to have the appropriate orientation. The fluorescent light emitted from such molecules is also polarized, but the average polarization is decreased by the extent to which the molecules have rotated during the time between absorbing and emitting the light; the fluorescence lifetime is generally of the order of 10^{-9}–10^{-8} s.

Depolarization of the emitted light due to protein rotation can be measured in the steady state, when the depolarization is determined by both the rate of rotation and the fluorescence lifetime. To separate the two phenomena, measurements are made as a function of the temperature and viscosity of the solvent, which directly affect only the rate of rotation. Alternatively, the fluorescence lifetime can be measured directly, as can the

Table 7.2 *Examples of Translational and Rotational Diffusion Rates*

Molecule	Translational diffusion coefficient (10^{-7} cm²/s)	Rotational relaxation time
H_2O	200	10^{-2} ns
Glycine	106[a]	
Alanine	91[a]	
Ala-Gly	72[a]	
Tryptophan		8.7 ns[b]
Globular proteins		
Myoglobin		30 ns[b]
Ribonuclease A	12.6[c]	22 ns[d]
Lysozyme	10.6[c]	30 ns[e]
Chymotrypsin		45 ns[e]
Immunoglobulin G	3.8[c]	504 ns[f]
Serum albumin	6.7[a]	125 ns[f]
Unfolded proteins		
Serum albumin	1.9[g]	
Pepsinogen	2.5[g]	
Chymotrypsinogen	3.2[g]	
Tropomyosin	2.2[a]	
Fibrinogen	2.0[h]	3.5 ms[h]
Myosin	0.84[c]	
Collagen		0.5 ms[h]
Poly(benzyl-Glu) ($M_w = 3.4 \times 10^5$)		
α-Helix	0.85[i]	
Random coil	1.30[i]	
Tobacco mosaic virus	0.3–0.4[c]	1.2–1.6 ms[c]

[a] From L. J. Gosting, *Adv. Protein Chem.* 11:429–554 (1956).

[b] From S. R. Anderson et al., *Biochemistry* 9:4723–4729 (1970).

[c] From F. D. Carlson, *Ann. Rev. Biophys. Bioeng.* 4:243–264 (1975).

[d] From S. Krause and C. T. O'Konski, *Biopolymers* 1:503–515 (1963).

[e] From D. R. Bauer et al., *J. Am. Chem. Soc.* 99:2580–2582 (1975).

[f] From J. Yguerabide et al., *J. Mol. Biol.* 51:573–590 (1970).

[g] From M. E. McDonnell and A. M. Jamieson, *Biopolymers* 15:1283–1299 (1976).

[h] From V. A. Bloomfield, *Ann. Rev. Biophys. Bioeng.* 10:421–450 (1981).

[i] From N. C. Ford et al., *J. Chem. Phys.* 50:3098–3100 (1969).

rate of depolarization after a very short pulse of polarized exciting light.

Depolarization measurements give the rate of rotation of the entire protein molecule only when the fluorescent groups are rigid parts of the molecule. In reality, individual groups in a protein molecule also rotate; for example, the side chains of the residues on the surface of a protein rotate about their single bonds independently and at a more rapid rate than the entire molecule moves. There may also be varying degrees of flexibility between different parts of a protein, especially between independent domains.

Other techniques that give information about both translational and rotational motions of macromolecules are flow birefringence, NMR, and quasi-elastic light scattering, also known as intensity fluctuation spectroscopy.

Reference

Rotational Brownian motion and polarization of the fluorescence of solutions. G. Weber. *Adv. Protein Chem.* 8:415–459 (1953).

7.1.3 Spectral Properties

The various environments of the chromophores of a folded protein and the unique stereochemistry of the polypeptide chain affect their spectral properties in various ways. These can be used to characterize and to follow changes in the folded conformation in solution.

References

Optical spectroscopy of proteins. C. R. Cantor and S. N. Timasheff. In *The Proteins*, 3rd ed., H. Neurath and R. L. Hill, eds., vol. 5, pp. 145–306. Academic Press, New York, 1982.

Spectral methods of characterizing protein conformation and conformational changes. F. X. Schmid. In *Protein Structure: A Practical Approach*, T. E. Creighton, ed., pp. 251–285. IRL Press, Oxford, 1989.

a. Absorbance

Absorbance of UV light by proteins is not very sensitive to their conformations or environments, except for that by the aromatic rings of Phe, Tyr, and Trp residues (Fig. 1.4). The spectral properties of the aromatic residues reflect their environments. Their absorbance spectra are shifted somewhat to longer wavelengths (redshifted) in a nonpolar environment such as the interior of a protein. For example, the phenolic side chain of Tyr has its λ_{max} increased by 3 nm when the solvent is changed from water to CCl_4. The absorbance spectra of the aromatic groups consequently can be used to determine their average exposure to water. The exposure of aromatic residues of a protein to solvent can also be measured by **solvent perturbation.** The absorbance of aromatic groups buried in protein interiors is not sensitive to changes in the solvent, whereas that of aromatic groups accessible to solvent is perturbed significantly by addition of reagents such as glycerol, ethylene glycol, or sucrose. Proteins have a variety of environments of aromatic side chains, however, and the value obtained in this way usually reflects the average exposure of all the aromatic rings.

The degrees of solvent exposure measured from the absorbance spectrum of the protein and from its perturbation by solvent are complementary. They can give similar results, as indicated in Table 7.3 for a series of conformational states of bovine pancreatic trypsin inhibitor (BPTI), ranging from the unfolded reduced form, with at least 85% exposure of the Tyr rings to solvent (relative to that in a small model peptide), to the native form with 25–36% exposure. The crystal structure of BPTI indicates 31% exposure of the aromatic residues, primarily of residues Tyr 10 and Tyr 21.

Reference

Calculation of protein extinction coefficients from amino acid sequence data. S. C. Gill and P. H. von Hippel. *Anal. Biochem.* 182:319–326 (1989).

b. Fluorescence

Fluorescence by the aromatic side chains (see Table 1.3) is much more sensitive to their environment than is absorbance, but it varies in an unpredictable manner. The quantum yield may be either increased or decreased by folding, so a folded protein can have either greater or less fluorescence than the unfolded form. The magnitude of the fluorescence is not very informative in itself, but it can serve as a sensitive probe of any perturbations of the folded state. The wavelength of the emitted light is a better indication of the environment of the emitting group. For example, Trp residues that are exposed to water fluoresce maximally at a wavelength of about 350 nm, whereas totally buried residues emit at about 330 nm.

Fluorescence by a protein is especially complex when there is more than one aromatic side chain. The close proximity of aromatic groups in a folded protein usually results in very efficient energy transfer between them (Sec. 5.2.2.b). In this process, light absorbed by one chromophore is transferred to another that absorbs at a longer wavelength, which may then emit the energy as fluorescence. Because the absorbance wavelengths of the aromatic amino acids are in the order Phe < Tyr < Trp (see Fig. 1.4), proteins containing all three types of residues generally emit fluorescent light typical of Trp; Tyr fluorescence is observed only in the absence of Trp; that of Phe, only in the absence of both Tyr and Trp.

Reference

Phase-resolved spectral measurements with several two tryptophan containing proteins. M. R. Eftink et al. *Biochemistry* 26:8338–8346 (1987).

Table 7.3 *Exposure of Tyrosine Residues of Various Conformational States of Bovine Pancreatic Trypsin Inhibitor (BPTI)*

Form of BPTI[a]	Fractional Exposure of Tyr Residues (%)			
	Compared with R		Compared with Gly-Tyr-Gly	
	Comparison spectra[b]	Perturbation spectra[c]	Comparison spectra	Perturbation spectra
R	100	100	84	86
(5–30)	73	80	59	69
(30–51)	64	67	51	57
(30–51, 5–14) + (30–51, 5–38)	60	63	47	53
(30–51, 14–38)	49	49	37	42
(30–51, 5–55)	27	41	16	35
Refolded + (5–55, 14–38)	36	37	25	32
Native	36	35	25	30

[a] The various forms of BPTI were isolated as trapped intermediates in refolding of the reduced protein R (Fig. 7.31). They are designated by the numbers of the Cys residues paired in disulfide bonds; native and refolded BPTI are (30–51, 5–55, 14–38). BPTI has four Phe and four Tyr residues, but the latter dominate the absorbance.

[b] The redshift of the absorbance spectrum was used to quantify the fraction of residues buried.

[c] The perturbant was 20% ethylene glycol.

From P. Kosen et al., *Biochemistry* 19:4936–4944 (1980).

c. Circular Dichroism

The circular dichroism (CD) and optical rotary dispersion (ORD) spectra of a protein are very sensitive to its conformation. In the far-UV region (below 250 nm), these spectral characteristics are determined primarily by the polypeptide backbone conformation, especially its secondary structure. The spectrum of a protein of known structure is usually close to that expected from the average of the spectra of α-helices, β-sheets, and irregular conformations of model polypeptides (Fig. 5.12), weighted by the fraction of the polypeptide chain in each conformation. Consequently, CD spectra can be used to estimate the relative proportions of the various types of secondary structure in a protein. Early methods interpreted the CD spectrum in terms of the model spectra of α-helix, β-sheet, and irregular confirmations (Fig. 5.12); more recent procedures use spectra of a number of proteins of known structure to fit the spectrum being analyzed. As long as the unknown spectrum does not have any unique features, fitting it with actual protein spectra usually gives the most meaningful interpretation. The agreement is not always satisfactory, however, and it is clear that other chromophores, especially aromatic rings, can contribute significantly to the far-UV spectrum of a protein. Caution must therefore be used in interpreting CD and ORD spectra. Infrared and Raman spectroscopy are being developed to measure protein secondary structure in solution (Sec. 6.4.2.b).

Most folded proteins also have significant optical activity in the near-UV spectral region (250–300 nm), owing primarily to the presence of the aromatic side chains in asymmetric environments and to the chirality of disulfide bonds. Such spectra cannot be interpreted in terms of protein structure but are sensitive probes of changes in the tertiary structure.

References

Protein secondary structure analysis using Raman amide I and amide III spectra. R. W. Williams. *Methods Enzymol.* 130:311–338 (1986).

Secondary structure of proteins through circular dichroism spectroscopy. W. C. Johnson, Jr. *Ann. Rev. Biophys. Biophys. Chem.* 17:145–166 (1988).

Protein secondary structure and circular dichroism: a practical guide. W. C. Johnson, Jr. *Proteins: Struct. Funct. Genet.* 7:205–214 (1990).

7.1.4 Ionization

The folded conformations of proteins have a variety of effects on the ionization of their polar groups. Many charged groups are brought into close proximity on the surface of a folded protein, so ionization of groups that would increase the net charge may be hindered. This general electrostatic effect influences the ionization of all the groups. Specific interactions, such as hydrogen bonding or salt bridging (Fig. 4.1), also occur and primarily affect the ionization of particular groups. The pK_a values of groups can be influenced by many environmental and electrostatic effects even in small molecules (Tables 4.5 and 4.6). The variety of environments in folded proteins can produce very unusual ionization properties. The pK_a values of residues of one type can vary widely within a single protein, often over a range of 3–4 pH units, because of their different environments. For example, the observed pK_a values of the His residues in sperm whale myoglobin range from 5.5 to 8.1, compared with the values of 6.0–7.0 observed in isolated His residues (Table 1.2).

Understanding and simulating electrostatic effects in the heterogeneous environments of a folded protein immersed in water or a membrane are much more complex than in a homogeneous liquid, where a simple dielectric constant can describe the effect of the environment (Sec. 4.1.2). Detailed modeling of electrostatic effects in proteins requires consideration of all the atoms and charges of both the protein and the solvent, plus their atomic polarizabilities. The complexity of folded protein structures prevents such analysis, and simpler approximate models are usually used. Electrostatic effects in proteins are also complicated by the presence of counterions in the aqueous solvent and by binding of ions by the protein. The ionization of each group on a protein is affected by its environment: by the protein, by the solvent, and by the ionization of other groups on the protein. Consequently, it is impossible at present to predict accurately the ionization behavior of any one group or the titration curve of the total protein, but progress is being made.

The effects on the ionization of a particular residue by neighboring charged groups can be determined by mutating the neighboring charged groups one at a time. The results of such a series of mutations on the ionization of His 64 at the active site of subtilisin are given in Table 7.4. Removing a positive charge from a neighboring residue increased the pK_a of His 64 by 0.1 pH unit, but removing a single negative charge decreased its pK_a by up to 0.4 pH unit. Replacing a negatively charged neighboring group by a positively charged group decreased the pK_a of His 64 by 0.6 pH unit; making the same substitution simultaneously for two groups decreased the pK_a by 1 pH unit. The effective dielectric constant for each of these electrostatic interactions was calculated by using the known distances between His 64 and the mutated charged groups in the crystal structure of the unmutated protein (Eq. 4.2) and was found to range between 45 and 173. The relatively high values result from these groups being at the surface of the protein, accessible to water. The effective dielectric constants greater than that of water (80) result from the polar groups on the protein being in appropriate positions and orientations to solvate the charges more effectively than does water. Probably for that reason, the occasional isolated charged groups found in protein interiors are usually surrounded by appropriate protein dipoles: $C{=}O$ dipoles for positive charges and $N{-}H$ for negative charges.

References

Experimental evaluation of the effective dielectric constant of proteins. D. C. Rees. *J. Mol. Biol.* 141:323–326 (1980).

Electrostatic effects in proteins. J. B. Matthew. *Ann. Rev. Biophys. Biophys. Chem.* 14:387–417 (1985).

Electrostatic effects on modification of charged groups in the active site cleft of subtilisin by protein engineering. A. J. Russell et al. *J. Mol. Biol.* 193:803–813 (1987).

Stabilization of charges on isolated ionic groups by polarized peptide units. F. A. Quiocho et al. *Nature* 329:561–564 (1987).

Electrostatic energy and macromolecular function. A. Warshel and J. Åqvist. *Ann. Rev. Biophys. Biophys. Chem.* 20:267–298 (1991).

7.1.5 Chemical Properties

The unique environments of reactive groups in folded proteins can substantially affect their chemical properties. Before the availability of protein crystal structures and site-directed mutagenesis, surveying the chemical reactivities of a protein with various reagents was the preferred method for inferring its architecture and functional design. The protein was treated with various reagents, such as those described in Section 1.3, to determine how many groups and which groups of the protein had reacted and to determine the consequent effect on the biological activity. Such studies uncovered a bewildering range of reactivities of various groups in different proteins, from unreactivity to hyperreactivity (in which the reaction was much more rapid than in a typical group). Two factors were thought to be involved: the effect of the environment on the electronic state of a group (i.e., its intrinsic reactivity) and the steric effects on accessibility to the modifying reagent. Explanation of such results in the light of a protein's crystal structure was sometimes successful and sometimes not. Groups on the surface of a protein can be unreactive, and those buried in the interior can be the most reactive; an exam-

Table 7.4 *Effects of Various Mutations of Ionized Residues on the Apparent pK_a Value of His 64 of Subtilisin at Low Ionic Strength[a]*

Mutant	Measured ΔpK_a[b]	Mean distance from charge to His 64 nitrogen atoms (Å)[c]	Effective dielectric constant, D_{eff}[d]
Asp 99 → Ser	−0.40	12.6	48
Glu 156 → Ser	−0.38	14.4	45
Ser 99 → Lys	(−0.25)	15.0	65
Ser 156 → Lys	(−0.25)	16.5	59
Lys 213 → Thr	+0.08	17.6	173
Asp 36 → Gln	−0.18	15.1	90
Asp 99 → Lys	−0.64	(13.8)	55
Gly 156 → Lys	−0.63	(15.5)	50
Asp 99 → Ser and Glu 156 → Ser	−0.63	(13.5)	57
Asp 99 → Lys and Glu 156 → Lys	−1.00	(14.7)	66

[a] Numbers in parentheses denote experimental values calculated from two or more other values rather than determined directly.

[b] Values for the pK_a shifts are the mean values of the two histidine imidazole nitrogens; the normal pK_a value of His 64 is 7.0–7.1, depending on the ionic strength.

[c] Mean distances are the average from the side-chain nitrogen or oxygens to the two histidine imidazole nitrogens; mean distances for mutants were obtained assuming the side chain to be fully extended.

[d] The effective dielectric constant was calculated using the equation

$$D_{eff} = \frac{244}{(\Delta q) r (\Delta pK_a)}$$

where Δq is the change in number of charges and r is the distance in Å.

From M. J. E. Sternberg et al., *Nature* 330:86–88 (1987).

ple is cytochrome *c*, in which the two internal Tyr residues (48 and 67) are much more reactive toward tetranitromethane than are the two at the surface of the molecule (residues 74 and 97). In contrast, iodination modifies primarily residues 67 and 74.

That the relative reactivity of a residue often depends on the nature of the reagent suggests that it is the local concentration of the reagent, determined by its interaction with neighboring parts of the protein, that is often the crucial factor. Steric or electrostatic repulsions may reduce the local concentration of reagent to far below that of the bulk solvent, leading to unreactivity of a nearby group. In contrast, tight binding of a reagent to the protein can produce extremely high local concentrations, perhaps up to the equivalent of about 10^{10} *M* (see Sec. 4.4). This could lead to apparent hyperreactivity of a group that is nearby and in appropriate proximity to react.

One of the classic examples of such apparent hyperreactivity involves the Ser residue that is characteristic of the so-called serine proteases (e.g., trypsin, chymotrypsin, and elastase; see Sec. 9.3.2.a). These related enzymes are inactivated by treatment with acylating reagents, such as diisopropyl fluorophosphate (DFP), which react with a unique Ser residue (number 195 in the usual numbering system based upon chymotrypsinogen):

$$[(CH_3)_2CH{-}O{-}]_2\overset{\overset{\textstyle O}{\|}}{P}{-}F + HOCH_2{-}$$
$$\text{DFP} \qquad\qquad \text{Ser 195}$$

$$\downarrow$$

$$[(CH_3)_2CH{-}O{-}]_2\overset{\overset{\textstyle O}{\|}}{P}{-}OCH_2{-} + F^- + H^+ \qquad (7.7)$$

Table 7.5 *Relative Rates of Alkylation of Histidine and of Two His Residues of Ribonuclease A*

Alkylating reagent	Second-Order Rate Constant[a] $(10^{-4} \text{ s}^{-1} M^{-1})$		
		Ribonuclease A[b]	
	L-Histidine	His 12	His 119
Iodoacetate		7.3	51.1
Iodoacetamide	0.012	1.1	0
Bromoacetate	0.086	20.5	184.5
L-α-Bromopropionate	0.0027	0.19	0.66
D-α-Bromopropionate	0.0028	4.16	1.84
D-α-Bromo-n-butyrate		3.60	1.11
β-Bromopyruvate		0	911
β-Bromopropionate	0.0229	0	6.33

[a] Reactions were carried out at 25°C and pH 5.3–5.5.

[b] His 12 is always alkylated at atom $N^{\epsilon 2}$, His 119 at $N^{\delta 1}$; reaction of one atom inhibits reaction at the other.

From R. L. Heinrickson et al., *J. Biol. Chem.* 140:2921–2934 (1965); R. G. Fruchter and A. M. Crestfield, *J. Biol. Chem.* 242:5807–5812 (1967).

A Ser hydroxyl normally does not react with such acylating reagents, and Ser 195 of the serine proteases does not react when the native conformation is disrupted. Therefore, the occurrence of this reaction in the native protein was attributed to a greatly enhanced nucleophilicity of this particular group by the folded conformation. It is now clear, however, that reagents such as DFP bind in the active sites of serine proteases, much like the substrates of these enzymes, and produce very high local concentrations and apparently very great reactivities (see Chap. 9).

Another example of hyperreactivity of residues is the reaction of His 12 and His 119 of ribonuclease A with iodoacetate, iodoacetamide, and other alkyl halides (Table 7.5). These reagents are usually most reactive with thiols (Eq. 1.47), but they also react with His residues, although much less rapidly. The rates of reaction of both His 12 and His 119 of ribonuclease A are considerably enhanced. The rate also depends on the nature of the reagent, indicating that interactions between the reagent and the surrounding protein are important. The orientation of the His residues in the folded conformation is also important for determining their accessibility to reagent, because His 12 invariably is alkylated on the $N^{\epsilon 2}$ atom and His 119 on the $N^{\delta 1}$ atom.

A method for quantitatively characterizing the chemical properties of specific groups in proteins is **competitive labeling**. A small amount of a reactive, radioactively labeled reagent is incubated with the protein under conditions that permit only a small fraction of the protein molecules to react. This limitation minimizes complications of the modification altering the properties of the protein. The relative extents to which the various groups are modified should reflect their relative reactivities in the unmodified protein. The modification of the protein is subsequently completed with a different isotopic form of the same reagent. The modified residues are identified by chemical procedures, and the incorporation of radioactivity from the first reagent is measured. The reactivity of each group is compared with that of an internal standard of a suitable model compound.

For example, only the nonionized form of amino groups reacts with most reagents (Sec. 1.3.7), so the variation of reactivity with pH can be used to determine the apparent pK_a value of the group in a protein and the relative reactivity of its nonionized form relative to the standard (Fig. 7.5A). The reactivities of different residues of a particular type often vary uniformly with their pK_a values because a group's affinity for protons can also tend to reflect its affinity for other reagents. This is usually illustrated with a **Brønsted plot** of model compounds (Fig. 7.5B). Departures of protein groups from this relationship reflect the effect of protein structure on the reactivity. In the case of the three α-amino groups of α-chymotrypsin (which has three polypeptide chains, Fig. 2.13), the pK_a and reactivity of Cys 1 are normal (Fig. 7.5), consistent with its exposed situation in the

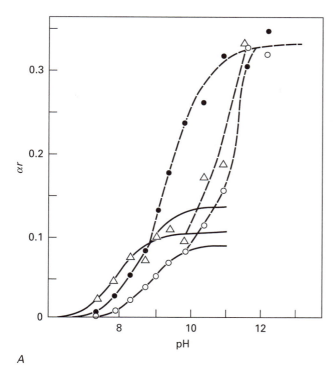

A

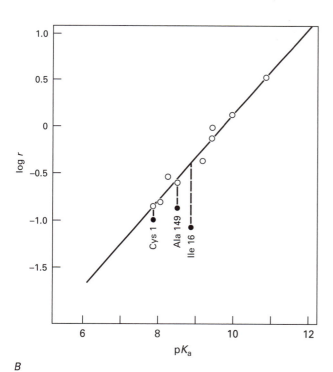

B

protein's crystal structure. The high pK_a value and low reactivity of the α-amino group of Ile 16 are consistent with its buried position in the folded protein, where it interacts in a salt bridge with Asp 194. The increased reactivities of these groups above pH 10 are due to unfolding of the protein at the high pH. The α-amino group of Ala 149 is partly buried in the folded protein and has a somewhat elevated pK_a value and a diminished reactivity. Its greatly increased reactivity at about pH 9 is attributed to the effect of ionization of the nearby Ile 16 α-amino group.

The proximities of other groups in a folded protein can also lead to unexpected intramolecular reactions. In hen lysozyme, iodine unexpectedly attacks preferentially one of the least accessible Trp residues (Trp 108). Other such reagents tend to react with other Trp residues, so presumably the reactive iodine species, the I^+ ion, must bind to the protein near Trp 108. In solution, oxidation of an indole ring is believed to occur from attack by the I^+ species, followed by displacement of HI by a hydroxyl ion or water molecule, to yield a residue of oxindole alanine:

$$(7.8)$$

FIGURE 7.5

Competitive labeling of the three α-amino groups, of residues 1, 16, and 149, of α-chymotrypsin. *A:* Reactivities of the groups with acetic anhydride as a function of pH. The reactivities are relative to the nonionized standard and are expressed as αr, where α is the fraction of nonionized α-amino group and r is the relative reactivity of the nonionized form. The solid lines are the theoretical curves for the following pK_a and r values, respectively: 7.9 and 0.10 for Cys 1, 8.9 and 0.087 for Ile 16, and 8.5 and 0.13 for Ala 149. The dashed curves are attributed to the increased reactivities of the groups that result from pH-dependent changes in the structure of the protein. *B:* Comparison of the pK_a values and reactivities of the α-amino groups of folded α-chymotrypsin ($\bullet$) with those of a series of model compounds ($\circ$) using a Brønsted plot. The extent to which the α-chymotrypsin α-amino groups have lower reactivities than otherwise expected reflects the effects of the protein conformation. (From H. Kaplan, *J. Mol. Biol.* 72:153–162, 1972.)

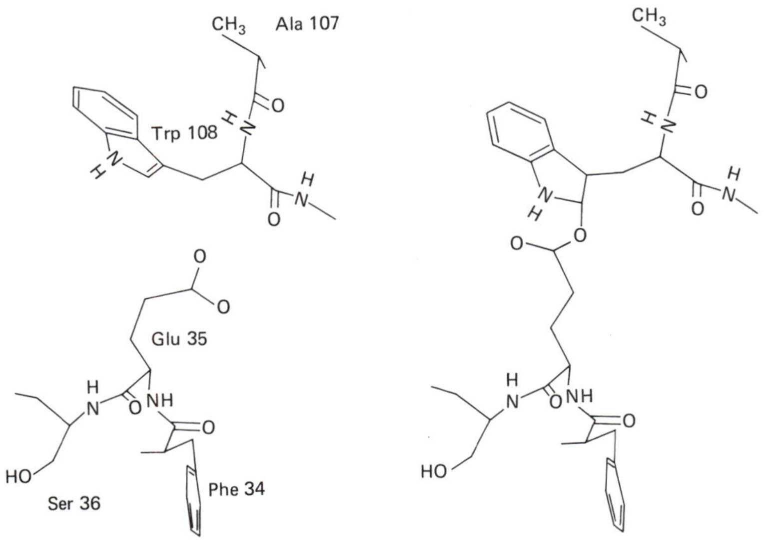

FIGURE 7.6

Covalent cross-link between Glu 35 and Trp 108 of hen lysozyme produced by iodine treatment. The positions of these two residues in native lysozyme are shown at *left*. Iodine presumably reacts initially with Trp 108, but then the adduct reacts preferentially with Glu 35 rather than with water, owing to the proximity of the Glu side chain. The structure of the cross-linked protein is shown at *right*. (Adapted from C. R. Beddell et al., *J. Mol. Biol.* 97:643–654, 1975.)

In the case of lysozyme Trp 108, however, the indole ring is buried in the folded structure, and a hydroxyl or water molecule is not available. The neighboring Glu 35 carboxyl group is in good proximity, however, and reacts with the intermediate to yield an internal ester cross-link (Fig. 7.6).

In general, groups on a protein that are held in close proximity and appropriate orientation by the folded conformation can undergo reactions that are not significant between separate molecules in solution.

References

Chemical modification as a probe of structure and function. L. A. Cohen. In *The Enzymes*, 3rd ed., P. D. Boyer, ed. vol. 1, pp. 147–211. Academic Press, New York, 1970.

The chemical modification of proteins by group-specific and site-specific reagents. A. N. Glazer. In *The Proteins*, 3rd ed., H. Neurath and R. L. Hill, eds., vol. 2, pp. 1–103. Academic Press, New York, 1976.

Chemical characterization of functional groups in proteins by competitive labelling. N. M. Young and H. Kaplan. In *Protein Function: A Practical Approach*, T. E. Creighton, ed., pp. 225–245. IRL Press, Oxford, 1989.

Chemical modification. T. Imoto and H. Yamada. In *Protein Function: A Practical Approach*, T. E. Creighton, ed., pp. 247–277. IRL Press, Oxford, 1989.

7.2   Proteins in Membranes

Membranes provide a physical and insulating barrier between the cell interior and its environment; they also divide eukaryotic cells into compartments. The basic structure of a membrane is the well-known lipid bilayer, in which the nonpolar "tails" of lipids aggregate side-by-side and tail-to-tail to avoid contact with water. The membrane interacts with water primarily through the polar, usually charged, head groups of the lipids. The basis of membrane structure is the amphiphilic structure of the lipid molecules. Natural membranes vary in their lipid compositions, as do the two layers of their bilayer in some cases.

The most important physical property of membranes is that they are essentially impermeable to charged molecules. The energetics for a charged group to be in the nonpolar interior of a membrane are so unfavorable that a charged molecule cannot remain in the bilayer long enough to traverse it. Polar but uncharged molecules can traverse membranes but only at low rates. The energetics of polar molecules in the lipid bilayer are not so unfavorable if they are not charged.

Membranes are essentially two-dimensional fluids. Above a certain temperature, the lipid molecules are free to diffuse in the plane of the bilayer, but they do not generally flip between the two layers of the bilayer.

Proteins typically compose 50% of the mass of most natural membranes, but this can vary between as little as 25% and as great as 75%. The proteins mediate various functions of the membrane such as transport of appropriate molecules into or out of the cell, catalysis of chemical reactions, receiving and transducing chemical signals from the cell environment, and maintaining the membrane structure. Membrane proteins are no less important biologically than those that are water soluble, but they have not been as thoroughly studied for simple technical reasons. Membrane proteins have amphipathic structures that reflect the membrane in which they reside. They have both polar surfaces that interact with the aqueous solution and with the lipid head groups, and nonpolar surfaces that interact with the nonpolar interior of the lipid bilayer. Consequently, they are soluble neither in aqueous solution nor in nonpolar solvents. They can be manipulated and studied only when immersed in a lipid bilayer or a detergent micelle. Detergents, which are amphipathic, mimic lipids and can be used to replace lipids to produce a more well defined bilayer. They can, however, denature proteins, as exemplified by sodium dodecyl sulfate, SDS (Sec. 1.5.3).

References

Physical principles of membrane organization. J. N. Israelachvili et al. *Quart. Rev. Biophys.* 13:121–200 (1980).

Lipid conformation in model membranes and biological membranes. J. Seelig and A. Seelig. *Quart. Rev. Biophys.* 13:19–61 (1980).

The dynamics of membrane structure. P. J. Quinn and D. Chapman. *Crit. Rev. Biochem.* 8:1–117 (1980).

New biophysical techniques and their application to the study of membranes. D. Chapman and J. A. Hayward. *Biochem. J.* 228:281–295 (1985).

Conformation and mode of organization of amphiphilic membrane components: a conformational analysis. R. Brasseur and J. M. Ruysschaert. *Biochem. J.* 238:1–11 (1986).

Polar lipids of thermophilic prokaryotic organisms: chemical and physical structure. *Ann. Rev. Biophys. Biophys. Chem.* 16:25–47 (1987).

The role of unsaturated lipids in membrane structure and stability. P. J. Quinn et al. *Prog. Biophys. Mol. Biol.* 53:71–103 (1989).

7.2.1 Association with Membranes

Different proteins associate with membranes to varying extents, depending on what fraction of the polypeptide chain is immersed in the membrane bilayer. Integral membrane proteins tend to be most immersed in the lipid bilayer because their polypeptide chains generally traverse the membrane completely and are in contact with the aqueous solvent on both sides. The polypeptide chains of some integral membrane proteins are almost entirely within the membrane, with only a few residues exposed to the aqueous solvent. In other membrane proteins, the segments accessible to the solvent may be extensive and may even correspond to one or more domains, like those of water-soluble globular proteins. Multiple transmembrane segments of some integral membrane proteins are connected by loops of varying size and aqueous solvent exposure; in others, a single segment of polypeptide chain spans the membrane.

Nonintegral membrane proteins are essentially water soluble but are anchored to the membrane only by fatty acid chains attached covalently by their polar ends to the protein; the nonpolar ends of the fatty acids are incorporated into the membrane bilayer (Sec. 2.4.4). Other proteins, also essentially water soluble, are associated with membranes by noncovalent interactions with the exposed surfaces of integral membrane proteins or with other components of the membrane surface. Only integral membrane proteins will be addressed further in this section.

The **topography** of an integral membrane protein —that is, which segments of its polypeptide chain are in the membrane and which are exposed to solvent—can be determined experimentally in a number of ways. The solvent-exposed segments on one side of a membrane can be identified by using a sealed, uniform membrane preparation in which the same side of the membrane is always enclosed. Such preparations can be made of intact cells or organelles, of plasma membrane vesicles, or of other vesicles. The exposed side of the membrane is reacted with a chemical reagent that cannot penetrate the sealed membrane. Only amino acid residues of the proteins that are accessible on the exposed side of the membrane react with the reagent, and they can be identified by chemical peptide mapping techniques (Sec. 1.6). It is often useful to compare the results obtained with two related reagents, one that can cross the membrane and one that cannot. Residues labeled by the permeable reagent, but not the other, should be those in the membrane or on the other side. A complementary approach is to use hydrophobic reagents that accumulate in the membrane and react preferentially with the interior segments of the protein. The difficulty here is the paucity of reactive groups on the nonpolar amino acid residues that tend to be in membranes. Other reagents that can detect segments outside the membrane are proteases (Sec. 1.6.2.e) and antibodies (Sec 8.3.1). None of these techniques is perfect, however, and conflicting information has been obtained for virtually every membrane protein studied in this way.

How membrane proteins get into membranes was discussed in Section 2.3.4.

References

Amphitropic proteins: a new class of membrane proteins. P. Burn. *Trends Biochem. Sci.* 13:79–83 (1988).

Membrane-impermeant cross-linking reagents: probes of the structure and dynamics of membrane proteins. J. V. Staros. *Acc. Chem. Res.* 21:435–441 (1988).

Topography of membrane proteins. M. L. Jennings. *Ann. Rev. Biophys.* 58:999–1027 (1989).

On the microassembly of integral membrane proteins. J. L. Popot and D. de Vitry. *Ann. Rev. Biophys. Biophys. Chem.* 19:369–403 (1990).

Membrane protein folding and oligomerization: the two-stage model. J. L. Popot and D. M. Engelman. *Biochemistry* 29:4031–4037 (1990).

7.2.2 Structures of Integral Membrane Proteins

Membrane proteins do not readily form three-dimensional crystals, primarily because of the membrane lipids or detergents that are necessarily bound to their nonpolar surfaces. These technical problems have been solved so recently, primarily by using detergents with short chains, that the three-dimensional structures of only three membrane proteins are known in detail. Two are photosynthetic reaction centers from related bacteria, and their structures are closely similar (Fig. 7.7). The other is the bacterial outer membrane protein, porin, which differs markedly from the reaction center proteins.

a. Photosynthetic Reaction Centers

Both photosynthetic reaction centers are large complex assemblies, undoubtedly due to their complex functions in photosynthesis. That from *Rhodopseudomonas viridis* is a complex of four different polypeptide chains containing 1187 amino acid residues and 14 cofactors. The detergent molecules that had replaced the membrane before crystallization were not apparent in the electron density map, so they are assumed to be disordered in the crystal. Their positions could, however, be determined at low resolution by neutron diffraction. The scattering of neutrons by water was made insignificant by using an appropriate mixture of ^1H_2O and ^2H_2O;

the two isotopes of hydrogen have opposite scattering factors for neutrons (Table 6.1), so the net contribution of water can be made zero.

Three of the polypeptide chains, designated H, M, and L, traverse the membrane; L and M five times each and H just once (Fig. 7.7). L and M are largely confined to the membrane, whereas subunit H has a large globular domain that is on one surface of the membrane, exposed to the aqueous solvent, and typical of water-soluble proteins. The fourth subunit, a cytochrome, is attached to the other side of the membrane only by two fatty acids that are esterified to a glycerol moiety attached to the N-terminal Cys thiol group. The two fatty acids are incorporated into the membrane; otherwise, the cytochrome is a typical globular protein. The cytochrome in the reaction center complex of *Rhodobacter sphaeroides* is not fixed in this way and is not an integral part of the complex.

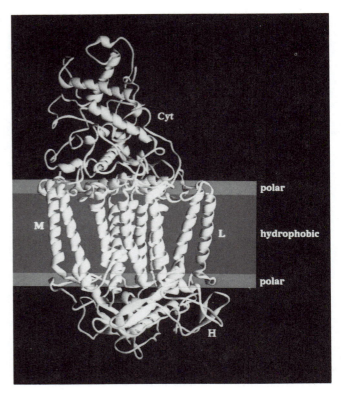

FIGURE 7.7

The photosynthetic reaction center protein from *Rhodopseudomonas viridis* embedded within a membrane. H, L, M, and Cyt refer to the four polypeptide chains. The backbone of each polypeptide chain is shown, with α-helices depicted as coils. The approximate position of the membrane bilayer is shown, with the hydrophobic interior and the polar surface groups indicated. (Kindly provided by J. Deisenhofer.)

The polypeptide segments of the H, M, and L chains that traverse the lipid bilayer are all in the α-helical conformation (Fig. 7.7). The 11 helical segments are 19–30 residues in length, packed together tightly, and tilted to varying extents from the perpendicular to the membrane plane. The average tilt is 22°, the greatest is 35°. The segments connecting the helices are loops of various lengths, comprising mainly hydrophilic residues. Some of the connecting loops adopt the α-helical conformation and lie parallel to the membrane surface. These interact with the rest of the protein, with the solvent, and with the polar groups of the membrane lipids.

The photosynthetic reaction center differs from water-soluble proteins only in its nonpolar interface with the membrane. Otherwise, it has the usual internal packing density and the usual surface area and roughness. The interior of the reaction center structure is not significantly different from that of water-soluble proteins in its amino acid composition or its packing density. It is no more or no less hydrophobic. At one time it was thought that membrane proteins would be "inside-out" versions of water-soluble proteins, with more polar interiors, but this has not been confirmed.

Although water-soluble proteins have mostly polar surfaces, the surfaces of the membrane proteins that are in contact with the lipid bilayer are extremely nonpolar and consist primarily of the side chains of Leu, Ile, Val, and Phe residues. This surface is more hydrophobic than the protein interior. Although the surfaces of membrane and water-soluble proteins differ in their physical natures, both have diverged rapidly during evolution. Only the polar/nonpolar nature of much of the protein surface, not the specific residues, seems generally important for both types of protein.

Charged residues ring the nonpolar surface of the reaction center protein where it penetrates the membrane. These residues interact with both the solvent and the polar surface of the membrane and apparently help to position the helices in the bilayer. They also may help to orient the protein because the residues on the cytoplasmic side of the membrane tend to be positively charged and those on the other side negatively charged. The cytoplasm of the bacterium is negatively charged due to the pumping of ions across the membrane, so the topology of the reaction center protein causes it to interact favorably with this electrostatic potential.

Most integral membrane proteins are thought to traverse the membrane as α-helical segments similar to those of the photosynthetic reaction center. The primary structures of most membrane proteins have one or more hydrophobic segments that could form hydrophobic helices of the required length. Many of these helices are amphipathic (Fig. 5.8), with polar residues along one side.

References

The photosynthetic reaction centre from the purple bacterium *Rhodopseudomonas viridis* J. Deisenhofer and H. Michel. *EMBO J.* 8:2149–2169 (1989).

Hydrophobic organization of membrane proteins. D. C. Rees et al. *Science* 245:510–513 (1989).

The bacterial photosynthetic reaction center as a model for membrane proteins. D. C. Rees et al. *Ann. Rev. Biochem.* 58:607–633 (1989).

Model for the structure of bacteriorhodopsin based on high-resolution electron cryo-microscopy. R. Henderson et al. *J. Mol. Biol.* 213:899–929 (1990).

Three-dimensional structure of plant light-harvesting complex determined by electron crystallography. W. Kühlbrandt and D. N. Wang. *Nature* 350:130–134 (1991).

High-resolution structure of photosynthetic reaction centers. J. Deisenhofer and H. Michel. *Ann. Rev. Biophys. Biophys. Chem.* 20:247–266 (1991).

b. Porin

Not all integral membrane proteins are helical. The best established exception is the bacterial outer membrane protein, porin. Its amino acid content is not especially hydrophobic, and it has a 16-stranded β-barrel structure. The barrel is perpendicular to the plane of the membrane, and its interior is a pore through the membrane that probably functions as a weakly selective channel for polar solutes. Half of the side chains of the β-barrel extend into the interior of the barrel, and the other half are on the outside, interacting with the membrane. The side chains within porin β-barrels are very polar, and those on the outside are very nonpolar.

Porins are unusual membrane proteins in not being very hydrophobic and in being composed of β-structure, which may be because of their occurrence in the outer membrane of bacteria. After their biosynthesis, proteins in the outer membrane must first be exported through the inner membrane. The fact that strongly hydrophobic segments are not exported efficiently through the inner membrane may be the reason that the porins have their particular β-barrel structure. Whatever the reason for its existence, the β-barrel structure of porin demonstrates that alternatives to hydrophobic transmembrane helices are feasible.

Reference

Molecular architecture and electrostatic properties of a bacterial porin. M. S. Weiss et al. *Science* 254:1627–1630 (1991).

7.2.3 Identifying Amino Acid Sequences Likely to Traverse Membranes

The integral membrane proteins of known structure are not markedly different in structure or amino acid composition from water-soluble proteins, except that they are slightly more hydrophobic. They differ mainly in the nature of the amino acid side chains that are on part of their surfaces. Those side chains of membrane proteins that are on the surface in contact with the membrane bilayer are less polar than the protein interior, whereas the surfaces of water-soluble proteins are much more polar than the interior. These observations make it likely that segments of polypeptide chains that traverse membranes could be identified from their amino acid sequences alone. In fact, the transmembrane segments of proteins of unknown tertiary structure have been identified simply by searching their primary structures for hydrophobic stretches of amino acid residues like those in known membrane proteins (Fig. 2.8). This effort has been most successful with very nonpolar segments that lack hydrophilic residues. It is not yet possible, however, to discriminate between these segments and the hydrophobic segments that occur in the interiors of globular proteins. Also, some membrane-spanning segments may escape identification because they appear to contain substantial numbers of polar residues, which are probably functional, for example, in pumping ions through the membrane. It is not known how these polar groups are accommodated within the membrane.

References

Identifying nonpolar transbilayer helices in amino acid sequences of membrane proteins. D. M. Engelman et al. *Ann. Rev. Biophys. Biophys. Chem.* 15:321–353 (1986).

Amphipathic helix motif: classes and properties. J. P. Segrest et al. *Proteins: Struct. Funct. Genet.* 8:103–117 (1990).

The prediction of transmembrane proteins sequences and their conformation: an evaluation. G. D. Fasman and W. A. Gilbert. *Trends Biochem. Sci.* 15:89–92 (1990).

Structure predictions of membrane proteins are not that bad. F. Jähnig. *Trends Biochem. Sci.* 15:93–95 (1990).

Membrane proteins: from sequence to structure. G. von Heijne and C. Manoil. *Protein Eng.* 4:109–112 (1990).

7.2.4 Dynamic Behavior in Membranes

Membrane proteins generally diffuse rapidly in the two-dimensional plane of the membrane, with diffusion coefficients of about 10^{-10} cm²/s, unless they are interacting with other molecules inside or outside the membrane. They usually retain their vertical orientation in the membrane, however, and do not flip between the two surfaces. The membrane lipids move even more rapidly in the membrane plane, with diffusion coefficients of 10^{-8} cm²/s, and only very infrequently do they move from one side of the bilayer to the other. Proteins in a membrane generally induce disorder in the lipid bilayer and restrict the diffusion of neighboring lipid molecules. The restricted lipids exchange positions rapidly with others, however, indicating that the interactions between the lipids and the proteins are weak and nonspecific. Similarly, neither ordered detergent nor lipid molecules are strongly evident in the crystal structures of membrane proteins. The physical state of the membrane also affects the functional properties of its proteins, but in widely varying ways. Interactions between proteins and membranes are complicated by the usual heterogeneity of the lipids in natural membranes.

Proteins in membranes tend to interact with each other much more than do proteins in solution. The large sizes of proteins and their high concentrations in most membranes, typically at least 25% of the membrane volume, produce a large excluded volume effect. There is not much empty space in a membrane for a protein molecule to move into. Also, the orientations of the proteins are fixed relative to the membrane and to each other; fewer degrees of freedom need to be lost for them to interact specifically. Perhaps partly for these reasons, many proteins in membranes are oligomeric.

References

Surface dynamics of the integral membrane protein bacteriorhodopsin. M. A. Keniry et al. *Nature* 307:383–386 (1984).

Interactions between components in biological membranes and their implications for membrane function. G. Benga and R. P. Holmes. *Prog. Biophys. Mol. Biol.* 43:195–257 (1984).

How bilayer lipids affect membrane protein activity. A. Carruthers and D. L. Melchior. *Trends Biochem. Sci.* 11:331–335 (1986).

Interactions between proteins localized in membranes. B. Grasberger et al. *Proc. Natl. Acad. Sci. USA* 83:6258–6262 (1986).

Lipid fluidity and membrane protein dynamics. G. Lenaz. *Biosci. Rep.* 7:823–837 (1987).

Spectroscopic studies of lipids and biological membranes. E. Oldfield. *Biochem. Soc. Trans.* 16:1–10 (1988).

Transcending the impenetrable: how proteins come to terms with membranes. G. von Heijne. *Biochim. Biophys. Acta* 947:307–344 (1988).

7.3 Flexibility of Protein Structure

The structures of proteins in crystals demonstrate varying degrees of conformational flexibility in that the electron density of any particular atom in the calculated electron density map may be spread out to varying extents. In part, this spreading reflects the existence of populations of alternative conformations (Sec. 6.2.7). Even greater flexibility would be expected in solution, without the constraint of the crystal lattice. Indeed, it is a thermodynamic requirement that molecules the size of proteins have substantial transient fluctuations.

The most prevalent and best understood movements of atoms in molecules are the small-scale vibrations of bond lengths and angles that are detectable by infrared and Raman spectroscopy techniques (Sec. 5.4.2.b). These vibrations in proteins are similar to those observed in small molecules, and they occur at frequencies between 6×10^{12}/s and 10^{14}/s. On a longer time scale, larger movements occur, such as those of domains of large proteins that are linked together by relatively flexible "hinge" segments. In antibody molecules (Sec. 8.3.1), for example, the domains rotate relative to each other during intervals of 10^{-8}–10^{-7} s. On the longest time scale, folded conformations are only marginally stable (Sec. 7.4) and therefore spontaneously undergo transient but complete unfolding with a frequency of 10^{-4}–10^{-12}/s, even under conditions that are optimal for stability. Protein flexibility therefore involves movements of widely varying magnitudes on a time scale that spans perhaps 26 orders of magnitude.

Describing protein flexibility is not straightforward, except for that of the side chains on the protein surface, which usually can move to extents similar to those observed in small molecules and unfolded proteins (Table 5.1). The close packing of atoms in the protein interior requires coordination of the movements of neighboring atoms. The complexity of such coordinated movements means that they are described most easily by computer simulation. Molecular dynamics calculations have been used extensively, but only on time scales as long as 10^{-10} s. The classical equations of motion for protein atoms are solved by using expressions for the energy as a function of the conformation. The validity of molecular dynamics simulations depends on the validity of the energy parameters used and the method of including the solvent. The mean vibrations of atoms calculated in this way are comparable to those observed crystallographically, but a variety of effects contribute to vibrations in crystals (Sec. 6.2.7).

The rate at which conformational changes occur is only one aspect of protein flexibility; another is the energetics of the various conformations. In some cases, such as the rotation of a symmetrical side chain, one conformation has the same energy as another and each is equally likely to occur. In other cases, a perturbed conformation has a much higher free energy and is encountered only infrequently and briefly. Only the low-energy conformations are normally present to a substantial extent.

There are severe constraints on the extent to which a folded protein conformation normally varies. These constraints are indicated by the slight variations that are produced by crystallization in different crystal lattices and by varying the temperature and pressure (Sec. 6.2.7). For example, proteins in solution and in a crystal are compressed by high pressure no more than is solid ice. Nevertheless, the structure of the small, rigid protein BPTI is altered by an average of 0.4–0.5 Å in the relative positions of the backbone C^α atoms in three different crystal lattices. Proteins can be thought of as existing normally in a range of distinct but closely related microstate conformations that are usually interconverted rapidly at room temperature. At very low temperatures, however, molecules can become trapped in different microstates.

Integral membrane proteins have varying degrees of internal flexibility, comparable to those of soluble proteins. Amino acid side chains that extend into the membrane, however, are much more restricted in their flexibility than are those of water-soluble proteins that extend into the aqueous solution.

References

Dynamics of proteins: elements and function. M. Karplus and J. A. McCammon. *Ann. Rev. Biochem.* 53:263–300 (1983).

Characterization of the distribution of internal motions in the basic pancreatic trypsin inhibitor using a large number of internal NMR probes. G. Wagner. *Quart. Rev. Biophys.* 16:1–57 (1983).

Protein fluctuations and the thermodynamic uncertainty principle. A. Cooper. *Prog. Biophys. Mol. Biol.* 44:181–214 (1984).

Structural and functional aspects of domain motions in proteins. W. S. Bennett and R. Huber. *Crit. Rev. Biochem.* 15:291–394 (1984).

Compressibility-structure relationship of globular proteins. K. Gekko and Y. Hasegawa. *Biochemistry* 25:6563–6571 (1986).

Dynamics of Proteins and Nucleic Acids. A. McCammon and S. C. Harvey. Cambridge Univ. Press, New York, 1987.

Conformational substates in proteins. H. Frauenfelder et al. *Ann. Rev. Biophys. Biophys. Chem.* 17:451–479 (1988).

Molecular dynamics simulations in biology. M. Karplus and
G. A. Petsko. *Nature* 347:631–639 (1990).

7.3.1 Hydrogen Exchange

The best evidence for extensive motility of protein structure is that internal groups of proteins do react at a finite rate with appropriate reagents in solution, even if only very slowly. Either the usually buried protein group must occasionally be at the surface, accessible to the reagent, or the reagent must permeate the protein interior; either condition would require disruption of the normal protein conformation. The interpretation of such reactions, however, is often complicated by the tendencies of reagents to bind to the protein, reacting rapidly with nearby groups (Sec. 7.1.5) and perhaps perturbing the conformation. The most useful reagent, therefore, is one that is normally present—namely, water. It can be used in its isotopic forms (^1H_2O, ^2H_2O, and ^3H_2O) to measure the tendencies of the various hydrogen atoms of the protein to exchange with the solvent.

To obtain these measurements, a protein with hydrogen atoms of one isotope is transferred to water with hydrogen atoms of a different isotope, and the exchange of isotopes is measured. Hydrogen atoms bonded cova-lently to various atoms exchange with solvent at different intrinsic rates, depending on the tendency of that atom to ionize (Fig. 7.8). Hydrogen atoms on oxygen, nitrogen, or sulfur atoms exchange rapidly, whereas those attached to carbon atoms exchange at very low rates that are significant only in certain instances.

Amide hydrogens of the polypeptide backbone are most often studied in exchange measurements of proteins because these hydrogens exchange on a convenient time scale. Also, they are often buried in proteins, which further slows exchange. Both acid and base catalysis of amide hydrogen exchange occurs in model peptides. Catalysis by acid is thought to occur by transient protonation of the peptide $C\!=\!O$ oxygen, followed by transient loss of the adjacent —NH hydrogen and its replacement from the solvent. Hydroxide ion transiently removes the —NH hydrogen directly; again, exchange occurs when this hydrogen is replaced from the solvent. The minimum rate of exchange of an —NH group occurs at about pH 3, where the acid- and base-catalyzed processes are equal in rate; the rate increases tenfold for each unit change in pH away from the minimum value (Fig. 7.8). The intrinsic rate of exchange is temperature dependent, generally increasing about threefold with a 10°C increase in temperature, which corresponds to an activation energy of 17–20 kcal/mol.

The rate of exchange is also influenced by the environment and by inductive and charge effects on the amide, so it is more complex in large molecules such as proteins than in model compounds. Even in an unfolded protein, such as performic acid–oxidized ribonuclease A, the rate of exchange of individual amide hydrogens varies 100-fold, apparently because of inductive effects of the neighboring amino acid side chains. Also, the rate is not as pH dependent as would be expected from model compounds. Perhaps for the same reason, some solvent-exposed groups on the surface of a protein have been found to exchange only 10^{-3} times as rapidly as expected from the model compounds. Nevertheless, exchange of buried groups is even slower.

Classical methods of measuring hydrogen exchange determine, from the average content of hydrogen isotope, only the average number of hydrogen atoms exchanged as a function of time. A fraction of the hydrogen atoms are usually found to exchange rapidly, approximately at the rate observed for model amides. This is consistent with these groups being on the surface of the protein. Other hydrogen atoms exchange more slowly, with a broad distribution of rates, presumably because the groups are buried and varying degrees of conformational motility are required for exchange to occur. The number of slowly exchanging hydrogen atoms and their rates of exchange vary with the protein and with the conditions. A rapidly exchanging, presum-

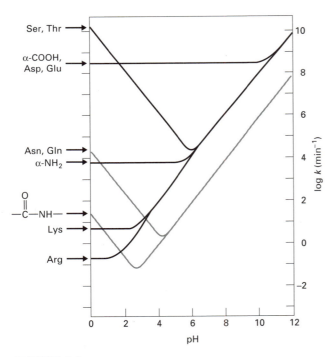

FIGURE 7.8
Dependence on pH of hydrogen exchange rates (k) of model groups.

ably flexible protein (insulin) is compared with one that is slowly exchanging and relatively inflexible (BPTI) in Figure 7.9. In the latter case, the most slowly exchanging hydrogen atoms do so at about 10^{-8} times the rate observed in model amides.

The exchange of individual hydrogen atoms can be followed using ^{1}H-NMR because of its high resolution, the ability to assign resonances to individual protons, and its discrimination between ^{1}H and ^{2}H atoms (Fig. 7.10). ^{1}H atoms give NMR signals whereas ^{2}H atoms do not. With this procedure, it has been found generally that the atoms that exchange least readily from folded proteins are those that are in the interior of the molecule and involved in hydrogen bonding in β-sheets; exchange is slower in the center of the β-strands than at either end. The —NH groups of α-helices also ex-

change slowly, and the interior side of an α-helix exchanges more slowly than the exterior side. The rates of exchange are so slow in the interiors of proteins because the —NH groups are inaccessible to solvent and because they are involved in hydrogen bonding, but it is not possible to separate the contributions of the two effects. The exchange rates of the —NH groups of proteins are usually both acid- and base-catalyzed, but not always to the extent expected theoretically. The rates of exchange generally increase with increasing temperatures, but in a complex manner: the mechanism of exchange seems to change.

Neutron diffraction readily distinguishes between ^{1}H and ^{2}H atoms (Sec. 6.1.8), permitting measurement of the exchange of individual atoms in the crystalline state (see Fig. 6.11). Crystallographic measurements,

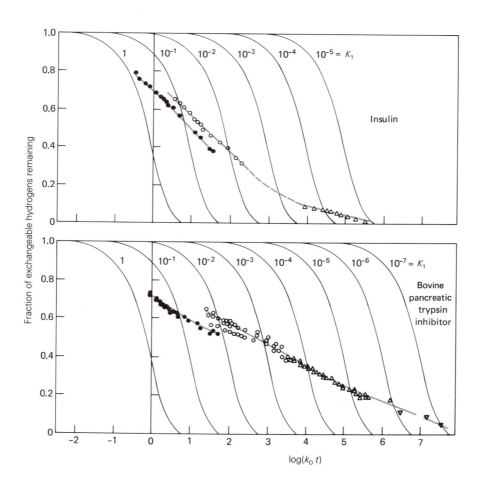

FIGURE 7.9

Exchange-rate curves of insulin and bovine pancreatic trypsin inhibitor (BPTI) at 25°C. k_0 is the exchange rate of an average solvent-exposed peptide group, and t is time. The measurements were made in the time interval from about 5 min to 10 h after dissolution of the protein, and at the following values of pH: insulin, (●) pH 3.0, (○) pH 4.1, (△) pH 7.7; BPTI, (●) pH 3.4, (○) pH 5.1, (△) pH 7.3, (▽) pH 9.2. The solid curves are calculated for the various probabilities of exposure to solvent of the peptide groups (K_1) indicated. (Adapted from A. Hvidt and E. J. Pedersen, *Eur. J. Biochem.* 48:333–338, 1974.)

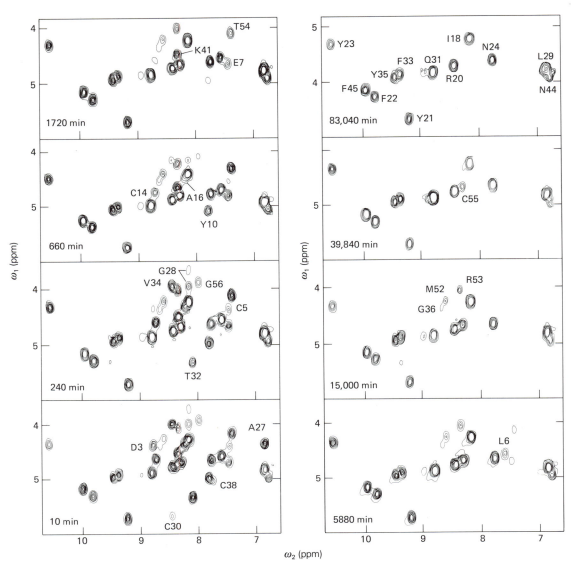

FIGURE 7.10

Hydrogen exchange of individual backbone amide protons in BPTI followed by the disappearance of the cross-peaks between the NH and C$^\alpha$H hydrogens of each residue in two-dimensional COSY NMR spectra. The ^{1}H-labeled protein was dissolved in ^{2}H$_2$O and kept at 36°C for the indicated periods of time before the spectra were measured. When the —N^{1}H— group becomes —N^{2}H— as a result of exchange with the solvent, the cross-peak disappears from the spectrum. The cross-peaks that disappear completely are identified on the last spectrum in which they are apparent, using the one-letter abbreviation of the amino acid followed by the residue number. The assignments of the most slowly exchanging amides are given on the last spectrum. In the COSY spectra, ω_1 corresponds to the chemical shift of the C$^\alpha$H hydrogen, ω_2 to that of the NH. (From G. Wagner and K. Wüthrich, *J. Mol. Biol.* 160:343–361, 1982.)

however, are not very amenable to rate measurements. The few studies have generally confirmed the conclusions of solution studies and have demonstrated that exchange occurs at many sites in the interior of the protein molecule, even in the crystal lattice.

In spite of many studies of hydrogen exchange using many different proteins, there is no consensus on how it occurs and what dynamic fluctuations are responsible for exchange of interior groups. Does the water penetrate the protein to reach the exchanging groups, or are the groups exposed to the solvent by unfolding of the protein? It is known that the rare transient unfolding of the entire protein is not usually responsible because there is a wide variation in exchange rates of different atoms in a protein, and even the slowest exchange rates are too fast to occur by total unfolding. Exchange from the unfolded state becomes significant only if the conditions favor unfolding, as at high temperatures, when the rates of exchange of different sites converge to comparable values.

Local unfolding, or "breathing," is often invoked to explain exchange of interior hydrogen atoms. This classical interpretation of hydrogen exchange, due initially to Linderstrøm–Lang, invoked a variety of transient "open" conformations of the protein, with the exchanging groups exposed locally to the solvent:

$$\text{folded} \underset{k_{-1}}{\overset{k_1}{\rightleftharpoons}} \text{open} \xrightarrow{k_{ex}} \text{hydrogen exchanged} \qquad (7.9)$$

The hypothetical open form would be unstable and transient; consequently k_1 must be much less than k_{-1}, so the equilibrium constant for the opening, K_1 ($\equiv k_1/k_{-1}$), is much less than unity. It is necessary to postulate a variety of open states to account for the different rates of exchange of various sites.

The rates of exchange are generally considered in terms of whether the rate is determined by k_1 (opening of the structure) or by k_{ex} (the exchange reaction). The former is known as the **EX$_1$ mechanism** and the latter as the **EX$_2$ mechanism**. An EX$_1$ mechanism requires that $k_{ex} > k_{-1}$, and the observed rate of exchange directly gives the value of k_1, the rate of opening of the protein structure. An EX$_2$ mechanism applies if $k_{ex} < k_{-1}$; the observed rate of exchange is $K_1 k_{ex}$. The rate of exchange is proportional to only the putative equilibrium constant for the local unfolding process, K_1, not its rate constant. In the EX$_2$ mechanism, the rate of exchange should be sensitive to factors known to affect the intrinsic exchange reaction in model compounds, especially the pH. Proteins demonstrate EX$_2$ types of exchange under most conditions; the data in Figure 7.9 are of this type and are interpreted there as giving the equilibrium constant for opening of the proteins. The EX$_1$ mechanism is observed only under certain conditions, especially at high pH, where hydrogen exchange is intrinsically very rapid and no longer rate-limiting.

An alternative hypothesis to local unfolding is that hydrogen exchange occurs in the protein interior upon rare instances of diffusion of solvent molecules to various sites. Quenchers of fluorescence are thought to be able to diffuse rapidly into protein interiors (see next section); therefore, it is not implausible for H_2O molecules to do so, although ^+H_3O and ^-OH might be expected to be excluded because of their charges. The rate of exchange for each site would depend in a complex manner on the general flexibility of the protein in creating the necessary channels, on the probability of diffusion of solvent molecules to the appropriate location, and on the stability of the hydrogen bond in which the amide hydrogen atom is usually involved.

The available evidence is not sufficient to decide which mechanism of hydrogen exchange occurs in any particular case. The evidence most indicative of the classical breathing process is the observation that the rate of exchange correlates with the solvent accessibility of the residue and of its neighbors but not with the distance of its —NH from the surface. Also, substantial effects on the rate of exchange of all slowly exchanging atoms in BPTI are produced by localized covalent modifications of the protein. The magnitudes of these effects are roughly proportional to the extent to which the stability of the folded conformation is decreased by the modification. Also, binding a ligand at one site on a protein almost invariably diminishes the rate of hydrogen exchange in all the interior groups throughout a folded protein. Observations more consistent with the solvent permeation mechanism are that (1) exchange occurs in the crystalline state of at least some proteins at rates very similar to those in solution, and unfolding would be expected to be greatly diminished by the crystal lattice; (2) the rate of exchange in solution is not increased by low concentrations of denaturing agents, such as urea, that would be expected to favor unfolding; and (3) the rate is decreased by increased pressure, suggesting a need to create channels into the protein. It seems most likely that hydrogen exchange occurs at different sites in folded proteins by a wide range of different processes, depending on the protein and the conditions.

References

Hydrogen exchange in RNase A: neutron diffraction study. A. Wlodawer and L. Sjolin. *Proc. Natl. Acad. Sci. USA* 79:1418–1422, 1982.

Protein dynamics investigated by the neutron diffraction–hydrogen exchange technique. A. A. Kossiakoff. *Nature* 296:713–721, 1982.

Hydrogen exchange and structural dynamics of proteins and nucleic acids. S. W. Englander and N. R. Kallenbach. *Quart. Rev. Biophys.* 16:521–655 (1984).

Mechanism of surface peptide proton exchange in bovine pancreatic trypsin inhibitor. Salt effects and O-protonation. E. Tüchsen and C. Woodward. *J. Mol. Biol.* 185:421–430 (1985).

Hydrogen exchange kinetics of bovine pancreatic trypsin inhibitor β-sheet protons in trypsin-BPTI, trypsinogen-BPTI, and trypsinogen-Ile-Val-BPTI. P. Brandt and C. Woodward. *Biochemistry* 26:3156–3162 (1987).

Correlation between calculated local stability and hydrogen exchange rates in proteins. A. A. Rashin. *J. Mol. Biol.* 198:339–349 (1987).

Electrostatic effects and hydrogen exchange behaviour in proteins. The pH dependence of exchange rates in lysozyme. M. Delepierre et al. *J. Mol. Biol.* 197:11–130 (1987).

Proton exchange in amides: surprises from simple systems. C. L. Perrin. *Acc. Chem. Res.* 22:268–275 (1989).

7.3.2 Fluorescence Quenching and Depolarization

The fluorescence of aromatic groups is instantly quenched by close physical encounter of the excited molecule with some small molecules such as O_2, I^-, and acrylamide. Fluorescence emission usually occurs spontaneously within nanoseconds of excitation, so the encounter with a quencher must take place within this time. As expected, aromatic side chains on the surface of a protein are quenched by diffusion-controlled encounters with such small molecules. Somewhat more surprisingly, many internal residues are quenched only slightly less efficiently by O_2, indicating that this quencher can diffuse through the interiors of proteins within the lifetime of the excited state, at diffusion rates that would have to be 20–50% of those it has in water. Charged and polar quenchers, such as I^- and acrylamide, are less efficient and are likely to act only when at least part of the fluorescent side chain is accessible to the solvent. Detailed interpretations of such studies are not possible, however, because they are complicated by energy transfer between fluorescent groups in the protein (Sec. 5.2.2.b), by the varying quantum yields of different groups, by possible binding of the quenchers to sites on or within the protein, and by perturbations of the protein structure upon excitation of the aromatic groups; for example, the pK_a of the Tyr side-chain hydroxyl decreases markedly upon excitation by UV light.

When polarized light is used to excite fluorescence, the polarization is diminished to varying extents, depending on the extent to which the fluorescent group rotates before it emits the fluorescence (Sec. 7.1.2.d). In some proteins, this indicates substantial flexibility of the aromatic side chains, whereas in others the only rotation is that of the entire protein molecule. Again, detailed interpretation is complicated by changes in the structural and physical properties of the aromatic side chains by excitation.

References

Time-resolved fluorescence of proteins. J. M. Beechem and L. Brand. *Ann. Rev. Biochem.* 54:43–71 (1985).

Fluorescence analysis of protein dynamics. A. P. Demchenko. *Essays Biochem.* 22:120–157 (1986).

Protein fluorescence quenching by small molecules: protein penetration versus solvent exposure. D. B. Calhoun et al. *Proteins: Struct. Funct. Genet.* 1:109–115 (1986).

Pressure dependence of fluorescence quenching reactions in proteins. M. R. Eftink and Z. Wasylewski. *Biophys. Chem.* 32:121–130 (1988).

7.3.3 Rotations of Side Chains

Side chains on protein surfaces, as well as terminal methyl groups of side chains in the interiors, are observed to have mobilities comparable to those in unfolded proteins or in small-molecule analogues, rotating on time scales of 10^{-11}–10^{-8} s. Such rapid motions can be measured by fluorescence depolarization (Sec. 7.1.2.d) or by ^{13}C-NMR (Sec. 5.2.3.b). The slower motions of internal groups, however, are masked in many techniques by rotation of the entire protein molecule. Nevertheless, that such slower motions do occur can be inferred from ^{1}H-NMR spectra of aromatic rings. The two *ortho* ($C^{\delta 1}$ and $C^{\delta 2}$) and *meta* ($C^{\epsilon 1}$ and $C^{\epsilon 2}$) hydrogen atoms of Tyr and Phe residues in the interiors of folded proteins are in unique environments and would be expected to give separate resonances in ^{1}H-NMR spectra. If the rings are flipping by 180° rotations about the C^{β}—C^{γ} bond rapidly (on the NMR time scale), however, the two atoms of each ring spend equal amounts of time in both environments and would then give only a single, averaged resonance. The ^{1}H-NMR time scale depends on the spectral region, but generally lies in the range of 10^{-5}–1 s.

Most proteins give averaged spectra for Phe and Tyr residues, suggesting that the rings are rotating by 180° flips with a frequency of at least 10^4/s, even when the rings are fully buried. Of course, such apparently averaged ^{1}H-NMR spectra could arise by coincidence of the separate resonances, but this is unlikely in general. It has been shown not to be the case in BPTI, where the two Tyr and the three Phe residues that are most buried give immobilized spectra at low temperatures (Table 7.6). As the temperature is increased, averaged spectra are obtained, indicating that rotations are occurring more frequently. Even rapidly rotating rings appear fixed on the ^{13}C-NMR time scale, indicating that the flips occur with frequencies no greater than 5×10^7/s.

The flipping of aromatic rings in the close-packed interior of a protein requires movement of the surrounding atoms and is a measure of protein flexibility. The movements required, however, are perhaps not as great as might be expected with the usual representation of Tyr and Phe side chains as flat planar rings. These

Table 7.6 *Rotation of Aromatic Rings in BPTI*

Residue	Frequency of 180° Rotations (s⁻¹) at Temperature of			Activation Parameters		
	4°C	40°C	80°C	Enthalpy $\Delta H^{\ddagger}$ (kcal/mol)	Entropy $\Delta S^{\ddagger}$ [cal/(mol · °C)]	Volume $\Delta V^{\ddagger}$ (Å³)
Tyr 10	Rotating rapidly at all temperatures					
Tyr 21	Rotating rapidly at all temperatures					
Tyr 23	<5	3×10^2	5×10^4	26	35	
Tyr 35	<1	50	5×10^4	37	68	60
Phe 4	Rotating rapidly at all temperatures					
Phe 22	Rotating rapidly at all temperatures					
Phe 33	Rotating rapidly at all temperatures					
Phe 45	30	1.7×10^3	5×10^4	17	11	50

From G. Wagner et al., *Biophys. Struct. Mech.* 2:139–159 (1976); *J. Mol. Biol.* 196:227–231 (1987).

rings are actually more like flattened spheres or oblate ellipsoids, with a thickness of 3.4 Å and a diameter of 6.8 Å. The perturbations required for flipping of the aromatic rings have been studied extensively with BPTI; computer simulations indicate that adjacent atoms need to move only by small bond rotations, no greater than 17°.

It is likely that flipping is an infrequent process that occurs rapidly once initiated, because the half-rotated intermediate state should be very unstable. The frequencies of flipping of Phe 45 and Tyr 35 of BPTI are decreased by elevated pressures, indicating that the transition states of the process have substantially increased volumes.

In contrast to Tyr and Phe residues, buried Trp and His side chains do not detectably undergo such ring flipping, undoubtedly because of the much greater size of the indole ring and the absence of symmetry in both His and Trp side chains. Full 360° rotations would consequently be required to occur in both cases, with much greater conformational adjustments. Buried Trp residues generally do undergo much more restricted vibrations about their mean positions on the 10^{-8} s time scale.

References

Carbon-13 nuclear magnetic resonance relaxation studies of internal mobility of the polypeptide chain in basic pancreatic trypsin inhibitor and a selectively reduced analogue. R. Richarz et al. *Biochemistry* 19:5189–5196 (1980).

Activation volumes for the rotational motion of interior aromatic rings in globular proteins determined by high resolution ¹H NMR at variable pressure. G. Wagner. *FEBS Letters* 112:280–284 (1980).

Pressure dependence of aromatic ring rotations in proteins: a collisional interpretation. M. Karplus and J. A. McCammon. *FEBS Letters* 131:34–36 (1981).

Side-chain rotational isomerization in proteins: a mechanism involving gating and transient packing defects. J. A. McCammon et al. *J. Amer. Chem. Soc.* 105:2232–2237 (1983).

Rates and energetics of tyrosine ring flips in yeast iso-2-cytochrome *c*. B. T. Nall and E. H. Zuniga. *Biochemistry* 29:7576–7584 (1990).

7.4 Stability of the Folded Conformation

The folded comformations of proteins are only marginally stable under the best of conditions and can often be disrupted by an environmental change, such as a rise in temperature, variation of pH, increase in pressure, or the addition of a variety of denaturants. The protein is then said to be *denatured*. Denaturation need not involve changes in covalent structure and is usually reversible, when it is clearly due to unfolding. In some cases, proteins can be unfolded by breaking any disulfide bonds present, removing any essential cofactors, mutating certain crucial residues, or deleting residues from the primary structure.

7.4.1 Reversible Unfolding Transitions

As the environment of a small, single-domain protein is gradually altered toward conditions that favor unfolding, the folded conformation initially changes very little, if at all (Fig. 7.11). There may be increases in flexibility

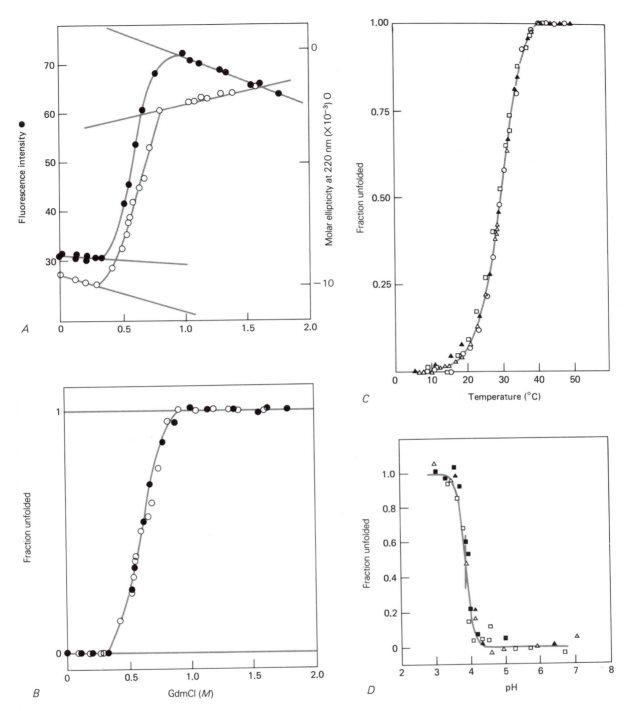

FIGURE 7.11

Equilibrium measurements of unfolding transitions. *A, B:* Guanidinium chloride (GdmCl) unfolding of phosphoglycerate kinase. The experimental observations of unfolding using fluorescence and circular dichroism are illustrated in *A*. The straight lines show the effect of
GdmCl on the spectral properties of the folded state at low GdmCl concentrations and on
the unfolded state at high concentrations. The same effects are assumed to apply throughout
the transition region. The fraction of unfolding indicated by the spectral measurements is
plotted in *B*, illustrating that the two methods give the same smooth unfolding curve, consistent with a two-state unfolding transition. (Adapted from H. Nojima et al., *J. Mol. Biol.*
116:429–442, 1977.) *C:* The temperature-induced unfolding of bovine ribonuclease A in
HCl–KCl at pH 2.1 and 0.019 ionic strength, measured by the increase in viscosity (□) and
by the decreases in optical rotation at 365 nm (○) and in UV absorbance at 287 nm (△). The

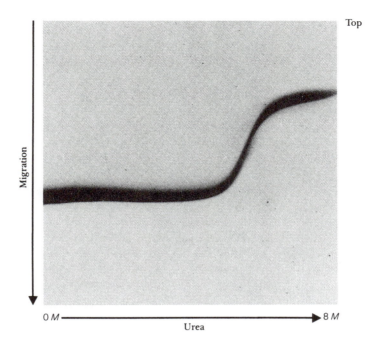

FIGURE 7.12

Transverse urea-gradient electrophoresis of cytochrome *c*. The folded protein was layered on the top of the polyacrylamide gel, which contained a linear gradient of urea from left to right. Electrophoresis at pH 4.0 was from top to bottom. At low urea concentrations, the protein remains folded and migrates rapidly; at high urea concentrations, it is unfolded and migrates more slowly. The same pattern is obtained starting with unfolded protein. This and the continuous band of protein through the abrupt unfolding transition indicate that unfolding and refolding were rapid relative to the time of electrophoresis. Therefore, the fraction of unfolding at equilibrium determined the rate of migration. The smooth shape of the transition, with a single inflexion point, indicates that only two conformational states with different electrophoretic mobilities were present to significant extents. (From T. E. Creighton, *J. Mol. Biol.* 129:235–264, 1979.)

and localized conformational alterations, but the average structure is unchanged. The protein then unfolds completely within a limited range of conditions. For example, staphylococcal nuclease changes from being totally folded to being totally unfolded if the pH is changed by only 0.3 pH unit (Fig. 7.11D). Such an abrupt transition cannot be caused by the ionization of a single group, which should require 2 pH units to go from 9% to 91% ionization. The abruptness of the unfolding transition is indicative of a very cooperative transition; for example, each of a number of groups might be ionized only if all of them are ionized simulta-

neously. The unfolding of many proteins occurs in the acidic pH range because His residues buried in the folded conformation can ionize only when the protein is unfolded (Sec. 7.4.3.a).

Unfolding transitions can be observed by any method that is sensitive to the protein conformation, but the most dramatic change that occurs upon unfolding is the increase in dimensions of the polypeptide chain. Unfolding can be visualized easily by electrophoresis in polyacrylamide gel slabs in which there is a gradient of urea perpendicular to the direction of migration (Fig. 7.12). The unfolded protein migrates much less rapidly

▲ show measurements of a second melting after cooling from 41°C for 16 h. The somewhat altered values at low temperatures indicate a slight irreversibility of the thermal unfolding transition. (Adapted from A. Ginsburg and W. R. Carroll, *Biochemistry* 4:2159–2174, 1965.) *D:* The acid-induced unfolding of staphylococcal nuclease A, measured by viscosity (squares) and by circular dichroism at 220 nm (triangles); the open symbols were measurements made during acidification and the solid symbols upon raising the pH. (Adapted from C. B. Anfinsen, *Biochem. J.* 128:737–749, 1972.)

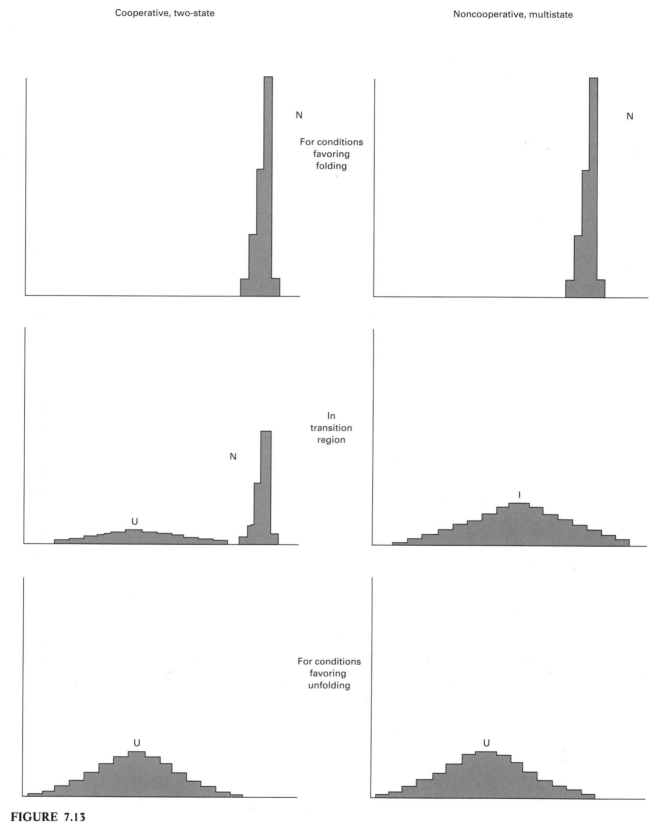

FIGURE 7.13

Illustration of a cooperative, two-state folding transition *(left)*, as usually observed with single-domain proteins, compared with a noncooperative, multistate transition *(right)*. Histograms of the numbers of molecules (vertical axis) with different degrees of folding (horizontal axis)

than the compact folded form, owing to its greater hydrodynamic volume.

The unfolding of most small proteins is reversible, and equilibrium can be attained. It is important to distinguish reversible unfolding transitions from those in which a protein is irreversibly inactivated, which usually involves chemical alterations of the primary structure (Sec. 10.1). Measurements of irreversible denaturation are usually made by subjecting a protein to high temperature or extremes of pH for various lengths of time and then assaying for the decrease in biological activity under physiological conditions. If the unfolding is reversible, no decrease in biological activity is observed because the protein refolds under the assay conditions.

When the degree of reversible unfolding is measured by a number of different methods, all usually give the same curve (Fig. 7.11), which suggests that unfolding is a two-state phenomenon, with only the fully folded (N) and fully unfolded (U) states present. At 50% unfolding, 50% of the molecules are N and 50% are U, rather than all or any significant fraction of the molecules being partially folded (Fig. 7.13). Partially folded structures are unstable relative to U or N. Exceptions to two-state transitions usually occur in large multidomain proteins, in which the domains undergo two-state transitions independently.

For a two-state transition, the equilibrium constant between the N and U states can be measured directly from the average fraction of unfolding (α) in the transition region:

$$K_{eq} = \frac{[N]}{[U]} = \frac{1 - \alpha}{\alpha} \qquad (7.10)$$

Where the value of α is significantly different from 0 or 1, the value of K_{eq} is known. This gives the free energy of N relative to that of U, ΔG_{fold}, under each set of conditions:

$$\Delta G_{fold} = G_N - G_U = -RT \ln K_{eq} \qquad (7.11)$$

The most pertinent value of ΔG_{fold} is that under normal conditions—for example, in the absence of denaturant and at room temperature—but this value can only be estimated by extrapolation. Through the transition region, where it can be measured, the value of ΔG_{fold} is

generally found to vary linearly with the denaturant concentration:

$$\Delta G_{fold} = \Delta G_{fold}^{H_2O} + m \, [\text{denaturant}] \qquad (7.12)$$

The parameter m reflects the dependence of the free energy on the denaturant concentration; typical values of m are 1 and 3 kcal/mol · M for urea and guanidinium chloride, respectively. The values of ΔG_{fold} are usually extrapolated linearly to determine its value in the absence of denaturant, $\Delta G_{fold}^{H_2O}$, although there is no theoretical basis for doing so. There can be substantial uncertainty in such lengthy extrapolations, but they usually give optimal values of $\Delta G_{fold}^{H_2O}$ in the region of -5 to -10 kcal/mol. These values give maximum values for K_{eq} for folding (Eq. 7.10) of $10^4 - 10^7$. Therefore, these proteins have probabilities of $10^{-4} - 10^{-7}$ of being fully unfolded, under even optimal conditions.

References

Conformational stability of globular proteins. C. N. Pace. *Trends Biochem. Sci.* 15:14–17 (1990).

pH dependence of the urea and guanidine hydrochloride denaturation of ribonuclease A and ribonuclease T1. C. N. Pace et al. *Biochemistry* 29:2564–2572 (1990).

7.4.2 Nature of the Unfolded State

Many unfolded proteins, in strong denaturants such as 6 M guanidinium chloride or 8 M urea or at extremes of pH, have been shown to have the average hydrodynamic, physical, and thermodynamic properties expected of random-coil polypeptides. A wide variety of evidence, however, suggests that at least some unfolded proteins are not true random coils under less extreme conditions. This is perhaps not too surprising, because in a truly random coil the energetics of interactions among parts of the polypeptide chain must be exactly balanced by interactions with the solvent. This is virtually impossible for a polypeptide chain composed of 20 different amino acid side chains with a diversity of physical properties. If interactions among different parts of a polypeptide chain are energetically favored over those with the solvent, the polypeptide chain will tend to be

are illustrated for conditions favoring folding and unfolding, and for the transition region. The folded state N is represented as a narrow distribution of folded conformations, the unfolded state U as a broad distribution of less compact conformations. In the transition region, a two-state transition will have two distinct populations in equilibrium, similar to those at the two limiting conditions of folding and unfolding, whereas a noncooperative transition will have the distribution of the entire population shifted so that most molecules have intermediate, partially folded conformations (I). (From T. E. Creighton, *Biochem. J.* 270:1–16. 1990.)

more compact and less disordered than expected for a random coil. The opposite occurs if there are especially favorable interactions between solvent and polypeptide or electrostatic repulsions in the polypeptide chain. Unfolded states produced under various unfolding conditions often have different physical properties. Nevertheless, they are indistinguishable thermodynamically, having the same enthalpy and heat capacity; they are probably different subsets of the truly random spectrum of nonnative conformations. Most importantly, unfolded proteins do not usually contain cooperative folded structures.

Unfolded proteins that retain disulfide bonds or other cross-links are not as expanded as those without them. Any such cross-link decreases the conformational flexibility of the unfolded polypeptide chain and decreases its free energy. This is one way in which disulfide bonds contribute to the net stability of the folded state (see Sec. 5.2.2.c).

A variety of proteins have been observed to exist, under certain conditions, in conformations that are neither fully folded nor fully unfolded, the so-called **molten globule state**. The most common properties of this state are

1. The overall dimensions of the polypeptide chain are much less than those of a random coil and only marginally greater than those of the fully folded state.

2. The average content of secondary structure, as measured by far-UV CD spectroscopy, is similar to that of the folded state.

3. The side chains are in homogeneous surroundings, as judged by near-UV CD and NMR spectroscopy, in contrast to the different and asymmetric environments they have in the interior of the fully folded state.

4. Many amide groups exchange hydrogen atoms with the solvent much more rapidly than they do in the folded state but more slowly than in the fully unfolded state.

5. The enthalpy of the molten globule state is very nearly the same as that of the fully unfolded state and substantially different from that of the folded state.

6. Interconversions of the molten globule state with the fully unfolded state are rapid and noncooperative, whereas those with the full folded state are slow and cooperative.

The precise structural features of this conformational state are not yet known. As explained in Chapter 5, it is difficult to characterize in detail the unfolded state of a protein because so many conformations are possible. The molten globule state appears to be the preferred conformational state of the unfolded protein under refolding conditions, in which it is usually only transient (Sec. 7.5.3.c). Unfolded proteins may show a continuum of conformations, with the fully unfolded state at one extreme and the collapsed molten globule at the other.

References

Protein denaturation. C. Tanford. *Adv. Protein Chem.* 23:122–282 (1968), 24:1–95 (1970).

Thermodynamic study of the apomyoglobin structure. Y. V. Griko et al. *J. Mol. Biol.* 202:127–138 (1988).

The molten globule state as a clue for understanding the folding and cooperativity of globular-protein structure. K. Kuwajima. *Proteins: Struct. Funct. Genet.* 6:87–103 (1989).

Characterization of a partly folded protein by NMR methods: studies on the molten globule state of guinea pig α-lactalbumin. J. Baum et al. *Biochemistry* 28:7–13 (1989).

Residual structure in large fragments of staphylococcal nuclease: effects of amino acid substitutions. D. Shortle and A. K. Meeker. *Biochemistry* 28:936–944 (1989).

Heat capacity and conformation of proteins in the denatured state. P. L. Privalov et al. *J. Mol. Biol.* 205:737–750 (1989).

Heat capacity of proteins. II. Partial molar heat capacity of the unfolded polypeptide chain of proteins: protein unfolding effects. P. L. Privalov and G. I. Makhatadze. *J. Mol. Biol.* 213:385–391 (1990).

The "molten globule" protein conformation probed by disulphide bonds. J. J. Ewbank and T. E. Creighton. *Nature* 350:518–520 (1991).

7.4.3 Physical Basis for Protein Denaturation

Many studies have been made of the effects on protein stability of varying the environment. The results are expected to give insight into the physical basis for the stability of the folded state, but the complexity of the process has prevented simple, unequivocal conclusions. Nevertheless, some generalities are possible, although the unfolding of any particular protein is probably caused by a number of factors.

Reference

The thermodynamic stability of proteins. J. A. Schellman. *Ann. Rev. Biophys. Biophys. Chem.* 16:115–137 (1987).

a. Extremes of pH

Many proteins unfold at pH values less than about 5 or greater than 10. Unfolding at such extremes of pH usually occurs because the folded protein has groups buried in nonionized form that can ionize only after unfolding. Most prevalent are His and Tyr residues, which tend to cause unfolding at acid and alkaline pH values, respectively. The general electrostatic repulsion between the ionized groups on the surface of a protein might also tend to cause unfolding when the protein has a substantial net charge, because such repulsions would be minimized in the unfolded state. There is little evidence for such repulsions being sufficient to produce unfolding by themselves, however, although proteins do tend to be most stable near their isoelectric point, where they have no net charge.

Salt bridges between ionizing groups can contribute to the stability of the folded state of some proteins and can be disrupted by extreme pH values, at which one of the interacting groups is no longer ionized. The stability of such salt bridges can be estimated from the extent to which the pK_a values of the two interacting groups are perturbed. An extreme example is the salt bridge between the side chains of Asp 70 and His 31 in T4 lysozyme, which alters the apparent pK_a values of these two residues from the normal values of 3.5–4.0 and 6.8, respectively, in the unfolded state to 0.5 and 9.1, respectively, in the folded state. These pK_a values imply that the salt bridge is stable by 3–5 kcal/mol. This further implies that the salt bridge contributes to the stability of the folded state to the same degree because ionization and conformational stability are linked functions (see Eqs. 7.16–7.18 and Fig. 7.32). The greater the stabilizing influence of a salt bridge, the less susceptible it is to disruption by extremes of pH.

References

The equilibrium unfolding parameters of horse and sperm whale myoglobin. Effects of guanidine hydrochloride, urea, and acid. D. Puett. *J. Biol. Chem.* 248:4623–4634 (1973).

pH-induced denaturation of proteins: a single salt bridge contributes 3–5 kcal/mol to the free energy of folding of T4 lysozyme. D. E. Anderson et al. *Biochemistry* 29:2403–2408 (1990).

b. Denaturants

A large number of reagents affect protein stability when added to the aqueous solvent. Those that decrease protein stability are known as *denaturants,* the best known of which are urea and guanidinium chloride:

$$
\begin{array}{cc}
\underset{\text{urea}}{H_2N-\overset{\overset{\textstyle O}{\|}}{C}-NH_2} & \underset{\text{guanidinium ion}}{H_2N-\overset{\overset{\textstyle NH_2}{|}}{C}=NH_2{}^+}
\end{array} \qquad (7.13)
$$

Initial explanations of the effects of these denaturants focused on their obvious potential for hydrogen bonding. They were considered to act by breaking protein hydrogen bonds, and they undoubtedly do interact with peptide groups in unfolded proteins by hydrogen bonding. Further reflection and experiment, however, indicated that these denaturants are probably no more potent in hydrogen bonding than is water.

The effects of additives on protein stability can be interpreted largely in terms of the preferential interactions of the additives with aqueous interfaces and with protein surfaces, which are similar to those interactions that determine protein solubility (Sec. 7.1.1). Additives that increase the stability of the folded state of proteins also tend to decrease their solubilities. Such additives are excluded from the protein surface in that their concentration near the protein is lower than that in the bulk solvent. There is apparently negative binding of stabilizing additives, and the protein is said to be "preferentially hydrated" (Fig. 7.14). In contrast, denaturants tend to increase protein solubility and to interact preferentially with the protein surface, thus appearing to be bound. The pertinent connection of this phenomenon with protein stability is that unfolded proteins have much greater surface area exposed to solvent than do folded proteins. Therefore, an additive that interacts more favorably with protein surfaces, especially those nonpolar surfaces that tend to be in the interiors of folded conformations, than with the bulk solvent tends to be a denaturant, whereas any additive that is excluded from the protein surface is a protein stabilizer. Many of these effects occur indirectly through the solvent, and the protein is involved indirectly; in this case, the additive is excluded from any aqueous interface, and it increases the surface tension of water (Sec. 7.1.1).

These general effects can be overwhelmed, however, by specific interactions of an additive with the protein. For example, divalent cations such as Ca^{2+} and Mg^{2+} would be expected to stabilize proteins because of their effects on water. But they are often destabilizing because they interact directly with protein surfaces, more of which are exposed in the unfolded state. Nonpolar additives are repelled by the charged groups on the surface of a folded protein and consequently decrease its solubility, but they tend to interact strongly with the nonpolar surface exposed by unfolding, thereby decreasing the stability of the folded state (Fig. 7.15).

Both urea and guanidinium chloride increase the solubilities of both polar and nonpolar molecules in

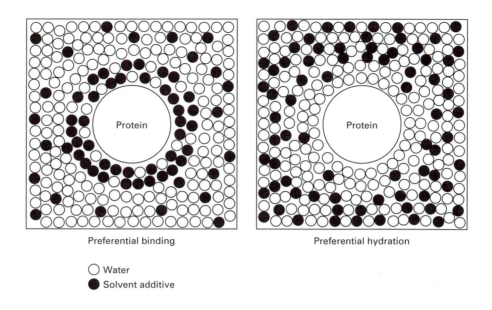

Preferential binding   Preferential hydration

○ Water
● Solvent additive

FIGURE 7.14

Schematic illustration of preferential binding and preferential hydration by solvent additives. In preferential binding, the additive occurs in the solvation shell of the protein at a greater local concentration than in the bulk solvent. Preferential hydration results from exclusion of the additive from the surface of the protein. (From S. N. Timasheff and T. Arakawa, in *Protein Structure: A Practical Approach,* T. E. Creighton, ed., pp. 331–345. IRL Press, Oxford, 1989.)

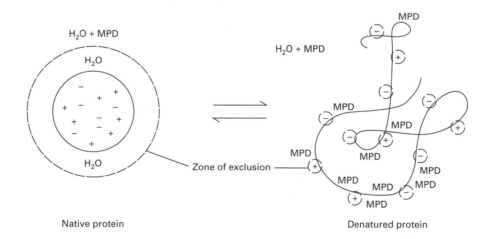

Native protein   Denatured protein

FIGURE 7.15

Schematic illustration of why a nonpolar additive, such as 2-methyl-2,4-pentanediol (MPD), decreases the solubility of a protein but destabilizes its folded conformation. In the folded state, the MPD is repelled by the high charge density on the protein surface, producing preferential hydration. This decreases the solubility of the folded conformation, and MPD is a potent agent for inducing crystallization of proteins. MPD decreases the stability of the folded state because the electrostatic repulsions are minimized in the unfolded state and because the MPD interacts favorably with the nonpolar surfaces that are exposed by unfolding. (From S. N. Timasheff and T. Arakawa, in *Protein Structure: A Practical Approach,* T. E. Creighton, ed., pp. 331–345. IRL Press, Oxford, 1989.)

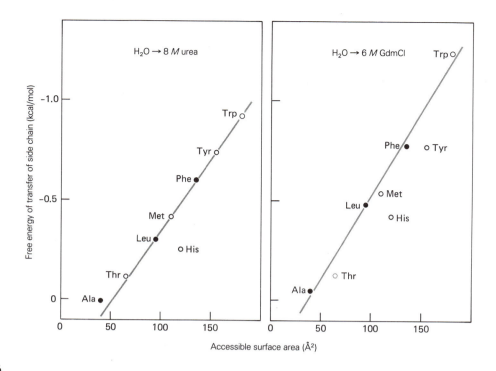

FIGURE 7.16

The denaturants urea and guanidinium chloride (GdmCl) increase the solubilities of both polar and nonpolar amino acid side chains, as measured by the free energy of transfer from water to either denaturant solution (Y. Nozaki and C. Tanford, *J. Biol. Chem.* 238:4074–4081, 1963; 245:1648–1652, 1970). There is a linear correlation of this effect with their accessible surface areas (Table 4.4), although the curves do not extrapolate through the origin. The solid lines have slopes of 7.1 and 8.3 cal/(mol · Å²) for 8 *M* urea and 6 *M* GdmCl, respectively. Residues indicated by open circles have polar groups on side chains. (From T. E. Creighton, *J. Mol. Biol.* 129:235–264, 1979.)

rough proportion to their accessible surface areas. These two denaturants decrease the magnitude of the hydrophobic interaction by up to one-third (Fig. 7.16). Urea and guanidinium ion affect the structure of water because they have hydrogen-bonding capabilities comparable to that of water but different geometries (Sec. 4.2.2). Their effects on water, however, increase the surface tension, which should increase protein stability. It must be concluded, therefore, that urea and guanidinium ion interact with both nonpolar and polar surfaces more favorably than does water. The physical nature of the interaction of these denaturants with nonpolar surfaces is not known; the interaction may simply not be as weak as is their solvation by water. Crystallographic studies of α-chymotrypsin in the presence of guanidinium chloride and urea have demonstrated directly that both denaturants bind to the surface of the folded protein. Urea molecules, but not the charged guanidinium ion, also permeated the interior, occupying small cavi-

ties and somewhat perturbing the close-packed interior. These denaturants are therefore acting as ligands, and the binding of any ligand preferentially to one conformational state increases the relative stability of that conformation (Chap. 8), even if the ligand is a denaturant. An example is the enzyme urease, which hydrolyzes urea and consequently has a specific substrate binding site for it; urease is not denatured by urea. These complications of specific denaturant binding make uncertain any quantitative interpretation of denaturant-induced unfolding curves.

The denaturation potencies of guanidinium salts are affected by the nature of their anions according to the Hofmeister series (Sec. 4.2.3.b). Guanidinium thiocyanate is more potent than the chloride whereas guanidinium sulfate actually stabilizes proteins (Fig. 7.17). Exclusion of the sulfate ion from the protein surface overcomes the destabilizing effect of the guanidinium ion.

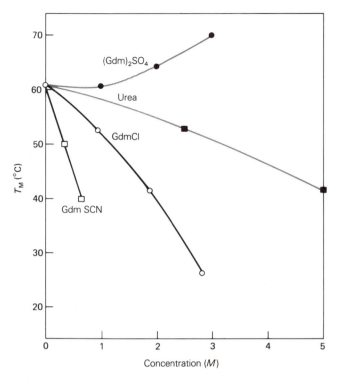

FIGURE 7.17
Thermal stability of ribonuclease A as a function of the concentration of urea and various guanidinium (Gdm⁺) salts. The temperature at the midpoint of the thermal unfolding transition, T_m, is given. (Adapted from P. H. Von Hippel and K. Y. Wong, *J. Biol. Chem.* 240:3909–3923, 1965.)

References

Expression of functionality of α-chymotrypsin. Effects of guanidine hydrochloride and urea in the onset of denaturation. L. S. Hibbard and A. Tulinsky. *Biochemistry* 17:5460–5468 (1978).

Selective binding and solvent denaturation. J. A. Schellman. *Biopolymers* 26:549–559 (1987).

Stabilization of protein structure by solvents. S. N. Timasheff and T. Arakawa. In *Protein Structure: A Practical Approach*, T. E. Creighton, ed., pp. 331–335. IRL Press, Oxford, 1989.

Why preferential hydration does not always stabilize the native structure of globular proteins. T. Arakawa et al. *Biochemistry* 29:1924–1931 (1990).

Surface tension measurements show that chaotropic salting-in denaturants are not just water-structure breakers. R. Breslow and T. Guo. *Proc. Natl. Acad. Sci. USA* 87:167–169 (1990).

c. Thermodynamics of Unfolding

The temperatures at which various proteins unfold vary enormously. Most proteins unfold at elevated tempera-tures, and some unfold at very low temperatures. Many proteins unfold at temperatures only a few degrees higher than those at which they function. Others are stable to much higher temperatures. Proteins from thermophilic organisms tend to be more heat stable than those from other organisms (Sec. 7.4.4.a).

The temperature dependence of protein unfolding is of intrinsic thermodynamic importance. Calorimetric measurements of the effects of heating solutions of lysozyme at three pH values are shown in Figure 7.18. The quantity measured is the energy required to raise the temperature of the entire solution, which is its heat capacity at constant pressure, C_p. The contribution of the protein, its partial C_p, is determined by subtracting the corresponding measurements of the aqueous solvent. The partial C_p is not just that of the protein but includes any effects it has on the surrounding solvent. The partial C_p of the folded protein initially changes only very slightly as the temperature is increased. Unfolding is apparent as a peak in the heat capacity curve, indicating a large absorption of heat. When unfolding is complete, at higher temperatures, the partial C_p of the protein becomes more constant and is greater than its original value.

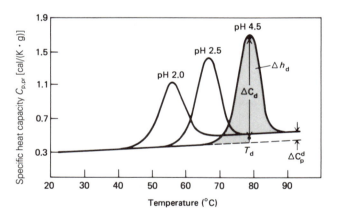

FIGURE 7.18
Calorimetric measurement of thermal unfolding of lysozyme at three pH values. The calorimeter measures directly the amount of heat required to raise the temperature of the protein solution by a specified amount (i.e., the heat capacity). The specific heat capacity is that of the solvent and the protein relative to the solvent alone. At low temperatures, the measurement gives the specific heat capacity of the folded protein; at high temperatures, it is the specific heat capacity of the unfolded protein; both are found to be independent of pH. The difference between the two is ΔC_p of unfolding. The heat absorbed during unfolding, when integrated over the entire transition (the shaded area at pH 4.5), gives the specific enthalpy change upon unfolding, Δh_d. (Adapted from P. L. Privalov and N. N. Khechinashvili, *J. Mol. Biol.* 86:665–684, 1974.)

The area under the transition peak gives the enthalpy change upon unfolding. A most important result of these measurements for a number of small proteins is that the enthalpy change measured calorimetrically in this way, ΔH_{cal}, is very nearly the same as that measured from the temperature dependence of the equilibrium constant, the van't Hoff enthalpy change, ΔH_{VH} (Sec. 5.2.3.a):

$$\Delta H_{VH} = \Delta H_{cal} \qquad (7.14)$$

This must be the case for a simple one-step reaction. Its occurrence with protein unfolding confirms that these unfolding transitions are cooperative and two-state, with only the fully folded and the fully unfolded states present at equilibrium (Fig. 7.13). The ratio of the two enthalpy changes measured for several proteins averages 1.05 ± 0.03, indicating that generally at most 5% of the molecules at the transition midpoint can be in intermediate conformations, with enthalpies different from both the fully folded and the fully unfolded molecules. In contrast, multidomain proteins give ratios much greater than unity. For example, papain and a dimer of immunoglobulin light chains (a Bence Jones protein), both with two domains per polypeptide chain, gave values of 1.80 and 1.90, respectively. These observations indicate that the domains tend to unfold independently. Thermodynamic measurements have concentrated on the two-state transitions of single-domain proteins because of their relative simplicity.

The partial heat capacity C_p is a measure of the temperature dependence of both the enthalpy and the entropy of the protein in solution (see Eq. 4.19). The partial heat capacities of the pre- and posttransition regions are those of the folded and unfolded states in water, respectively, and give the temperature dependence of their enthalpies and entropies. The measured partial heat capacity of the folded state is close to that expected from the measured values (Fig. 4.13) of the polypeptide backbone and of the amino acid side chains that are exposed to solvent in the crystal structure of the folded protein. The partial C_p of the unfolded state is approximately that expected if the unfolded protein is fully unfolded and fully exposed to solvent. The partial heat capacities of proteins are dominated by the nonpolar surface area exposed to water (Fig. 4.13). Their large values are thought to result from the tendency of water to order itself around such surfaces. The unfolded state has a greater heat capacity than the folded state; the difference between the two is roughly proportional to the nonpolar surface area buried in the interior of the folded protein and assumed to be exposed by unfolding (Fig. 7.19).

The complete thermodynamic characterization of the protein lysozyme (Fig. 7.20) illustrates the complex-

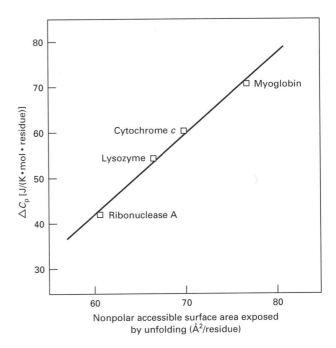

FIGURE 7.19
Relationship between the measured change in heat capacity upon unfolding of several proteins and the nonpolar surface area that is buried in the interior of the protein and is assumed to be exposed to solvent upon unfolding. Note that the relationship is not one of direct proportionality, in that it does not extrapolate to the origin. (Adapted from P. L. Privalov and G. I. Makhatadze, *J. Mol. Biol.* 213:385–391, 1990.)

ity of the physical basis of protein structure. The observed net stability of the folded state is the result of a very small difference between very large but compensating factors. For example, the enthalpies and entropies of both the folded and the unfolded states are strongly temperature dependent, as shown by the substantial heat capacities of the protein states. The enthalpic and entropic contributions to the free energy vary by up to 700 kcal/mol over the range 0°–100°C (Fig. 7.20A,B). But the enthalpies and entropies vary similarly and compensate each other, so the free energy is a relatively small difference between the two and varies only one-tenth as much (Fig. 7.20C). The free energies of the folded and unfolded states also vary similarly; thus the difference between them, the observed net stability, is no greater than 16 kcal/mol (Fig. 7.20D). Consequently, accounting for the net stabilities of proteins in terms of the primary interactions that stabilize them is a hazardous accounting procedure, in which the net result is minuscule relative to the individual terms. An error of more than 2% in a term of 700 kcal/mol would obliterate the net difference of only 16 kcal/mol. In spite of this, the small single-domain proteins studied have

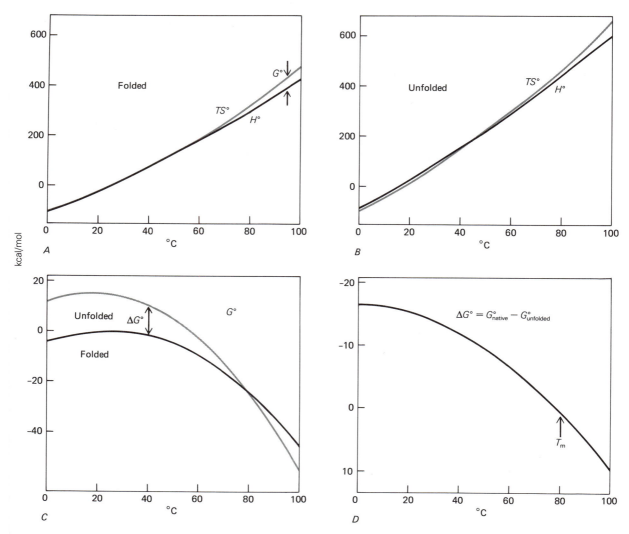

FIGURE 7.20
Thermodynamic parameters for the folded and unfolded forms of lysozyme at pH 7.0 and various temperatures. The enthalpic *H* and entropic *TS* contributions to the free energies of the folded and unfolded states are plotted as a function of temperature in *A* and *B*, respectively. The Gibbs free energy *G* of each state is the difference between the enthalpic and entropic contributions, as indicated in *A*. These values are plotted in *C* as a function of temperature for both the folded and unfolded states. The net stability of the folded state, ΔG, is the difference between the free energies of the two states, as indicated in *C*; this value is plotted in *D* as a function of temperature. Note the change in energy scale from *A* and *B* to *C* and then to *D*; the final ΔG illustrated in *D* is a very small difference between the individual enthalpy and entropy contributions of *A* and *B*. (Data from W. Pfeil and P. L. Privalov, *Biophys. Chem.* 4:41–50, 1976.)

similar marginal net stabilities in their folded structures (Fig. 7.21).

The large ΔC_p of protein unfolding causes there to be a temperature at which stability of the folded state is at a maximum (Fig. 7.21). The net stability decreases at both higher and lower temperatures. Proteins therefore would be expected to unfold at both high and low tem-

peratures. Proteins can almost always be unfolded by raising the temperature sufficiently, but unfolding at low temperatures can be observed only under circumstances in which it occurs in an accessible temperature range, above the freezing point of water. Low-temperature unfolding has thermodynamic characteristics opposite to those at high temperatures, in that heat

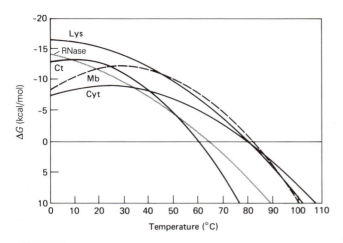

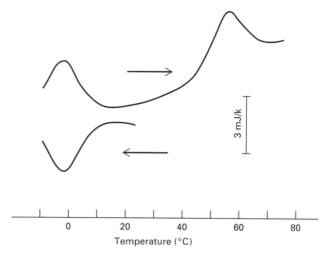

FIGURE 7.21

Temperature dependence of the difference in free energy between the folded and unfolded states of several proteins, expressed per mole of protein. Lys, hen lysozyme; RNase, ribonuclease A; Mb, metmyoglobin; Ct, α-chymotrypsin; Cyt, cytochrome c. The pH of each solution was that for which the protein is most stable. (Adapted from P. L. Privalov and N. N. Khechinashvili, *J. Mol. Biol.* 86:665–684, 1974.)

FIGURE 7.22

Unfolding of apomyoglobin at high and low temperatures measured calorimetrically. In the lower trace, folded apomyoglobin at room temperature was cooled to −10°C; the trough in the heat capacity is caused by the release of heat upon unfolding at −6°C. The cooled solution was then warmed, to produce the upper trace. The peak at −6°C corresponds to the uptake of heat as the apomyoglobin refolds; this is followed by a second peak of heat uptake, above 50°C, as the protein unfolds. (From Y. Griko et al., *J. Mol. Biol.* 202:127–138, 1988.)

is released by unfolding at low temperatures but is taken up by unfolding at high temperatures (Fig. 7.22). The heat capacity change is the same, however, and the unfolded states are indistinguishable thermodynamically. The opposite values of ΔH and ΔS upon unfolding at high and low temperatures are simply a consequence of the large heat capacity change. Physically, cold unfolding can be thought of as resulting from the increasing propensity of ordered water to solvate the nonpolar surface areas of the protein that are exposed by unfolding (see Sec. 4.3.1).

The stabilities of various small single-domain proteins can be compared by normalizing their thermodynamic quantities to correct for their different sizes. When this is done, the measured entropy and enthalpy changes of unfolding of the various proteins differ somewhat, but on extrapolation they converge at a high temperature (Fig. 7.23). When ΔC_p was believed to be independent of temperature, this temperature of convergence was thought to be about 110°C. ΔC_p is now known to decrease at higher temperatures, however, which causes the curves of Figure 7.23 to curve downward at high temperatures, and the convergence temperature is generally taken to be approximately 140°C. On the other hand, it is a reasonable approximation, and much more convenient, to take ΔC_p as constant at lower temperatures, in which case the stabilities of the various small proteins studied thus far can be described at most

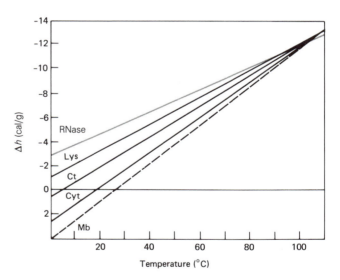

FIGURE 7.23

The specific enthalpy difference, Δh (per gram of protein), between the folded and unfolded states of five proteins: RNase, ribonuclease A; Lys, hen lysozyme; Ct, bovine α-chymotrypsin; Cyt, cytochrome c; Mb, metmyoglobin. The pH of each solution was that at which the protein is most stable. (Adapted from P. L. Privalov and N. N. Khechinashvili, *J. Mol. Biol.* 86:665–684, 1974.)

temperatures T by a simple equation in which the only variable is the ΔC_p of unfolding:

$$\Delta G_{unfold} = \Delta H^* - T\,\Delta S^*$$
$$+ \Delta C_p\left[(T - T^*) - T \ln \frac{T}{T^*}\right] \quad (7.15)$$

where $\Delta H^* = 1.54$ kcal/(mol · residue) or 6.4 kJ/(mol · residue) and $\Delta S^* = 4.35$ cal/(mol · residue) or 18.1 J/(mol · residue), which are the common values at the convergence temperature T^*. The value of T^* is taken to be 112°C with the approximation that the ΔC_p of unfolding is constant.

It is remarkable that the convergence temperature T^* is virtually the same as T_s, the temperature at which the entropy of transfer of nonpolar molecules from liquid to water is zero and at which the enthalpy of transfer from the gas to water is zero (Sec. 4.3.1, Fig. 4.9). It is tempting to conclude that all the proteins have the same enthalpy and entropy of unfolding at that temperature because the nonpolar surfaces that are exposed to water by unfolding have no net interactions with the water. In this case, their common enthalpy and entropy values would reflect the other common features of protein structure.

The thermodynamics of protein unfolding share many features of the hydrophobic interaction measured in model systems (Sec. 4.3.1), including the large change in heat capacity, the large temperature dependence of the changes in enthalpy and entropy, the occurrence of a temperature at which the free-energy change is a maximum, and the convergence of thermodynamic parameters. These similarities are undoubtedly due to the exposure of nonpolar surfaces to water that occurs in both. There are substantial differences, however, in that there are additional enthalpic factors that stabilize the folded state of proteins and entropic factors that stabilize the unfolded state. The thermodynamics of protein unfolding are much more similar to the dissolution of crystals of cyclic dipeptides of nonpolar amino acids than to hydrophobic interactions. This was not unexpected in view of the close packing of the protein interior, which approximates a crystal much more than a liquid. What is most surprising is that Equation (7.15) indicates that the more hydrophobic the interior of a protein (i.e., the greater the value of ΔC_p of unfolding, Fig. 7.19), the lower its net stability. This is not what would be expected if the hydrophobic interaction were the major force stabilizing the folded conformation.

References

Stability of proteins. Small globular proteins. P. L. Privalov. *Adv. Protein Chem.* 33:167–241 (1979).

Stability of proteins. Proteins which do not present a single cooperative system. P. L. Privalov. *Adv. Protein Chem.* 35:1–104 (1982).

Stability of protein structure and hydrophobic interaction. P. L. Privalov and S. J. Gill. *Adv. Protein Chem.* 39:191–234 (1988).

Thermodynamic problems of protein structure. P. L. Privalov. *Ann. Rev. Biophys. Biophys. Chem.* 18:47–69 (1989).

Common features of protein unfolding and dissolution of hydrophobic compounds. K. P. Murphy et al. *Science* 247:559–561 (1990).

Cold denaturation of proteins. P. L. Privalov. *CRC Crit. Rev. Biochem.* 25:281–305 (1990).

d. Rationalizing Stabilities of Folded Conformations

The folded structures of proteins indicate that they are likely to be stabilized by a combination of hydrogen bonds, van der Waals interactions, electrostatic interactions, and the hydrophobic interaction. In each case, of course, the net contribution to stability of the folded structure is the difference between the strength of the interaction in the folded state and that of all the interactions of the same groups in the unfolded state, including their solvation by water. The polar groups of the unfolded protein undoubtedly form hydrogen bonds and have favorable electrostatic interactions with water (Tables 4.8 and 4.9). Consequently, the hydrophobic interaction has long been considered to be the only interaction unique to the folded state and the primary driving force for folding. The thermodynamics of unfolding, however, indicate that this assumption may not be correct. The ΔC_p term in Equation (7.15) is proportional to the nonpolar surface area exposed to water by unfolding; that is, to the hydrophobicity of the protein interior (Fig. 7.19). The other term by which the ΔC_p term is multiplied, involving only the temperature, is positive for all temperatures less than 100°C. The term reflecting the hydrophobicity of the protein, therefore, decreases the stability of the folded protein; the more hydrophobic the protein, the less its net stability at lower temperatures.

This observation is the opposite of what would be expected if hydrophobicity were the predominant factor in protein stability. It implies that the other types of interactions in proteins, such as hydrogen bonding and electrostatic interactions between polar groups, contribute more to the stability of the folded state than do the van der Waals interactions between nonpolar surfaces and the hydrophobic effect. This conclusion is confirmed by the thermodynamic similarity of protein unfolding to the dissolution of crystals of cyclic dipeptides of nonpolar residues. In these crystals, the peptide

Table 7.7 *Free-Energy Contributions of Various Groups to the Stability of Cyclic Dipeptide Crystals in Water, Compared with Their Free Energy of Transfer to a Nonpolar Liquid*

	Transfer from Water to	
Groups	Cyclic dipeptide crystal[a] (kcal/mol)	Nonpolar liquid (kcal/mol)
$\overset{\displaystyle O}{\overset{\displaystyle \|}{-C-NH-}}$	-0.38 ± 0.29	$+6.12^{b}$ $+0.55$ (hydrogen bonded)[b]
Apolar hydrogen, — CH	-0.31 ± 0.05	-0.45^{c}
Phenyl ring	-1.37 ± 0.43	-2.58^{c}
—OH	-0.07 ± 0.26	$+2.23^{c}$

[a] From K. P. Murphy and S. J. Gill, *Thermochim. Acta* 172:11–20 (1990).

[b] From M. A. Roseman, *J. Mol. Biol.* 201:621–623 (1988).

[c] From D. J. Abraham and A. J. Leo, *Proteins: Struct. Funct. Genet.* 2:130–152 (1987).

groups are involved in chains of hydrogen bonds between molecules, with channels in which the nonpolar side chains interact. Analysis of a series of such dipeptides enables assignment of the contribution of each group to the stability of the crystal (Table 7.7). As is apparent, the greatest contribution to the stability of the crystals comes from the peptide group, which is involved in hydrogen bonding between molecules.

The only way that hydrogen bonds could contribute to the net stability of the folded state of a protein would be if those bonds in a folded protein were much stronger than the hydrogen bonds between the unfolded protein and water. This situation is possible because the hydrogen bonds in a protein are intramolecular whereas those between the unfolded protein and water are intermolecular. Such intramolecular hydrogen bonds should be especially stable if a number of them in a folded protein comprise a cooperative system. In this case, the simultaneous presence of numerous intrinsically weak interactions gives a much greater contribution to net stability than is possible with individual interactions (Sec. 4.4).

The cooperativity of protein folded structures is confirmed by the two-state nature of the unfolding transitions (Figs. 7.11–7.13). Partially folded conformations are unstable relative to the fully folded and unfolded conformations. Therefore, weakening one or a few interactions by partial unfolding weakens the other interactions so that their contributions to stability are decreased, and the free energy rises substantially (see Fig. 4.14). The unfolding transition is much more abrupt than would be expected from the disruption of a single interaction (Fig. 7.11).

There are probably two major reasons for the cooperativity of folding transitions. The first is that unfavorable interactions may occur in the partially folded states that do not occur in either the fully folded or the fully unfolded states. Two plausible examples of such interactions are the increase in free energy produced (1) by breaking an internal hydrogen bond without supplying comparable hydrogen-bonding partners to the acceptor or donor and (2) by pulling two nonpolar surfaces sufficiently far apart to diminish greatly the van der Waals interactions but not far enough apart to establish comparable interactions with other surfaces or the solvent. Any such conformational strain in partially folded structures, but not in U or N, should contribute to the cooperativity of folding but not to the net stability of the folded state.

The second reason for cooperativity of folding transitions is likely to arise from the simultaneous presence of many interactions in a single conformation, as described in Section 4.4. This entropic cooperativity should contribute to the net stability of the folded state. Although basically entropic, this effect can also contribute to the enthalpy if the intramolecular interactions are as a consequence also more favorable enthalpically. For example, most hydrogen bonds within water and between a protein and water are usually present only a fraction of the time, whereas those in folded proteins are present essentially all of the time; consequently, hydrogen bonds in folded proteins should have the more negative enthalpy.

It therefore seems reasonable to conclude that most of the interactions in the folded protein are more favorable energetically, in both enthalpy and free energy, than the corresponding interactions of the unfolded state with itself and with water. The stabilizing effects of all the interactions in the fully folded state are consequently greater than the interactions in any other conformation, making the fully folded state the preferred conformation.

The various contributions to the stability of a folded protein can be understood on the basis of Equation (7.15), if it is assumed that the significance of the convergence temperature T^* is that there are no net interactions between the exposed nonpolar surfaces of the protein and the solvent at that temperature. The observed value of ΔH^* would then represent all the net stabilizing interactions in the folded state, such as the van der Waals interactions, hydrogen bonds, and electrostatic interactions, that are disrupted by unfolding. Table 7.8 gives the magnitudes of these values for a protein the size of hen egg-white lysozyme, 129 residues (Fig. 7.20). The value of ΔS^* would represent primarily the greater conformational entropy of the unfolded state; its value corresponds to a reasonable eightfold increase in the average number of conformations per residue upon unfolding. According to this analysis, a folded conformation like that of lysozyme is held together by the -198-kcal/mol contribution of the stabilizing interactions, but this is balanced by the increased entropy of the unfolded polypeptide chain and by the solvation of the nonpolar surfaces by water at lower temperatures. The contribution of the entropy of unfolding is $+167$ kcal/mol at 25°C. This destabilizing factor, which increases in magnitude at higher temperatures, is responsible for thermal unfolding. The contribution of the solvation of nonpolar surfaces to the free energy is only $+17$ kcal/mol at 25°C, but this destabilizing contribution increases in magnitude at lower temperatures and is responsible for cold unfolding.

If hydrophobicity is not the primary force contributing to the stability of folded conformations, the contributions of the van der Waals interactions among nonpolar atoms in the protein interior must be considerably less than those in a liquid (see Fig. 4.9). Otherwise, the surface area buried in the protein interior could provide a free energy of stabilization from the hydrophobic effect that would be adequate to account for all the interactions thought to stabilize the folded state. Nevertheless, the thermodynamic analysis just given indicates that the hydrophobic contribution is less than half of this (Table 7.8). Therefore, the van der Waals interactions among the atoms in a protein interior must be less favorable than those in a nonpolar liquid, even though the interiors of proteins appear to be densely packed. This analysis is confirmed by the contribution of nonpolar groups to the stability of the crystals of cyclic dipeptides (Table 7.7) being that expected from their transfer from

Table 7.8 *Estimated Contributions to the Free Energy of Folding at 25°C of a Typical Protein the Size of Hen Lysozyme*

Contribution	$G^N - G^U$ (kcal/mol)
Greater conformational entropy of U[a]	$+167$
Net stabilizing interactions[b]	-198
Solvation of nonpolar surface in U[c]	$+17$
Net stability	-14

[a] $T\,\Delta S_{conf}$; $\Delta S_{conf} = 4.35$ cal/(K · mol · residue)

[b] Sum of van der Waals interactions in N, net greater stability of hydrogen bonds and other polar interactions in N relative to U, minus any conformational strain. Calculated from $\Delta H^* = 1.54$ kcal/(mol · residue).

[c] Favorable interactions of nonpolar surface with water at 25°C, calculated from $\Delta C_p[T - T^* - T \ln (T/T^*)]$, where $T^* = 112$°C and $\Delta C_p = 12.5$ cal/(K · mol · residue), the measured value for hen lysozyme.

water to nonpolar liquid. The dependence of the van der Waals attraction on the sixth power of the distance (see Fig. 4.2) implies that there could be a substantial decrease in the strength of the interaction in proteins with only a slight decrease in packing interactions. Although the interiors of proteins are densely packed, they are not closely packed with all adjacent atoms in van der Waals contact simultaneously (Figs. 6.20 and 6.21). Imperfect van der Waals interactions imply that there must be constraints preventing them from improving. In the case of the dipeptide crystals discussed above, the constraint may be the crystal lattice; with folded proteins it is most likely to be the covalent structure. There may then be a dynamic balance in the folded state between the tendency to maximize the van der Waals interactions and the energetic cost of the conformational strain due to the covalent constraints.

The concept of hydrophobicity now appears to have less relevance to the stability of folded protein structures than had previously been thought. Although interactions among nonpolar atoms are important, it is perhaps more meaningful to consider the van der Waals interactions involved. Van der Waals interactions in proteins appear to be much less than optimal and considerably weaker than those that occur in a liquid. Hydrogen bonds and other specific interactions appear to contribute most to the stability of protein folded conformations, probably because of the entropic cooperativity between them (Sec. 4.4). Interactions with directional requirements, such as hydrogen bonding, should gain most from entropic effects. Packing of the protein interior is especially important in determining the strengths of the van der Waals interactions and the cooperativity of hydrogen bonding. The role of the aqueous solvent is primarily to interact favorably with the charged and polar groups on the protein surface, which it can do equally well in both the folded and unfolded states. The interaction of water with nonpolar groups is increasingly favorable at low temperatures, which decreases the stability of the folded conformation and can lead to cold-induced unfolding.

Even though the folded state is stabilized by many interactions, it has only marginal net stability because of the large conformational entropy of the unfolded state. Consequently, the net stability of the folded state, ΔG_N, is a relatively small difference between the substantial compensating interactions stabilizing the folded and unfolded states (Fig. 7.20).

References

An empirical approach to protein conformation stability and flexibility. T. E. Creighton. *Biopolymers* 22:49–58 (1983).

Energetics of complementary side-chain packing in a hydrophobic core. J. T. Kellis, Jr., et al. *Biochemistry* 28:4914–4922 (1989).

Influence of interior packing and hydrophobicity on the stability of a protein. W. S. Sandberg and T. C. Terwilliger. *Science* 245:54–57 (1989).

Hydrophobic packing in T4 lysozyme probed by cavity-filling mutants. M. Karpusas et al. *Proc. Natl. Acad. Sci. USA* 86:8237–8241 (1989).

Coupling between local structure and global stability of a protein: mutants of staphylococcal nuclease. A. T. Alexandrescu et al. *Biochemistry* 29:4516–4525 (1990).

Thermodynamics of dissolution of solid cyclic dipeptides containing hydrophobic side groups. K. P. Murphy and S. J. Gill. *J. Chem. Thermodyn.* 21:903–913 (1989).

Stability of folded conformations. T. E. Creighton. *Curr. Opinion Struct. Biol.* 1:5–16 (1991).

The role of internal packing interactions in determining the structure and stability of a protein. W. A. Lim and R. T. Sauer. *J. Mol. Biol.* 219:359–376 (1991).

7.4.4 Effects on Stability of Variation of the Primary Structure

The stability of the folded conformation of a protein depends on its primary structure, but in a complex manner. The amino acid sequences of proteins have changed during evolutionary divergence, but their folded conformations have changed relatively little (see Fig. 6.33), nor have their net stabilities (Fig. 7.21). Yet single amino acid replacements can alter the stability of a folded protein quite drastically. Selective pressures must have acted during protein evolution to maintain the stability of a protein near its optimum level. Net stability appears not to be maximized, because proteins are generally only marginally stable under the conditions in which they function, even though increased stability would be feasible. The natural variation among proteins and the variations introduced by chemical modification, genetic mutation, or site-directed mutagenesis provide a plethora of data concerning the physical basis of protein stability.

a. Natural Variation in Structures and Environments

Proteins that function under normal (physiological) conditions tend to have similar stabilities to unfolding (Eq. 7.15), even though they have different folds and different amino acid sequences. Most of these proteins come from organisms known as *mesophiles*. Many organisms are adapted to other environments, such as low or high temperature (psychrophiles and thermophiles, respectively) or high salt concentrations (halophiles), in which mesophilic proteins would unfold or become insoluble, so it is not surprising that the proteins from these organisms also have adapted to their environments. How they have done so continues to intrigue protein chemists.

Most thermophilic proteins are intrinsically more resistant to high temperatures than are their mesophilic counterparts. Even when purified, they generally resist unfolding at the elevated temperatures (up to 90°C) at which they normally function, and they are often impervious to denaturants. Nevertheless, the structures of thermophilic proteins that have been determined are essentially the same as those of their mesophilic counterparts. It is usually impossible to distinguish between thermophilic and mesophilic proteins on the basis of their crystal structures alone. The only apparently significant structural differences between them seem to be that thermophilic proteins have more salt bridges on their surfaces or incorporate ions as part of their folded structures. They appear to rely on additional polar interactions for their greater stability. It might have been thought that thermophilic proteins would be more hydrophobic because the hydrophobic effect in model systems increases energetically with increasing temperature (Sec. 4.3), but that they are not is also consistent with the tendency of more hydrophobic proteins to be less stable (Eq. 7.15).

Proteins in halophiles often exist in ionic environments equivalent to saturated KCl solution. When isolated, they tend to require high ionic strengths for stability of their folded conformations. The few halophilic proteins that have been studied bind anomalously large amounts of water and salt and appear to incorporate ions into their folded conformations, although the structural details remain to be elucidated.

These observations are based on a few studies of a few proteins, so their generality is not established.

References

Mechanism of thermophily for thermolabile glyceraldehyde-3-phosphate dehydrogenase from the facultative thermophile *Bacillus coagulans* KU. J. McLinden et al. *Biochim. Biophys. Acta* 871:207–216 (1986).

Thermitase, a thermostable subtilisin: comparison of predicted and experimental structures and the molecular cause of thermostability. C. Frömmel and C. Sander. *Proteins: Struct. Funct. Genet.* 5:22–37 (1989).

Extremely thermostable D-glyceraldehyde-3-phosphate dehydrogenase from the eubacterium *Thermotoga maritima.* A. Wrba et al. *Biochemistry* 29:7584–7592 (1990).

Halophilic proteins and the influence of solvent on protein stabilization. G. Zaccai and H. Eisenberg. *Trends Biochem. Sci.* 15:333–337 (1990).

b. Mutagenic Studies

The principles of protein stability have been examined extensively by measuring the effects of altering the primary structure either chemically or mutationally. The latter approach has recently used site-directed mutagenesis (Fig. 2.6) to make specific alterations. Alternatively, mutations can be introduced randomly, either to the entire gene or to just a segment of it, and those mutations that produce the desired effect can be selected. For example, the in vivo biological function of a protein depends on its having a stable folded conformation. Mutations that decrease its stability sufficiently to inhibit its biological function in vivo can often be selected on the basis of the absence of the biological function in the mutant form. The method of selection and the changes in stability that are detectable in this way depend on the particular protein, but the principles are general. After inactive mutants have been isolated, further mutagenesis and selection can identify amino acid replacements that reverse the particular destabilization or that increase stability generally. If random mutagenesis is targeted to a certain part of the gene and protein and is sufficiently drastic to have produced all possible amino acid replacements, those mutations that are not identified by the selection process can be assumed to have passed the selection criterion. Negative observations thereby become significant.

The effects on stability of any particular mutation depend on the role that the original residue played in the folded structure, the role of the introduced residue, and any alterations of conformation caused by the replacement. The folded structures of proteins are generally not altered substantially by a single replacement. One of the largest structural changes observed upon mutagenesis involved extension of a helix after replacing a Pro residue. The original Pro residue presumably terminates the α-helix prematurely in that protein. The structures of proteins seem to be sufficiently plastic to accommodate small replacements, but this plasticity varies throughout the protein structure. The packing of atoms is observed to change locally around most mutations, but the extent of the changes depends on both the magnitude of the covalent change and the site in the protein. Partly for these reasons, it is difficult to predict or to rationalize the structural and energetic consequences of any particular mutation.

The effects of single mutations on the stability of the folded structure vary widely, both in and between proteins. Only a few generalities can be made. Replacements on the surface of a protein generally have little or no effect on the stability (Fig. 7.24) unless either the original or the introduced side chains have specific roles. Charged residues on the protein surface can be interchanged or replaced by nonpolar residues, and vice versa, with only very small effects on stability, unless the charged groups are involved in particular salt bridges. The interior of the protein is generally the most sensitive to mutation. Large decreases in net stability are usually produced by replacing buried hydrophobic side chains by even larger side chains that do not fit, by smaller side chains that leave cavities or cause substantial repacking of the interior, and by polar or charged side chains that then exist in unfavorable environments. The net stability of the folded state can be decreased by up to 9 kcal/mol by a single interior mutation, comparable to the overall net stability of the folded state, so a single such mutation can cause a protein to unfold. The greatest energetic effects generally occur at sites in the best packed, least flexible part of the protein structure. Other parts of the protein appear more able to accommodate mutations by altering the local packing of adjacent atoms. For example, Ala 90 of the λ repressor protein is totally buried, yet it can be replaced by a number of different residues, including Trp (Fig. 7.24). Although totally buried, the Ala 90 side chain is sufficiently near the surface that small conformational rearrangements can permit the larger side chains to be accommodated. Inserting extra residues in α-helices or β-strands usually causes a substantially greater decrease in conformational stability than insertions in less regular parts of the

A

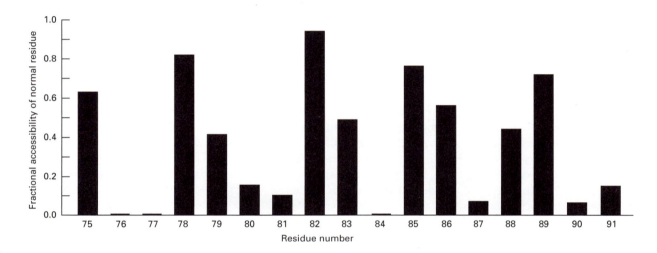

B

FIGURE 7.24

Amino acid substitutions in an α-helical segment that do not substantially lower the net stability of the N-terminal domain of the λ repressor (J. U. Bowie et al., *Science* 247:1306–1310, 1990). *A:* The normal sequence is shown in boldface, and the permitted amino acids are listed above the normal sequence in order of their relative hydrophobicities. *B:* The fractional accessibility of the normal side chain in the folded λ repressor. The most accessible residues generally are most tolerant of amino acid substitutions, although exceptions do occur, such as Ala 90. (Kindly provided by R. Sauer.)

polypeptide chain. The effects of insertion and deletion are more variable because of the number of ways in which the change can be accommodated by the protein structure.

The energetic effects of two simultaneous alterations on the stability of a folded protein are frequently additive. Exceptions occur primarily when the replacements occur in adjacent parts of the folded conformation. In these cases, the residue that is present at one site influences the effect of a replacement at the adjacent site.

There is often a rough correlation between the hydrophobicity of the side chain introduced into a protein interior and the net stability of the protein. The effect on stability can be up to two to three times greater than that expected from the free energy of transfer from water to a nonpolar liquid (see Table 4.8). The protein interior is not comparable to a liquid because the correlation breaks down when the side chain introduced is significantly larger than the original one. There appears to be an energetic cost to a folded protein rearranging its interior packing substantially. Multiple replacements that maintain the hydrophobicities and volumes of interior residues can still have substantial effects on net stability, indicating that it is not just hydrophobicity that is important. The correlations with hydrophobicity may be fortuitous and may actually reflect the more important aspects of the extent of the van der Waals interactions between nonpolar atoms.

Altering particular interactions or aspects of the structure can also affect protein stability. Replacing a charged residue involved in a particularly stabilizing salt bridge can decrease the stability by up to 5 kcal/mol. The backbone conformational properties of Gly and Pro residues are often important, so replacing or introducing them can have substantial effects on stability. Residue substitutions in an α-helix that increase the helical propensity, due to either the intrinsic helical propensity of the residue (Table 5.3) or the electrostatic interaction with the helix dipole (Sec. 5.3.1), can increase protein stability, as long as other consequences of the mutation do not predominate.

Most mutations are found to have little effect on the stability of the natural folded state. Natural selection seems to have optimized the stabilizing effect of the amino acid sequence. Only a few mutations are found to increase protein stability; they seem to introduce new stabilizing interactions (van der Waals interactions, salt bridges, disulfide bonds, etc.) to improve the packing of internal side chains and to relieve slightly unfavorable aspects of the native structure. In almost all instances, however, the additional stability is less than might be expected from the stabilizing interaction introduced, so there is generally an energetic cost to be paid upon altering a protein structure, possibly due to inadvertent unfavorable interactions. Native proteins do not have the greatest possible stability, but alterations to their covalent structures are much more likely to be destabilizing than stabilizing. Nevertheless, protein stability has been improved by natural selection to produce thermophilic proteins (Sec. 7.4.4.a).

Disulfide bonds and Gly and Pro residues alter the conformational entropy of the unfolded state, and their introduction or replacement is one of the ways in which protein stability can be altered. As in this case, the consequences of many mutations are attributed to their effects on either the folded or the unfolded state. This is not straightforward, however, because the free energies of proteins with different covalent structures cannot be compared directly. The best that can be done is to consider unfolding ($N \rightarrow U$) and the mutational event ($A \rightarrow B$) as linked functions (see also Eq. 4.26 and Fig. 7.32):

$$
\begin{array}{ccc}
N_A & \xrightarrow{K^A_{unfold}} & U_A \\
\left\downarrow{\scriptstyle K^N_{mut}}\right. & & \left\downarrow{\scriptstyle K^U_{mut}}\right. \\
N_B & \xrightarrow{K^B_{unfold}} & U_B
\end{array}
\tag{7.16}
$$

$$
\frac{K^A_{unfold}}{K^B_{unfold}} = \frac{K^N_{mut}}{K^U_{mut}}
\tag{7.17}
$$

$$
\Delta G_{unfold} = \Delta G_{mut}
\tag{7.18}
$$

Therefore, the difference in stability caused by the mutation $A \rightarrow B$ should be the same as the difference in free energies used in making the mutation in the folded and the unfolded states. The latter factor, ΔG_{mut}, cannot be measured experimentally but can be simulated computationally by **free-energy perturbation** methods. In this procedure, the mutation is made gradually to a model of the protein. Its free energy can be computed if there is no substantial change in the structure. This provides a feasible method of computing the effects of mutations on conformational stability. Computing the free energies of the unfolding process directly would be impractical due to the enormous conformational changes that occur; instead, the equivalent value of ΔG_{mut} is calculated. The linkage relationships of Equations (7.17) and (7.18) are also useful for rationalizing the effects of mutations on stability, by considering the difference in the energetic consequences of making the mutation in the folded and in the unfolded states.

Mutations are generally considered to affect primarily the folded state, unless conformational effects on the unfolded protein can be shown. Such effects on the unfolded state are most apparent with staphylococcal nuclease, for reasons that are not clear.

References

Genetic and structural analysis of the protein stability problem. B. W. Matthews. *Biochemistry* 26:6885–6888 (1987).

Probing the determinants of protein folding and stability with amino acid substitutions. D. Shortle. *J. Biol. Chem.* 264:5315–5318 (1989).

Genetic analysis of protein stability and function. A. A. Pakula and R. T. Sauer. *Ann. Rev. Genetics* 23:289–310 (1989).

Substantial increase of protein stability by multiple disulphide bonds. M. Matsumura et al. *Nature* 342:291–293 (1989).

Amino acid substitutions that increase the thermal stability of the λ Cro protein. A. A. Pakula and R. T. Sauer. *Proteins: Struct. Funct. Genet.* 5:202–210 (1989).

Structural studies of mutants of T4 lysozyme that alter hydrophobic stabilization. M. Matsumura et al. *J. Biol. Chem.* 264:16059–16066 (1989).

Residual structure in large fragments of staphylococcal nuclease: effects of amino acid substitutions. D. Shortle and A. K. Meeker. *Biochemistry* 28:936–944 (1989).

Deciphering the message in protein sequences: tolerance to amino acid substitutions. J. U. Bowie et al. *Science* 247:1306–1310 (1990).

Contributions of the large hydrophobic amino acids to the stability of staphylococcal nuclease. D. Shortle et al. *Biochemistry* 29:8033–8041 (1990).

Accommodation of single amino acid insertions by the native state of staphylococcal nuclease. J. Sondek and D. Shortle. *Proteins: Struct. Funct. Genet.* 7:299–305 (1990).

Additivity of mutational effects in proteins. J. A. Wells. *Biochemistry* 29:8509–8517 (1990).

c. Cleavage of the Polypeptide Chain

Most fully folded structural domains of proteins are resistant to proteolytic enzymes. The polypeptide segments linking independent structural domains are often flexible and susceptible to cleavage, however, which provides a diagnostic test for the presence of such domains and a method for their separation and isolation. The separated but intact domains usually remain folded, and they often unfold and refold like small single-domain proteins.

Exposed flexible polypeptide loops on the surfaces of folded proteins can also be cleaved proteolytically; the folded conformation is often still stable. The classic example is ribonuclease S, in which the peptide bond between residues 20 and 21 is cleaved; the two fragments are held together by noncovalent forces, with only local alterations of the conformation. The cleaved protein unfolds more readily than the normal intact protein, and the two fragments then dissociate. The equilibrium constant for reversible unfolding therefore becomes concentration dependent.

The separated fragments of single-domain proteins usually have either no stable conformation or a native-like conformation that is less stable than that of the original protein. For example, the S-peptide fragment of ribonuclease S, residues 1–20, is mostly unfolded, with only a limited tendency to adopt the helical conformation that it has in the folded complex. The other fragment, the S-protein, has the four original disulfide bonds and retains a folded conformation with some similarities to native ribonuclease. It unfolds, however, if interchange of the four disulfides is permitted (Fig. 7.25) when the disulfides become "scrambled." The four native disulfide bonds are determined by folding of the entire polypeptide chain including the S-peptide. In the S-protein alone, these four disulfide bonds constrain the polypeptide chain and stabilize the nativelike conformation, but alternative disulfides and conformations have lower free energies.

Many such pairs of fragments can reassemble to generate the original noncovalent complex, in a complicated, little understood process involving mutual recognition and folding of the two parts. For example, adding S-peptide to scrambled S-protein can regenerate

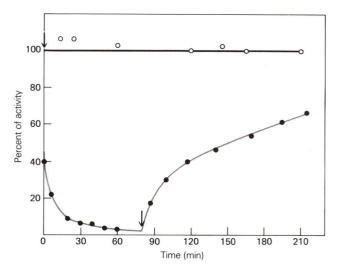

FIGURE 7.25

Inactivation and disulfide interchange of ribonuclease S-protein by the enzyme protein disulfide isomerase. The isomerase was added to the S-protein at $t = 0$. At the times indicated by the arrows, 1.3 equivalents of S-peptide were added. The S-peptide prevents inactivation if present initially, by forming the stable ribonuclease S complex. The S-peptide can reverse the disulfide rearrangement of S-protein, presumably by pulling the disulfide equilibrium toward the form of the S-protein with the folded conformation and the correct disulfides. (Adapted from I. Kato and C. B. Anfinsen, *J. Biol. Chem.* 244:1004–1007, 1969.)

ribonuclease S and the four native disulfides (Fig. 7.25). Virtually all the original residues must be represented at least once in the two fragments for a stable complex to be generated. Folded complexes can also be formed if a segment with the same sequence of residues is present in both fragments. The redundant segment can be removed proteolytically from the folded complex, indicating that only one of the two repeated segments, often from either of the two fragments, is involved in the folded conformation (Fig. 7.26). Not all such pairs of fragments, however, yield stable conformations, even if they contain all the original residues. The various covalent bonds of the polypeptide chains apparently differ widely in their energetic importance for stability of the folded conformation.

The newly generated amino and carboxyl ends may still be sufficiently proximate within the folded conformation of a cleaved polypeptide chain to be able to regenerate the original peptide bond. As a consequence, the equilibrium constant may favor bond formation rather than hydrolysis, and proteolytic enzymes can be used to catalyze peptide bond formation (the reverse of their normal reaction). Such a phenomenon is used by protease inhibitors, which bind tightly at the

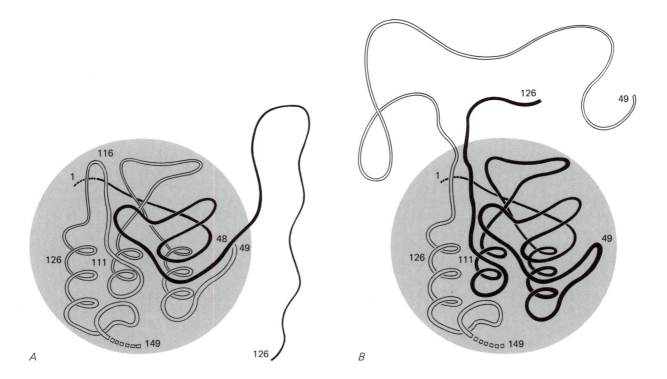

FIGURE 7.26

Schematic representation of the two types of complementation formed by fragments of residues 1–126 (dark ribbon) and 49–149 (light ribbon) of staphylococcal nuclease. The native enzyme consists of 149 residues. In *A*, residues 49–116 of the folded structure are provided by fragment (49–149), whereas in *B* they are provided by fragment (1–126). In both cases, the redundant fragments can be cleaved off by proteases, leaving a stable complex. The conformation of the ordered complex, enclosed by the shaded circle, is assumed to be like that of the native enzyme. (Adapted from H. Taniuchi and C. B. Anfinsen, *J. Biol. Chem.* 246:2291–2301, 1971.)

active sites of proteolytic enzymes without being degraded. Hydrolysis of the target peptide bond is readily reversible, with equilibrium constants in the region of unity (Sec. 9.3.2.a). Peptide bond reformation can also be induced chemically; for example, the homoserine lactone generated at the carboxyl end of a polypeptide chain after cyanogen bromide cleavage at Met residues (Sec. 1.6.2.e) is sufficiently reactive that a newly generated amino group still held in the correct proximity can react with it. Such procedures make possible the preparation of semisynthetic proteins, in which one fragment has been modified specifically or synthesized de novo.

In many cases, a few residues can be removed from the ends of a polypeptide chain with retention of a stable folded conformation. At some point, however, depending on the protein, the stability of the folded conformation disappears, and the remaining polypeptide chain is unfolded. Removal of carboxyl-terminal residues seems most deleterious for the few proteins studied in this way.

Removing cofactors from proteins has varying effects; for example, apomyoglobin has a nativelike conformation whereas apocytochrome *c* is unfolded.

References

Formation of randomly paired disulfide bonds in des-(121–124) ribonuclease after reduction and reoxidation. H. Taniuchi. *J. Biol. Chem.* 245:5459–5468 (1970).

The dynamic equilibrium of folding and unfolding of nuclease T'. H. Taniuchi. *J. Biol. Chem.* 248:5164–5174 (1973).

Spontaneous re-formation of a broken peptide chain. D. F. Dykes et al. *Nature* 247:202–204 (1974).

Protein complementation. I. Zabin and M. R. Villarejo. *Ann. Rev. Biochem.* 44:295–314 (1975).

Enzymatic resynthesis of the hydrolyzed peptide bonds in ribonuclease S. G. A. Homandberg and M. Laskowski, Jr. *Biochemistry* 18:586–592 (1979).

An 8-fold $\beta\alpha$ barrel protein with redundant folding possibilities. K. Luger et al. *Protein Eng.* 3:249–258 (1990).

7.5 Mechanism of Protein Folding

The observation of reversible unfolding transitions that attain equilibrium implies that the protein is both unfolding and refolding at equilibrium. The net stability of protein folding is so small, even under optimal conditions, that most proteins must unfold spontaneously and completely at finite frequencies; this spontaneous unfolding must be rapidly reversed. Refolding of most proteins can be demonstrated directly by placing the fully unfolded protein under conditions in which the folded state is stable.

Refolding of most proteins is a self-assembly process in that all the information required is present in the amino acid sequence, and refolding occurs spontaneously under the appropriate conditions. Proteins that do not refold in vitro generally have undergone some interfering covalent modification, have lost some cofactor required for the folded state (Chap. 8), or have precipitated. Some proteins do, however, require biological factors in vivo for folding (Sec. 7.5.6), primarily to prevent aggregation.

That proteins fold spontaneously should be somewhat surprising. It was shown in Chapter 5 that polypeptide chains can adopt an astronomical number of conformations, a number so large that it would not be feasible for any protein to try out all of its conformations on a practical time scale. For example, if a polypeptide chain of 100 residues has 10^{100} possible conformations (i.e., an average of 10 conformations per residue) and if it converts one conformation into another in the shortest possible time (i.e., perhaps 10^{-13} s), the average time required to sample all possible conformations would be 10^{77} years (10^{85} s). Nevertheless, proteins are observed to fold in 10^{-1}–10^3 s both in vivo and in vitro. The inescapable conclusion is that proteins do not fold by sampling all possible conformations randomly until the one with the lowest free energy is encountered. Instead, to occur on a short time scale, the folding process must be directed in some way.

If kinetic factors are so important, the possibility arises that the observed folded conformation is not the one with the lowest possible free energy but rather the most stable of those conformations kinetically accessible. Other unobserved, and probably unobservable, conformations might have even lower free energies.

This very basic question has been almost philosophical in nature, and only recently have clear-cut examples been found.

The potential for existing in a stable folded conformation is not sufficient to guarantee a pathway to it. This is indicated by proteins that cannot refold, such as those that normally fold as a precursor or after translocation through a membrane. The normal folding pathway is presumably not accessible after unfolding of the mature protein because part of the original protein or some other aspect of the biosynthetic machinery is missing. Further indirect evidence comes from mutations that block folding to a dramatic extent but do not alter the stability of the folded state. Not surprisingly, the experimentally determined pathways of folding are far from random.

References

Energetics of protein structure and folding. D. P. Goldenberg and T. E. Creighton. *Biopolymers* 24:167–182 (1985).

Protein folding: hypotheses and experiments. O. B. Ptitsyn. *J. Protein Chem.* 6:273–293 (1987).

Genetic studies of protein stability and mechanisms of folding. D. P. Goldenberg. *Ann. Rev. Biophys. Biophys. Chem.* 17:481–507 (1988).

Protein folding. T. E. Creighton. *Biochem. J.* 270:1–16 (1990).

Intermediates in the folding reactions of small proteins. P. S. Kim and R. L. Baldwin. *Ann. Rev. Biochem.* 59:631–660 (1990).

Up the kinetic pathway. T. E. Creighton. *Nature* 356:194–195 (1992).

7.5.1 Kinetic Analysis of Complex Reactions

To determine the mechanism and pathway of unfolding and refolding, the intermediates that define and direct the pathway must be identified, but these are usually unstable thermodynamically (Fig. 7.13). They might be detectable as kinetic intermediates, but only if they occur on the pathway before the rate-limiting step and if their free energies are comparable to or lower than that of the initial state. No other kinetic intermediates are populated to substantial levels, even transiently.

With a simple one-step reaction followed as a function of time t, there is expected to be a single kinetic phase, characterized by a single rate constant k:

$$\text{Fraction folded conformation} = 1 - \exp(-kt) \quad (7.19)$$

More complex kinetic behavior would be observed either if there were multiple rate-limiting steps in the

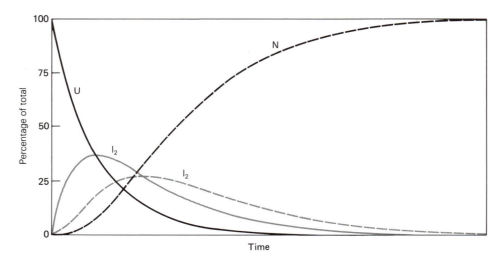

FIGURE 7.27
Kinetics of an obligatorily sequential pathway with multiple rate-limiting steps. The progress of the reaction $U \rightarrow I_1 \rightarrow I_2 \rightarrow N$ was simulated with each step having the same rate. The proportion of molecules in each species is plotted as a function of time. Note that U disappears and I_1 appears without a lag, but that I_2 and N appear only after lag periods. The lag period for I_2 corresponds to the time during which the concentration of the precursor I_1 is increasing. The lag period in formation of N is longer due to the need to build up the concentration of I_2 also.

reaction or if the starting material were heterogeneous, with different populations having different rates of reaction. Discriminating between these possibilities with a multiphase reaction is not always straightforward. It is too often assumed that one kinetic phase of a folding reaction represents formation of an obligatory intermediate I and that a second phase represents its conversion to another intermediate or to N:

$$U \underset{k_{-1}}{\overset{k_1}{\rightleftharpoons}} I \underset{}{\overset{k_2}{\rightleftharpoons}} N \qquad (7.20)$$

If this were the case, there would be a lag period in the appearance of N of approximate magnitude $(k_2 + k_1)^{-1}$, during which the steady-state concentration of I would be generated (Fig. 7.27). This effect is cumulative with additional steps. Therefore, the magnitude of the lag period in formation of the final folded conformation should be correspondingly longer with an increasing number of obligatory, sequential intermediates along a pathway. Yet very few claims of obligatory intermediates in protein folding are supported by the observation of an appropriate lag period. Most kinetic complexities of protein folding arise from heterogeneity of the unfolded state (Sec. 7.5.3.a).

Because of the great conformational heterogeneity of the unfolded state, protein folding is a special kinetic phenomenon in which every molecule of a typical population is likely to have a unique conformation at every instant. For example, 1 mg of a protein with a molecular

weight of 10^4 comprises 0.1 μmol, or 6×10^{16} molecules, whereas many more conformations are likely when it is unfolded. How is this conformational heterogeneity apparent in the kinetics of refolding? Does each molecule refold at its own rate, determined by its conformation at time zero, or do molecules somehow fold by a common mechanism at a common rate? If each molecule does not fold uniquely, how do different molecules manage to follow the same rate-limiting step?

It is clearly unrealistic to expect to elucidate all the details of a complex reaction like protein folding. Although it occurs much more rapidly (on the second to minute time scale) than expected for a random search, this time is long enough for each molecule to undergo perhaps some 10^{11}–10^{13} conformational changes. Because each molecule of a substantial population starts out with a different conformation, it might be feasible to determine at what stage different molecules start to follow the same pathway. At best it may be possible only to characterize the slowest transitions and the conformations and energetics of the most stable intermediates, to identify the overall rate-limiting step, and to characterize the transition state.

7.5.2 Kinetics of Unfolding

Unfolding of proteins is almost universally observed to be an all-or-none process, with little or no partial unfolding preceding complete unfolding. When a native, covalently homogeneous protein is placed in unfolding conditions at time zero, unfolding almost always occurs with a single kinetic phase and a single rate constant, as in Equation (7.19). There is no lag period, and all probes of unfolding give the same rate constant. Therefore, there is a single rate-limiting step in unfolding, and all folded molecules have the same probability of unfolding. Exceptions generally result from exceptions to the usual homogeneous nature of the folded state.

The rate of unfolding usually changes uniformly with variation of the unfolding conditions. In particular, logarithmic plots of unfolding rates versus denaturant concentration or temperature (Fig. 7.28) are generally linear, suggesting that the mechanism of unfolding is not changing. There appears to be a single transition state for unfolding under these conditions.

7.5.3 Kinetics of Refolding

Kinetic complexities are encountered almost universally in protein refolding. These complexities usually result from conformational heterogeneity of the unfolded state, with slow- and fast-refolding molecules:

$$U_S \underset{slow}{\rightleftharpoons} U_F \underset{fast}{\rightleftharpoons} N \tag{7.21}$$

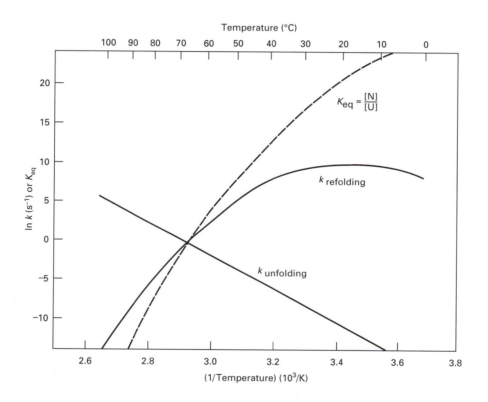

FIGURE 7.28

Typical temperature dependence of the rates and equilibria of protein folding transitions not involving intrinsically slow isomerizations. The natural logarithms of the rate constants for unfolding and refolding are plotted as a function of (temperature)$^{-1}$ in an Arrhenius plot. A similar plot of the equilibrium constant (K_{eq}) between the folded (N) and unfolded (U) states is a van't Hoff plot. The curvature of the van't Hoff plot is due to the greater apparent heat capacity of U than of N. The linear Arrhenius plot for the rate of unfolding indicates that the transition state for the folding transition has the same heat capacity as N. The greater heat capacity of U is reflected entirely in the curvature of the Arrhenius plot for the rate of refolding, because ln K_{eq} = ln $k_{refolding}$ − ln $k_{unfolding}$. The data used to construct this diagram are for hen egg-white lysozyme at pH 3. The data for stability were from W. Pfeil and P. L. Privalov, *Biophys. Chem.* 4:23–50 (1976); the rates of folding were from S. I. Segawa and M. Sugihara, *Biopolymers* 23:2473–2488 (1984), extrapolated to the absence of GdmCl. Although it is necessary that $k_{unfolding} = k_{refolding}$ at $K_{eq} = 1$, it is a coincidence that the rate constants had the value 1/s at this temperature, so that all three curves intersect at a common point. (From T. E. Creighton, *Biochem. J.* 270:1–16, 1990.)

In virtually all characterized cases, the heterogeneity arises from *cis–trans* isomerization of peptide bonds preceding Pro residues (Sec. 5.2.1).

a. Peptide Bond Isomerization

Cis peptide bonds are often found in folded proteins, but almost only when the next residue is Pro (Sec. 6.2). A peptide bond is usually *cis* or *trans* in essentially all the folded molecules because the folded conformation generally favors one over the other (although there are exceptions). When a protein is unfolded, however, the constraints favoring one form over the other are released, and an equilibrium between *cis* and *trans* isomers is attained at each peptide bond. When the protein is refolded, a fraction of the molecules, U_F, have all the necessary peptide bonds as the correct isomer whereas the others, U_S, have one or more as an incorrect isomer.

Cis–trans isomerization of Pro peptide bonds is intrinsically slow (Sec. 5.2.3.b). When the rate of refolding of the U_F molecules is faster than *cis–trans* isomerization, U_F and U_S molecules have different rates of refolding (Eq. 7.21). If all the peptide bonds must be of the correct isomer for refolding to occur, the greater the number of Pro residues the greater the fraction of U_S molecules and the slower their refolding. The actual situation is more complex, however, because some proteins can refold to a nativelike conformation with an incorrect isomer of one or more peptide bonds. Also, the rates of isomerization can be either increased or decreased by the conformation of the protein.

For example, two of the four Pro residues in bovine ribonuclease A have *cis* peptide bonds in the folded conformation. Unfolded ribonuclease A (with the four disulfide bonds intact) refolds in three different kinetic phases, corresponding to at least three different unfolded species. One accounts for 15% of the molecules and refolds within less than a second under optimal conditions; it is thought to have all correct peptide bond isomers. The remaining 85% of the molecules are thought to have one or more nonnative peptide bond isomers that must slow refolding. A second kinetic species represents 65% of the molecules and refolds on a time scale of seconds. Under conditions strongly favoring folding, this species folds more rapidly into a nativelike conformation, retaining the incorrect peptide bond isomer. The remaining 20% of the molecules make up the third kinetic species and refold even more slowly. The second and third kinetic species are believed to result from *trans* isomers predominating in the unfolded protein of the two peptide bonds that are *cis* in native ribonuclease A.

Replacing Pro residues with other amino acids, especially those Pro residues with *cis* peptide bonds in the folded state, can abolish the slow refolding transitions. This is often complicated, however, by destabilizing effects on the folded conformation because a *cis* peptide bond is unfavorable without the Pro residue (Eq. 5.5).

References

Consideration of the possibility that the slow step in protein denaturation reactions is due to *cis–trans* isomerism of proline residues. J. F. Brandts et al. *Biochemistry* 14:4953–4963 (1975).

Replacement of proline-76 with alanine eliminates the slowest kinetic phase in thioredoxin folding. R. F. Kelley and F. M. Richards. *Biochemistry* 26:6765–6774 (1987).

Separation of the nativelike intermediate from unfolded forms during refolding of ribonuclease A. L. N. Lin and J. F. Brandts. *Biochemistry* 27:9037–9042 (1988).

b. Refolding in the Absence of Slow Peptide Bond Isomerization

In a population of unfolded molecules with the same *cis–trans* isomers as the native state, the refolded protein generally appears with a single rate constant and without a significant lag period. The absence of an observable lag period indicates that there is a single rate-limiting step in refolding and that all preceding and subsequent steps are more rapid. Consequently, refolding can be simplified to three stages (Fig. 7.29): (1) the nature of the unfolded protein under refolding conditions, the "prefolded" conformation; (2) the nature of the rate-limiting step and the overall transition state for folding; and (3) the nature of the folded conformation under refolding conditions, especially its flexibility.

Considering the conformational heterogeneity of the unfolded state (but excluding intrinsically slow isomerizations), it is noteworthy that all the molecules with the same covalent structure are usually observed to fold with the same rate constant. A single rate constant is consistent with all the molecules folding by the same rate-determining step. The folding of many conformationally heterogeneous molecules by the same rate-limiting step requires that there be a rapid conformational equilibration prior to the rate-limiting step (Fig. 7.29C). That this occurs is also indicated by the general observation that the rate of refolding depends only on the final folding conditions, not the initial unfolding conditions. Proteins unfolded in different ways generally have different average physical properties. Nevertheless, they refold at indistinguishable rates under the same final folding conditions. The rate of folding is determined not by the nature of the initial unfolded protein but by the properties it rapidly adopts when placed under the final folding conditions.

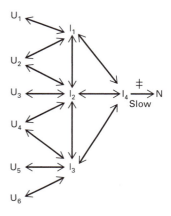

$$U \underset{\underset{Slow}{\longleftrightarrow}}{\overset{\ddagger}{}} I_1 \longrightarrow I_2 \longrightarrow I_3 \longrightarrow N$$

A

B

C

FIGURE 7.29

Examples of models for protein folding in the absence of intrinsically slow isomerizations. U_i are various unfolded molecules with different conformations at the start of folding, I_i are partially folded molecules, and N is the fully folded protein. All kinetic steps indicated by arrows are rapid except for those labeled *slow*; $\ddagger$ indicates the occurrence of the overall transition state. Single-headed arrows indicate steps that effectively occur only in the indicated direction under conditions strongly favoring folding. *A:* The nucleation, rapid-growth model (D. B. Wetlaufer, *Proc. Natl. Acad. Sci. USA* 70:697–701, 1973), in which a nucleation event in the unfolded protein, here indicated as formation of I_1, is the rate-limiting step. The nucleation event is very local and occurs randomly, so it could occur in all the unfolded molecules at a reasonable rate. Subsequent steps through various intermediates are rapid and essentially irreversible under strongly folding conditions. *B:* The jigsaw puzzle model (S. C. Harrison and R. Durbin *Proc. Natl. Acad. Sci. USA* 82:4028–4030, 1985), in which every unfolded molecule refolds by a unique sequence of events. The pathways converge only at the fully folded conformation. Each pathway will occur at a unique rate. *C:* The general model indicated by the experimental data. All the unfolded molecules rapidly equilibrate under folding conditions with a few partially folded, marginally stable in-

How do all unfolded molecules equilibrate rapidly prior to refolding if sampling of all conformations by a random coil requires such a long period of time? The answer undoubtedly is that an unfolded protein under refolding conditions does not behave as a random coil but adopts a limited set of energetically favored nonrandom conformations, as described in the next section. In this way, all the molecules converge to follow a common subsequent pathway and have the same rate-limiting step. This convergence is in contrast to the proposal that each protein molecule folds by a unique pathway (Fig. 7.29B).

The rate of direct refolding generally varies with temperature in a complex manner, giving a nonlinear Arrhenius plot (Fig. 7.28). At low temperatures, the rate of refolding increases with temperature, as do most chemical reactions. The increase in rate diminishes, however, and the rate reaches a maximum and then decreases dramatically at high temperatures. This temperature dependence is unusual for chemical reactions, but it might be expected for a complex reaction like protein folding that is dependent on the presence of metastable, partially folded intermediates. Such metastable intermediates would be destabilized at high temperatures, and the rate of refolding would decrease accordingly. This explanation, although simple and appealing, is not currently accepted, however. Instead, the temperature dependence of the rate of refolding is held to be a consequence of the difference in heat capacities between the unfolded state and the transition state. The difference in heat capacity of the folded and unfolded states (Fig. 7.18) must be reflected in nonlinear Arrhenius plots of the rates of either unfolding or refolding or both, depending on the relative heat capacity of the transition state. The data of Figure 7.28 demonstrate that the transition state has the same heat capacity as the fully folded state in this instance.

References

Identification and characterization of the direct folding process of hen egg-white lysozyme. S. Kato et al. *Biochemistry* 21:38–43 (1982).

Toward a better understanding of protein folding pathways. T. E. Creighton. *Proc. Natl. Acad. Sci. USA* 85:5082–5086 (1988).

termediates, which are also in rapid equilibrium. All the molecules pass through a common slow step, which involves going through a transition state that is a distorted form of the nativelike conformation. Any intermediates that occur after the rate-limiting step are probably very unstable relative to N. (From T. E. Creighton, *Biochem. J.* 270:1–16, 1990.)

c. The Prefolded State

The prefolded state is the unfolded protein under refolding conditions, prior to the rate-limiting step and complete refolding. The prefolded state is intrinsically unstable and is populated only transiently. Nevertheless, a variety of evidence indicates that it has considerable nonrandom conformation in many proteins. Evidence is accumulating that the prefolded state generally is similar to the so-called molten globule state (Sec. 7.4.2). The molten globule state of α-lactalbumin is stable under certain conditions, whereas that of the homologous protein hen egg-white lysozyme is not stable under any known conditions. Yet both proteins during refolding rapidly adopt similar prefolded conformations that are like the stable molten globule state of α-lactalbumin. The nature of the prefolded state is being investigated very actively, using its spectral properties, its susceptibility to proteases, protection from exchange of its labile hydrogens, and the effects of mutagenesis.

References

Evidence for identity between the equilibrium unfolding intermediate and a transient folding intermediate: a comparative study of the folding reactions of α-lactalbumin and lysozyme. M. Ikeguchi et al. *Biochemistry* 25:6965–6972 (1986).

Use of a trypsin-pulse method to study the refolding pathway of ribonuclease. K. Lang and F. X. Schmid. *Eur. J. Biochem.* 159:275–281 (1986).

A hydrophobic cluster forms early in the folding of dihydrofolate reductase. E. P. Garvey et al. *Proteins: Struct. Funct. Genet.* 6:259–266 (1989).

Evidence for a molten globule state as a general intermediate in protein folding. O. B. Ptitsyn et al. *FEBS Letters* 262:2024 (1990).

Transient folding intermediates characterized by protein engineering. A Matouschek et al. *Nature* 346:440–445 (1990).

Characterizing protein folding intermediates. R. L. Baldwin and H. Röder. *Curr. Biol.* 1:218–220 (1991).

d. Conformational Equilibria in Polypeptide Fragments

Other information pertinent to understanding the prefolded state comes from protein fragments. Any nonrandom conformation in such fragments is also likely to be present in the prefolded protein unless the other parts of the intact protein actively interfere. The occurrence of nonrandom conformation in protein fragments has become widely recognized only recently. Previously, short peptides were thought always to be totally unstructured in water, but it is now recognized that α-helices have marginal stability in many cases (Sec. 5.3.1). Reverse turns have also been claimed to be populated to some extent in short peptides. Proteins missing only a few residues from one end of the polypeptide chain were thought to approximate random coils. Although proteins lacking residues from the C-terminus are often unfolded, at least some are far from being random coils. The physical basis for the marginal stability of these conformations is not yet understood.

From its average properties a polypeptide chain might appear to be a random coil simply because no conformation is populated by a substantial fraction of the molecules. Consequently, a seemingly random-coil polypeptide could have a conformation that is populated by, say, 10^{-2} of its molecules. In a truly random-coil polypeptide this value would be expected to be much less, perhaps 10^{-n}, where n is the number of amino acid residues. The actual value is important for understanding the energetics of polypeptide conformation. Such conformational equilibria are usually expressed by K_{conf}, the equilibrium constant between the particular folded conformation of interest F and all other conformations U:

$$U \xrightleftharpoons{K_{conf}} F$$

$$K_{conf} = \frac{[F]}{[U]} \qquad (7.22)$$

Relatively small values of K_{conf} can be measured immunochemically, using the conformational specificity of antibodies against proteins (Sec. 8.3.1.b). Antibodies generated against a native protein are specific for the native conformation, at least for the unique configuration of groups on its surface that is recognized by the antibodies. Antibodies directed against unfolded proteins are quite different and recognize many linear segments of the polypeptide chain in many different conformations. The cross-reaction between the two sets of antibodies can be used to measure the probability that an unfolded polypeptide is in the nativelike conformation and therefore binds to antibodies against that conformation, or that a folded protein is unfolded transiently and binds to antibodies against the unfolded protein.

Consider a polypeptide chain that possesses all the parts recognized by antibodies against the folded protein N but that is in this conformation only infrequently and in equilibrium with other conformations, U. A reasonable assumption is that when this polypeptide is in the N-like conformation, it binds to the anti-N antibodies with the same affinity K_N as the folded protein:

$$U \underset{}{\overset{K_{conf}}{\rightleftharpoons}} N \underset{Ab}{\overset{K_N}{\rightleftharpoons}} Ab \cdot N$$

$$K_N = \frac{[Ab \cdot N]}{[N][Ab]} \tag{7.23}$$

where Ab is an antibody-combining site. The observed affinity for the polypeptide involved in the conformational equilibrium is

$$K_{app} = \frac{[Ab \cdot N]}{([U] + [N])[Ab]} = \frac{[Ab \cdot N]}{\left(1 + \dfrac{1}{K_{conf}}\right)[N][Ab]}$$

$$= \frac{K_N}{\left(1 + \dfrac{1}{K_{conf}}\right)} \tag{7.24}$$

Therefore, the affinity of the polypeptide for the anti-N antibodies is lower by the factor $[1 + (1/K_{conf})]$. If K_{conf} is very small, this factor becomes $1/K_{conf}$ (Fig. 7.30).

Values of K_{conf} measured with a few unfolded proteins or fragments are in the range of $10^{-3}-10^{-4}$. These values are consistent with the generally unfolded state of these protein fragments but are considerably larger than might be expected for random occurrence of a unique conformation in a fully unfolded polypeptide chain.

References

An immunologic approach to the conformational equilibria of polypeptides. D. H. Sachs et al. *Proc. Natl. Acad. Sci. USA* 69:3790–3794 (1972).

An immunological approach to the conformation equilibrium of staphylococcal nuclease. B. Furie et al. *J. Mol. Biol.* 92:497–506 (1975).

Seeding protein folding. R. L. Baldwin. *Trends Biochem. Sci.* 11:6–9 (1986).

Conformation of peptide fragments of proteins in aqueous solution: implications for initiation of protein folding. P. E. Wright et al. *Biochemistry* 27:7167–7175 (1988).

Residual structure in large fragments of staphylococcal nuclease: effects of amino acid substitutions. D. Shortle and A. K. Meeker. *Biochemistry* 28:936–944 (1989).

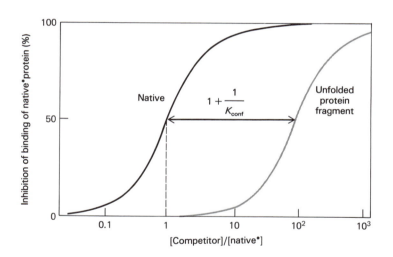

FIGURE 7.30

Immunochemical measurement of the equilibrium between U and N conformations in a protein fragment using antibodies that recognize only the N conformation. A constant amount of radioactive folded protein, designated *native**, is mixed with varying amounts of nonradioactive protein or fragment to be tested for their ability to bind to the antibody and compete with the labeled protein. This mixture is added to a limiting amount of antibodies that recognize the N conformation. Binding of radioactivity to the antibodies is measured. Binding of the unlabeled, competing protein to the antibodies is reflected in an inhibition of binding of radioactive protein. The black curve is that expected when the competitor unlabeled protein is in the folded conformation and indistinguishable from the radioactive protein; an equivalent amount of radioactive and nonradioactive proteins should produce 50% inhibition. The gray curve is that expected for a polypeptide that is in the N conformation only 1% of the time (i.e., $K_{conf} = 10^{-2}$). In general, such a curve is offset to the right by the factor $(1 + 1/K_{conf})$.

e. The Transition State for Folding

The transition state in protein folding is that species along the reaction pathway with the highest free energy, which is encountered in the rate-limiting step (Sec. 5.2.3.a). Transition states cannot be characterized directly but only by measuring the effect on the rates of unfolding and refolding of varying either the conditions or the protein. The rate constant according to transition state theory is inversely related to the relative free energy of the transition state (see Eq. 5.18). The same transition state should be encountered in both directions of the reaction under the same conditions; that is, the ratio of the rate constants for unfolding and refolding should be the same as the measured equilibrium constant. This is usually observed with direct unfolding and refolding reactions of proteins (Fig. 7.28), which indicates that the transition state is a useful concept for a complex reaction like protein folding.

The transition state for protein folding appears to be much closer to the fully folded state than to the unfolded state (Fig. 7.29C). In refolding, very substantial conformational changes often precede the rate-limiting step, whereas in unfolding there is little or no partial unfolding prior to complete unfolding. The transition state for folding appears from the available data to be a distorted high-energy form of the native conformation. The free-energy barrier to unfolding is substantial, presumably as a result of the cooperativity of the fully folded state. Distorting the native conformation raises the free energy of all the interactions stabilizing it (see Fig. 4.14). This high-energy distorted form of the native conformation is also the highest energy barrier overall for refolding. This contrasts with nucleation–rapid growth models of folding, in which the overall transition state would involve a nucleation event in the unfolded conformation (Fig. 7.29A); in this case, the folded conformation would be very flexible, with many partially unfolded forms preceding complete unfolding. High pressure is usually observed to slow folding reactions dramatically, indicating that the transition state has an expanded volume relative to both U and N, which barely differ in volume.

Altering the protein at specific sites and measuring the effect on the relative stability of N and on the rates of unfolding and refolding are currently being used to characterize the folding transition state in greater detail.

References

Characterization of the transition state of lysozyme unfolding. S. I. Segawa and M. Sugihara. *Biopolymers* 23:2473–2488, 2489–2498 (1984).

Low-temperature unfolding of a mutant of phage T4 lysozyme. 2. Kinetic investigations. B. L. Chen et al. *Biochemistry* 28:691–699 (1989).

Mapping the transition state and pathway of protein folding by protein engineering. A. Matouschek et al. *Nature* 340:122–126 (1989).

7.5.4 Folding Pathways

Elucidating the mechanism of protein folding requires characterization of the initial, intermediate, and final conformational states, plus determination of the steps by which they are interconverted. The kinetic roles of the various states can be determined most readily if there is some means of control over the rates and equilibria of the various steps. This control would also make it possible to ensure that unstable intermediates accumulate to substantial levels, at least transiently. Ideally, the unstable intermediates would be trapped in a stable form so that they could be characterized. To control the rates of formation and breakage of hydrogen bonds would be almost ideal because every protein structure includes hydrogen bonds. During folding, protein molecules with 1, 2, 3, . . . intramolecular hydrogen bonds might accumulate kinetically; if they could be trapped and identified, a pathway could be defined in terms of hydrogen bonding. Unfortunately, it is not possible to trap hydrogen bonds, but disulfide bonds can be trapped, due to the reduction–oxidation nature of the covalent disulfide interaction between thiol groups (Sec. 1.6.2.h).

a. Trapping Intermediates with Disulfides

Some proteins that contain disulfide bonds in their folded conformations require these disulfides for stability of their folded conformations. In this case, the reduced protein is unfolded, even in the absence of denaturants, and folding and disulfide bond formation are coupled. Protein species with different numbers of disulfide bonds that accumulate during unfolding and refolding can be trapped and separated (Fig. 7.31) and their disulfide bonds identified (Fig. 1.14).

The kinetic roles of the intermediates can often be determined unambiguously due to the ability to control the kinetics and thermodynamics of the disulfide interaction (Sec. 1.6.2.h). Under appropriate conditions, the disulfide interaction can be very dynamic, with disulfides being formed, broken, and rearranged on time scales as short as 10^{-5} s. The rates of the intermolecular steps in disulfide formation reflect the protein conformational transitions involved. The approach is useful

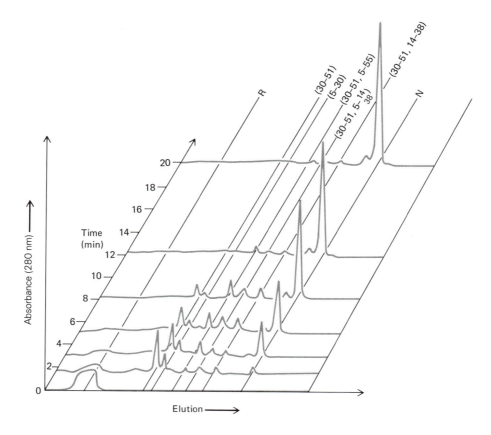

FIGURE 7.31

Isolation of intermediates trapped during refolding of reduced BPTI. Refolding and disulfide formation in the reduced protein were initiated by addition of glutathione disulfide. The process was stopped at the indicated times by addition of 0.1 M iodoacetate, which rapidly reacts with all thiol groups, converting free Cys residues to acidic carboxymethyl-Cys residues. Protein molecules with different numbers of disulfide bonds then differ in their net charges and are readily separated by ion-exchange chromatography. BPTI molecules are eluted roughly in order of their content of disulfide bonds, although molecules with the same number, but different pairings, of disulfides are also resolved because the charge distribution on the protein molecule is also important in its binding to the resin. Fully reduced BPTI, R, was present initially and has largely disappeared after 1.5 min, when one-disulfide intermediates are near their maximum levels. Two-disulfide intermediates accumulate more slowly, reaching their maximum levels after about 5 min. The major intermediates are identified by the disulfides they contain. The peak labeled N eventually predominates; it contains fully refolded BPTI plus a quasi-native species lacking the 30–51 disulfide bond. The small peak preceding N contains incorrectly folded molecules, with nonnative disulfide bonds. (Adapted from T. E. Creighton, *Methods Enzymol.* 107:305–329, 1984.)

only with proteins that unfold when their disulfides are broken; unfolding and refolding of the protein consequently can be controlled by varying just the intrinsic disulfide stability. There is no need to use denaturants, and the strengths of all other types of interactions that stabilize proteins are not affected.

Although only the disulfide bonds are trapped, the conformations that direct disulfide bond formation are effectively trapped also. The stabilities of protein disulfides and of the conformations that specify them are linked functions. It is thus a thermodynamic requirement that whatever conformation stabilized a particular disulfide bond must be stabilized to the same extent by the presence of that disulfide (Fig. 7.32). Therefore, the conformational basis of folding should be evident from the conformations of the trapped intermediates as long

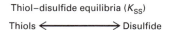

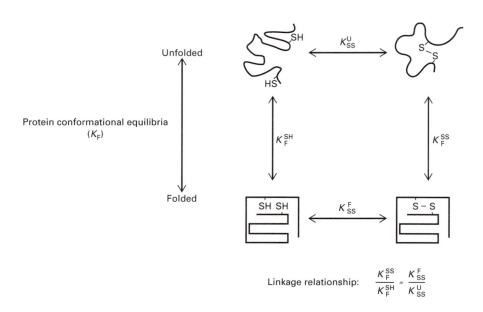

FIGURE 7.32

An example of linked functions, in this case of protein disulfide bond stability and conformational stability. A protein with two Cys residues that can form a disulfide bond is illustrated in the unfolded and folded conformations. The folded conformation holds the two Cys residues in proximity to form a disulfide bond. The indicated equilibrium constants represent the stabilities of the disulfide bonds (K_{SS}) and of the folded conformation (K_F). The linkage relationship results from the general requirement that the free-energy change around any cycle be zero. In this case, it states that whatever effect the folded conformation has on the stability of the disulfide bond, the disulfide bond must have the same effect on the stability of the folded conformation. Any two phenomena that affect each other can be represented in this way, and they are subject to the same linkage relationship. In particular, comparable linkage relationships pertain to all interactions in the folded conformation, not just disulfide bonds. (From T. E. Creighton, *Biochem. J.* 270:1–16, 1990.)

as the conformations are not affected by the trapping procedure.

b. Disulfide Folding Pathway of BPTI

The most detailed and informative folding pathway elucidated thus far is that of bovine pancreatic trypsin inhibitor (BPTI) (Fig. 7.33), which appears to have most of the properties observed for the folding of proteins not involving disulfides.

The three disulfide bonds of native BPTI (between Cys 5–55, 14–38, and 30–51) are required for stability of the folded conformation. The fully reduced protein is unfolded, even under physiological conditions, and nativelike elements of conformation are barely detectable. Whatever the predominant conformations of the re-

duced protein, they do not lead to folding. Even under the most productive folding conditions, formation of the first disulfide bond involves all six Cys residues in at least approximately random pairings. The rate is close to that expected for a random coil and is almost the same as in the presence of 8 M urea, where random one-disulfide intermediates are generated.

In contrast, the one-disulfide intermediates that actually accumulate under folding conditions are far from random. The intermediate with the nativelike disulfide between Cys 30 and Cys 51 [designated as (30–51)] usually accounts for 60% of the one-disulfide molecules, whereas the other 14 possible disulfides comprise roughly equal fractions of the remaining 40%. Whatever disulfide is formed initially is rapidly rearranged intramolecularly by thiol–disulfide interchange. The

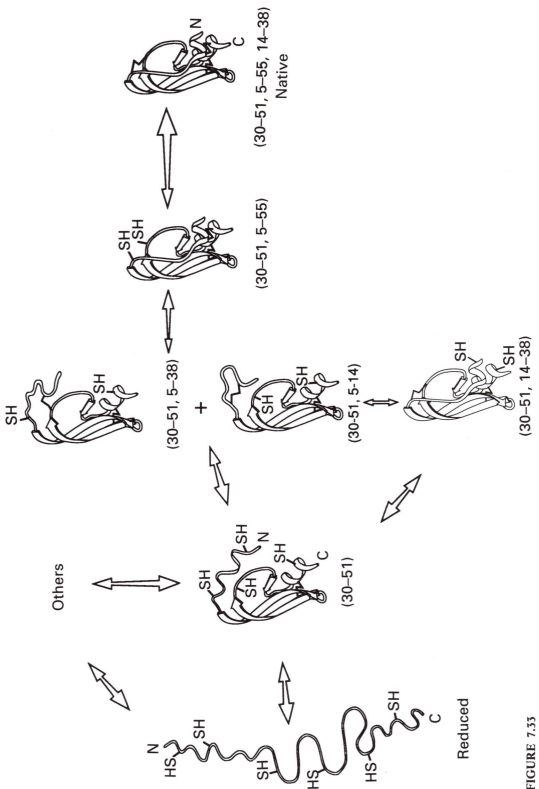

FIGURE 7.33

The disulfide folding pathway of BPTI. The polypeptide backbone of the protein is depicted by a smooth open line when its conformation is not regular or well defined, by arrows for β-strand conformation, and by a coil for α-helix. The approximate conformations of the intermediates are indicated. The positions of the six Cys residues are depicted, and the intermediates are designated by the disulfides (solid crosslinks) they contain. The relative rates of the intramolecular steps at alkaline pH are indicated semiquantitatively by the thickness of the appropriate arrowhead; the wider the arrowhead, the greater the rate in that direction. The fully reduced protein R is unfolded under the normal conditions of folding. Consequently, formation of the initial disulfide bonds is nearly random. The (30–51) one-disulfide intermediate predominates. It is in rapid equilibrium with the other one-disulfide intermediates, which seem to be largely unfolded. Three different second disulfides, 14–38, 5–14, and 5–38, are formed readily in intermediate (30–51), 10^4 times more rapidly than disulfide 5–55. These three second disulfides are rearranged intramolecularly to the nativelike intermediate (30–51, 5–55), which readily forms the 14–38 disulfide bond to complete disulfide formation and refolding. The + between intermediates (30–51, 5–14) and (30–51, 5–38) indicates that they have comparable kinetic roles. A quasi-native (5–55, 14–38) intermediate that is formed directly from minor intermediate (5–55) is not included in this diagram because it is not on the productive pathway. Unfolding and disulfide breakage occur by the reverse of this pathway. (Adapted from T. E. Creighton, *Biochem. J.* 270:1–16, 1990.)

intermediates that accumulate reflect the equilibrium mixture of one-disulfide species, and those that accumulate to the greatest extent are those with the lowest free energies. This is a result of their favorable conformational properties, which in turn depend on both the folding conditions and the sequence of the protein.

The preferential accumulation of the most stable intermediate (30–51) has important kinetic consequences because the productive pathway for refolding leads from this intermediate, and all further productive intermediates retain the 30–51 disulfide bond. The rapid equilibration of the one-disulfide species demonstrates how unfolded proteins can equilibrate rapidly prior to the rate-limiting step in refolding, how all unfolded molecules can follow the same pathway, and why the initial state of the unfolded protein is not important in determining the rate or pathway of folding.

The major α-helix and much of the β-sheet of the fully folded conformation appear to be present and moderately stable in intermediate (30–51). The interaction between these elements of secondary structure involves much of the hydrophobic interior of the protein. This nonrandom conformation is not populated substantially in the absence of the disulfide bond and is probably present in reduced BPTI in no more than 0.1% of the molecules. The predominant stability of (30–51) relative to the other one-disulfide intermediates is not a result of the preexistence of its favorable conformation in the reduced protein but is due to its reciprocal stabilization by the disulfide bond (Fig. 7.32).

Of the three disulfides in native BPTI, the only one well populated at the single-disulfide stage is 30–51, even though 5–55 is more stable in the fully folded conformation. This demonstrates that the most stable parts of a fully folded protein are not necessarily those that are initially formed in folding. The corresponding one-disulfide intermediate, (5–55), is present as only about 3% of the one-disulfide intermediates under normal folding conditions. The other native disulfide, 14–38, is considerably less stable in native BPTI and is not present at a detectable level in the one-disulfide intermediates.

The presence of nativelike β-sheet and α-helix in intermediate (30–51), with the remainder of the polypeptide chain disordered, is consistent with the tendency of the second disulfide bond to be formed between Cys residues 5, 14, and 38 at a total rate similar to that of forming the first disulfide. That the disulfide bond between Cys 14 and 38 is formed at such a low rate indicates that the nativelike conformation in (30–51) does not extend to both Cys 14 and Cys 38; otherwise, this disulfide should be formed at least 200-fold more rapidly. Cys 55 of intermediate (30–51) does not readily form a disulfide with any of the other three Cys residues.

It is often claimed that this could be due to inaccessibility of the Cys 55 thiol group, but this thiol is observed experimentally to be accessible and normally reactive. Formation of a disulfide between Cys 5 and 55 would produce the nativelike two-disulfide species (30–51, 5–55), and this step probably is so slow because it involves traversing the high-energy barrier that separates the more unfolded species from the native conformation. Probably for the same reason, intermediate (30–51, 14–38) also does not readily complete refolding by forming the 5–55 disulfide directly.

The rate-limiting step in BPTI refolding occurs just before reaching the stable native conformation. The energetically most favorable pathway into the native conformation of BPTI is by intramolecular disulfide rearrangement of the nonnative second disulfides of intermediates (30–51, 5–14) and (30–51, 5–38) to that of the nativelike (30–51, 5–55). The two remaining free Cys residues are held in proximity by the native conformation of (30–51, 5–55) and are on the surface of the molecule, where conformational distortions in forming disulfide bonds are not so energetically unfavorable. Consequently, the 14–38 disulfide bond is now formed rapidly.

Unfolding and disulfide breakage of these proteins occur by the reverse of this process, upon merely destabilizing the disulfide interaction. The height of the free-energy barrier to unfolding is observed to be inversely proportional to the stability of the folded state.

Similar energetics of disulfide bond formation and breakage have been observed in other proteins that have been examined in this way. In general, forming disulfide bonds that produce the stable native conformation in which the disulfide bond being formed will be buried is energetically unfavorable because the transition state involves a distorted form of the native conformation.

The disulfide intermediates of BPTI are less stable than either the fully reduced or fully folded states under all known conditions and have less stable folded conformations than N, so the BPTI disulfide folding transition demonstrates the usual cooperativity of folding. All but the rate-limiting steps are readily reversible, so the initial one- and two-disulfide intermediates rapidly equilibrate with the fully reduced protein prior to refolding and are collectively the equivalent of the prefolded state (Sec. 7.5.3.c). Intermediate (30–51, 5–55) rapidly equilibrates with native protein prior to complete unfolding but is much less stable. This intermediate barely differs from the native conformation, which typifies the limits to conformational flexibility of the folded state.

That the energetically most favorable pathway into and out of the native conformation of BPTI is by these disulfide rearrangements is believed to reflect the exceptionally high stability of the native conformation of BPTI; distorting it is also exceptionally difficult. The disulfide rearrangement pathway is not most favorable energetically with less stable homologues of BPTI, which may be more typical of small proteins. With these less stable proteins, formation of the 5–55 disulfide in intermediate (30–51) is the energetically preferred rate-limiting step in unfolding and refolding, although it is still slow. Consequently, the disulfides of these less stable proteins are formed in a seemingly simple, sequential manner, first 30–51, then 5–55, and finally 14–38. The slowness of the rate-limiting step indicates that forming the 5–55 disulfide probably still involves going through a distortion of the nativelike conformation. This might be analogous to the disulfide rearrangement pathway, although conformationally on a smaller scale, but in this case there are no Cys residues in appropriate positions to detect these conformational distortions by disulfides. The disulfide rearrangements of BPTI demonstrate vividly, if to a somewhat exaggerated extent, the importance of the high free-energy barrier of the distorted nativelike conformation.

References

The problem of how and why proteins adopt folded conformations. T. E. Creighton. *J. Phys. Chem.* 89:2452–2459 (1985).

Conformations of intermediates in the folding of the pancreatic trypsin inhibitor. D. J. States et al. *J. Mol. Biol.* 195:731–739 (1987).

Disulphide bonds and protein stability. T. E. Creighton. *BioEssays* 8:57–63 (1988).

A peptide model of a protein folding intermediate. T. G. Oas and P. S. Kim. *Nature* 336:42–48 (1988).

Mutational analysis of a protein-folding pathway. D. P. Goldenberg et al. *Nature* 338:127–132 (1989).

The kinetic roles and conformational properties of the nonnative two-disulphide intermediates in the refolding of bovine pancreatic trypsin inhibitor. N. J. Darby et al. *J. Mol. Biol.* 224:905–911 (1992).

7.5.5 *Folding of Large Proteins*

Large proteins are composed of multiple structural domains, multiple subunits, or both (Sec. 6.2.6). Individual domains can often be excised proteolytically from a protein, or the corresponding fragment can be produced by protein engineering. In many cases, the isolated domains are as stable as when they are in the intact protein, and they are independent structural units in the intact protein (Fig. 7.34). The independent domains un-

fold and refold like single-domain proteins, which can lead to complex unfolding curves for a protein when its domains unfold under different conditions. There can also be varying degrees of interaction between the domains when they are part of the same molecule. If these interactions are mutually stabilizing, the isolated domains are correspondingly less stable. In extreme cases, domains can be so interdependent as to become a single cooperative unit.

Where the isolated domains are stable, folding of an intact multidomain protein appears to occur by folding of individual domains, followed by their association. Somewhat surprisingly, the domains linked by a polypeptide chain often fold more slowly than when they are isolated. The various segments of a polypeptide chain seem to interfere with the folding of each other. Association of the folded chains is often the slowest step in the overall folding process, either because the domains are not folded entirely correctly or because the small adjustments required for their interaction are energetically unfavorable. It appears to be a general observation that altering the conformation of a protein becomes energetically more difficult the closer it approaches the fully folded conformation.

When association of folded domains is slow, an intermediate state accumulates during folding in which the individual domains are folded but unpaired. These domains apparently can interact with the complementary domains of other molecules, which often leads to indefinite aggregation of the protein and to its precipitation. For this reason, productive folding of large proteins usually must be carried out at very low protein concentrations.

The folding of oligomeric proteins is subject to similar considerations because their subunits often consist of multiple domains. With oligomers, however, specific interactions between molecules are necessary. The monomers generally fold to nearly their final conformations before any association steps occur; specific association presumably requires a folded conformation to provide the interaction site. Nevertheless, further folding generally occurs after association. The rate-limiting step in regenerating a native oligomeric protein can be either intramolecular folding (of a subunit or a higher order subassembly) or association of two particles (again, a monomer or a subassembly). Which is rate-limiting often depends on the protein concentration.

No scheme for folding and assembly is general to all oligomeric proteins. This should not be surprising in view of the many different quaternary structures that are encountered in proteins (Fig. 6.24). But even homologous proteins with essentially the same quaternary structure can use apparently different assembly mechanisms.

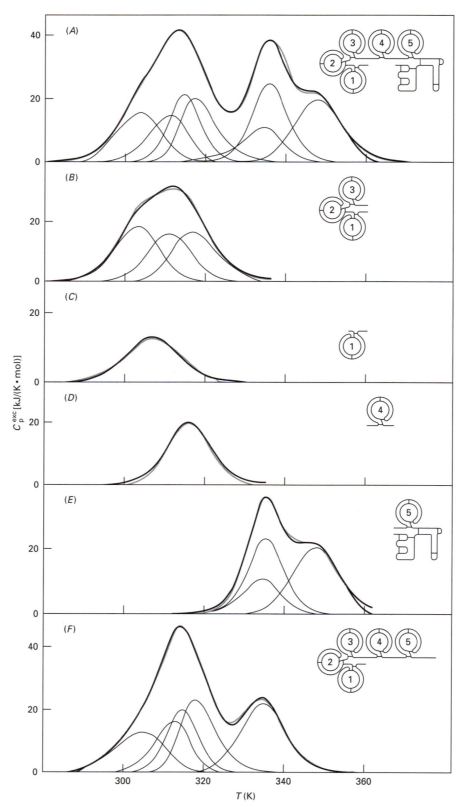

FIGURE 7.34

Calorimetric measurement of thermal unfolding of human plasminogen, which consists of 790 amino acid residues and at least six independent domains. The three-dimensional structure of this protein is not known, but the presence of six domains is apparent from the pri-

Association of folded monomers would seem to be a straightforward process, but it is often observed to be relatively slow, with association constants of only 10^3–10^5 $M^{-1}s^{-1}$, compared with diffusion-controlled rate constants of 10^8 $M^{-1}s^{-1}$ (Sec. 8.2.3). The final adjustments of the structure upon association seem to involve a significant energetic barrier. A somewhat different situation must occur for oligomeric proteins when the subunits are not separate structural units, such as the *trp* repressor dimer, in which the two monomers are intimately entwined (Fig. 6.24C). In this case, the rates of both unfolding and refolding are concentration dependent, indicating that folding and dimerization occur together. The details of how partially folded monomers recognize each other are not yet known.

With folded domains and subunits preceding the rate-limiting step in overall folding, ligands can affect the rate of refolding of multidomain and multisubunit proteins. Binding of a ligand specifically to a folded protein invariably stabilizes that conformation for simple reasons of mass action (Sec. 8.2.1). If a ligand binds to a folded domain or to a subunit in an intermediate in folding, that intermediate is stabilized, and this stabilization can alter the rate of refolding. This situation contrasts with the folding of single-domain proteins, in which the presence of the ligand does not usually alter the rate of refolding, only the rate of unfolding, because partially folded domains generally do not bind ligands specifically.

References

Stability of proteins. Proteins which do not present a single cooperative system. P. L. Privalov. *Adv. Protein Chem.* 35:1–104 (1982).

Folding and association of proteins. R. Jaenicke. *Prog. Biophys. Mol. Biol.* 49:117–237 (1987).

A simple model for proteins with interacting domains. Applications to scanning calorimetry data. J. F. Brandts et al. *Biochemistry* 28:8588–8596 (1989).

Alternate succession of steps can lead to the folding of a multidomain oligomeric protein. A. Murry-Brelier and M. E. Goldberg. *Proteins: Struct. Funct. Genet.* 6:395–404 (1989).

Folding and stability of *trp* aporepressor from *Escherichia coli*. M. S. Gittelman and C. R. Matthews. *Biochemistry* 29:7011–7020 (1990).

7.5.6 Biosynthetic Folding

Proteins are synthesized on ribosomes in vivo as linear polypeptide chains (Chap. 2), but they rapidly fold to their final conformations either during or after biosynthesis. Very little is known directly about how or when this happens in vivo. A polypeptide chain is assembled unidirectionally starting from the amino end, so it is attractive to envisage the initial amino-terminal fragment folding into a nucleus of fixed conformation, which then directs folding of the remainder of the polypeptide chain as it is assembled. Any such asymmetric folding cannot be a general kinetic requirement, however, because most intact, complete polypeptides can refold readily. It is also unlikely that an incomplete nascent polypeptide adopts a stable conformation because isolated fragments of such domains are generally observed to be disordered (see Sec. 7.4.4.c). Removal of just a few carboxyl-terminal residues, the last to be added to a nascent chain, is often sufficient to produce unfolding of small single-domain proteins. Only complete domains generally adopt stable folded conformations.

With multidomain proteins, there are a number of indications that each individual domain folds as it is assembled, without waiting for completion of the polypeptide chain. Folding in vitro of the intact polypeptide chain cannot, therefore, mimic the in vivo folding. This may be one of the reasons that the in vitro refolding of large proteins is generally inefficient, usually because the protein aggregates instead. In vitro, the domains are

mary structure and the disulfide bonds, as indicated schematically on the figure. Starting from the N-terminus, there are five homologous regions, labeled 1–5, which are known as *kringle* domains from their disulfide bond pattern. Kringle domains are also found in other proteins (Fig. 3.19). At the other end of the polypeptide chain is a serine protease domain that appears to unfold as two structural units. The polypeptide chain can be cleaved into various fragments containing different combinations of the domains. *A* shows the calorimetric trace of unfolding of the entire protein and its deconvolution into the contributions of the seven structural domains. The sum of these seven contributions virtually coincides with the experimentally determined curve. *B, C, D,* and *F* show the curves measured for fragments containing, respectively, kringle domains 1–3, 1, 4, and 1–5. *E* shows the results for a fragment with the fifth kringle domain plus the serine protease domain. The first four kringle domains unfold independently, but the stabilities of the fifth kringle and of the serine protease domain are mutually dependent. (From V. V. Novokhatny et al., *J. Mol. Biol.* 179:215–232, 1984.)

folding at the same time and often are observed to interfere with the folding of each other. The N-terminal domains may never fold in vivo in the presence of other domains from the carboxyl end of the polypeptide chain.

Oligomeric proteins often have the problem of subunits not being stable and needing to associate with another polypeptide chain to complete folding. Such partially folded monomers frequently precipitate in vitro. This complication is minimized in vivo by the presence of **molecular chaperones,** which bind transiently to certain incomplete proteins and prevent them from engaging in illicit associations until their proper partners present themselves. Examples are GroE from *E. coli* and related proteins from mitochondria and chloroplasts (also known as Hsp 60), plus the protein BiP (also known as Hsp 70) from the endoplasmic reticulum. There are indications that these proteins participate in an ordered sequence of events to facilitate the folding of a newly synthesized polypeptide chain.

Release of the protein from a chaperone is often energy dependent, requiring ATP hydrolysis. If the released protein is still unfolded, it tends to bind again. Multiple cycles of binding and release occur until the protein is properly folded and assembled, when it no longer binds to the chaperone. The few chaperones tend to assist in the folding and assembly of numerous different proteins, so they seem to bind unfolded proteins fairly nonspecifically. There are no indications that they specifically catalyze correct folding of the bound proteins. Instead, they seem to act simply by sequestering the unfolded protein and keeping it from aggregating nonspecifically. Binding by the chaperone may also have an "editing" function if proteins that cannot fold readily are retained by the chaperone and, perhaps, targeted for degradation.

Other in vivo factors tend to prevent completion of folding. This can be important when the polypeptide chain is to be translocated across a membrane, which appears to be impossible when the protein is folded. Special steps are incorporated to inhibit the folding of such proteins. In the case of proteins translocated into the endoplasmic reticulum (ER), assembly of the bulk of the polypeptide chain is interrupted by binding of the nascent chain to SRP until the ribosome and nascent chain are docked onto the ER membrane. The growing unfolded chain can then be extruded directly across the membrane (see Fig. 2.7). Proteins destined for other organelles, such as mitochondria, are synthesized on free ribosomes and are released into the cytoplasm. They are prevented from folding completely, however, and bind reversibly to certain proteins. The binding is reversible, so the unfolded protein is probably released when it encounters the site where it is to be translocated through a membrane.

Cis-trans isomerization of peptide bonds preceding Pro residues is an intrinsically slow process that often limits the rate of protein folding in vitro (Sec. 7.5.3.a). It is not surprising, therefore, that proteins capable of catalyzing such *cis-trans* isomerization have been detected and are known as **prolyl peptide isomerase.** Nevertheless, it was surprising that the two classes of prolyl isomerase are both proteins that had been identified as the receptors for two immunosuppressive drugs, cyclosporin A and FK506. The relationship between these functions is being elucidated.

Disulfide bond formation between Cys residues has been studied extensively in vitro (see Sec. 7.5.4), but how it occurs naturally after biosynthesis is not known. Not even the nature of the oxidizing agent is known. Disulfide bond formation may occur by thiol–disulfide exchange with small-molecule disulfides, of which glutathione is the most prevalent in most cells (Sec. 2.4.10). Most glutathione is present intracellularly in the reduced thiol form, rather than as the disulfide. Other such disulfides in the same environment are likely to be subject to the same redox equilibrium because chemical thiol–disulfide exchange is rapid, and enzymes are also present to catalyze such reactions. Consequently, the predominance of the reduced form of glutathione in the cell cytoplasm is often considered to indicate that the inside of a cell is too reducing to permit protein disulfide formation. This is not the case, however, when the protein conformation helps to stabilize the disulfide by bringing two Cys residues into proximity. The proteins that have been studied fold and form disulfides in vitro under thiol redox conditions like those in vivo. On the other hand, where the protein conformation keeps Cys residues apart, the in vivo redox conditions are such that these Cys residues should remain largely in the thiol form.

Relatively few intracellular proteins are known to form disulfides; most proteins that form disulfides are normally extracellular. This state of affairs is not unreasonable because a disulfide bond provides more stability to a folded conformation in a more oxidizing environment. Proteins with disulfides are usually secreted during biosynthesis; disulfide formation, therefore, occurs outside the cytoplasm, probably within the endoplasmic reticulum (see Fig. 2.9).

Rearrangement of disulfide bonds is often important and rate-limiting during folding of such proteins in vitro (Sec. 7.5.4). Not surprisingly, in all cells examined there is an enzyme, **protein disulfide isomerase,** that is capable of catalyzing such rearrangements. When isolated, this enzyme increases the rate of all steps that involve both disulfide interchange and substantial conformational alterations of the protein. The protein is localized within the endoplasmic reticulum, which is the logical place for it to be to function in vivo.

The protein disulfide isomerase is a dimer of identical monomers. Every monomer contains two homologous domains that each contain a pair of Cys residues involved in interchange of disulfides and that can form a disulfide bond reversibly. This disulfide bond can be formed and reduced by another enzyme, thioredoxin reductase, using the cofactors NADP and NADPH. It is possible, therefore, that the isomerase can also catalyze net disulfide bond formation or reduction, in addition to catalyzing disulfide interchange. Intriguingly, the isomerase appears to serve additional functions as a subunit of several protein complexes. The best established example is of prolyl-4-hydroxylase, the enzyme that hydroxylates the Pro residues of collagen (Sec. 5.5.3). There may be multifunctional ordered assemblies of the enzyme in the endoplasmic reticulum that are involved in several posttranslational modifications.

Disulfide bonds are stable indefinitely under the appropriate conditions. They may, therefore, stabilize a folded conformation that is no longer thermodynamically stable, perhaps owing to covalent modification of the protein after biosynthesis and folding (Fig. 7.25). For example, many hormones and enzymes, such as insulin and chymotrypsin, are synthesized initially as inactive precursors that fold and form disulfide bonds and are then activated by proteolytic cleavage (Sec. 2.4.1). The active forms of these proteins are stable under most conditions unless disulfide interchange is permitted. These proteins then rearrange their disulfides, unfold, and become inactive. In these particular instances, the original disulfides are stabilizing an inherently unstable folded conformation.

Additional complexities are introduced by the many instances of covalent modifications of the polypeptide chain after its biosynthesis (Chap. 2). A few of these, such as glycosylation, probably occur on the nascent chain before folding and might even be required for folding to occur. Most modifications, however, occur only after folding, and the folded conformation is usually important for the specificity of the modifications. In some cases, the covalent modification of the folded protein prevents refolding of the polypeptide chain, even if the native conformation is stable. Why refolding does not occur in such cases remains to be determined.

References

Molecular chaperones: proteins essential for the biogenesis of some macromolecular structures. R. J. Ellis and S. M. Hemmingsen. *Trends Biochem. Sci.* 14:339–342 (1989).
Protein disulfide isomerase: multiple roles in the modification of nascent secretory proteins. R. B. Freedman. *Cell* 57:1069–1072 (1989).

Protein oligomerization in the endoplasmic reticulum. S. M. Hurtley and A. Helenius. *Ann. Rev. Cell Biol.* 5:277–307 (1989).

Physiological role during export for the retardation of folding by the leader peptide of maltose binding protein. G. Liu et al. *Proc. Natl. Acad. Sci. USA* 86:9213–9217 (1989).

Protein sorting to mitochondria: evolutionary conservation of folding and assembly. F. U. Hartl and W. Neupert. *Science* 247:930–938 (1990).

Molecular chaperones. R. J. Ellis, ed. *Seminars Cell Biol.* 1:1–72 (1990).

Heat shock proteins. M. J. Schlesinger. *J. Biol. Chem.* 265:12111–12114 (1990).

The mechanism of protein folding. Implications of in vitro refolding models for *de novo* protein folding and translocation in the cell. G. Fischer and F. X. Schmid. *Biochemistry* 29:2205–2212 (1990).

Protein-catalysed protein folding. A. L. Horwich et al. *Trends Biotechnol.* 8:126–131 (1990).

Protein disulfide-isomerase is a substrate for thioredoxin reductase and has thioredoxin-like activity. J. Lundström and A. Holmgren. *J. Biol. Chem.* 265:9114–9120 (1990).

Chaperonin-mediated protein folding at the surface of groEL through a "molten-globule"-like intermediate. J. Martin et al. *Nature* 352:36–42 (1991).

Protein folding in the cell. M. J. Gething and J. Sambrook. *Nature* 355:33–45 (1992).

Successive action of DnaK, DnaJ and GroEL along the pathway of chaperone-mediated protein folding. T. Langer et al. Nature 356:683–689 (1992).

Exercises

1. There have been many suggestions that all the enzymes of a metabolic pathway, such as glycolysis, would ideally exist associated in an ordered structure in which the product of one enzyme could be transferred directly to the next enzyme in the pathway. Some investigators have reported extensive associations of proteins (e.g., F. M. Clarke and C. J. Masters, *Biochim. Biophys. Acta* 381:37–46, 1975) whereas others have found no evidence for such association (e.g., C. DeDuve, in *Structure & Function of Oxidation-Reduction Enzymes*, A. Åkeson and A. Ehrenberg, eds., pp. 715–728. Pergamon, New York, 1972). Find a plausible explanation for the differences in these observations.

ANSWER
A. B. Fulton, *Cell* 30:345–347 (1982).

2. When polyethylene glycol is added to the buffer used for gel filtration, proteins have been observed to elute later, as if they were of smaller size (S. C. B. Yan et al., *Anal.*

Biochem. 138:137–140, 1984; C. L. DeLigny et al. *J. Chromatography* 194:223–233, 1984). How can this be explained?

ANSWER
T. Arakawa, *Anal. Biochem.* 144:267–268 (1985).

3. There are many recipes in the literature for estimating the molecular weight M_W of a protein from its sedimentation coefficient *s*. Usually they are of the form

$$M_W = (6.7 \times 10^3)s^{3/2}$$

where the *s* value is expressed in Svedberg units. This equation can be derived from Equations (5.28) and (7.5); what assumptions are necessary?

   You are studying an enzyme composed of two different proteins designated A and B, which are available individually and as a complex of the two. You observe that their sedimentation coefficients are 2.7 S for the A protein, 5.1 S for the B protein, and 6.4 S for the complex (T. E. Creighton and C. Yanofsky, *J. Biol. Chem.* 241:980–990, 1966). What would you conclude from the above recipe about the stoichiometry of the A and B proteins in the complex?

   The A protein is a monomer with a molecular weight of 29,000, and the B protein is reported to be a dimer of two identical polypeptide chains with a molecular weight of 54,000 (D. W. Wilson and I. P. Crawford, *J. Biol. Chem.* 240:4801, 1965). Would that affect your conclusion about the structure of the complex? What experiments would you perform to determine the structure of the complex? What are the implications of the sedimentation properties of the various forms of the protein for the three-dimensional structure of the complex?

ANSWERS
M. Goldberg et al., *J. Mol. Biol.* 21:71–82 (1966); C. C. Hyde et al., *J. Biol. Chem.* 263:17857–17871 (1988).

4. Some bacterial proteins are found to increase the probability that supercooled water will freeze (R. L. Green and G. J. Warren, *Nature* 317:645–648, 1985), whereas many polar fish contain proteins that inhibit ice formation (R. E. Feeney, *Amer. Scient.* 62:712–719, 1974). How might these proteins act to cause such opposite effects?

ANSWER
R. E. Feeney et al., *Ann. Rev. Biophys. Biophys. Chem.* 15:59–78 (1986); P. Wolber and G. Warren, *Trends Biochem. Sci.* 14:179–182 (1989); H. Mizuno, *Proteins: Struct. Funct. Genet.* 5:47–65 (1989); C. L. Hew and D. S. C. Yang, *Eur. J. Biochem.* 203:33–42 (1992).

5. The temperatures at which proteins unfold increase by as much as 40°C when the water is removed, either by drying or by transferring the protein to a nonaqueous solvent (Y. Fujita and Y. Noda, *Intl. J. Peptide Protein Res.* 18:12–17, 1981). What are the implications of this for the roles of the aqueous solvent and of the hydrophobic interaction in stability of protein folded conformations?

ANSWER
T. E. Creighton, *Curr. Opinion Struct. Biol.* 1:5–16 (1991).

6. The protein you are studying appears to unfold reversibly at about 3 *M* urea, with a redshift in the absorbance of the aromatic residues, a decrease in the absolute value of the circular dichroism ellipticities at 222 nm and 280 nm, an increase in the fluorescence intensity, and a shift of the wavelength of maximum emission. After normalizing the data to represent the fraction of total change, all of these parameters give the same simple unfolding transition, except that the wavelength of maximum fluorescence intensity follows a different curve. Is this evidence that the unfolding transition is not two-state?

7. Chemical modifications of charged side chains generally appear to produce relatively small electrostatic effects in folded proteins. For example, changing the net charge of chymotrypsin by 28 units, by succinylation (Sec. 1.3.7) of the 14 Lys residues, altered the pK_a value of His 57 by only 1.0 pH unit (P. Valenzuela and M. L. Bender, *Biochim. Biophys. Acta* 250:538–548, 1971). A slightly smaller effect was obtained by converting 13 carboxyl groups to positively charged groups, by coupling with ethylenediamine. Acetylation of all the Lys residues of trypsin lowered the pK_a of His 57 by only 0.2 pH unit (W. E. Spooner and J. F. Wootton, *Biochim. Biophys. Acta* 235:164–171, 1971). Changing ribonuclease A from a very basic to a very acidic protein, by succinylation of its 11 amino groups, had no significant effect on the stability of its folded conformation (M. Hollecker and T. E. Creighton, *Biochim. Biophys. Acta* 701:395–404, 1982). Why are substantially greater effects observed by mutagenesis in Table 7.4?

ANSWER
S. Dao-pin et al., *J. Mol. Biol.* 221:873–887 (1991).

8. The protein that you are studying normally has a Glu residue that interacts in a complex with a Lys residue of a second protein, A. Another protein, B, differs from A only in having this Lys residue replaced by Glu. Not surprisingly, it interacts much less effectively with your protein. You would like to modify your protein by genetic engineering to interact with B instead of A. What is the likelihood that replacing the Glu residue in your protein with Lys will produce a protein that interacts as effectively with protein A?

ANSWER
J. K. Hwang and A. Warshel, *Nature* 334:270–272 (1988).

What would you expect if you replaced the Glu residue with Arg instead of Lys?

ANSWER

D. B. Wigley et al., *Biochim. Biophys. Res. Commun.* 149:927–929 (1987).

9. Kinetic criteria were established to distinguish between kinetic intermediates that were either on or off the productive pathway of protein folding (A. Ikai and C. Tanford, *Nature* 230:100–102, 1971). What unstated assumption did this analysis make?

ANSWER

P. J. Hagerman, *Biopolymers* 16:731–747 (1977).

10. The refolding of a hypothetical protein depends on the formation of a crucial intermediate state. Once formed, this intermediate completes folding with a constant rate and a constant decrease in free energy. What would be the effect on the rate of refolding of increasing the stability of this intermediate relative to the fully unfolded state? If nature has evolved proteins to fold at the most rapid rate possible, what are the implications for characterizing the productive intermediates in protein folding?

ANSWER

T. E. Creighton, *Biochem. J.* 270:1–16 (1990).

11. Gly and Pro residues affect the conformational entropy of an unfolded polypeptide chain. What effects on stability of the unfolded state would be expected upon replacing Gly and Pro residues with other residues, or upon replacing other residues with Gly or Pro?

ANSWER

B. W. Matthews et al. *Proc. Natl. Acad. Sci. USA* 84:6633–6667 (1987).

12. The following two statements can correctly be made about the roles of disulfide bonds in proteins:

   a. Disulfide bonds are covalent cross-links that "increase the protein's stability mainly by constraining the unfolded states, thereby decreasing their conformational entropy and increasing their free energy" (C. N. Pace, *Trends Biotechnol.* 8:93–98, 1990).

   b. Disulfide bonds are a weak, reversible interaction between Cys residues that "is similar in many ways to hydrogen bonding" (T. E. Creighton, *Biochem. J.* 270:1–16, 1990).

   How can both of these statements be true? What qualifications should be added to each statement?

13. Reduced ribonuclease A, with the four disulfide bonds broken, has been reported to adopt the nativelike conformation some 0.04–6% of the time (J. R. Garel, *J. Mol. Biol.* 118:331–345, 1978; L. G. Chavez and H. A. Scheraga, *Biochemistry.* 19:996–1004, 1980). If any one of the four native disulfide bonds were to be formed, how stable should the nativelike conformation be on the basis of the stabilizing effects of single disulfide bonds (Eq. 5.13)?

ANSWER

T. E. Creighton, *Biophys. Chem.* 31:155–162 (1988).

14. The kinetics of rebinding of carbon monoxide by myoglobin after photodissociation has been taken to indicate that folded proteins exist in a number of distinct microstates. The rebinding process is rapid and heterogeneous in rate at low temperatures (R. H. Austin et al. *Biochemistry* 14:5255–5373, 1975). The nonexponential kinetics of rebinding could indicate that the protein molecules are heterogeneous, existing in different microstate conformations; alternatively, the carbon monoxide might have dissociated to different sites in otherwise homogeneous protein molecules. How could you distinguish between these possibilities?

ANSWER

H. Frauenfelder et al., *Ann. Rev. Biophys. Biophys. Chem.* 17:451–479 (1988).

15. On what time scales might you expect the —NH_2 group of Asn and Gln side chains to be rotating when in various positions in a folded protein? How might you measure this experimentally?

ANSWER

C. L. Perrin, *Acc. Chem. Res.* 22:268–275 (1989); E. Tüchsen and C. Woodward, *Biochemistry* 26:8073–8078 (1987).

16. Temperature-sensitive mutations in bacteria that prevent growth without added histidine at elevated temperatures, as a result of inactivating one of the enzymes required for histidine biosynthesis, were observed to be "corrected" by the addition of neutral salts to the growth medium (T. Kohno and J. Roth, *Biochemistry* 18:1386–1392, 1979). Suggest some possible explanations for how salt might have this effect.

POSSIBLE CLUES

T. K. Van Dyk et al., *Nature* 342:451–453 (1989).

17. What are the likely structural and energetic consequences of deleting by mutagenesis nonpolar side chains that occupy substantial volumes in the interior of a folded protein?

ANSWER

A. E. Eriksson et al., *Science* 255:178–183 (1992).

How might such effects be alleviated in the mutant protein with the cavity?

ANSWER

A. E. Eriksson et al., *Nature* 355:371–373 (1992).

18. The transient kinetic intermediates in the refolding of BPTI with nonnative disulfide bonds (Fig. 7.31) were detected in lower quantities after trapping by acid than after trapping by reaction with iodoacetate (J. S. Weissman and P. S. Kim, *Science* 253:1386–1393, 1991). On this basis, the pathway of Figure 7.33 was revised to omit the nonnative intermediates. Was this warranted by the experimental data?

ANSWER
T. E. Creighton, *BioEssays* 14:195–199 (1992); *Science* 256:111–112 (1992).

19. The presence of nonpolar side chains on the surface of a folded protein usually is thought to have no substantial effect on stability, because the nonpolar side chain will be equally exposed to solvent in the unfolded protein. On the other hand, the stability of folded λ repressor protein was increased by replacing a Tyr residue that was exceptionally exposed to solvent on the surface of the protein by Val, Phe, Leu, Gln, His, Cys, and Asp residues; replacing it by Trp decreased the stability (A. A. Pakula and R. T. Sauer, *Nature* 344:363–364, 1990). What are the implications of these observations for the role of hydrophobicity in stabilizing protein structures?

ANSWER
A. J. Doig et al., *Nature* 348:397 (1990).

20. Gly residues commonly adopt the local left-handed α-helical conformation, while the other 19 amino acids rarely do so (Figs. 5.2 and 5.7). How might the stability of the folded state of a protein be expected to change if non-Gly residues in this conformation are replaced by Gly?

ANSWER
H. Nicholson et al. *J. Mol. Biol.* 210:181–193 (1989).

Interactions with
Other Molecules

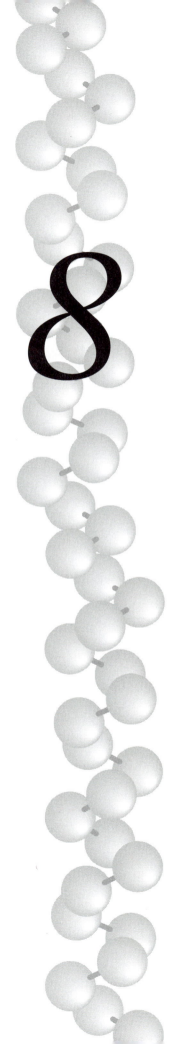

*T*he biological functions of proteins almost invariably depend on their direct, physical interaction with other molecules. The biological importance of this fact should not be underestimated. All organisms and cells survive only because they interact effectively with their environments. Both useful and dangerous molecules must be distinguished and dealt with. Organisms have sophisticated sensory organs for being aware of their environments. At a lower level, all cells have a variety of receptors for this purpose. Virtually every molecule in a cell was first bound by the enzyme that produced it or by the receptor on the cell surface that enabled it to be taken up. Proteins may bind very tightly and specifically to other proteins, generating large complexes; to nucleic acids, especially when controlling their replication and expression; to polysaccharides, especially to those on the surfaces of cell membranes; and to lipids, often becoming incorporated within membranes. Every aspect of the structure, growth, and replication of an organism depends on such interactions. Of crucial importance is the specificity of such interactions. In the crowded interior of a cell, each protein must interact only with the appropriate molecules and not with any of the others that are present, often in extremely high concentrations.

Proteins are generally classified according to the purpose and consequences of their binding; examples are structural proteins, enzymes, repressors, lectins, toxins, immunoglobulins, hormones, receptors, membrane transport proteins, and proteins of motility. The physical principles of the interactions are similar in all these cases and are the subject of this chapter. (Systems involving chemical changes, such as enzyme catalysis, are left to the next chapter). The following discussion focuses on the protein; the molecule that it interacts with, even if it is another protein, is designated the **ligand**. A protein with its ligand bound is known as the **holo** form; without the ligand a protein is in the **apo** form.

This chapter emphasizes specific, reversible interactions and gives little attention to interactions with components of the solvent, such as water molecules and hydrogen ions (for water-soluble proteins) or lipids (for membrane proteins). These interactions are not fundamentally different, but they occur at many sites on the protein and are generally weak, occurring only because the solvent molecules are present in such high concentration. They have been described in Chapter 7.

The interactions described here are distinguished by being specific for the appropriate ligand and occurring at a very limited number of sites on the protein, most often just one site per polypeptide chain.

8.1 Structures of Protein–Ligand Complexes

The structures of complexes of most ligands and proteins can generally be determined in the same way as the structures of the proteins themselves, using the techniques described in Chapter 6. Special approaches are required, however, if the ligand is especially large, such as DNA or a polysaccharide, or especially small, such as an electron. Such atypical ligands are not discussed in this general section but will be treated in a later discussion of particular ligands. The present discussion is confined to typical ligands with dimensions not much greater than those of the protein to which they bind.

One special aspect of experimental studies of the complex of a protein with a ligand is that the two components can be varied individually, in both their concentrations and their structures. The chemical structure of the ligand can be varied, even to the extent of adding groups that react with the protein. Similarly, the covalent structure of the protein can be varied. The two components can differ in isotopic composition; for example, one can have ^2H atoms, the other ^1H, ^{13}C instead of ^{12}C, ^{15}N instead of ^{14}N. This can be very useful with certain techniques, especially NMR.

8.1.1 The Difference Fourier Crystallographic Technique

The most detailed information about the physical nature of the interactions between a protein and its ligands comes from crystallographic determination of the structure of the complex (Fig. 8.1). In some cases, the ligands are prosthetic groups that are integral parts of the protein and normally present in its crystal structure. Other complexes have been prepared by mixing the components and crystallizing them together. This is the only feasible course when the ligand is large, such as another protein molecule or a nucleic acid. In this case, crystals of the complex are necessarily different from those of the protein alone, so the crystal structure of the complex must be determined from first principles. If the structure of either the protein or the ligand is known, is not substantially altered in the complex, and makes up a significant part of the structure of the complex, its position in the crystal of the complex can be determined crystallographically by the method of molecular replacement. The structure of the entire complex can then be determined by refinement procedures.

Many small ligands can be diffused into an existing protein crystal; the substantial volume of solvent in most protein crystals provides channels through which such ligands can diffuse. A crystalline protein is usually able to bind its appropriate ligand, as in solution, unless the binding site is sterically blocked by neighboring molecules in the crystal lattice or unless large conformational rearrangements occur in the protein that destroy the crystal lattice (Sec. 8.4). Diffusing a ligand into a protein crystal of known structure is usually the easiest method of visualizing the structure of the complex. The advantage is that the crystal of the complex is almost certain to be isomorphous with that of the original protein, in which case the amplitudes and phases of each reflection are only slightly different for the two crystals. The amplitudes of all the reflections for the crystal of the complex can be measured directly but not their phases. Nevertheless, an approximate electron density map of the complex can be calculated (Sec. 6.1.4) by using, for each reflection, the measured amplitude for the crystal of the complex and the phase for the crystal of the protein without the ligand. The phases are only approximately correct, but a map calculated in this way usually demonstrates the bound ligand as added electron density. The extra electron density is not at the expected magnitude, owing to the use of somewhat incorrect phases. This initial structure of the complex can be refined to produce a more accurate map, as with protein structures themselves (Sec. 6.1.6).

The usual procedure with such crystallographic data is to calculate a **difference Fourier map**. The Fourier calculation is exactly analogous to that in Equation (6.2) but uses for each reflection the *difference* between the amplitude of the protein crystal and the amplitude of the protein–ligand complex. Again, the phase of the protein crystal is used, and the resulting difference Fourier map gives the difference in electron density between the two crystal structures. Electron density of the added ligand is represented by positive values; negative values mean that something has disappeared or moved

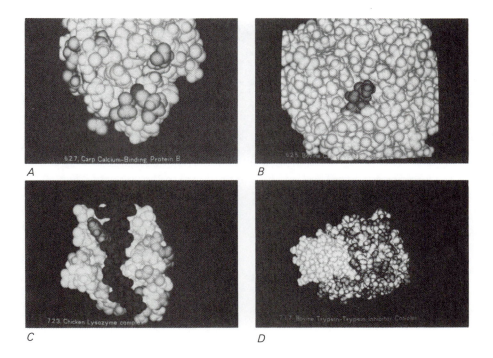

FIGURE 8.1

Examples of some protein – ligand complexes, illustrated as space-filling models viewed from the protein surface. *A:* Ca^{2+} ion (dark) bound to parvalbumin; the Asp and Glu residues of the protein are shaded. *B:* An inhibitor, glycyl tyrosine (dark), bound to carboxypeptidase A; a Zn^{2+} ion, also bound tightly to the protein, is barely visible behind the inhibitor. *C:* A hexasaccharide, (*N*-acetylglucosamine)$_6$ (black), bound to hen lysozyme; Trp, Phe, Arg, Lys, Asn, and Gln residues are shaded. *D:* The complex of bovine pancreatic trypsin inhibitor, BPTI (white), with bovine trypsin (some atoms are shaded). (Adapted from R. J. Feldman and D. H. Bing, *Teaching Aids for Macromolecular Structure,* Division of Computer Research and Technology, National Institute of Health, Bethesda, Md. 1980.)

upon ligand binding. Positive and negative values may compensate, and no feature appear, if the ligand displaces some other ligand, a part of the protein, or solvent molecules.

The great advantage of difference Fourier maps is their sensitivity to small changes that might not be apparent in comparison of the entire structures of the protein and of the complex. If part of the protein moves slightly upon binding the ligand, the difference map shows positive density on the side to which that part of the protein has moved, with comparable negative density on the other side. With a large movement of atoms, there is large negative density in their original positions and corresponding large positive density in their new positions (Fig. 8.2).

The difference Fourier technique is analogous to using isomorphous heavy atoms as ligands to determine the phases of the reflections of the original crystal (Sec. 6.1.3); it differs, however, in that very small ligands can be used. The difference Fourier technique is very sensitive and accurate because many of the systematic errors

in Fourier maps are the same for the crystals with and without the ligand and consequently cancel in the difference between them. Moreover, all of the many reflections from the entire protein asymmetric unit contribute to the structure of a much smaller ligand. The changes in the intensities of individual reflections need not even be apparent to give an accurate and interpretable difference Fourier map. On the other hand, the assumptions made, particularly that the phase of each reflection is unaltered by the presence of the ligand, results in values for the difference peaks of no more than half the intensity expected under ideal conditions, when the protein structure is totally unchanged by the ligand. Nevertheless, ligands as small as O_2 molecules can usually be observed readily using difference Fourier maps.

Reference

The difference Fourier technique in protein crystallography: errors and their treatment. R. C. Henderson and J. K. Moffat. *Acta Cryst.* B27:1414 – 1420 (1971).

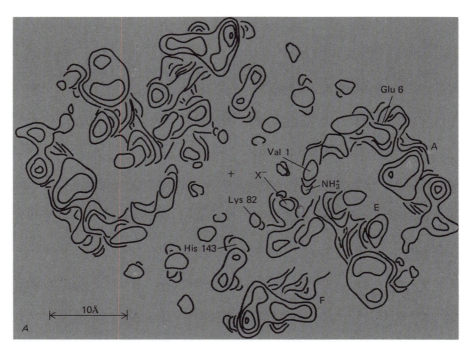

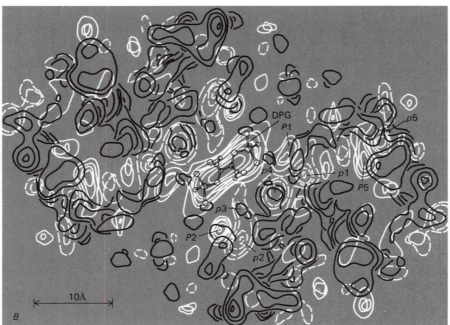

FIGURE 8.2

Difference Fourier map due to binding of diphosphoglycerate (DPG) between the two α-amino groups of the β chains of deoxyhemoglobin. *A:* The electron density map of this portion of native human deoxyhemoglobin; the $^+$ indicates the symmetry axis relating the two β chains. Helices A, E, and F are labeled, as are residues Val 1, Glu 6 (the residue that is changed to Val in sickle-cell hemoglobin, Sec. 3.2.2), Lys 82, and His 143; peak X^- is thought to be a sulfate or phosphate ion. *B:* The difference Fourier map (in white contours) with bound DPG is superimposed on the native map; positive differences are in solid contours, negative in dashed. The large positive peak in the center is due to the bound DPG molecule, and a symmetry-averaged molecule is superimposed. DPG is not a symmetric molecule, but it is bound on a twofold symmetry axis, where it is probably bound in the two

8.1.2 NMR

The entire structure of a protein–ligand complex can be determined by using the NMR methods outlined previously (Sec. 6.3) for determining protein structures in solution, as long as the complex is not too large in size. The structures of the ligand and of the protein are usually also determined separately. Which groups of the two components are in close proximity in the complex can be determined by NOE interactions or by magnetization transfer between their H atoms. Alternatively, when the structure of the ligand or the protein is not altered in the complex, the NMR chemical shifts or the rates of hydrogen exchange with the solvent will change most for those groups in the complex that are in contact with the other component.

Many ligands or proteins are too large for a complete NMR structure determination, but the structure of the smaller component in the complex can be determined by using natural or incorporated isotopes in the smaller and not in the larger component of the complex. Examples of isotopes that can be observed by NMR are ^1H, ^{13}C, ^{15}N, ^{19}F, and ^{31}P. Such "isotope editing" of the NMR spectra can be used to select only those NMR signals that arise from just one component or the other, or only those that arise between the two.

References

NMR studies of interactions of ligands with dihydrofolate reductase. J. Feeney. *Biochem. Pharmacol.* 40:141–152 (1990).

An antibody binding site on cytochrome *c* defined by hydrogen exchange and two-dimensional NMR. Y. Paterson et al. *Science* 249:755–759 (1990).

The NMR structure of cyclosporin A bound to cyclophilin in aqueous solution. C. Weber et al. *Biochemistry* 30:6563–6574 (1991).

NMR studies of [U-^{13}C]cyclosporin A bound to cyclophilin: bound conformation and portions of cyclosporin involved in binding. S. W. Fesik et al. *Biochemistry* 30:6574–6583 (1991).

8.1.3 Chemical Methods of Determining Binding Sites

Chemical methods of determining where ligands are bound to a protein can be very useful. Most of these methods involve covalent modification of the protein, using the chemical reactions described in Chapter 1, followed by determination of which part of the covalent structure has reacted. They directly relate binding only to the primary structure of the protein.

The classical method has been to examine the reactivity of various groups on a protein in the presence and absence of the ligand. A bound ligand should make any groups involved in its binding site inaccessible and should protect them from reaction with an added reagent. In another approach, chemical modification of various groups on the free protein or their replacement by other residues can be correlated with the ability of the modified protein to bind the ligand. If modification or replacement of a certain group produces a large decrease in affinity, that group is likely to be involved directly in binding.

Results obtained by chemical modification can be misleading if the structure of the protein is altered by binding of the ligand or by modification of the protein. A technique that minimizes such limitations is **affinity labeling**, which uses the specificity of the ligand–protein interaction to label preferentially the binding site by incorporating a reactive group into the ligand. The reaction between the ligand and the protein is essentially intramolecular in the complex between the two, when very high effective concentrations are possible with two groups in correct position and proximity (see Table 4.11). Such a reaction should be much more rapid than the usual bimolecular reaction of other groups of the protein with reagent that is not bound and is present in solution at only low concentrations. Unfortunately, the ligand may bind in such a way that no group of the protein capable of reacting is in proximity. Therefore, negative results with this technique are not significant.

The constraint of the requirement for close proximity of an appropriate group on the protein can be

---

possible orientations with equal probability. The other positive and negative peaks indicate slight movements of various parts of the hemoglobin molecule upon DPG binding. The positive and negative peaks $P1$ and $p1$ indicate that the α-amino groups move toward the DPG; the pair of peaks $P2$ and $p2$ show similar movements of the His 143 side chains. Negative peak $p3$ indicates a similar movement of Lys 82, but no compensating positive peak is apparent. Negative peak $p4$ indicates that the inorganic anion X^- has been displaced by the DPG. Other positive and negative peaks such as $P5$ and $p5$ indicate that the A helix moved slightly toward the center of the molecule. (From A. Arnone, *Nature* 237:146–149, 1972.)

relaxed by attaching the reactive group to the ligand by a flexible "arm" that can come into contact with a number of groups on the protein that are within its radius. The price for this greater scope for reaction is that the effective concentrations of the protein and ligand groups are correspondingly lower, so that the reaction with the bound ligand is not so favored.

Bifunctional reagents have two reactive groups. One group can be used to react with the ligand, to generate the affinity label. This is then added to the protein, and the second group is encouraged to react with nearby portions of the protein. The reactivities of the two groups must be controlled so that each reacts only when required. Photoactivated groups are especially useful for the second step because they are reactive only in the presence of light. Also, they are then highly reactive with a variety of groups, making it likely that reaction with the protein will occur. Most commonly used are azide groups, which are totally inert until activated to the nitrene:

$$-N_3 \xrightarrow[N_2]{} -\overset{\cdot\cdot}{\underset{\cdot\cdot}{N}}$$

$$\text{azide} \qquad \text{nitrene} \qquad (8.1)$$

Nitrenes are reactive even toward methyl and methylene ($-CH_2-$) groups.

Bifunctional reagents with the same reactive group at both ends can be used to cross-link the protein and the ligand, which is very useful when the ligand is large or is also a protein, as in complicated macromolecular structures such as ribosomes, viruses, chromatin, and enzyme complexes. Whether or not the components of a complex are in close proximity can often be inferred by whether or not bifunctional reagents will cross-link them. When the components are polypeptide chains, those that have been cross-linked can usually be determined most readily by SDS electrophoresis (Sec. 1.5.3).

The versatility of the cross-linking approach lies in the ability to vary the length of the cross-linking group to serve as a molecular "ruler" for measuring the distances between two molecules in a complex. The nature of the bifunctional reagent can also be varied. For example, membrane proteins that are in proximity at the surface of the membrane can be cross-linked with a polar reagent that is confined to the aqueous solvent; in contrast, nonpolar reagents that permeate the membrane can cross-link protein molecules that interact in the membrane. Cross-linking with bifunctional reagents has been most useful with very large complexes, composed of numerous molecules, too large or too nonsymmetric to be studied crystallographically.

References

Chemical cross-linking: reagents and problems in studies of membrane structure. K. Peters and F. M. Richards. *Ann. Rev. Biochem.* 46:523–551 (1977).

Photoaffinity labelling of biological systems. V. Chowdhry and F. H. Westheimer. *Ann. Rev. Biochem.* 48:293–325 (1979).

Chemical cross-linking in biology. M. Das and C. F. Fox. *Ann. Rev. Biophys. Bioeng.* 8:165–193 (1979).

Mapping of contact areas in protein–nucleic acid and protein–protein complexes by differential chemical modification. H. R. Bosshard. *Methods Biochem. Anal.* 25:273–301 (1979).

Reaction of (bromoacetamido)nucleoside affinity labels with ribonuclease A: evidence for steric control of reaction specificity and alkylation rate. C. F. Hummel et al. *Biochemistry* 26:135–146 (1987).

Site-directed cross-linking. Establishing the dimeric structure of the aspartate receptor of bacterial chemotaxis. D. L. Milligan and D. E. Koshland, Jr. *J. Biol. Chem.* 263:6268–6275 (1988).

8.1.4 General Properties of Ligand Binding Interactions

In most cases, there is a single unique binding site for a particular ligand on each molecule of a polypeptide chain. If the polypeptide chain has internal symmetry, however, as when it has arisen by gene duplication (Sec. 3.4), each of the structural units may have a binding site. For example, gene-duplicated ferredoxins contain two similar iron–sulfur complexes related by the twofold symmetry of the molecule (see Fig. 6.36).

One structural domain can bind more than one ligand at separate binding sites, but it is unusual for there to be more than two or three such binding sites on any one domain. A protein that binds a number of different ligands often binds them on separate domains. These domains are often designated by their binding properties; an example is in Figure 8.3.

There are a few spectacular exceptions to the generalization of one binding site for a particular ligand per protein structural unit: Cytochrome c_3, a single polypeptide chain of only 118 residues, binds four identical heme groups in different environments. Bacteriochlorophyll protein from a green photosynthetic bacterium binds seven chlorophyll molecules, each in a different position within a "string bag" of a 15-stranded β-sheet closed to form a flattened barrel.

The earlier description of protein structures (Chap. 6) noted their tendency to be spherical. Binding sites for

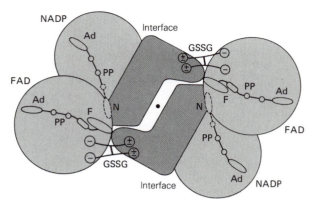

FIGURE 8.3

Schematic diagram of the domain structure of the dimeric enzyme glutathione reductase. The twofold axis relating the two polypeptide chains is indicated by the dot in the center. The two interface domains bind to each other and determine the dimeric structure. In the NADP-binding domain, the adenine (Ad), ribose (circles), diphosphate (PP), and nicotinamide (N) moieties are indicated; only the nicotinamide portion is completely buried. In the FAD-binding domain, the coenzyme is indicated similarly, except that F is the flavin moiety; only the adenine group is exposed to the solvent. The substrate, oxidized glutathione (GSSG), binds between the two subunits. (Adapted from G. E. Schulz et al., *Nature* 273:120–124, 1978.)

cules are not basically different from the interior of either protein. The interface between protein and ligand is usually as closely packed as is the protein interior. All polar groups in the interface are paired in hydrogen bonds, and electrostatic charges are generally neutralized. Hydrogen bonds are especially prominent in pairing polar groups, and water molecules frequently act as intermediaries. Ionic interactions are generally not buried in the interface unless it is necessary for charged groups on the ligand to be buried upon association; they then tend to be neutralized by groups of opposite charge. In other cases, however, there is no neutralizing charge, and the charge on the ligand appears to be effectively "solvated" by multiple hydrogen bonds to other groups of the protein (Fig. 8.4). Such hydrogen-bonding groups of the protein probably have fewer degrees of freedom than water molecules do, and they may offer a more stable solvation shell for the charged groups. The hydrogen-bonding groups are usually part of hydrogen-bond arrays that lead to the solvent or to groups with the opposite charge; the latter are in some

ligands are often sizable depressions on the surface and represent the greatest exceptions to this generality. It is probably more correct to state that the complex of a protein and its natural ligands tends to be spherical and that the interacting surface between them tends to be maximized. A very small ligand does not perturb the dimensions of a protein much and tends to be bound inside a relatively spherical protein molecule. Larger ligands tend to bind in depressions on the surface, where they can associate and dissociate. Those ligands that do not need to dissociate often or at all, such as hemes and some other prosthetic groups, tend to be bound deep in the protein interior and to be integral parts of the protein structure. Long, linear ligands such as polysaccharides tend to be bound in clefts on the surface. If the protein and its ligand are of similar size (e.g., two associating protein molecules), their interface tends to be flat and large. With very large ligands, such as nucleic acids, the protein tends to bind to depressions on the surface of the ligand.

Interactions between proteins and ligands demonstrate both steric and physical complementarity between the two. These interactions follow structural rules similar to those in the proteins themselves; indeed, many interfaces between two interacting protein mole-

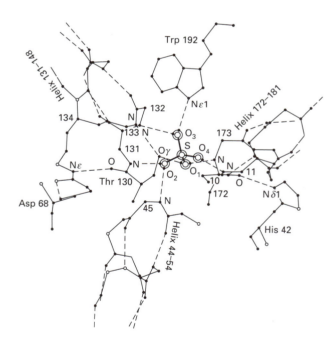

FIGURE 8.4

Binding of the SO_4^{2-} dianion to the sulfate-binding protein involved in bacterial active transport. The SO_4^{2-} molecule (depicted with double-circle atoms) is inaccessible to solvent, and there are no positively charged groups nearby. The crystal structure is consistent with there being seven hydrogen bonds between the ligand and the protein; all hydrogen bonds in the vicinity of the anion are indicated by dashed lines. (From F. A. Quiocho et al., *Nature* 329:561–564, 1987.)

Table 8.1 *Conformational Changes Observed upon Ligand Binding to Some Nonallosteric Proteins*

Protein	Ligand	Average Relative Movement (Å)	
		All atoms	Main chain
Trypsin	Bovine pancreatic trypsin inhibitor (BPTI)		0.26[a]
Trypsinogen	BPTI		0.28[b]
Lysozyme	Gd^{3+}	0.48[c]	
Myoglobin	O$_2$	0.92	0.61[d]
Concanavalin A	Ca^{2+} + Mn^{2+}	1.1	1.0[e]
Glyceraldehyde-3-phosphate dehydrogenase	NAD	1.2	1.2[f]
Carboxypeptidase A	Protein inhibitor		0.42[g]
Streptomyces griseus protease A	Ac-Pro-Ala-Pro-Phe-OH	0.11	0.10[h]
	Ac-Pro-Ala-Pro-Tyr-OH	0.10	0.09
	Ac-Pro-Ala-Pro-Phe-H	0.14	0.13

[a] From W. Bode and P. Schwyger, *J. Mol. Biol.* 98:693–717 (1975).

[b] From W. Bode et al., *J. Mol. Biol.* 118:99–112 (1978).

[c] From S. J. Perkins et al., *Biochem. J.* 173:607–616 (1978).

[d] From S. E. V. Phillips, *J. Mol. Biol.* 142:531–554 (1980).

[e] From M. Shoham et al., *J. Mol. Biol.* 131:137–155 (1979).

[f] From M. R. N. Murthy et al., *J. Mol. Biol.* 138:859–872 (1980).

[g] From D. C. Rees and W. N. Lipscomb, *Proc. Natl. Acad. Sci. USA* 78:5455–5459 (1981).

[h] M. N. G. James et al., *J. Mol. Biol.* 144:43–88, (1980).

cases the ends of α-helices, which have partial charge as a result of the helix dipole (Sec. 5.3.1). The hydrogen-bond arrays seem to have the role of effectively dispersing the formal buried charge.

The structure of a protein domain generally does not change substantially when it binds a ligand; exceptions are usually of functional importance. Small movements of atoms of the protein do occur in every case (Table 8.1), but they are often comparable to the experimental errors in crystallographic structure analysis. The most extreme changes in domains generally involve movements of flexible loops on the protein surface. On the other hand, some small adjustments are probably important in general to permit rapid rates of association and dissociation; totally rigid complex structures in which atoms interlock and interdigitate would be unlikely to be able to come together readily.

The analogy is often made with a key fitting into a lock. Although the rigidity implied by this analogy is too extreme for proteins, it conveys the correct message that a defined protein structure is probably necessary for specificity in ligand binding (Fig. 8.5). A very malleable protein would adopt its shape to match that of many ligands and would bind many of them with similar affinities. The difference in the affinities for two ligands is limited by the energy required to distort the normal conformation that is complementary to the high-affinity ligand to a conformation that is complementary to the low-affinity ligand. In the few known cases, binding of ligands with low affinity to a protein is not observed to produce large changes in conformation of the protein. The low affinities of such ligands primarily reflect their noncomplementarity to the preexisting binding site. The difference in binding energies for low- and high-affinity ligands is not sufficient to distort the protein binding site to make it complementary to the low-affinity ligand (Fig. 8.5).

Substantial changes in protein structure upon ligand binding are most frequently limited to motions of rigid domains or subunits relative to each other. Many ligands bind between domains that move together to engulf the ligand. This may help to maximize the inter-

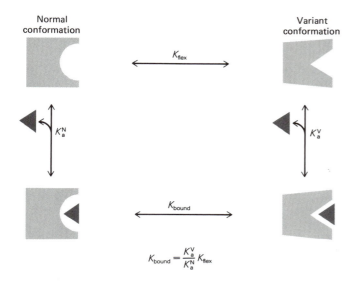

$$K_{bound} = \frac{K_a^V}{K_a^N} K_{flex}$$

FIGURE 8.5

Estimation of protein flexibility by the conformational distortions produced by binding a noncomplementary ligand. The normal conformation of a protein is imagined not to be complementary to a ligand (designated by the triangle) and consequently has poor affinity for that ligand. A variant conformation of the protein resulting from its flexibility might be complementary to the ligand and bind it tightly, so $K_a^V \gg K_a^N$, but this conformation would not normally be populated substantially (i.e., $K_{flex} \ll 1$). When the ligand is bound, however, the variant conformation should be more stable by the ratio of the two binding affinities, so it might be populated in the liganded state. For example, if $K_{flex} = 10^{-3}$ and the variant conformation binds the ligand 10^4 times more tightly than the normal conformation, K_{bound} will be 10, so the variant conformation should be present in the complex 91% of the time. On the other hand, if no substantial changes in the protein are observed upon binding a poor ligand, the protein must be relatively inflexible, so K_{flex} is extremely small.

actions between the protein and the ligand and to minimize interactions with other components of the solvent, while permitting the ligand to associate and to dissociate. The other role of domain or subunit movements upon ligand binding is to produce functional alterations at other sites on the protein, as in allosteric proteins (Sec. 8.4). Binding in this case is frequently regulated by changes elsewhere in the protein. Binding sites between domains or subunits seem to be dynamic locations, probably owing to the much greater flexibility of the protein's quaternary structure than of its tertiary structure.

In a few cases, binding of a ligand does produce substantial changes in the structure of protein domains. An example is the effects of Ca^{2+}-binding by regulatory proteins such as calmodulin (Sec. 8.3.4.a).

From these general principles, it is often possible to guess correctly the structure of a protein–ligand complex if the structures of the two components are known. Nevertheless, do not underestimate the difficulty of fitting together two molecules of known structure to make the most stable complex. There are an enormous number of ways that two molecules can associate, especially if at least one is the size of a protein, and only one of the ways is likely to be the correct one.

References

Structural and functional aspects of domain motions in proteins. W. S. Bennett and R. Huber. *Crit. Rev. Biochem.* 15:291–384 (1984).

Computer studies of interactions between macromolecules. S. J. Wodak et al. *Prog. Biophys. Mol. Biol.* 49:29–63 (1987).

Stabilization of charges on isolated ionic groups sequestered in proteins by polarized peptide units. F. A. Quiocho et al. *Nature* 329:561–564 (1987).

Substrate specificity and affinity of a protein modulated by bound water molecules. F. A. Quiocho et al. *Nature* 340:404–407 (1989).

High specificity of a phosphate transport protein determined by hydrogen bonds. H. Luecke and F. A. Quiocho. *Nature* 347:402–406 (1990).

The structure of protein–protein recognition sites. J. Janin and C. Chothia. *J. Biol. Chem.* 265:16027–16030 (1990).

8.2 *Energetics and Dynamics of Binding*

A fundamental aspect of the interaction of a protein with a ligand is the affinity of the two for each other, which is a measure of the overall free energy of the interaction. The magnitude of the affinity determines whether a particular interaction is relevant under a given set of conditions. The observed affinities of proteins for ligands vary enormously, ranging from very high values, for which dissociation is immeasurably small, to very low values, for which the concentration of free ligand required for a significant degree of binding is so great as to cast doubt on its relevance. Whether or not any particular affinity of a protein for a ligand is significant depends on the concentration of the ligand that the protein is likely to encounter; no other generalizations are possible.

If the affinity is very high, the protein is likely to be found and isolated as the complex; if such a ligand is relatively small, it is designated a **prosthetic group**. Examples are the heme groups of the globins and cytochromes, some coenzymes that bind tightly to enzymes, and metal ions that are integral parts of the protein structure. With lower affinities, ligands that are originally bound to a protein are likely to be lost during purification, unless they are added to the protein solution.

8.2.1 Binding Affinities

The affinity between a protein P and a ligand A is measured by the **association constant K_a** for the binding reaction at equilibrium:

$$P + A \xrightleftharpoons{K_a} P \cdot A \qquad (8.2)$$

$$K_a = \frac{[P \cdot A]}{[P][A]} \qquad (8.3)$$

All species are presumed to be present at sufficiently low concentrations for thermodynamic ideality to apply; if not, activities rather than concentrations must be measured. K_a is a constant under a given set of conditions and is measured experimentally by the dependence of binding on the free ligand concentration. Several commonly used graphic methods of analyzing binding data are illustrated in Figure 8.6.

The ratio of bound to free protein should be, according to Equation (8.3), directly proportional to the free-ligand concentration:

$$\frac{[P \cdot A]}{[P]} = K_a[A] \qquad (8.4)$$

An experimentally more useful measure of binding is the fraction y of protein molecules with bound ligand:

$$y = \frac{[P \cdot A]}{[P] + [P \cdot A]} = \frac{K_a[A]}{1 + K_a[A]} \qquad (8.5)$$

The greater the value of K_a, the greater the affinity. The value of K_a has units of $(\text{concentration})^{-1}$, however, and it is often intuitively easier to consider the **dissociation constant K_d**, which is simply the reciprocal of K_a and has units of concentration. With concentrations of free ligand below K_d, little binding to the protein occurs. With a concentration equal to K_d, half the protein molecules have bound ligand. An occupancy of 90% requires a nine times greater concentration of free ligand, whereas 99% occupancy requires that the concentration be 99 times K_d. Binding equilibria are simplest when the ligand is present at a concentration much greater than that of the protein binding sites. Uptake of the ligand by the protein does not then significantly alter the concentration of free ligand.

Specific binding by a protein of one ligand, and not another, depends on their relative affinities, their concentrations, and whether they bind at the same site. If two ligands are present at a concentration of 10^{-5} M but have different values of K_d—say, 10^{-3} M and 10^{-6} M—only the ligand with the lower K_d is bound significantly. If both are present at much higher concentrations—say, 10^{-2} M—both are bound to the protein to the maximum extent if they bind at separate sites. In this case, the higher affinity of one ligand is almost immaterial. If the two ligands compete for the same site, however,

$$P \cdot A \underset{K_d^A}{\overset{A}{\xrightleftharpoons{}}} P \underset{K_d^B}{\overset{B}{\xrightleftharpoons{}}} P \cdot B \qquad (8.6)$$

$$[P] = \frac{[P \cdot A]K_d^A}{[A]} = \frac{[P \cdot B]K_d^B}{[B]} \qquad (8.7)$$

$$\frac{[P \cdot A]}{[P \cdot B]} = \frac{K_d^B}{K_d^A}\frac{[A]}{[B]} \qquad (8.8)$$

the ligand with the higher affinity is bound to a correspondingly greater extent when the ligands are present at the same concentration. Weaker affinity can always be overcome by a higher concentration of that ligand, however, so binding affinities should always be considered relative to the concentration of the ligand.

The energetics of binding are often expressed by the Gibbs free energy of binding, ΔG_{bind}:

$$\Delta G_{bind} = -RT \ln K_a = RT \ln K_d \qquad (8.9)$$

It must be kept in mind, however, that K_a and K_d have units of concentration and that the value of ΔG_{bind} depends on which units are used (i.e., the standard state). If the units are moles per liter, the standard state is 1 M, and the calculated value of ΔG_{bind} applies only under the rather arbitrary situation when the concentration of free ligand is 1 M. In many instances a "unitary" free energy of binding is used as a measure of the intrinsic affinity; this is the free energy of binding that would occur with ligand at a hypothetical concentration of 55 M, the normal concentration of water. This parameter is not of any special significance, however, except when the ligand is water, and it does not represent the free energy of interaction that would occur in a unimolecular interaction (see Sec. 8.2.2).

The energetics of binding are defined more explicitly as the difference in free energies of the free and liganded protein, ΔG_b:

$$\Delta G_b = -RT \ln (K_a[A]) = -RT \ln \left(\frac{[A]}{K_d}\right) \qquad (8.10)$$

In this case, the concentration of free ligand must be specified. In a similar way, the enthalpy and entropy of

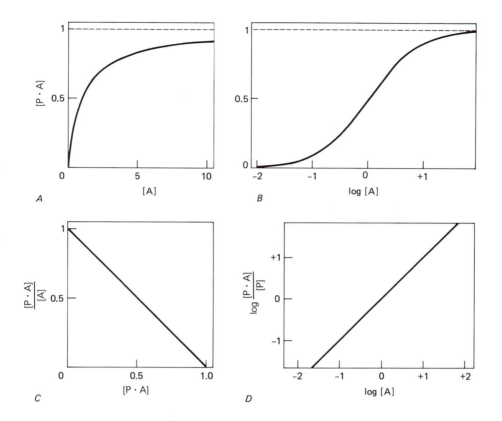

FIGURE 8.6

Some common methods of plotting binding data, using theoretical curves for the simple binding reaction $P + A \rightleftharpoons P \cdot A$. The concentration of free ligand is expressed relative to its dissociation constant, which is that concentration of free ligand that gives half-maximal binding. The concentrations of free and liganded forms of the proteins are given relative to the concentration of total protein. *A:* The normal hyperbolic relationship between binding and free-ligand concentration, demonstrating that a free-ligand concentration 9-fold greater than its dissociation constant produces only 90% of maximal binding (indicated by the dashed line); a 99-fold greater concentration is required for 99% saturation. *B:* A logarithmic scale emphasizes the wide range of free-ligand concentrations required for a complete binding curve. *C:* Scatchard plot. The negative slope gives the value of the association constant (the reciprocal of the dissociation constant). The horizontal intercept gives the extrapolated extent of the maximal binding. *D:* Hill plot. An accurate value for the maximum binding is required for this plot because both the liganded and the free protein concentrations are required. The value of the dissociation constant is given by the value of the free-ligand concentration where the vertical axis is zero (i.e., at half-maximal binding). This plot is used primarily for analyzing cooperative binding (see Fig. 8.24C).

binding are defined by the temperature dependence of the binding affinity.

A general consequence of ligand binding is that the protein is stabilized against unfolding and is less flexible. Neither of these observations need imply that the ligand has altered the structure of the protein. Instead, they are simply a consequence of the ligand binding more tightly to the fully folded conformation (N) than to the fully unfolded state (U) and any distorted or partially unfolded forms that result from flexibility of the structure. This can be illustrated very simply for the case of

unfolding of the protein when the ligand L binds solely to the folded state N:

$$N \cdot L \underset{K_d^L}{\overset{L}{\rightleftharpoons}} N \overset{K_u}{\longleftrightarrow} U \qquad (8.11)$$

$$K_{app} = \frac{[N \cdot L] + [N]}{[U]} = K_u \left(1 + \frac{[L]}{K_d^L}\right) \qquad (8.12)$$

The protein is stabilized against unfolding by the presence of the ligand. Even at very high ligand concentrations, above those at which the folded protein is fully saturated, the apparent stability of the protein is increased in proportion to the concentration of free ligand.

Ligand binding is simple in dilute solutions, but proteins often function in extremely concentrated aqueous solutions, as in the cytosol. For example, the interior of the red blood cell is about 35% hemoglobin by weight. Such solutions are very nonideal. The pertinent equilibria must be expressed in terms of the thermodynamic activities of the protein and of the ligand, which can be very different from their concentrations. Even though a particular protein might not be present in high concentration, the presence of molecules other than water in the environment can lead to substantial excluded volume effects (see Fig. 7.2). Added molecules favor any conformational or binding reaction that leads to a more spherical shape of a protein molecule, with less surface area exposed to solvent (Sec. 7.1.1). Consequently, binding of a ligand to a protein is often considerably greater in a concentrated solution than might be expected. It is possible that most proteins in the cytosol usually exist bound to each other, to membranes, to cytoskeleton, or to some other organized structure.

References

The meaning of Scatchard and Hill plots. F. W. Dahlquist. *Methods Enzymol* 48:270–299 (1978).

Protein affinities for small molecules: conceptions and misconceptions. I. M. Klotz and D. L. Hunston. *Arch. Biochem. Biophys.* 193:314–328 (1979).

Excluded volume as a determinant of macromolecular structure and reactivity. A. P. Minton. *Biopolymers* 20:2093–2120 (1981).

Calorimetric approaches to protein–ligand interactions. H. J. Hinz. *Topics Mol. Pharmacol.* 71–122 (1983).

Characterization of binding equilibrium data by a variety of fitted isotherms. R. Broderson et al. *Eur. J. Biochem.* 169:487–495 (1987).

An accurate method for determination of receptor–ligand and enzyme–inhibitor dissociation constants from displacement curves. A. Horovitz and A. Levitzki. *Proc. Natl. Acad. Sci. USA* 84:6654–6658 (1987).

Study of strong to ultratight protein interactions using differential scanning calorimetry. J. F. Brandts and L. N. Lin. *Biochemistry* 29:6927–6940 (1990).

8.2.2 Accounting for Relative Affinities

Can we account for the particular affinity of a protein for a given ligand? Do the structures of the two give any

insight into why the affinity is high or low? A qualified yes is permissible in answer to these questions in certain cases. Generally, however, only qualitative conclusions are feasible. Yet it should be possible eventually to account for, or even to predict, binding affinities for ligands and to design ligands or proteins with useful binding properties. The implications are immense for chemotherapy and for drug design.

It is currently impossible to rationalize the values of K_a or K_d of any protein for any ligand, even when the structures of the complex and of the free components are known. The practical difficulties in rationalizing ligand affinities arise from our poor understanding of the energetics of protein structure (Chap. 7) and of the strengths of the basic interactions in aqueous solvent (Chap. 4). The observed affinity depends on the relative free energies of the complex and of the components. This includes not only the interactions between the two in the complex but also any changes in their average conformations and their flexibilities produced by complex formation, any differences in their various interactions with solvent, the loss of translational and rotational freedom of each component, plus the displacement of solvent and any other ligands present in the binding sites before formation of the complex. Many of these factors compensate each other, and the net observed effect is a small difference between several terms of large and uncertain magnitude. It is not yet possible to calculate these quantities accurately enough to predict the values of K_a and K_d.

The relative affinities of two related ligands for the same protein (e.g., A and B), or of two closely related proteins for the same ligand, are more easily analyzed because many of the factors are the same in the two cases; differences in affinities can often be related to just one or two factors. Also, the ratio of their affinities is dimensionless:

$$\Delta(\Delta G^\circ)_{A-B} = -RT \ln \frac{K_a^A}{K_a^B} = +RT \ln \frac{K_d^A}{K_d^B} \quad (8.13)$$

There is then no complication in defining standard states. The most successful method for analyzing differences in binding interactions between closely related ligands or proteins is the free-energy perturbation method (Sec. 7.4.4.b) for simulating the effects of differences in chemical structure of the ligand or the protein, where the free energy is calculated in both the complex and the free molecule as the group that differs is gradually "mutated" during the calculation.

Interactions between a protein and a ligand always involve a substantial number of groups. The general approach to understanding the observed affinity has been to dissect it into the contributions of each group by measuring the effect on the affinity of removing them individually. Varying the ligand has traditionally been easiest, but even closely related ligands are occasionally observed to bind to the same protein in very different ways; in this case, a comparison of their measured binding affinities is largely meaningless. It is now more acceptable to use the same ligand but to vary the protein, using site-directed mutagenesis (Sec. 2.2), because the structures of the variant proteins tend to remain more constant. It is still advisable, however, to determine that the ligand binds in the same way to the variant proteins.

Given a series of binding affinities of related ligands for the same protein or of variant proteins for the same ligand, how is the binding energy dissected? It might be thought that the total binding energy (Eq. 8.9) is simply the sum of the contributions of each group, but it is not that straightforward. This can be illustrated in a manner first presented by Jencks.

Consider a ligand composed of two parts A and B; A might be capable of hydrogen bonding, and B might be hydrophobic. The affinity of ligand AB is compared with the affinities of A and B separately:

$$\xrightleftharpoons{K^{AB}} \tag{8.14}$$

$$\xrightleftharpoons{K^{A}} \tag{8.15}$$

$$\xrightleftharpoons{K^{B}} \tag{8.16}$$

where the equilibrium constants K^{AB}, K^A, and K^B are for either association or dissociation. In general, there is no simple relationship among these constants, and the classical binding energies calculated from them using Equation (8.9) are generally not additive:

$$-RT \ln K_a^{AB} \neq -RT \ln K_a^A - RT \ln K_a^B \tag{8.17}$$

even if the standard state is taken as 55 M so that unitary binding energies are calculated (Sec. 8.2.1). This nonadditivity is illustrated by the binding of biotin and some derivatives to the protein avidin (Table 8.2).

Table 8.2 *Binding of Biotin Derivatives to Avidin*

Derivative	Dissociation constant (M)	Free-energy contribution to binding (kcal/mol)
Biotin	1.3×10^{-15}	
Desthiobiotin	5×10^{-13}	
	3.4×10^{-5}	-13.3
$CH_3-(CH_2)_4-CO_2^-$	3×10^{-3}	-10.7

Data from N. M. Green, *Adv. Protein Chem.* 29:85–133 (1975).

The reason for the nonadditive nature of specific binding affinities becomes clear if the binding of AB is dissected into steps:

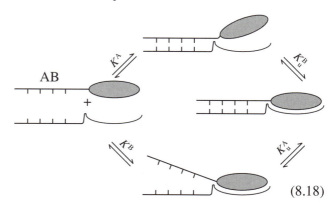

$$\tag{8.18}$$

The first step in the binding of the entire ligand should be analogous to the binding of each part when present alone (Eqs. 8.15 and 8.16), so the first two steps are assigned the bimolecular binding constants of each part.

The second step, binding of the second part of the ligand, however, is now a unimolecular step, rather than bimolecular. Consequently, the bimolecular binding constants K^A and K^B do not apply to the second step; instead, these second steps are assigned the unimolecular equilibrium constants K_u^A and K_u^B. The values of these two constants are not independent because they are linked functions:

$$K_a^{AB} = K_a^A K_u^B = K_a^B K_u^A \qquad (8.19)$$

From Equation (8.19), it is apparent that the constants K_u^A and K_u^B are the ratios of the binding constants of ligands with and without each of the respective moieties:

$$K_u^A = \frac{K_a^{AB}}{K_a^B} \qquad K_u^B = \frac{K_a^{AB}}{K_a^A} \qquad (8.20)$$

The contribution to the free energy of binding of each moiety can then be calculated:

$$\Delta\Delta G_i^A = -RT \ln K_u^A = -RT \ln \frac{K_a^{AB}}{K_a^B} \qquad (8.21)$$

$$\Delta\Delta G_i^B = -RT \ln K_u^B = -RT \ln \frac{K_a^{AB}}{K_a^A} \qquad (8.22)$$

For example, the free-energy contribution to binding of the sulfur atom of biotin can be estimated from the relative affinities of the first two compounds of Table 8.2 to be -3.5 kcal/mol:

$$\Delta\Delta G_i^S = -RT \ln \frac{5 \times 10^{-13} M}{1.3 \times 10^{-15} M}$$
$$= -3.5 \text{ kcal/mol} \qquad (8.23)$$

Similarly, the contributions of the remaining five-membered ring and of the acidic hydrocarbon group can be estimated to be -13.3 and -10.7 kcal/mol, respectively (Table 8.2).

The incremental binding energy contributions calculated in this way give a measure of the increased affinity caused by the presence of each group of the ligand. Their values depend critically on the relationship between the two parts of the ligand during binding, that is, on the effective concentration of the second part of the ligand when the first is bound, which can be designated as [A/B].

$$K_u^A = K_a^A [A/B] = \frac{[A/B]}{K_d^A} \qquad (8.24)$$

$$K_u^B = K_a^B [A/B] = \frac{[A/B]}{K_d^B} \qquad (8.25)$$

The same effective concentration applies to both parts because these are linked functions (see Eq. 8.19).

The effective concentration of either part in an intermediate complex can conceivably be zero, when it is kept away from the binding site and so provides no contribution to binding. Or, it might have a value of up to 10^{10} M (see Table 4.11), when parts A and B of the ligand are always in optimal orientation for simultaneous binding to a perfectly complementary binding site. The large values of effective concentrations in intramolecular reactions result from the entropic effect of the covalent linkage of the two parts. A ligand must lose a substantial amount of translational and rotational entropy upon binding; this is one of the factors determining the values of both K^A and K^B. In the case of ligand AB, however, at least some of this entropy is lost when the first part is bound; the second part of the ligand is then fixed to some extent and need not lose as much entropy upon completion of the binding as if that part were binding by itself. Consequently, the greater the rigidity between the two parts of a ligand, the greater the entropic contribution to the effective concentration is likely to be. This entropic contribution is the primary reason that the binding contributions of individual parts of a ligand do not add to give the observed affinity. For example, the contributions to the binding of biotin by the sulfur atom, the five-membered ring, and the acidic hydrocarbon (Table 8.2) total -27.5 kcal/mol. In contrast, the free energy of binding that would be calculated from Equation (8.9) is only -20.2 kcal/mol. The difference between these values reflects primarily the greater entropy that must be lost when parts of a ligand bind as separate molecules relative to the entropy that must be lost when they bind as parts of the same molecule.

If neither the ligand nor the protein is strained by binding, very high effective concentrations and free-energy contributions to binding may be observed. For example, the data of Table 8.2 for the two halves of desthiobiotin imply that their effective concentrations in the hypothetical intermediate complex are 2×10^5 M because

$$[A/B] = \frac{K_a^{AB}}{K_a^A K_a^B} = \frac{K_d^A K_d^B}{K_d^{AB}} \qquad (8.26)$$

With such high effective concentrations, ionic and hydrogen-bond interactions between ligand and protein may contribute substantially to binding, even in aqueous solution, where they must compete with intermolecular interactions between the solvent and the free protein and free ligand.

Because effective concentrations of the different parts of ligands are likely to vary substantially in different ligands and different binding situations, it is unrealistic to expect a constant contribution to binding of a hydrogen bond, a van der Waals interaction, and so on, in all ligand-binding interactions.

Table 8.3 *Large Contributions to Ligand Affinities for Proteins*

Group of ligand	Free-energy contribution to binding to protein[a] (kcal/mol)	Free energy of transfer from water to nonpolar liquid[b] (kcal/mol)
$-CH_3$	-2.0 to -3.9	-0.5
$-CH_2CH_3$	-6.5	-1.0
$-CH-(CH_3)_2$	-9.6	-1.5
$-CH_2-CH_2-CH_2-CH_3$	-7 to -8	-2.6
$-SCH_3$	-4.9	
$-CH_2-CH_2-S-CH_3$	-10 to -11	-1.3
$-SH$	-5.4 to -9.1	
$-OH$	-8	
$-NH_2$	-4.5	
$-NH_3^+$	-6.7	
$-CO_2^-$	-4.3	

[a] Contributions to binding were measured by the difference in affinities of ligands that differ only in the presence or absence of the indicated group; the free-energy contribution was calculated using Equation (8.21). Data from W. P. Jencks, *Proc. Natl. Acad. Sci. USA* 78:4046–4050 (1981); A. R. Fersht, *Proc. Roy. Soc. Lond. [Biol.]* 212:351–379 (1981).

[b] Data from Y. Nozaki and C. Tanford, *J. Biol. Chem.* 246:2211–2217 (1971).

Some examples of large incremental contributions to binding by various groups, measured by the relative affinities of ligands that differ only in that group, are tabulated in Table 8.3. The values for the nonpolar groups are considerably greater than their free energies of transfer from water to nonpolar liquids, which often is considered an analogous process. This discrepancy is further evidence that a protein—at least, its binding site—is not equivalent to an organic liquid. Instead, the folded protein has a higher concentration of atoms, and a binding site for nonpolar groups probably presents a more rigidly defined cavity with greater van der Waals interactions than is possible for a liquid. If part of the ligand or the protein normally involved in binding is missing, there might be a void at the interface between protein and ligand. Such a void could be filled by an isolated solvent molecule, or the protein and ligand could adapt to attain complementarity; but both are energetically costly. If a polar group normally involved in hydrogen bonding is deleted, its partner can be left in an energetically unfavorable situation without an alternative group to hydrogen-bond to. Consequently, interpreting such binding data in terms of individual interactions is not straightforward.

The data of Table 8.3 demonstrate that a protein can discriminate very effectively between its proper ligand and a ligand that lacks just one small part. Discrimination of ligands containing extra groups can be even more powerful because additional groups can interfere sterically with the complementarity between ligand and binding site.

Nevertheless, there are limits to the specificity of binding that is possible, set by the energetics of the interactions between groups. These limits are exceeded in some instances for which extreme specificity is necessary: for example, in the replication, transcription, and translation of genetic information. DNA replication occurs with an error frequency of only 10^{-10} even though the tautomerization of the nucleic acid bases, which will cause incorrect base-pairing, occurs with a frequency of 10^{-5}. Amino acids are also incorporated into proteins with considerably greater fidelity than expected, even from the data of Table 8.3. Much of that data comes from binding of amino acids to tRNA synthetases, which carry out the most crucial step of attaching the correct amino acid to the correct tRNA molecule. For example, how does a tRNA synthetase discriminate effectively against Gly when adding Ala to its tRNA, Val in the case of Ile, and Ser in the case of Thr? These pairs differ only by one $-CH_2-$ group and might be expected (Table 8.3) to differ in affinity by only a factor of 10^2. The answer in this case is that the enzyme seems to check the amino acid twice, discriminating at the first binding step and then subjecting the selected amino acid to a second check designed to detect the most likely fraudulent amino acid. Any caught by the second step are hydrolyzed from the tRNA and expelled (see Sec. 9.3.1.b). In such a *double-sieve* editing mechanism, the probability that an incorrect amino acid will be missed by both steps is the product of the two separate probabilities (e.g., $10^{-5} \times 10^{-5} = 10^{-10}$). In this way, biological specificities can be greatly enhanced over those possible with simple physical principles. Comparable multiple checks on specificity appear to be used in DNA replication. Such methods are used only when absolutely necessary, however, because there is a cost involved, in that a certain fraction of correct molecules are also removed at the subsequent recognition steps due to the intrinsic limitations on binding specificity.

References

Binding energy, specificity, and enzymic catalysis. W. P. Jencks. *Adv. Enzymol.* 43:219–410 (1975).

Entropy, binding energy, and enzymic catalysis. M. I. Page. *Angew. Chem. Int. Ed. Engl.* 16:449–459 (1977).

On the attribution and additivity of binding energies. W. P. Jencks. *Proc. Natl. Acad. Sci. USA* 78:4046–4050 (1981).

Effects of engineering complementary charged residues into the hydrophobic subunit interface of tyrosyl-tRNA synthetase. W. H. J. Ward et al. *Biochemistry* 26:4131–4138 (1987).

Relationships between apparent binding energies measured in site-directed mutagenesis experiments and energetics of binding and catalysis. A. R. Fersht. *Biochemistry* 27:1577–1580 (1988).

The hydrogen bond in molecular recognition. A. R. Fersht. *Trends Biochem. Sci.* 12:301–304 (1987).

Computer modeling of the interactions of complex molecules. P. A. Kollman and K. M. Merz, Jr. *Acc. Chem. Res.* 23:246–252 (1990).

8.2.3 Rates of Binding and Dissociation

The rates of binding and of dissociation of a ligand from a protein are determined by the respective rate constants k_a and k_d:

$$P + A \underset{k_d}{\overset{k_a}{\rightleftarrows}} P \cdot A \qquad (8.27)$$

Their ratio gives the association constant:

$$K_a = \frac{k_a}{k_d} \qquad (8.28)$$

The rate constants for binding ligands to proteins vary considerably, depending on the sizes of both and on any conformational changes that must take place in each upon binding. Many small ligands are found to bind very rapidly, at rates approaching those expected for diffusion control, k_D. This expected rate can be estimated from the diffusion coefficients of the protein and ligand, D_P and D_A, respectively, treating them as small spherical molecules that must approach within a distance r_{PA} for binding to occur:

$$k_D = 4\pi N_A(D_P + D_A)r_{PA} \qquad (8.29)$$

where N_A is Avogadro's number. For molecules with typical diffusion coefficients under normal circumstances (see Table 7.2), values of k_D in the region of 10^9 M^{-1} s^{-1} are expected. Larger molecules have smaller diffusion coefficients, but the value of k_D does not decrease accordingly because the value of r_{PA} is correspondingly larger. If the two molecules attract or repel each other at a distance, the term r_{PA} in Equation (8.29) should be replaced by a term containing the energy of interaction as a function of distance. For example, electrostatic interactions are significant over substantial distances and, when favorable, can increase rate constants for association to 10^{11} M^{-1} s^{-1}.

Rates of binding that are observed to be lower than k_D imply either that the two molecules must be in defined orientations for productive binding to occur or that changes occur during binding to produce a multistep association reaction. Both phenomena are undoubtedly important with proteins. The binding sites on proteins usually bind ligands only in defined orientations, and they generally comprise only small fractions of the protein surface. Consequently, most encounters between ligand and protein would be expected to be unproductive, and association to be relatively slow, but there are exceptions. For example, cytochromes c are thought to transfer electrons to and from other proteins through only 0.6% of their surfaces, where the heme group is accessible (see Fig. 6.31), and only when the two proteins interact in very specific orientations. On this basis, the rate of their interaction would be expected to be lower than that for diffusion-controlled encounters by a factor of at least 1000, but it is not. The reason is thought to be that asymmetric distributions of charges on the proteins orient them so that they tend to approach each other rapidly in a productive manner. The charge distribution of horse cytochrome c indicates a large dipole moment of just over 300 Debye units, and the dipole axis passes through the presumed binding site. Electrostatic interactions have also been shown to guide charged ligands to their binding sites on other proteins.

Table 8.4 *Rate Constants for Formation of Complexes of tRNA Synthetase (P) and tRNA*[a]

$$P + tRNA \underset{k_{-1}}{\overset{k_1}{\rightleftharpoons}} P \cdot tRNA \underset{k_{-2}}{\overset{k_2}{\rightleftharpoons}} P * tRNA$$

tRNA synthetase	tRNA	Initial Binding and Dissociation		Isomerization of Complex	
		$k_1 (M \cdot s)^{-1}$	$k_{-1} (s^{-1})$	$k_2 (s^{-1})$	$k_{-2} (s^{-1})$
Yeast Ser[b]	Yeast Ser	2.7×10^8	220	760	330
Yeast Phe[c]	Yeast Phe	2×10^8	250	420	750
Yeast Phe[c]	E. coli Tyr	8×10^8	1600	—	—

[a] The source and amino acid specificity of each tRNA synthetase and tRNA are indicated. The difference between the two complexes P · tRNA and P * tRNA is not known.

[b] Data from D. Riesner et al., *Eur. J. Biochem.* 68:71–80 (1976).

[c] Data from G. Krauss et al., *Eur. J. Biochem.* 68:81–93 (1976).

Binding of a ligand to a protein probably occurs generally via diffusion-controlled formation of an unstable *encounter* complex, followed by its rearrangement to the final complex:

$$P + A \underset{k_{-1}}{\overset{k_1}{\rightleftharpoons}} \{P \cdots A\} \underset{k_{-2}}{\overset{k_2}{\rightleftharpoons}} P \cdot A \qquad (8.30)$$

Such encounter complexes are usually not observed, and their structures are not known. The observed rate of binding is still proportional to the free-ligand concentration but is slower than diffusion-controlled k_1 because the encounter complex is unstable (i.e., $k_{-1} > k_1$ [A]), and it dissociates more rapidly than it completes binding (i.e., $k_{-1} > k_2$). Such encounter complexes can be observed only if high ligand concentrations are used so that k_2 becomes rate-limiting. Their occurrence may be important for permitting the two reactants to find their appropriate orientations for tight binding, rather than relying on the two molecules being in exactly the correct orientation in the initial encounter.

Even if the observed rate of association is apparently diffusion-controlled, additional steps may follow the initial association. For example, the initial interaction between a tRNA molecule and a tRNA synthetase enzyme occurs at nearly the diffusion-controlled rate, about $10^8/(M \cdot s)$, but rearrangements occur subsequently on the millisecond time scale (Table 8.4). Very little is known about what occurs structurally during the course of binding of ligands to proteins. There is considerable scope for conformational changes in both protein and ligand in most cases, but techniques for following them have yet to be devised.

Diffusion-limited encounters occur between all molecules in a solution; therefore, stable and specific binding must be reflected primarily in slow rates of dissociation, k_d (Eq. 8.28). Energetically favorable rearrangements of the complex after initial association have the effect of decreasing the apparent rate of dissociation. For example, binding of a yeast tRNA synthetase to a bacterial tRNA is not productive and is weaker than with the homologous pairs, but the initial association is somewhat faster. The weaker binding results from a greater rate of dissociation of the initial complex and from the apparent absence of a second isomerization step (Table 8.4).

The analysis just presented has been addressed to solutions in which the molecules are free to diffuse in three dimensions. In this case, the rate of diffusion-limited association depends on the size of the target, but this is no longer true when diffusion is confined to a space of lower dimensions, such as the two-dimensional plane of a membrane. For example, the mean diffusion times τ to reach a small target of radius a in the middle of a space of radius R ($R \gg a$) are given by

$$\tau_3 = \left(\frac{R^2}{3D_3}\right)\left(\frac{R}{a}\right) \qquad \text{in three dimensions} \quad (8.31)$$

$$\tau_2 = \left(\frac{R^2}{2D_2}\right) \ln \frac{R}{a} \qquad \text{in two dimensions} \quad (8.32)$$

$$\tau_1 = \left(\frac{R^2}{3D_1}\right) \qquad \text{in one dimension} \quad (8.33)$$

where D_i ($i = 1, 2, 3$) are the diffusion coefficients for the indicated dimensions. These times can be substantially shorter in one and two dimensions than in three. For example, a ligand could be imagined to bind rapidly to a protein in a membrane by first binding rapidly and nonspecifically to the membrane, then diffusing through the plane of the membrane to the protein. Although this is an attractive and popular idea, there are no well-characterized examples of this phenomenon.

One-dimensional systems might not be expected in biological systems, but they are probably approximated by long, linear macromolecules. For example, DNA is a long, extended double-helical structure with a high density of ionized phosphate groups. Specific DNA-binding proteins find short, specific nucleotide sequences in such molecules much more rapidly than would be expected for normal diffusion through solution in three dimensions. Also, the rate of finding specific sequences is observed to be increased by extending the length of the molecule with nonspecific sequences, whereas the opposite might be expected. The proteins appear initially to bind nonspecifically anywhere along the DNA molecule and then to diffuse one-dimensionally along the linear molecule until the correct sequence is found. There may also be "jumping" by the protein molecules between segments of the same DNA molecule that happen to come into proximity in solution. The initial, nonspecific binding of such a protein to DNA is caused by electrostatic interactions between the two and by the release of loosely bound counterions (Sec. 8.3.2). Upon reaching the correct sequence, the binding interactions are much more specific and tighter. Similar considerations may apply to the binding of proteins to long protein aggregates, such as the muscle protein actin.

References

The asymmetric distribution of charges on the surface of horse cytochrome c: functional implications. W. H. Koppenol and E. Margoliash. *J. Biol. Chem.* 257:4426–4437 (1982).

Diffusion-controlled macromolecular interactions. O. G. Berg and P. H. von Hippel. *Ann. Rev. Biophys. Biophys. Chem.* 14:131–160 (1985).

Reflections on the kinetics of substrate binding. H. Gutfreund. *Biophys. Chem.* 26:117–121 (1987).

Diffusion effects on rapid bimolecular chemical reactions. J. Keizer. *Chem. Rev.* 87:167–180 (1987).

Real-time spectroscopic analysis of ligand–receptor dynamics. L. A. Sklar. *Ann. Rev. Biophys. Biophys. Chem.* 16:479–506 (1987).

Dynamical simulation of rate constants in protein–ligand interactions. D. A. Case. *Prog. Biophys. Mol. Biol.* 52:39–70 (1988).

Facilitated target location in biological systems. P. H. von Hippel and O. G. Berg. *J. Biol. Chem.* 264:675–678 (1988).

8.2.4 Affinity Chromatography

The interactions of proteins with specific ligands makes possible one of the most powerful methods of protein purification, **affinity chromatography**. The ligand is chemically attached to an insoluble and porous solid support that is suitable for chromatography, in such a way that the ligand is still available for the protein to bind to it. To ensure that the ligand is sufficiently distant from the support that binding of the protein is not physically obstructed, a spacer group is usually inserted between the ligand and the chromatographic support. Obviously, the site on the ligand that is appropriate for its attachment to the solid support is one that is not involved in binding to the protein, but such details of the binding interaction are frequently not known. If the parts of the ligand required for binding to the protein are not known, various linkages between ligand and support must be tried until one is found to work.

In the ideal case, only the desired protein binds to the ligand resin, and all other proteins pass straight through the column. After the column is washed to remove all other proteins, those molecules that are bound tightly to the affinity resin can be eluted by adding soluble ligand to compete with the support or by changing the conditions to decrease the affinity of protein for the bound ligand; in extreme cases, the protein may have to be unfolded before it can be eluted. The use of different concentrations of soluble ligand permits quantitative measurements of the protein's affinities for both the free and the support-bound ligand, which are not necessarily the same. When multiple proteins bind to the same ligand, they can often be eluted separately by different ligands (Fig. 8.7) or by different concentrations of the same ligand (Fig. 8.8).

The principles of affinity chromatography are similar to those of the other types of column chromatography used with proteins, such as ion-exchange, hydrophobic, or reverse-phase chromatography; it differs only in the specificity of the interactions between the protein and the column resin.

References

Affinity chromatography of macromolecules. P. Cuatrecasas. *Adv. Enzymol.* 36:29–89 (1972).

Biospecific affinity chromatography and related methods. J. Porath and T. Kristiansen. In *The Proteins,* 3rd ed., H.

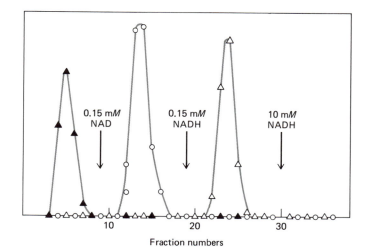

FIGURE 8.7

Separation of three proteins by affinity chromatography. The column matrix was Sepharose with the cofactor NAD linked covalently to it by a diaminohexane spacer. A mixture of bovine serum albumin, glyceraldehyde 3-phosphate dehydrogenase, and lactate dehydrogenase was applied to the column. The two dehydrogenases bind NAD and were bound by the column, whereas serum albumin (▲) does not bind NAD and was not retarded. Lactate dehydrogenase (△) binds NADH more tightly than NAD and was eluted by low concentrations of NADH. In contrast, glyceraldehyde-P dehydrogenase (○) binds NAD more tightly than NADH and was eluted from the column by low concentrations of free NAD. (Adapted from K. Mosbach et al., *Biochem. J.* 127:625–631, 1972.)

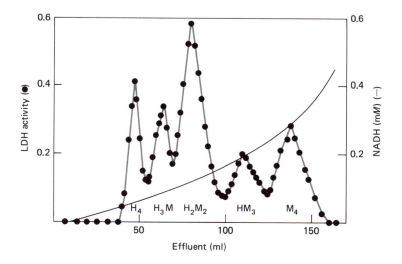

FIGURE 8.8

Separation of the five isozymes of lactate dehydrogenase by affinity chromatography. This enzyme is a tetramer that can be formed from two different but related polypeptide chains, usually designated H and M. The two chains can associate nearly randomly, so five tetramers are possible: the H_4, H_3M, H_2M_2, HM_3, and M_4 isozymes. The individual subunits of the tetramer bind NADH independently, with H chains having a fivefold greater affinity than M chains. Consequently, H_4 is eluted from an affinity column with lower concentrations of free NADH in the buffer and the others in order of their average affinities. The affinity column ligand was AMP, which corresponds to half an NADH molecule, bound to Sepharose. (Adapted from P. Brodelius and K. Mosbach, *FEBS Letters* 35:223, 1973.)

Neurath and R. L. Hill, eds., vol. 1, pp. 95–178. Academic Press, New York, 1975.

The purification of biologically active compounds by affinity chromatography. M. Wilchek and C. S. Hexter. *Methods Biochem. Anal.* 23:345–385 (1976).

The role of protein structure in chromatographic behavior. F. E. Regnier. *Science* 238:319–323 (1987).

Receptor-affinity chromatography. P. Bailon and D. V. Weber. *Nature* 335:839–840 (1988).

8.3 Relationship between Protein Conformation and Binding

Binding sites comprise relatively little of the structure of most proteins and often are just a small patch on the protein surface; small localized alterations of the protein, therefore, can produce large changes in its affinities for ligands, without changes in the overall protein conformation. Consequently, homologous proteins can bind very different ligands. Examples are the bacterial binding proteins present in the periplasm and involved in uptake of nutrients from the medium. All of these proteins that have been studied have very similar three-dimensional structures, consisting of two domains with a single binding site between them. Yet the various proteins bind a variety of small molecules, often with extreme specificity, including monosaccharides, oligosaccharides, amino acids, oligopeptides, sulfate (Fig. 8.4), and phosphate. Other examples are the immunoglobulins (Sec. 8.3.1), in which the same structural framework is used to bind an immense variety of antigens.

Conversely, the same ligand can be bound by unrelated proteins with different three-dimensional structures. This is illustrated most spectacularly by the binding of the heme group (iron-protoporphyrin IX) by a variety of proteins, especially the globins and various cytochromes. The globins (e.g., hemoglobins, myoglobin, erythrocruorin, and leghemoglobin) are all homologous and bind the heme group similarly (see Fig. 6.30). Very different are the structures of a large number of cytochrome c-like proteins (see Fig. 6.31) and of the unrelated cytochrome b_5, cytochrome b_{562}, and cytochrome c_3. This last binds four heme groups simultaneously in different ways at four different sites. Thus, a number of proteins appear to have acquired independently the ability to bind heme groups. The ability to bind the same ligand, therefore, cannot be used to imply that two proteins are related.

Nevertheless, the members of a number of different protein families do tend to bind similar ligands. Examples are the globins (Sec. 8.4.3) and the various protein families that bind DNA (Sec. 8.3.2) and Ca^{2+} ions (Sec. 8.3.4.a). The proteins in each of these examples are known to be evolutionarily related, and they obviously have retained their ligand-binding functions. In other cases, an evolutionary relationship is not obvious, and there is the possibility that similar three-dimensional structures have evolved to bind similar ligands in similar ways, perhaps for physical reasons. For example, the numerous 8-fold $\alpha\beta$ barrel proteins (Sec. 6.4.2) are the most likely candidates for examples of convergence to the same three-dimensional structure, yet they all seem to bind negatively charged ligands in similar positions at the same end of the barrel structure. Similarly, nucleotides are bound in different proteins to similar supersecondary structures, known collectively as the nucleotide-binding fold or the Rossmann fold (Sec. 8.3.3), that have no detectable amino acid sequence similarities. Do these similar structures and interactions with ligands reflect common ancestry? Or do they reflect similar physical principles of structure and binding? The verdict is not yet in.

The question of the relationship between protein structure and ligand binding is important because the biological functions of newly discovered proteins are often inferred from their homology to proteins of known structure and function. Rather than just give rules, some of the most informative examples are described.

References

The evolution of dehydrogenases and kinases. W. Eventoff and M. G. Rossmann. *Crit. Rev. Biochem.* 3:111–140 (1975).

Relation between structure and function of α/β proteins. C. I. Brändén. *Quart. Rev. Biophys.* 13:317–338 (1980).

The structural motif of β-lactoglobulin and retinol-binding protein: a basic framework for binding and transport of small hydrophobic molecules? J. Godovac-Zimmerman. *Trends Biochem. Sci.* 13:64–66 (1988).

Crystal structures of two intensely sweet proteins. S. H. Kim et al. *Trends Biochem. Sci.* 13:13–15 (1988).

Atomic structures of periplasmic binding proteins and the high-affinity active transport systems in bacteria. F. A. Quiocho. *Phil. Trans. Roy. Soc. Lond.* B 326:341–351 (1990).

8.3.1 Immunoglobulins

Antibody molecules are capable of prodigious diversity; individually they bind a few antigens very specifically,

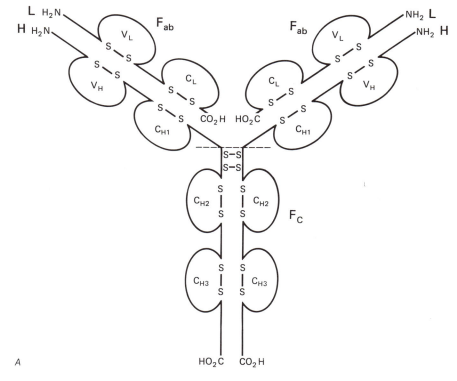

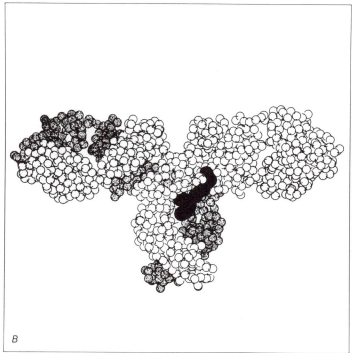

FIGURE 8.9

Schematic representation of a typical immunoglobulin structure *(A)* and a space-filling model determined crystallographically *(B)*. *A* shows the L and H polypeptide chains as solid lines, with the intramolecular disulfides linking Cys residues, about 60 residues apart in the primary structure, that are characteristic of each immunoglobulin domain. The site of cleavage by papain is shown by the dashed line; this cleavage yields two F_{ab} fragments and one F_c. If cleavage occurs on the carboxyl side of the disulfide linking the H chains, as occurs with pepsin, the two F_{ab}-like fragments are linked by the disulfide and are usually designated as F'_{ab}. In *B*, each sphere represents one amino acid residue. One complete heavy chain is white, the other heavily shaded; both light chains are white. The carbohydrate attached to the C_{H2} domain of each heavy chain is black. The antigen-binding sites are at the tips of the F_{ab} arms, at the far left and far right, where the V_H and V_L domains meet. (From E. W. Silverton, et al., *Proc. Natl. Acad. Sci. USA* 74:5140–5144, 1977.)

while collectively they are able to recognize virtually any molecule. Despite this binding diversity, antibody molecules have common structural and functional features. The consequences of ligand binding include elimination of the antigen–antibody complex from the bloodstream, complement-induced lysis of cells, histamine release, and stimulation of secretion of antibodies by lymphocytes, depending on the class of antibody. All immunoglobulin molecules of a given class must have these common functions, which are then combined with different specificities for various antigens. How these different functions are combined is a major question that is only now being answered in molecular terms.

The basic structure of an intact immunoglobulin is a Y-shaped molecule composed of two H and two L chains (Fig. 8.9); in immunoglobulins A and M, these molecules are assembled further into larger complexes. The L chain consists of V_L and C_L domains, and the H chains consist of V_H, C_{H1}, C_{H2}, and C_{H3} domains. All these domains consist of about 100 residues, are homologous in their primary structures, and are independent, stable structural units. All immunoglobulin domains have essentially the same conformation, designated the **immunoglobulin fold**, consisting of two layers of antiparallel β-sheets that are usually linked by a disulfide bond (Fig. 8.10). The arrangement of the various domains in the Y-shaped molecule is shown schematically in Figure 8.9A. The polypeptide chains between the domains are susceptible to proteases. Most susceptible is the hinge region linking the two arms to the base of the Y. After cleavage at this site, the two arms are released individually and are known as F_{ab} fragments; the base is known as the F_c fragment. Each F_{ab} fragment contains the V_L, C_L, V_H, and C_{H1} domains; the F_c fragment has two copies of each of the C_{H2} and C_{H3} domains. F_V fragments are produced in other ways and consist of only the V_H and V_L domains. The connecting segments between the domains have varying degrees of flexibility, and the individual domains undergo considerable motion relative to each other. The Y-shaped molecule shown in Figure 8.9 is just one of many shapes that an immunoglobulin can adopt in solution.

The immunoglobulin domains interact with each other in a variety of ways in an intact H_2L_2 molecule (Fig. 8.9B). The two C_{H2} domains interact with each other in the F_c portion, as do the two C_{H3} domains. Each arm of the molecule is composed of one C_{H1} domain interacting with one C_L domain, plus the interacting V_H and V_L domains. All pairs of C domains associate in a similar manner in which the members of one of their pairs of β-sheets associate isologously (see Fig. 6.25); in contrast, the V domains associate by means of the other β-sheet. Less extensive, and presumably less stable, interactions also take place between domains adjacent in

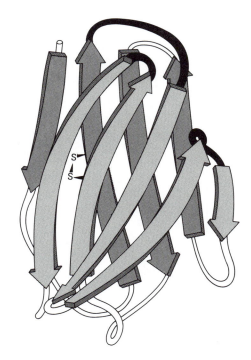

FIGURE 8.10

The immunoglobulin fold. Two layers of antiparallel β-sheet are folded on top of each other to form a sandwichlike structure. Between the two layers are hydrophobic side chains and the indicated disulfide bond linking two Cys residues about 60 residues apart; one Cys residue is in the middle of the second strand from the left of the top sheet; the other is in the middle of the second strand of the bottom sheet. In V_L domains, as illustrated here, the top β-sheet has five strands, the bottom four. The two strands at the right edge of the five-strand sheet are missing in C domains. The loops containing the hypervariable regions in V domains are shaded dark. (From J. Richardson, *Adv. Protein Chem.* 34:167–339, 1981.)

the primary structure. Nevertheless, the segments of polypeptide chain linking the domains have varying degrees of flexibility; especially flexible is the hinge region linking the F_{ab} and F_c portions, which makes the two F_{ab} arms particularly mobile.

References

Three-dimensional structure of antibodies. P. M. Alzari et al. *Ann. Rev. Immunol.* 6:555–580 (1988).

Elbow motion in the immunoglobulins involves a molecular ball-and-socket joint. A. M. Lesk and C. Chothia. *Nature* 335:188–190 (1988).

Antibody: the flexible adaptor molecule. D. R. Burton. *Trends Biochem. Sci.* 15:64–69 (1990).

a. Recognition of Antigens

The antigen-binding sites are at the tips of the two F_{ab} arms, situated between the V_L and V_H domains. Both F_{ab} and F_V fragments retain the antigen-binding sites. Each site is composed of the residues of the three irregular loops between β-strands of both the L and H chains (Fig. 8.11). Different binding sites are generated with different amino acid side chains in these positions, which are known as the **complementarity determining regions, or CDRs.** Their conformations depend largely on the conserved structure of the remainder of the immunoglobulin fold, which serves as a scaffold (Fig. 8.10). Consequently, the conformations of the CDR regions of many antibodies can be predicted with some degree of success from just their amino acid sequences. Also, the CDR regions from one immunoglobulin can be grafted onto the scaffold of another, and it is now becoming possible to design antibodies to order.

An enormous number of antibody molecules with different antigen specificities are made by complex organisms, and this diversity of immunoglobulin primary structures is generated by a special mechanism during their biosynthesis. The variable domains are encoded by separate gene segments, designated as variable (V_H), diversity (D), and joining (J) segments for the heavy chain and V_L and J_L segments for the light chain. For the heavy chain in the mouse, there are $100-1000$ V_H elements, approximately 12 D elements, and 4 J elements. Similarly, the light chain is encoded by more than 100 V and 5 J_L elements. Different antibody molecules are generated by joining these elements in different combinations. Further genetic variation is introduced at the sites where these gene segments are joined together by the genetic fusion mechanism. In this way, 10^8-10^{10} different antibody molecules can be generated from a limited number of gene segments. Most of this variation is produced in the residues comprising the three CDR regions of each polypeptide chain. Accordingly, the CDR regions are hypervariable in different antibody molecules, and they are also known as the **hypervariable regions.**

Each antibody-producing cell normally produces a single antibody molecule. When such a cell encounters an antigen that its antibody recognizes sufficiently (at this stage known as an *immunogen*), the cell is induced to synthesize the antibody in large quantities, to undergo cell division, and to proliferate. In the process, the genetic segment coding for the variable region of the antibody molecule undergoes mutation at a rate much greater than normal. Progeny cells producing antibodies with greater affinity for the immunogen are then selected. Consequently, the immunological response to an immunogen changes with time. Initially, many low-affinity antibodies are produced, with $K_d = 10^{-5}-10^{-7}$ M, but with time, antibodies of increasing affinity are produced. In an organism, many different antibody molecules are produced by many different cells, so normal antisera are very heterogeneous. Individual antibody-producing cells can be selected and cloned, however, and homogeneous **monoclonal antibodies** produced by them. Furthermore, the methods of protein engineering (Sec. 2.2) make it possible to manipulate the genes for antibodies, to express them in microorganisms, and to select for antibodies of the desired specificity.

There is one important restriction in the number of different antibody molecules that the normal immune system produces naturally against various antigens. Antibodies are not usually produced against molecules that are normally present in the host organism, for obvious reasons.

The interactions of antibody-combining sites with antigens seem to be no different from the interactions of other proteins with ligands. They occur with comparable rates and binding constants. Crystal structures of several immunoglobulin–antigen complexes demonstrate the usual, but still remarkable, complementarity of shapes of the CDR regions and of the antigen, with the usual van der Waals interactions and hydrogen bonds. No large conformational changes take place in the antibody molecules upon binding antigens. The six CDR loops play varying roles in the various complexes, and not all necessarily interact with each antigen. Generally, the heavy-chain CDR regions seem to play the major role, and single V_H domains often have significant affinities for ligands.

A major question to be answered is how the binding of antigens to the binding sites of intact immunoglobulins at the tips of the F_{ab} arms triggers the effector functions, such as complement activation, which are a property of the F_c portion of the molecule. Different classes of immunoglobulins have different C domains and different effector functions. There is little evidence for specific conformational changes in immunoglobulins upon antigen binding, although the immunoglobulins undergo a wide variety of motions of the various domains relative to each other. Effector functions appear to be triggered primarily by the formation of large antigen-immunoglobulin aggregates. In particular, the first protein of the classical complement activation pathway, C1q, is a complex structure resembling a bunch of six tulips, with six globular heads joined by six collagen-like stems that are held together in the lower half. Each head binds to the C_{H2} domain of an immunoglobulin; the simultaneous binding of many heads to aggregated immunoglobulins and antigens appears to be the trigger that sets off complement activation.

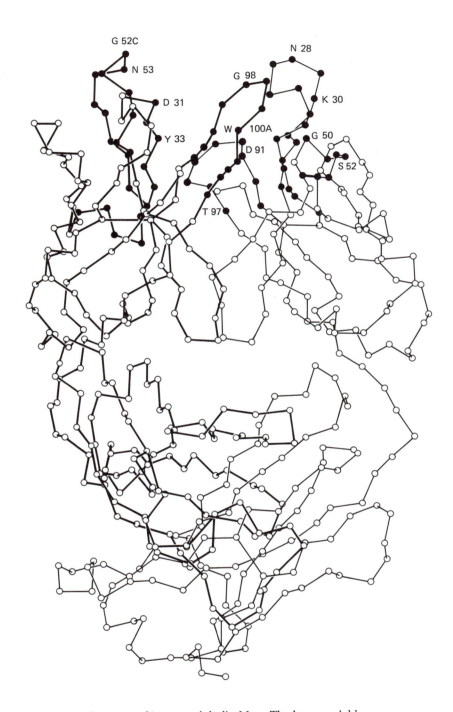

FIGURE 8.11

The α-carbon backbone of the F_{ab} fragment of immunoglobulin M_{603}. The hypervariable residues comprising the antigen-binding site are darkened. Two residues of each hypervariable segment are identified using the one-letter code for the amino acids. The heavy chain is shown with thicker lines than the light chain. The top half of the structure comprises the V_H and V_L domains, the bottom half the C_L and C_{H1} domains. (Kindly provided by D. R. Davies and H. Metzger.)

References

Thermodynamics of hapten–antibody interactions. T. K. S. Mukkur. *Crit. Rev. Biochem.* 16:133–167 (1987).

Diversity and the genesis of high affinity antibodies. C. Milstein. *Biochem. Soc. Trans.* 15:779–787 (1987).

Reshaping human antibodies for therapy. L. Riechmann et al. *Nature* 332:323–327 (1988).

Binding activities of a repertoire of single immunoglobulin variable domains secreted from *Escherichia coli*. E. S. Ward et al. *Nature* 341:544–546 (1989).

Conformations of immunoglobulin hypervariable domains. C. Chothia et al. *Nature* 342:877–883 (1989).

The role of somatic hypermutation in the generation of antibody diversity. D. L. French et al. *Science* 244:1152–1157 (1989).

Antibodies: a paradigm for the biology of molecular recognition. C. Milstein. *Proc. Roy. Soc. Lond. B.* 239:1–16 (1990).

Antibody–antigen complexes. D. R. Davies et al. *Ann. Rev. Biochem.* 59:439–473 (1990).

Three-dimensional structure determination of an anti-2-phenyloxazolone antibody: the role of somatic mutation and heavy/light chain pairing in the maturation of an immune response. P. M. Alzari et al. *EMBO J.* 9:3807–3814 (1990).

Man-made antibodies. G. Winter and C. Milstein. *Nature* 349:293–299 (1991).

b. Proteins as Antigens

If no ligands are known for a given protein, they can be made to order by preparing antibodies against it. This can be extremely important for identifying and quantifying a protein. For example, a new protein known only from its gene sequence can usually be identified by synthesizing a peptide corresponding to a short segment of its primary structure, preparing antibodies against this peptide, and using these antibodies to identify the protein. This procedure is most successful if this peptide corresponds to a part of the primary structure that is accessible to antibodies, even in the folded native conformation.

Specific proteins are often identified immunochemically by a procedure known as **Western blotting.** Proteins in samples separated by electrophoresis in polyacrylamide gels (Sec. 1.5.3) are transferred electrophoretically, in a way that retains the electrophoretic pattern, to a membrane to which the proteins stick tightly. The membrane is then treated with antibodies specific for the protein to be detected, and the positions of the bound antibodies are visualized in various ways. Using Western blotting, a single protein can be detected specifically among a mass of other proteins. The technique of Western blotting need not employ antibodies but can use other ligands that bind specifically to certain proteins. At least some proteins seem, perhaps surprisingly, to refold after or during transfer to the membrane because they exhibit ligand-binding properties that are dependent on the folded conformation of the protein.

There are numerous other immunochemical methods for quantifying specific proteins, some of which are of clinical importance. All of these immunochemical procedures are dependent on the conformational specificities of antibodies directed against proteins and peptides, and their cross-reactivities, so there has been substantial interest in proteins as immunogens. After much controversy, the determination of several crystal structures of complexes of proteins bound to antibodies directed against them (Fig. 8.12) has substantially clarified the situation. These crystal structures demonstrate that neither the antibody nor the protein antigen change their conformations substantially upon interacting. The interface between them is similar to those between other interacting protein molecules. Antibodies generally recognize an area on the surface of the protein determined largely by the size of the binding site on the antibody; this binding invariably involves atoms from residues of the protein that are distant in the sequence but are brought into close proximity by the protein's conformation. All antibodies against proteins are specific in some way for their conformations, which is the basis of their use in measuring conformational equilibria (Sec. 7.5.3.d).

Native proteins used as immunogens generally elicit at least some antibodies specific for the folded conformation, depending on the stability of the protein to the immunization procedure. Antibodies recognizing the folded conformation may be directed against any portion of the protein surface unless the host organism contains a protein with the same surface. If the protein is not very stable, unfolded molecules will also be present or will be produced by the immunogenic manipulations, and antibodies recognizing the unfolded state will also be produced. Proteins unfolded irreversibly and peptide fragments of a protein can also be used to produce antibodies recognizing the unfolded state. The specificity of such antibodies remains to be determined, but it is most likely that each antibody is specific for one of the many conformations that disordered polypeptides can adopt. Accordingly, unfolded proteins and peptides have substantially lower affinities for their antibodies than do folded proteins.

Antibodies against folded proteins cross-react with unfolded proteins or peptide fragments, and vice versa, to varying extents, depending on the conformational flexibilities of the two conformational states of the protein (Sec. 7.5.3.d). Antibodies against a peptide frag-

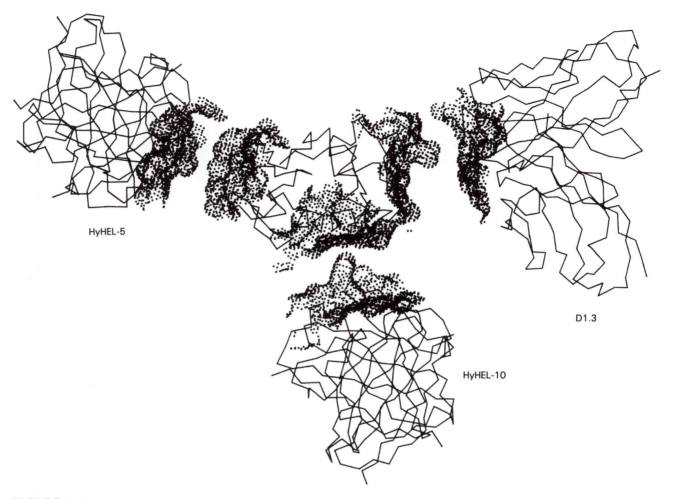

FIGURE 8.12

A composite of hen lysozyme binding to the F_V regions of three different antilysozyme antibodies, D1.3, HyHEL-5, and HyHEL-10. For simplicity of illustration, the three F_V fragments have been pulled away somewhat from the lysozyme molecule in the center. C^α representations are used for the F_V fragments and for lysozyme, and a dot representation is used for their interacting surfaces. Note that the three epitopes on lysozyme recognized by these three antibodies do not overlap, except for a small overlap between those of HyHEL-10 and D1.3. (From D. R. Davies et al., *Ann. Rev. Biochem.* 59:439–473, 1990.)

ment recognize the native protein to the extent that it unfolds spontaneously or if it is induced to do so by the manipulations used in the immunochemical measurements. The possibility that the peptide fragment tends to generate antibodies against a particular conformation must also be considered.

Antibody molecules can also be used, in a different species, as immunogens. Of special interest are the antibodies that are directed against their antigen-binding regions, known as **anti-idiotopes.** If such anti-idiotopes are complementary to the immunogenic antibody binding site, they should be equivalent to the antigen to which the immunogenic antibody was raised (Fig. 8.13).

Anti-idiotopes should be useful for binding studies in place of that antigen, and this expectation has been fulfilled in some instances. Although the antigen and anti-idiotope surfaces are equivalent to at least some extent, sufficient to permit their binding to the same antibody molecule, they need not be structurally equivalent.

References

The antigenic structure of proteins: a reappraisal. D. C. Benjamin et al. *Ann. Rev. Immunol.* 2:67–101 (1984).

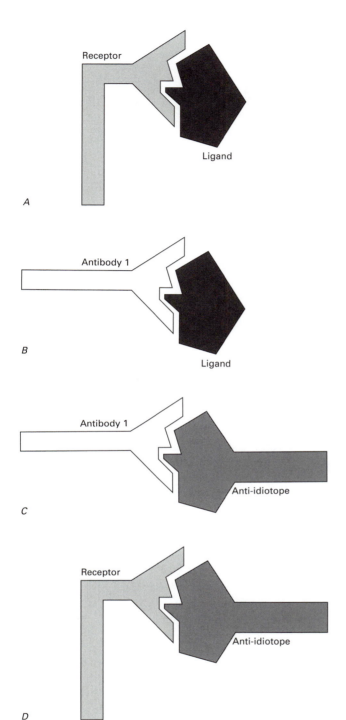

FIGURE 8.13
Schematic illustration of anti-idiotope antibodies and their uses. *A:* A sought-after receptor for a specific ligand bound to the ligand. *B:* Antibodies directed against the functional region of the ligand should have a combining site that mimics the receptor. *C:* An anti-idiotopic antibody that is raised against the combining site of the first antibody should have a combining site that mimics the ligand. *D:* Such anti-idiotopes should be able to identify the sought-after receptor.

Continuous and discontinuous protein antigenic determinants. D. J. Barlow et al. *Nature* 322:747–748 (1986).

The structural basis of antigen–antibody recognition. R. A. Mariuzza et al. *Ann. Rev. Biophys. Biophys. Chem.* 16:139–159 (1987).

Antibody–antigen complexes. D. R. Davies et al. *J. Biol. Chem.* 263:10541–10544 (1988).

On the attribution of binding energy in antigen–antibody complexes McPC 603, D1.3, and HyHEL-5. J. Novotny et al. *Biochemistry* 28:4735–4749 (1989).

Antibody geometry and form: three-dimensional relationships between anti-idiotypic antibodies and external antigens. W. V. Williams et al. *Trends Biotechnol.* 8:256–263 (1990).

Three-dimensional structure of an idiotope–anti-idiotope complex. G. A. Bentley et al. *Nature* 348:254–257 (1990).

8.3.2 DNA-Binding Proteins

Proteins that bind specifically to DNA are of great biological importance because they are usually involved in replication or expression of genetic information. Of greatest interest have been those proteins that bind to very specific sites on the DNA, defined by specific sequences of the four nucleotides A, T, C, and G at a few adjacent positions. For a protein to distinguish among different nucleotide sequences in double-stranded DNA is not straightforward because the nucleotides of the two antiparallel strands are base-paired (A-T, T-A, C-G, and G-C) in the interior of the double helix (Fig. 8.14). The exterior surface of the double helix is almost independent of its nucleotide sequence, being composed primarily of the constant phosphate–sugar backbone. Only the edges of the nucleotides are accessible to the solvent and to the protein, primarily in the major groove of the DNA double helix. The nucleotides are distinguished primarily by the polar groups that are accessible.

If a protein is to discriminate among DNA base-pairs by interacting with their edges in the major groove, it needs to have interacting groups that protrude substantially from its surface, to be able to contact the nucleotides at the base of the groove. The best characterized structural motif that accomplishes this is the **helix-turn-helix,** which protrudes from the protein surface. It is observed in a number of proteins that have no other structural similarities (Fig. 8.14). This structural motif seems to have sufficient intrinsic stability to be able to exist as a protuberance, with few interactions with the rest of the protein structure, in order to penetrate the DNA major groove. A number of hydrophobic

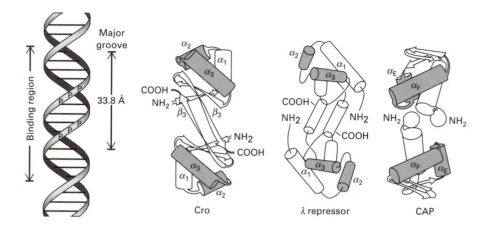

FIGURE 8.14

Comparisons of the dimeric structures of three DNA-binding proteins with the helix-turn-helix motif, relative to the structure of double-stranded DNA *(left)*. The three proteins are the repressor Cro, the amino-terminal domain of λ repressor, and the carboxyl-terminal domain of catabolite gene activator protein (CAP). The two α-helices of each helix-turn-helix motif are shaded. In the foreground of each protein, labeled α_3 or α_F, is the recognition helix, which fits most closely into the major groove of the DNA. Note that the spacing between the two recognition helices of the dimer is very close to the 33.8-Å spacing between the DNA major grooves, but that the orientations of the helices relative to the major groove differ. (Kindly provided by B. W. Matthews.)

side chains of this motif pack between the helices and provide a structural framework for it (Fig. 8.15). A number of these residues are similar among different proteins, presumably for such structural reasons, even when the rest of these proteins are not detectably similar in sequence or in structure. This conserved pattern of residues often makes it possible to detect the helix-turn-helix motif from just the primary structures of suspected DNA-binding proteins.

The specificities of the various helix-turn-helix motifs for binding to different DNA sequences arise primarily from the different amino acid side chains that emanate from the amino end of the second helix, known as the recognition helix, and that enter the major groove of the DNA. The other α-helix lies across the major groove and makes nonspecific contacts with the DNA (Fig. 8.16). The amino acid side chains make hydrogen bonds and van der Waals contacts with the exposed edges of the nucleotides in the DNA major groove. Hydrogen bonding to the nucleotides is especially important, and water molecules are frequently involved in networks of hydrogen bonds. Because of the importance of hydrogen bonding for discriminating among nucleotides, the residues that interact with the DNA are primarily polar, especially those with multiple hydrogen-bonding side chains, such as Asn, Gln, Arg, Asp, and Glu. These direct interactions involve flexible side chains of the protein, although their flexibility is usually

limited somewhat by interactions with neighboring residues. Consequently, the different helix-turn-helix motifs interact with DNA in a variety of geometries (Fig. 8.14). With this type of recognition, there is no simple code relating the amino acid sequence of the protein to the nucleotide sequence of DNA that it recognizes.

Many helix-turn-helix DNA-binding proteins bind as dimers, with both equivalent binding sites making the same interactions to DNA with the same sequence. The DNA sites at which they bind are said to be **palindromic,** in that the nucleotide sequence on one DNA strand is repeated in complementary fashion and in reverse order: For example, the binding site for the *lac* repressor has the following sequence of its two complementary DNA strands:

5′ **TGTGTGGAATTGTX$_9$ACAATTTCACACA**
3′ **ACACACCTTAACAY$_9$TGTTAAAGTGTGT**

$$(8.34)$$

All but one of thirteen base-pairs of each half obey the palindromic nature of this binding site; they are indicated in boldface. The nine nucleotides between the palindromic sequences are indicated by X$_9$ and Y$_9$. The palindromic sequences define a local twofold symmetry axis at right angles to the DNA helix that relates the two halves of the DNA binding site. The dimeric protein binds with its twofold axis coinciding with that of the DNA (Fig. 8.14). Having a double binding site undoubt-

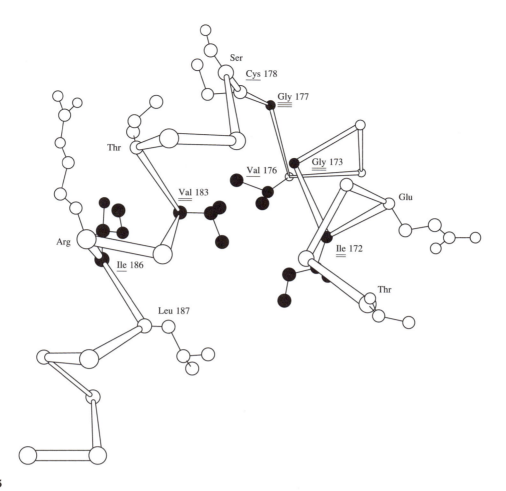

FIGURE 8.15

Detailed structure of the helix-turn-helix motif of CAP. The first α-helix is to the right; the second is the recognition helix and is to the left. The degree of conservation of the various residues in other motifs is indicated by the number of underlines. The Gly 177 residue is generally conserved in different proteins because it is important for the bend. Residues Ile 172, Val 176, Val 183, and Ile 186 are usually hydrophobic in other such motifs because they pack between the two helices and stabilize their conformations. (Adapted from T. A. Steitz, *Quart. Rev. Biophys.* 23:205–280, 1990.)

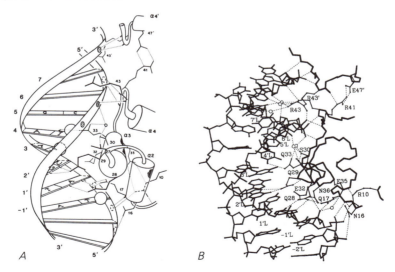

FIGURE 8.16

Interactions between one helix-turn-helix motif of the 434 repressor and its DNA binding site. *A:* A schematic drawing. *B:* A more detailed view. Dashed lines represent likely hydrogen bonds; circles are water molecules involved in such hydrogen bonds. The amino acid residues are identified by the one-letter code; residue numbers with primes come from the other subunit of the dimer. The labels 1′L, 2′L, and so on, refer to the nucleotides of the DNA. (From A. K. Aggarwal et al., *Science* 242:899–907, 1988.)

edly increases markedly the strength and specificity of the binding interaction (Sec. 8.2.2).

Another structural means of providing for an α-helix to protrude from the protein surface sufficiently to interact with the nucleotide bases in the major groove of DNA is provided by the **zinc fingers.** The most conspicuous feature of this structure is a Zn^{2+} ion chelated by two Cys and two His residues (Fig. 8.17A). These residues occur in a characteristic sequence motif, $-X_3-Cys-X_{2\text{-}4}-Cys-X_{12}-His-X_{3\text{-}4}-His-X_4-$, where X is any amino acid. Hundreds of such sequence motifs have been found in proteins involved in many aspects of gene regulation in eukaryotes. The four residues that chelate the zinc ion are held in the appropriate positions by the three-dimensional structure of the finger polypeptide chain, which is composed of a two-stranded β-hairpin and a single α-helix (Fig. 8.17). This fold is apparently stabilized by several hydrophobic side chains that make up the core of the structure. Each zinc finger seems to be a stable, autonomous structural unit, even though the fingers generally consist of only 28–31 residues.

Multiple zinc fingers are usually present in tandem along the polypeptide chain, linked by only a few residues; frequently -Thr-Gly-Glu-Lys-. An individual

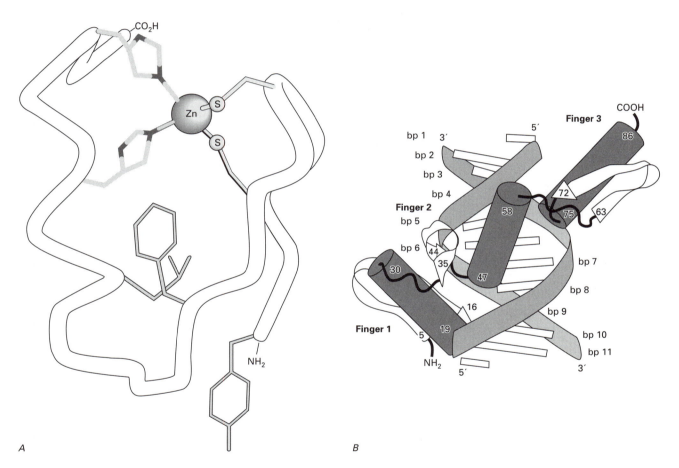

A

B

FIGURE 8.17

A: The structure of a single 2Cys, 2His zinc finger. *B:* The interaction of a three-finger protein with DNA. *A* shows the three-dimensional structure of the zinc finger *Xfin* determined by NMR analysis. The polypeptide backbone is represented by the thick cylinder; the single segment of α-helix is on the left, the β-hairpin is on the right. The two His and two Cys side chains coordinating the Zn^{2+} ion are shown. Three hydrophobic side chains that comprise the core of the three-dimensional structure are included. (Kindly provided by P. Wright.) *B* shows the crystal structure of a polypeptide chain comprising three zinc fingers from protein *Zif268.* The α-helix and β-hairpin of each finger are depicted, but the Zn^{2+} ions are omitted. Amino acid side chains from residues at the amino end of each helix contact the edges of three sequential base-pairs in the major groove of the DNA. (From N. P. Pavletich and C. O. Pabo, *Science* 252:809–817, 1991.)

finger binds to DNA only weakly, but simultaneous interactions involving a number of zinc fingers increase the binding substantially. Each zinc finger contacts three adjacent nucleotide base-pairs in the major groove of the DNA (Fig. 8.17B). Adjacent zinc fingers interact with adjacent triplets of the DNA. Each different zinc finger can have a different specificity for the triplet that it recognizes, using amino acid side chains from residues at the amino end of the α-helix. The proteins containing multiple zinc fingers are therefore of modular construction, and nature seems to have devised a variety of such proteins with specificities for different DNA sequences by mixing-and-matching various individual fingers with specificities for different triplets.

Another class of zinc-finger proteins has four Cys residues chelated to a single Zn^{2+} ion. The members of this class, as typified by the glucocorticoid receptor, have similar conformations that differ from the first class. A further type of Zn^{2+}-stabilized structure is found in proteins related to GAL4. Both of these types of DNA recognition domains bind as dimers to palindromic sequences, in contrast to the zinc fingers of Figure 8.17.

Other DNA-binding motifs are known, and others await characterization. For example, in the case of the *Met* repressor, a two-stranded β-sheet is involved in interacting with the DNA.

Direct interaction between the protein and the DNA nucleotides is a major factor in the specificity of binding, but another factor is thought to be the deformability of the DNA double helix. DNA is observed to bind to some proteins in conformations that are distorted to varying extents from the classical linear double helix (Fig. 8.14). The DNA structure, not that of the protein, is perturbed in these cases, indicating that the DNA is more pliable structurally than is the protein (Fig. 8.5). An exception is the Cro protein, in which binding to DNA causes the two monomers to rotate 40° relative to each other by a twisting of the two β-strands that connect one monomer with the other (Fig. 8.14); the basis for this exception remains to be determined. There is considerable other evidence for the plasticity of the DNA double-helix structure, and the extent and nature of this plasticity vary with the nucleotide sequence. Some DNA-binding proteins are thought to discriminate among different nucleotide sequences by binding their specific sequence in a distorted conformation that is energetically favorable for that sequence but not for others to which the protein might otherwise bind.

Proteins that bind to DNA irrespective of its nucleotide sequence, or with little sequence specificity, recognize primarily the DNA backbone of phosphate and sugar groups. Electrostatic interactions and the release of counterions and bound water molecules appear to provide the major driving force for binding proteins to the polyanionic DNA molecule, which has one ionized phosphate group for each nucleotide. Even with proteins that bind specifically, the initial binding of a protein to DNA is often nonspecific and is driven electrostatically and by the release of mobile cations bound loosely to the DNA (Sec. 4.4.3). Such attractions are thought to keep the protein bound loosely to the DNA while permitting it to search along the essentially one-dimensional molecule for its specific binding site (Sec. 8.2.3), to which it binds much more tightly. For example, the *lac* repressor binds nonspecifically to any DNA with a K_a of 10^6 M^{-1} at 0.1 M salt, but binding to the correct sequence is 10^8-fold tighter. The K_a for nonspecific binding decreases at higher salt concentrations, but the strength of the specific binding is not so sensitive to the ionic strength.

Interactions between proteins and DNA are usually measured experimentally by band-shift electrophoresis and by "footprinting" in which the bound protein protects the DNA from chemical modification. The high affinities of many DNA-binding proteins for specific DNA segments mean that their rates of dissociation are extremely slow. Consequently, the complexes between them can be stable during electrophoresis. The presence of DNA-binding proteins causes the electrophoretic mobility of the DNA segment to be shifted, which is a useful assay for such proteins.

Which nucleotides of the DNA segment are interacting directly with the protein can be determined by which are protected from chemical reagents that cleave the DNA. The DNA segment is radioactively labeled at one end initially. After partial cleavage in the presence and the absence of the protein, the resulting mixture of DNA fragments is subjected to electrophoresis under denaturing conditions, in which the mobility is determined only by the length of the DNA strand. Only the fragments retaining the original end with the radioactive groups are visualized, using autoradiography. The normal cleavage pattern is modified by any bound protein molecule, in that segments that arose from cleavages that are protected by the protein are missing from the pattern. Such a footprint can localize binding of a protein to specific nucleotides of the DNA.

Most DNA-binding proteins have their specific affinities for DNA modulated by another ligand. This ligand is frequently bound to a protein domain other than that which binds to the DNA. For the one case in which the mechanism is clearly known, the ligand alters the affinity of the protein for DNA by altering the dimeric protein structure so that the spacing of the two DNA binding sites does or does not coincide with the spacing

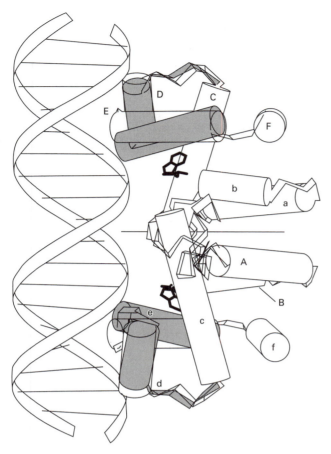

FIGURE 8.18

Structural change in the *trp* repressor induced by binding of the corepressor, L-tryptophan. Where the structure is altered by binding of L-tryptophan, the shaded model is that without the ligand. Helices A–F are from one polypeptide chain of the dimer, a–f are from the other. The ligand bound to each of the two sites of the dimer is shown as a skeletal model. The ligand binds between helices B, C, and E of each half of the dimeric protein. In doing so, it pushes helices D and E of the helix-turn-helix motif away from the center of the molecule and places the two equivalent recognition helices E and e the correct distance apart to bind in adjacent major grooves of normal double-helical DNA. The binding affinity for DNA is increased 10^3-fold by the presence of the ligand. (From R. G. Zhang et al., *Nature* 327:591–597, 1987.)

of the palindromic sequences of the DNA (Fig. 8.18). How other ligands affect DNA binding and how binding of a protein to DNA affects the expression of the genetic information are topics of very great interest at the present time.

References

Structural basis of protein–nucleic acid interactions. D. L. Ollis and S. W. White. *Chem. Rev.* 87:981–995 (1987).

Selection of DNA binding sites by regulatory proteins. O. G. Berg and P. H. von Hippel. *Trends Biochem. Sci.* 13:207–211 (1988).

Helix-turn-helix, zinc-finger, and leucine-zipper motifs for eukaryotic transcriptional regulatory proteins. K. Struhl. *Trends Biochem. Sci.* 14:137–140 (1989).

The helix-turn-helix DNA binding motif. R. G. Brennan and B. W. Matthews. *J. Biol. Chem.* 264:1903–1906 (1989).

λ-Repressor: a model system for understanding protein–DNA interactions and protein stability. R. T. Sauer et al. *Adv. Protein Chem.* 40:1–61 (1990).

The stereochemistry and biochemistry of the *trp* repressor-operator complex. B. F. Luisi and P. B. Sigler. *Biochim. Biophys. Acta* 1048:113–126 (1990).

Structural studies of protein–nucleic acid interaction: the sources of sequence-specific binding. T. A. Steitz. *Quart. Rev. Biophys.* 23:205–280 (1990).

Protein–DNA conformational changes in the crystal structure of a λ Cro-operator complex. R. G. Brennan et al. *Proc. Natl. Acad. Sci. USA* 87:8165–8169 (1990).

Zinc finger domains: hypotheses and current knowledge. J. M. Berg. *Ann. Rev. Biophys. Biophys. Chem.* 19:405–421 (1990).

Zinc finger–DNA recognition: crystal structure of a Zif268–DNA complex at 2.1 Å. N. K. Pavletich and C. O. Pabo. *Science* 252:809–817 (1991).

A structural taxonomy of DNA-binding domains. S. C. Harrison. *Nature* 353:715–719 (1991).

DNA recognition by GAL4: structure of a protein–DNA complex. R. Marmorstein et al. *Nature* 356:408–414 (1992).

8.3.3 Nucleotide Binding

Many proteins that bind any of the dinucleotide coenzymes, such as nicotinamide adenine dinucleotide (NAD, plus NADP) and flavin adenine dinucleotide (FAD), or the single nucleotides such as adenosine mono-, di-, and triphosphate (respectively, AMP, ADP, and ATP) and flavin mononucleotide (FMN), do so in remarkably similar ways. The binding site for each nucleotide is composed of loops at the carboxyl ends of two parallel β-strands linked by an intervening α-helix (see Fig. 6.35). The positive end of the macrodipole of the helix (Sec. 5.3.1) appears to be interacting with the phosphate group of the nucleotide. In some cases, additional βα units are present. For example, the NAD-binding domains of dehydrogenases consist of a six-stranded parallel β-sheet with four linking α-helices (see Fig. 6.34). The first three β-strands and two helices bind the AMP portion of NAD; the other half binds the nicotinamide portion. In other proteins, binding of the AMP portion is the more similar, whereas the position of the second part of the coenzyme tends to vary much more.

The $\beta\alpha\beta$ unit involved in nucleotide binding is known as the **mononucleotide binding domain** or the **Rossmann fold,** after its discoverer. Not all such units have sufficient sequence similarity to indicate homology, leaving the evolutionary origin of this structural and functional motif uncertain. The similarities in their sequences, however, usually appear to be for structural and functional reasons (see Fig. 6.35). For these reasons, it is often possible to use this pattern of similarities in sequence to detect the likely presence of a mononucleotide binding domain in the sequences of proteins of unknown structure.

The mononucleotide binding domain binds the nucleotides at the C-terminal ends of the β-strands. Other parallel β-sheets are also found generally to bind ligands at the C-terminal ends of the β-strands. Whether this is due to evolutionary or physical reasons, or whether it reflects interactions with the β-sheet or the intervening α-helices, is not yet clear.

References

Interaction of pyrophosphate moieties with α-helixes in dinucleotide binding proteins. R. K. Wierenga et al. *Biochemistry* 24:1346–1357 (1985).

Prediction of the occurrence of the ADP-binding $\beta\alpha\beta$-fold in proteins, using an amino acid sequence fingerprint. R. K. Wierenga. *J. Mol. Biol.* 187:101–107 (1986).

Rubredoxin reductase of *Pseudomonas oleovorans.* Structural relationship to other flavoprotein oxidoreductases based on one NAD and two FAD fingerprints. G. Eggink et al. *J. Mol. Biol.* 212:135–142 (1990).

The P-loop—a common motif in ATP- and GTP-binding proteins. M. Saraste et al. *Trends Biochem. Sci.* 15:430–434 (1990).

Binding of nucleotides by proteins. G. E. Schulz. *Curr. Opinion Struct. Biol.* 2:61–67 (1992).

8.3.4 Very Small Ligands

When the ligand is very small, such as an ion or an electron, there might seem to be little possibility of steric complementarity between it and the binding site of a protein. Considerable specificity, however, is still possible. In general, ion-binding sites are in the interior of the protein and are composed of a number of groups of the protein that surround and interact simultaneously with the ion, producing a cooperative effect. The affinities for ions of individual groups on proteins are too low to produce the required affinities and specificities of ion binding. Nevertheless, chemical considerations are also involved because ions tend to bind to protein groups for which they have some intrinsic affinity. For example,

Ca^{2+} ions tend to be bound to oxygen atoms, Zn^{2+} to sulfur atoms and to the imidazole nitrogen atoms of His residues, and Fe^{2+} and Fe^{3+} to the sulfur atoms of Cys residues or to sulfide ions, as in iron–sulfur proteins. Copper ions tend to interact with thiol or imidazole groups; Mg^{2+} ions are bound along with the phosphate groups of ligands.

The most detailed structural information about metal-binding sites in proteins comes from protein crystal structures, but the **EXAFS** (extended X-ray absorption fine structure) technique can be used to determine the type, number, and distances of atoms liganding the metal ion. Some metal-binding sites observed in proteins are illustrated in Figure 8.19. All of these ions are bound spontaneously to the apoprotein, even in the case of the seemingly complex iron–sulfur complexes of ferredoxins.

References

The symbiosis of metal and protein functions. R. J. P. Williams. *Eur. J. Biochem.* 150:231–248 (1985).

Characterization of metal centres in biological systems by X-ray absorption spectroscopy. S. S. Hasnain and C. D. Garner. *Prog. Biophys. Mol. Biol.* 50:47–65 (1987).

Surface topography of histidine residues: a facile probe by immobilized metal ion affinity chromatography. E. S. Hemdan et al. *Proc. Natl. Acad. Sci. USA* 86:1811–1815 (1989).

Carboxylate–histidine–zinc interactions in protein structure and function. D. W. Christianson and R. S. Alexander. *J. Amer. Chem. Soc.* 111:6412–6419 (1989).

Geometry of interaction of metal ions with sulfur-containing ligands in protein structures. P. Chakrabarti. *Biochemistry* 28:6081–6085 (1989).

Systematics in the interaction of metal ions with the main-chain carbonyl groups in protein structures. P. Chakrabarti. *Biochemistry* 29:651–658 (1990).

Where metal ions bind in proteins. M. M. Yamashita et al. *Proc. Natl. Acad. Sci. USA* 87:5648–5652 (1990).

a. Ca^{2+}-Binding

The Ca^{2+} ion has a wide variety of biological roles and binds to a variety of proteins. It is especially important as a "second messenger" to trigger muscle contraction and other types of motility and to control the release of neurotransmitters and hormones. These effects are mediated by the specific binding of Ca^{2+} ions to important regulatory proteins, such as calmodulin and troponin C, that have similar conformations.

In each of these regulatory proteins, Ca^{2+} binds to a common structural motif designated the **EF hand** from the two α-helices that were called the E and F helices in

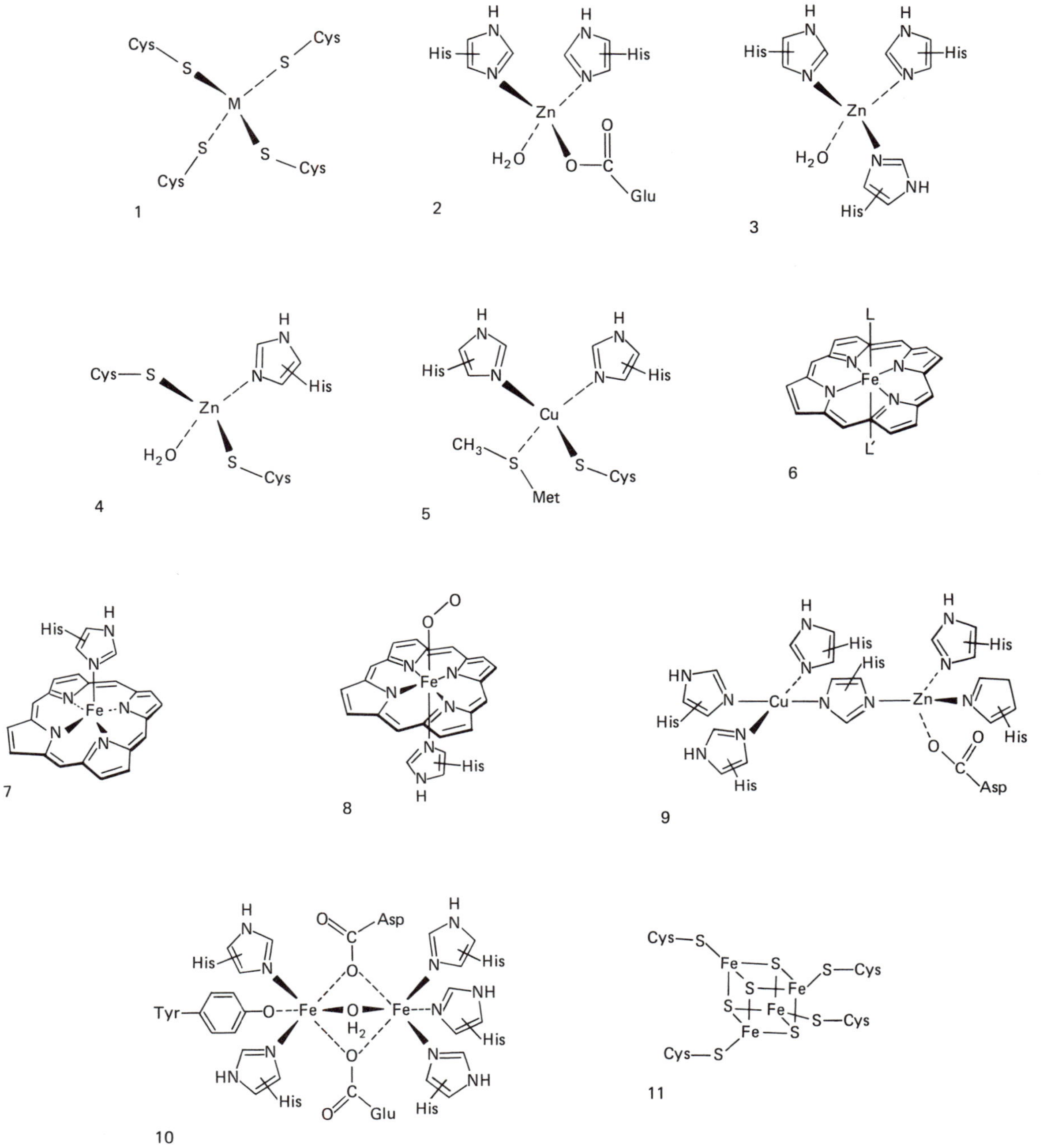

FIGURE 8.19

Examples of metal ions bound tightly to proteins, with the partial structures of the amino acid side chains to which they are attached. (1) M is either Fe, as in rubredoxin, or Zn, as in the structural ions of aspartate transcarbamylase and liver alcohol dehydrogenase. (2) Carboxypeptidase A. (3) Carbonic anhydrase and insulin. (4) Catalytic Zn^{2+} of liver alcohol dehydrogenase. (5) Azurin and plastocyanin. (6) Heme group of cytochromes; L is usually His and L' is His in cytochrome b_5 or Met in cytochrome c. (7) Deoxy heme groups in myoglobin and hemoglobin. (8) Oxy form of (7). (9) Superoxide dismutase. (10) Methemerythrin. (11) Some ferredoxins and high-potential iron protein; planar (2Fe–2S) and (3Fe–3S) complexes are also found in other ferredoxins. (Adapted from J. A. Ibers and R. H. Holm, *Science* 209:223–235, 1980.)

with the ligand are part of the segment of 12 residues that comprise the polypeptide loop between the helices; these residues are generally at positions 1, 3, 5, 7, 9, and 12 of the loop. The EF hand can be recognized from just the primary structure of a protein because of the frequent occurrence of Asp, Asn, Glu, Thr, and Ser residues at these positions, plus other structural constraints. It is likely that all of the EF-hand structures arose evolutionarily from a common ancestor by gene duplication and divergence.

The regulatory proteins calmodulin and troponin C have a domain with two Ca^{2+}-binding EF hands at each end of a long α-helix (Fig. 8.21). Both of these proteins interact reversibly with other proteins depending on whether or not Ca^{2+} is bound at the two sites that have lowest affinities. The structure of the EF hand appears to be quite different when the Ca^{2+} ion is not bound (Fig. 8.21), and the conformational changes that occur upon binding are probably used to signal changes in Ca^{2+} concentration. The long, central α-helix appears to serve as a flexible tether permitting the domains at each end to move as semirigid bodies to form single binding sites for other proteins, depending on whether or not Ca^{2+} is bound. Binding of Ca^{2+} to the regulatory proteins affects their affinities for the other proteins, so binding to these other proteins must also affect the affinity for Ca^{2+} of the regulatory protein (see Fig. 7.32).

Other types of Ca^{2+} binding sites are present in other proteins. The only property they share is the participation of multiple oxygen atoms of the protein, but in no common orientation. The Ca^{2+} ion does not prefer any particular geometry in its binding interactions but, being large, forms many long electrostatic bonds.

References

Structure and evolution of calcium-modulated proteins. R. H. Kretsinger. *Crit. Rev. Biochem.* 8:119–174 (1980).

Calmodulin. C. B. Klee and T. C. Vanaman. *Adv. Protein Chem.* 35:213–321 (1982).

A model for the Ca^{2+}-induced conformational transition of troponin C. A trigger for muscle contraction. O. Herzberg et al. *J. Biol. Chem.* 261:2638–2644 (1986).

Interactive properties of calmodulin. J. A. Cox. *Biochem. J.* 249:621–629 (1988).

Crystal structures of the helix-loop-helix calcium-binding proteins. N. C. J. Strydnaka and M. N. E. James. *Ann. Rev. Biochem.* 58:951–998 (1989).

How calmodulin binds its targets: sequence independent recognition of amphiphilic α-helices. K. T. O'Neil and W. F. DeGrado. *Trends Biochem. Sci.* 15:59–64 (1990).

Calcium binding induces conformational changes in muscle regulatory proteins. A. C. R. da Silva and F. C. Reinach. *Trends Biochem. Sci.* 16:53–57 (1991).

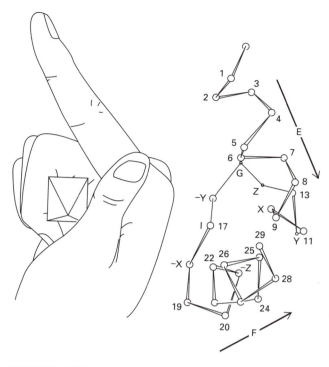

FIGURE 8.20

The EF hand that binds one Ca^{2+} ion in parvalbumin. The hand illustrates the octahedral arrangement of amino acid residues that provide the oxygen atoms chelating the central Ca^{2+} ion. The forefinger of the hand represents the E helix, the thumb the F helix. The positions of the α-carbon atoms in parvalbumin are illustrated on the right. The Ca^{2+} ion is bound within the turn connecting the E and F α-helixes. Oxygen atoms that chelate the ion at the vertices of the octahedron come from the residues designated X, Y, Z, −X, −Y, and −Z by their octahedral positions. The Ca^{2+} ion is seven-coordinated; the Glu residue at position −Z is a bidentate ligand that uses both oxygen atoms of its side chain. The Gly and Ile residues at positions 15 and 17, respectively, are highly conserved in different Ca^{2+} binding proteins. Helices E and F have hydrophobic side chains at residues 2, 5, 6, 9, 22, 25, 26, and 29 that pack against another EF hand related by an approximate twofold axis of rotation (see Fig. 8.21). (Kindly provided by R. H. Kretsinger.)

the protein in which this motif was discovered. In this structural motif, the two helices are roughly perpendicular to each other and are connected by a loop (Fig. 8.20). There are often two or four such motifs in each protein, each binding one Ca^{2+} ion. Two Ca^{2+} ions generally bind near each other in neighboring motifs, and the binding of one ion probably increases the affinity for the other. Each ion binds to a specific constellation of seven oxygen atoms from Asp, Asn, Glu, Thr, and Ser side chains, from backbone carbonyls, or from intermediary water molecules. The protein groups that interact

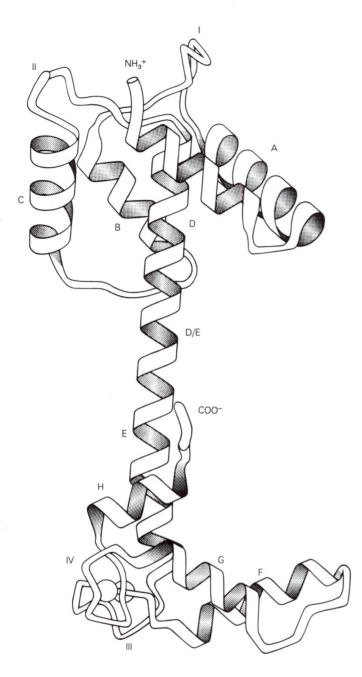

FIGURE 8.21
The crystal structure of troponin C, with Ca^{2+} ions bound at two of the EF-hand binding sites (III and IV) but not at sites I and II. Each of the four Ca^{2+} binding sites is similar to that of the EF hand of parvalbumin (see Fig. 8.20). In troponin C, each of the four sites is defined by two α-helices: site: I, helices A and B; II, C and D; III, E and F; IV, G and H. The three-dimensional structure of the upper domain, without Ca^{2+}, is substantially different from that of the lower domain. The long D/E α-helix is probably flexible in solution. (From I. Herzberg and M. N. G. James, *J. Mol. Biol.* 203:761–779, 1988.)

Intracellular calcium-binding proteins: more sites than insights. C. W. Heizmann and W. Hunziker. *Trends Biochem. Sci.* 16:98–103 (1991).

b. O₂-Binding

Sufficient discrimination of some small ligands appears not to be possible with a protein alone, so prosthetic groups bound to the protein serve as physical intermediaries in binding such small ligands. The most numerous examples occur in the physiologically vital binding of O_2 and electrons. Oxygen molecules can be bound in different ways: by the ferrous heme groups of the globins (Fig. 8.22); by two Fe^{2+} atoms held by His, Glu, and Asp side chains and bridged by an oxide anion O^{2-}, as in hemerythrin (Fig. 8.19); or by two Cu^+ ions held in close proximity by six His side chains in the hemocyanins. Fe^{2+} and Cu^+ ions and ferrous heme groups free in solution do not bind O_2 reversibly; instead, they are oxidized to the Fe^{3+} and Cu^{2+} forms. Accordingly, one of the functions of the protein must be to provide the appropriate environment for the binding of O_2 by the ions without their being oxidized.

The protein environment also determines which ligands bind and the consequences of their binding. For example, heme groups are used not only for O_2 binding in proteins like myoglobin and hemoglobin but also in other proteins to accept electrons (e.g., in cytochromes) and to catalyze redox reactions (e.g., in cytochrome P450 and horseradish peroxidase). These differences in function are due primarily to the environment of the central iron atom, particularly the groups liganded to it. In the case of the O_2-binding globins, a His side chain is usually liganded to the iron atom on one side of the heme, known as the proximal side (Fig. 8.22). No comparable ligand exists on the other, distal side in deoxymyoglobin and deoxyhemoglobin; a His residue is usually nearby, but it is held too far from the heme iron to serve as a ligand to it. It is here on the distal side of the heme group that O_2 binds reversibly to the iron atom and also forms a hydrogen bond to the side chain of the distal His residue (Fig. 8.22).

The related ligand carbon monoxide, CO, binds to isolated heme groups several thousand-fold more tightly than O_2 does. This intrinsic affinity is reduced in myoglobin and hemoglobin by a factor of about 100. The discrimination against CO has obvious physiological advantages because this poison is present under normal circumstances, being produced naturally in cells. O_2 binds at an angle to the heme plane whereas CO binds best to porphyrin when it is perpendicular. This configuration is made unfavorable in the protein, however, by steric clashes with the distal His residue

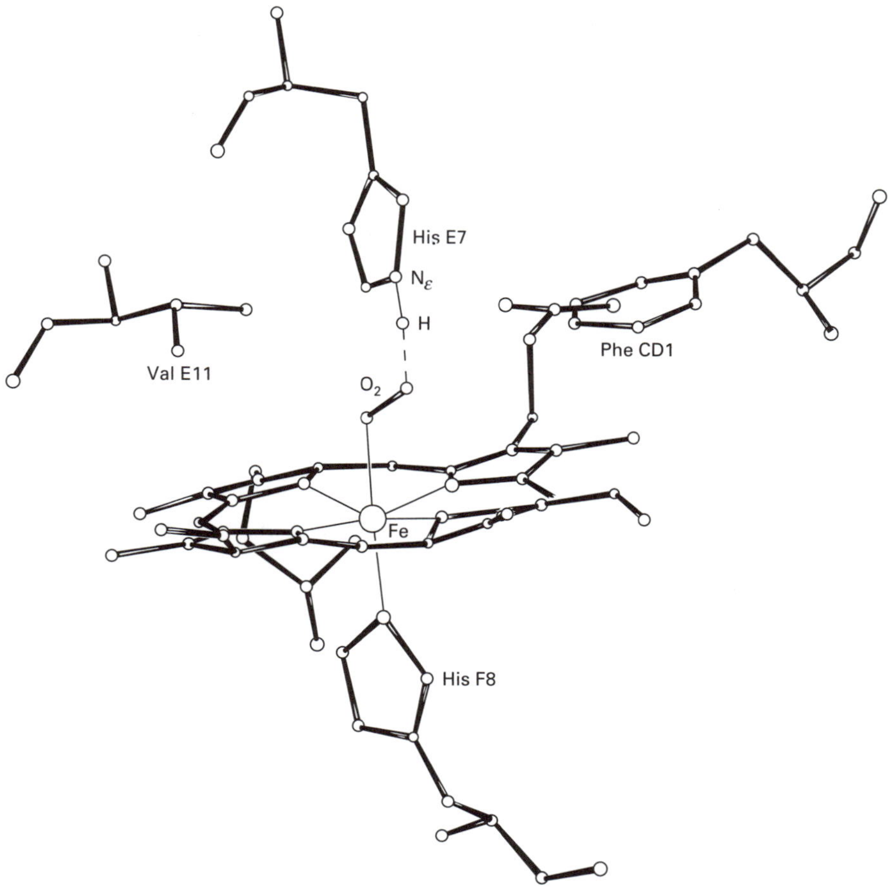

FIGURE 8.22

O_2 bound to the heme group of myoglobin. The iron atom of the heme group is firmly liganded on its proximal side chain to His F8 (so designated because it is the eighth residue of
the F helix). The O_2 molecule is bound on the distal side to the iron atom and is also hydrogen-
bonded to the distal His E7. Two other protein side chains that line the heme pocket are
also illustrated. This drawing depicts the individual atoms almost as points, but their van der
Waals volumes would largely fill the space. (From M. F. Perutz, *Trends Biochem. Sci.*
14:42–44, 1989.)

and a Val residue of the heme pocket. Consequently,
CO is found to bind to myoglobin in the bent configuration, like that of O_2. Removing the His and Val side
chains by mutagenesis decreases the discrimination
against CO by myoglobin.

The heme pocket is buried in the interiors of both
myoglobin and hemoglobin, with no channel to the solvent apparent in their crystal structures. Yet O_2 and CO
associate and dissociate rapidly. Fluctuations of the
protein structure are presumed to produce a channel of
sufficient size between the solvent and the heme pocket.
Binding of either O_2 or CO to the heme group produces
substantial changes in its absorbance spectrum, so the
kinetics of ligand binding have been studied exten-

sively. In particular, bound O_2 or CO can be dissociated
very quickly from the heme group by a flash of intense
light, and the rebinding of the ligand can be followed
experimentally. Many such studies have uncovered a
number of steps in rebinding, but of uncertain nature.
These studies demonstrate vividly what a complex process ligand binding to a protein can be when studied on
short time scales.

References

Binuclear oxygen carriers: hemerythrin. I. M. Klotz and D. M.
    Kurtz, Jr. *Acc. Chem. Res.* 17:16–22 (1984).

Reaction of myoglobin with phenylhydrazine: a molecular doorstop. D. Ringe et al. *Biochemistry* 23:2–4 (1984).

Rate theories and puzzles of hemeprotein kinetics. H. Frauenfelder and P. G. Wolynes. *Science* 229:337–345 (1985).

X-ray structure and refinement of carbonmonoxy (Fe^{II})-myoglobin at 1.5 Å resolution. J. Kuriyan et al. *J. Mol. Biol.* 192:133–154 (1986).

Control of the catalytic activity of prosthetic heme by the structure of hemoproteins. P. R. Ortiz de Montellano. *Acc. Chem. Res.* 20:289–294 (1987).

The structure of myoglobin-ethyl isocyanide: histidine as a swinging door for ligand entry. K. A. Johnson et al. *J. Mol. Biol.* 207:459–463 (1989).

Myoglobin and haemoglobin: role of distal residues in reactions with haem ligands. M. F. Perutz. *Trends Biochem. Sci.* 14:42–44 (1989).

c. Electrons and Redox Proteins

A variety of proteins bind electrons reversibly; they are especially important in photosynthesis and respiration. Because they are being reduced and oxidized, such proteins are termed **redox** proteins. Free electrons are not usually present in solution, so the electrons are usually transferred to and from other molecules, often another protein. These other molecules are specifically bound as ligands by the redox protein, and the electrons are transferred in the complex. The specificity in electron binding is not for the electron itself but for its donor and acceptor. How the actual electron transfer occurs is not known, but the site where the electron is held is usually obvious.

Perhaps the simplest redox proteins, in which no prosthetic groups are present, are those that use reversible formation of a disulfide bond between two Cys residues (Sec. 2.4.10), as in thioredoxin. Such proteins seem to be relatively rare, however, and most redox proteins use prosthetic groups such as flavins or NAD. Many use the ability of iron and copper ions to exist in two different redox states; for instance, Fe^{2+}/Fe^{3+} and Cu^+/Cu^{2+}. In some cases, such as the cytochromes, the iron atom is incorporated into a heme group whereas in others it is part of an iron–sulfur complex (Fig. 8.17). The heme groups of cytochromes bind electrons, rather than O_2 molecules as the globins do, because the sixth coordinating position of the iron atom, where O_2 would be bound, is occupied by a Met side chain in the cytochrome.

The tightness of binding of very small ligands such as electrons by proteins via prosthetic groups is determined both by the intrinsic properties of the prosthetic group and by modulation of its properties by the protein. For example, the redox potentials of heme-containing proteins vary enormously, from -1100 mV to $+300$ mV, covering some 24 orders of magnitude in affinity for an electron. Understanding how this modulation of affinity occurs is still not complete, owing to our insufficient understanding both of protein structure and of the properties of the prosthetic groups. Indeed, these particular groups may be used by proteins because they are sufficiently complex that binding an electron produces substantial changes in them that can be transmitted to the rest of the protein, and vice versa. As a very simple example, the Cu^+ ion prefers to be tetrahedrally coordinated whereas Cu^{2+} prefers planar or octahedral coordination. Consequently, a copper ion held in a tetrahedral coordination should have a stronger preference for being Cu^+ (i.e., have a higher electron affinity) than a copper ion held in planar or octahedral coordination. Therefore, the orientations of the coordinating side chains in a protein should alter the redox potential of the copper ion. Furthermore, the binding or release of an electron in this case would be expected to produce a tendency to alter the protein structure through the side chains coordinating the copper ions. The effects of ligand binding on the protein and the effect of the protein on ligand affinity are linked functions (see Fig. 7.32).

All such perturbations of the intrinsic affinity of a prosthetic group, including the 10^{24}-fold variation in electron affinities of heme groups, imply a suitably rigid protein conformation. Otherwise, the protein would simply adapt to the consequences of ligand binding (Fig. 8.5) and would have little power to modulate the affinity of the prosthetic group. Alternatively, the redox state that is disfavored by the protein geometry would be energetically unstable, and the protein would tend to unfold, or the prosthetic group would dissociate. Very little is known about how a redox protein modulates the affinity of the prosthetic group, but the covalent linkage between the two probably serves to prevent their dissociation. The structures of both redox states of known redox proteins demonstrate very few, if any, geometrical differences; large energetic effects on binding, therefore, do not require large conformational changes in proteins.

A major physical problem to be solved is how two redox proteins transfer their bound electrons so rapidly from one to the other. The redox centers are generally buried in the interiors of proteins, where their electrons are inaccessible to the solvent or to other molecules. Yet the electrons are transferred very rapidly to the appropriate acceptor protein over substantial distances between their redox centers. The electron transfer process appears to involve electron tunneling, a quantum mechanical phenomenon that is outside the scope of this volume.

References

Proteins containing 4Fe–4S clusters: an overview. V. W. Sweaney and J. C. Rabinowitz. *Ann. Rev. Biochem.* 49:139–161 (1980).

The crystal structure of poplar apoplastocyanin at 1.8 Å resolution. The geometry of the copper-binding site is created by the polypeptide. T. P. J. Garrett et al. *J. Biol. Chem.* 259:2822–2825 (1984).

The structure, function and evolution of cytochromes. F. S. Mathews. *Prog. Biophys. Mol. Biol.* 45:1–56 (1985).

Control of the redox potential of cytochrome *c* and microscopic dielectric effects in proteins. A. K. Churg and A. Warshel. *Biochemistry* 25:1675–1681 (1986).

Crystal structure analyses of reduced (Cu^I) poplar plastocyanin at six pH values. J. M. Guss et al. *J. Mol. Biol.* 192:361–387 (1986).

Long-distance electron transfer in proteins and model systems. G. McLendon. *Acc. Chem. Res.* 21:160–167 (1988).

Long-range electron transfer in multisite metalloproteins. H. B. Gray and B. G. Malmström. *Biochemistry* 28:7499–7505 (1989).

Electrical wiring of redox enzymes. A. Heller. *Acc. Chem. Res.* 23:128–134 (1990).

Mechanisms of long-distance electron transfer in proteins: lessons from photosynthetic reaction centers. S. G. Boxer. *Ann. Rev. Biophys. Biophys. Chem.* 19:267–299 (1990).

d. Metal Ion Storage and Transport

In contrast to the earlier examples, in which one or a few ions are bound very specifically by a protein molecule, other proteins bind many ions simultaneously. Large amounts of some essential ions, such as iron, must be available for use when required and must be stored in a benign form to avoid toxicity. Such a function is served by the protein **ferritin,** one molecule of which can bind about 4500 iron atoms. The protein alone, apoferritin, provides a spherical coat consisting of 24 identical polypeptide chains of 163 amino acid residues each; this coat encloses a microcrystalline inorganic matrix with the approximate composition $(FeOOH)_{8n}(FeO:OPO_3H_2)_n$. The protein does not provide specific binding sites for each stored ferric ion; instead, it probably serves primarily to take up specifically Fe^{2+} ions, oxidize them to Fe^{3+}, and precipitate them in the internal inorganic matrix.

Iron is transported in serum by being bound very tightly ($K_d = 10^{-20}\,M$) to the proteins of the **transferrin** family. These proteins have two Fe^{3+} binding sites, one in the interior of each of two domains. A $CO_3{}^{2-}$ anion is bound along with each Fe^{3+} ion in the same site, and two of its oxygen atoms participate in ligating the iron.

Another protein believed to be important for binding metal ions is **metallothionein,** which binds cadmium, copper, and zinc ions. The synthesis of this protein is induced by the presence of cadmium salts and other metals and by other types of stress, suggesting that its biological role is to sequester harmful metal ions. Metallothioneins generally consist of 61–64 residues of which 20 are Cys. The thiol groups of these Cys residues are the ligands for seven metal ions, which are bound in two clusters of three and four metal atoms in the two domains of the protein. Eight of the Cys thiols serve as bridging ligands to two different metal atoms.

References

Biochemistry of metallothionein. J. H. R. Kägi and A. Schäffer. *Biochemistry* 27:8509–8515 (1988).

Apolactoferrin structure demonstrates ligand-induced conformational change in transferrins. B. F. Anderson et al. *Nature* 344:784–787 (1990).

Proteins of iron storage and transport. R. R. Crichton. *Adv. Protein Chem.* 40:281–363 (1990).

The ferritin family of iron storage proteins. E. C. Theil. *Adv. Enzymol.* 63:421–449 (1990).

8.4 Allostery: Interactions between Binding Sites

Association between a single ligand molecule and a single site on a protein is relatively simple, but binding of multiple ligands at multiple sites can lead to very complex behavior. The multiple sites may be identical and have the same intrinsic binding properties, as when each subunit of an oligomeric protein contains the same binding site, or there may be different sites with different affinities. Binding at one site may have no effect on the affinities of the other sites or may increase or decrease them. Two ligand molecules that interact in this way may be identical **(homotropic interactions)** or may differ **(heterotropic interactions).**

Examples of interactions between ligands have been encountered already. Two ligands can compete for the same binding site (Eq. 8.6). Binding of the corepressor L-tryptophan increases the affinity of the *trp* repressor for its DNA binding site (Fig. 8.18). Ferric ions bind to transferrins only if carbonate ion also binds at an adjacent site (Sec. 8.3.4.d). In such cases, the natures of the interactions between the sites are apparent because the sites and the bound ligands are adjacent to each other. In most instances of interactions between ligands, however, the ligands are not related to each other, and the sites to which they bind are distant in the protein structure. The term **allostery** was introduced to describe this biologically important type of interaction; it is one of the primary ways in which physiological processes are regulated. Allosteric systems will be shown to have an additional level of complexity in that they invariably involve oligomeric or multidomain proteins, and the allosteric phenomenon involves changes in that quaternary structure.

Reference

Allosteric proteins and cellular control systems. J. Monod et al. *J. Mol. Biol.* 6:306–329 (1963).

8.4.1 Multiple Binding Sites and Interactions between Them

Most information about interactions between ligands comes from complexities in their binding behavior. Unfortunately, there is often considerable confusion about the interpretation and significance of complex binding curves, partly because the various methods for measuring binding can give different types of information. The most commonly used methods measure binding physically; an example is equilibrium dialysis, in which the free ligand, but not the protein or the complex, equilibrates across a semipermeable membrane. Such methods can measure the number of ligands bound per protein molecule, but averaged over all the binding sites of the protein population; they do not distinguish between different sites on each protein molecule. Some procedures, such as radioimmunoassay, measure a protein antigen as bound so long as it is bound to at least one antibody molecule, irrespective of the actual number. Spectral measurements of binding, which use spectral changes in the signals from either the protein or the ligand upon binding, have the potential to distinguish among classes of binding sites, but they have the disadvantage of usually determining only the fraction of the maximum binding that is possible, not the actual stoichiometries of ligand molecules per protein molecule. The first crucial step in analyzing binding data is to make certain exactly what is being measured.

References

Ligand-receptor interactions. I. M. Klotz. *Quart. Rev. Biophys.* 18:227–259 (1985).

The origin and use of the terms competitive and non-competitive in interactions among chemical substances in biological systems. H. McIlwain. *Essays Biochem.* 22:158–186 (1986).

Cooperative binding to macromolecules. A formal approach. B. Perlmutter-Hayman. *Acc. Chem. Res.* 19:90–96 (1986).

a. Identical and Independent Sites

Multiple identical and independent binding sites on a protein molecule usually occur when the protein is a symmetrical oligomer and each subunit has the same binding site. The binding curves in this case are similar in shape to those with a single ligand, but the maximum stoichiometry of binding should be an integer n equal to the number of sites per molecule (Fig. 8.23A). The intrinsic K_d of a site is given by the concentration of free ligand that produces half of the maximum binding, as in the case of a single site. Such binding curves are often interpreted, however, in terms of association or dissociation constants for the 1st, 2nd, . . . , nth ligand molecules to bind; the use of such binding constants introduces a statistical factor. For example, the first ligand to bind to the protein has n empty binding sites available, so the rate constant for association is n times that for an individual site, k_a. The rate constant for dissociating is the same as that for an individual site, k_d. The association constant for binding the first ligand, K_a^1, is then

$$K_a^1 = \frac{nk_a}{k_d} = nK_a \qquad (8.35)$$

or n times that of an individual site. Conversely, the nth ligand to bind has only one site available for binding, but there are n ways of losing one ligand from the fully saturated molecule, so

$$K_a^n = \frac{k_a}{nk_d} = \frac{1}{n}K_a \qquad (8.36)$$

or $1/n$ that of an individual site. Statistical factors also apply in similar ways to the intervening association constants.

b. Independent and Nonidentical Sites

In this case, the sites with highest affinity become occupied at the lowest ligand concentrations, followed by those sites of decreasing affinity. Unless site-specific binding is measured, the curves of total binding are complex. In particular, Scatchard plots (Fig. 8.6C) become concave, the curvature depending on the differences in affinities of the sites. The curve may have two or more linear portions if the affinities of the sites are very different, and the stoichiometries and affinities of

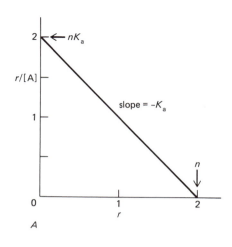

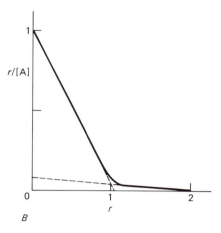

FIGURE 8.23

Scatchard plots of binding of ligand A to two sites on a protein P. The degree of binding is expressed as r, the moles of ligand bound per mole of protein. *A:* Two identical but noninteracting sites, each with an intrinsic association constant K_a of 1.0, expressed in the same units as the concentration of ligand. The slope is $-K_a$, and the x intercept is the extrapolated maximum number of ligand molecules bound, n. The y intercept is nK_a. *B:* Two different and independent sites with intrinsic association constants that differ 50-fold: $K_a^1 = 1.00$ and $K_a^2 = 0.02$. The two linear segments of the curve correspond approximately but not exactly to the binding to the two sites. The y intercept of the steep portion gives an apparent value of 1.02 for the binding constant for the first ligand to bind, which is $K_a^1 + K_a^2$.

the various classes of sites can then be estimated (Fig. 8.23B).

c. Cooperative Binding of the Same Ligand to Multiple Sites

Binding of ligand to one site on a protein molecule can increase or decrease the affinities of the other sites; that is, cooperativity can be either positive or negative. When the sites are intrinsically identical, as in an oligomeric protein of identical subunits, the interactions between the sites are said to be **homotropic.**

Positive cooperativity produces a sigmoidal relationship between degree of binding and free-ligand concentration (Fig. 8.24A) because the affinity for successive ligand molecules bound to one protein molecule increases. A convex Scatchard plot results (Fig. 8.24B); this plot is most sensitive to low degrees of cooperativity. It is not widely used, however, because its interpretation is not straightforward. Hill plots are most often used for positive cooperativity (Fig. 8.24C) because the slope in the middle of the curve, the **Hill coefficient,** gives a quantitative measure of the degree of cooperativity. It corresponds to the hypothetical number of ligand molecules that would have to be bound fully cooperatively, in an all-or-none fashion, to give such a slope. In this hypothetical case, only empty and fully occupied protein molecules would be present:

$$P + nA \rightleftharpoons P \cdot A_n \qquad (8.37)$$

$$K_a = \frac{[P \cdot A_n]}{[P][A]^n} \qquad (8.38)$$

$$\frac{y}{1 - y} = \frac{[P \cdot A_n]}{[P]} = K_a [A]^n \qquad (8.39)$$

$$\log\left(\frac{y}{1 - y}\right) = \log K_a + n \log A \qquad (8.40)$$

where y is the fraction of maximum binding (Eq. 8.5). A totally linear Hill plot with a slope of n would be expected in this hypothetical case.

In reality, binding is never totally cooperative, and the Hill plot has a slope greater than unity only in the middle. The ends of the curves have slopes of unity; these linear segments correspond to the binding to the protein molecule of the first and the last ligand molecules, and they give the association constants of the first and last ligands to bind. The Hill coefficient is given by the maximum slope of the curve. It may have any value, not necessarily an integer, depending on the degree of cooperativity. Its value could be as large as the number of interacting binding sites, but only if there were complete cooperativity. In reality, partially occupied protein molecules are always present even if there is positive cooperativity, but at lower concentrations than with independent binding (Fig. 8.24D).

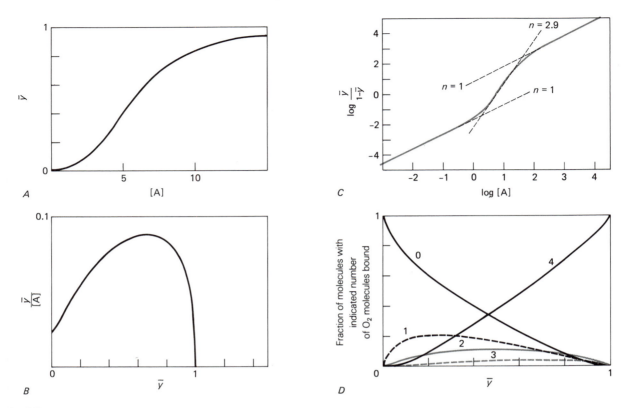

FIGURE 8.24

Positive cooperativity in ligand binding. The binding curve in each case is that for binding of O_2 by normal hemoglobin at 25°C in 0.1 M NaCl at pH 7.4, where the apparent association constants for the first, second, third, and fourth O_2 molecules bound to the four heme groups are, respectively, 0.024, 0.074, 0.086, and 7.4 mm^{-1}; the O_2 concentration is expressed as its partial pressure in millimeters of mercury. A: Plot of $\bar{y}$, the average fraction of complete binding, versus the ligand concentration [A]. B: Scatchard plot. C: Hill plot. (From Y. Tyuma et al., *Biochemistry* 12:1491–1498, 1973.) D: The fractions of hemoglobin molecules with zero to four O_2 molecules bound are plotted as a function of the fractional saturation. As a result of the cooperativity of binding, the molecules with zero and four bound O_2 molecules predominate at all stages of binding, with partially saturated molecules being relatively rare. With no cooperativity, in contrast, partially saturated molecules would predominate from approximately $\bar{y} = 0.16$ to $\bar{y} = 0.84$.

Negative cooperativity produces Hill plots with slopes of less than unity, owing to the decreased affinity of the protein molecule for successive ligand molecules to be bound. Scatchard plots are used more frequently to analyze such binding data because they give concave curves, often biphasic, like those with sites of different affinities (Fig. 8.23B). Curved Scatchard plots resulting from negative cooperativity are not, however, readily interpretable in terms of the affinities for the successive ligands or of the interactions between them.

Binding curves alone cannot distinguish negative cooperativity between otherwise identical sites from independent but different sites on the same protein. Apparent negative cooperativity can even result simply from unsuspected heterogeneity of the protein molecules present. For these reasons, there are few well-documented cases of negative cooperativity. The effect of added ligand on the rate of dissociation of radioactive ligand from partially occupied protein molecules is useful for distinguishing the various possible explanations; the rate of dissociation should be increased by added ligand only if there is negative cooperativity between the sites.

References

Ligand competition curves as a diagnostic tool for delineating the nature of site–site interactions: theory. Y. I. Henis and A. Levitzki. *Eur. J. Biochem.* 102:449–465 (1979).

Table 8.5 *Allosteric Effectors of O_2 Affinity of Hemoglobin*

	Major Binding Site on Human T State	
Decrease O_2 affinity	α Chain	β Chain
Organic phosphates (e.g., diphosphoglycerate; see Fig. 8.2)		α-NH_3^+ His 2 Lys 82 His 143
CO_2	α-NH_2	α-NH_2
Anions (e.g., Cl^-)	α-NH_3^+ Arg 141	Lys 82
H^+ (Bohr effect)	α-NH_2	His 146
	Binding Site on R State	
Increase O_2 affinity	α Chain	β Chain
O_2 and other heme ligands	Heme Fe	Heme Fe

Data from A. Arnone and M. F. Perutz, *Nature* 249:34–36 (1974); J. V. Kilmartin, *Trends Biochem. Sci.* 2:247–250 (1977); S. O'Donnell et al., *J. Biol. Chem.* 254:12204–12208 (1974); J. V. Kilmartin and L. Rossi-Bernardi, *Biochem. J.* 124:31–45 (1971); A. Arnone, *Nature* 237:146–149 (1972).

Free energy coupling within macromolecules. The chemical work of ligand binding at the individual sites in cooperative systems. G. K. Ackers et al. *J. Mol. Biol.* 17:223–242 (1983).

d. Interactions between Different Ligands

The binding of one ligand can affect the binding of different ligands at other sites on the same protein molecule. These **heterotropic** effects are easier to analyze than homotropic effects because the concentrations of the different ligands can usually be varied independently, and the interactions between them can be observed directly.

Interactions between sites are **linked functions** (see Fig. 7.32). Consider the binding of two ligands A and B to different sites on a protein P, with the indicated association constants:

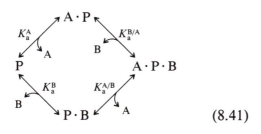

$$(8.41)$$

Because the free-energy change around such a cycle must be zero,

$$\frac{K_a^A}{K_a^{A/B}} = \frac{K_a^B}{K_a^{B/A}} \tag{8.42}$$

In other words, whatever effect the binding of ligand A has on the affinity for ligand B, the binding of B must have exactly the same effect on the affinity for A. For example, binding of organic phosphates, hydrogen ions, and CO_2 affects the oxygen affinity of hemoglobin (Table 8.5); likewise, the binding of oxygen must affect the affinity of hemoglobin for these ligands. If an inducer affects the binding of a repressor molecule to the operator region of DNA (Fig. 8.18), binding to the DNA must affect the repressor's affinity for the inducer.

References

Linked functions and reciprocal effects in hemoglobin: a second look. J. Wyman, Jr. *Adv. Protein Chem.* 19:223–286 (1964).

Linkage graphs: a study in the thermodynamics of macromolecules. J. Wyman. *Quart. Rev. Biophys.* 17:453–488 (1984).

The structure of the saccharide-binding site of concanavalin A. A. Derewenda et al. *EMBO J.* 8:2189–2193 (1989).

8.4.2 Allosteric Models

The problem of how different binding sites on a protein interact is very closely related to the phenomenon of protein flexibility (Fig. 8.5). The relationship between these two problems has been clarified by the proposal of two opposing, extreme structural models for allostery: the sequential and the concerted models.

Reference

Mechanisms of cooperativity and allosteric regulation in proteins. M. F. Perutz. *Quart. Rev. Biophys.* 22:139–236 (1989).

a. The Sequential Model

One school of thought holds that a protein is sufficiently flexible that it is easy for the binding of a ligand at one site to alter directly the protein conformation at another site, thereby affecting the affinity of the second site for its ligand. Such arguments are plausible on energetic terms because binding affinities in allosteric systems are usually changed by no more than a few orders of magnitude, which corresponds to differences of only a few kilocalories per mole in binding interaction. Such changes could probably be produced by small conformational perturbations of the binding site. If so, virtually any type of allosteric interactions would be possible, depending only on what conformational changes are induced at one site upon ligand binding at another. The binding of any ligand to any site on a protein could affect independently every other binding site, increasing or decreasing its affinity for its particular ligand. Binding different ligands at the same site could also have different effects on the other sites. Some relatively simple models, proposed by Koshland, Nemethy, and Filmer, can account for the known allosteric properties of proteins. Their general scheme has come to be known as the **KNF** or **sequential model**, the latter because the conformational and affinity effects occur sequentially as a consequence of binding each ligand.

With the sequential model view of ligand binding, it is almost surprising that so many proteins exhibit normal, nonallosteric binding behavior. Binding a ligand to one site on a protein might be expected invariably to alter significantly the ligand affinity of another binding site. This does not appear to be the usual situation, although there have been few systematic studies of binding to separate sites on monomeric proteins. The changes in conformation upon binding ligand are usually small (Table 8.1), although only a small change in structure might be needed to produce a significant

change in affinity. One of the largest movements detected upon ligand binding is a 12-Å movement of a peptide loop upon binding of the cofactor NAD to each of the four sites on tetrameric lactate dehydrogenase, yet there is no effect of this on the NAD affinities of the other subunits.

References

Comparison of experimental binding data and theoretical models in proteins containing subunits. D. E. Koshland et al. *Biochemistry* 5:365–385 (1966).
Energetics of ligand binding to proteins. G. Weber. *Adv. Protein Chem.* 19:1–83 (1975).

b. The Concerted Model

Proteins demonstrating interactions between nonoverlapping binding sites are invariably oligomeric, consisting of at least several identical polypeptide chains and often also of different polypeptides. Upon binding an appropriate ligand, they demonstrate very large conformational changes, sufficient to disrupt a crystal lattice; in contrast, nonallosteric proteins generally bind ligands in an unaltered crystal lattice unless the lattice contacts are directly perturbed by the ligand. The large conformational changes in allosteric proteins involve rearrangements of relatively unaltered subunits (i.e., they are primarily quaternary structure changes).

The archetypal allosteric protein is the well-known hemoglobin of vertebrates. Its deoxy and oxy (or other liganded) forms differ by a 15° rotation of one pair of α and β subunits relative to the other two in the $\alpha_2\beta_2$ tetramer (Fig. 8.25). Owing to this large change, crystals of the two forms of the protein are different, and both crack and dissolve upon converting the protein to the other form.

These properties of allosteric proteins led Monod, Wyman, and Changeux to propose a model for allosteric interactions that envisages the other extreme of protein flexibility to the sequential model. In their scheme, the binding of ligand at one site has no direct effect on the affinities of the other sites, but it alters the conformational equilibrium between two alternative quaternary conformations of the protein. One conformation, having low intrinsic affinity for the ligand at all its sites, is designated T (for *tense*) because it is imagined to be constrained in some manner. The other conformation, having high affinity for the ligand, is designated R (for *relaxed*). The two forms with i ligand molecules bound are often referred to as T_i and R_i. According to the model, these two conformations coexist even in the absence of ligand, with an equilibrium constant L between

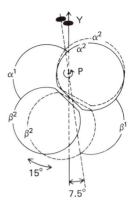

FIGURE 8.25

The quaternary structures of deoxy (T) and liganded (R, oxy) hemoglobins. The $\alpha^1\beta^1$ dimers of both are superimposed, and the relative positions of the $\alpha^2\beta^2$ dimer are depicted in solid lines for T and dashed lines for R. The $\alpha^2\beta^2$ unit rotates 15° relative to $\alpha^1\beta^1$ about the axis P perpendicular to the plane of the paper and to the twofold symmetry axis Y. The only direct contacts between the subunits are between α^1 and β^1 and between α^1 and β^2, plus the symmetric contacts between α^2 and β^2 and between α^2 and β^1. The binding site for the organic phosphate allosteric effectors DPG, IPP, and IHP in the T state (Fig. 8.2) is at the bottom of the molecule straddling the twofold axis between the two β chains. In the R state, the two chains move closer together, essentially destroying this binding site. (From J. M. Baldwin, *Trends Biochem. Sci.* 5:224–228, 1980.)

T_0 and R_0. The T_0 form would normally be favored ($L < 1$), and the protein would have relatively low affinity for the first ligand molecule. The R conformation has the higher affinity by a factor c, so ligand molecules will be bound to it preferentially, which will pull the conformational equilibrium toward the R state because the conformational change and ligand binding are linked functions. The conformational equilibrium between the two conformations with one ligand molecule bound will be cL. The other vacant sites on all the R_1 molecules will then be in the high-affinity form; therefore, the average affinity of the vacant sites of the entire population will be increased. The same phenomenon occurs upon binding subsequent ligand molecules, giving positive cooperative homotropic interactions.

In this model, heterotropic interactions involving other ligands arise because these ligands bind preferentially to either the R or the T states. Affinity for the original ligand is decreased by other molecules that bind preferentially to the T state because the other molecules pull the conformational equilibrium back toward the T state. Affinity for the ligand is increased by any molecule that binds preferentially to the R state. In general, each ligand controls the apparent affinity of the protein for

other ligands simply by shifting the equilibrium between the R and T states.

The concerted model is also known as the **MWC**, **symmetric**, or **two-state allosteric model** because it envisages only two symmetric quaternary structures. It is illustrated in its simplest form and compared with the sequential model in Figure 8.26. The two models differ in terms of protein conformation primarily with respect to the conformations and ligand affinities of the partially liganded states. Both models imply that ligand binding has effects on the protein conformation. In the sequential model, such effects extend directly to the other binding sites and affect their ligand affinities. In the concerted model, the conformational effects need extend only to the interface between the subunits to alter the conformational equilibrium between the R and T quaternary states. The two models also differ in that the concerted model envisages the two conformations to be present even in the absence of ligand, R_0 and T_0, whereas in the sequential model the R conformation is induced only upon ligand binding. The sequential model predicts that the conformational change upon ligand binding should parallel the extent of ligand binding, whereas with the concerted model the two need not coincide because the conformational change should tend to occur at one particular stage of ligand binding of each molecule: the stage when Lc^i becomes greater than unity (Fig. 8.26).

The concerted model is much more restrictive than the sequential model. The only parameters that can be varied are L, the conformational equilibrium constant in the absence of any ligand, and the affinities of the two states for each ligand; the relative affinities of the two states specify the allosteric parameter c. Moreover, the two parameters L and c are not observed to be independent in the case of hemoglobin (Sec. 8.4.3): alteration of one also changes the other. This interdependence probably reflects linkage between the conformational change and the change in affinity for the ligand in state T.

The greatest restriction of the concerted model is that it does not predict negative cooperativity because a ligand can pull the conformational equilibrium only toward the form with high affinity for it. This phenomenon will be considered further in the next section.

References

On the nature of allosteric transitions: a plausible model. J. Monod et al. *J. Mol. Biol.* 12:88–118 (1965).

On the nature of allosteric transitions: implications of non-exclusive ligand binding. M. M. Rubin and J. P. Changeux. *J. Mol. Biol.* 21:265–274 (1966).

Close correlation between Monod-Wyman-Changeux param-

Sequential

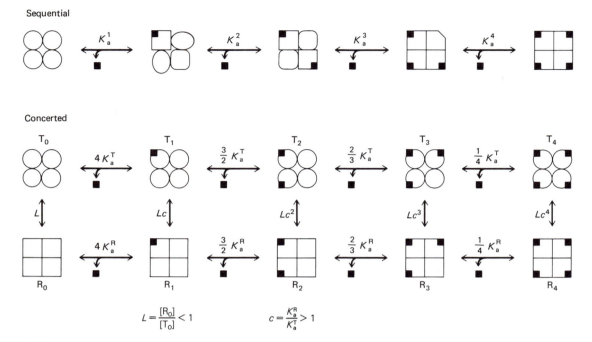

Concerted

$$L = \frac{[R_0]}{[T_0]} < 1 \qquad c = \frac{K_a^R}{K_a^T} > 1$$

FIGURE 8.26

Schematic illustration of the sequential and concerted models for positive allosteric cooperativity of ligand (■) binding, using a hypothetical tetrameric protein. In the sequential model, binding of each ligand induces a conformational change in the subunit to which it is bound and in the other subunits, thereby increasing their affinities for ligand, so that $K_a^1 < K_a^2 < K_a^3 < K_a^4$. The conformations depicted and the sequence of binding by the individual subunits are arbitrary. In the concerted model, there are only two conformational states, T and R, with intrinsic affinities of individual sites for ligand of K_a^T and K_a^R, respectively, which are not altered by ligand binding to other subunits. Instead, binding of each ligand shifts the equilibrium from T to R by the factor c, which is the ratio of the two intrinsic affinities, K_a^R/K_a^T. The association constants for binding of the first, second, and so on, ligand molecules include statistical factors to take into account the numbers of free and occupied sites on the protein molecule. The intrinsic affinities of all sites in the T state are postulated to be identical, as are those in the R state.

eters, L and c, and its implication for the stereochemical mechanism of haemoglobin allostery. S. Matsukawa et al. *J. Mol. Biol.* 150:615–621 (1981).

An analytic solution to the Monod-Wyman-Changeux model and all parameters in this model. A. Zhou et al. *Biophys. J.* 55:275–280 (1989).

Is symmetry conservation an unessential feature of allosteric theory? C. Debru. *Biophys. Chem.* 37:15–23 (1990).

8.4.3 The Allosteric Properties of Hemoglobin

The most extensively studied allosteric protein is vertebrate hemoglobin because its allosteric properties are so obviously of physiological importance in optimizing O_2 transport by erythrocytes. Homotropic interactions occur between the four O_2 molecules bound to the four heme groups of the $\alpha_2\beta_2$ structure (Fig. 8.24C,D), and there are heterotropic interactions with other allosteric effectors — namely, organic phosphates, CO_2, Cl^- ions, and H^+. The well-known sigmoidal curve of O_2 binding as a function of O_2 pressure (the equivalent of O_2 concentration in solution), and its modulation by the other allosteric effectors, doubles the efficiency with which hemoglobin delivers O_2 from the lungs to the tissues and makes it very responsive to changes in oxygen pressure. Hemoglobin served as the prototype for the concerted allosteric model and still most closely fits that model, even though it has been modified somewhat. Nevertheless, many details remain to be elucidated.

Hemoglobin has an $\alpha_2\beta_2$ quaternary structure (Fig. 8.25) that is close to a tetrahedral tetramer of identical polypeptide chains. The α and β chains are not identical, however, although they are homologous (see Fig. 3.12) and have similar functional properties. The $\alpha_2\beta_2$ structure is often approximated as a tetramer with four identical binding sites. However, the structure is really a dimer of $\alpha\beta$ units. The α and β subunits of the same dimer unit are designated with the same superscripts (i.e., $\alpha^1\beta^1$ and $\alpha^2\beta^2$). The $\alpha^1\beta^1$ and $\alpha^2\beta^2$ units are identical and are related by a twofold symmetry axis (Fig. 8.25). The distinction between the α and β subunits complicates analysis of the O_2 binding properties of hemoglobin. Instead of there being five ligation states in a symmetrical tetramer, with zero to four O_2 molecules bound, there are ten distinguishable liganded forms of the $\alpha_2\beta_2$ structure. There are numerous uncertainties about the O_2 affinities of the α and β chains in the different forms of the protein, and the consequences of O_2 binding at the heme groups of the α and β chains are also somewhat different. Because of these uncertainties, the α and β chains will not be distinguished in this discussion.

Hemoglobin is found to exist in either of only two quaternary structures, R and T (Fig. 8.25), in accord with the concerted model. Only these two quaternary structures have been observed in numerous crystal structures of numerous forms of the protein. The individual $\alpha\beta$ dimers differ only slightly in the R and the T structures. The main difference between the two structures is that the $\alpha^2\beta^2$ dimer is rotated and translated relative to the $\alpha^1\beta^1$ dimer (Fig. 8.25). The two β subunits are about 5 Å farther apart in the T quaternary structure than they are in the R structure. This difference is due primarily to differences at the $\alpha^1\beta^2$ interface (and the symmetry-related $\alpha^2\beta^1$ interface) that involve alternative interdigitations of groups that must be either one way or the other, not in between (Fig. 8.27). In the T structure, Thr 41α of the C helix is dovetailed into a groove of the other subunit, whereas in the R structure it is Thr 38α that is in this position. Thr 41α is located one turn along the C helix from Thr 38α. Because of this interdigitation, intermediate quaternary structures are unlikely to be stable, so it is not surprising that no other quaternary structures have been observed. Which quaternary structure is adopted depends on which ligands are present. O_2 or other heme ligands favor the R state whereas the known heterotropic allosteric effectors bind preferentially to the T state (Table 8.5).

The most dramatic of the heterotropic allosteric effectors are 2,3-diphosphoglycerate (DPG) and the inositol pentaphosphate and hexaphosphate (IPP and IHP, respectively). They lower the O_2 affinity of hemoglobin in vivo by binding tightly to the T structure. All bind at the same single site per $\alpha_2\beta_2$ tetramer; only one site is present because it straddles the β subunits across the twofold symmetry axis (Fig. 8.2). This binding site is destroyed when the quaternary structure changes to the R state because the β chains move much closer together (Fig. 8.25). Consequently, the T state has much higher affinity for these effectors than does the R state. DPG is present in most mammalian erythrocytes but is replaced by IPP in birds and by HCO_3^- in crocodiles. Bird hemoglobins have two additional Arg residues on each β chain at positions favorable for participation in binding IPP, presumably to neutralize and interact with its greater number of phosphate groups. Three amino acid replacements in this region of the β chains of crocodile hemoglobins also can explain their preference for HCO_3^- as the allosteric effector. These instances of evolutionary replacements of amino acids seem to have occurred for functional reasons; they do not seem to be selectively neutral, as most of the other replacements of residues apparently are (Sec. 3.3.1.d).

The oxygen affinity of hemoglobin is physiologically coupled in two ways to the CO_2 content of the blood, with increased CO_2 causing release of O_2 from oxyhemoglobin. Direct interactions occur because CO_2 binds as the carbamate to the α-amino groups of both α and β chains in the T state:

$$-NH_2 + CO_2 \rightleftharpoons -NH-CO_2H \quad (8.43)$$

Binding of CO_2 to the T structure is favored because the ionized carbamate has electrostatic interactions with positively charged groups that are not in the same position in R.

Elevated concentrations of CO_2 in the blood also cause the pH to decrease, owing to formation of H^+ and HCO_3^-. This also lowers the O_2 affinity of hemoglobin because more protons happen to be bound by the T state than by the R state structure. Known as the **Bohr effect,** this phenomenon may result primarily from the elevated pK_a values of the His 146 side chain of β chains (8.09 in T_0 and 7.1 in R_4) and of the α-amino groups of the α chain (about 7.8 in T_0 and 7.0 in R_4) that are produced by their involvement in salt bridges in the T state but not in R. Anions such as Cl^-, DPG, and IPP also bind preferentially to the T structure and influence the Bohr effect. In particular, Cl^- binds between the α-amino group of one α chain and the guanidinium group of Arg 141 of the other α chain. This salt bridge is disrupted in the R state, and the Cl^- is released. Binding of this and other allosteric effectors such as DPG affects the ionization of the groups involved. Ionization of one group can also affect ionization of a neighboring group. Consequently, analysis of the Bohr effect is complex, and there is considerable uncertainty as to its assignment to individual groups.

The concerted allosteric model is very attractive for characterization of hemoglobin because of the occurrence of only two well-defined quaternary struc-

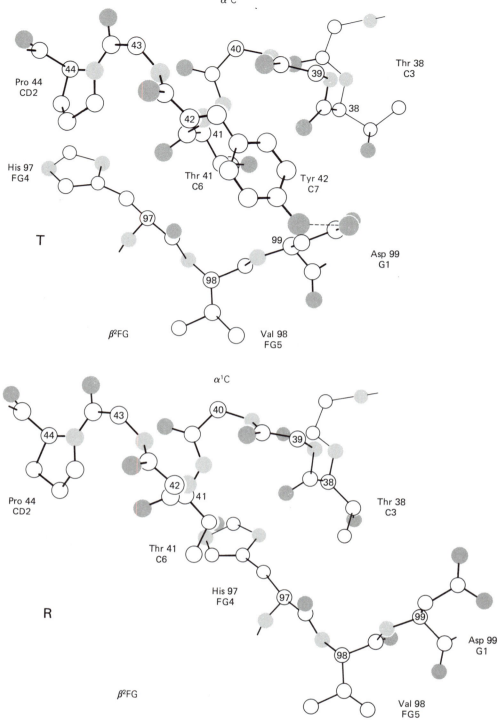

FIGURE 8.27

Part of the $\alpha^1\beta^2$ interface of hemoglobin, illustrating the large differences that occur there in the R (oxy) and T (deoxy) states. At the top of each structure is the C helix of the α^1 chain; at the bottom, the irregular corner linking the F and G helices of the β^2 chain. The conformations of both subunits are virtually the same individually in the R and T structures, but the contacts differ markedly owing to a shift of one subunit relative to the other. His 97 of the β chain is in contact with Thr 41 of the α chain in T, but with Thr 38 in R. Thr 38α and Thr 41α are on adjacent turns of the C helix. Intermediate positions of the two subunits would be unstable because His 97 and Thr 41 would be too close together. Therefore, hemoglobin $\alpha_2\beta_2$ molecules must be in either the R or the T quaternary structures. (From J. M. Baldwin, *Trends Biochem. Sci.* 5:224–228, 1980.)

tures. Moreover, the two quaternary structures can be shown to coexist, in the absence of any ligand, by modifying the protein at the $\alpha^1\beta^2$ interface or by disrupting the salt bridges that preferentially stabilize the T state. In this way, the value of L can be decreased from its normal value of approximately 10^{-4}, favoring the T state, to a value close to unity, so that R becomes populated substantially even in the deoxy form.

Likewise, liganded hemoglobin can be pulled into the T state by high concentrations of DPG, IPP, or IHP. In the absence of an $R \leftrightarrow T$ quaternary structure change, cooperativity of O_2 binding is much diminished.

The $\alpha_2\beta_2$ tetramer with O_2 or any other comparable ligand bound to all four heme groups is considered to be the R state because the binding properties of its four subunits are considered normal—similar to those of monomeric myoglobin, individual α chains, or $\alpha\beta$ dimers, although the affinity for O_2 is actually somewhat greater in the R state. In contrast, the ferrous deoxy tetramer is considered to be in the constrained T state because the oxygen affinities of its heme groups are decreased, usually about 500-fold. The R state appears to be unique because its O_2 affinity is not very sensitive to changes in conditions. The affinity of the T structure is not constant, however, and varies substantially depending on which allosteric effectors are present; mutations that affect the relative stability of R and T also affect the O_2 affinity of the T state. The greater the stability of T relative to R—that is, the smaller the value of L (Fig. 8.26)—the lower is the O_2 affinity of T. The two parameters L and c of the concerted model are not independent in hemoglobin but are related by the equation

$$\log \frac{K_a^R}{K_a^T} = \log c = A - 0.25 \log L \qquad (8.44)$$

where A is a constant. The greater the constraints stabilizing T, the lower is its affinity for O_2.

Although the T state is considered to be constrained, it is more stable than the R state when all the hemes are deoxy. Its quaternary structure has somewhat more extensive interactions between the subunits and extra salt bridges involving certain C-terminal residues of both chains that are not possible in the R state. The T state is also much less prone to dissociate into $\alpha^1\beta^1$ and $\alpha^2\beta^2$ units than is the R state.

The difference in O_2 affinities of the R and T structures appears to be a result of each quaternary structure preferentially stabilizing the liganded and deoxy forms of the heme group, respectively. The properties of the heme group change significantly upon ligand binding, even though the tertiary structures of the individual subunits are almost unchanged (Fig. 8.28). In the ferrous, deoxy form, the iron atom is somewhat too large to fit in the center of the heme group and is forced 0.5–0.6 Å out of the mean plane of the heme. Partly as a consequence, the heme group is domed in the unliganded T state. In the liganded R state, the iron atom lies in the heme plane, and the heme group is flat. Upon binding O_2 in the T state, the residues surrounding the heme group tend to prevent it from adopting the planar configuration, with the iron atom in its plane. Therefore, the O_2 affinity of the T state is low. These constraints are largely absent in the R structure, so the heme can take up the geometry preferred by either the liganded or the deoxy heme group. The O_2 affinity of the R state is therefore high and similar to that of myoglobin.

How does the quaternary structure affect the heme group affinity for ligands, and how does O_2 binding affect the quaternary structure equilibrium? The movements of the iron atom appear to be crucial because it is also attached to the proximal His side chain, and it is here that the largest changes in tertiary structure of the individual chains are observed (Fig. 8.29). The movement of the iron atom in and out of the plane of the heme group causes movements of the proximal His residue and of the F helix of which it is a part. The low oxygen affinity of the T state is partly due to this segment of the protein preventing the iron atom from moving toward the heme group, as should occur upon O_2 binding. Indeed, when the very strong ligand nitrous oxide is bound tightly to the iron atom, pulling the tetramer into the T state by a heterotropic effector such as IHP actually breaks the iron–His bond. This constraint normally just lowers the affinity of O_2 in the T state. When O_2 is bound, however, the constraint is reversed, and the Fe and His groups are pulled toward the heme, which also pulls the F helix in that direction. This movement is transmitted to the ends of a rigid α-helix, and the carboxyl end of the F helix is in contact with the other subunit across the $\alpha^1\beta^2$ interface (Fig. 8.27). This is the interface that changes markedly in the $T \leftrightarrow R$ quaternary structure change; this movement of the F helix, therefore, may be a major route by which ligand binding and the quaternary structure change are structurally linked. Another route may be the tendency of the liganded heme group to flatten and to push on the side chains of Leu FG3 and Val FG5 at the turn between the F and G helices (Fig. 8.28). These two residues also form part of the $\alpha^1\beta^2$ interface that changes between the R and T quaternary structures (Fig. 8.27).

Many questions remain about the binding of O_2 by hemoglobin, but the following picture emerges. Deoxy hemoglobin exists predominantly in the T structure, with L approximately 10^{-4}, but this value depends on the concentrations of any heterotropic effectors that might be present to stabilize the T state. O_2 binds only poorly to the heme groups in T_0 because the tertiary

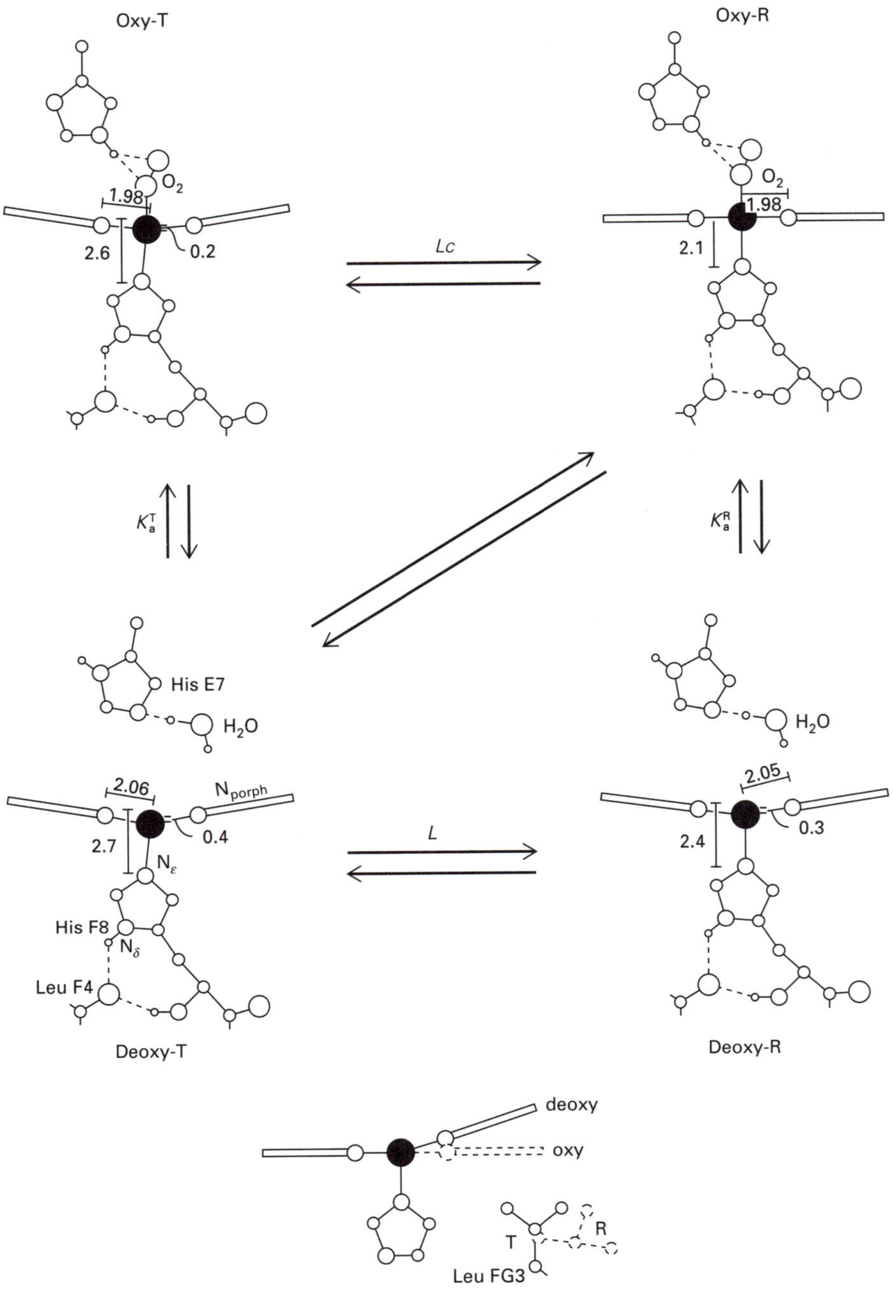

FIGURE 8.28

Schematic diagram of changes in heme stereochemistry on binding of O_2 by the α-subunits in the R and T structures of hemoglobin. The bottom diagram shows the change in conformation of the heme that ideally should take place upon binding O_2. The heme tends to flatten, but this is prevented in the T structure by the side chains of Leu FG3 and Val FG5 (Val FG5 is not depicted here because it lies behind Leu FG3). Consequently, upon uptake of oxygen by the T structure, the heme remains domed and the iron remains displaced from the porphyrin plane in the oxy-T state; the T state, therefore, has low O_2 affinity. This unfavorable stereochemistry is alleviated by the change to the R quaternary structure. The R structure does not constrain the heme group to be either flat or domed; therefore the heme can alternate between the oxy and deoxy conformations, and the R state has normal O_2 affinity. The numbers in the diagram indicate the distances (in angstroms) of N_ϵ of His F8 from the mean plane of the porphyrin carbons and nitrogens, the mean distance between the iron and the porphyrin nitrogens (N_{porph}), and the displacement of the iron from the plane of the porphyrin nitrogens. (From M. F. Perutz et al., *Acc. Chem. Res.* 20:309–321, 1987.)

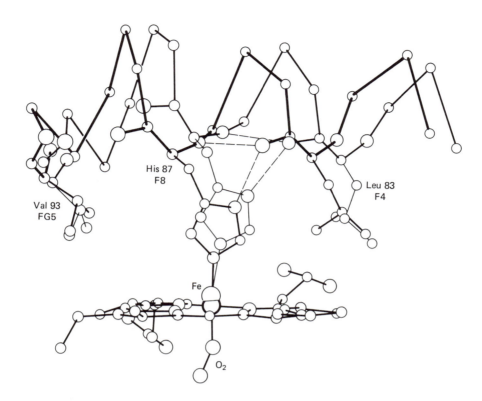

FIGURE 8.29

Changes at the proximal side of the heme group upon binding of O_2 to the α chains of hemoglobin. Similar changes occur in the β chains. The average positions of the heme groups of the deoxy (T) and oxy (R) structures are superimposed to illustrate the relative movements of the protein atoms. Some of the neighboring residues of the F helix and the FG corner between the F and G helices are shown in thick lines for oxy, thin lines for deoxy. Bound O_2 is shown below the heme group. Upon binding O_2 (or CO), the iron atom moves about 0.6 Å, from above the plane of the heme group into its plane. The iron atom pulls with it the proximal His 87, also known as F8 (the 8th residue of the F helix), and the F helix of which it is a part. This produces changes at the FG corner, on the left, which is in contact with the β^2 chain across the $\alpha^1\beta^2$ interface (see Fig. 8.27). It is this interface that changes in the T $\leftrightarrow$ R quaternary structure change, and the preceding sequence of events describes a plausible but unproved mechanism by which O_2 binding can affect the quaternary structure, and vice versa. The dashed lines illustrate hydrogen bonds that are thought to be important for orienting the His F8 side chain. (Kindly provided by B. Shaanan.)

structures of the individual α and β subunits inhibit the changes in structure of the heme group that should take place upon O_2 binding (Fig. 8.28). The greater the stability of the T state, the greater the inhibition of these changes and the lower the O_2 affinity. When O_2 is bound to the T state, the unfavorable interactions are transmitted to the $\alpha^1\beta^2$ interface, and the T state is weakened. Consequently, it has a greater tendency to dissociate into $\alpha\beta$ dimers and to change to the R quaternary structure because in both cases the constraints on O_2 binding are released and the affinity for O_2 is increased. The weakening of the T-state constraints may also cause the affinities of the other, unoccupied heme groups in the tetramer to increase somewhat. Upon binding a second and a third O_2 molecule, the T state is further destabi-

lized, and it is even more likely to dissociate or to change to the R state. The equilibrium between the T and R quaternary structures reaches unity when two or three O_2 molecules are bound, so the R structure comes to predominate at that stage. Because the unoccupied heme groups of R have high affinities for O_2, the greater the proportion of R, the higher the average affinity of the population of hemoglobin molecules for O_2. With four O_2 molecules bound, the R structure predominates.

The allosteric properties of hemoglobin largely confirm the expectations of the concerted Monod–Wyman–Changeux model, in that changes in quaternary structure are largely responsible for the cooperativity of O_2 binding and for the effects of the heterotropic effectors. The major difference is that the hemoglobin T state does not have a unique affinity for its ligands, although it appears to be a unique structure. The strengths of the structural constraints of the quaternary structure that produce its low affinity for O_2 are variable, not fixed, so any factor that alters the stability of the T structure, whether it be binding of H^+, Cl^-, DPG, or O_2, affects its affinity for O_2 molecules. Consequently, some cooperativity is manifest in the T state without a change of quaternary structure. This is, of course, the sort of phenomenon envisaged by the other extreme model, the sequential model (Fig. 8.26). Hemoglobin appears, therefore, to have used aspects of both allosteric models.

References

Hemoglobin: the structural changes related to ligand binding and its allosteric mechanism. J. Baldwin and C. Chothia. *J. Mol. Biol.* 129:175–220 (1979).

Hemoglobin tertiary structural change on ligand binding. B. R. Gelin et al. *J. Mol. Biol.* 171:489–559 (1983).

Species adaptation in a protein molecule. M. F. Perutz. *Adv. Protein Chem.* 36:213–244 (1984).

Haemoglobin: the surface buried between the $\alpha_1\beta_1$ and $\alpha_2\beta_2$ dimers in the deoxy and oxy structures. A. M. Lesk et al. *J. Mol. Biol.* 183:267–270 (1985).

Stereochemistry of cooperative mechanisms in hemoglobin. M. F. Perutz et al. *Acc. Chem. Res.* 20:309–321 (1987).

How much do we know about the Bohr effect of hemoglobin? C. Ho and I. M. Russu. *Biochemistry* 26:6299–6305 (1987).

Evolution of haemoglobin studied by protein engineering. K. Nagai et al. *BioEssays* 8:79–82 (1988).

Structure of haemoglobin in the deoxy quaternary state with ligand bound at the α haems. B. Luisi and N. Shibayama. *J. Mol. Biol.* 206:723–736 (1989).

Structure of deoxy-quaternary haemoglobin with liganded β subunits. B. Luisi et al. *J. Mol. Biol.* 214:7–14 (1990).

Oxygen transport in extreme environments. G. diPrisco et al. *Trends Biochem. Sci.* 16:471–474 (1991).

Molecular code for cooperativity in hemoglobin. G. K. Ackers et al. *Science* 255:54–63 (1992).

8.4.4 Other Allosteric O_2-Binding Proteins

Although vertebrate hemoglobin is most typical of allosteric systems, other types of cooperativity are known. Lamprey hemoglobin is one of the simplest allosteric proteins. Although the tertiary structure of its single polypeptide chain is like those of the myoglobin and hemoglobin chains (see Fig. 6.30), its cooperative binding of O_2 is based on reversible dissociation of identical subunits. The deoxy form of the protein tends to exist as dimers and tetramers, with relatively low O_2 affinity. Binding O_2 causes them to dissociate into monomers because the monomers have higher affinity. The O_2 affinity also depends on pH, and a substantial Bohr effect results from the release of protons upon binding O_2.

This simpler system has many parallels with vertebrate hemoglobin, which is not surprising in view of the likely common evolutionary origin of the two proteins (Sec. 3.4). The vertebrate globins have diverged into different polypeptide chains (e.g., α and β) whereas lamprey globin has retained only one. The low O_2 affinity of the lamprey dimers and tetramers indicates that constraints between the subunits in the quaternary structure inhibit the binding of O_2, as in the vertebrate T state. Binding O_2 causes the interactions between the subunits to diminish and to favor dissociation into monomers. Binding of O_2 to the T state of vertebrate $\alpha_2\beta_2$ hemoglobin also tends to induce its dissociation into $\alpha\beta$ dimers, which have high affinity. The major difference between lamprey and vertebrate hemoglobins is that in the latter the liganded $\alpha\beta$ dimers tend to associate into the R quaternary structure.

Certain clam hemoglobins have evolved another allosteric system from the same ancestral polypeptide chains. The clam globin chains adopt the usual globin tertiary structure (see Fig. 6.30), but they adopt a very different dimeric structure in which the two hemes are in close contact. Binding of O_2 or CO to the subunits causes the usual changes in structure of the heme group (Fig. 8.28), which cause the phenyl side chain of Phe residue F10 to be expelled from the heme pocket. The heme group moves in the protein tertiary structure, and the phenyl side chain moves to the interface between the subunits. These changes do not produce a substantial change in quaternary structure, but they probably directly alter the O_2 affinity of the other heme group, which lies directly across the dimeric interface. In this protein, the cooperativity is not substantial because the O_2 affinity increases only about eightfold, and the Hill coefficient for O_2 binding is approximately 1.5.

Clearly, the hemoglobin type of quaternary structure change is not a necessity, and other types of cooperativity are possible. Nevertheless, it is striking that cooperative systems almost always involve multiple sites on separate subunits or domains. Enzymes can also exhibit allosteric behavior and are described in Section 9.4.2.

References

The self-association and oxygen equilibrium of hemoglobin from the lamprey *Entosphenus japonicus*. D. Dohi et al. *J. Biol. Chem.* 248:2354–2362 (1973).

Linkage between ligand binding and the dimer–tetramer equilibrium in the Monod–Wyman–Changeux model of hemoglobin. S. J. Edelstein and J. T. Edsall. *Proc. Natl. Acad. Sci. USA* 83:3796–3800 (1986).

Structural transitions upon ligand binding in a cooperative dimeric hemoglobin. W. E. Royer, Jr., et al. *Science* 249:518–521 (1990).

8.4.5 Negative Cooperativity

Homotropic allosteric interactions in which the binding of one molecule of ligand decreases the affinity of other sites for the same ligand are unlikely with the concerted model. According to this model, binding of ligand can pull the T ↔ R transition only toward the high-affinity state, which can produce only positive cooperativity. Nevertheless, negative cooperativity is observed with some proteins, although binding evidence for this phenomenon is often found subsequently to be a result of artifacts (Sec. 8.4.1.c). In extreme cases, proteins even exhibit **half-of-the sites reactivity,** in which only half the expected sites of an oligomer bind a ligand, or half the individual groups react with a modifying reagent; the other half of the sites bind weakly or react slowly, or not at all.

One explanation for such behavior would be that initially identical and equivalent binding sites on an oligomeric protein are made nonidentical, and of lower affinity or reactivity, by binding of ligand or reagent to another site, as in the sequential allosteric model. An alternative explanation would be that the binding sites on the oligomeric protein were not equivalent initially, even though the subunits are identical. This is most likely if the proteins are constructed by heterologous association (see Fig. 6.24) using different binding interactions (Fig. 8.30A). Oligomeric proteins assembled by isologous association are expected to have exact symmetry, with each subunit being equivalent. Nevertheless, preexistent asymmetry could occur if only two alternative, complementary conformations were energetically favorable for each subunit and if the same conformation in all subunits in an oligomer were not possible for packing reasons (Fig. 8.30B). In this case, a binding site on one subunit might have one conformation and affinity whereas the comparable site on another subunit would have a different conformation and affinity. Both preexistent and induced asymmetry could, but need not, disappear upon binding ligand to all the sites, but no convincing examples of such asymmetric oligomers are known from the crystal structures of proteins. Slight degrees of asymmetry are apparent in the crystal structures of insulin, α-chymotrypsin, lactate dehydrogenase, and malate dehydrogenase, but the

nonequivalent subunits are in different crystal lattice environments, which could be responsible for the slight asymmetry. If, on the other hand, the preexisting asymmetry of an oligomer did not affect its orientation in the crystal lattice, only a symmetrical, averaged electron density map would result, so this phenomenon could be missed.

Another explanation for apparent negative cooperativity is that the equivalent binding sites on a symmetric protein oligomer either overlap or are sufficiently close to interact sterically or electrostatically, which could occur because they include, or are near, the symmetry axes of the oligomer. Binding of the first ligand molecule would block, or perhaps just inhibit, binding of a second molecule at the overlapping or adjacent site. For example, the binding site for organic phosphates on hemoglobin straddles the twofold axis (Fig. 8.2), and only one molecule is bound per $\alpha_2\beta_2$ hemoglobin molecule, whereas two would be expected for a symmetric dimer. Effects due to direct interactions between adjacent or overlapping sites or due to non-

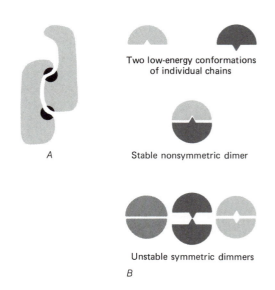

FIGURE 8.30

Possible explanations for nonsymmetrical dimers. *A:* Two-dimensional example of a heterologous dimer with two nonequivalent binding sites (dark shaded areas). *B:* Possible basis for deviations from symmetry in an isologous dimer. The individual chains are assumed to have two alternative low-energy conformations *(top)* that are complementary so that they can form a stable isologous dimer, with close-packing between the two interfaces but with deviations from symmetry *(middle)*. Symmetric dimers are not stable *(bottom),* either because the intermediate conformation that would produce a close-packed interface has high energy *(left)* or because the two low-energy conformations do not give close-packing when associated *(center and right)*.

identical sites are not allosteric in the original sense of the term.

All these phenomena appear to occur in various instances of negative cooperativity, although in no case is the structural explanation known. For example, tyrosyl tRNA synthetase exhibits half-of-the-sites reactivity in solution by binding tightly only one molecule of tyrosine or tyrosine adenylate per dimer molecule. It was concluded to be asymmetric in solution even though it crystallizes as a symmetrical dimer with each subunit having a complete binding site for tyrosine and tyrosine adenylate. The symmetry in solution of the molecule without ligand was investigated by constructing heterodimers in which one polypeptide chain was truncated slightly and by mutating the binding site of one or the other polypeptide. Which site bound ligand initially was found to be random, but the same site was used to bind successive ligand molecules. This dimeric enzyme appears to be intrinsically asymmetric even in the absence of ligands, but the structural basis is not known.

In some instances, glyceraldehyde-3-phosphate dehydrogenase (GPD) binds the cofactor NAD to the four sites on its tetramer with negative cooperativity. This has been concluded to be due to ligand-induced asymmetry. The crystal structures of the apo- and holo-GPD show the four subunits to be equivalent in both instances. Local rearrangements occur upon NAD binding, and the crystal structure of the tetramer with one NAD molecule bound shows it to be asymmetric, with the liganded subunit in the holo conformation and the other three subunits in the apo form. Only very slight changes were observed in the unliganded subunits, however, so how the affinities of these sites are lowered remains to be determined.

References

Enzymes with asymmetrically arranged subunits. Y. Degani and C. Degani. *Trends Biochem. Sci.* 5:337–341 (1980).

Molecular symmetry and metastable states of enzymes exhibiting half-of-the-sites reactivity. J. Herzfeld et al. *Biochemistry* 20:4936–4941 (1981).

Structural evidence for ligand-induced sequential conformational changes in glyceraldehyde 3-phosphate dehydrogenase. A. G. W. Leslie and A. J. Wonacott. *J. Mol. Biol.* 178:743–772 (1984).

Asymmetry of tyrosyl-tRNA synthetase in solution. W. H. J. Ward and A. R. Fersht. *Biochemistry* 27:1041–1049 (1988).

Exercises

1. Electrostatic interactions between charges on a protein surface and on a ligand can guide the ligand to its binding site on the protein and increase substantially the rate of binding over that expected by just diffusion (e.g., E. D. Getzoff et al., *Nature* 306:287–290, 1983). Consider the properties of the other types of interactions that can occur between proteins and ligands (Chap. 4) and decide whether any of them could also be responsible for increasing the rate of ligand binding.

POSSIBLE ANSWERS
K. Brady and R. H. Abeles, *Biochemistry* 29:7608–7617 (1990).

2. Your protein is observed to bind a ligand added at time zero only after an initial lag period. What are the possible explanations?

ANSWER
R. M. Schultz et al., *J. Biol. Chem.* 264:1497–1507 (1989).

3. Standard methods of analyzing binding data assume that binding sites not occupied by the ligand being studied are free and available for binding. What would be the consequences if there were a second ligand in the preparation of ligand being used for binding studies with your tetrameric protein containing four identical and noninteracting binding sites, one that was not detected by the binding assay you are using? What if the second ligand were present in the buffer being used for the ligand assays?

It is also generally assumed that the ligand being studied is free in solution when not bound to the protein. Suppose that the free ligand has a tendency to dimerize. What would be the consequences if the protein bound only the monomer or only the dimer? Consider the situation when the dimer is a major and a minor form of the ligand.

4. The SO_4^{2-} and CrO_4^{2-} ions bind to sulfate-binding protein (Fig. 8.4) at least 10^5-fold more tightly than does the very similar HPO_4^{2-} ion. The natural phosphate-binding protein similarly discriminates against SO_4^{2-}. How might a protein discriminate between these similar ions?

ANSWER
B. L. Jacobson and F. A. Quiocho, *J. Mol. Biol.* 204:783–787 (1988); H. Luecke and F. A. Quiocho, *Nature* 347:402–406 (1990).

5. The phosphate-binding protein of bacteria is homologous to the sulfate-binding protein (Fig. 8.4) and is able to bind both HPO_4^{2-} and $H_2PO_4^-$ ions with similar affinities, even though the ions are completely buried. How could it accommodate both of these anions, with either one or two hydrogen atoms and two or one net charges?

ANSWER
H. Luecke and F. A. Quiocho, *Nature* 347:402–406 (1990).

6. Would you expect to be able to imprint a protein with a ligand in such a way that the protein retained a memory of

the ligand and subsequently recognized it (e.g., A. J. Russell and A. M. Klibanov, *J. Biol. Chem.* 263:3194–3201, 1988)? What methods of imprinting would be most likely to succeed?

ANSWER

L. Braco et al. *Proc. Natl. Acad. Sci. USA* 87:274–277 (1990).

7. An early plausible theory for the appearance of antibodies specific for certain antigens was that the immunogen served as a template to induce a complementary shape in initially nonspecific antibody molecules (A. M. Silverstein, *A History of Immunology*, Academic Press, San Diego, 1989). What would be the simplest way to test this theory?

ANSWER

E. Haber, *Proc. Natl. Acad. Sci. USA* 52:1099–1106 (1964).

A similar process has been proposed for the major histocompatibility molecules that present processed antigens to T-cell receptors (A. Townsend et al., *Nature* 340:443–448, 1989; P. Parham, *Nature* 340:426–428, 1989). Is such a process plausible in this case?

DEBATE

J. M. Claverie; P. Parham, *Nature* 345:121–122 (1990).

8. Antigenic sites (epitopes) on the surfaces of folded proteins were found to be those parts of the surface that were most flexible, and such flexibility was proposed to be essential for immunogenicity (E. Westhoff et al., *Nature* 311:123–126, 1984; J. A. Tainer et al., *Nature* 312:127–133, 1984; H. M. Geysen et al., *Science* 235:1184–1190, 1987). What other explanations are plausible?

ANSWERS

D. C. Benjamin et al., *Ann. Rev. Immunol.* 2:67–101 (1984); J. Novotny et al., *Proc. Natl. Acad. Sci. USA* 83:226–230 (1986), *Science* 238:1584 (1987); J. M. Thornton et al., *EMBO J.* 5:409–413 (1986).

9. The gene for an important biological function has recently been identified on the basis that mutations in the gene inactivated that biological function. You have recently cloned and sequenced this gene and need to identify the protein it encodes, which performs this important biological function. You decide to prepare antibodies against peptide fragments and to use these antibodies in Western blots. How do you decide which segments of the primary structure to use in the peptides?

ANSWERS

T. P. Hopp and K. R. Woods, *Proc. Natl. Acad. Sci. USA* 78:3824–3829 (1981), *Mol. Immunol.* 20:483–489 (1983); P. A. Karplus and G. Schulz, *Naturwis.* 72:212–213 (1985); G. W. Welling et al., *FEBS Letters* 188:215–218 (1985).

10. Antibodies against short peptides were considered to cross-react with the intact protein at least four orders of magnitude more frequently than expected (H. L. Niman et al., *Proc. Natl. Acad. Sci. USA* 80:4949–4953, 1983). This was proposed to demonstrate that short peptides are less flexible than thought and that folded proteins are more flexible. Is there any need to change our view of these matters?

ANSWERS

L. Djavadi-Ohaniance et al., *Biochemistry* 23:97–104 (1984); R. Jemmerson, *Proc. Natl. Acad. Sci. USA* 84:9180–9184 (1987).

11. The red-colored heme group can be removed from myoglobin to produce apomyoglobin, which has a folded conformation that approximates that of myoglobin but is less stable and much more flexible (e.g., Y. V. Griko et al., *J. Mol. Biol.* 202:127–138, 1988). If precipitating antibodies prepared against apomyoglobin were to be mixed with myoglobin, what color would you expect the precipitate to be?

ANSWER

M. J. Crumpton, *Biochem. J.* 100:223–232 (1966).

Does this observation invalidate the use of antibodies to measure conformational equilibria in proteins (Sec. 7.5.3.d)?

12. The complex of the *trp* repressor with the DNA fragment that it is thought to recognize was somewhat surprising in that there were very few direct interactions between the two (Z. Otwinowsky et al., *Nature* 335:321–329, 1988). The specificity of the interaction was concluded to arise primarily from the sequence dependence of the distortion of the DNA that occurred on binding and from hydrogen bond networks involving water molecules. What other explanations are plausible?

SUGGESTIONS

S. E. V. Phillips et al., *Nature* 341:711–715 (1989); B. F. Luisi and P. B. Sigler, *Biochim. Biophys. Acta* 1048:113–126 (1990); D. Staacke et al., *EMBO J.* 9:1963–1967 (1990).

13. Members of a major class of DNA-binding proteins are made up of dimers of two-domain polypeptides. One domain dimerizes via the so-called leucine-zipper structure, actually a parallel coiled coil of α-helices from the two polypeptides (T. G. Oas et al., *Biochemistry* 29:2891–2894, 1990) that holds the two polypeptide chains together. The other domains of each polypeptide are thought to be structurally independent; they contain a preponderance of basic residues and are thought to bind directly to DNA. In isolation, these domains appear to be largely unfolded in the absence of the DNA (K. T. O'Neil et al., *Science* 249:774–778, 1990; L. Patel et al., *Nature* 347:572–575, 1990), and the dimer at physiological concentrations may even be dissociated into unfolded mono-

mers (M. A. Weiss et al., *Nature* 347:572–575, 1990). How could such a disordered protein be capable of binding specifically to DNA, and why might this type of system be useful in this case?

ANSWER
R. T. Sauer, *Nature* 347:514–515 (1990).

14. The zinc-finger motifs (Sec. 8.3.2) were attractive subjects for predicting the folded conformation from just the sequence: they are small, consisting of only about 30 residues; the two Cys and two His residues must be constrained to interact with a single Zn^{2+} atom; about 150 homologous sequences were known; and the structure was certain to be determined without undue delay. Evaluate the predicted structures (J. M. Berg, *Proc. Natl. Acad. Sci. USA* 85:99–102, 1988; T. Gibson et al., *Protein Eng.* 2:209–218, 1988), and the procedures used to generate them, in the light of the structure determined by NMR (M. S. Lee et al., *Science* 245:635–637, 1989) and by X-ray crystallography (N. P. Pavletich and C. O. Pabo, *Science* 252:809–817, 1991).

15. The oxidized and reduced forms of cytochrome *c* exhibit a variety of physical differences that have been taken to indicate that the conformations of the two proteins are significantly different (e.g., E. Margoliash and A. Schejter, *Adv. Protein Chem.* 21:113–286, 1966; D. Eden et al., *Proc. Natl. Acad. Sci. USA* 79:815–819, 1982; D. D. Ulmer and J. H. Kägi, *Biochemistry* 7:2710–2717, 1968; N. Osherhoff et al., *Proc. Natl. Acad. Sci. USA* 77:4439–4443, 1980). Yet the crystal structures of the two forms of the protein are virtually identical (T. Takano and R. E. Dickerson, *J. Mol. Biol.* 153:79–94, 95–115, 1981). Is it possible to reconcile these observations?

ANSWER
Y. Feng and S. W. Englander, *Biochemistry* 29:3505–3509 (1990).

16. The cooperativity of O_2 binding to hemoglobin means that the equilibrium distributions of molecules with O_2 bound at 1, 2, and 3 of the four subunits is less than if binding were random (Fig. 8.24D). The distribution of species can be determined directly only if some way is found to separate them on a time scale faster than the O_2 molecules associate and dissociate. Rapid freezing techniques indicated that about 6% of the hemoglobin tetramers had three O_2 molecules bound (M. Perrella et al., *J. Biol.*

Chem. 261:8391–8396, 1986), but analysis of the binding curves indicated that negligible amounts of the three O_2 species were present (S. J. Gill et al., *Trends Biochem. Sci.* 13:465–467, 1988). Can these two conclusions be reconciled?

ANSWER
M. C. Marden et al., *J. Mol. Biol.* 208:341–345 (1989).

17. Binding of O_2 by hemoglobin and dissociation of the $\alpha_2\beta_2$ tetramer into $\alpha\beta$ dimers are linked functions (see Fig. 7.32). The $\alpha\beta$ dimers bind O_2 noncooperatively, with an affinity like that of the R state. The T state of deoxy $\alpha_2\beta_2$ dissociates into dimers much less than the fully oxygenated R state, with respective dissociation constants of $2 \times 10^{-11}\,M$ and $1.2 \times 10^{-6}\,M$. What are the implications of this for the mechanism of cooperativity of O_2 binding?

ANSWERS
F. R. Smith and G. K. Ackers, *Proc. Natl. Acad. Sci. USA* 82:5347–5351 (1985); F. A. Ferrone, *Proc. Natl. Acad. Sci. USA* 83:6412–6414 (1986); S. J. Edelstein and J. T. Edsall, *Proc. Natl. Acad. Sci. USA* 83:3796–3800 (1986); M. L. Johnson, *Biochemistry* 25:791–797 (1986); G. Weber, *Biochemistry* 26:331–332 (1987); G. K. Ackers and F. R. Smith, *Ann. Rev. Biophys. Biophys. Chem.* 16:583–609 (1987); F. R. Smith et al., *Proc. Natl. Acad. Sci. USA* 84:7089–7093 (1987); M. Straume and M. L. Johnson, *Biochemistry* 27:1302–1310 (1988); M. L. Johnson, *Biochemistry* 27:833–837 (1988); V. J. LiCata et al., *Biochemistry* 29:9771–9783 (1990).

18. Zone electrophoresis in nondenaturing polyacrylamide or agarose gels is frequently used to assay for proteins that bind to specific regions of DNA (Sec. 8.3.2). A radioactive fragment of the DNA is mixed with possible sources of binding proteins, and the mixture is subjected to electrophoresis. The presence of proteins that bind to the DNA causes the electrophoretic mobility of the DNA fragment to decrease (band-shift). In some cases, multiple complexes can also be apparent from multiple new bands of the DNA, with different numbers and combinations of proteins bound. Most macromolecular complexes dissociate during zone electrophoresis and cannot be detected in this way. What conditions must be fulfilled for a DNA–protein complex to be detected?

ANSWER
J. R. Cann, *J. Biol. Chem.* 264:17032–17040 (1989).

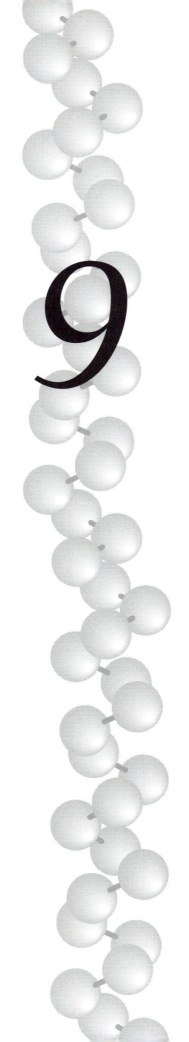

Enzyme Catalysis

9

*T*he interactions of a protein with ligands are very often followed by covalent changes in the ligands. In this case, the protein is known as an *enzyme,* and the altered ligands are its *substrates.* The enzyme is a catalyst of the chemical reaction undergone by its substrates and is not altered. In some cases, the chemical reaction can be coupled energetically to other processes, such as mechanical movements of muscle or flagella or the transport of molecules through membranes, often against concentration gradients.

Enzyme catalysis has been studied extensively for over a century, so a detailed description would require much more space than can be provided here. In the classical early studies, much was learned about the kinetics of enzyme action but little about the enzymes themselves; emphasis was on the substrates and products, and the enzyme was represented either as E or as an ill-defined black box. All this changed in the past three decades, when the structures and properties of enzymes and their complexes with ligands were elucidated, as described in previous chapters.

In this chapter, we will review the classical kinetics of enzyme action only briefly, because it is an extensive topic that is described adequately in many other books. Instead, we will concentrate on the structural properties of enzymes and the physical principles by which they catalyze reactions at apparently extraordinary rates. It is now believed that the kinetic properties of enzymes can be at least rationalized on the basis of established physical and chemical principles. Because of the subject's history, many different terms have been used to describe similar concepts of how enzymes function, including *approximation, orientation, entropy loss, propinquity, rotamer distribution, anchimeric assistance, proximity, orbital steering, stereopopulation control, distance distribu-*

tion function, togetherness, and *freezing at reactive centers of enzymes* (FARCE). Such jargon, which is more often misunderstood than useful, is avoided in the following description of the general principles involved. The principles are illustrated with a few of the best understood enzymes. The somewhat less certain subject of the regulation of enzyme activity is also described briefly.

References

Enzyme Structure and Mechanism, 2nd ed. A. Fersht. W. H. Freeman, New York, 1985.

Structure and catalysis of enzymes. W. N. Lipscomb. *Ann. Rev. Biochem.* 52:17–34 (1983).

9.1 The Kinetics of Enzyme Action

A catalyst, such as an enzyme, only increases the rate of a reaction and is not altered at the end of the reaction. Although a catalyst may participate intimately in the reaction, it is regenerated in its original form. Consequently, a catalyst cannot alter the equilibrium between reactants and products that would apply in its absence; it can only increase the rate of attainment of that equilibrium.

In classical studies of enzymes, the enzyme was used at very low concentrations relative to the substrates, and only the net effect of its presence on the rate of conversion of substrate to product was measured. The enzyme was usually catalyzing the reaction very rapidly on the time scale of the measurements, so it was in a steady state in which the concentrations of its various forms were constant and too low to affect the measurements. Because the enzyme was not observed directly, emphasis was placed on the substrates, products, inhibitors, and so forth. Nevertheless, these studies were pursued rigorously and often ingeniously, permitting the elucidation of complex reaction schemes and some significant insight into enzyme mechanisms. They also provided the framework for studying directly the reactions that take place on the enzyme.

Here we will consider briefly the general principles and observations derived from such studies. The following discussion is limited to "simple" enzymes; allosteric enzymes will be described in a subsequent section.

9.1.1 Steady-State Kinetics

When the enzyme is present only at low concentrations relative to its substrates, it catalyzes the same reaction numerous times. The concentrations of free substrates and products change slowly, and the enzyme is present in a steady state.

a. Single Substrate and Product

Most enzymes catalyze reactions that occur at negligible rates in their absence. The velocity of the catalyzed reaction v is the rate at which a substrate S disappears or a product P appears:

$$v = \frac{-d[\text{S}]}{dt} = \frac{d[\text{P}]}{dt} \tag{9.1}$$

The rate of catalysis by an enzyme is proportional to substrate concentration at low levels and becomes independent at high levels (Fig. 9.1A). The dependence of the reaction velocity on the substrate concentration is analogous to that of ligand binding (see Fig. 8.6), which indicates that catalysis occurs only after binding of the substrate to the enzyme E:

$$\text{E} + \text{S} \underset{k_{-1}}{\overset{k_1}{\rightleftharpoons}} \text{ES} \overset{k_2}{\rightarrow} \text{E} + \text{P} \tag{9.2}$$

ES is known as the **Michaelis complex**. The substrate concentration at which the reaction rate is half-maximal is known as the **Michaelis constant** or K_m. K_m is analogous to the dissociation constant of the Michaelis complex, but the values of the two constants are the same only if dissociation is more rapid than is conversion to product (i.e., $k_{-1} \gg k_2$). In general,

$$K_m = \frac{k_2 + k_{-1}}{k_1} \tag{9.3}$$

The value of K_m is therefore usually greater than or equal to the equilibrium constant for dissociation of substrate from the enzyme; it can be smaller, however, if intermediate forms of the Michaelis complex along the reaction pathway are present in significant amounts and have smaller dissociation constants. Observed values of K_m for different substrates and enzymes vary widely in the same way that ligand affinities vary (Sec. 8.2.1).

The velocity of an enzyme-catalyzed reaction becomes independent of substrate concentration when the enzyme is fully saturated with substrate in the steady state and when the concentration of free enzyme is negligible. That maximal velocity, known as V_{max}, is directly proportional to the total enzyme concentration:

$$V_{max} = k_{cat}[\text{E}_\text{T}] \tag{9.4}$$

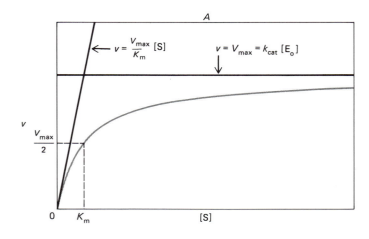

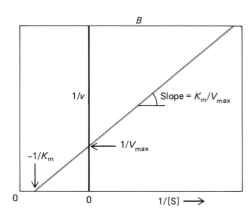

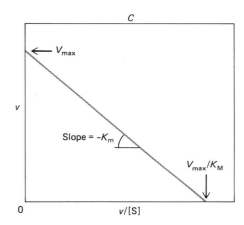

FIGURE 9.1

The dependence of the velocity v of an enzyme-catalyzed reaction on substrate concentration [S] when the enzyme is present at very low concentration. A shows the normal hyperbolic relationship between v and [S]. B: The double reciprocal Lineweaver–Burke plot of the same data is widely used because it provides easy estimation of the values of V_{max} and K_m. C: The Eadie–Hofstee plot is the equivalent of the Scatchard plot of binding data (see Figs. 8.6 and 8.24). (Adapted from A. Fersht, *Enzyme Structure and Mechanism,* W. H. Freeman, Reading, 1977.)

The value of k_{cat} is the rate of breakdown of the ES complex, k_2 in Equation (9.2). When expressed as moles of substrate consumed per unit of time per mole of enzyme, k_{cat} is often referred to as the **turnover number** of the enzyme and represents the number of times each enzyme molecule catalyzes the reaction per unit of time. Turnover numbers vary widely; the highest values observed are 4×10^7 s^{-1} for catalase and 10^6 s^{-1} for carbonic anhydrase.

The velocity of the enzyme-catalyzed reaction, v, is expressed as a function of substrate and enzyme concentration by the well-known **Michaelis–Menten equation:**

$$v = \frac{[S]}{K_m + [S]} V_{max} \qquad (9.5)$$

Enzyme kinetic data can be treated graphically in the same manner as those of ligand binding. **Eadie–Hofstee plots** (Fig. 9.1C) are exactly comparable to Scatchard plots of binding data (see Fig. 8.6C), plotting v versus $v/$[S]. Most widely used in enzyme kinetics are the **double-reciprocal Lineweaver–Burke plots** of $1/v$

versus $1/$[S] (Fig. 9.1B). Inverting the Michaelis–Menten equation (9.5) gives

$$\frac{1}{v} = \left(1 + \frac{K_m}{[S]}\right)\frac{1}{V_{max}} \qquad (9.6)$$

Consequently, the y intercept of the Lineweaver–Burke plot gives $1/V_{max}$, the x intercept is $-1/K_m$, and the slope of the straight line gives K_m/V_{max}.

Classical studies use the initial enzyme velocity, measured before the rate of decrease in concentration of substrate diminishes or the amount of products formed becomes significant. The complete time course of the approach to equilibrium can also be used to extract information about the effects of both substrate and product concentrations on the rate of the reaction, but substantial mathematical analysis is necessary.

At very low substrate concentrations, the combination of the substrate with the enzyme becomes rate-limiting. Under these conditions, the Michaelis–Menten equation becomes

$$v = \frac{k_{cat}}{K_m}[E_T][S] \qquad ([S] \ll K_m) \qquad (9.7)$$

Most of the enzyme is in the free form under these conditions, so Equation (9.7) becomes

$$v = \frac{k_{cat}}{K_m}[E][S] \qquad (9.8)$$

Expressed in terms of free enzyme, this equation is valid for all substrate concentrations. Consequently, k_{cat}/K_m represents the apparent rate constant for combination of a substrate with the free enzyme and is the most critical parameter in determining the specificity of an enzyme for a substrate. Because an enzyme and a substrate cannot combine more rapidly than diffusion permits (Sec. 8.2.3), there is an upper limit on enzyme catalysis: the value of k_{cat}/K_m cannot be greater than the diffusion limit of about $10^9 \ s^{-1} \ M^{-1}$. This value is lower if the rate of the catalytic step is less than k_{-1}, so that some substrate molecules dissociate rather than being converted to product, and if not all encounters between enzyme and substrate result in binding, as with any ligand-binding interaction. Nevertheless, some enzymes have values of k_{cat}/K_m that approach the diffusion limit, indicating extreme efficiency in binding the substrate and converting it to product.

As stated earlier, an enzyme does not alter the equilibrium of a reaction, so it must also catalyze the reaction in the reverse direction. An enzyme–product complex must exist, and there must be a Michaelis constant for the product, K_m^P, and a value of k_{cat} for the reverse reaction, k_{cat}^r.

$$E + S \xrightleftharpoons[\ ]{K_m^S} E \cdot S \xrightleftharpoons[k_{cat}^r]{k_{cat}^f} E \cdot P \xrightleftharpoons[K_m^P]{} E + P \quad (9.9)$$

The values of K_m^P and k_{cat}^r are not independent of the corresponding values for the forward reaction. They are related by the equilibrium constant for the reaction, because at equilibrium the enzyme must be catalyzing the reaction equally in both directions. The **Haldane relationship** expresses this:

$$\frac{k_{cat}^f/K_m^S}{k_{cat}^r/K_m^P} = K_{eq} = \frac{[P]_{eq}}{[S]_{eq}} \qquad (9.10)$$

where eq designates equilibrium concentrations and constants. The substrate and product compete for free enzyme; the outcome depends on their respective values of k_{cat}/K_m. The equilibrium ratio of product to substrate on the enzyme is given by k_{cat}^f/k_{cat}^r, which is different from the equilibrium in solution (K_{eq}) when the substrate and product have different K_m values.

References

Two rules of enzyme kinetics for reversible Michaelis–Menten mechanisms. T. Keleti. *FEBS Letters* 208:109–112 (1986).

The analysis of enzyme progress curves by numerical differentiation, including competitive product inhibition and enzyme reactivation. S. C. Koerber and A. L. Fink. *Anal. Biochem.* 165:75–87 (1987).

b. Multiple Substrates and Products

Many reactions catalyzed by enzymes have multiple substrates, or products, or both. A number of reaction mechanisms are possible in each case, and they can usually be distinguished by steady-state kinetic measurements. In brief, one measures the effects on the observed rate of the catalyzed reaction of independently varying the concentrations of the substrates and the products. Varying the concentration of one substrate, keeping all others constant, usually yields normal Michaelis–Menten kinetics, but different reaction mechanisms predict various effects of the other substrates and products on the apparent values of K_m and V_{max} for each substrate. The steady-state kinetic behavior expected for any particular reaction scheme can be predicted, most readily by using the ingenious graphical procedure of King and Altman. Reaction schemes inconsistent with the kinetic results can be excluded. Remember, however, that kinetic analysis cannot prove a mechanism, only disprove alternative mechanisms.

Sequential reactions are those in which all the substrates bind to the enzyme before the first product is formed. The binding of substrates or release of products may be essentially random,

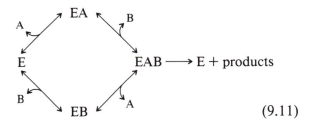

$$(9.11)$$

although the binding of one substrate or product may alter the affinity for others. An extreme situation in which there are very large differences in affinities for the substrates results in essentially ordered substrate binding; for instance,

$$E \xleftrightarrow{A} EA \xleftrightarrow{B} EAB \longrightarrow E + \text{products} \quad (9.12)$$

Similar considerations apply to product release.

A quite different type of reaction scheme is that in which one (or more) products are formed and released from the enzyme before the other substrates bind:

$$E \xleftrightarrow{A} EA \xleftrightarrow{P} E' \xleftrightarrow{B} E'B \longrightarrow E + Q \quad (9.13)$$

This **substituted-enzyme** type of mechanism is also referred to as the **ping-pong** or **double-displacement** mechanism. It usually occurs when part of substrate A is transferred to substrate B. The groups to be transferred in this way are usually held by the intermediate form of the enzyme E′ as covalent adducts, such as acyl, phosphoryl, or Schiff base forms of the enzyme.

The substituted-enzyme type of reaction often offers the first direct clue to the role of the enzyme in catalysis because the modified enzyme E′ can often be prepared by adding substrate A in the absence of the other substrates. In many cases, however, one of the substrates or products may be water. The concentration of water is difficult to vary, so other approaches must be used to determine the reaction scheme. For example, the occurrence of an acyl-enzyme intermediate of the serine proteases (described in Sec. 9.3.2.a) was inferred from the observation that substrates with different leaving groups gave the same value of k_{cat}. Such substrates were hydrolyzed nonenzymatically at different rates. A common enzymic rate, therefore, suggested that the rate-determining step was breakdown of a common acyl intermediate after release of the different leaving groups:

$$
\underset{\substack{\| \\ RC-OR'}}{O} + E \xrightarrow{R'OH} \underset{\substack{\| \\ RC-E}}{O}
$$

$$
k_{cat} \diagdown -H_2O
$$

$$
RCO_2H + E \qquad (9.14)
$$

Similarly, the addition of other acceptors that can compete with water, such as NH_2OH,

$$
\underset{\substack{O \\ \| \\ R-C-E}}{} \diagup \overset{\substack{O \\ \|}}{RCNHOH} + E
$$

$$
\underset{\substack{\\ H_2O}}{} \diagdown \overset{NH_2OH}{}
$$

$$
RCO_2H + E \qquad (9.15)
$$

gave product ratios that were independent of the R′ group of the substrate, whereas different ratios were obtained with the nonenzymatic reactions.

References

A systematic method for deriving the rate laws for enzyme-catalyzed reactions. E. L. King and C. Altman. *J. Phys. Chem.* 60:1375–1378 (1956).

Graphic rules in steady and non-steady state enzyme kinetics. K. C. Chou. *J. Biol. Chem.* 264:12074–12079 (1989).

Kinetic competence of enzymic intermediates: fact or fiction? W. W. Cleland. *Biochemistry* 29:3194–3197 (1990).

c. Inhibitors

Much indirect information about the catalytic activity of an enzyme can be gained from its inhibition, especially if the enzyme has a limited range of substrates. This discussion excludes inhibitors that react covalently with the enzyme and refers only to those that act by binding reversibly to the enzyme at specific sites. They are usually structural analogues of one of the substrates, but they can also be the products of the reaction or ligands that bind at some other specific functional site on the enzyme. An inhibitor might also be a substrate used in a different way from the normal substrate, but the simplest cases are those in which there is only reversible binding of the inhibitor. Each inhibitor is characterized by its inhibition constant K_i, which is analogous to K_m for a substrate. The K_i is usually the same as the dissociation constant of the inhibitor from the enzyme if the inhibitor binds to and dissociates from the enzyme rapidly and is not a substrate of the enzyme.

The modes of action of inhibitors are usually elucidated by their effects, at varying concentrations, on the utilization of various concentrations of a substrate by the enzyme. The inhibition generally can be designated as being competitive, uncompetitive, or noncompetitive. Each classification refers to only one substrate; if there is more than one substrate, a different type of inhibition is usually exhibited with respect to each substrate by any one inhibitor.

The simplest inhibitors are those, such as substrate analogues, that bind at the same site as the substrate and compete with it. This **competitive inhibition** has the effect of increasing the apparent K_m for the substrate. V_{max} is not altered, however, because high concentrations of the substrate displace the inhibitor from the enzyme (Fig. 9.2A).

Uncompetitive inhibition is expected if the inhibitor binds only to the enzyme–substrate complex, blocking its catalytic activity. The values of K_m and V_{max} are affected in parallel (Fig. 9.2B). This type of inhibition is rare with single-substrate enzymes but occurs more frequently when there are multiple substrates. For example, a competitive inhibitor of one substrate can give uncompetitive inhibition with respect to the other in and ordered, sequential reaction scheme (Eq. 9.12).

An inhibitor that binds to both the free enzyme and the enzyme–substrate complex might affect only V_{max}, giving pure **noncompetitive inhibition** (Fig. 9.2C). More commonly, the K_m of the substrate is also affected by such an inhibitor, so that reciprocal plots at various inhibitor concentrations intersect somewhere other

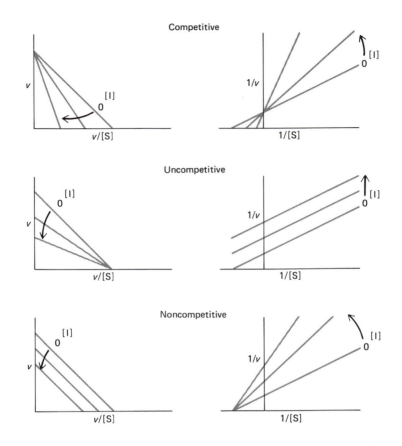

FIGURE 9.2

Steady-state kinetic analysis of enzyme inhibition. The velocity v of the reaction catalyzed by an enzyme present at very low concentration is measured at varying concentrations of substrate S and of inhibitor I. The relationship between v and [S] is plotted at various values of [I] in Eadie–Hofstee plots *(left)* and in Lineweaver–Burke plots *(right)*. The line labeled 0 is in the absence of added inhibitor and the arrow shows the changes produced by adding increasing amounts of inhibitor. (Adapted from A. Fersht, *Enzyme Structure and Mechanism,* W. H. Freeman, Reading, 1977.)

than on the x axis. This combination of effects is known as **mixed inhibition.**

Inhibition patterns can be complex, and any proposed reaction mechanism should be shown to be quantitatively consistent with all experimental observations. Inhibition patterns can, however, usually be predicted qualitatively from a few useful rules. An inhibitor affects the slope of a Lineweaver–Burke plot (Fig. 9.2) if it and the substrate whose concentration is being varied compete directly for the same form of the enzyme or for different forms that are in reversible equilibrium with each other. An inhibitor binding to any other form of the enzyme affects the y intercept (i.e., V_{max}). These two effects can occur separately, when they produce competitive or uncompetitive patterns, respectively; when they occur jointly, they produce noncompetitive inhibition. If the substrate or inhibitor combines with more

than one form of the enzyme, the resulting inhibition pattern is the sum of the different individual patterns.

For example, in the sequential ordered scheme

$$E \xleftrightarrow{\ A\ } EA \xleftrightarrow{\ B\ } EAB$$
$$\downarrow$$
$$E \xleftrightarrow[\ Q\ ]{} EQ \xleftrightarrow[\ P\ ]{} EPQ \tag{9.16}$$

an analogue of B that combines only with EA is competitive with respect to B but uncompetitive with respect to A. An analogue of A that binds only to E is competitive with respect to A but noncompetitive with respect to B. Product Q gives the same pattern of inhibition of the forward reaction as the analogue of A because both the analogue and Q compete with A for the free enzyme. Product P is noncompetitive with both A and B.

This analysis is for inhibitors that both associate with the enzyme and dissociate from it very rapidly compared with the rate at which the reaction is catalyzed. Many of the most interesting and useful inhibitors equilibrate with the enzyme very slowly, often on the second to hour time scale, primarily because they dissociate so slowly. Such inhibitors are often approximating an intermediate or transition state along the reaction pathway (Sec. 9.2.2).

Prediction of the inhibitor binding site on an enzyme from steady-state kinetic measurements alone is dangerous. Certain noncompetitive inhibitors have been found to bind at the same site as that for substrate; they give noncompetitive inhibition rather than the expected competitive type because they dissociate very slowly from the enzyme. Analysis of the inhibition of steady-state kinetics, therefore, must be supplemented by direct studies of the interaction between inhibitor and enzyme. Such studies of these slowly dissociating inhibitors can be extremely valuable because the complexes formed can be stable, in contrast to those in enzyme–substrate interactions.

References

Enzyme kinetics and mechanism. Part A. Initial rate and inhibitor methods. D. L. Purich, ed. *Methods Enzymol.* 63 (1979).

The behavior and significance of slow-binding enzyme inhibitors. J. F. Morrison and C. T. Walsh. *Adv. Enzymol.* 61:201–301 (1988).

d. Isotopically Labeled Substrates

To dissect the kinetic scheme of a complex enzymatic reaction, one must know the origin of each atom of each product: that is, whether each atom of the product arises from the solvent or from the substrate, and from which atom. This origin can be determined most readily by using specifically labeled substrates or solvent and following the fate of each atom. Which groups are transferred in each reaction step can also be inferred from the effect on the rate and on the equilibrium constant of substituting a heavier or a lighter isotope; lighter isotopes are generally transferred in chemical reactions more rapidly than are heavier ones. This difference is most dramatic with hydrogen atoms, for which the reaction rate for transfer of the deuterium and the tritium isotopes may be 1/24 and 1/79, respectively, of that for the naturally occurring isotope. Smaller isotope effects occur with other atoms, but even in these cases modern

sensitive methods of measuring the relative levels of two isotopes can reveal small kinetic differences between two isotopically labeled forms of the substrate.

Other techniques that use isotopes do not use the effect on rates; identification of isotopes in the products determines the origin of the atoms. The most widely used technique is isotope exchange between substrates or products. Enzymes with two or more substrates or products often catalyze partial reactions that can be detected by isotope exchange. For example, group-transfer reactions of the type

$$A\text{---}G + B \rightleftharpoons A + B\text{---}G \qquad (9.17)$$

where group G is transferred from substrate A to substrate B, can occur either by direct exchange on a ternary complex of the two substrates on the enzyme ($E \cdot A\text{---}G \cdot B$) or by a substituted-enzyme scheme in which the group is transiently transferred to the enzyme:

$$E + A\text{---}G \underset{\overset{A}{\frown}}{\rightleftharpoons} E\text{---}G \underset{\overset{B}{\frown}}{\rightleftharpoons} E + B\text{---}G \quad (9.18)$$

In the latter case, the enzyme catalyzes an exchange between A—G and A in the absence of B so that no net reaction occurs. This exchange can be detected if either A—G or A is isotopically labeled. A similar isotope exchange reaction between B and B—G is catalyzed by the enzyme in the absence of A. No such exchange should occur if the group is transferred directly from one substrate to the other.

This technique has been widely used to measure amino acid activation by amino-acyl tRNA synthetases (Sec. 9.3.1). In the absence of tRNA, to which the correct amino acid becomes attached in the full reaction, the enzyme catalyzes the partial reaction

$$E + ATP + \text{amino acid} \rightleftharpoons$$
$$\text{aminoacyl adenylate} \cdot E + PP_i \quad (9.19)$$

where the aminoacyl adenylate remains firmly bound to the enzyme. This reaction is readily measured by ^{32}P isotope exchange between ATP and pyrophosphate (PP_i).

Another approach is to follow isotope exchange between a substrate and product when all the substrates are present and the reaction is at equilibrium. No net reaction occurs at equilibrium, but the forward and reverse rates can be followed individually by isotope exchange. For example, the enzyme aldolase was found at equilibrium to catalyze isotope exchange between the substrate fructose 1,6-diphosphate and one of its products, glyceraldehyde-P, much more rapidly than with its other product, dihydroxyacetone-P. This was one piece of evidence that glyceraldehyde-P is released first from the enzyme, leaving the other product attached to the enzyme by a Schiff base.

$$E + \text{fructose 1,6-P}_2$$
$$\downarrow \nearrow \text{glyceraldehyde-P}$$
$$E \cdot \text{dihydroxyacetone-P}$$
$$\Updownarrow$$
$$E + \text{dihydroxyacetone-P} \qquad (9.20)$$

References

Determining the chemical mechanisms of enzyme-catalyzed reactions by kinetic studies. W. W. Cleland. *Adv. Enzymol.* 45:273–387 (1977).

Enzyme kinetics and mechanism. Part B. Isotopic probes and complex enzyme systems. D. L. Purich, ed. *Methods Enzymol.* 64 (1980).

The expression of isotope effects on enzyme-catalyzed reactions. D. B. Northrop. *Ann. Rev. Biochem.* 50:103–131 (1981).

Use of isotope effects to elucidate enzyme mechanisms. W. W. Cleland. *Crit. Rev. Biochem.* 13:385–428 (1982).

Positional isotope exchange. F. M. Raushel and J. F. Villafranca. *Crit. Rev. Biochem.* 23:1–26 (1988).

Hydrogen tunneling in enzyme reactions. Y. Cha et al. *Science* 243:1325–1330 (1989).

9.1.2 Reactions on the Enzyme

Steady-state kinetics provide only tantalizing hints of what processes take place on the enzyme, with postulated complexes of the enzyme with substrates and products, and with intermediate, substituted enzymes in ping-pong schemes. To verify the existence of these complexes and to learn more about what happens between the time when the first substrate is bound and when the last product is released, the enzyme itself must be studied; this requires substrate-level quantities of enzyme and rapid techniques that permit measurements within the turnover time of the enzyme. The turnover time is given by $1/k_{cat}$; its value can be as short as 10^{-6} s although longer times are more usual.

The existence of enzyme–substrate complexes has been amply demonstrated using primarily spectral techniques such as those used to measure ligand binding in general (Chap. 8), with results comparable to those for binding other ligands. Often, further chemical changes in the bound substrate can also be detected in this way.

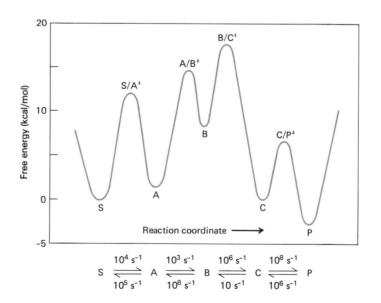

FIGURE 9.3

Free-energy profile for a hypothetical reaction S → P proceeding sequentially through intermediates A, B, and C. The rate constants for all the interconversions are given on the reaction scheme shown. The free energy of each transition state relative to its precursor was calculated at 25°C from transition-state theory (Sec. 5.2.3.a). During the reaction S → P, only intermediate A would accumulate to detectable levels, at most 10% of the molecules present. The overall rate of the reaction is determined by the free energy of the transition state B/C‡ between intermediates B and C, relative to the free energy of S.

A basic difficulty with kinetic studies of multistep reactions, however, is that only a few intermediates are likely to accumulate to detectable levels (Fig. 9.3). Any correlation between the importance of an intermediate and its probability of accumulating is likely to be inverse; that is, the most important intermediates are those least likely to accumulate. For example, the transition state is most important for determining the rate of a reaction, but it is the species that accumulates least (Sec. 5.2.3.a).

Intermediates can be detected most readily if many different substrates are available for the enzyme, because the various substrates may differ in the rates of individual steps; this can lead to the accumulation of different intermediates along the reaction pathway. Some substrates might undergo partial reaction very rapidly but then not react further at a comparable rate. If a product is released in the first step, that step will not be readily reversed, and intermediate EX will accumulate:

$$E + S \xrightarrow[P_1]{\text{fast}} EX \xrightarrow{\text{slow}} E + P_2 \qquad (9.21)$$

This kind of kinetic behavior with certain substrates was further evidence for an acyl intermediate of serine proteases (Sec. 9.3.2.a). Such substrates have practical uses as **active site titrants;** after mixing enzyme with excess substrate there is an initial burst of formation of product P_1 — one mole per mole of enzyme active site (Fig. 9.4). Subsequent turnovers of the enzyme are limited by the slow second step. In many cases, the intermediate has the second product covalently attached to the enzyme, so the complex can be characterized chemically. In such techniques, intrinsically labile linkages between enzyme and substrate may need to be trapped chemically — for example, by irreversibly reducing a Schiff base linkage (Sec. 1.3.7) — whereas others can be stabilized merely by disrupting the enzyme's structure.

If a substrate normally makes no known covalent intermediate, it may be induced to do so — thus providing the enzymologist with a convenient probe — by adding a reactive group to it. Such **active-site-directed irreversible inhibitors** are analogous to affinity labels used to react covalently with ligand binding sites (Sec. 8.1.3). Further specificity can be introduced by altering the substrate so that a very reactive group is generated during the catalytic cycle, which then reacts with an appropriate group nearby in the enzyme active site. Such **suicide substrates,** also called **enzyme-activated** or **mechanism-based inhibitors,** become reactive only after being subjected to catalysis. Most suicide inhibitors are based on the generation of an intermediate that has reactive conjugated double bonds. The classic

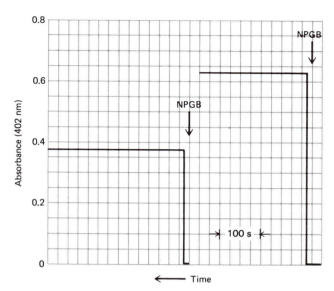

FIGURE 9.4

Active-site titration of 0.05 ml *(left)* and 0.08 ml *(right)* of a solution of β-trypsin, using the titrant NPGB:

At the times indicated by the arrows in the figure, the NPGB was added to a 1.0-ml solution containing the trypsin. The almost instantaneous increase in absorbance is due to the release of *p*-nitrophenol upon acylation of all the trypsin molecules. The subsequent very slow increase is limited by the slow rate of deacylation of the enzyme. Extrapolation to the time of adding NPGB indicates that the concentrations of trypsin active sites in the cuvette were 2.22×10^{-5} M and 3.76×10^{-5} M on the left and right, respectively. Therefore, the stock solution was $4.57 \pm 0.13 \times 10^{-4}$ M in active trypsin. (Adapted from T. Chase and E. Shaw, *Methods Enzymol.* 19:20–27, 1970.)

example is that for the enzyme β-hydroxy-decanoyl-dehydrase, which normally catalyzes the reactions

$$C_6H_{13}-CH=CH-CH_2-CO-NAC$$

$$\updownarrow$$

$$C_6H_{13}-CH_2-CH=CH-CO-NAC$$

$$\downarrow H_2O$$

$$C_6H_{13}-CH_2-CHOH-CH_2-CO-NAC \quad (9.22)$$

where —NAC is —S—CH$_2$—CH$_2$—NH—CO—CH$_3$. The suicide inhibitor has a triple bond, so it is isomerized by the enzyme to a form with two adjacent double bonds. This species is chemically reactive and reacts with a His residue in the enzyme active site:

$$C_6H_{13}-C\equiv C-CH_2-CO-NAC + enzyme$$

$$\downarrow$$

$$C_6H_{13}-HC=C=CH-CO-NAC \cdot enzyme$$

$$\downarrow$$

$$C_6H_{13}-HC=\underset{|}{C}-CH_2-CO-NAC$$
$$enzyme \qquad (9.23)$$

Some natural toxins seem to act in just this way and irreversibly inactivate crucial enzymes. Acting irreversibly, they are much more potent than reversible inhibitors, which can be displaced by the substrate.

The pH dependence of each step in an enzyme-catalyzed reaction can be used to measure the pK_a values of crucial acids or bases in the free enzyme, the substrate, and the enzyme–substrate complex. The small, relatively simple enzymes that have been studied most extensively usually demonstrate straightforward pH dependence of the strength of substrate binding or the rate of catalysis; both binding and catalysis often appear to depend principally on the ionization of only one or two groups. Care must be exercised, however, in assigning such pK_a values to specific groups. The pK_a values of groups in enzyme active sites are often substantially perturbed from their normal values, and they frequently are even more perturbed in the enzyme–substrate complex. Some examples are given in Table 9.1. The identities of these and other groups in the active site can be determined by the abolition of substrate binding or of catalysis upon their specific covalent modification or replacement by site-directed mutation.

The most direct and detailed information about enzyme action would come from crystallographic determination of enzyme–substrate complexes at various stages of reaction, but this determination is hampered by the rapid rates at which enzyme-catalyzed reactions occur. It is much more feasible to determine the structures of stable complexes of the enzyme with incomplete sets of the substrates, of substrate analogues that react very slowly, and of inhibitors; many such reactions have been studied crystallographically. The binding of substrates to enzymes is found not to be fundamentally different, at least to a first approximation, from the binding of ligands to specific sites on a protein. This type of crystallographic information has provided great insight into how substrates are held in precise orientations at enzyme active sites. Such studies do not observe catalysis directly, however, and extrapolation of the results is necessary to imagine what occurs during actual catalysis, when significant changes in both the enzyme and the substrate could be occurring. Inhibitors can be misleading because they may be inactive as substrates simply because they bind in a nonproductive mode that is different from the binding of a true substrate.

An experimental approach that can alleviate these difficulties involves following the kinetics of enzyme action at very low temperatures, down to −70°C, where all reactions become very much slower. The en-

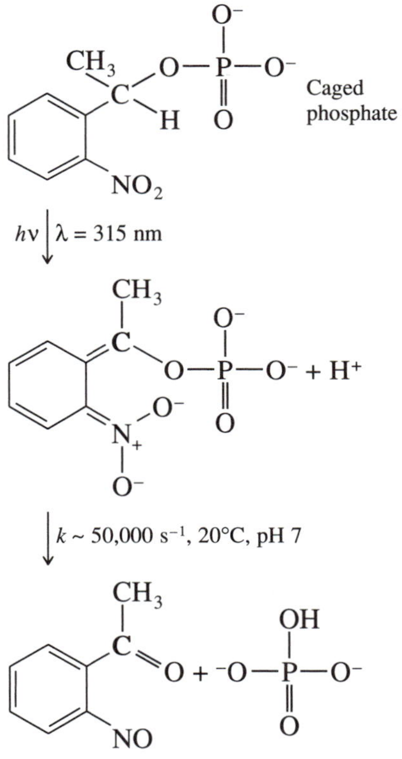

FIGURE 9.5
Light-induced release of phosphate from caged 2-nitro-phenyl phosphate. Upon illumination (*hv*) with light at 315 nm, the phosphate is released with a half-time of 14 μs. (From J. Hajdu and L. N. Johnson, *Biochemistry* 29:1669–1678, 1990.)

Table 9.1 *Anomalous* pK_a *Values of Ionizing Groups in Enzyme Active Sites*

Enzyme	Ionizing group	Observed pK_a	Normal pK_a[a]
Acetoacetate decarboxylase[b]	Lys	6.0	10.4–11.1
Carboxypeptidase A[c]	Glu 27	7.0	4.3–4.5
α-Chymotrypsin[d]	α-NH$_2$	10.0	6.8–8.0
Lysozyme[e]	Glu 35	6.5	4.3–4.5
Papain[f]	His 159	8.5	6.0–7.0
	Cys 25	3.3	9.0–9.5
Pepsin[g]	Asp 32	1.5	3.9–4.0
Rhodanese[h]	Cys 247	6.5	9.0–9.5

[a] Table 1.2.

[b] F. C. Kokesh and F. H. Westheimer, *J. Amer. Chem. Soc.* 93:7270–7274 (1971).

[c] P. H. Petra and H. Neurath, *Biochemistry* 10:3171–3177 (1971).

[d] A. R. Fersht, *Enzyme Structure and Mechanism,* 2nd ed., p. 174. W. H. Freeman, New York, 1985.

[e] C. C. F. Blake et al., in *Ciba Foundation Symposium,* vol. 60, pp. 137–172. Excerpta Medica, Amsterdam, 1978.

[f] S. D. Lewis et al., *Biochemistry* 20:48–51 (1981).

[g] J. A. Hartsuck and J. Tang, *J. Biol. Chem.* 247:2575–2580 (1972).

[h] J. H. Ploegman et al., *J. Mol. Biol.* 127:149–162 (1979).

ergy necessary to overcome an activation barrier comes from the kinetic energy of ordinary thermal fluctuations; the rate, therefore, is decreased exponentially as the temperature is lowered to an extent determined by the enthalpy and entropy of the transition state (Sec. 5.2.3.a). Intermediates may accumulate for much longer times under these conditions, and these productive species can be studied by the most powerful methods, including X-ray crystallography. New intermediates may also be detected at low temperatures; the rates of the various steps are slowed to different extents, and new steps may become rate-determining. Drawbacks of this approach include the need to use antifreeze solvents such as 50–80% methanol, dimethyl sulfoxide, or ethylene glycol; there is always the possibility that the antifreeze, as well as the low temperature, will change the reaction mechanism.

Recent advances in X-ray crystallography make it possible to measure diffraction patterns within 10^{-10} s (Sec. 6.1.7) and thus make it feasible to follow the course of any reaction on an enzyme in the crystalline state. There are numerous technical difficulties, however. One is to synchronize the reaction on all the enzyme molecules that make up the crystal. Mixing the enzyme and substrates rapidly, as is usually done in solution kinetic studies, is not feasible in a crystal, where it is necessary for the substrate to diffuse through the channels of the crystal lattice. Instead, the substrate must first be diffused into the crystal in such a way that the enzymatic reaction does not take place. After the substrate is in place, the enzymatic reaction must be initiated simultaneously in all the molecules. One of the best methods is to use a **caged compound.** The substrate is made inert by covalent attachment of a photolabile protecting group (Fig. 9.5). Flash illumination dissociates the protecting group, liberating the substrate within 10^{-2}–10^{-5} s.

This technology is only now being developed, but experimental results are beginning to be forthcoming. The caged substrate has been found in some instances to be bound by the enzyme in an unusual mode, but removal of the protecting group permits the liberated substrate to bind rapidly in the correct manner. A technical difficulty is that the released protecting group often tends to react with the protein. A more fundamental problem is that only particularly stable intermediate states will be visible by this technique. If such intermediates do not accumulate on a substantial fraction of the enzyme molecules in the crystal, all that can be observed crystallographically will be a shift with time in the proportion of enzyme molecules that have either substrate or product bound.

References

Transients and relaxation kinetics of enzyme reactions. H. Gutfreund. *Ann. Rev. Biochem.* 40:315–344 (1971).

Chemical studies of enzyme active sites. D. S. Sigman and G. Mooser. *Ann. Rev. Biochem.* 44:899–931 (1975).

X-ray cryoenzymology. A. L. Fink and G. A. Petsko. *Adv. Enzymol.* 52:177–246 (1981).

Suicide substrates, mechanism-based enzyme inactivators: recent developments. C. T. Walsh. *Ann. Rev. Biochem.* 53:493–535 (1984).

Characterization of transient enzyme-substrate bonds by resonance Raman spectroscopy. P. R. Carey and A. C. Storer. *Ann. Rev. Biophys. Bioeng.* 13:25–49 (1984).

Enzyme-activated/mechanism-based inhibitors. M. G. Palfreyman et al. *Essays Biochem.* 23:28–80 (1987).

Catalysis in enzyme crystals. J. Hajdu et al. *Trends Biochem. Sci.* 13:104–109 (1988).

Analysis and prediction of the location of catalytic residues in enzymes. M. J. J. M. Zvelebil and M. J. E. Sternberg. *Protein Eng.* 2:127–138 (1988).

Properties and uses of photoreactive caged compounds. J. A. McCray and D. R. Trentham. *Ann. Rev. Biophys. Biophys. Chem.* 18:239–270 (1989).

Time-resolved X-ray crystallographic study of the conformational change in Ha-Ras p21 protein on GTP hydrolysis. I. Schlichting et al. *Nature* 345:309–315 (1990).

9.2 Theories of Enzyme Catalysis

The rate of a chemical reaction depends on the relative free energies of the initial free reactants, of any stable intermediates in the reaction, and of the transition states along the reaction (Fig. 9.3). A catalyst can increase the rate of a reaction by any of several mechanisms: by destabilizing the initial reactants, by stabilizing the transition state (Fig. 9.6), or by altering the mechanism of the reaction. Enzymes use all of these mechanisms. An additional consideration with enzymes is to account for the specificity of their catalysis.

References

How do enzymes work? J. Kraut. *Science* 242:533–540 (1988).

The interplay between chemistry and biology in the design of enzymatic catalysts. P. G. Schultz. *Science* 240:426–433 (1988).

9.2.1 Rate Enhancements

The magnitude of enzyme catalysis is given by the relative rates at which the reaction converting the substrate S to product P occurs on the enzyme, k_{cat}, and normally in solution in its absence, k_n:

$$ES \xrightarrow{k_{cat}} EP \qquad (9.24)$$

$$S \xrightarrow{k_n} P \qquad (9.25)$$

This comparison is straightforward for unimolecular reactions. Some examples of the rate enhancements produced by enzymes under comparable conditions are given in Table 9.2; the values range between 10^6 and 10^{14}. Few rate constants for uncatalyzed reactions under normal conditions are available, however, because such reactions generally are so slow. The sample for which values have been measured, therefore, may

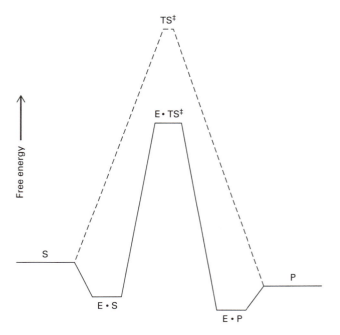

FIGURE 9.6

Simplified free-energy profile of an uncatalyzed reaction (dashed line) and a reaction catalyzed by an enzyme (solid line). The relative free energies of the substrate S, the transition state TS‡, and the product P are shown when free and when bound to the enzyme E. The reaction is catalyzed by the enzyme when the energy of the transition state is lowered more upon binding to the enzyme than is the energy of the substrate. The relative free energies of S, TS‡, and P when free and when bound depend on their respective affinities for the enzyme and on the concentration of the enzyme. The height of the free-energy barrier to the catalyzed reaction is given by the free energy of E · TS‡ relative to E · S; this difference determines the value of k_{cat} (see Eq. 5.18). The free energy of E · S relative to S is determined by the concentrations of E and S and their affinity; the same considerations apply to E · P and P. The reaction is catalyzed when the enzyme has greater affinity for the transition state than for the substrate.

Table 9.2 *Relative Rates of Enzyme-Catalyzed and Noncatalyzed Reactions, Under Conditions Optimal for the Enzyme*

Enzyme	Reaction	Ratio of enzyme-catalyzed to noncatalyzed rate
	Unimolecular Reactions	
Chorismate mutase		1.9×10^{6} [a]
Triose phosphate isomerase	$P_i-CH_2-\overset{OH}{\underset{\mid}{CH}}-\overset{O}{\overset{\parallel}{CH}} \longrightarrow P_i-CH_2-\overset{O}{\overset{\parallel}{C}}-CH_2OH$	3×10^{8} to 10^{9} [b]
Serine proteases	$R_1-\overset{O}{\overset{\parallel}{C}}-NH-R_2 \xrightarrow{H_2O} R_1-\overset{O}{\overset{\parallel}{C}}OH + H_2N-R_2$	$10^{5}-10^{10}$ [c]
Lysozyme	$(N\text{-acetylglucosamine})_2 \xrightarrow{H_2O} 2\ N\text{-acetyl glucosamine}$	$10^{7}-10^{9}$ [d]
Urease	$NH_2-\overset{O}{\overset{\parallel}{C}}-NH_2 \xrightarrow{H_2O} CO_2 + 2NH_3$	10^{14} [e]
Carbonic anhydrase		1.1×10^{8} [f]
	Bimolecular Reactions [g]	
Hexokinase	glucose + ATP $\longrightarrow$ glucose 6-P + ADP	$> 8 \times 10^{10}$ M
Phosphorylase	glucose 1-P + glycogen$_n$ $\longrightarrow$ glycogen$_{n+1}$ + P$_i$	$> 9 \times 10^{11}$ M
Alcohol dehydrogenase	ethanol + NAD $\longrightarrow$ acetaldehyde + NADH	$> 2 \times 10^{10}$ M
Creatine kinase	creatine + ATP $\longrightarrow$ creatine-P + ADP	$> 4 \times 10^{8}$ M

[a] From P. R. Andrews et al., *Biochemistry* 12:3492–3498 (1973).

[b] From R. Wolfenden, *Acc. Chem. Res.* 5:10–18 (1972).

[c] From A. Hall and J. R. Knowles, *Biochemistry* 14:4384–4352 (1975); G. Hess, in *The Enzymes*, 3rd ed. P. D. Boyer, ed., vol. 3, pp. 213–248. Academic Press, New York, 1971; J. Kraut, *Ann. Rev. Biochem.* 46:331–358 (1977).

[d] From D. Chipman et al., *Biochemistry* 10:1714–1722 (1971).

[e] From W. P. Jencks, *Catalysis in Chemistry and Enzymology*, p. 5. McGraw-Hill, New York, 1969.

[f] From Y. Pocker and J. E. Meany, *J. Am. Chem. Soc.* 89:631–636 (1967).

[g] From D. E. Koshland, Jr., *J. Cell. Comp. Physiol.* 47 (Suppl. 1):217–234 (1956).

Table 9.3 *Relative Rates of the Esterification Reaction*

$$-OH + HOC\!\!\!\overset{\displaystyle O}{\|} \longrightarrow -O-C\!\!\!\overset{\displaystyle O}{\|} + H_2O$$

Reactants	Rate constant	Effective concentration of intramolecular groups
(phenol) OH + CH_3COH	10^{-10} s^{-1}M^{-1}	—
(benzene ring with OH and side chain —OH, C=O)	3.2×10^{-6} s^{-1}	3.2×10^4 M
(H_3C, CH_3, CH_3 substituted ring, —OH, C=O)	3.3×10^{-6} s^{-1}	3.3×10^4 M
(benzene ring, —OH, C=O, H_3C CH_3)	3.6×10^{-5} s^{-1}	3.6×10^5 M
(H_3C, CH_3, CH_3 ring, —OH, C=O, H_3C CH_3)	8.5×10^{-2} s^{-1}	8.5×10^8 M

Rate constants for the uncatalyzed reactions in water from S. Milstien and L. A. Cohen, *Proc. Natl. Acad. Sci. USA* 67:1143–1147 (1970).

be biased in favor of rapid nonenzymatic rates and may give a conservative estimate of the magnitude of enzyme catalysis.

Comparison of chemical and enzymic reaction rates for reactions involving two or more substrates is not straightforward. The nonenzymatic rate constants are of second order, or even higher, whereas the enzymic rate k_{cat} is always of first order (Table 9.2). A chemical multimolecular reaction requires the simultaneous encounter of all the reactants, and consequently its rate is proportional to each of their concentrations in solution. In enzymatic reactions, the reactants are bound to the enzyme, and the reaction occurs in this complex in an essentially unimolecular process.

The unimolecular nature of enzyme catalysis points to a major way in which enzymes can catalyze reactions. In unimolecular reactions, the effective concentrations (Sec. 4.4) of reactants can far exceed those that are possible in bimolecular reactions (Table 9.3). The maximum effective concentration for bimolecular reactions is thought to be about 10^{10} M, which simply reflects the entropy that must be lost in a bimolecular encounter between reactants in solution that need not be lost in a unimolecular reaction (Sec. 4.4). The substrates in an enzyme–substrate complex have already lost this entropy in binding to the enzyme, and this factor is included in their binding affinities. Consequently, there is a large entropic advantage in a multimolecular reaction occurring on an enzyme. The greater the number of reactants binding simultaneously, the greater the advantage. For example, the probability of simultaneous encounters among three or four reactants in solution, when all are present at low concentrations, is essentially nil, whereas ternary and quaternary complexes on an enzyme are not uncommon. Of course, simultaneous binding of reactants to an enzyme is not sufficient; high effective concentrations can occur only if the reactive groups are held in the appropriate positions and orientations for reaction to occur (Table 9.3). Substrates held apart would be prevented from reacting and would have an effective concentration of zero.

This entropic factor might seem not to apply to unimolecular reactions with only a single substrate. Virtually every chemical reaction can be catalyzed in solution, however—for example, by nucleophiles or electrophiles in general acid or base catalysis (Fig. 9.7)—and the observed rate of the reaction depends on the concentrations of the catalysts.

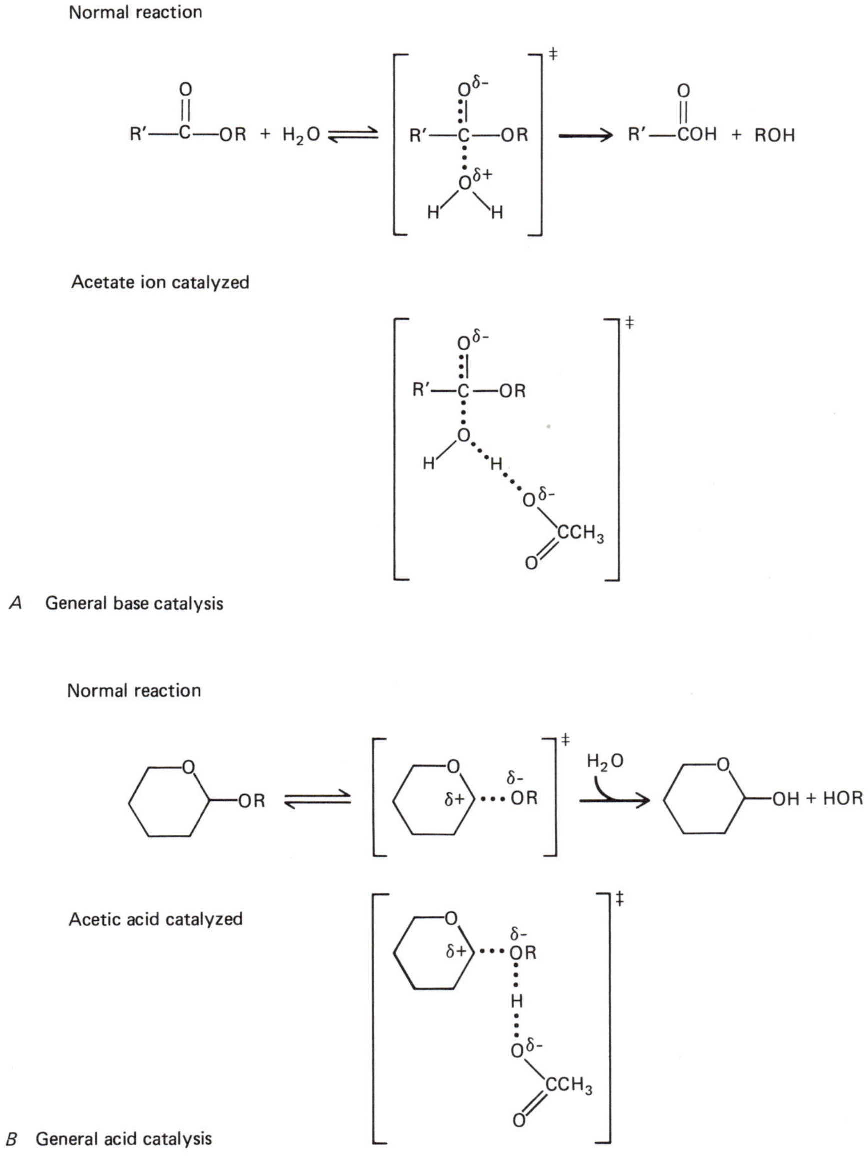

FIGURE 9.7
Examples of general base and general acid catalysis; the presumed transition states are enclosed in brackets. *A:* Acetate ion catalyzes hydrolysis of an ester, presumably by interacting with the transition state so as to make the charge separation more favorable and to lower its free energy. *B:* Acetic acid catalyzes hydrolysis of an acetal, presumably by stabilizing the transition state as indicated. (Adapted from A. Fersht, *Enzyme Structure and Mechanism,* W. H. Freeman, Reading, 1977.)

Enzymes have numerous functional groups that potentially could play a role as acid or base catalysts and could be present at very high effective concentrations to the substrate, if held in the correct position in the enzyme active site. For example, the reactions of Table 9.3 are catalyzed by acid, by hydroxide ion, and by imidazole, with second-order rate constants of between 10^{-6} and 10^6 M^{-1} s^{-1}. If just one of these catalysts were present in an enzyme active site in exactly the correct position to present an effective concentration of 10^6 M to the substrate, the reactions catalyzed on the enzyme would have first-order rate constants of between 1 and 10^{12} s^{-1} by this mechanism alone. These hypothetical reactions would occur between 10^6 and 10^{14} times more rapidly than the uncatalyzed reaction in solution. Consequently, unimolecular reactions, in theory, can also be catalyzed at very high rates on enzymes simply by incorporating a chemical acid or base catalyst into a unimolecular complex.

The presence of multiple substrate molecules or catalytic groups in an enzyme–substrate complex implies that the reactions that occur there may be somewhat different from those that occur in solution. The most favorable mechanism in solution might be limited to bimolecular encounters between reactants or substrates simply because higher order encounters are so improbable. On the enzyme, there is no such entropic restriction on encounters between multiple groups because of the unimolecular nature of the complex. Concerted reactions between multiple groups can readily occur on an enzyme; for example:

$$
\begin{array}{c}
\underset{|}{R} \\
B: \curvearrowright H-\underset{|}{N}: \curvearrowright C=O \\
H
\end{array}
$$

$$
\Updownarrow
$$

$$
\begin{array}{c}
\underset{|}{R} \quad | \\
BH^+ \quad N-C-O^- \\
| \quad | \\
H
\end{array} \tag{9.26}
$$

Enzymes may also speed reactions by producing intermediate states along the reaction pathway. In a single-step reaction, the geometry of the transition state is somewhere between substrate and product and therefore has a correspondingly higher free energy than either. With a relatively stable intermediate between reactant and product, each of the two transition states will be closer to either the substrate or the product. Consequently, neither transition state need differ from substrate or product so much in geometry or in free energy (Fig. 9.8), and the overall kinetic barrier would

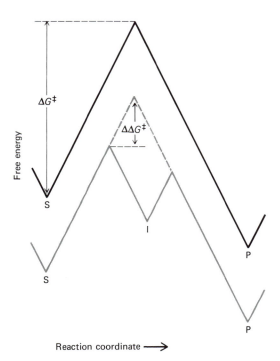

FIGURE 9.8

Naive illustration of how the occurrence of an intermediate might increase the rate of a reaction. The free energy of the molecule is postulated to be directly proportional to the distance along the reaction coordinate from the reactant S, the product P, or any intermediate I. The upper curve illustrates such a free-energy profile for a reaction that occurs in a single step, with the rate-determining free energy of the single transition state indicated. The lower curve shows the same reaction that has an unstable intermediate I midway along the reaction coordinate. The rate-determining transition state is that on the left; its free energy is lower than that in the absence of the intermediate by $\Delta\Delta G^\ddagger$.

not be expected to be so high. Therefore, reactions entailing a number of small changes should occur faster than those involving one large change. The intermediate need not, and should not, be more stable than either substrate or product for the overall rate to be increased. Being closer to the transition state, however, the intermediate is likely to be tightly bound by the enzyme (Sec. 9.2.2) and consequently to be more stable than when free in solution during the normal uncatalyzed reaction.

Interactions of the reactants with the solvent are extremely important for determining the rate of a chemical reaction. Charge separations almost invariably occur in any transition state because electrons are being redistributed in the process of covalent bond breakage and reformation. Polar solvents can greatly stabilize charges, so reactions involving increased charge separation in the transition state occur much more rapidly in such solvents. Conversely, reactions with less charge separation occur more rapidly in nonpolar solvents. For example, the reaction

$$(9.27)$$

occurs $10^4 - 10^5$ times faster in ethanol than in water. The intermediate in brackets is thought to approximate the transition state and has the charge separation of the original molecule greatly diminished. Enzymes could conceivably increase the rates of reactions simply by removing the substrates from aqueous solution and sequestering them in an active site environment that is appropriately either more or less polar.

Catalysts can be intimately involved in a chemical reaction, actually forming covalent bonds with the reactants in intermediate stages and thereby changing the reaction mechanism. An excellent example is the way in which carbonyl compounds reversibly form Schiff bases with amines, thereby altering the chemical properties of both:

$$R^1 \underset{R^2}{\diagdown} C{=}O + H_2N{-}R^3$$

$$\updownarrow H_2O$$

$$R^1 \underset{R^2}{\diagdown} C{=}N{-}R^3$$

Schiff base $$(9.28)$$

Protonation of the Schiff base occurs under normal conditions, and it then becomes an electron "sink," assist-

ing loss of a proton from groups R_1, R_2, or R_3. This makes these groups much more reactive as nucleophiles than they would be in the parent compound; for example:

$$H_3C \underset{R_2}{\diagdown} C{=}N{-}R_3$$

$$\downarrow H^+$$

$$H_3C \underset{R_2}{\diagdown} C{=}\overset{+}{N}H{-}R_3$$

$$\downarrow H^+$$

$$H_2C \underset{R_2}{\diagdown} C{-}NH{-}R_3$$

$$(9.29)$$

After reaction of one of the activated peripheral groups, the Schiff base can dissociate back to the carbonyl and the amine. Whichever of the original reactants was not altered by the reaction would be an apparent catalyst of the reaction of the other.

Enzymes engage in such tricks by using suitable groups as cofactors and by channeling the intrinsic catalytic capabilities of the cofactors. For example, many enzymes catalyzing reactions involving amino acids use pyridoxal phosphate as cofactor, which by itself readily forms a Schiff base with amino acids and catalyzes a number of their reactions:

pyridoxal-P

$$+ H_2N{-}\underset{R}{\overset{R}{\underset{|}{CH}}}{-}CO_2^-$$

$$\updownarrow H_2O$$

$$(9.30)$$

The pyridine ring of the Schiff base acts as an electron sink and stabilizes a negative charge very effectively. Therefore, each group on the C^α atom of the amino acid can be more readily cleaved off as cations — for example, the hydrogen atom:

(9.31)

The consequence of transient removal of the hydrogen atom can be racemization of the amino acid, alteration of the amino acid side chain, or loss of the carboxyl group as CO_2. In each case, unaltered pyridoxal phosphate is regenerated, and it has served only as a catalyst. In other enzymatic reactions, the pyridoxal phosphate is converted from the aldehyde to the amine by cleavage of the C^α—N bond, and the amino group is subsequently transferred to a second substrate to complete a catalytic cycle.

Numerous other cofactors, prosthetic groups, and metal ions are used in enzymes to provide catalytic capabilities beyond those possible with amino acid residues alone.

References

Catalysis in Chemistry and Enzymology. W. P. Jencks. McGraw-Hill, New York, 1969.

Schiff base intermediates in enzyme catalysis. E. E. Snell and S. J. DiMari. In *The Enzymes*, 3rd ed., P. D. Boyer ed., vol. 2, pp. 335–370. Academic Press, New York, 1970.

Some pertinent aspects of mechanism as determined with small molecules. T. C. Bruice. *Ann. Rev. Biochem.* 45:331–373 (1976).

Binding energy, specificity, and enzymic catalysis: the Circe effect. W. P. Jencks. *Adv. Enzymol.* 43:219–410 (1975).

Transition states, standard states and enzymic catalysis. M. I. Page. *Intl. J. Biochem.* 11:331–335 (1980).

Electrostatic basis of structure-function correlation in proteins. A. Warshel. *Acc. Chem. Res.* 14:284–290 (1981).

Miniature organic models of enzymes. V. T. D'Souza and M. L. Bender. *Acc. Chem. Res.* 20:146–152 (1987).

Nickel enzymes. C. T. Walsh and W. H. Orme-Johnson. *Biochemistry* 26:4901–4906 (1987).

Quantitative modeling of proximity effects on organic reactivity. K. N. Houk et al. *Acc. Chem. Res.* 23:107–113 (1990).

Zinc coordination, function, and structure of zinc enzymes and other proteins. B. L. Vallee and D. S. Auld. *Biochemistry* 29:5647–5659 (1990).

9.2.2 Transition-State Stabilization

In general terms, a reaction occurs more rapidly on an enzyme than in solution simply because the difference in the free energies of the substrate and the transition state $T^\ddagger$ are not so great when they are bound to the enzyme (Fig. 9.6). From transition-state theory (Sec. 5.2.3.a):

$$S \xrightleftharpoons{K_n^\ddagger} T^\ddagger \xrightarrow{v} P \qquad (9.32)$$

$$ES \xrightleftharpoons{K_E^\ddagger} ET^\ddagger \xrightarrow{v} P \qquad (9.33)$$

$$\frac{k_{cat}}{k_n} = \frac{K_E^\ddagger}{K_n^\ddagger} \qquad (9.34)$$

The decreased free-energy difference on the enzyme could be due to either the free energy of the substrate being increased or that of the transition state being decreased.

The free energy of the substrate could be increased by binding if the bound substrate is strained, either sterically with distorted geometry or by placing it in an otherwise unfavorable environment. The free-energy "cost" of introducing this strain would be "paid for" from binding energy, in that the affinity of the enzyme for the substrate would be lower than for an unstrained comparable ligand. Alternatively, the strain could be present in the enzyme, even before the substrate is bound. An example is the perturbed chemical properties of groups often found in the active sites of enzymes (Table 9.1). The energetic cost of this strain would be paid for from the energy of the folded conformation because any such strained properties of the folded conformation necessarily lower its stability. In both cases, the strain must be such as to push the enzyme–substrate complex along the reaction coordinate toward the transition state so that the strain is relieved there and the reaction occurs more rapidly.

It is now thought that the enzyme usually lowers the free energy of the transition state rather than raising that of the substrate. This decrease in free energy can be achieved simply by having the active site optimally complementary to the transition state rather than to the substrate or product. Besides being sterically complementary, appropriate acid and base catalysts would be in the appropriate positions to stabilize electrostatic charges on the transition state (Fig. 9.7). How such a mechanism works will be illustrated in the examples of specific enzymes given later in this chapter (Sec. 9.3).

A consequence of the enzyme being more complementary to the transition state than to the substrate is that the enzyme should bind the transition state more tightly than the substrate, by the factor by which the enzyme increases the rate of the reaction. This is a result of linkage between binding and the free energies of the transition states:

$$\begin{array}{ccc} S & \xrightarrow{\;K_n^{\ddagger}\;} T^{\ddagger} & \longrightarrow P \\ E \updownarrow K_a^S & E \updownarrow K_a^T & \\ ES & \xrightarrow{\;K_E^{\ddagger}\;} ET^{\ddagger} & \longrightarrow EP \end{array} \qquad (9.35)$$

It is a thermodynamic requirement that

$$\frac{K_a^T}{K_a^S} = \frac{K_E^{\ddagger}}{K_n^{\ddagger}} \qquad (9.36)$$

and from Equation (9.34) it follows that

$$\frac{K_a^T}{K_a^S} = \frac{k_{cat}}{k_n} \qquad (9.37)$$

Consequently, enzymes are not expected to have extremely high affinities for their substrates because they would not be catalysts in that case. Instead, they would be like immunoglobulins, which bind antigens tightly but usually do not alter them.

Reference

Enzymatic catalysis and transition-state theory. G. E. Lienhard. *Science* 180:149–154 (1973).

9.2.3 Transition-State Analogues

Analogues of a substrate that tend to mimic the transition state should in theory bind more tightly to the enzyme than does the normal substrate (Eqs. 9.35–9.37). Attempts to design transition-state analogues have produced some that do in fact bind substantially more

tightly than substrates (Table 9.4). It would be unrealistic to imagine that a perfect transition-state analogue could be designed to display the predicted maximum increase in affinity of 6–14 orders of magnitude, because transition states are by definition extremely unstable species, with only partially formed bonds. Also, transition states often have anomalous ionization properties that could not be expected to be maintained in solution. Nevertheless, the affinities predicted for transition states are so high that even approximate analogues should bind very tightly to the appropriate enzyme.

Tight binding alone is not sufficient evidence that a substrate analogue is related to the transition state. For example, methotrexate is related structurally to dihydrofolate and binds with a K_i of 5.8×10^{-11} M to the enzyme dihydrofolate reductase; the tight binding of methotrexate suggested that it is a transition-state analogue of dihydrofolate. Crystallographic analysis, however, revealed that methotrexate binds to the active site of dihydrofolate reductase in an orientation that is inverted with respect to the way dihydrofolate binds as a substrate. Clearly, methotrexate is not even a good substrate analogue.

The general validity of the transition-state analogue approach has been demonstrated best in the case of thermolysin. A series of substrates were compared with the corresponding transition-state analogues. A reasonable correlation was observed between the K_i values for the transition-state analogues and the ratio of the K_m and k_{cat} values for the substrates, but not with the K_m values alone (Fig. 9.9). The value of k_{cat}/K_m gives the free energy of the enzyme transition state relative to that of the free substrate and enzyme (Sec. 9.2.1), so the correlation of this value with K_i for the inhibitors indicates that they are truly transition-state analogues and not just analogues of the substrate. Nevertheless, none of the K_i values of the analogues approach those expected of true transition states.

Many transition-state analogues have been observed to bind and to dissociate more slowly than do substrates. Slow dissociation would be expected from the tight binding, but slow rates of binding would not. The slow binding may indicate that the enzyme active site is designed to combine with the substrate and to release the product, but not the transition state. Upon binding substrate, varying extents of structural change occur in the enzyme, and more extensive changes may take place during the reaction. As a result of such structural changes (Sec. 9.2.5), there may be kinetic barriers to binding and releasing the transition state.

Other inhibitors that generally bind more tightly to an enzyme than do the normal substrates are those that are covalently linked versions of two (or more) sub-

Table 9.4 *Examples of Transition-State Analogues*

Enzyme	Reaction[a]	Increased affinity of transition-state analogue
Triose phosphate isomerase[b]		7.5×10^2
Glucose 6-phosphate isomerase[c]		1.0×10^3
Oxaloacetate decarboxylase[d]		3.5×10^3

[a] The reaction catalyzed by the enzyme is given with the presumed structure of the transition state in brackets; below it is the structure of the transition-state analogue. The dissociation constants used to calculate the increased affinity of the transition-state analogue are given below the ligands.

[b] From R. Wolfenden, *Nature* 223:704–705 (1969); F. C. Hartman et al., *Biochemistry* 14:5274–5279 (1975).

[c] From J. M. Chirgwin and E. A. Noltmann, *J. Biol. Chem.* 250:7272–7276 (1975); P. J. Shaw and H. Muirhead. *FEBS Letters* 65:50–55 (1976).

[d] From A. Schmitt et al., *Hoppe-Seylers Z. Physiol. Chem.* 347:18–34 (1966).

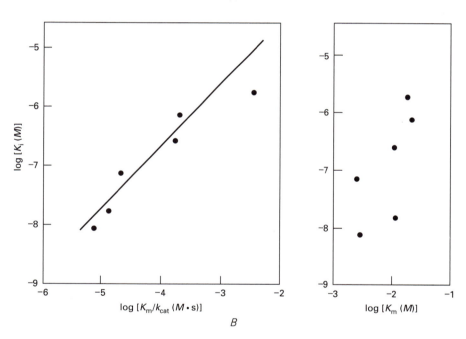

FIGURE 9.9

Comparison of a series of substrates and transition-state analogues of thermolysin. *A:* Structures of the substrate, the tetrahedral intermediate thought to occur during the reaction, and transition-state analogues. The series of substrates and analogues have various Y groups attached to the terminal carboxyl group of the blocked dipeptide Cbz-Gly-Leu-. Cbz is the benzyloxycarbonyl group blocking the terminal amino group. The transition-state analogues have a phosphonamidate replacing the peptide group of the substrate. This was modeled to mimic the tetrahedral intermediate, in which the carbonyl carbon also has tetrahedral geometry. *B:* Correlation between the inhibition constant K_i for the transition-state analogues of *A* and the catalytic parameters of the corresponding substrates. The measured values of K_i are correlated much more closely with the measured values of K_m/k_{cat} *(left)* than with their K_m values alone *(right)*. (From P. A. Bartlett and C. K. Marlowe, *Biochemistry* 22:4618–4624, 1983.)

strates that normally bind simultaneously to the enzyme in a ternary (or higher) complex (Eq. 9.10). For example, an enzyme that normally binds two substrates A and B, to accelerate reaction between them, will bind a covalently linked **bisubstrate analogue** A–B much tighter than either A or B alone (e.g., Fig. 9.30). Tight binding of such an analogue requires only that the covalent link between A and B be compatible with the two parts binding simultaneously to the enzyme in such a way that the effective concentrations of parts A and B relative to each other (Eq. 8.24) are greater than the concentrations of the individual substrates that are normally used.

References

Analog approaches to the structure of the transition state in enzyme reactions. R. Wolfenden. *Acc. Chem. Res.* 5:10–18 (1972).

Transition state analog inhibitors and enzyme catalysis. R. Wolfenden. *Ann. Rev. Biophys. Bioeng.* 5:271–306 (1976).

Transition-state analogues in protein crystallography: probes of the structural source of enzyme catalysis. E. Lolis and G. A. Petsko. *Ann. Rev. Biochem.* 59:597–630 (1990).

Transition-state characterization: a new approach combining inhibitor analogues and variation in enzyme structure. M. A. Phillips et al. *Biochemistry* 31:959–963 (1992).

9.2.4 Catalytic Antibodies

The preceding discussion should make it apparent why antibodies are not enzymes and why they do not catalyze reactions of their antigens. Antibodies have been selected for their affinity for the antigen (more correctly, the immunogen, Sec. 8.3.1) rather than for the transition state of any reaction the antigen might undergo. If the immunogen were a transition state or a transition-state analogue, however, antibodies should catalyze the appropriate reaction. If this were the case, it should be possible to make antibodies with catalytic activity to order.

Monoclonal antibodies directed against various transition-state analogues have been found to have some of the expected catalytic activities and specificities. Such antibodies have many of the characteristics of enzymes in that they accelerate reactions up to 10^5-fold over the noncatalyzed rate, show comparable substrate specificities, exhibit a Michaelis complex through saturation kinetics, have reasonable K_m values for substrates, and are subject to competitive inhibition. The catalytic activities of the antibodies generated thus far are still lower than those of natural enzymes, however,

probably as a result of deficiencies in the way the analogues mimic the true transition states.

Other approaches can be taken to generate catalytic antibodies. Bisubstrate inhibitors can generate antibodies that bind the two individual substrates and enhance reaction between them simply due to their proximity in the antibody combining sites. Reactive groups can be generated in combining sites by using the appropriate immunogen, by mutation of the antibody gene, or by chemical modification of the antibodies.

References

Antibodies as enzymes. R. A. Lerner and A. Tramontano. *Trends Biochem. Sci.* 12:427–430 (1987).

Catalytic antibodies. P. G. Schultz. *Acc. Chem. Res.* 22:287–294 (1989).

Catalytic antibodies. G. M. Blackburn et al. *Biochem. J.* 262:381–390 (1989).

At the crossroads of chemistry and immunology: catalytic antibodies. R. A. Lerner et al. *Science* 252:659–667 (1991).

9.2.5 Substrate Specificity and Induced Fit

The specificity of an enzyme for particular substrates could be imagined to be due simply to its specificity in binding ligands at its active site. An additional factor could be a requirement to bind substrates in a very specific orientation for catalysis, thereby discriminating further against ligands that bind nonproductively in other modes. The specificity exhibited by an enzyme for its substrate, however, can be even greater than that observed in ligand binding. For example, hexokinase catalyzes the phosphorylation of glucose,

$$\text{glucose} + \text{ATP} \rightleftharpoons \text{glucose-6-P} + \text{ADP} \quad (9.38)$$

4×10^4 times more rapidly than it catalyzes the phosphorylation of water,

$$\text{H}_2\text{O} + \text{ATP} \rightleftharpoons \text{ADP} + \text{P}_i \quad (9.39)$$

The enzyme is not particularly specific for glucose, since it readily phosphorylates other sugars; yet water is used as substrate very poorly, even though it is present in aqueous solution at a concentration of 55 M. Other phosphoryl-transferring enzymes are also very specific (e.g., phosphoglucomutase discriminates against H_2O by a factor of 3×10^{10} in favor of its normal substrate, glucose 6-P). It is imperative that enzymes of this type (i.e., kinases) discriminate effectively against water; otherwise they would also function as ATPases (Eq. 9.39) and would drain a cell of its supply of ATP.

Modest degrees of specificity are not difficult to account for. The 4×10^4-fold discrimination against bulk water observed in hexokinase conceivably could result from glucose and ATP binding in the hexokinase active site so that their effective concentration relative to each other (Sec. 9.2.1) is $4 \times 10^4 \times 55$ M ($= 2 \times 10^6$ M). Water could also be excluded from the active site so that its concentration there is less than 55 M. On the other hand, the more extreme examples, such as the 3×10^{10} discrimination against water by phosphoglucomutase, may not be explicable on this basis alone.

To explain such extreme substrate specificities, Koshland proposed that binding of a substrate induces the active conformation of the enzyme, that there is an **induced fit** on binding the substrate. That such a phenomenon occurs with hexokinase was evident from the observation that the binding to its active site of a sugar that cannot accept the phosphate group caused the K_m of the enzyme for ATP to decrease 40-fold and the rate of hydrolysis of the ATP to increase 18-fold. The binding of this substrate analogue activated the catalytic capability of the enzyme for ATP hydrolysis.

The structural basis of this phenomenon is apparent from the crystal structures of free hexokinase and of its complex with sugars. Upon binding glucose, the two domains of hexokinase rotate, essentially as rigid bodies, by 12° relative to each other, causing relative movements of atoms of as much as 8 Å (Fig. 9.10). Ligands that bind at the active site but do not produce this conformational change are not used as substrates; such movement, therefore, is probably essential for catalysis. After rotation of the two lobes, the glucose molecule is almost entirely engulfed by the enzyme, so it cannot enter or leave the active site; dissociation of product from the enzyme is quite slow and rate-limiting and is probably limited by the opening of the cleft.

As a consequence of glucose binding and the conformational change, the apparent affinity of the enzyme for ATP increases 50-fold. ATP binds in the cleft, interacting with both domains, in a way that is favored by the conformational change. The conformational change induced by glucose must also be important for activating the enzyme as a catalyst toward ATP hydrolysis. If sugars that cannot accept the phosphoryl group are added to the enzyme with ATP, the phosphoryl group of ATP tends to be slowly transferred to an adjacent Ser residue in the enzyme. Water does not cause this conformational change, so it is a poor acceptor of the phosphoryl group of ATP.

Further conformational changes probably take place in hexokinase after binding the second substrate, ATP, because the glucose 6-hydroxyl and ATP γ-phosphoryl groups are observed in the individual complexes to be 6 Å apart, too far for the reaction between them to occur in one step. The nature of these changes is not yet known in the case of hexokinase, but further changes are observed with the enzyme adenylate kinase, which catalyzes the transfer of a phosphoryl group from ATP to AMP:

$$ATP + AMP \rightleftharpoons 2ADP \qquad (9.40)$$

Upon binding AMP, the domain of this enzyme that binds AMP closes over the active site, with C^α atoms moving by up to 8.2 Å (Fig. 9.11). Binding of both AMP and ATP simultaneously has been mimicked by two adenosine moieties linked by five phosphate groups, rather than the four that are normally present. When this bisubstrate analogue binds, the AMP-binding domain closes farther down over the active site so that the substrates are removed from contact with water.

Many different kinase enzymes have similar bilobed structures of two or more domains, even though many of these kinases are not homologous in either primary or tertiary structure, other than having similar nucleotide-binding motifs related to their common use of ATP (see Sec. 8.3.3). This common structural feature of two domains is thought to indicate that all these kinases undergo similar movements of the domains relative to each other. That the conformational change involves rotation of otherwise unchanged domains, rather than rearrangements within them, is suggestive of a relatively rigid architecture for each domain.

Structural changes induced in the enzyme by binding a substrate can also account for another aspect of substrate specificity: it may be necessary for the enzyme to embrace the substrate intimately, bringing reactive groups into its proximity from all directions and preventing the transition state from dissociating, as well as excluding water. For example, a hydrophobic environment would be expected to facilitate the electrostatic interactions between enzyme and substrate thought to be of prime importance for catalysis. In the case of tyrosyl tRNA synthetase (Sec. 9.3.1), four basic side chains are induced to interact with two phosphoryl groups in the transition state by movements of a mobile polypeptide loop of the enzyme. Yet it is also necessary for an enzyme to bind substrates and to release products into the solvent; such movements would be a mechanism for giving access to an otherwise inaccessible active site. Indeed, the release of products is often observed to be the rate-determining step in an enzyme-catalyzed reaction. Structural changes in the enzyme during a reaction may also be a reason that transition-state analogues frequently bind so slowly to enzymes. There is experimental evidence in triose phosphate isomerase that large movements of loops to cover the substrates in the active site serve to prevent a labile intermediate from dissociating from the enzyme.

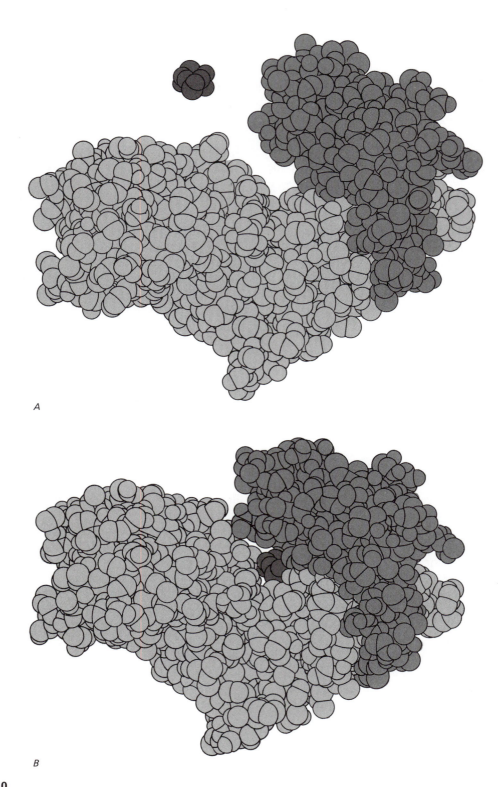

A

B

FIGURE 9.10

The conformational change that occurs upon glucose binding to hexokinase. *A:* A space-filling model of the free enzyme, with a glucose molecule approaching the active-site cleft. *B:* A model of the complex, showing the bound glucose in the closed cleft. The small lobe of the enzyme (darkly shaded) has rotated counterclockwise relative to the larger lobe (lightly shaded). (From W. S. Bennett and T. A. Steitz, *J. Mol. Biol.* 140:211–230, 1980.)

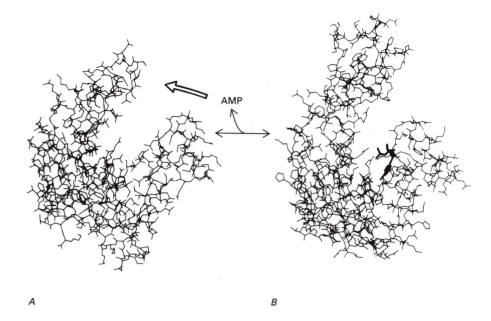

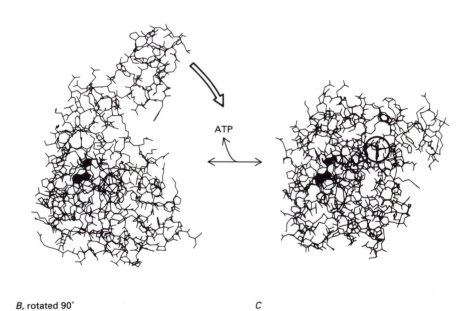

FIGURE 9.11

Domain movements (open arrows) correlated with substrate binding to adenylate kinase. The three protein structures were determined crystallographically, but are of three different forms of the enzyme: The porcine cytosolic form *(A)*, the bovine mitochondrial form with AMP bound *(B)*, and the *Escherichia coli* form with the bisubstrate analogue $A_{p5}A$ bound *(C)*. The three proteins are homologous, and it is assumed that they have similar conformations in each state, so that the comparisons reflect the consequences of ligand binding rather than the different sources of the enzyme. Structure *B* is rotated by 90° in the lower left representation to illustrate more clearly the cleft where the ATP binds. The bound substrates are in bold, and the purine and ribose rings of the AMP moiety are solid black. The $A_{p5}A$ analogue is presumed to approximate both substrates AMP and ATP bound simultaneously to the active site. The second adenosine moiety of $A_{p5}A$, presumed to correspond to the usual ATP substrate, is circled in *C*. (Adapted from G. E. Schulz et al., *J. Mol. Biol.* 213:627–630, 1990.)

Because of structural changes in the enzyme, describing the free enzyme after release of product as being the same as the enzyme that bound the substrate is probably an oversimplification. The state of an enzyme as it releases its product is necessarily different in some respect from the state that bound the substrate. This difference may not be large and may involve only the protonation state of catalytic residues or their precise positions, but the fact that products are chemically different from substrates demands differences in the forms of the enzyme that accept them. So long as the rate of interconversion of the two forms is faster than that of all other steps, the distinction is not critical, and the free enzyme can be considered to be a mixture of rapidly interconverted microstates. This interconversion can, however, be rate-limiting in some instances. It is more strictly correct to depict a simple enzyme mechanism as

$$E_1 \xrightarrow{\;S\;} E_1S \longleftrightarrow E_2P \xrightarrow{\;P\;} E_2 \longleftrightarrow E_1 \quad (9.41)$$

The phenomenon of induced fit might be expected to be general, but large conformational changes during the catalytic reaction are observed primarily in enzymes like kinases, where extreme substrate specificity is necessary. The reason that induced fit seems not to occur more generally may be that it does not contribute to increasing the rate of the reaction; also, structural changes upon binding substrate produce a greater K_m for the substrate than otherwise would be necessary, because some of the binding energy of the substrate must be used to pull the enzyme into the active conformation.

There is much discussion in the literature about the roles of the phenomenon of induced fit in enzyme catalysis. Much of this discussion results from the use of different definitions of the term. Induced fit can be taken to include any structural changes that occur when substrates are bound or only those changes that contribute to increasing the rate or the specificity of the reaction. For this reason, it may be preferable not to use the term at all.

References

Evidence for conformation changes induced by substrates of phosphoglucomutase. J. A. Yankeelov, Jr., and D. E. Koshland, Jr. *J. Biol. Chem.* 240:1593–1602 (1965).

Induced fit in yeast hexokinase. G. Dela Fuente et al. *Eur. J. Biochem.* 16:226–233 (1970).

What limits the rate of an enzyme-catalyzed reaction? W. W. Cleland. *Acc. Chem. Res.* 8:145–151 (1975).

Structure of a complex between yeast hexokinase A and glucose. W. S. Bennett and T. A. Steitz. *J. Mol. Biol.* 140:183–230 (1980).

Reconstruction by site-directed mutagenesis of the transition state for the activation of tyrosine by the tyrosyl-tRNA synthetase: a mobile loop envelopes the transition state in an induced-fit mechanism. A. R. Fersht et al. *Biochemistry* 27:1581–1587 (1988).

Induced-fit movements in adenylate kinases. G. E. Schulz et al. *J. Mol. Biol.* 213:627–630 (1990).

Stabilization of a reaction intermediate as a catalytic device: definition of the functional role of the flexible loop in triosephosphate isomerase. D. L. Pompliano et al. *Biochemistry* 29:3186–3194 (1990).

9.2.6 Testing Theories of Catalysis

The three-dimensional structures of an enzyme with substrates, products, transition-state analogues, and so on bound to the active site usually suggest a plausible mechanism by which the catalyzed reaction occurs on the enzyme. Polar groups that can act as acid or base catalysts or can simply stabilize the transition state are often positioned near groups in the substrate where the chemical reaction takes place. In many cases, reaction schemes involving the shifting of electrons can be proposed, using arrows in the way that organic chemists frequently describe chemical reactions. More appropriately, perhaps, a structure for the transition state of the reaction can be proposed by considering how the functional groups in the active site will tend to shift the electronic structure of the substrate.

Understanding enzyme catalysis involves understanding not only the structure of the transition state on the enzyme but also its energetics. It is necessary to explain why the free energy of the transition state is lower by the observed amount when it is on the enzyme than when it is in solution. Energy calculations should ideally be able to provide such answers, but they are of uncertain validity for complex proteins in aqueous solution. The most accurate are the free-energy perturbation methods, which can calculate free-energy differences between systems with small structural differences (Sec. 7.4.4.b). Consequently, they can simulate the differences among closely related substrates or enzymes, but not the free-energy difference between an enzyme–transition-state complex and the free enzyme and substrate. The differences between closely related enzymes or substrates can be simulated, but not the absolute magnitudes of k_{cat} or K_m.

What other approaches can be used to test the validity of a postulated enzyme mechanism? One method is to measure the pH dependence of the values of k_{cat} and K_m. Polar groups that act as acid or base catalysts are often involved in enzyme active sites. They should be ionized under at least some conditions, and the catalytic activity of the enzyme should vary with the state of ionization. The pH dependence of the reaction should correlate with the pK_a values of these groups, so long as changes in ionization of other groups are not producing large effects. A difficulty with this approach is that polar groups in enzyme active sites frequently have anomalous pK_a values (Table 9.1) substantially different from those that might be expected. The actual pK_a values can be determined in the free enzyme and in enzyme–substrate complexes, but these values can also be different in the transition state. Nevertheless, that this approach can be useful is well illustrated by α-chymotrypsin. The pH dependence of its value of k_{cat}/K_m for the hydrolysis of model substrates follows a bell-shaped curve, with apparent pK_a values of 6.8 and 8.8 for the two transitions between active and inactive forms of the enzyme. The value of 6.8 was shown to represent the ionization of the catalytically important base at the active site, whereas the value of 8.8 is due to the α-amino group of residue Ile 16, which when ionized maintains the enzyme in a catalytically active conformation (see Sec. 9.3.2.a).

The techniques most frequently used to identify the roles of particular groups in catalysis involve methodically varying the structure of the substrate or the enzyme and measuring the effects on k_{cat} and K_m. Until recently it was most practical to vary the substrate, but such studies can be bedeviled by the ability of substrate analogues to bind to the enzyme active site in remarkably different ways. For example, synthetic substrates for α-chymotrypsin with extraneous hydrophobic residues were found to bind with these groups in the hydrophobic binding site, rather than the Phe or Tyr side chain that usually occupies this site (Sec. 9.3.2.a). Comparing the affinity constants for such a series of substrates, or their values of k_{cat}, would be very misleading in the absence of structural information because different phenomena would be being compared.

The preferred method is to vary the protein structure, previously by covalent modification but now by using site-directed mutagenesis (see Fig. 2.6). Chemical alteration is less likely to change the structure of the protein than is a substrate to alter its mode of binding. Site-directed mutagenesis can specifically and effectively remove a single functional group—for example, by changing a particular Ser or Cys residue to Ala, or an Asp or Glu residue to Asn or Gln. Covalent modification has the disadvantage of usually adding extra groups to the enzyme; simply the extra bulk of these groups can disrupt enzyme activity, even when the functional group blocked plays no direct role in catalysis.

As an example of the dramatic effects that can be observed by using site-directed mutagenesis, the ϵ-amino group of Lys 258 of the pyridoxal phosphate-dependent enzyme aspartate amino transferase was postulated to be the base responsible for transferring the proton during the transamination of aspartate. Replacing this Lys residue by Ala caused the enzyme activity to decrease by more than six orders of magnitude even though other partial reactions of the enzyme were unaltered. This mutation demonstrated the major role that Lys 258 plays in this enzyme.

Replacing residues that are involved primarily in binding a substrate might be expected to change primarily K_m whereas changing those residues involved directly in catalysis should change only k_{cat}. Changing residues outside the active site might not be expected to alter the catalytic activity of the enzyme unless the mutation alters the structure of the enzyme. Classifying residues as being involved in catalysis, binding, or neither is much too simple, however; in general, binding is related to catalysis in that the strength of the binding of the transition state, relative to the substrate, determines the rate of the enzymatic reaction. Residues that are involved in binding are necessarily involved in catalysis.

Mutagenic data can be misleading if replacements in the active site affect the properties of other residues of the enzyme. Interactions between two residues in a protein can be uncovered by comparing the effects of single mutations of each of the two residues (e.g., X and Y) with the effect of mutating both simultaneously. The effect of the double mutation on the energy of an intermediate or transition state, $\Delta\Delta G_{(X,Y)}$, should be the sum of the energetic effects of the two individual mutations,

$$\Delta\Delta G_{(X,Y)} = \Delta\Delta G_{(X)} + \Delta\Delta G_{(Y)} \qquad (9.42)$$

unless the two residues interact in the enzyme or unless either causes a change in the reaction mechanism or in the reaction step that is rate-limiting.

The effects of mutations can be interpreted most confidently if the data exhibit **linear free-energy relationships;** in this case, a series of alterations of an enzyme or substrate produces a series of related changes in two or more variables of the reaction. Such an approach is well established in physical organic chemistry to elucidate the structure of the transition state of a reaction. For example, in a **Brønsted plot,** the logarithm of the rate constant for a reaction is plotted versus the pK_a of a reactant or a catalyst. The rate constant is related to the relative free energy of the transition state of the reaction, whereas the pK_a is a measure of the acidity or basicity of the reactant. If the reaction mechanism and rate-limiting step are unchanged for a series of reactants, a linear plot with a slope of β is generally obtained. When $\beta = 0$, the rate is independent of the basicity of the reactant, and its basicity is not evident in the transition state for the reaction. When $\beta = 1$, the

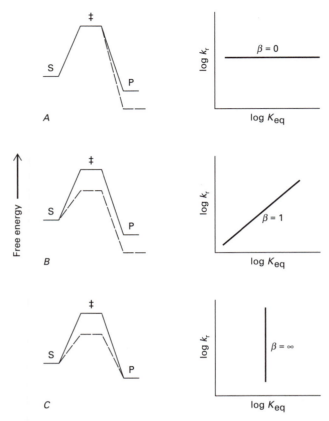

FIGURE 9.12
Schematic illustrations of the alterations in the energetics of a reaction that produce three particular types of linear free-energy relationships. The relative free energies of a reactant S, the product P, and the transition state by which they are interconverted (‡) are depicted on the left for two variants of the reactants. In each case, the free energies of ‡ and P are expressed relative to that of S. On the right are the linear free-energy relationships that would be apparent for a series of such variants of S when the logarithm of the rate constant k_r is plotted versus that of the equilibrium constant K_{eq}. The slope of this line is β. *A:* The differences between the variant reactants are apparent only in the relative free energies of the product, so $\alpha = 0$. *B:* The free-energy difference is apparent in both the transition state and in the product, so $\beta = 1$. *C:* The difference is apparent only in the energy of the transition state, and $\beta = \infty$. In the examples shown here, the magnitude of any free-energy difference is constant. Its magnitude will usually differ in ‡ and in P, so any value of β is possible.

rameter being varied. With $\beta = 0$ the transition state resembles the substrate; with $\beta = 1$ it resembles the product (Fig. 9.12).

Linear free-energy relationships are observed in some cases of enzyme catalysis (Fig. 9.13) but not in others. Such relationships are probably observed only for relatively simple cases in which catalysis is the result primarily of multiple independent binding interactions, as in the case of tyrosyl tRNA synthetase (Sec. 9.3.1). For cases of enzyme catalysis in which simple linear free-energy relationships are not apparent, the various interactions between enzyme and substrate are not independent. Alternatively, the reaction mechanism, the rate-determining step, or the structure of the enzyme may possibly have been changed significantly by the series of mutations.

When they are observed, linear free-energy relationships like those of Figure 9.12 suggest that neither the nature of the reaction nor the structure of the enzyme has changed, and they provide valuable evidence about the transition state. For example, when the slope β of the linear curve is close to zero, the series of mutations has altered the equilibrium constant but not the transition state. When $\beta = 1$, the energetic consequences of the mutations are the same for both parame-

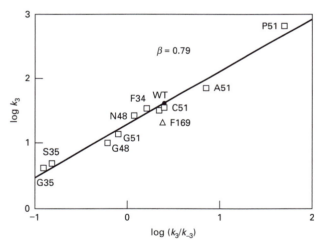

FIGURE 9.13
Free-energy relationship observed in formation of tyrosyl adenylate from tyrosine and ATP by various tyrosyl tRNA synthetases. The variant enzymes are designated by the mutation by which they differ from the normal enzyme (WT), using the one-letter abbreviation for the variant amino acid residue at the indicated residue number. The rate constants k_3 and k_{-3} are for the forward and reverse steps, respectively, in forming tyrosyl adenylate and pyrophosphate from tyrosine and ATP on the enzyme (see Fig. 9.14). (From A. R. Fersht et al., *Biochemistry* 26:6030–6038, 1987.)

rate is fully dependent on the basicity of the reactant, which must be expressed completely in the transition state. More generally, the rate of a reaction is compared with its equilibrium constant; the value of the slope of the line gives a measure of how much the transition state is like the substrate or the product in terms of the paters. Values of $\beta > 1$ are possible in enzymes, even to the extent that $\beta = \infty$, when the mutated residue affects only the free energy of the transition state, not the equilibrium constant (Fig. 9.12).

If a functional group is removed from an enzyme active site by mutagenesis, it should be possible to replace it by adding a similar small molecule to the solution or, perhaps, even to the substrate. As an example of the first possibility, the mutated Lys 258 side chain of aspartate amino transferase mentioned above could be replaced by added small molecule amines. In this case, the smaller the amine and the greater its pK_a value, the greater its efficiency in assisting to catalyze the reaction on the enzyme. The added amine presumably occupied the void left by the mutated Lys side chain. In the second case, removing a crucial His residue of subtilisin by mutation to Ala (Sec. 9.3.2.a) decreased k_{cat}/K_m by a factor of 10^6. This His residue could be replaced to some degree by a His residue at residue P2 in the polypeptide substrate, which occupies the same position in the enzyme–substrate complex. The mutant enzyme catalyzed the hydrolysis of substrates with this sequence much more rapidly than it did other substrates. Consequently, the mutant enzyme is much more specific for certain substrates than is the normal enzyme.

References

Tinkering with enzymes: what are we learning? J. R. Knowles. *Science* 236:1252–1258 (1987).

Relationships between enzymatic catalysis and active site structure revealed by applications of site-directed mutagenesis. J. A. Gerlt. *Chem. Rev.* 87:1079–1105 (1987).

Structure-activity relationships in engineered proteins: analysis of use of binding energy by linear free energy relationships. A. R. Fersht et al. *Biochemistry* 26:6030–6038 (1987).

Direct Brønsted analysis of the restoration of activity to a mutant enzyme by exogenous amines. M. D. Toney and J. F. Kirsch. *Science* 243:1485–1488 (1989).

Engineering subtilisin BPN' for site-specific proteolysis. P. Carter et al. *Proteins: Struct. Funct. Genet.* 6:240–248 (1989).

Site directed mutagenesis: a tool for enzyme mechanism dissection. C. R. Wagner and S. J. Benkovic. *Trends Biotechnol.* 8:263–270 (1990).

Additivity of mutational effects in proteins. J. A. Wells. *Biochemistry* 29:8509–8517 (1990).

Enzyme catalysis: not different, just better. J. R. Knowles. *Nature* 350:121–124 (1991).

Linear free energy relationships in enzyme binding interactions studied by protein engineering. A. R. Fersht and T. N. C. Wells. *Protein Eng.* 4:229–231 (1991).

9.3 Examples of Enzyme Mechanisms

In this section, the possible mechanisms of enzyme function are elaborated for a few particular enzymes, chosen for the amount of data available, for the degree to which they appear to be understood, and for their illustration of important principles.

The following descriptions of enzyme mechanisms should not be taken as dogma but merely as the current view, which may quickly change; the changes that have occurred since the first edition of this volume suggest that the mechanisms proposed here are likely to be refined or even replaced in the future.

Reference

Structures and catalysis of enzymes. W. N. Lipscomb. *Ann. Rev. Biochem.* 52:17–34 (1983).

9.3.1 Tyrosyl tRNA Synthetase

a. Catalytic Activity

This enzyme is said to be one of the simplest known, in that its catalytic powers arise almost entirely from specific binding interactions. There are no polar groups in the active site that are likely to be involved in general acid or base catalysis.

Only the first half of the reaction catalyzed by this enzyme has been characterized. It is the formation of the intermediate tyrosyl adenylate (Tyr-AMP) and pyrophosphate (PP_i) from the amino acid tyrosine and ATP:

$$E + \text{tyrosine} + \text{ATP} \rightleftharpoons$$
$$E \cdot \text{Tyr-AMP} + PP_i \quad (9.43)$$

The reaction is kinetically of random order in that either ATP or tyrosine can be bound first (Fig. 9.14). The affinity for ATP alone is low, however, and binding tends to be ordered, with tyrosine binding first and ATP second. The transition state relevant to catalysis is the one that occurs between the complex with tyrosine plus ATP and that with the Tyr-AMP plus PP_i. Most of the rate and equilibrium constants of the reaction can be measured directly by equilibrium binding studies, by presteady-state kinetic methods, and by steady-state kinetics. Consequently, the relative free energies of the relevant forms of the enzyme along the reaction pathway can be measured.

In the second half of the normal reaction catalyzed by this enzyme, the Tyr-AMP is transferred to its specific

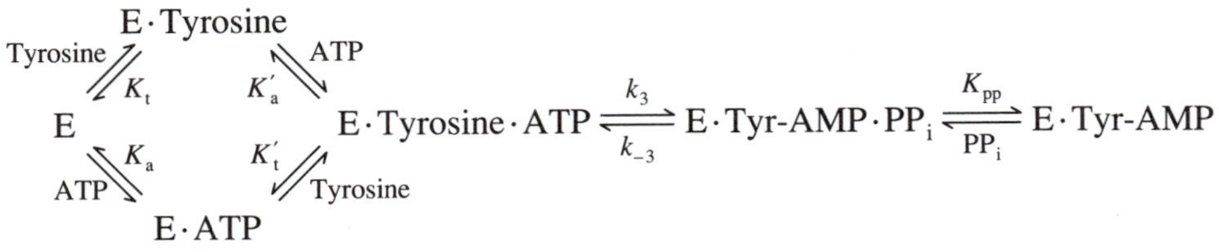

$$E \cdot Tyrosine$$

$$
\begin{array}{c}
Tyrosine \; / \; K_t \qquad K'_a \; \diagdown \; ATP \\
E \qquad\qquad\qquad E \cdot Tyrosine \cdot ATP \underset{k_{-3}}{\overset{k_3}{\rightleftharpoons}} E \cdot Tyr\text{-}AMP \cdot PP_i \underset{PP_i}{\overset{K_{pp}}{\rightleftharpoons}} E \cdot Tyr\text{-}AMP \\
ATP \; \diagdown \; K_a \qquad K'_t \; / \; Tyrosine \\
E \cdot ATP
\end{array}
$$

FIGURE 9.14

The kinetic scheme for the initial half-reaction of tyrosyl tRNA synthetase, the formation of tyrosyl adenylate (Tyr-AMP) and pyrophosphate (PP_i) from tyrosine and ATP. The substrate bound or product released at each step is depicted, along with its appropriate dissociation constant. The enzyme may strictly bind the substrates in either order to form the ternary complex E · Tyrosine · ATP. The bound ATP and tyrosine are converted to the tyrosyl adenylate (Tyr-AMP) and to pyrophosphate (PP_i). The pyrophosphate dissociates, but the amino acyl adenylate remains bound to the enzyme. The enzyme is now ready to transfer the tyrosyl moiety to the correct tRNA in the second stage of the overall reaction, which is not depicted.

tRNA, to generate the charged tRNA that is used in protein biosynthesis (Sec. 2.1.3.a). In the absence of the tRNA, the reactive and thermodynamically unstable Tyr-AMP is not released from the enzyme but is sequestered and stabilized in the enzyme active site. In the following discussion of the first half of the reaction, it must be kept in mind that Tyr-AMP is not the final product of the enzyme but an intermediate that is stabilized by the enzyme.

The enzyme from the thermophilic bacterium *Bacillus stearothermophilus,* which has been studied extensively, is a dimeric enzyme of two identical 418-residue polypeptides. The structures of the free enzyme and of the enzyme with either tyrosine or Tyr-AMP bound are known. In the crystals, only the 320-residue domain responsible for the first half-reaction is observed; the domain that binds the tRNA is disordered. The dimeric enzyme exhibits half-of-the-sites reactivity (Sec. 8.4.5), in that only one molecule of tyrosine is bound to the dimer and only one molecule of Tyr-AMP is formed. The dimer in solution is asymmetric, but the structural basis for this is not known; the crystalline enzyme appears to be symmetrical.

Eight hydrogen bonds between the enzyme and the bound Tyr-AMP are apparent in the complex elucidated crystallographically (Fig. 9.15). The roles of the enzyme groups involved in these interactions and in the interactions with PP_i have been determined by changing the various residues individually and in pairs, using site-directed mutagenesis. The replacements have been to those amino acids that lack the polar groups normally present, but with minimal other changes in overall structure. The changes in the relative free energies of the various complexes of the enzyme with substrates, with transition states, and with products are taken to represent the contributions of the group removed to the binding of that reactant. For example, residues Asp 78, Tyr 169, and Glu 173 of the enzyme form a binding site for the α-amino group of the bound substrate tyrosine. Replacement of any one of these residues weakens binding of tyrosine by about 3 kcal/mol, and similar effects are observed on the other complexes. This indicates that the function of these residues is simply to bind the amino group of the tyrosine substrate and that this function does not change appreciably as the reaction progresses. The specificity for tyrosine, rather than, say, phenylalanine, is contributed largely by interactions of the tyrosine phenolic hydroxyl with the side chains of Asp 176 and Tyr 34 of the enzyme. Replacing Tyr 34 with Phe decreased the affinity of the enzyme for tyrosine and the stabilities of the other complexes. The Phe 34 enzyme still favors tyrosine as substrate by a factor of 1×10^4 over phenylalanine, but this discrimination has been reduced by a factor of 15. The interactions of Tyr 34 with the substrate contribute to binding of tyrosine and are of similar magnitude throughout the reaction sequence. Like the interactions involved in binding the α-amino group just described, those between Tyr 34 and the substrate do not lower the free energy of the transition state relative to that of the substrates and do not, therefore, contribute directly to catalysis.

The crystal structure of the complex of the enzyme with Tyr-AMP indicated that residues Cys 35, Thr 51, and His 48 interact with the ribose ring of the adenylate moiety. Replacement of these residues demonstrated that they contribute substantially to stabilizing this complex and the enzyme–transition state, but not the complex of the substrate-ATP. Linear free-energy relationships observed for these mutations showed that, on average, 12% of the binding energy of these interactions

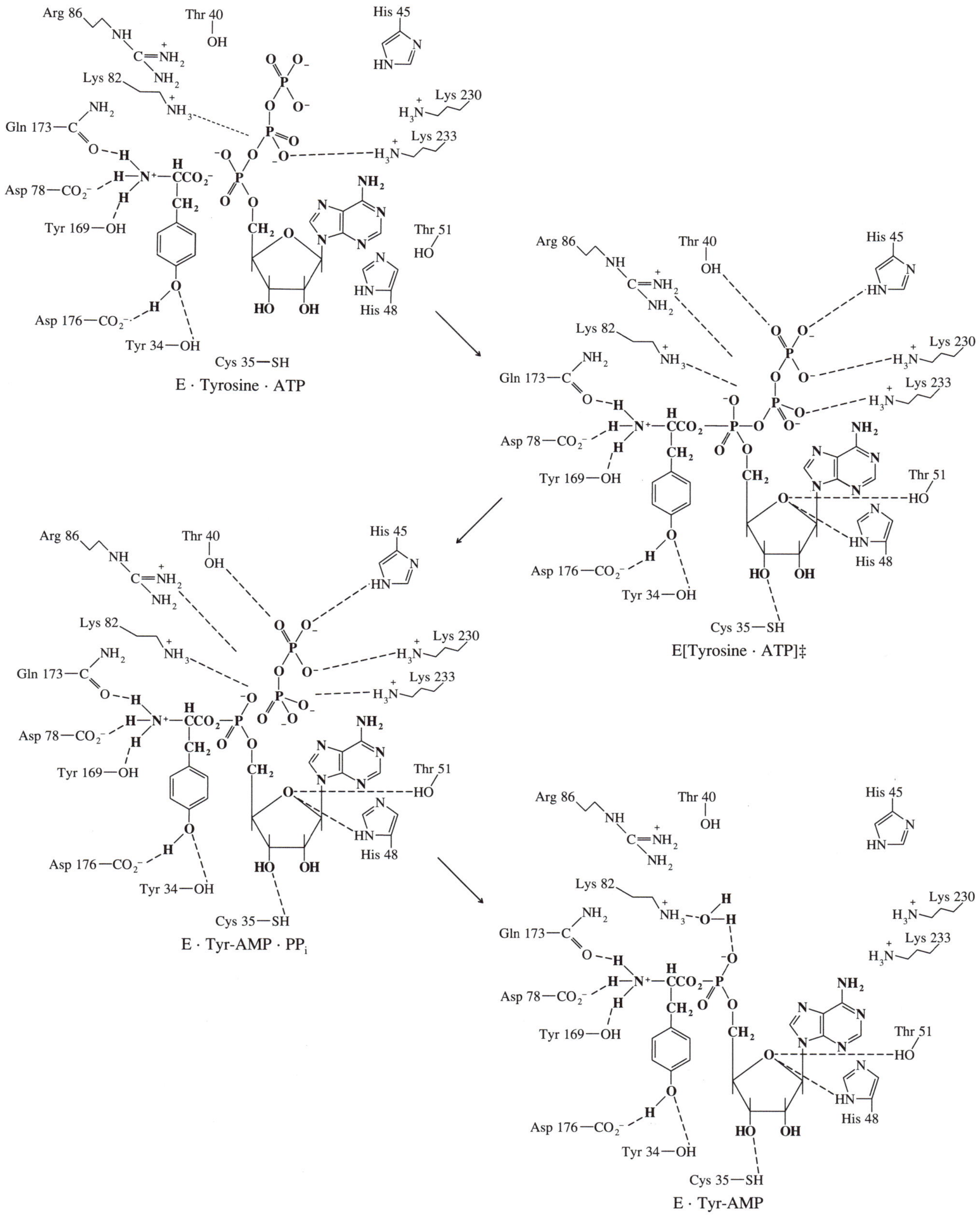

FIGURE 9.15

Interactions of tyrosyl tRNA synthetase (E) with the substrates tyrosine and ATP, with the transition state [Tyrosine · ATP]‡, and with the products tyrosyl adenylate (Tyr-AMP) and pyrophosphate (PP_i). The interactions with tyrosine and with Tyr-AMP have been determined crystallographically; the others are inferred from the results of mutagenesis studies in which the functional groups have been removed by mutation. (From A. R. Fersht et al., *Biochemistry* 27:1581–1587 1988.)

is realized on the binding of ATP to the E · tyrosine complex, 84% on binding the transition state, and 91% in the complex of Tyr-AMP and PP_i; the value of 100% pertains to the complex of the Tyr-AMP. These interactions of the enzyme stabilize the transition state more than they do the substrates and, therefore, contribute directly to catalysis.

Two residues, Thr 40 and His 45, do not interact directly with tyrosine, with ATP, or with the Tyr-AMP in complexes of the enzyme, and mutating them to Ala and Gly, respectively, had little effect on the affinities of the enzyme for these species. The mutations, however, decreased the rate of formation of Tyr-AMP by factors of 7000 and 200, respectively; the rate was decreased 3×10^5-fold in the double mutant. The stability of the E · Tyr-AMP · PP_i complex was also decreased, indicating that these groups interact with the pyrophosphate moiety. Similar observations were made for the basic side chains of residues Lys 82, Arg 86, Lys 230, and Lys 233. These residues are on flexible loops in the free enzyme and in the complex with Tyr-AMP, but they are thought to become fixed by interacting with the β and γ phosphoryl groups of bound ATP, which become the pyrophosphate product. All of these interactions occur only in the transition state and in the E · Tyr-AMP · PP_i product, not with the substrate ATP, because of the change in geometry that takes place in the reactants during the reaction. The reaction proceeds by inversion of the configuration of the α-phosphorous atom of ATP, probably through a 5-coordinate transition state.

The interactions that are thought, on the basis of the crystal structure of the enzyme and of the mutational studies, to occur between the enzyme and the substrates, the transition state, and the products are depicted in Figure 9.15.

The tyrosyl tRNA synthetase demonstrates how merely binding the transition state more tightly than the substrates increases the rate of a reaction. What is initially surprising in this case is that the transition state is not stabilized more than the Tyr-AMP and PP_i products. Such an enzyme usually would not be an efficient catalyst because it merely binds the products exceedingly tightly. This situation probably occurs in the case of tyrosyl tRNA synthetase because the tyrosyl adenylate is not the final product of the enzyme but is merely an intermediate that is to be transferred to the appropriate tRNA. The intermediate is intrinsically unstable, so tight binding on the enzyme protects it from hydrolysis and stabilizes it so that it is available for transfer to the tRNA. The equilibrium constant for the first half of the reaction—that is, the ratio of the equilibrium concentrations of Tyr-AMP and PP_i relative to those of tyrosine and ATP—is 2.3 on the enzyme but only 3.5×10^{-7} in solution.

References

Analysis of enzyme structure and activity by protein engineering. A. R. Fersht et al. *Ang. Chem. Intl. Ed. Engl.* 23:467–473 (1984).

Transition-state stabilization in the mechanism of tyrosyl-tRNA synthetase revealed by protein engineering. R. J. Leatherbarrow et al. *Proc. Natl. Acad. Sci. USA* 82:7840–7844 (1985).

Binding energy and catalysis: a lesson from protein engineering of the tyrosyl-tRNA synthetase. A. R. Fersht et al. *Trends Biochem. Sci.* 11:321–325 (1986).

Dissection of the structure and activity of the tyrosyl-tRNA synthetase by site-directed mutagenesis. A. R. Fersht. *Biochemistry* 26:8031–8037 (1987).

Correlations between kinetic and X-ray analyses of engineered enzymes: crystal structures of mutants Cys → Gly-35 and Tyr → Phe-34 of tyrosyl-tRNA synthetase. M. D. Fothergill and A. R. Fersht. *Biochemistry* 30:5157–5164 (1991).

b. Editing of Amino Acid Activation

The specificity with which tRNA synthetases attach amino acids to tRNA is crucial for protein biosynthesis (Chap. 2). The fidelity of assembling the correct sequence of amino acids into a protein depends on each tRNA synthetase recognizing the correct amino acid, activating it as the amino acyl adenylate, and transferring it to the correct tRNA. Much of the specificity comes from binding interactions; for example, tyrosyl tRNA synthetase binds tyrosine 1.5×10^5 times more tightly than it binds the similar phenylalanine. Much of this discrimination comes from the interactions of Asp 176 and Tyr 34 with the phenolic hydroxyl group of bound tyrosine (Fig. 9.15).

There are limits, however, to the degree of discrimination that is possible by just binding interactions (Sec. 8.2.2). For example, the extra methylene group of isoleucine relative to valine causes the isoleucyl tRNA synthetase to bind valine less tightly by a factor of only 100–200. There is a fivefold greater concentration of valine over isoleucine in vivo, so an error rate of 2–5% might be expected. Yet the observed error rate is only 0.03%.

The extra specificity is gained by incorporating an editing mechanism involving hydrolysis of incorrect structures, at least in those tRNA synthetases that require it. For example, addition of isoleucyl tRNA to the incorrect complex of valyl adenylate and isoleucyl tRNA synthetase results in hydrolysis of the valyl adenylate. In some cases, the hydrolysis may be of the adenylate directly, but in the cases that have been studied, hydrolysis occurs after transfer of the incorrect amino acid to the tRNA. With the correct amino acid, the rate of hydrolysis is much lower. As a consequence, the correctly charged tRNA usually predominates.

In the case of the valyl tRNA synthetase, there is evidence for a separate hydrolytic site at which editing occurs (Fig. 9.16). This hydrolytic site appears to be designed to recognize the closely related amino acid threonine, and it is imagined to have a donor or acceptor for a hydrogen bond to the hydroxyl group of threonine. If tight binding to this site occurs, the amino acid is hydrolyzed from the tRNA.

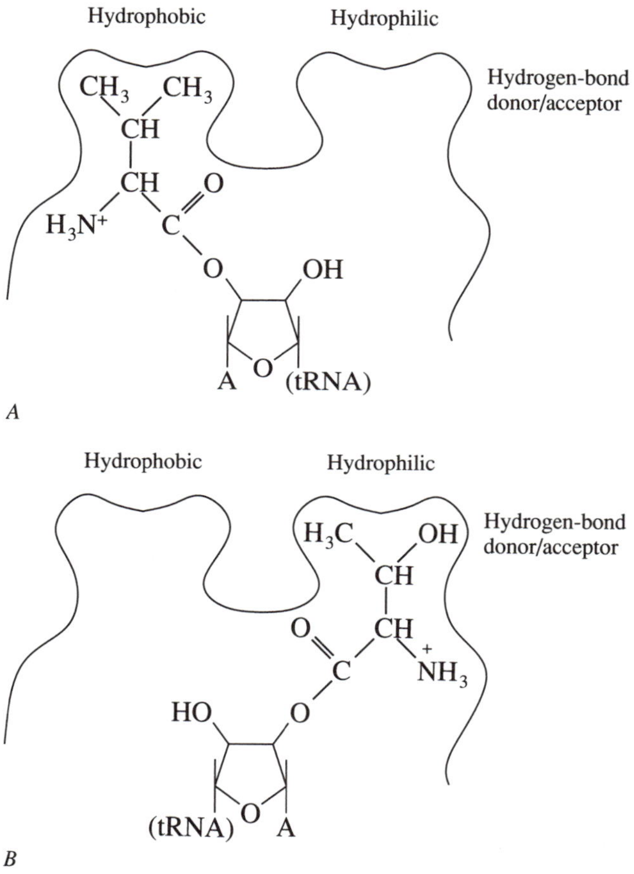

FIGURE 9.16
Proposed mechanism for the prevention of the misacylation of tRNA^Val with threonine. The valyl tRNA synthetase is thought to have two distinct sites: an acylation site where the amino acid is transferred from the adenylate to the tRNA and a hydrolysis site where amino acids that bind tightly are hydrolyzed from the tRNA. *A:* The acylation site is imagined to be specific for valine by being hydrophobic and complementary in shape to the valine side chain. *B:* The hydrolytic site is depicted as being specific for threonine by virtue of a hydrogen-bond donor or acceptor for its hydroxyl group; binding to this site produces hydrolysis of the charged tRNA. (From A. R. Fersht and M. Kaethner, *Biochemistry* 15:3342–3346, 1976.)

Such an editing process has been described as a *double sieve.* In the initial discrimination, upon binding to the tRNA synthetase, larger or markedly different amino acids do not bind with substantial affinity. Smaller or similar amino acids, however, may still bind to a significant extent. These incorrect amino acids are detected because they bind to the second, hydrolytic site more effectively than do the correct amino acids. Those that are accepted by the second sieve are hydrolyzed.

Such multiple editing mechanisms are used where extreme selection is required. Another example is DNA replication, in which error rates must be as low as possible. DNA replication occurs in bacteria with error rates as low as $10^{-8} - 10^{-10}$. Discriminating between the four nucleotides solely on the basis of binding affinities would not be sufficient to attain such accuracy, so the DNA polymerases also use further hydrolytic editing steps.

References

Enzymic editing mechanisms and the genetic code. A. R. Fersht. *Proc. Roy. Soc. London [Biol.]* 212:351–379 (1981).

Aminoacyl tRNA synthetases: general scheme of structure–function relationships in the polypeptides and recognition of transfer RNAs. P. Schimmel. *Ann. Rev. Biochem.* 56:125–158 (1987).

Mechanisms of aminoacyl-tRNA synthetases: a critical consideration of recent results. W. Freist. *Biochemistry* 28:6787–6795 (1989).

Fidelity mechanisms in DNA replication. H. Echols and M. F. Goodman. *Ann. Rev. Biochem.* 60:477–511 (1991).

9.3.2 Proteases

Four major classes of proteases are known and are designated by the principal functional group in their active site: serine, thiol, carboxyl, and metallo. As this classification implies, proteases do not rely simply on binding interactions to catalyze cleavage of peptide bonds; they also use crucial polar groups with functional roles. These groups have distinct roles to play in the four classes (Fig. 9.17), and four ways of catalyzing the same chemical reaction have evolved; there is no evidence that the four classes are evolutionarily related.

In spite of the differences in catalytic mechanisms, the enzymes of the four classes share a common property of going through an intermediate (or transition state) in which the normally trigonal carbonyl carbon of the peptide bond becomes tetrahedral due to the temporary addition of a nucleophile. In the case of the serine and thiol proteases, the nucleophile is the serine hy-

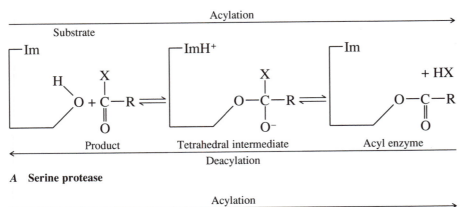

A   Serine protease

B   Thiol protease

C   Carboxyl protease

D   Metallo protease

FIGURE 9.17

Comparison of the mechanisms of action of proteases of the serine, thiol, carboxyl, and metallo classes. *A:* Serine proteases, with the hydroxyl group of the Ser residue and the imidazole (Im) group of a His residue indicated. In the substrate, X is R′—NH—. The nucleophilic attack by the Ser hydroxyl of the protein on the carbonyl carbon atom of the substrate is assisted by the His residue accepting the proton of the hydroxyl and leads to the tetrahedral intermediate. The His residue transfers the proton to the newly generated amino half of the peptide bond, which is now free to dissociate. The acylenzyme intermediate is hydro-

droxyl or cysteine thiol at the enzyme active site, whereas in the other two classes it is a water molecule that hydrolyzes the peptide bond. The reaction proceeds more directly in the latter classes, whereas the acyl-enzyme intermediate that occurs with the serine and thiol proteases must be hydrolyzed by water in a second step (Fig. 9.17).

The proteinases are some of the best characterized enzymes; in fact, the term *enzyme* was first used to describe trypsin. Their broad substrate specificities have made possible extensive kinetic studies of their catalytic powers to determine in detail the effects on k_{cat} and K_m of varying their substrates and their inhibitors. The intimate roles of unique amino acid side chains in catalysis were elucidated using protein chemistry approaches. The wealth of data about these enzymes makes it particularly instructive to examine their catalytic mechanisms in greater detail.

Reference

Common feature of the four types of protease mechanism. L. Polgár. *Biol. Chem. Hoppe-Seyler* 371 (Suppl.):327–331 (1990).

a. Serine Proteases

This class is characterized by the conspicuous presence in each enzyme of a unique Ser residue of apparently exceptional reactivity that forms covalent adducts with some substrates and inhibitors. Two major evolutionary families are represented in this class; they have very similar mechanisms of action, even though they are not otherwise detectably related, and provide one of the most striking examples at the molecular level of apparent evolutionary convergence. One family is represented by the bacterial protease subtilisin. The other family, which is more extensive and more thoroughly studied, is the trypsin family; it includes chymotrypsin,

trypsin, and elastase, plus the less well characterized but physiologically very important enzymes thrombin, plasmin, kallikrein, and acrosin; other members are involved in a range of diverse cellular functions including blood clotting, complement activation, hormone production, and fertilization. This discussion focuses on the trypsin family, designating the homologous residues at the active site by their residue numbers in chymotrypsinogen.

The normal, presumably physiological, reaction catalyzed by these enzymes is the hydrolysis of peptide bonds in proteins and peptides:

$$-NH-\underset{\underset{R^n}{|}}{CH}-\underset{\overset{O}{\|}}{C}-NH-\underset{\underset{R^{n+1}}{|}}{CH}-$$

$$\downarrow H_2O$$

$$-NH-\underset{\underset{R^n}{|}}{CH}-CO_2H + H_2N-\underset{\underset{R^{n+1}}{|}}{CH}- \quad (9.44)$$

The terminology generally used to describe the specificity of proteases for various polypeptide substrates designates the positions of the residues of the substrate relative to the peptide bond that is cleaved:

$$-P_4-P_3-P_2-P_1-P_1'-P_2'-$$

$$\downarrow H_2O$$

$$-P_4-P_3-P_2-P_1-OH + H-P_1'-P_2'- \quad (9.45)$$

The various serine proteases catalyze this reaction in very similar ways but differ most strikingly in their preference for amino acid side chains at position P_1. Trypsin cleaves bonds only after Lys and Arg residues, chymotrypsin after large hydrophobic residues; the other proteases of this family have less distinct preferences at position P_1 but also depend to varying extents on the residues at neighboring positions.

---

lyzed by reversal of this sequence, in which HX is now H_2O. *B:* Thiol proteases, with a Cys residue playing the role of the Ser of serine proteases *(A).* The His residue plays a similar role but tends to extract the proton from the thiol group, which is more readily ionized than Ser even in the absence of substrate. *C:* Carboxyl proteases with the two Asp residues of pepsin indicated. These two carboxyls interact directly in the enzyme to favor the situation in which only one of their carboxyl groups is ionized. During the reaction, one carboxyl group temporarily donates a proton to the carbonyl of the substrate whereas the other accepts a proton from a water molecule. Both actions facilitate formation of the tetrahedral intermediate, which breaks down to products by accepting and donating different protons. *D:* Metallo proteases, in this case containing a Zn^{2+} ion. The attack of a water molecule to generate the tetrahedral intermediate is facilitated by simultaneous interactions both with an enzyme carboxyl group and with the Zn^{2+} ion. The products are generated by transfer of the proton from the protein carboxyl group to the —NH— group of the peptide bond being hydrolyzed. (From L. Polgár, *Biol. Chem. Hoppe-Seyler* 371(Suppl.):327–331, 1990.)

The proteases usually catalyze related reactions of appropriate small amides and esters, especially their hydrolysis:

$$R^1-\overset{O}{\underset{\|}{C}}-NH-R^2 \xrightarrow{H_2O} R^1-\overset{O}{\underset{\|}{C}}OH + H_2N-R^2$$

$$R^1-\overset{O}{\underset{\|}{C}}-OR^2 \xrightarrow{H_2O} R^1-\overset{O}{\underset{\|}{C}}OH + HO-R^2 \quad (9.46)$$

Exchange of the oxygen atoms of the carboxylic acids with those of solvent can also be catalyzed:

$$R^1-\overset{O}{\underset{\|}{C}}OH + H_2{}^{18}O \rightleftharpoons R^1-\overset{O}{\underset{\|}{C}}{}^{18}OH + H_2O \quad (9.47)$$

Acyl acceptors other than water can be used, so the most general representation of the reaction catalyzed by these proteases is

$$R^1-\overset{O}{\underset{\|}{C}}-X + HY \rightleftharpoons R^1-\overset{O}{\underset{\|}{C}}-Y + HX \quad (9.48)$$

Which substrates are used by each enzyme depends primarily on the nature of the group R_1.

The nonenzymic cleavage of amides and esters is thought to proceed through a **tetrahedral intermediate,** or a transition state approximating such a structure, that is generated by nucleophilic attack by group Y on the acyl carbon:

$$
\begin{array}{c}
R^1-\overset{\overset{\curvearrowleft O}{\|}}{\underset{\underset{\curvearrowright}{YH}}{C}}-X \\
\Updownarrow \\
\left[\begin{array}{c} R^1 \quad O^- \quad X \\ \diagdown\;\vdots\;\diagup \\ C \\ | \\ Y \end{array}\right] + H^+ \\
\downarrow \\
R^1-\overset{O}{\underset{\|}{C}}-Y + HX \quad (9.49)
\end{array}
$$

The term *tetrahedral intermediate* is derived from the configuration about the acyl carbon atom, which contrasts with the trigonal geometry in the original amide or ester. Consequently, groups R_1 and X must move relative to the carbon and oxygen atoms to attain the tetrahedral configuration. The tetrahedral species is unstable and breaks down by expelling either group X or

group Y. Much effort has been made to detect this tetrahedral species during protease catalysis, but the evidence for its occurrence is still only indirect. It is now considered to be close to the transition state for the reaction and too unstable to accumulate to substantial levels. The enzymic reaction has been found to proceed through an **acylenzyme intermediate** in which the group

$$R^1-\overset{O}{\underset{\|}{C}}-$$

is transferred first to the uniquely reactive Ser residue of the enzyme that gives this class of proteases its name:

$$
\begin{array}{c}
E-CH_2OH + R^1-\overset{O}{\underset{\|}{C}}-X \\
\downarrow \\
E-CH_2O-\overset{O}{\underset{\|}{C}}-R^1 + HX \quad (9.50)
\end{array}
$$

$E-CH_2OH$ represents the enzyme and this Ser side chain, which is at residue 195 in the trypsin family and at position 221 in subtilisin. The acyl enzyme then reacts with the second substrate, usually water, to complete the reaction:

$$
\begin{array}{c}
E-CH_2O-\overset{O}{\underset{\|}{C}}-R^1 + HY \\
\downarrow \\
E-CH_2OH + R^1\overset{O}{\underset{\|}{C}}-Y \quad (9.51)
\end{array}
$$

The acyl enzyme can accumulate with reactive ester substrates and can lead to a burst of the first product (Fig. 9.4) because the acyl enzyme can be formed much more rapidly than it is hydrolyzed. It does not accumulate, however, with the less reactive amide substrates. That the rate-limiting step in ester hydrolysis does not depend on the nature of the group X was an early indication that group X has been released as a product prior to the rate-determining step, which would be hydrolysis of the acyl enzyme (Sec. 9.1.1.b). Since then, acylenzyme intermediates have been trapped, isolated, and characterized.

The enzymic reaction, then, is of the ping-pong or substituted-enzyme type (Eq. 9.13). This fact was not determined, however, by the usual steady-state kinetic analysis (Sec. 9.1.1) because the second substrate is H_2O, whose concentration is difficult to vary independently. In addition to dividing the reaction into two steps, this type of mechanism results in a symmetric reaction sequence in which hydrolysis of the acyl enzyme can occur by the reverse of its formation (Fig. 9.17A), with water replacing the product HX. There are then two acyl transfers, first from the substrate to the enzyme, forming an ester with the Ser side chain, and then to H_2O or other acceptor HX. In both transfers, it seems likely that a tetrahedral intermediate or transition state would be involved.

The unique reactivity with substrates and with inhibitors of the one particular Ser side chain of the serine proteases led to this residue being considered an extraordinarily potent nucleophile. A nonionized His residue (His 57 in the trypsin family) was also found to be involved in both catalysis and the reactivity of the Ser side chain; this His residue reacted with affinity labels, and any modification of it, including protonation with a pK_a of about 7, greatly diminished both the catalytic activity of the enzyme and the reactivity of the Ser side chain. The obvious chemical use of a His side chain would be to interact as a base with the Ser hydroxyl group and to promote its nucleophilicity.

Accordingly, the crystal structure of the first serine protease to be determined, that of α-chymotrypsin, showed the reactive Ser 195 to be in the active site, with the important His 57 nearby. Moreover, on the other side of His 57 was the buried side chain of Asp 102, which forms a hydrogen bond to both His 57 and another group. Buried carboxyl groups are very rare in proteins, presumably because such a situation is energetically unfavorable for folding. The burial of Asp 102 in this enzyme, although in a relatively polar environment, suggested an important function for this residue. The unfavorable negative charge of the Asp residue was imagined to be transferred to the Ser 195 side chain via His 57 to varying extents during catalysis:

$$\text{(9.52)}$$

This arrangement of Asp, His, and Ser residues has come to be known as the **charge-transfer relay system** or **catalytic triad.**

Very similar constellations of three such residues were later found in the other serine proteases, including subtilisin, where no homology to the trypsin family is detectable in its primary or tertiary structures. The three residues are Asp 32, His 64, and Ser 221 in subtilisin. Note that these three residues occur in different orders in the primary structures of subtilisin and the trypsin

proteases, making it very unlikely that the two families arose by divergence from a common ancestor. That the same catalytic triad arose independently in subtilisin, presumably by convergent evolution, is indicative of its catalytic importance.

The roles of the Asp, His, and Ser residues have been dissected in subtilisin by mutating them to Ala residues. The values of k_{cat} were diminished by the individual mutations by factors of 3×10^4, 2×10^6, and 2×10^6, respectively, with only very small changes in K_m. The simultaneous presence of all three mutations diminished the value of k_{cat} by only 2×10^6-fold, so the effects of the three mutations are not additive. This would be expected in this case because once any one of the three residues is mutated, the catalytic function of the others should be severely impaired. The catalytic activity of the triple mutant was still 10^3 times greater than the nonenzymatic rate, so other groups of the protein also contribute to catalysis.

The precise roles of the three groups in the catalytic triad are uncertain, however. A significant transfer of charge from the Asp to the Ser residue, via the His, seems unlikely because the Asp and His residues titrate with approximately normal pK_a values. The position of the proton between the Asp and His residues was a matter of contention for a long time, but it was found to be located on the His residue by neutron diffraction analysis and by ^{15}N-NMR analysis. The strength of the hydrogen-bond interaction between the His and Ser residues seems to be variable in the two serine protease families, as judged by the distance between the atoms in the protein crystal structures. Nevertheless, mutating the Asp residue to Asn decreases k_{cat} by a factor of 10^4 and also diminishes the reactivity of the Ser hydroxyl by a similar amount, without altering the reactivity of the His residue. It has been suggested that the role of the Asp residue is to stabilize (1) the conformation of the His side chain, (2) the appropriate protonation state of the His residue (the hydrogen must be on atom $N^{\delta 1}$ rather than $N^{\epsilon 2}$), and (3) the positive charge on the His side chain that is thought to form during the reaction.

The steps that occur on the enzyme during catalysis have been inferred from the crystal structures of the enzymes with various acyl groups, products, and inhibitors bound, but with few good substrates because of their rapid turnover. The only structure of an enzyme–substrate Michaelis complex at low temperature shows the substrate bound at the active site, moving the His 57 side chain toward Ser 195 and perhaps inducing a hydrogen-bond interaction between the two. The most detailed models of enzyme–substrate complexes come from the complexes of trypsin with bovine pancreatic trypsin inhibitor (BPTI), trypsin with soybean trypsin inhibitor, and subtilisin with its protein inhibitor. These

protease inhibitors bind tightly and specifically to the active sites of the respective proteases and thereby inhibit them. They are substrates of the proteases, being hydrolyzed very slowly at one peptide bond that is in the appropriate position in the enzyme active site. They are, however, tightly folded protein molecules that remain folded after cleavage of this peptide bond and still bind tightly to the protease. Consequently, the newly generated α-amino group of the P_1' residue remains approximately in its original position, inhibiting access of water to the cleavage site; it is also in position to reverse readily the hydrolysis step to resynthesize the peptide bond:

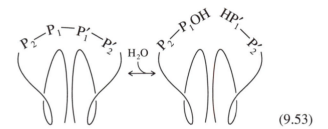

$$(9.53)$$

An equilibrium is generally reached in which roughly half the inhibitor molecules are cleaved, the precise value depending on the inhibitor and the conditions. The hydrolysis reaction is readily reversed because it is a unimolecular reaction, in contrast to the normal cleavage of a peptide into two products (Eq. 9.44). The latter reaction is normally irreversible with small concentrations of products, but high concentrations do lead to reversal and indicate an equilibrium constant for Equation (9.44) of approximately 1 M with model peptides. Therefore, the reversal of this reaction with the protease inhibitors requires only an effective concentration of about 1 M between the newly generated α-amino and α-carboxyl groups of residues P_1 and P_1' in the cleaved, but folded, inhibitor. On the enzyme, equilibrium greatly favors the original form with the intact peptide bond ($k_{cat}^r > k_{cat}^f$), and the cleaved form has the higher K_m as expected from the Haldane relationship (Eq. 9.10).

Even though they are not homologous, different protease inhibitors of this type bind very similarly to the serine proteases. The P_1 amino acid side chain is bound in the primary specificity pocket of the enzyme, designated subsite S_1. There are additional subsites S_2, S_3, and S_4, which interact to varying extents with residues P_2, P_3, and P_4, respectively, of the inhibitor; this part of the polypeptide backbone of the inhibitor is extended and forms an antiparallel β-sheet interaction with an extended strand of the enzyme. Further but less extensive interactions occur between residues P_1' and P_2', on the other side of the scissile bond, and subsites S_1' and S_2'. Substrates are thought to make similar interactions with the enzyme. Other classes of protein inhibitors of pro-

teases use different stratagems to block the active site of the target protease.

The molecular basis of the primary specificity of each serine protease for the residue at position P_1 is evident from the three-dimensional structures of the enzymes. The specificity of trypsin for Arg and Lys residues is due to its S_1 binding site being deep, with an Asp (residue 189) at the bottom to neutralize the charge of the Arg or Lys residue. Chymotrypsin is specific for aromatic residues (Phe, Tyr, and Trp) and has a large, narrow hydrophobic pocket to accommodate their planar aromatic rings. The Asp residue at the bottom of the pocket in trypsin is replaced by Ser in chymotrypsin. Elastase prefers short nonpolar side chains as substrates, and its binding site is less deep owing to the presence of large, bulky residues (e.g., Val 216 and Thr 226) at the bottom of the binding pocket; residues 216 and 226 are both Gly in trypsin and chymotrypsin (Fig. 9.18). The bacterial *Streptomyces griseus* protease A is similar to chymotrypsin in its specificity, preferring large hydrophobic residues, and it has a similar large hydrophobic binding site. The highly homologous α-lytic protease prefers small neutral residues and has its S_1 binding pocket shortened by the insertion of five residues into its polypeptide chain after residue 217.

The details of the binding interactions are not always straightforward, however. Arg P_1 side chains interact directly in hydrogen bonds with the Asp 189 of trypsin, but the shorter Lys residues interact via an intermediate water molecule. Replacing Asp 189 with Lys virtually abolished the catalytic activity of trypsin toward substrates with Arg and Lys at the P_1 position, as would be expected, but none of the expected activity toward substrates with Asp and Glu at the P_1 position was detectable. Instead, the low intrinsic activity of trypsin toward hydrophobic residues at the P_1 position was increased. The positively charged ϵ-amino group of the introduced Lys 189 residue is thought to be directed outside the substrate-binding pocket due to interactions with other groups of the enzyme. Replacing Asp 189 with a Ser residue, as in chymotrypsin, greatly diminished the catalytic activity of trypsin toward Lys and Arg substrates, but the 10- to 50-fold increase in activity toward hydrophobic substrates was much less than expected from the catalytic activity of α-chymotrypsin, even though there are no other major differences between the trypsin and chymotrypsin S_1 substrate-binding sites. Even more surprisingly, the differences in catalytic activity were due to changes in the value of k_{cat}, not K_m. Similar observations were made with trypsins in which Gly 216 and Gly 226 were replaced by Ala residues. The energetics of the interactions of the various substrates with the residue at position 189 are not apparent in the Michaelis complex, but only in the transition state. A detailed explanation for these in-

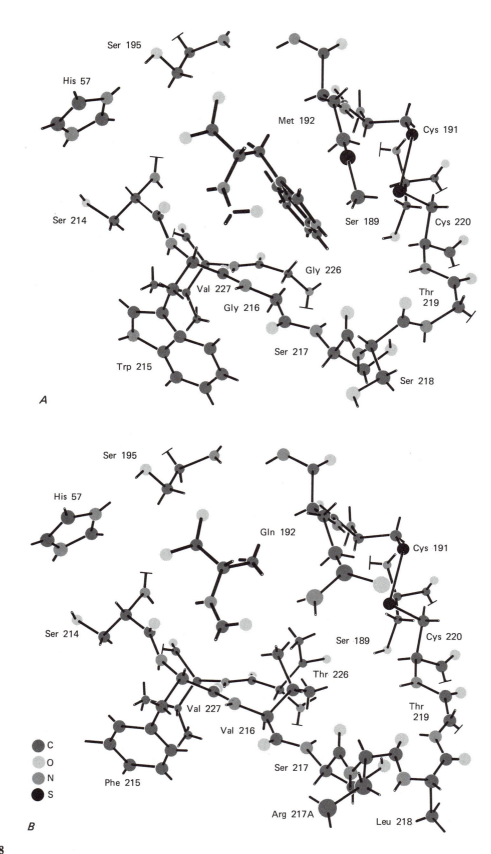

FIGURE 9.18

Differences in the primary binding sites of α-chymotrypsin and elastase that account for their different specificities. *A:* α-chymotrypsin, with formyl L-tryptophan bound in the center. The deep binding pocket adequately contains the Trp side chain. The large size of the pocket is due in part to the absence of side chains on residues Gly 216 and Gly 226. *B:* Elastase, with formyl L-alanine bound. The residues 216 and 226 are, respectively, Val and Thr, which block the pocket so that only small side chains can fit. (From B. S. Hartley and D. M. Shotton, in *The Enzymes,* 3rd ed., P. D. Boyer et al., eds., vol. 3, pp. 323–373. Academic Press, New York, 1970.)

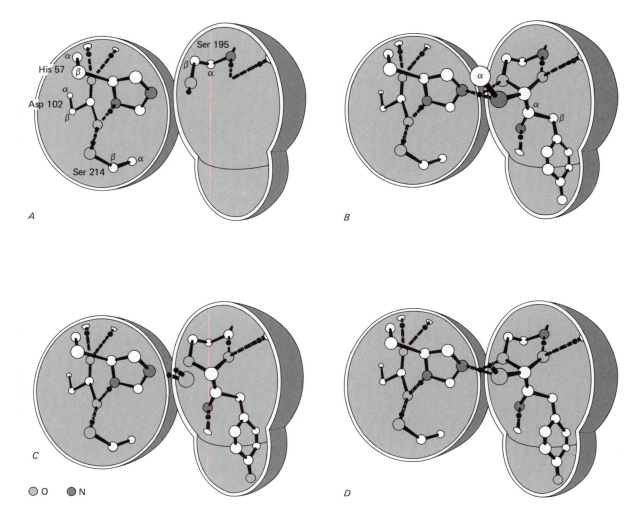

FIGURE 9.19

Schematic and stylized views of the active site of a serine protease during catalysis, based on the structure of *Streptomyces griseus* protease A. Atoms other than carbon are shaded; important hydrogen atoms are indicated by small black spheres; hydrogen bonds are broken cylinders. The superscript designations of some carbon atoms are indicated. *A:* The native enzyme before substrate is bound. The hollow disk on the left contains the side chains of His 57 and Asp 102 of the catalytic triad and Ser 214. Residue Ser 214 serves to orient Asp 102, as do two hydrogen bonds to the other oxygen atom of the Asp 102 side chain from the backbone NH groups of residues 56 and 57. The disk on the right shows the side chain of Ser 195, along with potential hydrogen bonds from the backbone NH groups of residues 193 and 195, which make up the oxyanion binding site. *B:* Part of a bound peptide substrate, approximating the presumed tetrahedral intermediate. Residue P_1 (Tyr) and atoms N and C^α of residue P_1' are shown. The side chain of residue P_1 is bound in the S_1 subsite of primary specificity (at lower part of right disk); its NH group is hydrogen-bonded to the backbone carbonyl oxygen of residue 214. The carbonyl oxygen of the substrate is bound in the oxyanion hole; the C′ atom is in the tetrahedral configuration and bonded, at least partially, to Ser 195. The proton originally on Ser 195 is in the process of being transferred by His 57 $N^{\epsilon 2}$ to the NH group of the leaving residue P_1'. *C:* The P_1' residue has dissociated as one of the products, and a water molecule has taken its place, ready to hydrogen-bond with His 57 $N^{\epsilon 2}$. The remaining part of the substrate is present as the ester with Ser 195; the C′ carbon is shown as still being tetrahedrally distorted due to the complementarity of the active site. *D:* The presumed tetrahedral intermediate in deacylation, with the proton of the water molecule being transferred by His 57 to Ser 195. The ester is subsequently hydrolyzed, and the carboxylic acid product dissociates to regenerate the native enzyme *(A)*. (From M. N. G. James et al., *J. Mol. Biol.* 144:43–88, 1980.)

Table 9.5 *Kinetic Parameters for the Hydrolysis of Peptide Substrates by Two Relatively Nonspecific Serine Proteases*[a]

	Elastase			α-Lytic protease		
Substrate[b]	k_{cat} (s^{-1})	K_m (mM)	k_{cat}/K_m (s^{-1}M^{-1})	k_{cat} (s^{-1})	K_m (mM)	k_{cat}/K_m (s^{-1}M^{-1})
P$_5$ -P$_4$ -P$_3$ -P$_2$ -P$_1$ -P$_1'$ -P$_2'$ $\downarrow$						
Ac-Ala-NH$_2$	<0.0008	160	<0.005	<0.0004	300	<0.0013
Ac-Pro-Ala-NH$_2$	0.007	100	0.07	0.07	300	0.23
Ac-Ala-Pro-Ala-NH$_2$	0.09	4.2	21	0.70	110	6.3
Ac-Pro-Ala-Pro-Ala-NH$_2$	8.5	3.9	2200	1.1	23	48
Ac-Ala-Pro-Ala-Pro-Ala-NH$_2$	5.3	3.9	1360	0.97	15	64
Ac-Pro-Ala-Pro-Gly-NH$_2$	0.1	22	5	0.12	42	2.9
Ac-Pro-Ala-Pro-Val-NH$_2$	6.0	35	208	0.42	28	15
Ac-Pro-Ala-Pro-Leu-NH$_2$	3.0	11	270	<0.023	>9.9	<3.2
Ac-Pro-Ala-Pro-Ala-Gly-NH$_2$	26	4.0	6500	3.0	12.7	236
Ac-Pro-Ala-Pro-Ala-Ala-NH$_2$	37	1.5	24700	2.67	5.5	500
Ac-Pro-Ala-Pro-Ala-Phe-NH$_2$	18	0.64	28800	1.52	5.2	288
Ac-Pro-Ala-Pro-Ala-Ala-Ala-NH$_2$	—	—	—	17.5	8.3	2100

[a] These data demonstrate that k_{cat}/K_m increases over 10^6-fold when the size of the substrate increases from a single residue to a hexapeptide, primarily due to increases in k_{cat}. Interactions with substrate residue P$_2$ are most important for α-lytic protease, but with elastase it is residue P$_4$. Elastase prefers an Ala residue at P$_1$, but α-lytic protease tolerates both smaller and slightly larger side chains. Interactions with residues P$_1'$ and P$_2'$ can increase k_{cat}/K_m, but there is little specificity for the amino acid side chain.

[b] Ac- is the $\overset{\displaystyle O}{\overset{\|}{CH_3C}}$— group blocking the α-amino group of the next residue; —NH$_2$ is the amide group on the α-carboxyl group.

Data from R. C. Thompson and E. R. Blout, *Biochemistry* 12:57–65 (1973); C. A. Bauer et al., *Biochemistry* 15:1296–1299 (1976); and C. A. Bauer et al., *Eur. J. Biochem.* 120:289–294 (1981).

triguing observations is not apparent, and the binding interactions probably act by aligning the substrate.

The interaction of residue P$_1$ of a substrate with the primary subsite S$_1$ of the enzyme is crucial for orienting the substrate properly for catalysis (Fig. 9.19). Where this interaction is extensive, as with the large Lys and Arg side chains in the case of trypsin and with Phe, Tyr, and Trp in chymotrypsin, it is of primary importance for catalysis, and the other sites have little effect on the rate of catalysis. Where the S$_1$ site is small and makes limited interactions with the P$_1$ residue, however, as with elastase, interactions at subsites S$_2$, S$_3$, and S$_4$ are much more important and substantially affect catalysis (Table 9.5). Strong interactions at these sites are exhibited primarily as increases in k_{cat} rather than as increased binding affinity, so they are used to enhance catalysis. They probably do this by assisting in the productive orienta-

tion of the substrate at the bond to be cleaved. Interactions with the substrate at the other side of the scissile bond, at residues P$_1'$, P$_2'$, and so on, are much less extensive, presumably because this portion of the substrate is to be expelled first as a product.

Three other interactions in the catalytic center serve to fix the scissile bond of the substrate in a specific orientation. One is a hydrogen bond between the backbone —NH— group of the P$_1$ residue and the backbone carbonyl group of residue 214 of the enzyme. The other two are hydrogen bonds from the carboxyl oxygen of the P$_1$ residue to the two —NH— groups of residues 193 and 195 of the enzyme, known as the **oxyanion binding site** or **hole**. With these three points of attachment, the position of the peptide bond to be cleaved is now fixed, within the limits imposed by the flexibility of the enzyme. Note, however, that this enzyme primarily uses groups of its backbone to secure the substrate, rather than the more flexible side chains. Any potential substrate that cannot participate simultaneously in these interactions tends to bind nonproductively, with a different geometry.

Optimum interactions between substrate and enzyme at these sites in the catalytic center and at the other subsites occur only if the carbonyl carbon of the scissile bond is distorted toward the tetrahedral geometry expected in the transition state (Eq. 9.49). In particular, the carbonyl oxygen can only then form geometrically favorable hydrogen bonds to the donors of the oxyanion binding site (Fig. 9.19B). At the same time, the dipole interaction with these groups can be imagined to induce polarization of the carbonyl group, producing an accumulation of negative charge on the oxygen atom and of positive charge on the carbonyl carbon. This carbon atom is close to the hydroxyl of Ser 195, and its developing positive charge favors nucleophilic attack by Ser 195. In the process, the proton from the Ser hydroxyl is transferred to His 57. As a result, the $C=O$ bond of the substrate lengthens to a single bond, and the oxygen atom acquires a net negative charge. These changes in the substrate increase its interactions with the oxyanion binding site. The site consists of backbone —NH— groups in the trypsin proteases, so their contribution to catalysis cannot be tested by the usual methods of site-directed mutagenesis or chemical modification. In subtilisin, however, one of the hydrogen-bonding groups in the oxyanion binding site is from the side chain of Asn 155. Site-directed mutation of this residue to Leu, which is similar in shape to Asn but nonpolar, decreased k_{cat} by a factor of 200, with little effect on K_m.

In the serine proteases, the active site is both sterically and electrostatically more complementary to the tetrahedral state of a substrate than to the normal geometry. Although the tetrahedral intermediate has not been observed directly, it is likely to be close to the transition state by analogy with many of the chemical inhibitors of serine proteases. These inhibitors produce tetrahedral adducts with the Ser residue and bind with the expected geometry (Fig. 9.20), so they may be considered transition-state analogues. One such analogue is the classical diisopropyl fluorophosphate (DFP), which reacts chemically with the unique Ser residue:

$$[(CH_3)_2CH-O-]_2 \overset{\overset{\displaystyle O}{\|}}{P}-F + HOCH_2-E$$
$$\text{DFP} \qquad\qquad \text{Ser 195}$$

$$\downarrow$$

$$[(CH_3)_2CH-O-]_2 \overset{\overset{\displaystyle O}{\|}}{P}-OCH_2-E + F^- + H^+$$

(9.54)

Inactivation of a protease by DFP is generally taken as an indication that the protease is of the serine class. Because a Ser hydroxyl group does not usually react with such acylating reagents, and because Ser 195 of the trypsin proteases does not react when the native conformation is disrupted, this reaction was attributed to a greatly enhanced nucleophilicity of the Ser 195 hydroxyl group in the native conformation. It is now clear, however, that the reaction of DFP with Ser 195 is due to the similarity of DFP to a substrate or a transition state. The adduct formed by this reaction does undergo further reaction, like a substrate, but slowly, and only one of the isopropyl groups is removed:

$$\begin{array}{c} E \\ | \\ CH_2 \\ | \\ O \\ | \\ (CH_3)_2CH-O-P-O-CH(CH_3)_2 \\ \| \\ O \end{array}$$

$$\downarrow\!\!\!\searrow \quad H_2O$$
$$\searrow \quad H^+$$

$$\begin{array}{c} E \\ | \\ CH_2 \\ | \\ O \\ | \\ HOCH(CH_3)_2 + {}^-O-P-O-CH(CH_3)_2 \\ \| \\ O \end{array}$$

(9.55)

In the case of the chloromethyl ketone affinity labels (Fig. 9.20), the sulfonyl portion forms a tetrahedral adduct with the Ser residue, and the reactive chloromethyl group then reacts further with the nearby His residue. The protein inhibitors also may be considered transition-state analogues because they appear to be strained in their normal conformation somewhat toward tetrahedral geometry.

The His 57 side chain in the tetrahedral intermediate is in good proximity to transfer the proton to the leaving-group of the substrate. The leaving-group cannot bind to the S_1' site in the acyl enzyme because this would force its amino group to be too close to the carbonyl group of the acyl enzyme. Also, there are few interactions between the leaving-groups and the enzyme at subsites S_1', S_2', and so forth, so the leaving-group dissociates from the acyl enzyme. The remainder of the substrate remains as the acyl intermediate, esterified to Ser 195.

Half the reaction is now complete, and there remains only to hydrolyze the acyl intermediate, presum-

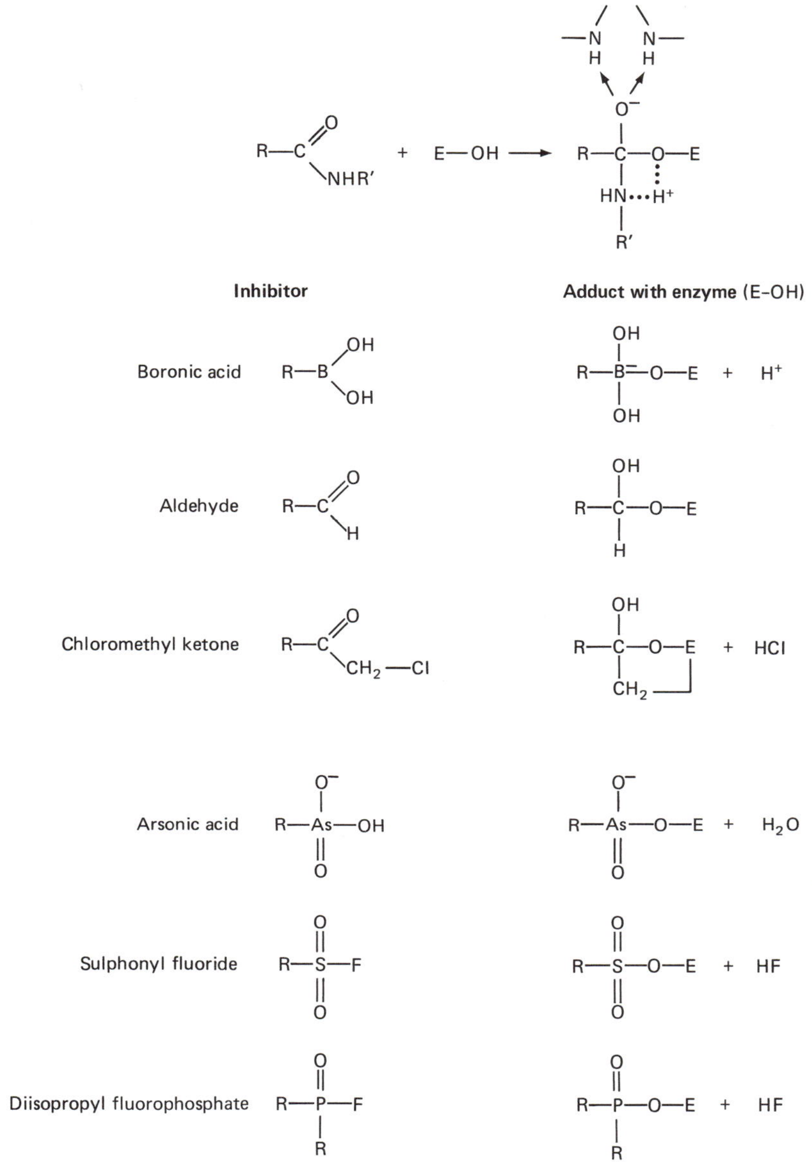

FIGURE 9.20

Tetrahedral transition-state analogues of serine proteases. At the top is shown a normal amide substrate, the enzyme (E—OH, the OH group being that of the active-site Ser residue), and the presumed tetrahedral intermediate. The carbonyl oxygen is shown interacting with the NH groups of the enzyme oxyanion binding site, and the proton is being transferred from the Ser to the substrate leaving group. Below are the inhibitors known to form covalent bonds with the active-site Ser residue, and the adducts. Note that, in each case, the adducts are tetrahedral, with an oxygen atom or hydroxyl group in position to interact with the oxyanion binding site. With the chloromethyl ketone inhibitors, the ClCH$_2$-group reacts covalently with the adjacent His residue of the catalytic triad. (Adapted from J. Kraut, *Ann. Rev. Biochem.* 46:331–358, 1977.)

ably by the reverse of the process just described. The acyl intermediate is observed to be less stable and more reactive than a normal ester. These properties may also be due to its being constrained toward a tetrahedral species by the nature of the catalytic site; consequently, the intermediate is activated toward attack by nucleophiles such as water. A water molecule is observed crystallographically to occupy the site vacated by the leaving-group, hydrogen-bonded by one of its protons to the N$^{\epsilon 2}$ atom of His 57 and possibly to another group on the protein. Its lone-pair electrons are oriented for nucleophilic attack on the acyl carbon. In the process, the water proton can be transferred transiently to His 57 to form the tetrahedral intermediate more fully. This proton can then be transferred to Ser 195, producing breakdown of the tetrahedral intermediate to generate the free-acid product and the intact enzyme. With dissociation of the product, the catalytic cycle is complete.

In summary, serine proteases appear to catalyze amide and ester hydrolysis by having an active site that is more complementary to the tetrahedral transition state than to the substrates, the products, or the acyl-enzyme intermediate. Energetically important interactions occur between enzyme and substrate throughout the binding subsites and in the oxyanion site to polarize the acyl group toward the tetrahedral geometry. The Ser 195 residue participates primarily because it is in the correct position to interact with the tetrahedral distorted carbon atom of the substrate, and because His 57 is nearby to accept its proton transiently. Ser 195 is apparently not especially potent as a nucleophile, but it appears to be especially reactive because bound ligands are activated toward the tetrahedral state. His 57 serves as both a base in transiently accepting a proton and as an acid in transferring it again, a role for which imidazole groups are admirably suited. The role of His 57 is undoubtedly augmented by its interaction with Asp 102, although the detailed energetics of this interaction are not clear.

Finally, note that no large movements of the enzyme need occur according to the mechanism of Figure 9.19, although His 57 could no doubt shuttle the proton between Ser 195 and the leaving-group more readily if it rocked between two conformations. Transient fluctuations of the enzyme and substrate may be important generally during catalysis, but there is little direct evidence in any instance. The primary catalytic function of the serine proteases appears to be stabilization of the tetrahedral transition state thought to participate in the nonenzymic reaction, both by favorable electrostatic interactions and by steric complementarity. The enzyme also provides a novel reaction intermediate, the acyl enzyme, thereby permitting the reaction to occur in two symmetric half-steps. A possible advantage of the acyl intermediate is that the leaving and attacking groups (HX of Fig. 9.17A) can share the same position, so the same catalytic groups can be involved in the two half-reactions.

References

Structure and mechanism of chymotrypsin. D. M. Blow. *Acc. Chem. Res.* 9:145–152 (1976).

Serine proteases: structure and mechanism of catalysis. J. Kraut. *Ann. Rev. Biochem.* 46:331–358 (1977).

Protein inhibitors of proteinases. M. Laskowski, Jr., and I. Kato. *Ann. Rev. Biochem.* 49:593–626 (1980).

Direct determination of the protonation states of aspartic acid-102 and histidine-57 in the tetrahedral intermediate of the serine proteases: neutron structure of trypsin. A. A. Kossiakoff and S. A. Spencer. *Biochemistry* 20:6462–6474 (1981).

Crystallographic and NMR studies of the serine proteases. T. A. Steitz and R. G. Shulman. *Ann. Rev. Biophys. Bioeng.* 11:419–444 (1982).

Confirmation of the assignment of the low-field proton resonance of serine proteases by using specifically nitrogen-15 labeled enzyme. W. W. Bachovchin. *Proc. Natl. Acad. Sci. USA* 82:7948–7951 (1985).

Structure of α-chymotrypsin refined at 1.68 Å resolution. H. Tsukada and D. M. Blow. *J. Mol. Biol.* 184:703–711 (1985).

Site-directed mutagenesis and the role of the oxyanion hole in subtilisin. B. Bryan et al. *Proc. Natl. Acad. Sci. USA* 83:3743–3745 (1986).

Selective alteration of substrate specificity by replacement of aspartic acid-189 with lysine in the binding pocket of trypsin. L. Graf et al. *Biochemistry* 26:2616–2623 (1987).

The catalytic role of the active site aspartic acid in serine proteases. C. S. Craik et al. *Science* 237:909–913 (1987).

Dissecting the catalytic triad of a serine protease. P. Carter and J. A. Wells. *Nature* 332:564–568 (1988).

Subtilisin—an enzyme designed to be engineered. J. A. Wells and D. A. Estell. *Trends Biochem. Sci.* 13:291–297 (1988).

Electrostatic complementarity within the substrate-binding pocket of trypsin. L. Graf et al. *Proc. Natl. Acad. Sci. USA* 85:4961–4965 (1988).

How do serine proteases really work? A. Warshel et al. *Biochemistry* 38:3629–3637 (1989).

Human leukocyte and porcine pancreatic elastase: X-ray crystal structures, mechanism, substrate specificity, and mechanism-based inhibitors. W. Bode et al. *Biochemistry* 28:1951–1963 (1989).

Structure and activity of two photoreversible cinnamates bound to chymotrypsin. B. L. Stoddard et al. *Biochemistry* 29:4871–4879 (1990).

b. Thiol Proteases

Thiol proteases are most comparable to the serine proteases, with a Cys side chain playing the role of the Ser. The known thiol proteases are not detectably homologous to the serine proteases. Until recently, there was no known evolutionary variant in which the Ser and Cys residues at the active site were interchanged. Proteins that are homologous to thiol proteases, but with a Ser residue at the position of the catalytic Cys residue, have recently been identified from gene sequences, but they have not yet been shown to be active proteases. The active-site Ser side chain of subtilisin has been converted chemically to Cys, but the enzyme is then no longer active proteolytically, only being able to hydrolyze active esters. Nevertheless, the thiol proteases are catalytically similar to the serine proteases, catalyzing similar reactions involving both a tetrahedral intermediate and an acylenzyme intermediate in which the acyl group is bonded to the Cys sulfur atom.

The structures of two homologous thiol proteases, papain and actinidin, are known in great detail. Both are composed of two structural domains of about 212 amino acid residues, each comprising one of the two halves of the polypeptide chain. The active site lies in a deep cleft between these two domains and consists of seven subsites (S_4, S_3, S_2, S_1, S_1', S_2', and S_3'), for seven residues of the peptide substrate. The substrate specificities of the proteases arise primarily from interactions at subsites S_2 and S_1', which are specific for hydrophobic side chains of polypeptide substrates. The catalytic site is defined by residues from both domains. In particular, the Cys 25 residue that forms the acyl enzyme interacts closely with His 159 of the other domain, which is thought to be analogous to His 57 of the serine protease catalytic triad. There is no obvious analogue to Asp 102; the nearest acidic group is 7.5 Å away, although an Asn residue may play a comparable role. There is an equivalent of the oxyanion binding site. The catalytic site is probably sterically and electrostatically complementary to the tetrahedral-like transition state, just as in the serine proteases.

Ionization of the Cys 25 thiol group and of the His 159 side chain accounts for the pH dependence of the catalytic activity. Their ionization is complex because close interaction between them results in marked interdependence among their ionization states. Ionization of one group is thought to affect the pK_a of the other by 4.2 pH units:

$$(9.56)$$

The active form of papain is believed to be that at the bottom in Equation (9.56), where His 129 and Cys 25

are both ionized and interacting electrostatically. In the enzyme, there may be an equilibrium between the two forms in the middle: the form where both the His and Cys residues are neutral and that where both are ionized.

The pK_a values of the His and Cys side chains are substantially different from those considered normal (see Table 1.2), not only as a result of the interaction between them but probably also as a result of electrostatic interactions with the dipole of the α-helix on which Cys 25 is located (Sec. 5.3.1), with an Asn and an Asp side chain, and with the solvent.

The differences between the serine and thiol proteases are primarily due to the much greater tendency of a thiol to ionize compared with a hydroxyl group. As a result, the thiol proton can be transferred to the His residue in the normal protein; perhaps for this reason, there is no need for an ionized carboxyl group to be part of the catalytic triad (Fig. 9.17B).

With a thiol group required for activity, the thiol proteases are extremely susceptible to inhibition by the many reagents that react with thiols (Sec. 1.3.10). The thiol proteases are often isolated in a form that is to a great extent inactive. Mixed disulfides with other thiol compounds, plus some other adducts, can be removed by treatment with reducing agents and the enzymes activated. Other derivatives of thiol proteases are inhibited irreversibly. An example is the oxidation product —SO_2^-, which appears to be formed readily owing to the adjacent oxyanion binding site, in which one of the oxygen atoms binds.

References

On the reactivity of the thiol group of thiolsubtilisin. L. Polgár et al. *Eur. J. Biochem.* 39:421–429 (1973).

Effect of cysteine-25 on the ionization of histidine-159 in papain as determined by proton nuclear magnetic resonance spectroscopy. Evidence for a His-159–Cys-25 ion pair and its possible role in catalysis. S. D. Lewis et al. *Biochemistry* 20:48–51 (1981).

The thiol proteases: structure and mechanism. E. N. Baker and J. Drenth. In *Biological Macromolecules and Assemblies*, F. A. Jurnak and A. McPherson, eds., vol. 3, pp. 314–368. Wiley, New York, 1987.

The active site of papain. All-atom study of interactions with protein matrix and solvent. J. A. C. Rullmann et al. *J. Mol. Biol.* 206:101–118 (1989).

Cysteinyl proteinases and their selective inactivation. E. Shaw. *Adv. Enzymol.* 63:271–347 (1990).

c. Carboxyl Proteases

The class of carboxyl proteases has long intrigued enzymologists because its best known representative, pepsin, is active only at extreme pH values of 1–5 (no doubt because it must function in the acidic environment of the stomach). It is also a very acidic protein, having about 39 carboxyl side chains and only 2 Arg, 1 Lys, and 1 His residue. The enzymes of this class are also known as the **acidic proteases** (though not all require low pH for activity) and as **aspartyl proteases** because the catalytic carboxyl groups are usually from Asp residues.

Like the members of the other protease families, carboxyl proteases perform many functions. The best known gastric enzymes — such as pepsin, gastricsin, and chymosin (formerly known as rennin, used in cheese making) — are involved in digestion and have only limited substrate specificities. Cathepsin D is a lysosomal enzyme that degrades proteins intracellularly. Other carboxyl proteases produce a very specific cleavage in a single protein; for example, renin catalyzes removal of the decapeptide angiotensin I, which plays a major role in the control of blood pressure, from the amino terminus of angiotensinogen. In the retroviruses, including human immunodeficiency virus HIV-1, many proteins are synthesized as polyproteins that are cleaved by a carboxyl protease during activation of the virus.

The structures of most of the known carboxyl proteases are very similar, and most of them are a single polypeptide chain of approximately 327 residues that is folded into a bilobed structure with two domains of similar structure. The two domains are related by an approximate twofold rotation axis, suggesting that the ancestral protein was a dimer of two identical chains. Just such a dimeric structure is present in the HIV protease, in which the polypeptide chain is only 99 residues long. The retroviral protease may be a molecular "fossil" that has retained the dimeric structure.

The active site of the carboxyl proteases is located in the cleft between the two domains, or subunits in the case of the retroviral protease. The two Asp residues for which this class of proteases is named are located in the active site and are related by the twofold axis that is approximate in the two-domain proteases and exact in the retroviral dimeric enzyme. The two carboxyl groups are connected to one another by a complex network of hydrogen bonds. They interact closely and share one proton in the active form of the enzyme, in which only one of the carboxyls is ionized (Fig. 9.21). Due to this interaction, the first carboxyl to ionize has an anomalously low pK_a value of about 1.5 whereas the other has an elevated pK_a of 4.7. This is similar to the ionization of maleic acid, $HO_2C—CH_2—CHOH—CO_2H$, which has pK_a values of 1.9 and 6.2. In the two-domain pepsin, Asp 32 has been assigned the lower pK_a value because it is modified specifically by epoxides, which react with ionized carboxyls, whereas Asp 215 reacts with diazo

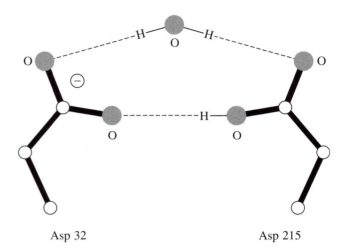

Asp 32   Asp 215

FIGURE 9.21
Interaction between the two carboxyl groups of the catalytic Asp residues of carboxyl proteases. Only one of the carboxyls is ionized. The positions of the protons are not known experimentally, but those shown are thought to be the most probable. A water molecule bridges two oxygen atoms of the carboxyl groups.

compounds that are specific for nonionized carboxyls (Sec. 1.3.5).

Inhibitors that are thought to mimic substrates or transition states bind very similarly in the active-site cleft, making a variety of interactions with the enzyme. One part of pepsin, known as the *flap* and consisting of residues 72–81 in a β-hairpin bend, is flexible in the free enzyme but folds over inhibitors bound to the active site. This folding has the effect of removing the peptide bond to be hydrolyzed from contact with the solvent and surrounding it with atoms of the protein. In the process, a number of water molecules are expelled from the cleft, including that bound to the catalytic carboxyl groups (Fig. 9.21), but the two carboxyl groups do not move substantially.

Substrates are thought to bind with the $C=O$ of the bond to be cleaved situated between the catalytic carboxyls, with the oxygen atom close to the position of the displaced water molecule that normally bridges them (Fig. 9.21). The nonionized carboxyl is thought to protonate the carbonyl oxygen of the substrate (Fig. 9.17C). A water molecule that is hydrogen-bonded to the external oxygen atom of Asp 32 is in position for nucleophilic attack on the carbonyl carbon, with transfer of one of its protons to Asp 32. This produces a tetrahedral intermediate that breaks down to product upon protonation of the —NH— group, either from a solvent molecule or from the catalytic carboxyl groups. An alternative mechanism, however, has the protonation states of the Asp residues reversed and the bridging water molecule as the nucleophile.

The importance of the tetrahedral intermediate is illustrated by inhibitors with comparable geometry that bind tightly to the carboxyl proteases. Notable is the naturally occurring inhibitor **pepstatin,** an unusual peptide that contains two residues of the amino acid statine:

$$
\begin{array}{c}
\text{HC(CH}_3)_2 \\
| \\
\text{CH}_2 \quad \text{OH} \quad\quad \text{O} \\
| \quad\quad | \quad\quad\quad || \\
-\text{NH}-\text{CH}-\text{CH}-\text{CH}_2-\text{C}- \\
\end{array}
\qquad (9.57)
$$

When bound to pepsin, one of the statine residues occupies the catalytic site, with the tetrahedral

$$
\begin{array}{c}
\text{OH} \\
| \\
-\text{CH}- \\
\end{array}
$$

group occupying the position normally occupied by the

$$
\begin{array}{c}
\text{O} \\
|| \\
-\text{C}- \\
\end{array}
$$

group of a substrate. Similarly, a pepstatin analogue with a carbonyl group in this position binds as the hydrated form $-\text{C(OH)}_2-$, which has tetrahedral geometry.

The carboxyl proteases are similar to the serine and thiol proteases (Table 9.5) in that extending the size of the substrate, so that it interacts with more subsites of the enzyme, alters primarily k_{cat} and not K_m. There is no covalent intermediate in the case of the carboxyl proteases, in contrast to the serine and thiol proteases. There is a tetrahedral intermediate or transition state, but it is obtained by attack of water on the carbonyl carbon rather than by attack of a Ser hydroxyl or a Cys thiol. The carboxyl protease mechanism is closer to that of the metalloproteases than to those of the serine or thiol class.

References

The mechanism of the catalytic action of pepsin and related acid proteases. J. S. Fruton. *Adv. Enzymol.* 44:1–36 (1976).

Stereochemical analysis of peptide bond hydrolysis catalyzed by the aspartic proteinase penicillopepsin. M. N. G. James and A. R. Sielecki. *Biochemistry* 24:3701–3713 (1985).

The rational design of renin inhibitors: X-ray studies of aspartic proteinases complexed with transition-state analogues. T. L. Blundell et al. *Biochemistry* 26:5585–5590 (1987).

The structure and function of the aspartic proteinases. D. R. Davies. *Ann. Rev. Biophys. Biophys. Chem.* 19:189–215 (1990).

Structure and function of retroviral proteases. P. M. D. Fitzgerald and J. P. Springer. *Ann. Rev. Biophys. Biophys. Chem.* 20:299–320 (1991).

d. Metalloproteases

Metalloproteases use bound metals in their active sites, usually Zn^{2+}, and include medically important enzymes such as angiotensin-converting enzyme, enkephalinase, and collagenase. Carboxypeptidases A and B and thermolysin are the most thoroughly studied representatives of this class. The carboxypeptidases catalyze the stepwise removal of single residues from the carboxyl terminus of a polypeptide substrate; that is, they are **exopeptidases** and are specific for the nature of the P_1' residue. Carboxypeptidase A is specific for large hydrophobic side chains and form B for basic residues. The two forms are homologous and have similar conformations, but carboxypeptidase A has a hydrophobic S_1' site whereas B has an Asp residue to interact with basic P_1' side chains. Therefore, these two exopeptidases are undoubtedly related evolutionarily in the same way as are the chymotrypsin and trypsin variants of the serine proteases. Thermolysin, which is not homologous with the carboxypeptidases in either primary or tertiary structures, is an **endopeptidase,** catalyzing hydrolysis of nonterminal peptide bonds, especially those with hydrophobic P_1' residues. Nevertheless, its active site shows many similarities to those of the carboxypeptidases, presumably as a result of convergent evolution, analogous to the relationship of subtilisin to the other serine proteases.

Although the difference between an exopeptidase and an endopeptidase is crucial from the functional point of view, it is not a fundamental consideration at the molecular level. Thermolysin is an endopeptidase, as are the other proteases described earlier, because it has S_2', S_3', . . . sites that interact favorably with P_2', P_3', . . . residues of the substrate or at least do not discriminate against them. In contrast, carboxypeptidase A has an S_1' site that is essentially a deep dead-end pocket for a terminal P_1' residue, and this structure is unable to accommodate residues at positions P_2', P_3', and so on. This exopeptidase is also specific for an α-carboxyl group on the P_1' residue, in the form of a salt bridge with residue Arg 145 and hydrogen bonds to the side chains of Tyr 248 and Asn 144 (Fig. 9.22).

The Zn^{2+} atom is bound to both thermolysin and carboxypeptidase by interactions with the imidazole side chains of two His residues and with the carboxyl side chain of a Glu residue. A further coordination position of each Zn^{2+} atom is occupied by a water molecule,

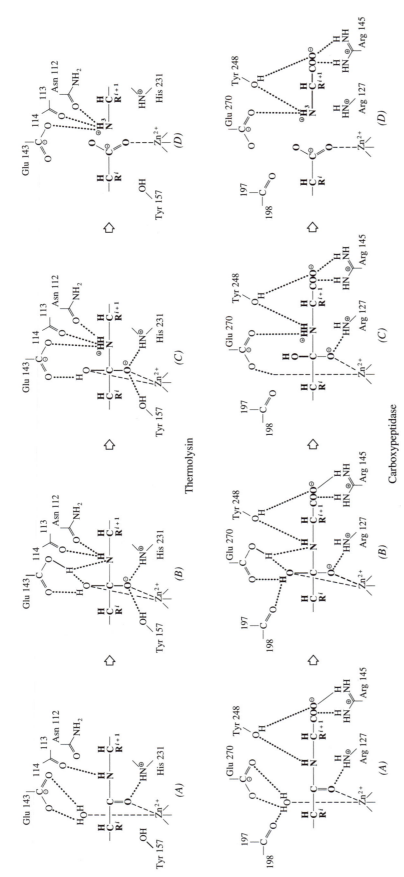

FIGURE 9.22

Comparison of the proposed mechanisms for thermolysin (*top*) and carboxypeptidase A (*bottom*). The common features involve equivalent Zn^{2+} ions and Glu residues (143 in thermolysin and 270 in carboxypeptidase A), plus water molecules interacting with both of them. *A*: The Michaelis complex; note the specificity in carboxypeptidase A for the α-carboxyl group of the P_1' residue, so this residue must be at the C-terminus. *B*: The water molecule has reacted with the carbonyl carbon to generate the tetrahedral intermediate. *C*: The proton has been transferred from the Glu carboxyl to the —NH— of the substrate. *D*: The peptide bond has been broken to generate the product α-carboxyl and α-amino groups. (From B. W. Matthews, *Acc. Chem. Res.* 21:333–340, 1988.)

which plays a crucial role in the catalytic activity of each enzyme.

The structures of a number of substrate analogues and inhibitors bound to carboxypeptidase A and thermolysin suggest that they function very similarly (Fig. 9.22). Superimposing one active site on the other, using the positions of the Zn^{2+} ion and a bound dipeptide substrate, demonstrates that the position of Glu 270 of carboxypeptidase A corresponds closely to that of Glu 143 of thermolysin. Both residues are implicated in having central roles in the catalytic function. Similarly, Arg 127 of carboxypeptidase A and His 231 of thermolysin have very similar positions and similar electrostatic interactions with the Zn^{2+} ion. Except for these pairs of residues and the Zn^{2+} atom, the other groups that are in the active site and involved in catalysis are not structurally equivalent (Fig. 9.22).

Early studies of carboxypeptidase led to the suggestion that the Glu 270 side chain acts as a nucleophile and reacts directly with the carbonyl carbon of the peptide bond to be cleaved of certain substrates, generating a mixed anhydride:

$$
\underset{\text{enzyme}}{E-CO_2^-} + \underset{\text{substrate}}{R^1-\overset{\overset{\textstyle O}{\|}}{C}-X}
$$

$$
\downarrow
$$

$$
\underset{\text{mixed anhydride intermediate}}{E-\overset{\overset{\textstyle O}{\|}}{C}-O-\overset{\overset{\textstyle O}{\|}}{C}-R^1 + X^-} \tag{9.58}
$$

In contrast, structural studies of thermolysin suggested that the nucleophile that reacts with the carbonyl carbon is the water molecule that is bound to both the Zn^{2+} ion and the Glu 143 carboxyl group (Figs. 9.17D, 9.22). This is now also believed to be the case with carboxypeptidase A, and there probably is not a covalent intermediate.

The carbonyl oxygen atom of a bound peptide substrate interacts with the Zn^{2+} atom, and the water molecule should be pushed toward the Glu carboxyl. Both the Glu residue and the Zn^{2+} ion promote the nucleophilicity of this water molecule, and it reacts with the carbonyl carbon to generate the tetrahedral intermediate or transition state, which leaves a proton on the Glu carboxyl. The tetrahedral species is further stabilized by electrostatic interactions of the carbonyl oxygen (which now has net negative charge) with the Zn^{2+} ion and in two hydrogen bonds to other groups that play a role similar to the oxyanion binding site of the serine proteases (Sec. 9.3.2.a). That the active site is more complementary to the tetrahedral form of the substrate

is shown by the binding of an inhibitor with a ketone group,

$$
-\overset{\overset{\textstyle O}{\|}}{C}-CH_2- \tag{9.59}
$$

which binds to carboxypeptidase A as the hydrated form

$$
-C(OH)_2-CH_2- \tag{9.60}
$$

The hydrated form occurs only rarely in solution but has a tetrahedral geometry and is thought to be bound more tightly by the enzyme. The other evidence for the tetrahedral intermediate comes from the transition-state analogues described in Figure 9.9.

In the complex with the tetrahedral species, the proton on the Glu carboxyl is transferred to the —NH— of the substrate to cleave the peptide bond and to generate the new amino group of the product. The Glu residue has the role of transferring a proton from the nucleophile to the —NH— leaving-group, the same role as the His residues of the serine and thiol proteases and the Asp residues of the carboxyl proteases (Fig. 9.17).

A substantial conformation change occurs in carboxypeptidase A, but not in thermolysin, in that a loop of five residues that is flexible in free carboxypeptidase A closes over the substrate when it is bound. This loop includes Tyr 248, whose phenolic side chain interacts with the substrate (Fig. 9.22). The role of this residue has evoked much controversy. Its phenolic hydroxyl was originally postulated to be involved in the transfer of a proton from the water nucleophile to the —NH— group because chemical modification of this Tyr residue decreased the protease activity of the enzyme. That Tyr 248 does not have a central role in catalysis, however, was demonstrated by mutating it to Phe; the mutant protein had k_{cat} decreased only 2.5-fold. Tyr 248 appears not to be important catalytically but to be involved primarily in binding the substrate throughout the catalytic cycle. The postulated catalytic role of Tyr 248 has now been attributed to Glu 270 (Fig. 9.22).

References

Binding of *N*-carboxymethyl dipeptide inhibitors to thermolysin determined by X-ray crystallography: a novel class of transition-state analogues for zinc peptidases. A. F. Monzingo and B. W. Matthews. *Biochemistry* 23:5724–5729 (1984).

Evidence against a crucial role for the phenolic hydroxyl of Tyr-248 in peptide and ester hydrolyses catalyzed by carboxypeptidase A: comparative studies of the pH dependencies of the native and Phe-248-mutant forms. D. Hilvert et al. *J. Amer. Chem. Soc.* 108:5298–5304 (1986).

Mechanism of carboxypeptidase A: hydration of a ketonic substrate analogue. D. W. Christianson et al. *Proc. Natl. Acad. Sci. USA* 84:1512–1515 (1987).

Structural basis of the action of thermolysin and related zinc peptidases. B. W. Matthews. *Acc. Chem. Res.* 21:333–340 (1988).

Carboxypeptidase A. D. W. Christianson and W. N. Lipscomb. *Acc. Chem. Res.* 22:62–69 (1989).

e. Zymogen Activation

Because most proteases are destructive, it is vital that their catalytic activities not be unleashed before they are required. Probably for that reason, proteases are generally synthesized as larger precursors known as **zymogens**, which are catalytically inactive. All are activated by proteolysis, but the mechanism of the activation and the basis for the inactivity of the precursor are different in each protease family.

Serine Proteases The mammalian trypsin proteases are synthesized as the zymogens trypsinogen, chymotrypsinogen, proelastase, and so on, which have polypeptide extensions at their N-termini. They are inactive as proteases; their low activities toward small substrates are no greater than that of a comparable solution of imidazole. Activation of these zymogens occurs by proteolytic cleavage of the peptide bond between residues 15 and 16. Other cleavages may also take place—for example, after residues 13, 146, and 148 in chymotrypsinogen A—to release the dipeptides 14–15 and 147–148 and to give the final disulfide-cross-linked, three-polypeptide-chain α-chymotrypsin (see Fig. 2.13). Only the cleavage after residue 15, however, is necessary to generate catalytic activity. The process of activating the serine proteases in the mammalian gut is initiated by a proteolytic enzyme, enterokinase, which is attached to the external side of the brush-border membrane. It is extremely specific for activating trypsinogen; the product trypsin then activates other molecules of trypsinogen, proelastase, and chymotrypsinogens A, B, and C.

The structural basis of this activation has been investigated by comparing the structures of chymotrypsinogen A and α-chymotrypsin and of trypsinogen and trypsin. The initially surprising finding was that the structures are so similar (Fig. 9.23). In particular, their active sites are largely intact, with a stereochemically acceptable catalytic triad of Asp 102, His 57, and Ser 195. In trypsinogen, four segments of residues 16–19, 142–152, 184–193, and 216–223 that border the active site are very flexible in the zymogen but become fixed on activation. They also become fixed in a trypsin-like conformation if the protein inhibitor BPTI (Sec. 9.3.2.a) is bound to the active site; the affinity of BPTI for trypsinogen is, however, only 10^{-7} that for trypsin.

Only the loop 142–152 is flexible in chymotrypsinogen, where it is usually cleaved proteolytically to release the dipeptide of residues 147–148 (Fig. 2.13). The other three segments of chymotrypsin adopt slightly different conformations in the zymogen and in the active enzyme. These small differences in the zymogen affect the binding pocket for the substrate and cause the effective absence of the oxyanion binding site, which consists of the backbone —NH— groups of residues 193 and 195. The —NH— of residue 193 points in the wrong direction in the zymogen, where it forms a hydrogen bond with the carbonyl oxygen of residue 180 in a type III turn.

The most important change on activation of the zymogen is thought to be the liberation of the new α-amino group of residue 16. This amino group forms a salt bridge with the side chain of Asp 194 in the active enzyme (Fig. 9.24), which involves a substantial rearrangement of residues 189–194. The α-amino group of residue 16 does not exist in the zymogen, and Asp 194 is in a polar environment with its side chain interacting with His 40. The Ile 16 side chain is accessible to solvent in the zymogen, but upon activation it displaces the buried side chain of Met 192, which moves into the solvent. As a result of these conformational rearrangements of residues 189–194, the —NH— group of residue 193 adopts its appropriate position for participating in the oxyanion hole. The changes extend through the middle of the protein molecule because the activating cleavage of the 15–16 peptide bond occurs on the surface of the molecule opposite to the active site (Fig. 9.24).

The importance of the salt bridge involving the α-amino group of residue 16 is demonstrated by the disappearance of activity of α-chymotrypsin if this group loses its proton. The bacterial trypsin proteases have no known zymogen precursors and no α-amino group of residue 16. They do, however, have the Arg side chain of residue 138 in these proteins, which is involved in a salt bridge to a buried Asp 194 side chain.

Serine proteases of the subtilisin class and bacterial examples of the trypsin class are also synthesized as precursors but with substantially longer extensions that can be at either end of the polypeptide chain. The terminal extensions are inhibitors of the protease, even after cleavage of the connecting segment. It is of interest that the N-terminal extension is required for the correct folding of the protein; the mature proteins do not unfold reversibly in the absence of the precursor peptide.

References

Mechanism of zymogen activation. R. M. Stroud et al. *Ann. Rev. Biophys. Bioeng.* 6:177–193 (1977).

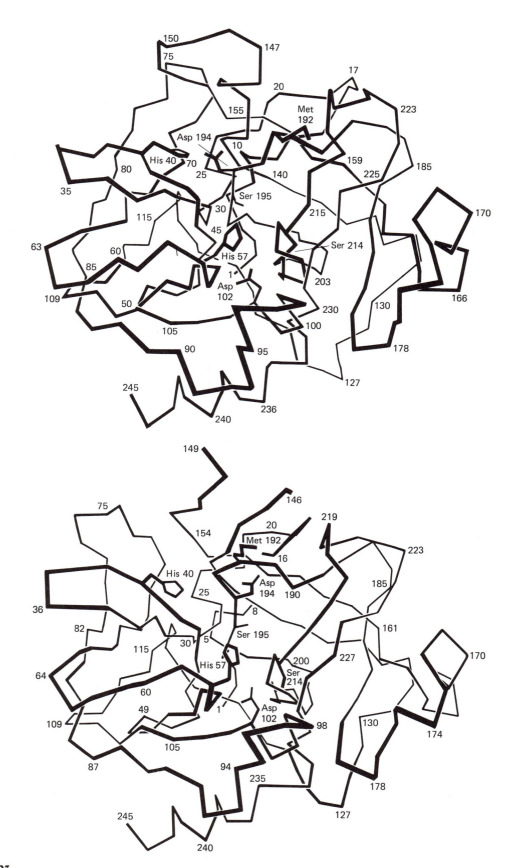

FIGURE 9.23

Comparison of the polypeptide backbone conformation of chymotrypsinogen A *(top)* and α-chymotrypsin *(bottom)*. The latter has had residues 14, 15, 147, and 148 removed proteolytically upon activation. The side chains of some residues important for catalysis and for activation are also shown. (From S. T. Freer et al., *Biochemistry* 9:1997–2009, 1970.)

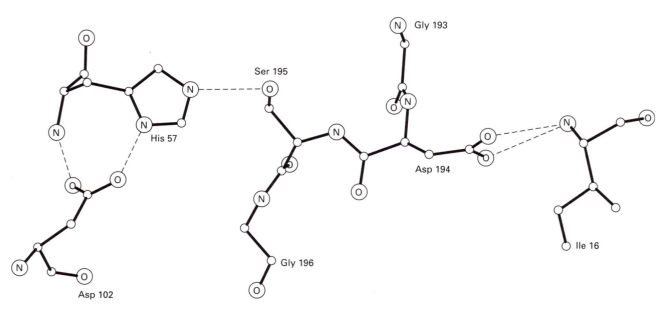

FIGURE 9.24

The position of the Ile 16 α-amino salt bridge to Asp 194 in the interior of the α-chymotrypsin molecule, relative to the catalytic triad of Asp 102, His 57, and Ser 195 at the active site. Presumed hydrogen bonds are shown as dashed lines. These residues virtually span the interior of the globular folded structure of the protein. The entrance to the active site is at the left, front; the rear surface of the protein is at the right (see Fig. 9.23). (Kindly provided by D. M. Blow, using coordinates from H. Tsukada and D. M. Blow, *J. Mol. Biol.* 184:703–711, 1985.)

Structural basis of the activation and action of trypsin. R. Huber and W. Bode. *Acc. Chem. Res.* 11:114–122 (1978).

Bovine chymotrypsinogen A. X-ray crystal structure analysis and refinement of a new crystal form at 1.8 Å resolution. D. Wang et al. *J. Mol. Biol.* 185:595–624 (1985).

A protein-folding reaction under kinetic control. D. Baker et al. *Nature* 356:263–265.

Carboxyl Proteases Pepsin is generated from the zymogen **pepsinogen** by the proteolytic removal of 44 residues from its N-terminus. The zymogen is stable at neutral pH but spontaneously activates when the pH is decreased to less than 5. The proteolytic cleavage can be catalyzed by another molecule of pepsin, or it can be a solely intramolecular process in which the active site of the zymogen cleaves its own polypeptide chain. This is apparent from the observation that the rate is independent of zymogen concentration and is not affected by the presence of a substrate for pepsin. Activation can also occur when pepsinogen is immobilized by covalent attachment to a resin, when different molecules cannot come into contact.

When pepsinogen is placed in acidic solution, a conformational change is observed spectrally to occur very rapidly, before formation of active pepsin, indicating an intermediate stage in the activation process. The peptide bond, between residues 16 and 17 of the pro segment is subsequently hydrolyzed more slowly, producing enzymatically active **pseudopepsin**. Further processing to remove the remainder of the pro segment and to generate mature pepsin occurs intermolecularly.

The structural basis of this phenomenon was suggested by the crystal structure of pepsinogen at pH 6. Residues 11–44 of the proenzyme segment cover the substrate-binding cleft of the pepsin portion of the zymogen and block access to the two catalytically active Asp residues. The pepsin portion of the zymogen is not detectably different from the mature, active enzyme. The pro segment is held in place by a number of electrostatic interactions between basic side chains of the pro segment and acidic residues of the pepsin portion. At least some of these interactions are thought to be weakened by lowering the pH and protonating the carboxyl groups, and electrostatic repulsions between the resulting unpaired basic residues are thought to alter the conformation. This permits the peptide bond between residues 16 and 17 to enter the active site and to be cleaved by the pepsin portion of the molecule. In pepsinogen, therefore, the proenzyme segment acts as an inhibitor of an otherwise active enzyme.

Reference

Molecular structure of an aspartic proteinase zymogen, porcine pepsinogen, at 1.8 Å resolution. M. N. G. James and A. R. Sielecki. *Nature* 319:33–38 (1986).

Metalloproteases The zymogens of this class of protease appear to be similar to those of the carboxyl proteases in that the metalloprotease zymogen contains an intrinsically active protease in which the active site is masked by the pro segment. Procarboxypeptidase B has a 95-residue pro segment that is folded and shields the active site of the protease by interacting with residues important for substrate recognition. Part of the pro segment occupies the substrate binding site, but does not fill it in the way that a substrate would. Consequently, the zymogen is inactive to large substrates, and the propeptide is not cleaved by the protease.

Activation is caused by trypsin cleaving the peptide bond between the pro and protease segments, which is very exposed in the zymogen. Cleavage of this connecting segment presumably lowers the affinity of the pro domain for the protease and favors its dissociation. The pro region remains folded in isolation, however, and can bind to and inhibit the protease.

Reference

Three-dimensional structure of porcine procarboxypeptidase B: a structural basis of its inactivity. M. Coll et al. *EMBO J.* 10:1–9 (1991).

9.3.3 Lysozyme

Hen egg-white lysozyme was the first enzyme to have its structure determined crystallographically, in the laboratory of D. C. Phillips, but until that time very little was known about its catalytic properties. Lysozymes hydrolyze various oligosaccharides, especially those of bacterial cell walls. The complex structures and chemical properties of the substrates have not encouraged large numbers of detailed enzymatic studies of these enzymes. Nevertheless, the structure of hen lysozyme and of its complexes with several saccharides provided Phillips and coworkers sufficient clues to propose a plausible

model of how catalysis might take place; this model has become the archetypal example in most biochemistry textbooks. Subsequent investigations have largely supported that model, although some unforeseen complexities and a change in emphasis have emerged, and many questions remain unanswered.

Briefly, the enzyme's active site is a cleft capable of binding simultaneously six sugar residues of an appropriate polysaccharide at subsites designated A–F (Figs. 8.1 and 9.25). Hydrolysis occurs between the sugar residues at sites D and E; thus, the six sites are comparable to subsites S_4–S_2' of proteases. Binding at all six subsites is important for catalysis because the enzyme hydrolyzes a hexasaccharide 10^7 times more rapidly than it does a disaccharide. Of its substrates, *N*-acetyl glucosamine (abbreviated here as GlcNAc but often as NAG) residues can be accommodated at all six subsites, whereas the additional lactyl group of an *N*-acetyl muramic acid (MurNAc, or NAM) residue interferes with its binding to subsites A, C, and E and limits its binding to the other sites. Consequently, the alternating oligosaccharide (—GlcNAc—MurNAc—) of bacterial cell walls is hydrolyzed only adjacent to the MurNAc residue bound to site D (Fig. 9.25). This substrate specificity for oligosaccharides is relatively straightforward, but lysozyme is much more active toward its natural substrate, bacterial cell walls, in which the polysaccharides are cross-linked by peptides. The natural substrate may undergo additional or different interactions with the enzyme.

Model-building a substrate into the active site of the free enzyme suggested that a normal, chair-shaped sugar residue could not be accommodated in site E unless it were sterically distorted into the "half-chair," or "sofa," conformation that was thought to occur in the transition state:

$$\tag{9.61}$$

The oxocarbonium ion form of the sugar residue, which would be in subsite D of the enzyme, is most stable with a planar conformation due to resonance, spreading the

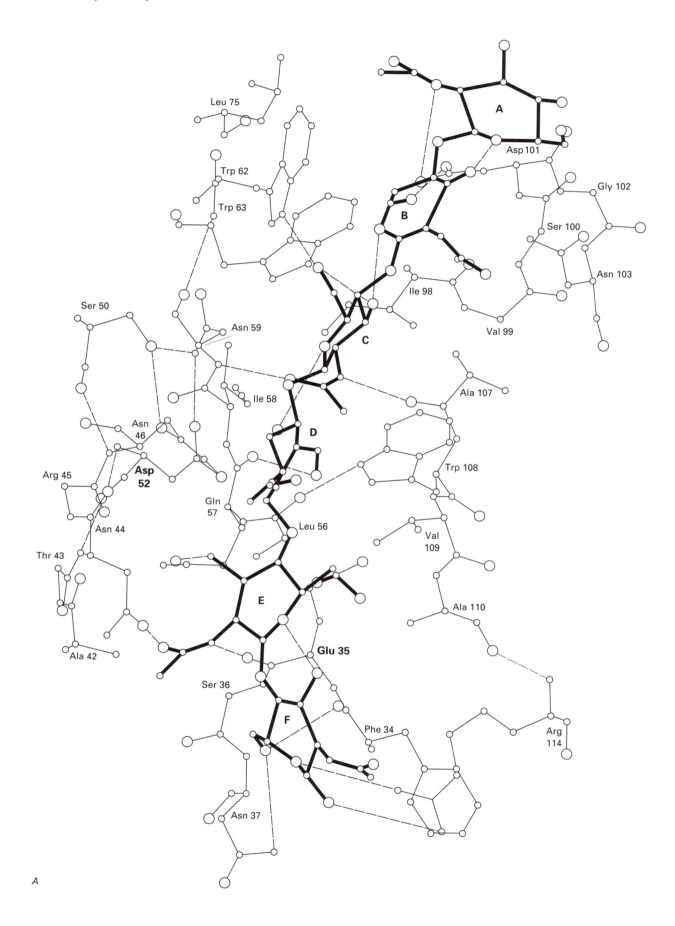

Leu 75

Trp 62

Trp 63

A

Asp 101

Gly 102

B

Ser 100

Ile 98

Asn 103

Val 99

Ser 50

Asn 59

C

Ala 107

Ile 58

Asn 46

D

Trp 108

Arg 45

Asp 52

Gln 57

Asn 44

Leu 56

Val 109

Thr 43

E

Ala 110

Ala 42

Glu 35

Ser 36

Arg 114

Phe 34

F

Asn 37

A

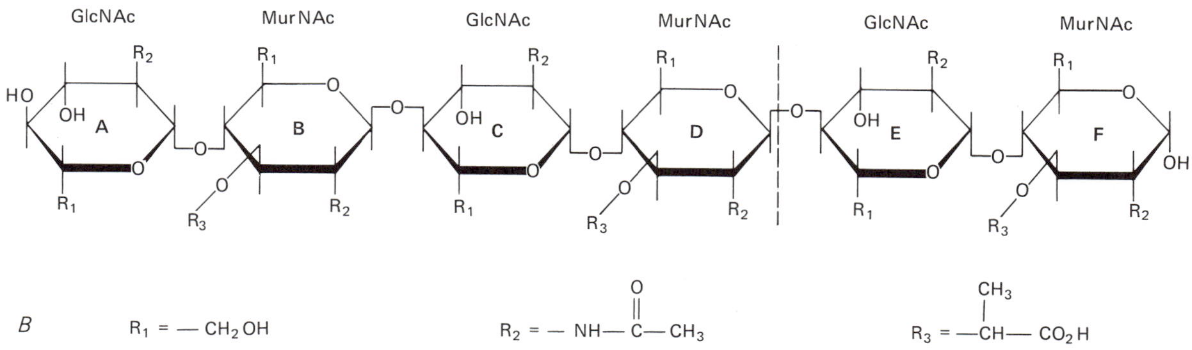

FIGURE 9.25

Proposed binding of a hexasaccharide to the active site of lysozyme. *A* shows six saccharide units (with thick bonds) bound to the subsites A–F of the protein. *B* depicts the structures of six saccharide residues of the normal substrate of lysozyme, an alternating polymer of *N*-acetyl-D-glucosamine (GlcNAc) and *N*-acetyl muramic acid (MurNAc). This substrate is normally hydrolyzed by lysozyme at the dashed line, when bound between subsites D and E. (Adapted from S. J. Perkins et al., *Biochem. J.* 193:553–572, 1981.)

positive charge onto the oxygen atom. In this way, lysozyme could catalyze hydrolysis of such a polysaccharide simply by being sterically complementary to the presumed sofa transition state. More detailed model-building analysis, however, did not support the suggestion that binding of the half-chair conformation would be more favorable, and the original proposal lost favor. Only recently has a detailed crystallographic study of substrate bound to site D found it to be distorted sterically in the way predicted by the original proposal. The history of lysozyme catalysis is one of the longest of any enzyme, and it makes fascinating reading as an example of how progress in understanding enzymes oscillates and converges only slowly to a consensus that may approximate the truth.

Studies of the binding of saccharides to the active site indicated that binding to subsite D is significantly less favorable than binding to the others (Table 9.6). Interactions with subsite C are the most extensive, and sugars have the greatest affinity for it. The change in overall binding affinity when adjacent residues are added to the saccharide gives the incremental binding energies of the additional group (see Eq. 8.11). As can be seen in Table 9.6, GlcNAc residues in subsite A, B, or C increase the affinity about 10^2-fold but only 3-fold when in subsite D. Binding of the alternative residue MurNAc to subsite D actually decreases the affinity 10^2-fold, indicating unfavorable interactions involving the lactyl side chain. Lactose and dehydro derivatives of GlcNAc, designed to mimic the transition state, bind only slightly more tightly in one case and less tightly in the other (Table 9.6). Close binding in subsite D is avoided unless very favorable interactions also occur between sugar residues in subsites E and F. In this case,

simultaneous binding to subsites A, B, C, E, and F seems to be sufficient to pull a residue into subsite D. Unfortunately, binding to subsites E and F has not been observed crystallographically because these sites are blocked by the crystal lattice. There are also no quantitative data for the incremental binding energies at these sites, although they are thought to be substantial.

The binding data indicate that binding of a normal substrate to subsite D is energetically unfavorable, as would be expected if the substrate had to be distorted by binding there (Eq. 9.61). The crystal structure of the complex shows some favorable interactions between the enzyme and the distorted substrate that should assist binding, but these seem not to compensate entirely for distorting the substrate.

In the active-site cleft are two acidic groups, Glu 35 and Asp 52, which should be in close proximity to the bond to be cleaved when a substrate is inserted fully into subsite D (Fig. 9.25). Replacing either of these residues with the corresponding amide virtually abolishes activity. In the free enzyme, Asp 52 is in a very polar environment and has a nearly normal pK_a value of 3.5 ± 0.5. In contrast, Glu 35 is in a hydrophobic environment and, probably as a consequence, has a substantially elevated pK_a value of 6.3 ± 0.2. Hen lysozyme is maximally active at pH 5, and its catalytic activity decreases in a manner consistent with a requirement that Asp 52 be ionized and Glu 35 not, comparable to the situation in the carboxyl proteases (Sec. 9.3.2.c).

Exactly how these two groups catalyze the reaction is a subject of contention, but it is generally agreed that the nonionized Glu 35 side chain tends to transfer its proton to the saccharide ring in subsite D, probably to the oxygen atom linking the rings in subsites D and E but

Table 9.6 *Energetics of Binding Saccharides to the Active Site of Hen Lysozyme[a]*

A	B	C	D	K_d (M)	Incremental binding contribution of group in italics (kcal/mol)
		GlcNAc		3.2×10^{-2}	
	GlcNAc -	GlcNAc		3.1×10^{-4}	-2.7
	GlcNAc -	GlcNAc -	*MurNAc*	7×10^{-4}	$+0.5$
GlcNAc -	GlcNAc -	GlcNAc		8.6×10^{-6}	-2.1
GlcNAc -	GlcNAc -	GlcNAc -	*(GlcNAc lactone)*	7.5×10^{-8}	-2.8
GlcNAc -	GlcNAc -	GlcNAc -	*XylNAc*	4.3×10^{-6}	-0.4
	MurNAc -	GlcNAc		1.0×10^{-4}	-3.4
GlcNAc -	MurNAc			9.0×10^{-5}	
GlcNAc -	MurNAc -	GlcNAc		3.2×10^{-6}	-2.0
GlcNAc -	MurNAc -	*GlcNAc*			-2.0
GlcNAc -	MurNAc -	GlcNAc -	*XylNAc*	7.1×10^{-7}	-0.9
GlcNAc -	MurNAc -	GlcNAc -	*GlcNAc*	1.1×10^{-6}	-0.6
GlcNAc -	MurNAc -	GlcNAc -	*MurNAc*	4.1×10^{-4}	$+2.9$
GlcNAc -	MurNAc -	GlcNAc -	$\Delta^{2,3}$*GlcNAc*	5.1×10^{-6}	$+0.3$

[a] Many discussions of the energetics of binding to the subsites of lysozyme use "unitary" free energies of binding, calculated from association constants and a saccharide concentration of 55 M. As discussed in Section 8.2, these values are not considered to have any special significance and should not be compared with the incremental free energies used here, which are calculated from the change in overall binding affinity produced by adding a single residue [i.e., $-RT \ln (K_d^A / K_d^{AB})$ for group B].

[b] Abbreviations: GlcNAc: N-acetyl-D-glucosamine; MurNAc: N-acetyl muramic acid; see Figure 9.25.

XylNAc:

GlcNAc lactone:

$\Delta^{2,3}$GlcNAc:

Data from M. Schindler et al., *Biochemistry* 16:423–431 (1977).

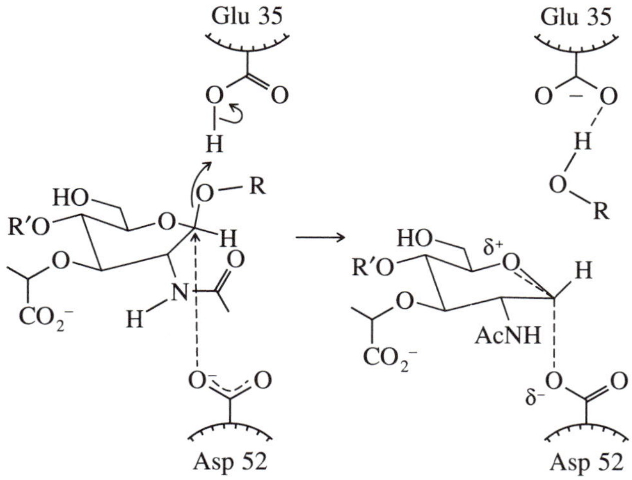

FIGURE 9.26
Plausible mechanism for lysozyme catalysis, based largely on the chemistry of model systems. Only residues Glu 35 and Asp 52 are shown, along with the one saccharide unit that occupies site D of the enzyme. R is the part of the oligosaccharide that occupies sites E and F and that dissociates first as the initial product, ROH. Only the first half of the reaction is shown. The reaction will be completed and the other half of the products generated if ROH is replaced by H_2O and the same reaction occurs in reverse. (From A. J. Kirby, *Crit. Rev. Biochem.* 22:283–315, 1987.)

perhaps to the ring oxygen instead. The ionized carboxyl group of Asp 52 would assist this process by stabilizing the resulting positive charge on the saccharide (Fig. 9.26). In this process, structural alterations to the saccharide ring must occur, and the enzyme active site seems to be much more complementary to the distorted substrate of the transition state than to the normal substrate (Eq. 9.61). On breaking the covalent bond linking the adjacent saccharide units, the half of the substrate held less tightly at the E and F subsites would dissociate. The overall reaction would be completed by the reverse of the first step, in which ROH would be HOH (Fig. 9.26).

Oligosaccharides have sufficient affinity for subsites E and F that they can compete with water in this last step and consequently be joined to the residual saccharide in subsite D, regenerating a glycosyl bond. This transglycosylation reaction, plus the many possible modes of substrate binding to the six subsites, complicates the kinetic study of lysozyme catalysis.

The lysozyme from phage T4 exhibits similar structural and binding properties; therefore, it may catalyze hydrolysis of the same substrates in a similar manner. This enzyme has no detectable amino acid sequence homology with the mammalian lysozymes, but it has barely sufficient structural similarities to suggest that they all have diverged from an ancient common ancestor.

References

Crystallographic studies of the activity of hen egg-white lysozyme. C. C. F. Blake et al. *Proc. Roy. Soc. Lond. [Biol.]* 167:378–388 (1967).

X-ray crystallography of the binding of bacterial cell wall trisaccharide NAM-NAG-NAM to lysozyme. J. A. Kelly et al. *Nature* 282:875–878 (1979).

Crystallographic determination of the mode of binding of oligosaccharides to T4 bacteriophage lysozyme: implications for the mechanism of catalysis. W. F. Anderson et al. *J. Mol. Biol.* 147:523–543 (1981).

Goose lysozyme structure: an evolutionary link between hen and bacteriophage lysozymes? M. G. Grütter et al. *Nature* 303:828–830 (1983).

Chemical mutations of the catalytic carboxyl groups in lysozyme to the corresponding amides. R. Kuroki et al. *J. Biol. Chem.* 261:13571–13574 (1986).

The roles of conserved aromatic amino-acid residues in the active site of human lysozyme. M. Muraki et al. *Biochim. Biophys. Acta* 916:66–75 (1987).

Does lysozyme follow the lysozyme pathway? An alternative based on dynamic, structural and stereoelectronic considerations. C. B. Post and M. Karplus. *J. Amer. Chem. Soc.* 108:1317–1319 (1986).

Mechanism and stereoelectronic effects in the lysozyme reaction. A. J. Kirby. *Crit. Rev. Biochem.* 22:283–315 (1987).

Site-directed mutagenesis of the catalytic residues Asp-52 and Glu-35 of chicken egg white lysozyme. B. A. Malcolm et al. *Proc. Natl. Acad. Sci. USA* 86:133–137 (1989).

Lysozyme revisited: crystallographic evidence for distortion of an *N*-acetylmuramic acid residue bound in site D. N. C. J. Strynadka and M. N. G. James. *J. Mol. Biol.* 220:401–424 (1991).

9.4 Regulation of Enzyme Activity

Enzymes are assumed to have evolved catalytic properties that are perhaps optimal, or at least suitable, for their physiological functions in vivo. This assumption raises the question of how the properties of an enzyme observed in vitro reflect its in vivo function. There is also the question of what catalytic parameters would be considered to be optimal evolutionarily for any particular enzyme. Living organisms do not ordinarily exist in a static environment, however, and they must be able to adapt to changes in their environment. For that reason, at least some enzymes, especially those catalyzing rate-determining steps in metabolism, must be able to respond to changes in availability of their substrates or in demand for their products.

Control of enzyme catalysis is accomplished by varying both the quantity of the enzyme present and its catalytic activity. Control by synthesis (Chap. 2) occurs most often in rapidly growing organisms such as bacteria; control via the rate of enzyme degradation is less common (Chap. 10). For most organisms in which the level of an enzyme is nearly constant, control is achieved primarily by varying the catalytic activity. The dependence of velocity on substrate concentration (Fig. 9.1) is one regulating factor for any enzyme, but the activity of an enzyme is usually modulated by the levels of metabolites other than its substrates and products.

References

The molecular basis for enzyme regulation, D. E. Koshland, Jr. Mechanisms of enzyme regulation in metabolism, E. R. Stadtman. Enzymes as control elements in metabolic regulation, D. E. Atkinson. In *The Enzymes*, 3rd ed., P. D. Boyer, ed., vol. 1, pp. 341–396, 398–459, and 461–489. Academic Press, New York, 1970.

The regulation of enzyme activity and allosteric transitions. E. Whitehead. *Prog. Biophys. Mol. Biol.* 21:321–397 (1970).

Regulation of enzyme activity. G. G. Hammes and G. W. Wu. *Science* 172:1205–1211 (1971).

Amino acid biosynthesis and its regulation. H. E. Umbarger. *Ann. Rev. Biochem.* 47:533–606 (1978).

Regulation of tryptophan biosynthesis. I. P. Crawford and G. V. Stauffer. *Ann. Rev. Biochem.* 49:163–197 (1980).

9.4.1 Enzyme Function in Vivo

It is straightforward to consider what properties the most efficient enzyme possible would have if it is assumed that the ultimate goal is to have the greatest catalytic capacity so that the smallest possible amount of enzyme is needed.

The impulsive conclusion is likely to be that such an enzyme would have the greatest conceivable value of k_{cat} and the lowest possible K_m. Such an enzyme, however, would be limited by the rate at which substrate molecules could diffuse to its active site, k_D (Sec. 9.1.1.a). In general, the value of k_{cat}/K_m cannot be greater than that of k_D. Increasing the value of k_{cat} beyond that at which binding of substrate becomes rate-limiting is unlikely to be beneficial because such increases in k_{cat} would have to be accompanied by corresponding increases in K_m. In general, an enzyme is unlikely to be more efficient than is functionally necessary; selective pressures would seem to be necessary to maintain the catalytic properties of an enzyme because most mutational events that affect its functional properties are likely to be detrimental.

On the basis of what is feasible, a "perfect" enzyme is considered to be one for which k_{cat}/K_m of the least stable substrate or product has the maximal value of k_D and in which no intermediate accumulates (Fig. 9.27). This requires that the diffusion-limited transition state be that with the highest free energy and that the free energies of all the enzyme complexes be somewhat greater than the free energy of the free enzyme and the most stable substrate. Considering the free enzyme E and the enzyme–substrate complex E · A, their relative free energies would depend on the concentration of the substrate and would be equal when the dissociation constant for the E · A complex is the same as the free substrate concentration. For E · A and E · B to have comparable free energies, it would be necessary for the enzyme to have different affinities for the substrate and the product to compensate for any difference in their relative free energies in solution, which determine the equilibrium constant for the reaction. The latter parameter is independent of the enzyme. These conditions are frequently found to be the case for actual enzymes, in that their K_m values for each substrate are similar to the normal in vivo concentration of that substrate, and the

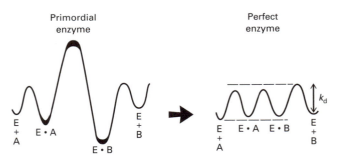

FIGURE 9.27

Hypothetical illustration of how enzymes would be expected to evolve to approach "perfection." The postulated primordial enzyme E is inefficient, in that it has a high free-energy barrier to catalyzing the reaction. It also has high affinities for the substrate A and product B, so the enzyme is usually present as a complex with one or the other, usually B in this case. In the perfect enzyme E, the transition state with the highest free energy would be limited by the rate of diffusion k_d of the less stable substrate, in this case B. The other transition states would be of lower free energy. Also, the free energies of the complexes at the physiological concentrations of A and B, and of any intermediates, would be higher than that of the free enzyme, so no complexed forms of the enzyme would be present at substantial levels. (Kindly provided by J. R. Knowles.)

most stable forms of the enzyme have similar free energies in such situations. At least some enzymes are thought to approach optimal efficiency; yeast triose phosphate isomerase is considered to be within 60% of being perfect, and evolutionary pressures are considered unlikely to be sufficient to produce a truly perfect enzyme.

Perfection in this sense applies only to one set of conditions, and virtually all enzymes are likely to encounter fluctuating concentrations of their substrates in vivo. Moreover, the substrate for one enzyme is usually the product of another enzyme, so the concentration of such a metabolite depends on the catalytic activity of both enzymes. A single enzyme should not be considered in isolation from the other components of metabolism.

Conditions in vivo are often very different from those usually used in studies of enzyme kinetics. For example, the enzymes of glycolysis, a central pathway of metabolism, are often present at high concentrations, with the concentration of each of the enzyme active sites in the range of 10^{-4} to 10^{-3} M. Such enzyme concentrations are greater even than the concentrations of some of their substrates, in particular the metabolic intermediates fructose-1,6-diphosphate, 1,3-diphosphoglycerate, and phosphoenol pyruvate, which are present at levels of only 10^{-4} M. Only the precursors and products of glycolysis, such as glucose-6-phosphate and ATP, occur at concentrations (from 10^{-3} to 10^{-2} M) that are greater than those of the enzymes. That the metabolic intermediates are present at lower concentrations than the enzymes that produce or use them suggests that the intermediates may not usually be free in solution but are present in enzyme-bound form.

The high in vivo concentrations of the enzymes makes it plausible that they interact physically and that metabolic intermediates may be passed directly from one enzyme to another without dissociating into the aqueous milieu. There is some experimental evidence that this can occur in vitro between certain pairs of enzymes at high concentrations, and the tendency for different forms of the enzyme to have similar stabilities (Fig. 9.27) can be explained as having evolved to maximize the probability of transfer of a metabolic intermediate between enzymes. On the other hand, the observed properties of individual enzymes in vitro have been found sufficient to simulate metabolism in vivo, assuming that the metabolites are free in solution and that there is no such direct transfer between enzymes.

The high concentrations of proteins inside cells and the existence of a molecular architecture in the cellular cytoplasm and organelles make it plausible that many enzymes exist as multimolecular aggregates that are not apparent after the cell is disrupted and the contents diluted. A number of multimolecular enzyme aggregates and polypeptides with multiple enzyme activities are known, and they may be simply the most stable examples. For example, the six enzymes involved in fatty acid synthesis are all part of a single large polypeptide chain in the mammalian enzyme system known as fatty acid synthase. In yeast, they are divided between two separate polypeptide chains, and in prokaryotes and plants the six enzymes occur on six individual polypeptide chains that can be separated by standard chromatographic procedures. Nevertheless, the individual enzymes are functionally very similar in all of these systems, and the mechanism of fatty acid chain elongation is the same. In each case, elongation occurs on a large multifunctional complex. During this process, the fatty acid is assembled while attached to the *acyl carrier protein*, which carries it from one active site to another.

Another notable example of variability in the association of the various enzyme activities of a biosynthetic pathway with different polypeptide chains is the pathway of tryptophan biosynthesis in microorganisms. Here, only one of the intermediates along the pathway, indole, is kept from dissociating into the aqueous solution. It is an intermediate between two consecutive enzyme activities that are part of the tryptophan synthase bifunctional complex. In some species, the two activities are part of the same polypeptide chain, whereas in others they reside on separate polypeptides, α and β, that aggregate into an $\alpha_2\beta_2$ complex. The most remarkable aspect is that the $\alpha_2\beta_2$ complex has a hydrophobic tunnel through which the indole intermediate is transferred 25–30 Å, from the active site on α where it is produced to the active site on β where it is used as a substrate.

It is probably necessary for indole to be channeled in this way because it is a very hydrophobic molecule that readily leaks through the cell membrane when released into the cytoplasm. The other intermediates in tryptophan biosynthesis are ionized or phosphorylated, as are most other metabolic intermediates, so they do not readily pass through membranes. The polar intermediates of tryptophan biosynthesis appear to be released into solution, even when produced and used by two sequential enzyme activities that are part of the same polypeptide chain. Why several enzyme activities of tryptophan biosynthesis occur on the same polypeptide chain or on different polypeptides that aggregate, and why this varies between species, is not known.

References

Perfection in enzyme catalysis: the energetics of triose phosphate isomerase. J. R. Knowles and W. J. Albery. *Acc. Chem. Res.* 10:105–111 (1977).

Metabolite transfer via enzyme–enzyme complexes. D. K. Srivastava and S. A. Bernhard. *Science* 234:1081–1086 (1986).

Biophysical chemistry of metabolic reaction sequences in concentrated enzyme solution and in the cell. D. K. Srivastava and S. A. Bernhard. *Ann. Rev. Biophys. Biophys. Chem.* 16:175–204 (1987).

Complexes of sequential metabolic enzymes. P. A. Srere. *Ann. Rev. Biochem.* 56:89–124 (1987).

Interpreting the behavior of enzymes — purpose or pedigree? S. Benner and A. D. Ellington. *Crit. Rev. Biochem.* 23:369–426 (1988).

Enzyme kinetics and molecular evolution. S. A. Benner. *Chem. Rev.* 89:789–806 (1989).

Why do many Michaelian enzymes exhibit an equilibrium constant close to unity for the interconversion of enzyme-bound substrate and product? G. Pettersson. *Eur. J. Biochem.* 195:663–670 (1991).

Structural basis for catalysis by tryptophan synthase. E. W. Miles. *Adv. Enzymol.* 64:93–172 (1991).

9.4.2 Allosteric Regulation

Regulation of enzyme activity could be imagined to occur by classical enzyme inhibition (Fig. 9.2) — that is, by a regulating metabolite competing directly with substrate for the active site — but this is not a common phenomenon. The two predominant ways in which enzyme activity is controlled are by allosteric interactions, with the regulatory metabolites binding to regulatory sites on the enzyme, and by covalent modification, which will be discussed later (Sec. 9.4.3).

In most metabolic pathways, the initial, rate-determining enzyme is inhibited allosterically by the end product of the pathway. This inhibition is exactly analogous to allosteric effects on binding affinity for ligands (Sec. 8.4.2), with the additional aspect that both K_m and k_{cat} may be affected. Allosteric interactions can be analyzed by their effects on enzyme activity in the same way as in ligand binding (see Fig. 8.24), but the additional complexity — that both affinity and catalytic activity may be affected — makes very tenuous any mechanistic deductions from analysis of enzyme activity alone. Physical and structural analysis is also required, and in no case is the mechanism of allosteric regulation of an enzyme fully elucidated. The two allosteric enzymes that are best understood are the bacterial phosphofructokinase and aspartate transcarbamoylase.

References

Kinetic co-operativity of monomeric mnemonical enzymes. The significance of the kinetic Hill coefficient. J. Ricard and G. Noat. *Eur. J. Biochem.* 152:557–564 (1985).

Co-operative and allosteric enzymes: 20 years on. J. Ricard and A. Cornish-Bowden. *Eur. J. Biochem.* 166:255–272 (1987).

a. Phosphofructokinsae

Phosphofructokinase (PFK) catalyzes the key control step of glycolysis:

$$\text{fructose 6-P} + \text{ATP} \longrightarrow \text{fructose 1,6-P}_2 + \text{ADP} \quad (9.62)$$

The PFK enzymes from eukaryotic sources are usually inhibited by ATP and by citrate and are activated by AMP, ADP, and cyclic AMP. The bacterial PFK enzymes are somewhat simpler, in that the eukaryotic proteins are twice the size as a result of gene elongation by duplication. The bacterial enzymes from various species, on the other hand, are more diverse in their regulatory properties. The most thoroughly studied PFKs from *Bacillus stearothermophilus* and *Escherichia coli* are inhibited by phosphoenolpyruvate (PEP), an end product of glycolysis, and are activated by ADP and GDP, by which they respond to the energy level of the cell. These regulatory effects are due to changes in the apparent K_m values for substrates, not to changes in k_{cat}. The allosteric nature of the regulation is illustrated by the sigmoidal dependence of the rate of the enzymatic reaction on the concentration of fructose 6-P, although not for the other substrate, ATP.

Phosphofructokinases are generally tetramers of identical polypeptide chains. The bacterial enzymes have approximately 320 residues in each chain, and each polypeptide chain is folded into two domains. Two polypeptide chains pack together with a large interface to form dimers (Fig. 9.28). Two dimers pack together with a smaller interface to form the tetramer.

Each active site is at the interface between the monomers of each dimer. ATP is bound predominantly to one of the domains of one monomer. Fructose 6-P, however, is bound by groups from both domains and from two different polypeptide chains across the smaller subunit interface. The allosteric effectors ADP and PEP bind to a common effector site that bridges two polypeptide chains in the same dimer, across the larger interface.

The allosteric effects on enzyme activity are consistent with the enzyme existing in either a T form with low affinity for fructose 6-P or an R form with high affinity, according to the concerted allosteric model (Sec. 8.4.2). The T state would have the lower affinity for fructose 6-P and for the activator ADP and the higher affinity for the inhibitor PEP. The crystal structures of the R and T forms of the enzyme demonstrate that they differ principally by a rotation of about 7° of one dimer relative to the other as a result of changes across the

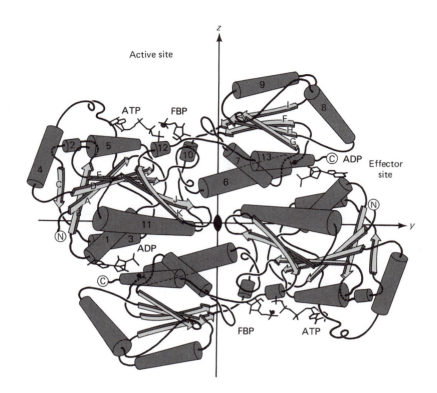

FIGURE 9.28

Schematic view of two subunits of the phosphofructokinase from *Bacillus stearothermophi-lus* that make up one dimer of the tetramer. The other dimer would be below, related by the *y* and *z* twofold symmetry axes. The dimers are similar in both the R and T states but differ primarily in their orientations relative to each other as a result of a rotation about the *x* axis (perpendicular to the page). Two active sites are shown with bound substrates ATP and fructose 6-P (F6P). The regulatory effector sites shown have bound activator ADP. The α-helices and β-strands of one subunit are labeled. (From P. R. Evans et al., *Philos. Trans. R. Soc. Lond. [Biol.]* 293:53–62, 1981.)

smaller interface. Less substantial changes take place in the individual dimers, but how these changes are related to ligand binding and to the changes in quaternary structure is not known. Nevertheless, the rotation of the dimers has a substantial effect on the binding site for fructose 6-P and explains why the affinity for this substrate is different in the two structures. In particular, one of the two Arg residues that interact with the phosphate group of fructose 6-P in the R state can no longer do so in the T state. Indeed, its three-dimensional position is taken in the T state by a Glu residue, which has the wrong charge to interact.

The allosteric activator ADP and the inhibitor PEP bind to the effector site in very similar ways, and this site is very similar in both the T and R states. One loop of the protein, however, moves to cover the bound effector, and its position depends on which effector is bound. This small structural difference is thought to determine whether the R state or the T state is stabilized, but the structural details are not clear.

The bacterial PFK is similar to hemoglobin (Sec. 8.4.3) in that the allosteric properties of both proteins

result primarily from the existence of two quaternary structures that have different affinities for various ligands. Only two quaternary states are thought to be possible in PFK, as in the case of hemoglobin (see Fig. 8.27), because an ordered layer of water molecules bridges two β-strands from different dimers in the R state. These water molecules are expelled in the T state, where the β-strands hydrogen-bond directly (Fig. 9.29). Such a layer of water molecules is thought to be either present or absent, although it conceivably could vary in size and in structure. With the PFK from the extreme thermophile *Thermus thermophilus*, the enzyme is inhibited by PEP due to dissociation of the tetramer into inactive dimers; the dissociation is reversed by fructose 6-P and by ADP.

References

Evolution of phosphofructokinase—gene duplication and creation of new effector sites. R. A. Poorman et al. *Nature* 309:467–469 (1984).

Structural basis of the allosteric behaviour of phosphofructo-

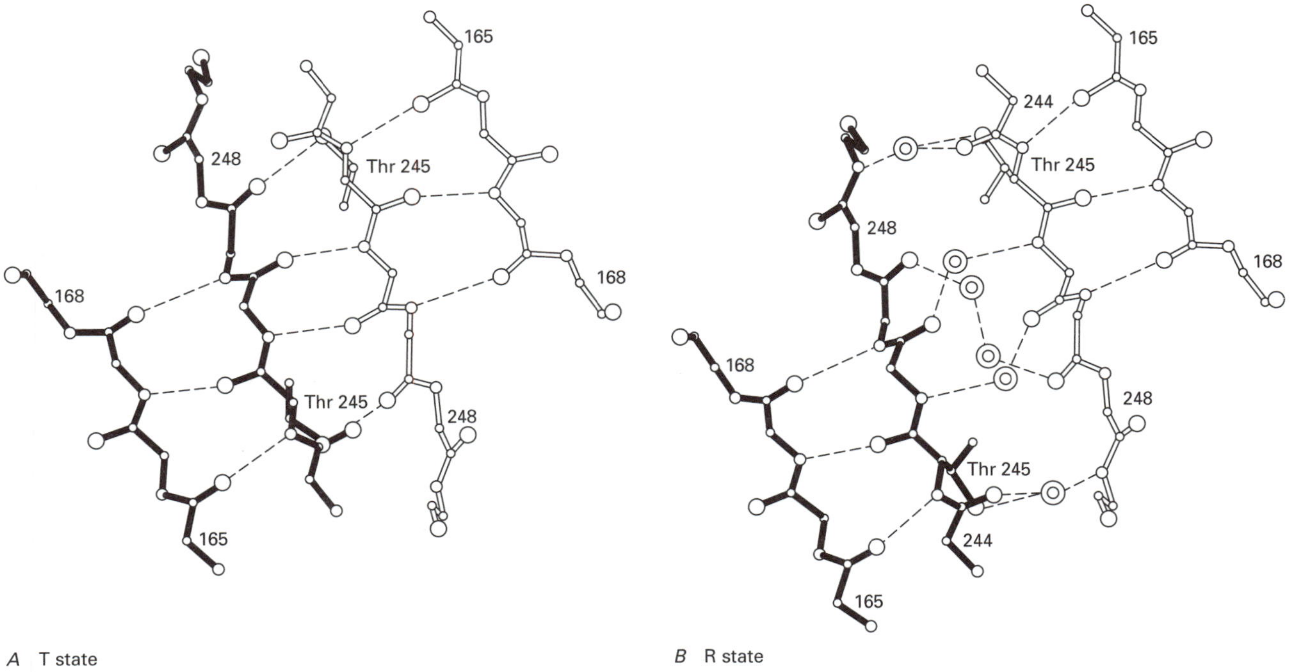

A T state   *B* R state

FIGURE 9.29

Interactions in the smaller interface between the two dimers of phosphofructokinase in the T state *(A)* and in the R state *(B)*. Two β-strands from opposite monomers are depicted, one with open bonds, the other with closed bonds. Normal hydrogen bonds are formed in the T state, so a continuous β-sheet is present. In the R state, however, the strands are separated and a layer of water molecules, depicted only by the oxygen atoms, hydrogen-bonds between the two sheets. (From T. Schirmer and P. R. Evans, *Nature* 343:140–145, 1990.)

kinase. T. Schirmer and P. R. Evans. *Nature* 343:140–145 (1990).

Tetramer–dimer conversion of phosphofructokinase from *Thermus thermophilus* induced by its allosteric effectors. J. Xu et al. *J. Mol. Biol.* 215:597–606 (1990).

b. Aspartate Transcarbamoylase

The aspartate transcarbamoylase (ATCase) of *E. coli* is the most thoroughly studied allosteric enzyme. It catalyzes the formation of carbamyl aspartate from carbamyl phosphate and aspartate:

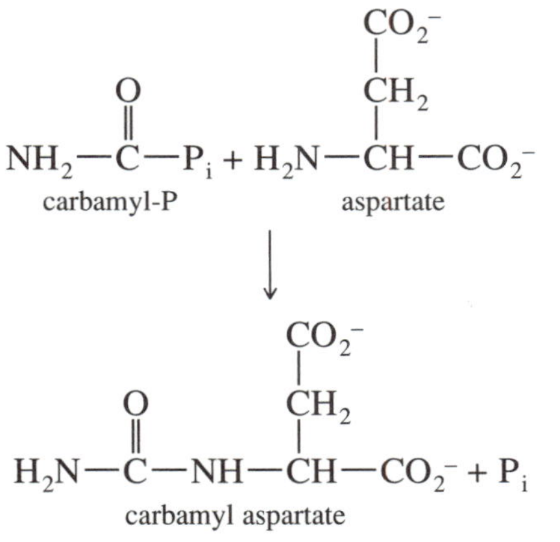

A

Carbamyl phosphate   L-aspartate

$$NH_2{-}\overset{\overset{\displaystyle O}{\|}}{C}{-}P_i + H_2N{-}\overset{\overset{\displaystyle CO_2^-}{|}\,\overset{\displaystyle CH_2}{|}}{CH}{-}CO_2^-$$

carbamyl-P   aspartate

$$H_2N{-}\overset{\overset{\displaystyle O}{\|}}{C}{-}NH{-}\overset{\overset{\displaystyle CO_2^-}{|}\,\overset{\displaystyle CH_2}{|}}{CH}{-}CO_2^- + P_i$$

carbamyl aspartate   (9.63)

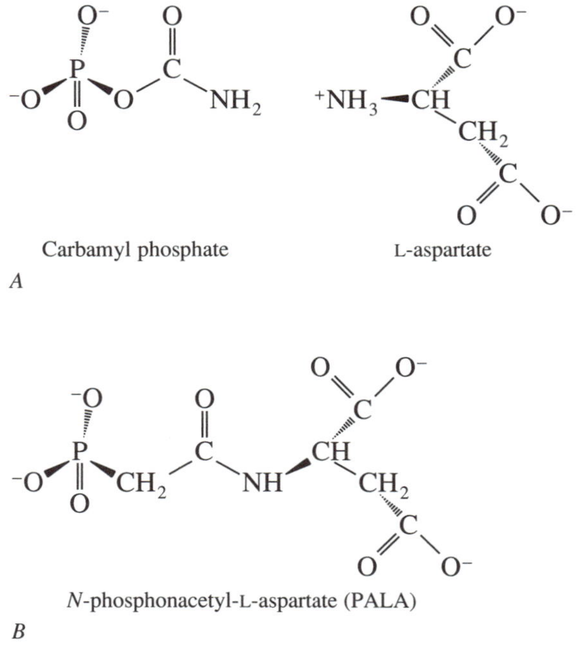

N-phosphonacetyl-L-aspartate (PALA)

B

FIGURE 9.30

The structures of the substrates of aspartate transcarbamoylase (carbamyl phosphate and aspartate) *(A)* compared with the structure of the bisubstrate analogue PALA *(B)*.

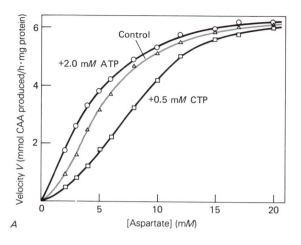

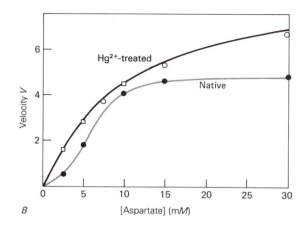

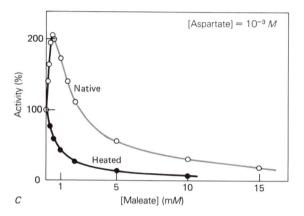

FIGURE 9.31

Steady-state kinetic behavior of aspartate transcarbamoylase. The velocity of the enzyme-catalyzed reaction is measured by the rate at which the product carbamylaspartate (CAA) is produced. *A:* The sigmoidal dependence of enzyme velocity V on the concentration of aspartate, at a fixed concentration of the other substrate, carbamyl-P (3.6 mM). CTP lowers the apparent affinity for aspartate and increases the cooperativity, whereas ATP has the opposite effect. *B:* The kinetic behavior of native and of mercuric ion–treated enzyme. The treated enzyme is dissociated into catalytic trimers and regulatory dimers. The kinetic response of the treated enzyme to aspartate concentration is hyperbolic (i.e., normal Michaelis–Menten), and the value of V_{max} is increased. *C:* The effect of the inhibitor maleate, which competes with aspartate, on the enzymatic activity of native and heat-dissociated aspartate transcarbamylase. With the dissociated enzyme, maleate acts as a normal competitive inhibitor, but it activates the native enzyme at low concentrations of both maleate and aspartate. The inhibitory effect of maleate binding at one or a few of the six active sites on the native enzyme must be more than compensated by an allosteric activating effect on the remaining active sites, increasing their affinity for aspartate. (From J. C. Gerhart, *Curr. Top. Cell Reg.* 2:275–325, 1970; J. C. Gerhart and A. B. Pardee, *Cold Spring Harbor Symp. Quant. Biol.* 28:491–496, 1963.)

The carbamyl group is transferred directly from carbamyl-P to aspartate on the enzyme. Binding of the substrates tends to be ordered, with carbamyl-P binding first 90% of the time. A bisubstrate analogue known as PALA (Fig. 9.30) binds at least 10^3 times more tightly than either single substrate.

Carbamyl aspartate is the first metabolite that is unique to the biosynthetic pathway for pyrimidine nucleotides. Consequently, the synthesis of ATCase is repressed by one product of this pathway, uracil, and the enzymic activity of ATCase is inhibited by the end products CTP and UTP. ATCase is activated by ATP, presumably to coordinate pyrimidine and purine nucleotide synthesis. These allosteric regulators affect primarily K_m, not V_{max} (Fig. 9.31A). The enzyme does not obey normal Michaelis–Menten kinetics in that the dependence of enzyme velocity on the concentration of either substrate is sigmoidal rather than hyperbolic. Homotropic interactions accompany heterotropic interactions, as occurs so often in allosteric systems (Sec. 8.4). CTP and ATP compete for the same site, but their binding produces opposite effects on homotropic cooperativity and on the apparent K_m values for the substrates (Fig. 9.31A). A striking consequence of the homotropic interactions is that substrate analogues that are competitive with the substrate aspartate and undoubtedly bind at the active site activate the enzyme at low concentrations in the presence of carbamyl phosphate (Fig. 9.31C). Such inhibitors at high concentrations display the usual net inhibition. The activation at low concentrations is a result of the allosteric effect on the other active sites of the same enzyme molecule produced by the inhibitor binding at just one or a few sites. This activating effect on the other sites more than compensates for the inhibiting effect on the sites to which the inhibitor is bound.

The enzyme consists of six copies of each of two polypeptide chains, noted here as c and r because one is primarily catalytic, the other regulatory. The c_6r_6 complex can be dissociated reversibly by mild treatments with heat, mercurials, or urea into two c_3 trimers (often designated C) and three r_2 dimers (or R):

$$c_6r_6 \rightleftharpoons 2c_3 + 3r_2 \quad \text{or} \quad C_2R_3 \rightleftharpoons 2C + 3R \tag{9.64}$$

The dissociated enzyme is somewhat more active than the original enzyme and has lost all its allosteric properties (Fig. 9.31B,C). The r_2 dimers have no catalytic activity but bind the allosteric effectors. Many studies with modified c and r chains and with various reconstituted complexes have demonstrated that both the homotropic and heterotropic allosteric interactions depend on the presence of both of the polypeptide chains — one catalytic, the other regulatory — assembled into the large quaternary structure. The catalytic trimer usually exhibits normal enzyme kinetics, although that of the Arg 105 → Ala mutant form exhibits cooperativity. The catalytic properties of reconstituted, hybrid complexes are determined largely by the c polypeptide chains and the regulatory properties by the r chains.

The crystal structures of ATCase show two rings of three c subunits associated face-to-face, with three r_2 dimers linking the two rings (Fig. 9.32). Both the c and r polypeptide chains are composed of two domains. Each of the domains of the c polypeptide chain binds one of the two substrates, carbamyl-P and aspartate, and the active site is situated between the two domains of adjacent polypeptide chains in the trimer. One of the two domains of the r polypeptide chain binds Zn^{2+}, the other binds either of the allosteric effectors ATP and CTP. The

Zn^{2+} ion is tetrahedrally chelated by the sulfhydryl groups of four Cys residues and is thought to play primarily a structural role. The Zn^{2+} domain plays a crucial role in linking the allosteric domain to the catalytic polypeptide chain. Mercurials dissociate the c_6r_6 complex (Fig. 9.31) because they displace the Zn^{2+} ion. The allosteric effectors bind to the r subunit on the periphery of the enzyme complex, some 60 Å from the active site.

The allosteric mechanism involves changes in quaternary structure in this enzyme also. The homotropic part of the allosteric mechanism of ATCase is usually treated as involving the usual R and T states of the concerted mechanism (Sec. 8.4.2). The unliganded enzyme is designated the low-affinity T state. On binding both substrates, or the bisubstrate analogue PALA, the enzyme increases its hydrodynamic volume substantially, due to a complex change in quaternary structure to that of the R state (Fig. 9.33). The X-ray diffraction studies show that the catalytic dimers reorient about the threefold axis by 10° and move apart; the regulatory dimers rotate about each twofold axis by 15°. As a result, the molecule elongates along its threefold axis by 11 Å. These conformational changes are almost entirely movements of rigid domains relative to each other.

A single molecule of PALA binding to just one of the six active sites on ATCase causes the quaternary structure to change from R to T when ATP is also present. As a result, the five other catalytic sites are converted to the high-affinity form, explaining the homotropic allosteric effects. That the transition from T to R is primarily concerted was shown by reconstituting ATCase from its dissociated polypeptide chains in which some of the c chains were inactivated by chemical modification. Hybrid forms of ATCase containing two native and four inactivated c chains were compared when the two active chains were part of the same c_3 trimer or different trimers. The two isomeric hybrids exhibited identical catalytic and regulatory behavior, indicating that the cooperative unit is the entire ATCase molecule, not some smaller part. Hybrid molecules could also be used to demonstrate that the structural alterations produced by substrate binding to some c subunits are experienced by other subunits. For this purpose, one of the c_3 trimers of the ATCase hybrid was chemically inactivated so that it could not bind aspartate or its analogues and nitrated (Sec. 1.3.9) to introduce a nitrotyrosyl chromophore that is sensitive to the protein conformation. The nitrotyrosyl spectrum of the hybrid ATCase changed when substrate analogues were bound at the other c_3 trimer.

The structures of the R and T states suggest a plausible stereochemical basis for the homotropic cooperativity, which is also supported by the properties of many mutant forms of the enzyme. Binding of carbamyl phos-

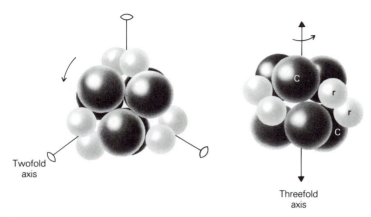

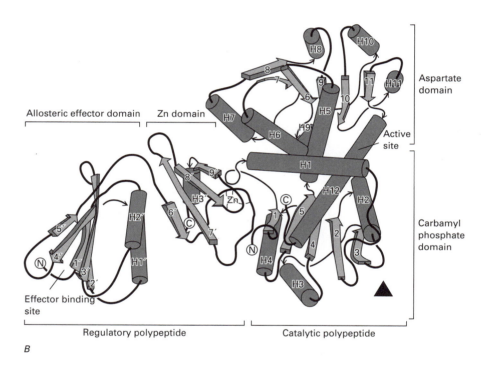

FIGURE 9.32

The structure of aspartate transcarbamoylase. *A:* Two views of the overall quaternary structure in the T state. *Left:* A view down the threefold axis, showing a trimer of three catalytic subunits (large spheres) lying above a second trimer just barely visible below it. The two trimers are linked on the outside by three dimers of regulatory subunits (small spheres). The arrow indicates the direction in which the top trimer is rotated upon conversion to the R state. Three of the six twofold axes relating the top and bottom halves are indicated. *Right:* A view down one twofold axis. The arrow indicates the direction in which the top trimer rotates upon conversion to the R state; the trimers also move apart along the threefold axis. *B:* The tertiary structure of one regulatory r and one catalytic c polypeptide chain. The view is down the threefold axis, indicated by the solid triangle. Each polypeptide chain is composed of two domains. The active site is between two c domains, each of which binds predominantly one of the substrates; the adjacent c chain of the trimer also contributes residues to the active site. The Zn^{2+} ion bound to the r chain is thought to have solely a structural role. The allosteric effectors ATP and CTP bind to the periphery of the r chain, some 60 Å from the active site. (From R. C. Stevens et al., *Biochemistry* 29:7691–7701, 1990.)

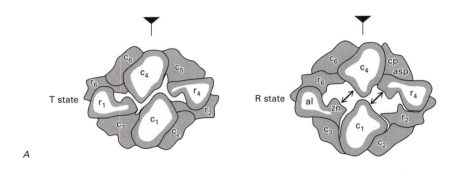

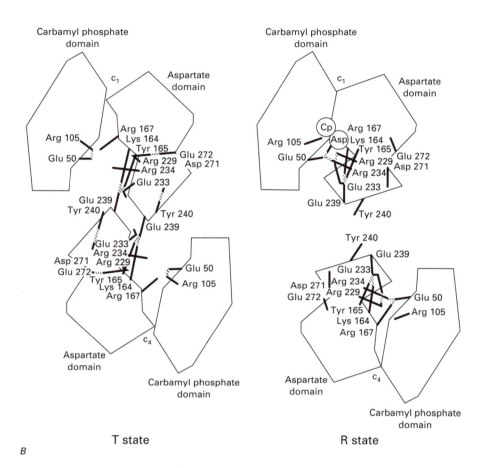

FIGURE 9.33

The differences in the quaternary structures of the T and R states of aspartate transcarbamoylase and its consequences for the active site. *A:* Schematic illustration of the quaternary structure. The view is down one of the twofold axes, with the threefold axis vertical. Catalytic subunits c_1 and c_4 are the pair from different trimers that are in direct contact in the enzyme complex (Fig. 9.32A); identical structures are present in the c_2–c_5 and c_3–c_6 pairs. In the other subunits that are visible, al and zn are the allosteric effector and zinc-binding domains of subunit r_1, and cp and asp are the carbamyl phosphate and aspartate domains of subunit c_5. *B:* Local conformational changes that are thought to link the change in quaternary structure and the catalytic activity of the active site. The substrates carbamyl phosphate (Cp) and aspartate (Asp) are bound to the active site of subunit c_1, and the quaternary structure change has altered the active site of subunit c_4. (From J. E. Gouaux et al., *Proc. Natl. Acad. Sci. USA* 86:8212–8216, 1989; E. R. Kantrowitz and W. N. Lipscomb, *Science* 241:669–674, 1988.)

phate to the active site induces a local conformational change that increases the affinity for aspartate. When both substrates are bound, the two domains of adjacent c subunits move closer together. This change is believed to be facilitated when the quaternary structure change also takes place, because the active site is connected to a loop at the interface between c subunits in different trimers (Fig. 9.33). Mixing two types of catalytic subunits that had been inactivated in different ways produced hybrid ATCase molecules that possessed some catalytic activity as a result of the generation of composite active sites with no modification (Fig. 9.34).

The structural basis of the opposing effects of the allosteric effectors CTP and ATP is less clear. Both bind to the effector site of an individual r subunit in very similar ways. Their triphosphate and ribose moieties make similar interactions; only their bases are bound somewhat differently. Perhaps most significantly, binding of ATP causes the base subsite to expand more than does binding of CTP. How this small difference might be communicated to the remainder of the ATCase molecule to produce opposite allosteric effects is not clear. There are structural differences in the r subunits of the R and T states, particularly in the relative orientations of their two domains. ATP and CTP probably affect primarily the R $\rightleftharpoons$ T equilibrium. Certainly, binding of ATP and CTP have the expected opposite effects on the affinities of substrates and analogues that bind at the active site; they also have opposite structural effects on the c subunits, as observed using the nitrated form. One observation that is difficult to explain on the basis of the two-state allosteric model is that both ATP and CTP bind to the two r effector sites with negative cooperativity. This is thought to result from direct changes in one r subunit upon effector binding to the other r subunit of the r_2 dimer unit, independent of any quaternary structure change. Furthermore, it is not certain that there are only two unique quaternary structures, as occurs in hemoglobin (see Fig. 8.27) and in phosphofructokinase (Fig. 9.29). There are reports of intermediate quaternary structures with ATCase, although the subject is controversial.

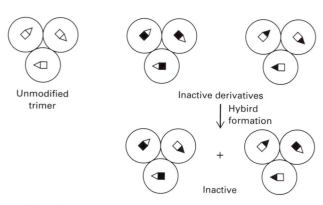

A   Active sites within monomers, unshared

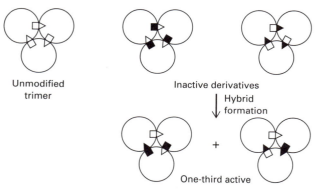

B   Active sites at monomer interfaces, shared

FIGURE 9.34

Rationale of the use of hybrid oligomers to determine whether pairs of residues contribute to the active site within (A) or between (B) polypeptide chains. Large circles designate individual c polypeptide chains that are associated into c_3 trimers. Squares and triangles indicate the two residues that are being tested; open symbols represent native residues, and closed symbols are residues that have been altered so as to inactivate any active site to which they contribute. Both homogeneous populations in which one of the pair of residues has been altered are inactive. Mixing the dissociated monomers and reconstituting these randomly into trimers generates active hybrids only if the active site uses the two residues from different polypeptide chains. In this way, Lys 84, Ser 52, and His 134 of ATCase were shown to contribute to an active site that is shared between two polypeptide chains (lower example, S. R. Wente and H. K. Schachman, *Proc. Natl. Acad. Sci. USA* 84:31–35, 1987.) (From E. A. Robey and H. K. Schachman, *Proc. Natl. Acad. Sci. USA* 82:361–365, 1985.)

References

Shared active sites in oligomeric enzymes: model studies with defective mutants of aspartate transcarbamoylase produced by site-directed mutagenesis. S. R. Wente and H. K. Schachman. *Proc. Natl. Acad. Sci. USA* 84:31–35 (1987).

Can a simple model account for the allosteric transition of aspartate transcarbamoylase? H. K. Schachman. *J. Biol. Chem.* 263:18583–18586 (1988).

Crystal structure of the Glu-239 → Gln mutant of aspartate carbamoyltransferase at 3.1 Å resolution: an intermediate quaternary structure. J. E. Gouaux et al. *Proc. Natl. Acad. Sci. USA* 86:8212–8216 (1989).

Heterotropic effectors promote a global conformational change in aspartate transcarbamoylase. E. Eisenstein et al. *Biochemistry* 29:3724–3731 (1990).

Crystal structures of aspartate carbamoyltransferase ligated with phosphonoacetamide, malonate and CTP or ATP at 2.8 Å resolution and neutral pH. J. E. Gouaux et al. *Biochemistry* 29:7702–7715 (1990).

Escherichia coli aspartate transcarbamoylase: the molecular basis for a concerted allosteric transition. E. R. Kantrowitz and W. N. Lipscomb. *Trends Biochem. Sci.* 15:53–59 (1990).

Escherichia coli aspartate carbamoyltransferase: the probing of crystal structure analysis via site-directed mutagenesis. R. C. Stevens et al. *Protein Eng.* 4:391–408 (1991).

9.4.3 Reversible Covalent Modification

Another level of control of enzyme activity, which acts upon various types of signals, uses reversible covalent modification of the enzyme. The modification most frequently used is phosphorylation (Sec. 2.4.8). In most cases, the modification is catalyzed by one enzyme and is removed by a second. These two reactions are not simply the reverse of each other. Instead, the sum of the two reactions involves a favorable net reaction, so both are energetically favorable with insignificant reversibility. For example, phosphorylation of an enzyme generally uses the γ phosphate group of ATP, whereas removal of the phosphate group is by simple hydrolysis:

$$\begin{aligned} \text{phosphorylation:} \quad & E \quad + ATP \longrightarrow E-P + ADP \\ \text{dephosphorylation:} \quad & E-P + H_2O \longrightarrow E \quad + P_i \\ \hline \text{net:} \quad & ATP + H_2O \longrightarrow ADP + P_i \end{aligned}$$

$$(9.65)$$

The net result is hydrolysis of ATP, so both reactions are favorable energetically. The two reactions must also be regulated, however, to prevent wasteful cycling. The potential for regulating the two reactions independently is undoubtedly the reason that this method is used throughout metabolism.

Reversible phosphorylation of proteins is frequently one step in the action of hormones. The hormone or other physiological effector usually binds to a receptor, which triggers a signaling system that increases or decreases the level of a second messenger such as cyclic AMP, Ca^{2+} ion, or inositol phosphates. The second messenger then activates or inhibits various protein kinases that phosphorylate certain target proteins. The target proteins are sometimes protein phosphatases, which remove the phosphoryl groups. Such a system involves a substantial amplification of the original physiological signal. Nanomolar levels of hormones produce micromolar concentrations of secondary messengers; each molecule of activated protein kinase or phosphatase catalyzes the phosphorylation or dephosphorylation of many molecules of the target protein. If this protein is an enzyme, changes in its activity can produce substantial changes in the levels of its substrates and products. Even greater effects can be obtained when there is a cascade in which one protein kinase or phosphatase acts on another kinase or phosphatase before modification of the final target protein.

Second messengers such as cyclic AMP generally bind directly to the protein kinase or phosphatase and either activate or inhibit it. The effects of Ca^{2+} ion, however, are often transmitted by calmodulin or other members of the same family of Ca^{2+}-binding proteins (Sec. 8.3.4.a). In the absence of Ca^{2+}, calmodulin is usually inactive. With Ca^{2+} bound, calmodulin interacts with a variety of receptor proteins; some of these proteins are specific calmodulin-dependent protein kinases.

Phosphorylation of an enzyme frequently changes its K_m for a substrate or its affinity for allosteric effectors. Substrates, activators, and inhibitors of the enzyme may also affect the rate at which it is phosphorylated or dephosphorylated. Although a bewildering array of phosphorylation–dephosphorylation reactions are involved in the control of metabolism, some generalities have emerged. Enzymes involved in degradation pathways are usually activated by phosphorylation, whereas most enzymes involved in biosynthetic processes are inactivated by this modification. There are relatively few protein kinases and phosphatases, and each of them usually modifies a number of target enzymes. In some cases, the same target protein can be phosphorylated at the same site or at different sites by various kinases. Phosphorylation of different sites of the same protein can have cooperative or antagonistic consequences. Some kinases appear to be specific for a single target protein. In such cases, the kinase often acts progressively and adds several phosphate groups to each target protein molecule.

The central role of protein kinases in cellular regulation is illustrated by many of them being encoded by proto-oncogenes, genes that when mutated in certain ways contribute to the transformation of a cell and the appearance of the cancerous condition. Such mutations in these genes alter the regulation of the kinase activity of the protein encoded by the gene.

Protein kinases are a structurally diverse group of proteins. They have modular structures that differ widely in size, subunit structure, subcellular localization, mechanism of activation, and substrate specificity. Two general classes exist; those that phosphorylate Ser and Thr residues and those that phosphorylate Tyr residues. Nevertheless, all eukaryotic protein kinases have catalytic domains that are related evolutionarily and are recognizable from just their amino acid sequence. The noncatalytic regions of various kinases show no homology and are important for the specific localization and regulation of each kinase. Regulatory domains may be part of the same polypeptide chain as the kinase domain, or the regulatory part may be on a different polypeptide chain that associates with the kinase catalytic subunit. In many cases, the regulatory domains seem to inhibit kinase activity by interacting with the kinase active site as a pseudosubstrate; they have short sequences of residues that resemble the specific sequence in the target proteins recognized by the kinase. When this sequence contains a phosphorylatable residue, this residue is often phosphorylated by the kinase active site, which alters the affinity of the regulatory domain for the catalytic domain. The autoinhibitory domain can be removed and the inhibition reversed by an activator of the kinase. A nonphysiological mechanism of reversing the inhibition is to remove the autoinhibitory domain proteolytically.

The protein phosphatases that remove phosphate groups from proteins are not as well characterized as the kinases. There seem to be fewer phosphatases than kinases, with wider specificities. On the other hand, these few phosphatases associate with various other proteins to generate a number of different forms. Their phosphatase activities can be regulated by phosphorylation and by the presence of cAMP and Ca^{2+}.

Besides phosphorylation, several other reversible regulatory modifications have been observed to alter enzyme activity: nucleotidylation of Tyr residues; ADP ribosylation, primarily of Arg residues; methylation of Glu or Asp carboxyl groups; and acetylation of Lys amino groups. The chemical natures of these reactions are described in Chapter 2 (Sec. 2.4).

Three of the most thoroughly studied systems will be described in some detail to illustrate some of the finer details.

References
The role of protein phosphorylation in the hormonal control of enzyme activity. P. Cohen. *Eur. J. Biochem.* 151:439–448 (1985).

Protein-tyrosine kinases. T. Hunter and J. A. Cooper. *Ann. Rev. Biochem.* 54:897–930 (1985).

Protein kinases in the brain. A. C. Nairn et al. *Ann. Rev. Biochem.* 54:931–976 (1985).

Control by phosphorylation. *The Enzymes*, 3rd ed., P. D. Boyer and E. G. Krebs, eds., vol. 17. Academic Press, New York, 1986.

Protein kinase activity of the insulin receptor. S. Gammeltoft and E. van Obberghen. *Biochem. J.* 235:1–11 (1986).

Protein serine/threonine kinases. A. M. Edelman et al. *Ann. Rev. Biochem.* 56:567–613 (1987).

A thousand and one protein kinases. T. Hunter. *Cell* 50:823–829 (1987).

Growth factor receptor tyrosine kinases. Y. Yarden and A. Ullrich. *Ann. Rev. Biochem.* 57:443–478 (1988).

The protein kinase C family: heterogeneity and its implications. U. Kikkawa et al. *Ann. Rev. Biochem.* 58:31–44 (1989).

The structure and regulation of protein phosphatases. P. Cohen. *Ann. Rev. Biochem.* 38:453–508 (1989).

cAMP-dependent protein kinase. Model for an enzyme family. S. S. Taylor. *J. Biol. Chem.* 264;8443–8446 (1989).

cAMP-dependent protein kinase: framework for a diverse family of regulatory enzymes. S. S. Taylor et al. *Ann. Rev. Biochem.* 59:971–1005 (1990).

Protein kinases. Regulation by autoinhibitory domains. T. R. Soderling. *J. Biol. Chem.* 265:1823–1826 (1990).

a. Glycogen Phosphorylase

Glycogen phosphorylase has the vital role of controlling the metabolism of the storage polysaccharide glycogen, or $(\alpha\text{-}1,4\text{-glucoside})_n$, by catalyzing the first step in its degradation:

$$(\alpha\text{-}1,4\text{-glucoside})_n + P_i \rightleftharpoons$$
$$(\alpha\text{-}1,4\text{-glucoside})_{n-1} + \text{glucose-1-P} \quad (9.66)$$

The glucose-1-P produced is used in muscle primarily to provide the energy needed for contraction and in the liver to maintain blood-sugar levels.

Phosphorylase normally occurs in two forms, designated *a* and *b*, which differ covalently only in that the former has Ser 14 of the 842-residue polypeptide chain phosphorylated. Yet the two forms of the enzyme differ markedly in their allosteric regulatory properties. The *b* form is activated by AMP or IMP and is inhibited by ATP, ADP, glucose, and glucose 6-P; in this way, it responds to the energy requirements of the cell. The *a* form escapes the controls by these metabolites but is inhibited by glucose; consequently, this form of the enzyme degrades glycogen even if the cell is not energy deficient.

In muscle, the conversion of phosphorylase from the *b* to the *a* form is caused by muscle contraction and by stimulation by the hormone adrenaline (also known as epinephrine). The peptide hormone glucagon plays a comparable role in the liver. These stimuli are signals that energy and glucose are going to be required in the near future and that phosphorylase should be activated, irrespective of the current energy state of the cell.

The hormones adrenaline and glucagon act by stimulating production of cyclic AMP, the common second messenger. Cyclic AMP is formed from ATP by adenylate cyclase, a membrane-bound complex of at least three proteins. One, on the outer surface of the membrane, is the receptor for the specific hormone. Another, on the inner side of the membrane, is the catalytic protein, which produces cyclic AMP from ATP in the cell, but only when hormone is bound to the receptor. Communication between these two proteins is mediated by the third, a G protein or GTPase, but only when it has GTP bound. The interactions among these proteins are intricate and only partly understood.

The increased level of cyclic AMP caused by hormone binding to the cell surface activates two different cAMP-dependent kinases. In the absence of cyclic AMP, both proteins are catalytically inactive. Both are composed of two regulatory and two catalytic subunits, R_2C_2; they differ only in the R chains. Cyclic AMP dissociates this complex by binding more tightly to the free regulatory dimer R_2 and thereby releases active catalytic subunits:

$$\underset{\text{inactive}}{R_2C_2} + 4cAMP \rightleftharpoons \underset{\text{active}}{[R(cAMP)_2]_2 + 2C} \qquad (9.67)$$

The R subunit has a substratelike sequence that is essential for interacting with the C subunit. It has two cAMP-binding domains that are not equivalent, and cAMP binds cooperatively to these two sites.

Activation of the cAMP-dependent protein kinase is also regulated by covalent modification and allosteric regulation. One type of R subunit can be phosphorylated by the catalytic subunit at its substratelike site, thereby increasing its tendency to dissociate. With the other type of regulatory subunit, ATP binds tightly and cooperatively to the R_2C_2 form, thereby inhibiting its tendency to be dissociated and activated by cyclic AMP.

The active cAMP-dependent kinases phosphorylate a number of different enzymes, each on accessible Ser (or Thr) residues occurring in the sequence -Lys-Arg-X-X-Ser or Arg-X-Arg-X-Ser. X can be almost any amino acid, and Thr can replace the Ser residue. As a consequence of these phosphorylations, degradative enzymes such as glycogen phosphorylase are activated, and biosynthetic enzymes such as glycogen synthetase are inhibited. These regulatory effects often are not direct but occur via intermediary enzymes in cascades. For example, cAMP-dependent kinases directly phosphorylate phosphorylase kinases, not glycogen phosphorylase. The phosphorylated phosphorylase kinase then phosphorylates phosphorylase *b*.

Phosphorylase kinase is a complex protein consisting of four copies of each of four different polypeptide chains. Two types of these chains can each be phosphorylated on a particular Ser side chain. Phosphorylation of one chain increases the rate of phosphorylation of the other. The kinase activity of the complex is apparent only if the first subunit is phosphorylated and if another subunit, calmodulin, has bound Ca^{2+} ion. This makes activation of glycogen phosphorylase sensitive to muscle contraction, which is triggered by Ca^{2+} ion release. The activated phosphorylase kinase then specifically phosphorylates Ser 14 of phosphorylase *b*.

All of these activation steps are balanced by reverse steps that are catalyzed by phosphatases. Under normal conditions, a steady-state condition of phosphorylation is probably balanced by dephosphorylation, even though ATP is hydrolyzed in the process. The adenylate cyclase is regulated by its middle component hydrolyzing its bound GTP and thereby temporarily stopping the hormone from causing cyclic AMP formation. Existent cyclic AMP is hydrolyzed to AMP by a specific phosphodiesterase, thereby favoring inhibition of the cAMP-dependent kinase by aggregation with its regulatory subunits (the reverse of Eq. 9.67). The two phosphoryl groups on phosphorylase kinase are removed by two different phosphatases; the phosphoryl group that was added first is also removed first. The responsible phosphatase also removes the phosphoryl group on Ser 14 of phosphorylase *a* and on glycogen synthetase. The rate at which it acts on Ser 14 probably depends on the conformational state of phosphorylase *a*. The phosphatase appears to bind tightly to phosphorylase *a* but to cleave the phosphoryl group only when it becomes accessible. Therefore, this covalent modification is also susceptible to allosteric control in the enzyme.

The activities of these two phosphatases are also regulated directly, one by calmodulin and Ca^{2+} binding and the other by two protein inhibitors. One of these inhibitors is active only if phosphorylated by cAMP-dependent protein kinase. This amplifies the effect of the hormone because the inhibitor is activated and then inhibits the phosphatase; consequently, the rate of inactivation of phosphorylase *a* is decreased, which tends to increase the rate of glycogen breakdown.

The structural effects of phosphorylation of Ser 14 of glycogen phosphorylase are known from the crystal structures of the proteins. Phosphorylases *a* and *b* have very similar dimeric structures (Fig. 9.35). In the absence of the phosphate group on Ser 14, the N-terminal 16 residues are disordered. When Ser 14 is phosphorylated, these residues fold into a distorted 3_{10}-helix that is bound to the rest of the protein. In doing so, the N-terminal residues displace four C-terminal residues, which become disordered.

The phosphate group of Ser 14 interacts in the ordered conformation with two Arg side chains, one from the other subunit of the dimer, and the movements undergone by these Arg residues cause slight rearrangements in the interface between the two subunits. There

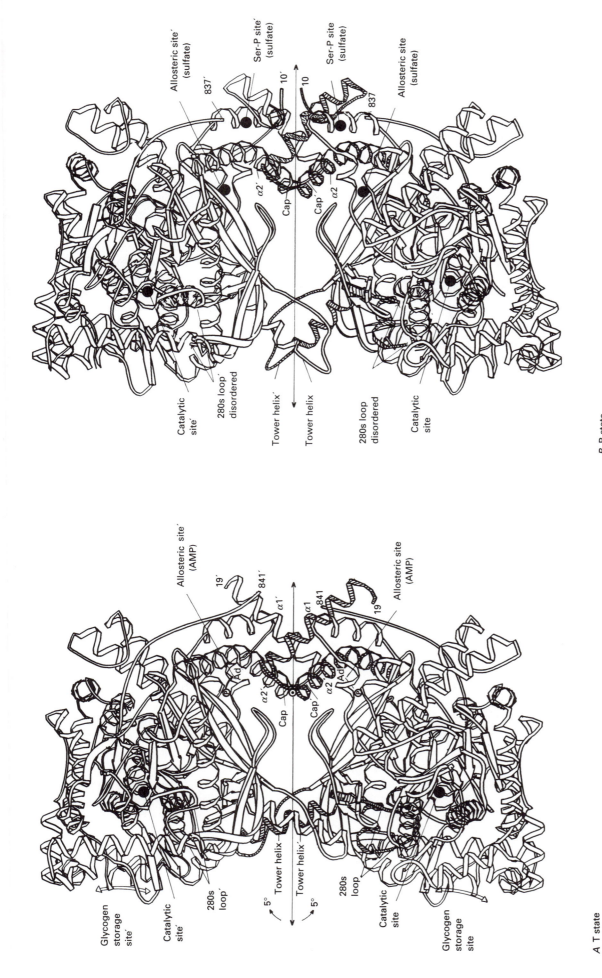

A T state

B R state

FIGURE 9.35

Ribbon representations of (*A*) the T-state and (*B*) the R-state structures of glycogen phosphorylase *b*. The catalytic site, the glycogen storage site, and the allosteric effector site are indicated, with any ligands present in the structure given in parentheses. Sulfate ions used to crystallize the R state occupy sites that normally are occupied by phosphate groups. The catalytic site is close to the 280s loop at the end of the tower helix. The allosteric effector site and Ser 14-phosphate site are situated between the α2 helix and the loop known as Cap that interacts with the other subunit. Features designated by ′ are present in the upper subunit. (From D. Barford and L. N. Johnson, *Nature* 340:609–616, 1989.)

are also slight structural changes that affect at least 64 residues within a 20 Å radius of the phosphate group, especially the binding sites for the allosteric effectors AMP and glucose 6-P.

The crucial structural aspect of phosphorylation of Ser 14 appears to be the introduction of the negative charge of the phosphate group. The acidic phosphate group induces the N-terminal segment to fold into the required helical conformation by compensating for the positive charge of other basic residues at positions 9, 10, 11, and 16. These basic residues comprise the specificity sequence of the phosphorylation site, but it is simply the excess of positive charge in this segment that is thought to be the reason this segment is disordered when Ser 14 is not phosphorylated.

The activity of glycogen phosphorylase is modulated allosterically as well as by reversible phosphorylation. The allosteric control is explained to at least some extent by the occurrence of different R and T quaternary structures (Figs. 9.35 and 9.36). When they are interconverted, one subunit is rotated relative to the other by 10° about an axis perpendicular to the twofold symmetry axis of the dimer (Fig. 9.35). In the process, a pair of α-helices that project away from the protein surface and form part of the interface between the subunits,

known as the *tower* helices, pull apart, tilt, and slide past each other to adopt an alternative mode of packing. The geometry of one helix is constrained by that of the symmetry-related helix, so that molecular symmetry is preserved, and the conformational change seems to be concerted. In this allosteric protein, as in hemoglobin (Fig. 8.27) and in phosphofructokinase (Fig. 9.29), the interface between subunits seems to be designed to permit only two possible modes of docking.

The complexity of the allosteric control of phosphorylase is due to the number of allosteric effectors and binding sites (Fig. 9.36). The allosteric site of each subunit binds cooperatively AMP, ATP, and glucose 6-P. AMP binds more tightly to the R state and activates the enzyme, and the other two bind more tightly to the T state and decrease the activity of the enzyme. Phosphorylation of Ser 14 stabilizes the R state preferentially; both the phosphate group and AMP interact with residues from both subunits that are available only in the R-state conformation. Phosphorylation of Ser 14 also causes small conformational changes that create a stronger binding site for the activator AMP and a weaker binding site for the inhibitors. In turn, the allosteric state of the protein affects the rate of phosphorylation of Ser 14 by the kinase and its dephosphorylation by the phosphatase.

The catalytic site is in a very narrow cleft between the two domains of each subunit. It is some 15 Å from the subunit interface but is connected to it by the tower helix. The movement of the tower helix in the quaternary structure change to the R state causes the opening of a tunnel to allow access for substrates and the movement of an Asp residue to permit its position to be occupied by an Arg residue, which is involved in binding phosphate groups of the substrates. Phosphorylated substrates consequently bind more tightly to the R state, but glucose binds preferentially to the T state (Fig. 9.36). The quaternary structure affects the binding of substrates and vice versa, explaining the cooperativity of their binding.

Near the catalytic site is a hydrophobic site that weakly binds AMP, other purine nucleosides, and other planar nonpolar molecules. They generally block access to the catalytic site and inhibit the enzyme. There is also a site for binding glycogen, known as the storage site; it is believed to function primarily to anchor the enzyme to its substrate, although it may also have a regulatory role.

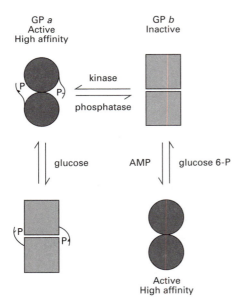

FIGURE 9.36
Simplified diagram of the allosteric and covalent regulation of glycogen phosphorylase (GP). Circles depict the R state, squares the T state. The *a* form has a phosphate group (P) on Ser 14, but the *b* form does not. Glucose binds more tightly to the T state of the *a* form, but glucose 6-P binds more tightly to the T state of the *b* form. AMP binds more tightly to the R state of the *b* form. (Adapted from D. Barford and L. N. Johnson, *Nature* 340:609–616, 1989.)

References

The structures and related functions of phosphorylase *a*. R. J. Fletterick and N. B. Madsen. *Ann. Rev. Biochem.* 49:31–61 (1980).

Glycogen phosphorylase structures and function. R. J. Fletterick and S. R. Sprang. *Acc. Chem. Res.* 15:361–369 (1982).

Structural changes in glycogen phosphorylase induced by phosphorylation. S. R. Sprang et al. *Nature* 336:215–221 (1988).

Glycogen phosphorylase. The structural basis of the allosteric response and comparison with other allosteric proteins. L. N. Johnson and D. Barford. *J. Biol. Chem.* 265:2409–2412 (1990).

Structural mechanism for glycogen phosphorylase control by phosphorylation and AMP. D. Barford et al. *J. Mol. Biol.* 218:233–260 (1991).

The GTPase superfamily: conserved structure and molecular mechanism. H. R. Bourne et al. *Nature* 349:117–127 (1991).

b. Isocitrate Dehydrogenase

The other enzyme for which the structural basis of its regulation by phosphorylation is known is the isocitrate dehydrogenase from bacteria. This enzyme catalyzes the conversion of isocitrate to α-ketoglutarate and CO_2 in an NADP-dependent reaction. Under certain metabolic conditions — such as when bacteria are growing on acetate, rather than glucose, as sole carbon source — the isocitrate must be used by a different metabolic pathway, the *glyoxylate bypass,* rather than by isocitrate dehydrogenase. This diversion to the other pathway is accomplished by phosphorylating Ser 113 of the 416-residue polypeptide chain of isocitrate dehydrogenase; its enzymatic activity is consequently inhibited 10^6-fold.

Phosphorylation of Ser 113 causes no substantial alteration in the tertiary structure of the dimeric enzyme, and only small rearrangements occur in the vicinity of the site to accommodate the phosphate group. The reason the phosphorylated isocitrate dehydrogenase is inactive is that the Ser 113 side chain is part of the enzyme active site and normally hydrogen-bonds to the γ-carboxyl group of the substrate. Phosphorylation of Ser 113 prevents this interaction and disrupts binding of isocitrate. The negative charge of the phosphate group alone is probably sufficient to inhibit the enzyme because mutating Ser 113 to Asp or Glu also totally inactivates the enzyme.

Phosphorylation acts on this enzyme by a direct effect on the enzyme's site. This contrasts with the situation in glycogen phosphorylase, described earlier, in which the modification acts through conformational changes. The different mechanisms may be related to glycogen phosphorylase being an allosteric protein and regulated by quaternary structure changes, whereas isocitrate dehydrogenase has no known allosteric properties even though it is a dimeric protein. Preliminary results with other enzymes suggest that a variety of structural mechanisms of regulation by phosphorylation await elucidation.

References

Regulation of isocitrate dehydrogenase by phosphorylation involves no long range conformational change in the free enzyme. J. H. Hurley et al. *J. Biol. Chem.* 265:3599–3602 (1990).

Regulation of an enzyme by phosphorylation at the active site. J. H. Hurley et al. *Science* 249:1012–1016 (1990).

Electrostatic and steric contributions to regulation at the active site of isocitrate dehydrogenase. A. M. Dean and D. K. Koshland, Jr. *Science* 249:1044–1046 (1990).

c. Glutamine Synthetase

Glutamine synthetase in bacteria catalyzes the synthesis of glutamine from ATP, glutamic acid, and ammonia. The ATP drives the reaction energetically and is hydrolyzed to ADP and P_i. The amide nitrogen of glutamine is used for the biosynthesis of virtually all amino acids, purine and pyrimidine nucleotides, glucosamine 6-P, and NAD. Thus, ammonia is the major source of nitrogen for bacteria under most conditions. Glutamine synthetase, therefore, links the uptake of nitrogen in the form of ammonia with the biosynthesis of proteins, nucleic acids, complex polysaccharides, and various coenzymes. Not surprisingly, this enzyme is subjected to a variety of regulatory mechanisms that include control of its rate of biosynthesis, feedback inhibition of its enzymatic activity, reversible covalent modification, and variation of its rate of degradation.

Glutamine synthetase from *E. coli* is a dodecamer of identical polypeptide chains of 468 residues. The 12 subunits in each molecule are arranged in a double-decked hexagon with sixfold dihedral symmetry and six twofold axes (Fig. 9.37). The 12 active sites are located between pairs of adjacent subunits in each hexameric ring. Two metal ions, either Mg^{2+} or Mn^{2+}, are bound tightly at each active site and are required for activity.

The enzymatic activity of glutamine synthetase is regulated primarily by adenylylation of Tyr 397 of each subunit. Under normal conditions, the adenylylated enzyme is inactive. The adenylyl group ($-AMP$) of ATP is transferred in a phosphodiester bond to the side-chain hydroxyl of Tyr 397:

$$ATP + E-OH \longrightarrow PP_i + E-AMP \quad (9.68)$$

The adenylyl group is removed by phosphorylation to produce ADP:

$$E-AMP + P_i \longrightarrow E-OH + ADP \quad (9.69)$$

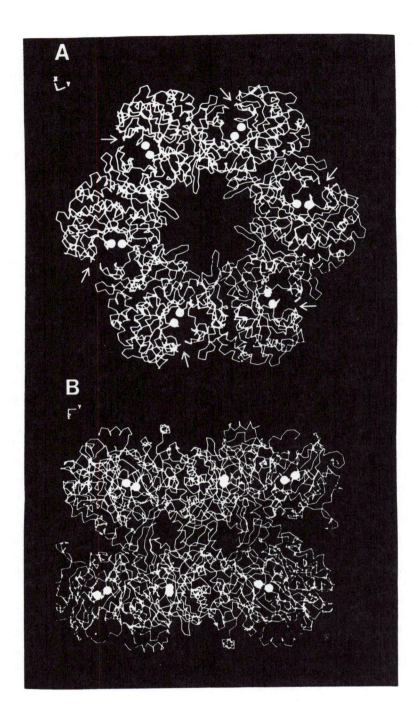

FIGURE 9.37

The quaternary structure of glutamine synthetase. The path of the polypeptide backbone is depicted as a line connecting successive C^α atoms. *A:* Six subunits of the top layer are shown in a view that looks down the sixfold axis. *B:* The six nearer subunits of two layers are shown in a view that is at right angles to the vertical sixfold axis. The location of each active site is indicated by the pair of Mn^{2+} ions, represented as spheres. The Tyr residue that is adenylylated in each subunit is indicated by the arrows in *A.* (From M. M. Yamashita et al., *J. Biol. Chem.* 264:17681–17690, 1989.)

The two reactions are not simply the reverse of each other, but they are catalyzed by the same bifunctional enzyme, adenylyl transferase, albeit at different catalytic sites.

Futile cycling of the two modification reactions, which would convert ATP and P_i to ADP and PP_i, is prevented by coupling the adenylylation cycle to the uridylylation cycle of another protein known as P_{II}. The two modification cycles are similar, differing mainly in the use of ATP in one, UTP in the other. The two cycles are coupled because unmodified P_{II} protein stimulates the adenylylation of glutamine synthetase whereas the uridylylated form of P_{II} is required for removal of the adenylyl groups. Ultimately, these two modification cycles are regulated by a variety of metabolites that influence the activities of the two modifying enzymes.

The P_{II} protein is also involved in regulating transcription of the gene for glutamine synthetase. Synthesis of glutamine synthetase is induced when the source of nitrogen is limited; this is when protein P_{II} is uridylylated. Synthesis of the enzyme is repressed when P_{II} is unmodified. Consequently, the nitrogen status of the cell affects both the synthesis and the activity of the glutamine synthetase.

The activity of *E. coli* glutamine synthetase is inhibited by nine end products of glutamine metabolism: CTP, AMP, glucosamine 6-P, histidine, tryptophan, carbamyl-P, alanine, glycine, and serine. The individual inhibitors cause only partial inhibition, but their effects are cumulative, and the enzyme activity can be completely inhibited by several inhibitors acting simultaneously. There appear to be multiple allosteric sites on the enzyme, but the structural details are not yet known. The sensitivity of the enzyme to feedback inhibition is dependent on its adenylylation state; adenylylation dramatically increases the sensitivity to inhibition.

The structural basis of the allosteric effects is not known, but the catalytic and adenylylation sites occur at the interface between subunits in the unusual quaternary structure of this protein (Fig. 9.37), which would be consistent with allosteric changes in quaternary structure being involved.

Finally, the amount of glutamine synthetase present in *E. coli* also depends on its rate of degradation. Under conditions of nitrogen starvation, the enzyme is inactivated by oxidation; this results in the irreversible covalent modification of His 269, which is involved in the metal-binding sites. The modified protein is subsequently degraded by proteolysis.

The susceptibility of the protein to oxidation depends on both its state of adenylylation and the levels of its substrates, ATP and glutamine. In the absence of the substrates, the unmodified protein is highly susceptible to oxidation whereas the adenylylated protein is not. In the presence of ATP and glutamate, however, the situation is reversed: the adenylylated protein is readily oxidized, whereas the unmodified protein is not. This situation is enigmatic but sensible. At low levels of ATP and glutamate, the enzyme tends not to be adenylylated. Even though potentially active, the protein is nonfunctional in the absence of its substrates. It can, therefore, be degraded to provide a supply of amino acids for the cell. When the levels of ATP and glutamine are high, the protein is not adenylylated and is resistant to oxidation so long as there is a need for glutamine. Otherwise, the enzyme is adenylylated, which makes it inactive; it is then susceptible to oxidation and degradation, because its catalytic activity is not needed.

References

Interconvertible enzyme cascades in cellular regulation. P. B. Chock et al. *Ann. Rev. Biochem.* 49:813–843 (1980).

Regulation of *Escherichia coli* glutamine synthetase. S. G. Rhee et al. *Adv. Enzymol.* 62:37–92 (1990).

Metal-catalyzed oxidation of proteins. E. R. Stadtman and C. N. Oliver. *J. Biol. Chem.* 266:2005–2008 (1991).

Exercises

1. The overall rate of the reaction of Figure 9.3 is said to be determined by the free energy of the transition state between intermediates B and C. Yet the conversion of B to C has a rate constant that is greater than those of most of the other steps along the pathway. In what way can this be said to be the rate-determining step?

   ANSWER
   W. J. Ray, Jr., *Biochemistry* 22:4625–4637 (1983); S. Yagisawa, *Biochem. J.* 263:958–988 (1989).

2. The K_m value for a substrate can be the same as the dissociation constant of the enzyme–substrate complex under certain conditions (Sec. 9.1). If this is the case in one direction for a simple one-substrate–one-product enzymatic reaction, can it also hold for the reverse direction?

   ANSWER
   T. Keleti, *FEBS Letters* 208:109–112 (1986).

3. Consider an enzyme operating as one step in a metabolic pathway. If an inhibitor of that enzyme is introduced, the rate of catalysis by the enzyme decreases and the steady-state concentration of its substrate increases. If the substrate competes with the inhibitor, the increase in substrate concentration decreases the inhibition, and a new steady state is attained with a similar flux but an elevated concentration of the substrate of the inhibited enzyme.

What would happen if the inhibition were uncompetitive (Sec. 9.1.1) relative to the substrate?

ANSWER
A. Cornish-Bowden, *FEBS Letters* 203:3–6 (1986).

4. Enzymes usually function in aqueous solution and are usually studied under such conditions. What would be the likely advantages and disadvantages of using enzymes in an anhydrous organic solvent rather than aqueous solution?

ANSWER
A. M. Klibanov, *Trends Biochem. Sci.* 14:141–144 (1989); *Acc. Chem. Res.* 23:114–120 (1990).

5. Many reactions in the gas phase have small activation barriers compared with the corresponding reaction in solution. It has been suggested that the large rates of reactions on enzymes are a result of the solvating water being squeezed out from between the enzyme and the substrate, so that the reaction between them takes place as it would in the gas phase (M. J. S. Dewar and D. M. Storch, *Proc. Natl. Acad. Sci. USA* 82:2225–2229, 1985). What objections might be raised to this suggestion?

SUGGESTIONS
A. Warshel et al., *Proc. Natl. Acad. Sci. USA* 86:5820–5824 (1989).

6. An enzyme is thought to function because its active site is complementary to the transition state for the reaction (Sec. 9.2.2). In this case, the enzyme need not bind the substrate with any substantial affinity. Why then is a Michaelis complex of enzyme with bound substrate always apparent?

ANSWER
J. Kraut, *Science* 242:533–540 (1988).

7. How would you determine which step in the reaction sequence of your enzyme is rate-limiting? What factors determine this?

ANSWER
W. J. Albery and J. R. Knowles, *Biochemistry* 25:2572–2577 (1986).

8. With a series of related reactions, a Brønsted-type analysis is often used (e.g., Fig. 9.12), in which the rate constant for the reaction (e.g., k_1) is plotted versus the equilibrium constant K_{eq}:

$$S \underset{k_{-1}}{\overset{k_1}{\rightleftharpoons}} P$$

In this case, $K_{eq} = k_1/k_{-1}$, so the parameter k_1 affects both the abscissa and the ordinate of the plot. In this case, is the linearity of data plotted in this way meaningful? Is a linear curve evidence for a linear free-energy relationship? Is the value of β defined by the slope of the plot a significant parameter?

DEBATE
D. A. Estell; A. R. Fersht, *Protein Eng.* 1:441–446 (1987).

9. The trypsin proteases have three-dimensional structures in which there is a twofold symmetry axis relating two similar halves of the polypeptide chain. The polypeptide chain of one half can be superimposed directly on that of the other half after a rotation through 180°. This situation is frequently observed with molecules that have evolved from a dimer of identical subunits (Sec. 6.4.3), but in the case of the trypsin proteases, the respective polypeptide chains of the superimposed domains are found to run in opposite directions (K. Nishikawa and T. Ooi, *J. Theor. Biol.* 43:351–374, 1974). What explanations for this observation are plausible?

SUGGESTIONS
A. D. McLachlan, *J. Mol. Biol.* 128:49–79 (1979); M. J. Dufton, *FEBS Letters* 271:9–13 (1990).

10. Peptide boronic acids are good inhibitors of serine proteases and appear to act by mimicking the tetrahedral intermediate or transition state in catalysis (Fig. 9.20). In this case, a series of peptide boronic acids would be expected to improve as inhibitors in parallel with increases in k_{cat}/K_m for a corresponding series of substrates (e.g., Fig. 9.9). This improvement is observed in a series of peptide boronic acids that have Val at the P_1 position and that differ in the number of residues in the peptide. Such a correlation is not observed, however, in a series of peptides differing only at the P_1 position (C. A. Kettner et al., *Biochemistry* 27:7682–7688, 1988). What is the most likely explanation for the discrepancy?

ANSWER
R. Bone et al., *Biochemistry* 28:7600–7609 (1989).

11. If trypsin is added to a cryosolvent at −30°C containing a chromogenic substrate, the *p*-nitroanilide of either N^α-carbobenzoxyl-L-lysine or -arginine, a "burst" of absorbance due to the release of product (Fig. 9.4) occurs prior to complete turnover of the enzyme (P. D. Compton and A. L. Fink, *Biochem. Biophys. Res. Commun.* 93:427–431, 1980). This turnover of the enzyme was attributed to the formation of a tetrahedral intermediate. What other explanations for this observation are possible?

ANSWER
P. D. Compton and A. L. Fink, *Biochemistry* 23:2989–2994 (1984).

12. A lag phase in the slow binding of transition-state analogues is frequently taken to indicate that binding is inherently slow and that the protein undergoes an induced-fit

conformational rearrangement to a tight-binding complex (e.g., A. Baici and M. Gyger-Marazzi, *Eur. J. Biochem.* 129:33–41, 1982; L. C. Kurz et al., *Biochemistry* 26:3027–3032, 1987). What other explanations are possible?

ANSWER

R. M. Schultz et al., *J. Biol. Chem.* 264:1497–1507 (1989); P. A. Bartlett and C. K. Marlowe, *Biochemistry* 26:8553–8561 (1987); H. M. Holden et al., *Biochemistry* 26:8542–8553 (1987).

13. The serine protease subtilisin is the active enzyme in many household laundry detergents. This enzyme is inactivated readily by oxidation of Met 222, a residue that is conserved in all known subtilisin sequences and occupies a partially buried position next to the catalytic Ser 221. It should be possible to produce a more stable and useful enzyme by mutating this residue to one that is not susceptible to oxidation. Which amino acid should you use to replace the Met residue?

ANSWER

D. A. Estell et al., *J. Biol. Chem.* 260:6518–6521 (1985).

14. A protein from *Plasmodium falciparum* identified from its gene sequence was proposed to be homologous to the thiol proteases (D. G. Higgins et al., *Nature* 340:604, 1989). What is the problem with this proposal?

ANSWER

A. E. Eakin et al.; J. C. Mottram et al., *Nature* 342:132 (1989).

15. The carboxyl proteases were previously thought to go through a covalent intermediate because they catalyze transpeptidation reactions. In such a reaction, pepsin-catalyzed hydrolysis of peptide Leu-Tyr-Leu gives as a product the peptide Leu-Leu (M. Takahashi and T. Hofmann, *Biochem. J.* 147:549, 1975; T. E. Wang and T. Hofmann, *Biochem. J.* 153:691, 1976). What other explanations are possible for this type of observation?

ANSWER

M. S. Silver and S. L. T. James, *Biochemistry* 20:3177–3182, 3183–3189 (1981); M. Blum et al., *J. Biol. Chem.* 266:9501–9507 (1991).

16. The catalytic mechanism of carboxypeptidase A was previously thought to occur via a covalent mixed-anhydride intermediate involving Glu 270 of the enzyme and the acyl portion of the substrate, comparable to the acylenzyme intermediate of the serine proteases (Fig. 9.17). The evidence for this mechanism included observation of the putative intermediate (M. W. Makinen et al., *J. Biol. Chem.* 254:356–366, 1979; L. C. Kuo and M. W. Makinen, *J. Biol. Chem.* 257:24–27, 1982; L. C. Kuo et al., *J. Mol. Biol.* 163:63–105, 1983; J. Suh et al., *J. Amer. Chem. Soc.* 107:4530–4535, 1985; M. E. Sander and H. Witzel, *Biochem. Biophys. Res. Commun.* 132:681–687, 1985). Are these observations compatible with the currently accepted mechanism of Figure 9.22?

POSSIBLE EXPLANATIONS

S. J. Hoffman et al., *J. Amer. Chem. Soc.* 105:6971–6973, (1983); D. W. Christianson and W. N. Lipscomb, *Acc. Chem. Res.* 22:62–69 (1989).

17. "Catalytic triads" of the Asp-His-Ser type first observed in the serine proteases (Sec. 9.3.2.a), but with the Ser residue replaced by a Zn^{2+} atom, have been found in many zinc-containing enzymes, including thermolysin and carboxypeptidase A (D. W. Christanson and R. S. Alexander. *J. Amer. Chem. Soc.* 111:6412–6419, 1989). How would the presence of such a catalytic triad structure affect the proposed catalytic mechanism of Figures 9.17D and 9.22?

18. An Asp-His-Ser catalytic triad involving a catalytic Ser residue has been described in each of two structures of lipases (L. Brady et al., *Nature* 343:767–770, 1990; F. K. Winkler et al., *Nature* 343:771–774, 1990). What are the implications of this triad for catalysis and for the evolution of these enzymes?

ANSWER

D. Blow, *Nature* 343:694–695 (1990).

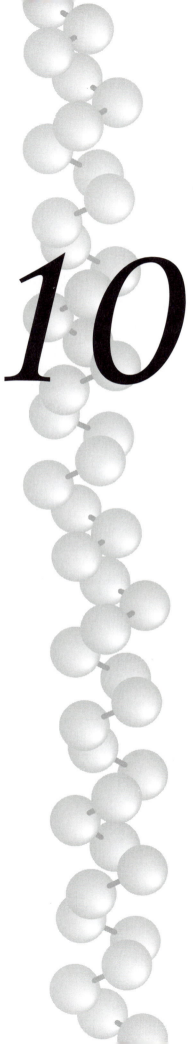

Degradation

10

A lthough proteins have many remarkable properties, they are not immortal. They eventually succumb to old age, brought on by gradual covalent modification in a variety of nonenzymatic chemical reactions. Moreover, in most cells there is also extensive turnover of normal, unmodified proteins. These proteins are degraded proteolytically to their constituent amino acids, which are used again in protein synthesis. This process of synthesis and degradation of proteins would seem to be costly energetically, so it is not surprising that it is regulated. Each protein has its own rate of degradation under any particular circumstances.

Turnover of proteins by degradation is physiologically important because it is often necessary for intracellular protein levels to change during development, during the cell cycle, and in response to changes in the environment. The level of a particular protein may be increased by accelerated synthesis, but it can decrease only if the cells are growing rapidly so that the existing protein molecules are diluted or if there is some mechanism for degrading the protein. In general, the greater the rate of degradation, the greater the possible degree of control of the level of the protein. Control is most efficient if there is regulation of the rates of both synthesis and degradation of individual proteins.

An extreme example of rapid but regulated degradation of a protein occurs during the cell cycle of eukaryotic cells. The transition from metaphase to anaphase, when the duplicated chromosomes of a dividing cell begin to separate, is induced by the specific proteolytic degradation of a regulatory protein known as **cyclin.** This protein normally activates a protein kinase (Sec. 9.4.3), which plays an important role in regulating cell metabolism. The protein kinase becomes inactive when the cyclin disappears at the start of chromosome separation. After completion of chromosome separation and cell division, the degradation of cyclin ceases, so its level increases as a result of its synthesis; the kinase consequently is activated, and another round of mitosis and cell division ensues. Similarly, the plant photoreceptor protein, phytochrome, is degraded rapidly in vivo after light converts it from a stable form that absorbs red light to one that absorbs far-red light. The turnover of other proteins by degradation is not so dramatic, but there is little doubt about the physiological importance of protein degradation.

Cells also need mechanisms to survive temporary conditions of starvation so that they can degrade "luxury" proteins in order to use their amino acids for biosynthesis of proteins that are more essential for survival. A mechanism is also needed for degrading abnormal proteins, either those that have succumbed to old age or those that were synthesized in an incorrect form due to mistakes in the biosynthetic machinery for assembly and folding of the polypeptide chain.

As a result of recent recognition of its importance, study of the phenomenon of protein degradation has become increasingly active in the past few years.

References

Control of enzyme levels in mammalian tissues. R. T. Schimke. *Adv. Enzymol.* 37:135–187 (1973).

Intracellular protein degradation in mammalian and bacterial cells. Part I. A. L. Goldberg and J. F. Dice. *Ann. Rev. Biochem.* 43:835–869 (1974); Part 2. A. L. Goldberg and A. C. St. John. *Ann. Rev. Biochem.* 45:747–803 (1976).

Intracellular protein degradation. F. J. Ballard. *Essays Biochem.* 13:1–37 (1977).

Intracellular protein degradation. J. Kay. *Biochem. Soc. Trans.* 6:789–797 (1978).

Mechanisms of intracellular protein breakdown. A. Hershko and A. Ciechanover. *Ann. Rev. Biochem.* 51:335–364 (1982).

Intracellular protein catabolism: state of the art. R. J. Mayer and F. Doherty. *FEBS Letters* 198:181–193 (1986).

Intracellular protein degradation. A. J. Rivett. *Essays Biochem.* 25:39–81 (1990).

Regulation of protein turnover in skeletal and cardiac muscle. P. H. Sugden and S. J. Fuller. *Biochem. J.* 273:21–37 (1991).

10.1   Chemical Aging

One factor that potentially limits the lifetime of a protein is its susceptibility to chemical change. The oxygen-rich, aqueous environments in which most proteins exist tend to produce a variety of chemical reactions in proteins. Their sulfur atoms are oxidized, their Gln and Asn residues deamidate, and they racemize about their C^α atoms to the D isomer and epimerize about any additional asymmetric centers of side chains (i.e., Thr and Ile). These reactions occur chemically and are largely unavoidable under physiological conditions, but living organisms have defensive mechanisms against some of them. For example, the enzymes superoxide dismutase and catalase are frequently present in cells, presumably to scavenge the small amounts of superoxide (O_2^-) and hydrogen peroxide that are invariably produced by oxidative metabolism. Most cells contain millimolar concentrations of small-molecule thiols, usually glutathione, to maintain the cellular thiol groups and to scavenge other reactive species such as the very reactive hydroxyl radical. An enzyme system is generally present in cells to maintain the glutathione, or other such compound, in the thiol form.

Many of the chemical reactions that occur most frequently with proteins are described in Chapter 1. The most common of these that are involved in chemical aging are hydrolysis of peptide bonds, especially at Asp residues (Sec. 1.6.2.e); deamidation of Asn residues, particularly those preceding Gly residues in the sequence (Sec. 1.3.6); oxidation of the sulfur atoms of Cys and Met residues (Sec. 1.3.10); and destruction of disulfide bonds between Cys residues, especially at elevated temperatures and high pH values. Enzymes are present that can reverse some of these modifications, in particular the oxidation of Met and Cys residues. Systems have also been proposed for reversing the formation of iso-Asp residues, which are generated during deamidation of Asn residues, but there is no evidence that they are significant in vivo.

Proteins are modified covalently by many of the chemicals present in our environment, many of which are man-made. One chemical reaction of particular significance to protein aging is the nonenzymatic reaction between protein amino groups and natural reducing sugars, such as glucose and glucose 6-P, to form a Schiff base. This Schiff base can then undergo an Amadori-type rearrangement to the more stable ketoamine, as shown at the top of the next page. The ketoamine product is subject to a variety of dehydrations to form yellow-brown fluorescent products and cross-links between protein molecules. These chemical reactions are largely responsible for many of the flavors and golden-brown colors that can be produced by cooking foods, but they occur even at low temperatures. The glycosylation of proteins occurs nonenzymatically under normal physiological conditions and is of clinical importance. This reaction is especially prevalent when glucose levels are high, as in the case of poorly controlled diabetes.

Only L-amino acids are incorporated into proteins during their biosynthesis (Chap. 2), but subsequently

$$
-NH_2 \; + \;
\begin{array}{c}
HC{=}O \\
| \\
HCOH \\
| \\
HOCH \\
| \\
HCOH \\
| \\
HCOH \\
| \\
CH_2OH
\end{array}
$$

glucose

H_2O

$$
\underset{\text{aldimine (Schiff base)}}{
\begin{array}{c}
HC{=}N{-} \\
| \\
HCOH \\
| \\
HOCH \\
| \\
HCOH \\
| \\
HCOH \\
| \\
CH_2OH
\end{array}}
\xrightarrow[\text{rearrangement}]{\text{Amadori}}
\underset{\text{ketoamine}}{
\begin{array}{c}
CH_2{-}NH{-} \\
| \\
C{=}O \\
| \\
HOCH \\
| \\
HCOH \\
| \\
HCOH \\
| \\
CH_2OH
\end{array}}
\qquad (10.1)
$$

they are spontaneously converted very slowly to the D-enantiomers. A similar phenomenon occurs in the asymmetric side chains of Thr and Ile residues. Most proteins are degraded before a significant amount of any racemized forms are generated, but Asp and Asn residues have been found to racemize at a rate of 0.1% per year in human enamel, dentine, and lens proteins. These residues are especially susceptible to racemization due to their tendency to form the succinimide derivative (Fig. 1.3), which racemizes relatively rapidly. Racemization reactions in other amino acid residues are slow and are of greatest significance for estimating the ages of fossils.

Chemical modifications that significantly alter the structure of a protein usually lead to its degradation, because most cells have degradation systems that recognize aberrant proteins. Modifications that do not trigger degradation tend to accumulate during the lifetime of a protein. Aging processes are especially evident in long-lived proteins such as those of the eye lens. Lens cells do not die, and there is little degradation of their proteins. Therefore, proteins synthesized at the embryonic stage are still present in the human eye lens 70 or more years later, which makes lens proteins excellent subjects for protein aging studies. Aging of lens proteins is also of physiological importance in that chemical modification results in decreased solubility of the proteins and increased pigmentation of the lens, with obvious deleterious effects on vision, including the formation of cataracts. Many structural proteins are also long-lived; other proteins exist for the life of the cell they

are in, such as hemoglobin, which is usually unchanged throughout the three-month average life span of an erythrocyte.

Chemical aging has frequently been proposed as the factor that determines the rate of turnover of a protein; that is, each protein is designed to have a primary or tertiary structure in which, for example, deamidation of an Asn residue or oxidation of a Met or Cys residue occurs at a particular rate. As a result of this modification, the protein is inactivated or degraded proteolytically. It is now recognized, however, that most proteins turn over at rates that are determined by enzymatic processes, not chemical reactions. The only type of chemical modification that appears to be the primary determinant of the rate of degradation of a protein is the oxidation of metals that destroys the amino acid side chains involved in binding them at specific sites on certain proteins. The best characterized example, glutamine synthetase, was described briefly in Section 9.4.3.b. Similar mechanisms might act to trigger the degradation of other proteins, but they have not yet been discovered.

References

Aging of protein molecules: lens crystallins as a model system. J. S. Zigler and J. Goosey. *Trends Biochem. Sci.* 6:133–136 (1981).

Lens proteins. H. Bloemendal. *Crit. Rev. Biochem.* 12:1–38 (1982).

Modifications of Proteins during Aging. R. C. Adelman. Alan R. Liss, New York, 1985.

Nonenzymatic covalent posttranslational modification of proteins in vivo. J. J. Harding. *Adv. Protein Chem.* 37:247–334 (1985).

Free radicals, lipids and protein degradation. S. P. Wolff et al. *Trends Biochem. Sci.* 11:27–31 (1986).

Aging of proteins and nucleic acids: what is the role of glucose? A. Cerami. *Trends Biochem. Sci.* 11:311–314 (1986).

Biochemistry of oxidative stress. H. Sies. *Ang. Chem. Intl. Ed. Engl.* 25:1058–1071 (1986).

Analysis of processes causing thermal inactivation of enzymes. T. J. Ahern and A. M. Klibanov. *Methods Biochem. Anal.* 33:91–127 (1988).

Post-translational Modifications of Proteins and Aging. V. Zappia et al. Plenum, New York, 1988.

10.2   Protein Turnover in Vivo

The rates of protein degradation need to be measured in vivo to obtain results that are physiologically significant. Such measurements are subject to many experimental complexities. The most prevalent methods use radioactive tracers; either a labeled protein is injected into the appropriate site, or it is synthesized in situ by giving a pulse of radioactive amino acids or precursors. The disappearance of the radioactivity in the relevant protein is then followed as a function of time. It is assumed that the radioactive protein is indistinguishable from the normal protein, but this may not be the case if the radioactive labeling was carried out by covalent modification. Pulse synthesis in situ of a protein from ^{3}H, ^{14}C, or ^{35}S isotopic forms of the normal amino acids is preferable for this reason but suffers from the tendency of the radioactive amino acids released from the degraded protein to be used for further rounds of protein synthesis. This tendency can be minimized by adding a large excess of unlabeled amino acid after the initial radioactive pulse, but this addition may alter the rate of degradation. With all such manipulations there is uncertainty about the size and nature of the intracellular pools of amino acids used for protein synthesis, as well as a limited ability to manipulate them. Nevertheless, this method has been used extensively. It has also been refined by using two pulses with different isotopes at different times (e.g., first ^{14}C, then ^{3}H later). All other factors being constant or normalized, the subsequent ratio of the two isotopes present in a protein will depend on the degree of degradation of the protein labeled by the first isotope that occurred during the time interval between the two pulses.

Nonradioactive methods are also useful for measuring rates of protein degradation. For example, the rate constant for degradation (k_D) of a protein P can be calculated from the difference between its measured rate of synthesis (k_S) and the rate of increase in its cellular level ($d[P]/dt$)

$$\text{amino acids} \xrightarrow{k_s} P \xrightarrow{k_D} \text{degradation} \quad (10.2)$$

$$\frac{d[P]}{dt} = k_s - k_D[P] \quad (10.3)$$

Some proteins have unique amino acids, generated by posttranslational covalent modification, that are not reused metabolically after degradation. For example, certain His residues in the muscle proteins actin and myosin are quantitatively methylated. This methyl-His amino acid does not occur in other proteins; it is not reused after degradation of actin and myosin but is excreted in the urine. Consequently, accurate measurements of the degradation of these two proteins can be made by measuring the rate of excretion, the total protein pool, and any dietary intake of this modified amino acid. Similarly, hydroxyproline in the urine has long been used clinically as a measure of the rate of degradation of collagen (Sec. 5.5.3).

When the conditions are kept constant, protein degradation is found to follow first-order kinetics (Sec. 5.2.3); the rate is described by a single apparent rate constant, and partially degraded proteins are usually not present in significant quantities. The occurrence of first-order kinetics of degradation has the important implication that degradation of all the molecules of any particular protein is random; newly synthesized molecules are just as likely to be degraded as old molecules of the same protein. A further implication is that a single event is sufficient to initiate degradation. The nature of this event is not specified, however, and the observed rate may be the sum of the rates of a number of different events that would be sufficient to initiate degradation independently. First-order kinetics of degradation rule out the possibility that a protein molecule is degraded rapidly only after suffering two or more such events, such as covalent modifications or proteolytic cleavages, two examples of events likely to trigger degradation.

A very wide range in the rates at which proteins are degraded has been observed under normal physiological conditions (Table 10.1), even in the same cell, cellular compartment, or tissue. The proteins that turn over most rapidly in normal cells tend to be those that catalyze rate-determining metabolic reactions. For example, one of the most rapidly degraded proteins known in rat liver, ornithine decarboxylase, has a half-life of only 11 min and catalyzes the rate-determining step in polyamine biosynthesis. Its enzymatic activity is not subject to any known regulation by allosteric control or by covalent modification. Instead, the rate of synthesis of the protein can vary up to 1000-fold. This variation in rate of synthesis and the rapid rate of degradation of this protein indicate that the activity of ornithine decarboxylase is regulated primarily by varying its levels in the liver. Other rapidly degraded proteins include the protein products of proto-oncogenes, which play central roles in the regulation of cell growth. The rate of protein degradation in a cell or tissue is not generally constant, however, but depends very much on the physiological state of the cell.

10.2.1 Factors That Determine the Rate of Degradation

The molecular basis for the varying rates of protein degradation has been sought by looking for correlations with molecular properties of the proteins. Seemingly significant correlations have been reported for the rate of degradation of a protein with the following: its susceptibility to thermal unfolding; the absence of stabilizing ligands; its susceptibility to protease digestion in

Table 10.1 *Degradation Rates of Various Enzymes and Proteins in Vivo*

Protein	Tissue	Half-life
Ornithine decarboxylase	Rat liver	11 min
RNA polymerase I	Rat liver	1.3 h
Tyrosine aminotransferase	Rat liver	1.5 h
Tryptophan oxygenase	Rat liver	2 h
Phosphoenolpyruvate carboxykinase	Rat liver	6 h
Hexokinase	Rat liver	1 day
Acetyl CoA carboxylase	Rat liver	2 days
Glyceraldehyde phosphate dehydrogenase	Rat liver	3–4 days
Arginase	Rat liver	4–5 days
α-Actinin	Rat cardiac muscle	5–6 days
Myosin heavy chain	Rat cardiac muscle	5–6 days
Myosin light chain	Rat cardiac muscle	9 days
Actin	Rat cardiac muscle	7–8 days
Troponin	Rabbit skeletal muscle	10–15 days
α-Actinin	Rabbit skeletal muscle	20–25 days
Tropomyosin	Rabbit skeletal muscle	20–25 days
Myosin	Rabbit skeletal muscle	30 days
Actin	Rabbit skeletal muscle	>50 days
Hemoglobin	Reticulocytes/erythrocytes	$\simeq \infty$[a]

[a] Limited by life of erythrocyte.
From J. Kay, *Biochem. Soc. Trans.* 6:789–797 (1978).

vitro; the intrinsic rate of deamidation of its Gln and Asn residues; the susceptibility of its Cys, His, and Met residues to oxidation; the presence of attached carbohydrates or phosphate groups; the presence of a free α-amino group; the net negative charge of the protein; increasing size of the polypeptide chain; and the flexibility of the folded conformation as measured by hydrogen exchange. Many, but not all, of these correlations seemed plausible if susceptibility to proteolysis were the limiting factor. None of these correlations, however, has been generally confirmed.

Any such correlations could hold only to the extent that the same mechanism is used for the degradation of all the proteins. That this is probably not the case is indicated by many observations that the rate of degradation of a particular protein changes under various conditions. For example, a number of yeast enzymes that are not required when glucose is present in the growth medium are "catabolite-inactivated" by specific degradation. In this case, the enzymes are taken up specifically by the vacuoles present in yeast and degraded there. Degradation of metabolically crucial enzymes is often regulated, as in the case of glutamine synthetase (Sec. 9.4.3.b). Maturation of reticulocytes into erythrocytes is accompanied by the selective degradation of mitochondria, ribosomes, and many proteins that are no longer required after the cell has synthesized its complement of hemoglobin. In diabetes or starvation, there are many proteins whose degradation rates are markedly enhanced but others that are little affected.

For these reasons, more specific signals, such as particular sequences, have been sought to explain the varying rates at which proteins are degraded. Rapidly degraded proteins have been proposed to have one or more segments of 12–60 residues in their primary structure that are rich in Pro, Glu, Ser, and Thr residues. Such regions have been designated **PEST regions,** after the single-letter abbreviations of these four amino acid residues. Another proposal is that the N-terminal residue determines the rate of degradation, at least in the ubiquitin-dependent system, which will be discussed later (Sec. 10.3.3).

The folded conformation of a protein is important, at least for minimizing its rate of degradation. Abnormal proteins — such as those resulting from chemical modification, mutation, chain termination, or incorporation of amino acid analogues — are almost invariably found to be rapidly degraded in cells by proteolysis. Excess, but normal, hemoglobin-α chains present as a result of a block in formation of β chains are usually also degraded, presumably because they cannot adopt their usual, more stable conformation in the $\alpha_2\beta_2$ hemoglobin tetramer (Sec. 8.4.3). Abnormal proteins are readily recognized by most cells and are degraded rapidly.

References

Protein structure and intracellular stability. M. Rechsteiner et al. *Trends Biochem. Sci.* 12:390–394 (1987).

Covalent modification reactions are marking steps in protein turnover. E. R. Stadtman. *Biochemistry* 29:6323–6331 (1990).

Regulated import and degradation of a cytosolic protein in the yeast vacuole. H. L. Chiang and R. Schekman. *Nature* 350:313–318 (1991).

10.3 Mechanisms of Protein Degradation

Many diverse mechanisms for protein degradation have been found thus far. Some use ATP hydrolysis, others are independent of ATP. Some involve conjugation to ubiquitin, others do not. Some degradation occurs in lysosomes, but most does not. All the degradation systems involve proteases, but there are a variety of proteases present in living systems.

10.3.1 Proteases Involved in Protein Turnover

Most studies of protein degradation have concentrated on proteolytic enzymes because hydrolysis of the protein to amino acids seems to be a universal aspect of this process. Cells and organisms have a variety of processes other than degradation, however, that are catalyzed by proteases, such as postbiosynthetic processing of pre and pro polypeptide chains (Chap. 2), activation of zymogen precursors of many proteins (Chap. 9), and digestion of exogenous proteins for their use as foodstuffs. Consequently, the identification of a proteolytic enzyme is not sufficient to conclude that it is involved in protein turnover.

A variety of proteases of varying specificities appear to be involved in the degradation of certain proteins and to be linked to some disease states such as muscular dystrophy, arthritis, degenerative skin disorders, respiratory and gastrointestinal diseases, and malignancy. Two such prominent proteases are collagenase and elastase, which degrade the structural proteins collagen and elastin, respectively. Many cells have Ca^{2+}-activated proteases, known as calpains, that once were thought to be involved in protein degradation but are now thought to catalyze regulatory cleavages of specific proteins. There are also reported to be proteases that are specific for pyridoxal phosphate-containing and NAD-containing proteins, but only the apoproteins are hydrolyzed. With the possibility of a large number of different proteases, with varying specificities for different proteins, it is easy to envision different rates of degradation with different regulatory features.

The most interesting protein degradation is that which occurs intracellularly, whereas most of the well-characterized proteases (Sec. 9.3.2) are normally present extracellularly. Intracellular proteases are distinguished from those that act extracellularly by size, the intracellular proteins being much larger than their extracellular counterparts. The large sizes of intracellular proteins usually arise from the presence of additional domains that do not have proteolytic activity but seem to have primarily regulatory roles. For example, calpains I and II are each composed of two subunits, a catalytic subunit of 80,000 M_W and a small subunit of 30,000 M_W. Their small subunits are identical, and the large subunits are different but homologous. Each catalytic subunit appears from its primary structure to have four domains, of which the second is apparently a thiol protease. The fourth domain contains four EF-hand structures involved in Ca^{2+} binding (see Fig. 8.20). The smaller subunit contains a similar Ca^{2+}-binding domain and a hydrophobic membrane-binding domain. Ca^{2+} is required for the calpains to exhibit their proteolytic activity, presumably by binding to the fourth domain with the EF-hand motifs. In addition, association with membranes activates calpain II.

One of the most striking aspects of protein degradation within the cell is that energy is usually required. Protein hydrolysis per se does not have this energy requirement, so its requirement in vivo was long presumed to be due to an indirect effect. Plausible candidates would be the requirement of ATP hydrolysis for the uptake of proteins by vesicles or the maintenance of low pH in lysosomes (Sec. 10.3.2). The energy is now known, however, to be required for the initiation of proteolytic degradation of at least some proteins. Several ATP-dependent proteolytic systems have been discovered in cell-free extracts from several sources. Some of these proteases are of the thiol type (Sec. 9.3.2.b) and are merely activated by ATP as an allosteric effector. Members of the other class are of the serine type; in this class, ATP hydrolysis is required for proteolytic activity. These proteins require Mg^{2+} ions (which are usually bound with ATP), are inhibited by vanadate (an ATPase inhibitor), and are not activated by ATP analogues that are not hydrolyzable. Their ATPase activity is stimulated by the presence of degradable protein substrates.

More than 20 proteases are known in *E. coli*, but only two require ATP; they are known as *La* and *Ti*. *La* is a tetramer of identical polypeptide chains of 87,000 M_W. This polypeptide is the product of the *lon* gene. Mutational inactivation of the *lon* gene causes the bacteria to have a number of unusual properties, including increased sensitivity to UV light and overproduction of capsular polysaccharide. These two phenotypic properties are due to the inability of the mutant *La* protease to degrade rapidly a UV-inducible protein that regulates filamentation and another protein that regulates the expression of genes for capsular polysaccharide synthesis. The synthesis of the *La* protease is induced by various stress situations including heat shock. Its role appears to be to degrade abnormal or denatured proteins generated by the stressful situation, as well as several short-lived regulatory proteins.

La is a serine protease in that it is inactivated by DFP (Eq. 9.54), but it is not detectably homologous to the trypsin family of eukaryotic serine proteases (Sec. 9.3.2.a). *La* differs from ordinary proteases in that it requires ATP for its catalytic activity. It is inhibited by ADP and has an ATPase activity that is stimulated by protein substrates. Under optimal conditions, two ATP molecules are used per peptide bond hydrolyzed. The ATP seems to play a regulatory role in the protease activity and is not used to modify or unfold the protein substrate. The protease activity is masked in the absence of ATP. When ATP is present, the protease cleaves a peptide bond of the substrate. Following this cleavage, the ATP is hydrolyzed to ADP, which is bound more tightly than is ATP; this inhibits the protease activity until the ADP is displaced by ATP. The rate-limiting step is thought to be release of the ADP, which is stimulated by the presence of a protein substrate. The four active sites of the tetramer appear to be subject to allosteric control.

Protease *Ti* is similar to *La* in many respects, but it is a complex of two polypeptide chains, one having intrinsic proteolytic activity that is active in isolation, the other having ATPase activity. ATP hydrolysis by one subunit activates the proteolytic subunit, and peptide bond cleavage by that subunit enhances ATP consumption by the other. Protease *Ti* is also a heat-shock protein that is induced when it is likely that there will be damaged proteins to degrade. Proteases comparable to *La* and to *Ti* are also found in mitochondria and chloroplasts.

The cytosol of eukaryotic cells contains a very large protease of $7 \times 10^5 M_W$ that is known as the *multicatalytic proteinase*, the *multicatalytic complex*, or the *proteasome*. It is usually composed of at least 10 different polypeptides, with individual molecular weights of between 2.2×10^4 and 3.4×10^4, that are arranged in a hollow cylindrical structure. It is able to cleave peptide bonds on the carboxyl side of basic, hydrophobic, and acidic residues, and it can catalyze the extensive breakdown of some protein substrates. Its physiological role is still being elucidated, but it is known to be a component of the very large ($1.5 \times 10^6 M_W$) complex that degrades proteins conjugated to ubiqitin (Sec. 10.3.3).

The molecular basis of the requirement for ATP hydrolysis generally seems to be to regulate the proteolytic activities of these enzyme complexes. Clearly it is important to ensure that the protease exists in an active form only after it has bound to an appropriate protein substrate, one that is meant to be degraded. Otherwise, the presence in the cytosol of a protease capable of degrading intact proteins would be expected to reduce the cell to a bag of amino acids.

References

Control of proteolysis. H. Holzer and P. C. Heinrich. *Ann. Rev. Biochem.* 49:63–91 (1980).

Intracellular proteases. J. S. Bond and P. E. Butler. *Ann. Rev. Biochem.* 56:333–364 (1987).

Genetics of proteolysis in *Escherichia coli*. S. Gottesman. *Ann. Rev. Genetics* 23:163–198 (1989).

High molecular mass intracellular proteases. A. J. Rivett. *Biochem. J.* 263:625–633 (1989).

The multicatalytic proteinase complex, a major extralysosomal proteolytic system. M. Orlowski. *Biochemistry* 29:10289–10297 (1990).

The mechanism and function of ATP-dependent proteases in bacterial and animal cells. A. L. Goldberg. *Eur. J. Biochem.* 203:9–23 (1992).

10.3.2 Lysosomes

Lysosomes are intracellular compartments that have long been considered to be involved in protein degradation because they contain a variety of proteases. The lysosomal proteases, known collectively as the **cathepsins,** are each related to the serine, carboxyl, or thiol proteases (Sec. 9.3.2). Cathepsins can be distinguished by their sensitivity to the unusual microbial inhibitors pepstatin and leupeptin and by their requirement for relatively acidic conditions, which are maintained in the lysosome by active transport of protons. Weakly basic reagents, such as ammonia and chloroquine, that can penetrate cell membranes raise the internal pH of the lysosome and inhibit the cathepsins. The partial sensitivity of protein degradation to these inhibitors in vivo suggests that some intracellular degradation, but not all, is catalyzed by the lysosomal proteases.

It has been well established that lysosomes are involved in the uptake and digestion of exogenous proteins. Such proteins are taken up from the extracellular medium by endocytosis, in which a region of the plasma membrane of a cell invaginates to form a closed vesicle entirely surrounded by cytoplasm. Extracellular materials are either trapped in the lumen of the vesicle or are bound to its lumenal surface. These vesicles fuse with lysosomes, which mixes the contents of the two compartments and leads to proteolytic degradation of any extracellular proteins that had been captured by the vesicle.

Cytoplasmic proteins are thought to enter lysosomes when a small volume of cytoplasm is sequestered by a cell membrane that then fuses with the lysosome, or when invagination of the lysosomal membrane occurs, followed by disruption of this invagination. The first pathway seems to be used primarily during starvation to obtain amino acids for synthesizing vital proteins; the second pathway is thought to be used for the normal turnover of principally long-lived proteins. Both pathways seem to be nonselective in what proteins of the cytoplasm are taken up. A third lysosomal pathway has recently been proposed: the uptake and degradation of cytosolic proteins in a highly selective manner that requires a specific peptide motif in the proteins to be degraded; this pathway is thought to be activated under certain conditions and to be specific to proteins that are dispensable.

References

Lysosomes. R. T. Dean and A. J. Barrett. *Essays Biochem.* 12;1–40 (1976).

Peptide sequences that target cytosolic proteins for lysosomal proteolysis. J. F. Dice. *Trends Biochem. Sci.* 15:305–309 (1990).

10.3.3 Ubiquitin-Mediated Pathway

The best defined mechanism of intracellular protein degradation, and the one that is increasingly found to be of great physiological significance, involves an intermediate in which the protein to be degraded is covalently attached to the protein **ubiquitin** (Fig. 10.1). This protein is abundant in all eukaryotes, as implied by its name. It also has the distinction of having the slowest rate of evolutionary divergence yet discovered (Chap. 3). For example, the known ubiquitins of animals are identical and differ at only three positions from those of yeast and plants. The function of this protein seems to

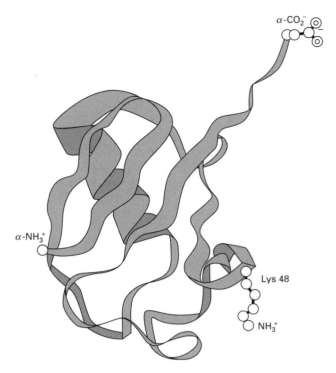

FIGURE 10.1
The structure of ubiquitin. The path of the polypeptide chain is illustrated, plus the side chain of Lys 48. Ubiquitin is conjugated to the amino groups of other proteins through its terminal carboxyl group, which is on a flexible segment. When ubiquitin is conjugated to another ubiquitin molecule, the α-carboxyl group of one molecule is linked in an amide bond to the ϵ-amino group of Lys 48 of the other ubiquitin molecule. (Figure kindly provided by G. Vriend.)

have been conserved throughout evolution, and virtually all the residues of this protein are likely to be involved in its function.

Ubiquitin is a small protein of only 76 residues that adopts a stable, compact globular conformation, with four strands of β-sheet and a single α-helix (Fig. 10.1). The three C-terminal residues, -Arg-Gly-Gly, are flexible and extend into the solvent. The terminal carboxyl group of ubiquitin can be covalently linked to other proteins through an amide bond to their α- and ϵ-amino groups.

The first step in linking ubiquitin to another protein is the activation of the ubiquitin α-carboxyl group and its linkage to a thiol group of the activating enzyme, known as E_1 (Fig. 10.2). The carboxyl is activated by its intermediate adenylation by ATP, as occurs in other such activation processes (Sec. 2.1.3.a, Fig. 9.14). The activated ubiquitin is transferred from E_1 to a second protein, E_2. Several forms of E_2 apparently transfer the ubiquitin to the α- and ϵ-amino groups of various proteins. Some proteins such as histones are known to be reversibly linked to ubiquitin, with various effects on their functions but without being degraded. For cases in which the protein is degraded, it is complexed first to one of a number of enzymes known collectively as E_3. Ubiquitin is transferred from one of the E_2 enzymes to the target protein in the complex with E_3. Multiple ubiquitin molecules are generally attached to the target protein to be degraded, often to the same site. These ubiquitin molecules are linked to each other in chains produced by an amide bond between the α-carboxyl group of one ubiquitin molecule and the side chain of Lys 48 of another. Such chains of ubiquitin molecules must be attached to the protein for it to be degraded by this system.

The target protein is degraded by a very large ATP-dependent protease complex, with a molecular weight of about 1.5×10^6, that is specific for ubiquitin–protein conjugates. The proteasome (Sec. 10.3.1) is part of this complex. ATP is necessary in this system for the attachment of ubiquitin to the target protein, for the assembly of the large protease complex, and for the proteolysis steps.

Ubiquitin is released from the protein being degraded and can be reused. It can be released from various proteins by several enzymes that cleave specifically the covalent bond between the ubiquitin C-terminus and the other protein. Such enzymes are probably important in removing ubiquitin molecules that are attached reversibly to other proteins, such as histones, and in the biosynthesis of ubiquitin. The genes for ubiquitin indicate that in many cases it is synthesized as a polyprotein in which multiple copies of ubiquitin are linked head-to-tail by peptide bonds, with a single additional residue at the C-terminus. In genes with a single ubiquitin sequence, another protein is linked at the carboxyl end; this other protein is incorporated into the structures of ribosomes. All of these proteins are released from the original polyprotein by specific proteolysis. The reason for this complex mechanism for the biosynthesis of ubiquitin is not known.

The question about the ubiquitin-dependent system that is most important for protein degradation is how the proteins to be degraded are selected. The E_3 enzymes seem to play a central role in this process because E_3 first binds the selected protein at specific

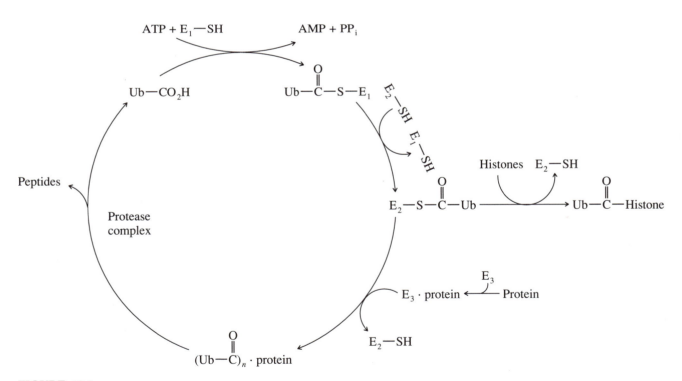

FIGURE 10.2
The ubiquitin-mediated pathway of protein degradation. Ub—CO_2H is ubiquitin, with its α-carboxyl group depicted. E_1, E_2, and E_3 are proteins involved in activating and transferring ubiquitin to the protein to be degraded. A protein complexed to E_3 has multiple (n) copies of ubiquitin conjugated to it. When protein E_3 dissociates from the protein being ubiquitinated and degraded is not known. Activated ubiquitin can also be transferred by specific E_2 enzymes to proteins that are not degraded, such as histones.

protein-binding sites. A free terminal α-amino group is necessary for such binding, and the identity of the terminal residue is important. In higher organisms, an N-terminal Met, Ile, Val, Gly, or Pro residue is thought to cause a low rate of degradation, whereas an Arg, Lys, His, Glu, Gln, Asp, Asn, Leu, Phe, Tyr, Trp, Ala, Ser, or Thr residue causes rapid degradation. The roles of the residues are somewhat different in yeast and other microorganisms. Three types of E_3 enzymes have been identified, with specificities for different N-terminal residues. Type I is specific for basic residues (Lys, Arg, His), type II for bulky hydrophobic residues (Phe, Tyr, Trp, Leu), and type III for small uncharged residues (Ala, Ser, Thr). These residues include many of the N-terminal residues that cause rapid degradation. Other residues that cause rapid degradation, such as Glu and Asp, are thought to function indirectly in that they favor the addition of an Arg residue to the N-terminus, which is the residue that most favors degradation. Arg residues can be added to the terminal α-amino group of Glu and Asp residues of proteins by transfer from Arg-tRNA catalyzed by specific enzymes (Sec. 2.4.2.b). This unusual posttranslational modification has been known for a long time, but this function for it has only recently been recognized. Terminal Asn and Gln residues are proposed to favor degradation of the protein only after their deamidation and the subsequent addition of a residue such as Arg.

The identity of the N-terminal residue of a protein is determined not only by its sequence but also by posttranslational modifications (Chap. 2). The N-terminal initiating Met residue may or may not be removed, and the α-amino group may or may not be acetylated or blocked in other ways. Whether or not these modifications occur depends on the identity of the next residue (Sec. 2.4.2). Intriguingly, the residues that favor removal of the initiator Met residue are those that seem to protect the protein from degradation by the ubiquitin-dependent system. In other cases, the initiating Met is not removed during biosynthesis, and an N-terminal Met residue also tends to inhibit degradation. Most proteins, therefore, tend to be synthesized in a form in which they are stabilized against degradation by the ubiquitin-dependent system. A new N-terminal residue, however, may be uncovered as a result of other proteolytic events. A modification to a protein that exposes a new N-terminus might then trigger its degradation. The specificity of this degradation system is such that one subunit of a tetrameric protein that acquires the degradative sequence can be degraded independently of the other, long-lived subunits of the same molecule.

It has been proposed that the presence of a Lys residue on a protein in a position, relative to the amino-terminal residue, that is suitable for the addition of ubi-quitin is another factor in the targeting of a protein by the ubiquitin-dependent system. It is likely that other aspects of the targeting of proteins for degradation by the ubiquitin system remain to be discovered.

The ubiquitin system has been shown to be responsible for the degradation of most of the short-lived proteins of cells such as cyclin and the products of the proto-oncogenes. It also is involved in the degradation of abnormal proteins. The synthesis of ubiquitin is induced as part of a cell's response to heat shock or other stressful conditions that can damage proteins, which then may have to be degraded.

References

The ubiquitin system: functions and mechanisms. D. Finley and A. Varshavsky. *Trends Biochem. Sci.* 10:343–347 (1985).

Ubiquitin-mediated pathways for intracellular proteolysis. M. Rechsteiner. *Ann. Rev. Cell Biol.* 3:1–30 (1987).

Ubiquitin-mediated protein degradation. A. Hershko. *J. Biol. Chem.* 263:15237–15240 (1988).

Cotranslational processing and protein turnover in eukaryotic cells. S. M. Arfin and R. A. Bradshaw. *Biochemistry* 27:7979–7984 (1988).

The degradation signal in a short-lived protein. A. Bachmair and A. Varshavsky. *Cell* 56:1019–1032 (1989).

Post-translational addition of an arginine moiety to acidic NH_2 termini of proteins is required for their recognition by ubiquitin-protein ligase. A. Elias and A. Ciechanover. *J. Biol. Chem.* 265:15511–15517 (1990).

New perspectives on the structure and function of ubiquitin. B. P. Monia et al. *Bio/Technology* 8:209–215 (1990).

Ubiquitin-conjugating enzymes: novel regulators of eukaryotic cells. S. Jentsch et al. *Trends Biochem. Sci.* 15:195–198 (1990).

Inhibition of the N-end rule pathway in living cells. R. T. Baker and A. Varshavsky. *Proc. Natl. Acad. Sci. USA* 88:1090–1094 (1991).

Cyclin is degraded by the ubiquitin pathway. M. Glotzer et al. *Nature* 349:132–138 (1991).

Ubiquitination. D. Finley and V. Chau. *Ann. Rev. Cell Biol.* 7:25–69 (1991).

Exercises

1. "Black smoker" bacteria were reported to have been isolated from deep-sea hydrothermal vents, where they were claimed to be growing at a temperature of 250°C (J. A. Baross and J. W. Deming, *Nature* 303:423–426, 1983). What is the greatest difficulty with this claim?

ANSWER
R. H. White, *Nature* 310:430–432 (1984).

2. The rates at which Asn and Gln residues in short peptides deamidate depend on the identities of their neighboring residues. It was proposed that the sequence around Asn and Gln residues in a protein could have been selected to produce a rate of deamidation that would be appropriate for the rate of turnover of that protein (A. B. Robinson et al., *Proc. Natl. Acad. Sci. USA* 66:753–757, 1970; A. B. Robinson, *Proc. Natl. Acad. Sci. USA* 71:885–888, 1974). What are the likely problems with this proposal?

ANSWERS
K. J. Dilley and J. J. Harding, *Biochim. Biophys. Acta* 386:391–408 (1975); J. J. Harding, *Colloq.-Inst. Natl. Sante Rech. Med.* 60:215–222 (1976).

3. Folded proteins are normally resistant to proteases, and it is plausible that local unfolding must precede proteolytic cleavage of the polypeptide chain. The flexibility and stability of a protein might then be expected to be a factor determining the rate at which a protein is normally degraded. How could you test this hypothesis?

ONE ANSWER
K. V. Rote and M. Rechsteiner, *J. Biol. Chem.* 261:15430–15436 (1986).

4. The carboxyl methyltransferase enzymes of eukaryotes were unexpectedly found to be specific for iso-Asp sequences, in which the peptide linkage occurs via the side chain β-carboxyl of the Asp residue rather than the α-carboxyl. Such residues are generated during the deamidation of Asn residues (see Fig. 1.3). This observation prompted the proposal that the carboxyl methyltransferase enzymes may be involved in the degradation or repair of proteins that have been modified by such deamidation (D. W. Aswad and B. A. Johnson, *Trends Biochem. Sci.* 12:155–158, 1987). How could this proposal be tested?

ANSWER
C. E. M. Voorter et al., *J. Biol. Chem.* 263:19020–19023 (1988).

5. The N-terminal amino acid residue of a protein appears to be especially important in determining the rate at which a protein is degraded (Sec. 10.3.3). How could you test whether a terminal amino group is necessary?

ONE ANSWER
K. V. Rote et al., *J. Biol. Chem.* 264:1156–1162 (1989).

Major Protein and DNA Sequence Data Banks

Data bank	Address
EMBL	European Molecular Biology Laboratory Meyerhofstrasse 1 6900 Heidelberg, Germany
GenBank	Los Alamos National Laboratory Los Alamos, New Mexico 87545
NEWAT	R. F. Doolittle Department of Chemistry University of California, San Diego La Jolla, California 92093
Protein Identification Resource (PIR)	National Biomedical Research Foundation Georgetown University Medical Center 3900 Reservoir Road NW Washington, DC 20007
PSeqlP	Computer Science Unit Institut Pasteur 28 rue du Dr. Roux 75724 Paris, Cedex 15, France
Swiss-Prot	A. Bairoch Departement de Biochemie Medicale Centre Medicale Universitaire 1211 Geneva 4, Switzerland

References to Protein Structures Determined Crystallographically to High Resolution

*P*roteins are listed in alphabetical order under their most common names. Related proteins that do not differ substantially are listed under the name of the most common member. Complexes with other macromolecules or in which the structure of the protein is substantially altered are designated by a / separating the names of the two components. Multiple references are given for structures that have been determined independently in different laboratories. Only the most recent or most comprehensive reference is given, so this list should not be taken to indicate the order or precedence of protein structure determinations.

Protein structure	Reference
Acetylcholinesterase	
Torpedo	J. L. Sussman et al., *Science* 253:872–879 (1991)
Aconitase	A. H. Robbins and C. D. Stout, *Proteins: Struct. Funct. Genet.* 5:289–312 (1989)
Actin	
/DNase I	W. Kabsch et al., *Nature* 347:37–44 (1990)
/profilin	C. E. Schutt et al., *J. Mol. Biol.* 209:735–746 (1989)
Actinidin	E. N. Baker, *J. Mol. Biol.* 141:441–484 (1980)
Acyl-CoA dehydrogenase	J. J. P. Kim and J. Wu, *Proc. Natl. Acad. Sci. USA* 85:6677–6681 (1988)
Adenosine deaminase	D. K. Wilson et al., *Science* 252:1278–1284 (1991)
Adenovirus major coat protein	M. M. Roberts et al., *Science* 232:1148–1151 (1986)

Protein structure

Reference

Adenylate kinase
  bacterial
  beef mitochondrial
  porcine cytosol
  yeast
Albumin, serum
Alcohol dehydrogenase
Aldolase
  KDPG
  muscle
Aldose reductase
  porcine
Alkaline phosphatase
  bacterial
α-Amylase
  pancreatic
  aspergillus
Amyloid β-protein precursor, protease inhibitor domain
Anaphylatoxin C3a
Annexin V
α_1-Antichymotrypsin
Anti-idiotopic Fab E225/antilysozyme Fab D1.3
Antithrombin III
α_1-Antitrypsin
Apolipoprotein
Apolipoprotein E receptor-binding domain
Arabinose-binding protein
Ascorbate oxidase
Asp tRNA synthetase/tRNA
Aspartate aminotransferase
  mitochondrial
  bacterial

Aspartate carbamoyl transferase
Aspartate receptor
  /aspartate
Azurin

Bacteriochlorophyll a protein
Barnase
Basic blue protein
  cucumber
Bean-pod mottle virus
Bence–Jones protein
  Au
  Loc
  Mcg
  Mcg-Weir hybrid
  Rei
  Rhe

C. W. Muller and G. E. Schulz, *J. Mol. Biol.* 202:909–912 (1988)
K. Diederichs and G. E. Schulz, *J. Mol. Biol.* 217:541–549 (1991)
D. Dreusicke et al., *J. Mol. Biol.* 199:359–371 (1988)
U. Egner et al., *J. Mol. Biol.* 195:649–658 (1987)
D. C. Carter and X. m. He, *Science* 249:302–303 (1990)
H. Eklund et al., *Biochemistry* 23:5982–5996 (1984)

I. M. Mavridis et al., *J. Mol. Biol.* 162:419–444 (1982)
J. Sygusch et al., *Proc. Natl. Acad. Sci. USA* 84:7846–7850 (1987)

J. M. Rondeau et al., *Nature* 355:469–472 (1992)

E. E. Kim and H. W. Wyckoff, *Clin. Chim. Acta* 186:175–188 (1989)

G. Buisson et al., *EMBO J.* 6:3909–3916 (1987)
E. Boel et al., *Biochemistry* 29:6244–6249 (1990)
T. R. Hynes et al., *Biochemistry* 29:10018–10022 (1990)

R. Huber et al., *Z. Physiol. Chem.* 361:1389–1399 (1980)
R. Huber et al., *EMBO J.* 9:3867–3874 (1990)
U. Baumann et al., *J. Mol. Biol.* 218:595–606 (1991)
G. A. Bentley et al., *Nature* 348:254–257 (1990)

M. Delarue et al., *Acta Crys.* B46:550–556 (1990)
H. Loebermann et al., *J. Mol. Biol.* 177:531–556 (1984)
D. R. Breiter et al., *Biochemistry* 30:603–608 (1991)
C. Wilson et al., *Science* 252:1817–1822 (1991)

F. A. Quiocho and N. K. Vyas, *Nature* 310:381–386 (1984)
A. Messerschmidt et al., *J. Mol. Biol.* 206:512–529 (1989)
M. Ruff et al., *Science* 252:1682–1686 (1991)

G. C. Ford et al., *Proc. Natl. Acad. Sci. USA* 77:2559–2563 (1980)
S. Kamitori et al., *J. Biochem.* 104:317–318 (1988)
D. L. Smith et al., *Biochemistry* 28:8161–8167 (1989)
H. Ke et al., *J. Mol. Biol.* 204:725–747 (1988)

M. V. Milburn et al., *Science* 254:1342–1347 (1991)
E. N. Baker, *J. Mol. Biol.* 203:1071–1095 (1988)
H. Nar et al., *J. Mol. Biol.* 221:765–772 (1991)

D. E. Tronrud et al., *J. Mol. Biol.* 188:443–454 (1986)
S. Baudet and J. Janin, *J. Mol. Biol.* 219:123–132 (1991)

J. M. Guss et al., *Science* 241:806–811 (1988)
Z. Chen et al., *Science* 245:154–159 (1989)

H. Fehlhammer et al., *Biophys. Struct. Mech.* 1:139–146 (1975)
M. Schiffer et al., *Biochemistry* 28:4066–4072 (1989)
K. R. Ely et al., *J. Mol. Biol.* 210:601–615 (1989)
K. R. Ely et al., *Mol. Immunol.* 27:101–114 (1990)
O. Epp et al., *Eur. J. Biochem.* 45:513–524 (1974)
W. Furey et al., *J. Mol. Biol.* 167:661–692 (1983)

Protein structure	Reference
Roy	P. M. Colman et al., *J. Mol. Biol.* 116:73–79 (1977)
Bilin binding protein	R. Huber et al., *J. Mol. Biol.* 198:499–513 (1987)
Black beetle virus	M. V. Hosur et al., *Proteins: Struct. Funct. Genet.* 2:167–176 (1987)
Bowman–Birk protease inhibitor	A. Suzuki et al., *J. Biochem.* 101:267–274 (1987)
/trypsin	Y. Tsunogae et al., *J. Biochem.* 100:1637–1646 (1986)
BPTI (bovine pancreatic trypsin	J. Deisenhofer and W. Steigemann, *Acta Cryst.* B31:238–250 (1975)
inhibitor)	A. Wlodawer et al., *J. Mol. Biol.* 198:469–480 (1987)
α-Bungarotoxin	R. A. Love and R. M. Stroud, *Protein Eng.* 1:37–46 (1986)
Calmodulin	Y. S. Babu et al., *J. Mol. Biol.* 204:191–204 (1988)
Canavalin	A. McPherson, *J. Biol. Chem.* 255:10472–10480 (1980)
Carbonic anhydrase	
I	K. K. Kannan et al., *Ann. N.Y. Acad. Sci.* 429:49–60 (1984)
II	A. E. Eriksson et al., *Proteins: Struct. Funct. Genet.* 4:274–282 (1988)
Carboxypeptidase	
A	D. C. Rees et al., *J. Mol. Biol.* 168:367–387 (1983)
A/inhibitor	D. C. Rees and W. N. Lipscomb, *J. Mol. Biol.* 160:475–498 (1982)
B	M. F. Schmid and J. R. Herriott, *J. Mol. Biol.* 103:175–190 (1976)
D,D	J. A. Kelly et al., *J. Mol. Biol.* 209:281–295 (1989)
serine, II	D. I. Liao and S. J. Remington, *J. Biol. Chem.* 265:6528–6531 (1990)
Cardiotoxin	
V_4^{II}	B. Rees et al., *J. Mol. Biol.* 214:281–297 (1990)
Catabolite gene activator	D. B. McKay et al., *J. Biol. Chem.* 257:9518–9524 (1982)
/cyclic AMP	I. T. Weber and T. A. Steitz, *J. Mol. Biol.* 198:311–326 (1987)
CAP91	M. C. Vaney et al., *Biochemistry* 28:4568–4574 (1989)
Catalase	W. R. Melik-Adamyan et al., *J. Mol. Biol.* 188:63–72 (1986)
Cathepsin B	D. Musil et al., *EMBO J.* 10:2321–2330 (1991)
CD4 cell-surface glycoprotein	
immunoglobulin-like domains	J. Wang et al., *Nature* 348:411–418 (1990)
	S. E. Ryu et al., *Nature* 348:419–426 (1990)
Cellobiohydrolase II	J. Rouvinen et al., *Science* 249:380–386 (1990)
Chaperone PapD	A. Holmgren and C. I. Brändén, *Nature* 342:248–251 (1989)
CheY	A. M. Stock et al., *Nature* 337:745–749 (1989)
	K. Volz and P. Matsumura, *J. Biol. Chem.* 266:15511–15519 (1991)
Chloramphenicol acetyltrans-	A. W. Leslie, *J. Mol. Biol.* 213:167–186 (1990)
ferase	
Cholesterol oxidase	A. Vrielink et al., *J. Mol. Biol.* 219:533–554 (1991)
Chymosin	G. L. Gilliland et al., *Proteins: Struct. Funct. Genet.* 8:82–101 (1990)
	P. Strop et al., *Biochemistry* 29:9863–9871 (1990)
α-Chymotrypsin	H. Tsukada and D. M. Blow, *J. Mol. Biol.* 184:703–711 (1985)
	R. A. Blevins and A. Tulinsky, *J. Biol. Chem.* 260:4264–4275 (1985)
/ovomucoid third domain	M. Fujinaga et al., *J. Mol. Biol.* 195:397–418 (1987)
/secretory leukocyte protease	M. G. Grütter et al., *EMBO J.* 7:345–351 (1988)
inhibitor	
γ-Chymotrypsin	G. H. Cohen et al., *J. Mol. Biol.* 148:449–479 (1981)
	M. M. Dixon and B. W. Matthews, *Biochemistry* 28:7033–7038 (1989)
Chymotrypsin inhibitor	C. A. McPhalen and M. N. G. James, *Biochemistry* 26:261–269 (1987)
/subtilisin novo	C. A. McPhalen and M. N. G. James, *Biochemistry* 27:6582–6598 (1988)
Chymotrypsinogen A	D. Wang et al., *J. Mol. Biol.* 185:595–624 (1985)
/pancreatic secretory trypsin	A. Hecht et al., *J. Mol. Biol.* 220:711–722 (1991)
inhibitor	
Citrate synthase	G. Wiegand and S. J. Remington, *Ann. Rev. Biophys. Biophys. Chem.* 15:97–117 (1986)
α-Cobratoxin	C. Betzel et al., *J. Biol. Chem.* 266:21530–21536 (1991)

Protein structure	Reference
ColE1 Rop protein	D. W. Banner et al., *J. Mol. Biol.* 196:657–675 (1987)
Colicin A fragment	M. W. Parker et al., *Nature* 337:93–96 (1989)
Concanavalin A	G. N. Reeke et al., *J. Biol. Chem.* 250:1525–1547 (1975)
Crambin	W. A. Hendrickson and M. M. Teeter, *Nature* 290:107–113 (1981)
Creatine aminohydrolase	M. Coll et al., *J. Mol. Biol.* 214:597–610 (1990)
p-Cresol methylhydroxylase (flavocytochrome *c*)	F. S. Matthews et al., *Biochemistry* 30:238–247 (1991)
Cro (λ) repressor	W. F. Anderson et al., *Nature* 290:754–758 (1981)
/operator	R. G. Brennan et al., *Proc. Natl. Acad. Sci. USA* 87:8165–8169 (1990)
Cro (434) repressor	A. Mondragón et al., *J. Mol. Biol.* 205:179–188 (1989)
/operator	A. Mondragón and S. C. Harrison, *J. Mol. Biol.* 219:321–334 (1991)
β-Crystallin B2	R. Lapatto et al., *J. Mol. Biol.* 222:1067–1083 (1991)
γ-Crystallin	
II	G. Wistow et al., *J. Mol. Biol.* 170:175–202 (1983)
IIIb	Y. N. Chirgadze et al., *FEBS Letters* 131:81–84 (1981)
IVa	H. E. White et al., *J. Mol. Biol.* 207:217–235 (1989)
Cupredoxin	E. T. Adman et al., *J. Biol. Chem.* 264:87–99 (1989)
Cyclodextrin glycosyltransferase	G. Klein and G. E. Schulz, *J. Mol. Biol.* 217:737–750 (1991)
Cystatin	W. Bode et al., *EMBO J.* 7:2593–2599 (1988)
Cytochrome	
*b*5	P. Argos and F. S. Mathews, *J. Biol. Chem.* 250:747–751 (1975)
*b*562	F. Lederer et al., *J. Mol. Biol.* 148:427–448 (1981)
c	T. Takano and R. E. Dickerson, *J. Mol. Biol.* 153:79–94 (1981)
	T. Tanaka et al., *J. Biochem.* 77:147–162 (1975)
	H. Ochi et al., *J. Mol. Biol.* 166:407–418 (1983)
*c*2	F. R. Salemme et al., *J. Biol. Chem.* 248:3910–3921 (1973)
	M. M. Benning et al., *J. Mol. Biol.* 220:673–685 (1991)
*c*3	M. Pierrot et al., *J. Biol. Chem.* 257:14341–14348 (1982)
	Y. Higuchi et al., *J. Mol. Biol.* 172:109–139 (1984)
*c*4	L. Sawyer et al., *J. Mol. Biol.* 153:831–835 (1981)
*c*5	D. C. Carter et al., *J. Mol. Biol.* 184:279–295 (1985)
*c*550	R. Timkovitch and R. E. Dickerson, *J. Biol. Chem.* 251:4033–4046 (1976)
*c*551	Y. Matsuura et al., *J. Mol. Biol.* 156:389–409 (1982)
*c*553	A. Nakagawa et al., *J. Biochem.* 108:701–703 (1990)
*c*554	M. L. Ludwig et al., in *Electron Transport and Oxygen Utilization*, C. Ho, ed., pp. 27–32. Elsevier, New York, 1982
*c*555	Z. R. Korszun and F. R. Salemme, *Proc. Natl. Acad. Sci. USA* 74:5244–5247 (1977)
c'	B. C. Finzel et al., *J. Mol. Biol.* 186:627–643 (1985)
P-450	R. Raag and T. L. Poulos, *Biochemistry* 28:7586–7592 (1989)
Cytochrome *c* peroxidase	J. Wang et al., *Biochemistry* 29:7160–7173 (1990)
Dienelactone hydrolase	D. Pathak et al., *J. Mol. Biol.* 204:435–445 (1988)
Dihydrofolate reductase	
E. coli	C. Bystroff and J. Kraut, *Biochemistry* 30:2227–2239 (1991)
human	J. F. Davies et al., *Biochemistry* 29:9467–9479 (1990)
lactobacillus	D. A. Matthews et al., *J. Biol. Chem.* 253:6940–6954 (1978)
R-plasmid	D. A. Matthews et al., *Biochemistry* 25:4194–4204 (1986)
DNA-binding protein II, bacteria	I. Tanaka et al., *Nature* 310:376–381 (1984)
DNA gyrase B protein	D. B. Wigley et al., *Nature* 351:624–629 (1991)
DNA polymerase I	
Klenow fragment	D. L. Ollis et al., *Nature* 313:762–766 (1985)
/DNA	P. S. Freemont et al., *Proc. Natl. Acad. Sci. USA* 85:8924–8928 (1988)
Deoxyribonuclease I	C. Oefner and D. Suck, *J. Mol. Biol.* 192:605–632 (1986)

Protein structure

/DNA
/actin
dUTPase
Eco R1 restriction endonuclease
  /DNA fragment

Eglin/subtilisin

Elastase
  porcine

  neutrophil
    /ovomucoid inhibitor
  pseudomonas
Elongation factor Tu
Endothiapepsin
δ-Endotoxin, bacillus
Enol-pyruvylshikimate-3-phos-
  phate synthase
Enolase
  yeast
Enterotoxin
Erabutoxin a
Erabutoxin b
Erythrocruorin
Exotoxin A
Fab antibody fragment
  AN02/DNP
  Anti-Lewis a
  B13l2/peptide
  D1.3/lysozyme
  D1.3/anti-idiotopic Fab E225
  HyHEL5/lysozyme
  Jel42
  J539
  Kol
  M603
  McPC603
  NC41/neuraminidase
  NQ10/12.5
  New
  R19.9
  4-4-20
  17/9/peptide
  36-71
Fab′ antibody fragment
  /peptide antigen
Fatty acid-binding protein

Favin
Fc antibody fragment
  /protein A

Reference

D. Suck et al., *Nature* 332:464–468 (1988)
W. Kabsch et al., *Nature* 347:37–44 (1990)
E. S. Cedergren-Zeppezauer et al., *Nature* 355:740–743 (1992)

C. A. Frederick et al., *Nature* 309:327–331 (1984)
Y. Kim et al., *Science* 249:1307–1309 (1990)
C. A. McPhalen and M. N. G. James, *Biochemistry* 27:6582–6598 (1988)
W. Bode et al., *Eur. J. Biochem.* 166:673–692 (1987)
D. W. Heinz et al., *J. Mol. Biol.* 217:353–371 (1991)

L. Sawyer et al., *J. Mol. Biol.* 118:137–208 (1978)
E. Meyer et al., *Acta Cryst.* B44:26–38 (1988)
M. A. Navia et al., *Proc. Natl. Acad. Sci. USA* 86:7–11 (1989)
W. Bode et al., *EMBO J.* 5:2453–2458 (1986)
M. M. Thayer et al., *J. Biol. Chem.* 266:2864–2871 (1991)
F. Jurnak, *Science* 230:32–36 (1985)
T. L. Blundell et al., *J. Mol. Biol.* 211:919–941 (1990)
J. Li et al., *Nature* 353:815–821 (1991)
W. C. Stallings et al., *Proc. Natl. Acad. Sci. USA* 88:5046–5050 (1991)

B. Stec and L. Lebioda, *J. Mol. Biol.* 211:235–248 (1990)
T. K. Sixma et al., *Nature* 351:371–377 (1991)
P. W. R. Corfield et al., *J. Biol. Chem.* 264:9239–9242 (1989)
P. E. Bourne et al., *Eur. J. Biochem.* 153:521–527 (1985)
W. Steigemann and E. Weber, *J. Mol. Biol.* 127:309–338 (1979)
V. S. Allured et al., *Proc. Natl. Acad. Sci. USA* 83:1320–1324 (1986)

A. T. Brünger et al., *J. Mol. Biol.* 221:239–256 (1991)
J. Vitali et al., *J. Mol. Biol.* 198:351–355 (1987)
R. L. Stanfield et al., *Science* 248:712–719 (1990)
A. G. Amit et al., *Science* 233:747–753 (1986)
G. A. Bentley et al., *Nature* 348:254–257 (1990)
E. A. Padlan et al., *Proc. Natl. Acad. Sci. USA* 86:5938–5942 (1989)
L. Prasad et al., *J. Biol. Chem.* 263:2571–2574 (1988)
S. W. Suh et al., *Proteins: Struct. Funct. Genet.* 1:74–80 (1986)
M. Matsushima et al., *J. Mol. Biol.* 121:441–459 (1978)
D. M. Segal et al., *Proc. Natl. Acad. Sci. USA* 71:4298–4302 (1974)
Y. Satow et al., *J. Mol. Biol.* 190:593–604 (1986)
P. M. Colman et al., *Nature* 326:358–363 (1987)
P. M. Alzari et al., *EMBO J.* 9:3807–3814 (1990)
F. A. Saul et al., *J. Biol. Chem.* 253:585–597 (1978)
M. B. Lascombe et al., *Proc. Natl. Acad. Sci. USA* 86:607–611 (1989)
J. N. Herron et al., *Proteins: Struct. Funct. Genet.* 5:271–280 (1989)
J. M. Rini et al., *Science* 255:959–965 (1992)
R. K. Strong et al., *Biochemistry* 30:3739–3748 (1991)

R. L. Stanfield et al., *Science* 248:712–719 (1990)
J. C. Sacchettini et al., *Proc. Natl. Acad. Sci. USA* 86:7736–7740 (1989)
G. Scapin et al., *Mol. Cell. Biochem.* 98:95–99 (1990)
G. N. Reeke, Jr., and J. W. Becker, *Science* 234:1108–1111 (1986)
J. Deisenhofer et al., *Z. Physiol. Chem.* 359:975 (1978)
J. Deisenhofer, *Biochemistry* 20:2361–2370 (1981)

Protein structure	Reference
Ferredoxin	
2-Fe	T. Tsukihara et al., *J. Mol. Biol.* 216:399–410 (1990)
	W. R. Rypniewski et al., *Biochemistry* 30:4126–4131 (1991)
3-Fe	C. R. Kissinger et al., *FEBS Letters* 244:447–450 (1989)
3-Fe/4-Fe	G. H. Stout et al., *Proc. Natl. Acad. Sci. USA* 85:1020–1022 (1988)
	C. D. Stout, *J. Mol. Biol.* 205:545–555 (1989)
	C. R. Kissinger et al., *J. Mol. Biol.* 219:693–715 (1991)
4-Fe	F. Fukuyama et al., *J. Mol. Biol.* 210:383–398 (1989)
8-Fe	E. T. Adman et al., *J. Biol. Chem.* 251:3801–3806 (1976)
	H. M. Kirshna Murthy et al., *J. Biol. Chem.* 263:18430–18436 (1988)
Ferredoxin NADP$^+$ reductase	P. A. Karplus et al., *Science* 251:60–66 (1991)
Ferredoxin reductase	S. Sheriff and J. R. Herriott, *J. Mol. Biol.* 145:441–451 (1981)
Ferritin	
L form	S. H. Banyard et al., *Nature* 271:282–284 (1978)
H form	D. M. Lawson et al., *Nature* 349:541–544 (1991)
Fibroblast growth factor	A. E. Eriksson et al., *Proc. Natl. Acad. Sci. USA* 88:3441–3445 (1991)
	J. Zhang et al., *Proc. Natl. Acad. Sci. USA* 88:3446–3450 (1991)
FIS (factor for inversion stimulation)	D. Kostrewa et al., *Nature* 349:178–180 (1991)
FK506 binding protein	G. D. van Duyne et al., *Science* 252:839–842 (1991)
Flavocytochrome b2	Z. X. Xia and F. S. Mathews, *J. Mol. Biol.* 212:837–863 (1990)
Flavodoxin	
bacterial	W. Watt et al., *J. Mol. Biol.* 218:195–208 (1991)
algal	K. Fukuyama et al., *J. Biol. Chem.* 265:15804–15812 (1990)
Foot and mouth disease virus	R. Acharya et al., *Nature* 337:709–716 (1989)
Fructose-1,6-bisphosphatase	H. Ke et al., *Proc. Natl. Acad. Sci. USA* 88:2989–2993 (1991)
Fv antibody fragment	
D1.3/lysosome	T. N. Bhat et al., *Nature* 347:483–485 (1990)
D-Galactose-binding protein	N. K. Vyas et al., *Science* 242:1290–1295 (1988)
	S. L. Mowbray and G. A. Petsko, *J. Biol. Chem.* 258:7991–7997 (1983)
Galactose oxidase	N. Ito et al., *Nature* 350:87–90 (1991)
Gene 5 DNA unwinding protein	G. D. Brayer and A. McPherson, *J. Mol. Biol.* 169:565–596 (1983)
Glucagon	K. Sasaki et al., *Nature* 257:751–757 (1975)
Glucocorticoid receptor/DNA	B. F. Luisi et al., *Nature* 352:497–505 (1991)
Glucose/galactose receptor, periplasmic	S. I. Mowbray et al., *Receptor* 1:41–53 (1990)
Glucose isomerase	K. Henrick et al., *Protein Eng.* 1:467–469 (1987)
Glucose-6-P isomerase	P. J. Shaw and H. Muirhead, *J. Mol. Biol.* 109:475–485 (1977)
Glutamate dehydrogenase	
bacterial	P. J. Baker et al., *Proteins: Struct. Funct. Genet.* 12:75–86 (1992)
Glutaminase-asparaginase	H. L. Ammon et al., *J. Biol. Chem.* 263:150–156 (1988)
Glutamine synthetase	M. Y. Yamashita et al., *J. Biol. Chem.* 264:17681–17690 (1989).
Glutaminyl tRNA synthetase/tRNA^{Gln}	J. J. Perona et al., *Proc. Natl. Acad. Sci. USA* 88:2903–2907 (1991)
Glutathione peroxidase	O. Epp et al., *Eur. J. Biochem.* 133:51–69 (1983)
Glutathione reductase	
erythrocyte	P. A. Karplus and G. E. Schulz, *J. Mol. Biol.* 210:163–180 (1989)
bacterial	U. Ermler and G. E. Schulz, *Proteins: Struct. Funct. Genet.* 9:174–179 (1991)
Glutathione S-transferase, π	P. Reinemer et al., *EMBO J.* 10:1997–2005 (1991)
Glyceraldehyde-P dehydrogenase	
lobster	M. Buehner et al., *J. Mol. Biol.* 90:25–49 (1974)

Protein structure

Reference

microbial	T. Skarzynski and A. J. Wonacott, *J. Mol. Biol.* 203:1097–1118 (1988)
Glycolate oxidase	Y. Lindqvist, *J. Mol. Biol.* 209:151–166 (1989)
Growth hormone	S. S. Abdel-Meguid et al., *Proc. Natl. Acad. Sci. USA* 84:6434–6437 (1987)
/receptor	A. M. de Vos et al., *Science* 255:306–312 (1992)
Guanylate kinase	T. Stehle and G. E. Schulz, *J. Mol. Biol.* 211:249–254 (1990)
Haloalkane dehalogenase	S. M. Franken et al., *EMBO J.* 10:1297–1302 (1991)
70k Heat shock protein	
ATPase fragment	K. M. Flaherty et al., *Nature* 346:623–628 (1990)
Hemagglutinin	W. Weis et al., *J. Mol. Biol.* 212:737–761 (1990)
Hemerythrin	
monomer	S. Sheriff et al., *J. Mol. Biol.* 197:273–296 (1987)
trimer	J. L. Smith et al., *Nature* 303:86–88 (1983)
octamer	M. A. Holmes et al., *J. Mol. Biol.* 218:583–593 (1991)
Hemocyanin	
arthropod	A. Volbeda and W. G. J. Hol, *J. Mol. Biol.* 209:249–279 (1989)
Hemoglobin	
mammalian	G. Fermi and M. F. Perutz, *Haemoglobin and Myoglobin*, Clarendon Press, Oxford, 1981
lamprey	R. B. Honzatko et al., *J. Mol. Biol.* 184:147–164 (1985)
annelid	G. Arents and W. E. Love, *J. Mol. Biol.* 210:149–161 (1989)
clam	W. E. Royer, Jr., et al., *Science* 249:518–521 (1990)
Hexokinase	C. M. Anderson et al., *J. Mol. Biol.* 123:1–13 (1978)
High-potential iron protein	C. W. Carter et al., *J. Biol. Chem.* 249:4212–4225 (1979)
Hirudin/α-thrombin	T. J. Rydel et al., *J. Mol. Biol.* 221:583–601 (1991)
	E. Skrzypczak-Jankun et al., *J. Mol. Biol.* 221:1379–1393 (1991)
Histidine decarboxylase	E. H. Parks et al., *J. Mol. Biol.* 182:455–465 (1985)
Histone octamer	G. Arenta et al., *Proc. Natl. Acad. Sci. USA* 88:10148–10152 (1991)
HIV-1 protease	A. Wlodawer et al., *Science* 245:616–621 (1989)
	M. A. Navia et al., *Nature* 337:615–620 (1989)
	R. Lapatto et al., *Nature* 342:299–302 (1989)
HLA histocompatibility antigen	
A2	M. A. Saper et al., *J. Mol. Biol.* 219:277–319 (1991)
Aw68	T. P. J. Garrett et al., *Nature* 342:692–696 (1989)
Homeodomain engrailed/DNA	C. R. Kissinger et al., *Cell*: 579–590 (1990)
HPr, phosphocarrier protein	O. A. L. El-Kabbani et al., *J. Biol. Chem.* 262:12926–12929 (1987)
Hydroxyacyl-CoA dehydrogenase	J. J. Birktoft et al., *Proc. Natl. Acad. Sci. USA* 84:8262–8266 (1987)
p-Hydroxybenzoate hydroxylase	H. A. Schreuder et al., *J. Mol. Biol.* 208:679–696 (1989)
3α,20β-Hydroxysteroid dehydrogenase	
bacterial	D. Ghosh et al., *Proc. Natl. Acad. Sci. USA* 88:10064–10068 (1991)
Immunoglobulin G	
Dob	E. W. Silverton et al., *Proc. Natl. Acad. Sci. USA* 74:5140–5144 (1977)
Kol	M. Marquart et al., *J. Mol. Biol.* 141:369–391 (1980)
Mcg	S. S. Rajan et al., *Mol. Immunol.* 20:787–799 (1983)
Insecticyanin	H. M. Holden et al., *EMBO J.* 6:1565–1570 (1987)
Insulin	T. Blundell et al., *Adv. Protein Chem.* 26:279–402 (1972)
	A. Wlodawer et al., *Acta Cryst.* B45:99–107 (1989)
Interferon β	T. Senda et al., *Proc. Japan. Acad.* 663:77–80 (1990)
Interferon γ	S. E. Ealick et al., *Science* 252:698–702 (1991)
	C. T. Samudzi et al., *J. Biol. Chem.* 266:21791–21797 (1991)
Interleukin	
1α	B. J. Graves et al., *Biochemistry* 29:2679–2684 (1990)

Protein structure	Reference
1β	B. C. Finzel et al., *J. Mol. Biol.* 209:779–791 (1989)
	J. P. Priestle et al., *Proc. Natl. Acad. Sci. USA* 86:9667–9671 (1989)
2	B. J. Brandhuber et al., *Science* 238:1707–1709 (1987)
8	E. T. Baldwin et al., *Proc. Natl. Acad. Sci. USA* 88:502–506 (1991)
Isocitrate dehydrogenase	J. H. Hurley et al., *J. Biol. Chem.* 265:3599–3602 (1990)
Isolectin I, *lathyrus ochrus*	Y. Bournet et al., *J. Biol. Chem.* 267:197–203 (1992)
/trisaccharide	Y. Bourne et al., *J. Biol. Chem.* 265:18161–18165 (1990)
Iso-1-cytochrome *c*	G. V. Louie et al., *J. Mol. Biol.* 199:295–314 (1988)
3-Isopropylmalate dehydrogenase	
bacterial	K. Imada et al., *J. Mol.. Biol.* 222:725–738 (1991)
Kallikrein A	W. Bode et al., *J. Mol. Biol.* 164:237–282 (1983)
/BPTI	Z. Chen and W. Bode, *J. Mol. Biol.* 164:283–311 (1983)
Kringle 4 domain, plasminogen	A. M. Mulichak et al., *Biochemistry* 30:10576–10588 (1991)
Kunitz-type trypsin inhibitor	
eruthrina	S. Onesti et al., *J. Mol. Biol.* 217:153–176 (1991)
α-Lactalbumin	K. R. Acharya et al., *J. Mol. Biol.* 221:571–581 (1991)
β-Lactamase	
bacillus	J. R. Knox and P. C. Moews, *J. Mol. Biol.* 220:435–455 (1991)
citrobacter	C. Oefner et al., *Nature* 343:284–288 (1990)
staphylococcus	O. Herzberg, *J. Mol. Biol.* 217:701–719 (1991)
streptomyces	O. Dideberg et al., *Biochem. J.* 245:911–913 (1987)
Lactate dehydrogenase	
bacterial	D. B. Wigley et al., *J. Mol. Biol.* 223:317–335 (1992)
heart, B4	J. J. Holbrook et al., *The Enzymes* 11:191–292 (1975)
muscle, A4	Abad-Zapatero et al., *J. Mol. Biol.* 198:445–467 (1987)
testicular, C4	H. H. Hogrefe et al., *J. Biol. Chem.* 262:13155–13162 (1987)
Lactoferrin	B. F. Anderson et al., *J. Mol. Biol.* 209:711–734 (1989)
/apo	B. F. Anderson et al., *Nature* 344:784–787 (1990)
β-Lactoglobulin	M. Z. Papiz et al., *Nature* 324:383–388 (1986)
Lambda (λ) repressor	C. O. Pabo and M. Lewis, *Nature* 298:443–445 (1982)
/operator	S. R. Jordan and C. O. Pabo, *Science* 242:893–899 (1988)
Lectin	
Ca^{2+}-dependent	W. I. Weis et al., *Science* 254:1608–1615 (1991)
/lactose	B. Shaanan et al., *Science* 254:862–866 (1991)
Leghemoglobin	B. K. Vainshtein et al., *Dokl. Akad. SSR* 223:238–241 (1977)
Leucine aminopeptidase	S. K. Burley et al., *Proc. Natl. Acad. Sci. USA* 88:6916–6920 (1991)
Leucine-binding protein	J. S. Sack et al., *J. Mol. Biol.* 206:193–207 (1989)
Leu, Ile, Val-binding protein	J. S. Sack et al., *J. Mol. Biol.* 206:171–191 (1989)
Lipase	
triacylglycerol	L. Brady et al., *Nature* 343:767–770 (1990)
	F. K. Winkler et al., *Nature* 343:771–774 (1990)
fungal	J. D. Schrag et al., *Nature* 351:761–764 (1991)
Lipoamide dehydrogenase	A. Takenaka et al., *J. Biochem.* 103:463–469 (1988)
	A. Matteri et al., *J. Mol. Biol.* 220:975–994 (1991)
Lysozyme	
hen	T. Imoto et al., *The Enzymes* 7:665–868 (1972)
human	P. J. Artymiuk and C. C. F. Blake, *J. Mol. Biol.* 152:737–762 (1981)
phage T4	L. H. Weaver and B. W. Matthews, *J. Mol. Biol.* 193:189–199 (1987)
streptomyces	S. Harada et al., *J. Biol. Chem.* 256:11600–11602 (1981)
α-Lytic protease	M. Fujinaga et al., *J. Mol. Biol.* 183:479–502 (1985)

Protein structure	Reference
Malate dehydrogenases	
cytoplasmic	J. J. Birktoft et al., *Biochemistry* 28:6065–6081 (1989)
mitochondrial	S. L. Roderick and L. J. Banaszak, *J. Biol. Chem.* 261:9461–9464 (1986)
Maltose-binding protein	J. C. Spurlino et al., *J. Biol. Chem.* 266:5202–5219 (1991)
Mandelate racemase	D. J. Neidhart et al., *Nature* 347:692–694 (1990)
Mast cell protease	S. J. Remington et al., *Biochemistry* 27:8097–8105 (1988)
Matα2 homeodomain	
/operator	C. Wolberger et al., *Cell* 67:517–528 (1991)
Mellitin	T. C. Terwilliger and D. Eisenberg, *J. Biol. Chem.* 257:6016–6022 (1982)
Mengo virus	S. Krishnaswamy and M. G. Rossmann, *J. Mol. Biol.* 211:803–844 (1990)
Mercuric ion reductase	N. Schiering et al., *Nature* 352:168–172 (1991)
Met repressor	J. B. Rafferty et al., *Nature* 341:705–710 (1989)
Met tRNA synthetase fragment	S. Brunie et al., *J. Mol. Biol.* 216:411–424 (1990)
Metallothionein	A. H. Robbins et al., *J. Mol. Biol.* 221:1269–1293 (1991)
Methylamine dehydrogenase	F. M. D. Vellieux et al., *EMBO J.* 8:2171–2178 (1989)
β2-Microglobulin	J. W. Becker and G. N. Reeke, Jr., *Proc. Natl. Acad. Sci. USA* 82:4225–4229 (1985)
Monellin	C. Ogata et al., *Nature* 328:739–742 (1987)
MS2 bacteriophage	K. Valegård et al., *Nature* 345:36–41 (1990)
Muconate lactonizing enzyme	A. Goldman et al., *J. Mol. Biol.* 194:143–153 (1987)
Muconolactone isomerase	S. K. Katti et al., *J. Mol. Biol.* 205:557–571 (1989)
Myoglobin	
horse	S. V. Evans and G. D. Brayer, *J. Mol. Biol.* 213:885–897 (1990)
sperm whale	S. E. V. Phillips, *J. Mol. Biol.* 142:531–554 (1980)
mollusc	M. Bolognesi et al., *J. Mol. Biol.* 205:529–544 (1989)
NADH oxidoreductase	I. G. Young et al., *Eur. J. Biochem.* 116:165–170 (1981)
NADH peroxidase	
streptococcus	T. Stehle et al., *J. Mol. Biol.* 221:1325–1344 (1991)
Nerve growth factor, murine	N. Q. McDonald et al., *Nature* 354:411–414 (1991)
Neuraminidase	
N2	J. N. Varghese et al., *Nature* 303:35–40 (1983)
N9	A. T. Baker et al., *Proteins: Struct. Funct. Genet.* 1:111–117 (1987)
A/Tokyo/3/67	J. N. Varghese and P. M. Colman, *J. Mol. Biol.* 221:473–486 (1991)
Neurophysin II	L. Chen et al., *Proc. Natl. Acad. Sci. USA* 88:4240–4244 (1991)
Neurotoxin	
scorpion	R. J. Almassy et al., *J. Mol. Biol.* 170:497–527 (1983)
snake, short	D. Tsernoglou and G. A. Petsko, *FEBS Letters* 68:1–4 (1976)
	P. W. R. Corfield et al., *J. Biol. Chem.* 264:9239–9242 (1989)
snake, long	M. D. Walkinshaw et al., *Proc. Natl. Acad. Sci. USA* 77:2400–2404 (1980)
Neutral protease	
bacillus	R. A. Pauptit et al., *J. Mol. Biol.* 199:525–537 (1988)
streptomyces	S. Harada et al., *J. Biochem.* 110:46–49 (1991)
Nitrite reductase	J. W. Godden et al., *Science* 253:438–442 (1991)
Oncomodulin	F. R. Ahmed et al., *J. Mol. Biol.* 216:127–140 (1990)
Ovalbumin	P. E. Stein et al., *J. Mol. Biol.* 221:941–959 (1991)
Ovomucoid 3rd domain	W. Bode et al., *Eur. J. Biochem.* 147:387–395 (1985)
cleaved	D. Musil et al., *J. Mol. Biol.* 220:739–755 (1991)
/protease B	R. J. Read et al., *Biochemistry* 22:4420–4433 (1983)
/α-chymotrypsin	M. Fujinaga et al., *J. Mol. Biol.* 195:397–418 (1987)
P1 nuclease	A. Volbeda et al., *EMBO J.* 10:1607–1618 (1991)
P2 myelin protein	T. A. Jones et al., *EMBO J.* 7:1597–1604 (1988)
Pancreatic polypeptide	T. L. Blundell et al., *Proc. Natl. Acad. Sci. USA* 78:4175–4179 (1981)

Protein structure

Reference

Pancreatic secretory trypsin inhibitor
  /chymotrypsinogen A — A. Hecht et al., *J. Mol. Biol.* 220:711–722 (1991)
  /trypsinogen — M. Bolognesi et al., *J. Mol. Biol.* 162:839–868 (1983)
Papain — I. G. Kamphuis et al., *J. Mol. Biol.* 179:233–256 (1984)
  /stefin B — M. T. Stubbs et al., *EMBO J.* 9:1939–1947 (1990)
Parvalbumin — A. L. Swain et al., *J. Biol. Chem.* 264:16620–16628 (1989)
  V. D. Kumar et al., *Biochemistry* 29:1404–1412 (1990)
  J. P. Declercq et al., *J. Mol. Biol.* 220:1017–1039 (1991)

Parvovirus — J. Tsao et al., *Science* 251:1456–1464 (1991)
Pea lectin — H. Einspahr et al., *J. Biol. Chem.* 261:16518–16527 (1986)
Penicillopepsin — M. N. G. James and A. R. Sielecki, *Biochemistry* 24:3701–3713 (1985)
Pepsin
  Endothia parasitica — T. L. Blundell et al., *J. Mol. Biol.* 211:919–941 (1990)
  porcine — A. R. Sielecki et al., *J. Mol. Biol.* 214:143–170 (1990)
  J. B. Cooper et al., *J. Mol. Biol.* 214:199–222 (1990)
  C. Abad-Zapatero et al., *Proteins: Struct. Funct. Genet.* 8:62–81 (1990)
  N. S. Andreeva et al., *J. Biol. Chem.* 259:11353–11356 (1984)
  Rhizopus chinensis — K. Suguna et al., *J. Mol. Biol.* 196:877–900 (1987)
Pepsinogen — A. R. Sielecki et al., *J. Mol. Biol.* 219:671–692 (1991)
Peptidases
  D-Ala-D-Ala — J. A. Kelly et al., *J. Biol. Chem.* 260:6449–6458 (1985)
  Zn^{+2} — O. Dideberg et al., *Nature* 299:469–470 (1982)
pFc′ antibody fragment — S. H. Bryant et al., *Acta Cryst.* B41:362–368 (1985)
Phage 434 repressor — A. Mondragon et al., *J. Mol. Biol.* 205:189–200 (1989)
  /operator — J. K. Aggarwal et al., *Science* 242:899–907 (1988)
Phaseolin — M. C. Lawrence et al., *EMBO J.* 9:9–15 (1990)
Phosphate-binding protein — H. Luecke and F. A. Quiocho, *Nature* 347:402–406 (1990)
Phosphocarrier protein III[glc]
  E. coli — D. Worthylake et al., *Proc. Natl. Acad. Sci. USA* 88:10382–10386 (1991)
Phosphofructokinase — W. R. Rypniewski and P. R. Evans, *J. Mol. Biol.* 207:805–821 (1989)
Phosphoglucomutase — Z. J. Lin et al., *J. Biol. Chem.* 261:264–274 (1986)
Phosphogluconate dehydro-
  genase — M. J. Adams et al., *EMBO J.* 2:1009–1014 (1983)
Phosphoglycerate kinase
  yeast — H. C. Watson et al., *EMBO J.* 1:1635–1640 (1982)
  horse — R. D. Banks et al., *Nature* 279:773–777 (1979)
Phosphoglycerate mutase — J. W. Campbell et al., *Nature* 250:301–303 (1974)
Phospholipase A2
  bee venom — D. I. Scott et al., *Science* 250:1563–1566 (1990)
  pancreatic — B. W. Dijkstra et al., *J. Mol. Biol.* 168:163–179 (1983)
  R. Renetseder et al., *J. Mol. Biol.* 200:181–188 (1988)
  secretory — J. P. Wery et al., *Nature* 352:79–82 (1991)
  snake venom — S. Brunie et al., *J. Biol. Chem.* 260:9742–9749 (1985)
  S. P. White et al., *Science* 250:1560–1563 (1990)
  D. R. Holland et al., *J. Biol. Chem.* 265:17649–17656 (1990)

Phospholipase C
  bacterial — E. Hough et al., *Nature* 338:357–360 (1989)
Phosphorylase
  a — S. Sprang and R. J. Fletterick, *J. Mol. Biol.* 131:523–551 (1979)
  b — L. N. Johnson et al., *J. Mol. Biol.* 140:565–580 (1980)
  /ligands — L. N. Johnson et al., *J. Mol. Biol.* 211:645–661 (1990)

Protein structure

Reference

Phosvitin/lipovitellin R. Raag et al., *J. Mol. Biol.* 200:553–569 (1988)
Photoreceptor protein D. E. McRee et al., *Proc. Natl. Acad. Sci. USA* 86:6533–6537 (1989)
Photosynthetic reaction center
   Rhodopseudomonas viridis J. Deisenhofer et al., *J. Mol. Biol.* 180:385–398 (1984)
   Rhodobacter sphaeroides H. Komiya et al., *Proc. Natl. Acad. Sci. USA* 85:9012–9016 (1988)
C-Phycocyanin M. Duerring et al., *J. Mol. Biol.* 217:577–592 (1991)
Phycoerythrocyanin M. Duerring et al., *J. Mol. Biol.* 211:633–644 (1990)
Plakalbumin H. T. Wright et al., *J. Mol. Biol.* 213:513–528 (1990)
Plasminogen activator inhibitor-1 J. Mottonen et al., *Nature* 355:270–273 (1992)
Plasminostreptin N. Kamiya et al., *Bull. Chem. Soc. Japan* 57:2075–2081 (1984)
Plastocyanin
   algal C. A. Collyer et al., *J. Mol. Biol.* 211:617–632 (1990)
   poplar W. B. Church et al., *J. Biol. Chem.* 261:234–237 (1986)
Platelet factor 4 R. St. Charles et al., *J. Biol. Chem.* 264:2092–2099 (1989)
Poliovirus J. M. Hogle et al., *Science* 229:1358–1367 (1985)
Porin, Rhodobacter M. W. Weiss et al., *Science* 254:1627–1630 (1991)
PRA isomerase-InGP synthase M. Wilmanns et al., *J. Mol. Biol.* 223:477–507 (1992)
   synthase domain M. Wilmanns et al., *Protein Eng.* 3:173–180 (1990)
Prealbumin C. C. F. Blake et al., *J. Mol. Biol.* 121:339–356 (1978)
Procarboxypeptidase B M. Coll et al., *EMBO J.* 10:1–9 (1991)
Profilin/actin C. E. Schutt et al., *J. Mol. Biol.* 209:735–746 (1989)
Protease
   A A. R. Sielecki et al., *J. Mol. Biol.* 134:781–804 (1979)
   A/chymostatin L. T. J. Delbaere and G. D. Brayer, *J. Mol. Biol.* 183:89–103 (1985)
   B L. T. J. Delbaere et al., *Can. J. Biochem.* 57:135–144 (1979)
   B/ovomucoid fragment R. J. Reed et al., *Biochemistry* 22:4420–4433 (1983)
   B/chymotrypsin inhibitor-1 H. M. Greenblatt et al., *J. Mol. Biol.* 205:201–228 (1989)
Protein A J. Deisenhofer, *Biochemistry* 20:2361–2370 (1981)
Protein kinase
   cAMP-dependent D. R. Knighton et al., *Science* 253:407–414 (1991)
α1-Proteinase inhibitor H. Loebermann et al., *J. Mol. Biol.* 177:531–556 (1984)
   S variant R. Engh et al., *Protein Eng.* 2:407–415 (1989)
Proteinase K Ch. Betzel et al., *Protein Eng.* 3:161–172 (1990)
Proteinase K & α-amylase inhibi- K. J. Zemke et al., *FEBS Letters* 279:240–242 (1991)
   tor
Prothrombin fragment 1 T. P. Seshadri et al., *J. Mol. Biol.* 220:481–494 (1991)
Protocatechuate 3,4-dioxygenase D. H. Ohlendorf et al., *Nature* 336:403–405 (1988)
Pseudoazurin K. Petratos et al., *Acta Cryst.* B44:628–636 (1988)
Purine nucleoside phosphorylase S. E. Ealick et al., *J. Biol. Chem.* 265:1812–1820 (1990)
α_1-Purothionin M. M. Teeter et al., *Proteins: Struct. Funct. Genet.* 8:118–132 (1990)
Pyruvate kinase H. Muirhead et al., *EMBO J.* 5:475–481 (1986)
ras protein
   oncogene p21 E. F. Pai et al., *EMBO J.* 9:2351–2355 (1990)
A. T. Brünger et al., *Proc. Natl. Acad. Sci. USA* 87:4849–4853 (1990)
   normal L. Tong et al., *J. Mol. Biol.* 217:503–516 (1991)
recA protein, *E. coli* R. M. Story et al., *Nature* 355:318–325 (1992)
Relaxin, human C. Eigenbrot et al., *J. Mol. Biol.* 221:15–21 (1991)
Renin A. R. Sielecki et al., *Science* 243:1346–1351 (1989)
$\gamma\delta$ Resolvase catalytic domain M. R. Sanderson et al., *Cell* 63:1323–1329 (1990)
Retinol-binding protein S. W. Cowan et al., *Proteins: Struct. Funct. Genet.* 8:44–61 (1990)
Reverse transcriptase
   HIV-1, ribonuclease H domain J. F. Davis et al., *Science* 252:88–95 (1991)

Protein structure

Reference

Rhinovirus
  1A
  14
Rhodanese
Riboflavin synthase

Ribonuclease
  A, bovine
  A, bovine, phosphate-free
  A, bovine, sulfate-free
  B, bovine
  bacillus Bi
  H

  Ms
  S, bovine

  St
  T1
Ribonucleotide reductase B2
Ribosomal protein
  L7/L12 C-terminal domain
  L30
Ribulose-1,5-bisphosphate car-
    boxylase (RuBisCO)
  tobacco
  spinach

  Rhodospirullum rubrum
Ricin
  A chain
  B chain

Rop, plasmid ColE1
Rous sarcoma virus protease
Rubredoxin
  desulfovibrio

  clostridium
Sarcoplasmic Ca²⁺-binding pro-
    tein
Satellite tobacco necrosis virus
Savinase proteinase, bacillus
Scorpion toxin II

Secretory leukocyte protease in-
    hibitor/α-chymotrypsin
Serine proteinase inhibitor 2
  barley
Serum albumin, human
Seryl tRNA synthetase
Sindbis virus coat protein

S. Kim et al., *J. Mol. Biol.* 210:91–111 (1989)
E. Arnold and M. G. Rossmann, *J. Mol. Biol.* 211:763–801 (1990)
J. H. Ploegman et al., *J. Mol. Biol.* 123:557–594 (1978)
R. Ladenstein et al., *J. Mol. Biol.* 203:1045–1070 (1988)

A. Wlodawer and L. Sjolin, *Biochemistry* 22:2720–2728 (1983)
A. Wlodawer et al., *Biochemistry* 27:2705–2717 (1988)
R. L. Campbell and G. A. Petsko, *Biochemistry* 26:8579–8584 (1987)
R. L. Williams et al., *J. Biol. Chem.* 262:16020–16031 (1987)
A. G. Pavlovsky et al., *FEBS Letters* 162:167–170 (1983)
K. Katayanagi et al., *Nature* 347:306–309 (1990)
W. Yang et al., *Science* 249:1398–1405 (1990)
T. Nonaka et al., *FEBS Letters* 283:207–209 (1991)
F. M. Richards and H. W. Wyckoff, *Ribonuclease-S.* Clarendon Press, Oxford
    (1973)
K. T. Nakamura et al., *Nature* 299:564–566 (1982)
D. Kostrewa et al., *Biochemistry* 28:7592–7600 (1989)
P. Nordlund et al., *Nature* 345:593–598 (1990)

M. Leijonmarck and A. Liljas, *J. Mol. Biol.* 195:555–580 (1987)
K. S. Wilson et al., *Proc. Natl. Acad. Sci. USA* 83:7251–7255 (1986)

M. S. Chapman et al., *Science* 241:71–74 (1988)
S. Knight et al., *J. Mol. Biol.* 215:113–160 (1990)
I. Andersson et al., *Nature* 337:229–234 (1989)
G. Schneider et al., *J. Mol. Biol.* 211:989–1008 (1990)
E. Rutenber et al., *Proteins: Struct. Funct. Genet.* 10:240–250 (1991)
B. J. Katzin et al., *Proteins: Struct. Funct. Genet.* 10:251–259 (1991)
E. Rutenber and J. D. Robertus, *Proteins: Struct. Funct. Genet.* 10:260–269
    (1991)
D. W. Banner et al., *J. Mol. Biol.* 196:657–675 (1987)
M. Jaskólski et al., *Biochemistry* 29:5889–5898 (1990)

M. Frey et al., *J. Mol. Biol.* 197:525–541 (1987)
E. T. Adman et al., *J. Mol. Biol.* 217:337–352 (1991)
K. D. Watenpaugh et al., *J. Mol. Biol.* 138:615–633 (1979)
W. J. Cook et al., *J. Biol. Chem.* 266:652–656 (1991)

T. A. Jones and L. Liljas, *J. Mol. Biol.* 177:735–767 (1984)
C. Betzel et al., *J. Mol. Biol.* 223:427–445 (1992)
J. C. Fontecilla-Camps et al., *Proc. Natl. Acad. Sci. USA* 85:7443–7447
    (1988)

M. G. Grütter et al., *EMBO J.* 7:345–351 (1988)

S. Ludvigsen et al., *J. Mol. Biol.* 222:621–635 (1991)
D. C. Carter and X. m. He, *Science* 249:302–303 (1990)
S. Cusack et al., *Nature* 347:249–255 (1990)
J. K. Choi et al., *Nature* 354:37–43 (1991)

Protein structure

Southern bean mosaic virus
  empty particle
Squash protease inhibitor
  /β-trypsin
Staphylococcal nuclease

Stefin B
  /papain
Streptavidin
Subtilisin
  Bacillus amyloliquefaciens
  BPN′ (or novo)

    /chymotrypsin inhibitor 2
    /eglin
    /subtilisin inhibitor
  Carlsberg
    /eglin

Subtilisin inhibitor
  /subtilisin BPN′
Sulfate-binding protein
Sulfilte reductase
Superoxide dismutase
  Cu, Zn
  Fe

  Mn

Taka-amylase A
Tendamistat
Thaumatin I
Thermitase

  /eglin
Thermolysin
Thioredoxin
  bacterial
  phage
Thioredoxin reductase
α-Thrombin
  /hirudin

Thymidine phosphorylase
Thymidylate synthase

Tobacco mosaic virus
  coat protein
Tomato bushy stunt virus
Tonin
Transferrin

Reference

A. M. Silva and M. G. Rossmann, *J. Mol. Biol.* 197:69–87 (1987)
J. W. Erickson et al., *Science* 229:625–629 (1985)

W. Bode et al., *FEBS Letters* 242:285–292 (1989)
F. A. Cotton and E. E. Hazen, *The Enzymes* 4:153–175 (1971)
P. J. Loll and E. A. Lattman, *Proteins: Struct. Funct. Genet.* 5:183–201 (1989)
T. R. Hynes and R. O. Fox, *Proteins: Struct. Funct. Genet.* 10:92–105 (1991)

M. T. Stubbs et al., *EMBO J.* 9:1939–1947 (1990)
W. A. Hendrickson et al., *Proc. Natl. Acad. Sci. USA* 86:2190–2194 (1989)

R. Bott et al., *J. Biol. Chem.* 263:7895–7906 (1988)
J. Drenth et al., *Eur. J. Biochem.* 25:177–181 (1972)
R. A. Alden et al., *Biochem, Biophys. Res. Commun.* 45:337–344 (1971)
C. A. McPhalen and M. N. G. James, *Biochemistry* 27:6582–6598 (1988)
D. W. Heinz et al., *J. Mol. Biol.* 217:353–371 (1991)
Y. Takeuchi et al. *J. Mol. Biol.* 221:309–325 (1991)
D. J. Neidhart and G. A. Petsko, *Protein Eng.* 2:271–276 (1988)
C. A. McPhalen and M. N. G. James, *Biochemistry* 27:6582–6598 (1988)
W. Bode et al., *Eur. J. Biochem.* 166:673–692 (1987)
Y. Mitsui et al., *J. Mol. Biol.* 131:697–724 (1979)
S. Hirono et al., *J. Mol. Biol.* 178:389–413 (1984)
J. W. Pflugrath and F. A. Quiocho, *J. Mol. Biol.* 200:163–180 (1988)
D. E. McRee et al., *J. Biol. Chem.* 261:10277–10281 (1986)

J. A. Tainer et al., *Nature* 306:284–287 (1983)
B. L. Stoddard et al., *Biochemistry* 29:8885–8893 (1990)
W. C. Stallings et al., *Proc. Natl. Acad. Sci. USA* 80:3884–3888 (1983)
M. L. Ludwig et al., *J. Mol. Biol.* 219:335–358 (1991)
M. W. Parker and C. C. F. Blake, *J. Mol. Biol.* 199:649–661 (1988)
Y. Matsuura et al., *J. Biochem.* 87:1555–1558 (1980)
J. W. Pflugrath et al., *J. Mol. Biol.* 189:383–386 (1986)
A. M. de Vos et al., *Proc. Natl. Acad. Sci. USA* 82:1406–1409 (1985)
Ch. Betzel et al., *Protein Eng.* 3:161–172 (1990)
A. V. Teplyakov et al., *J. Mol Biol.* 214:261–279 (1990)
P. Gros et al., *J. Mol. Biol.* 210:347–367 (1989)
M. A. Holmes and B. W. Matthews, *J. Mol. Biol.* 160:623–639 (1982)

S. K. Katti et al., *J. Mol. Biol.* 212:167–184 (1990)
B. O. Soderberg et al., *Proc. Natl. Acad. Sci. USA* 75:5827–5830 (1978)
J. Kuriyan et al., *Nature* 352:172–174 (1991)
W. Bode et al., *EMBO J.* 8:3467–3475 (1989)
T. J. Rydel et al., *Science* 249:277–280 (1990)
M. G. Grütter et al., *EMBO J.* 9:2361–2365 (1990)
M. R. Walter et al., *J. Biol. Chem.* 265:14016–14022 (1990)
K. M. Perry et al., *Proteins: Struct. Funct. Genet.* 8:315–333 (1990)
D. A. Matthews et al., *J. Mol. Biol.* 214:923-936 (1990)
K. Namba et al., *J. Mol. Biol.* 208:307–325 (1989)
A. C. Bloomer et al., *Nature* 276:362–373 (1983)
A. J. Olsen et al., *J. Mol. Biol.* 171:61–93 (1983)
M. Fujinaga and M. N. G. James, *J. Mol. Biol.* 195:373–396 (1987)
S. Bailey et al., *Biochemistry* 27:5804–5812 (1988)

Protein structure	Reference
Trimethylamine dehydrogenase	L. W. Lim et al., *J. Biol. Chem.* 261:15140–15146 (1986)
Triose phosphate isomerase	
chicken	D. C. Phillips et al., *J. Mol. Biol.* 119:319–351 (1978)
yeast	E. Lolis et al., *Biochemistry* 29:6609–6618 (1990)
glycosomal	R. K. Wierenga et al., *J. Mol. Biol.* 198:109–121 (1987)
trypanosomal	R. K. Wierranga et al., *J. Mol. Biol.* 220:995–1015 (1991)
Troponin C	O. Herzberg and M. N. G. James, *J. Mol. Biol.* 203:761–779 (1988)
	K. A. Satyshur et al., *J. Biol. Chem.* 263:1628–1647 (1988)
Trp repressor	R. W. Schevitz et al., *Nature* 317:782–786 (1985)
/tryptophan	R. G. Zhang et al., *Nature* 327:591–597 (1987)
/operator	A. Otwinowski et al., *Nature* 335:321–329 (1988)
Trypanothione reductase	J. Kuriyan et al., *Proc. Natl. Acad. Sci. USA* 88:8764–8768 (1991)
Trypsins	
bovine β	W. Bode and P. Schwager, *J. Mol. Biol.* 98:693–717 (1975)
	T. Earnest et al., *Proteins: Struct. Funct. Genet.* 10:171–187 (1991)
	H. D. Bartunik et al., *J. Mol. Biol.* 210:813–828 (1989)
/BPTI	R. Huber et al., *J. Mol. Biol.* 89:73–101 (1974)
/soybean inhibitor	R. M. Sweet et al., *Biochemistry* 13:4212–4228 (1974)
/squash protease inhibitor	W. Bode et al., *FEBS Letters* 242:285–292 (1989)
streptomyces	R. J. Read and M. N. G. James, *J. Mol. Biol.* 200:523–551 (1988)
Trypsinogen	H. Fehlhammer et al., *J. Mol. Biol.* 111:415–438 (1977)
	A. A. Kossiakoff et al., *Biochemistry* 16:654–664 (1977)
/BPTI	W. Bode et al., *J. Mol. Biol.* 118:99–112 (1978)
/pancreatic secretory inhibitor	M. Bolognesi et al., *J. Mol. Biol.* 162:839–868 (1983)
Tryptophan synthetase $\alpha_2\beta_2$	C. C. Hyde et al., *J. Biol. Chem.* 263:17857–17871 (1988)
Tumor necrosis factor	E. Y. Jones et al., *Nature* 338:225–228 (1989)
	M. J. Eck et al., *J. Biol. Chem.* 267:2119–2122 (1992)
Turnip crinkle virus	J. M. Hogle et al., *J. Mol. Biol.* 191:625–638 (1986)
Tyrosyl tRNA synthetase	P. Brick et al., *J. Mol. Biol.* 208:83–98 (1989)
U1 small nuclear ribonucleoprotein RNA-binding domain	K. Nagai et al., *Nature* 348:515–520 (1990)
Ubiquitin	
human	S. Vijay-Kumar et al., *J. Mol. Biol.* 194:531–544 (1987)
yeast, oat	S. Vijay-Kumar et al., *J. Biol. Chem.* 262:6396–6399 (1987)
Uteroglobulin	I. Morize et al., *J. Mol. Biol.* 194:725–739 (1987)
	R. Bally and J. Delettré, *J. Mol. Biol.* 206:153–170 (1989)
Variant surface glycoprotein (VSG) of trypanosomes	D. Freymann et al., *J. Mol. Biol.* 216:141–160 (1990)
Verotoxin-1	P. E. Stein et al., *Nature* 355:748–750 (1992)
Vitamin D-dependent intestinal Ca^{2+}-binding protein	D. M. E. Szebenyi and K. Moffat, *J. Biol. Chem.* 261:8761–8777 (1986)
Wheat germ agglutinin	C. S. Wright, *J. Mol. Biol.* 215:635–651 (1990)
Xylose isomerase	
arthrobacter	K. Henrick et al., *J. Mol. Biol.* 208:129–157 (1989)
actinoplanes	F. Rey et al., *Proteins: Struct. Funct. Genet.* 4:165–172 (1988)
streptomyces	H. L. Carrell et al., *Proc. Natl. Acad. Sci. USA* 86:4440–4444 (1989)
	G. K. Farber et al., *Protein Eng.* 1:459–466 (1987)
	M. Whitlow et al., *Proteins: Struct. Funct. Genet.* 9:153–173 (1991)
ϕX 174 virus	R. McKenna et al., *Nature* 355:137–143 (1992)
Zinc finger	
Zif268/DNA	N. P. Pavletich and C. O. Pabo, *Science* 252:809–817 (1991)
Zn^{2+} G peptidase	O. Dideberg et al., *Nature* 299:469–470 (1982)

Index

Page numbers for major entries, illustrations, and tables are in bold. Page numbers for items including a reference are in italics.

Absorption spectra
  aromatic residues, **14**–**17**, 20, *23, 31,* **177**–**179**
  hemes, 365
  polypeptides, 190
  proteins, 20, *23,* 29, *31,* **270**
Accessible surface
  area, **141**–**142,** *159,* **161, 162, 227**–**232, 295,** *328*
  definition, **227**–**229**
Acetic anhydride, reaction with
  amino groups, 11, 35, **275**
  carboxyl groups, 32, 35
  Tyr residues, 16
Acetoacetate decarboxylase, 395
Acetylation
  amino groups, *11*–*12,* 35, **198, 275,** 453, 472
  biosynthetic, 69, **87**–**88,** 472
  hydroxyl groups, 16
Acetyl CoA, 87
Acetyl CoA carboxylase, 467
N-Acetyl galactosamine, **92**–**94**
N-Acetylglucosamine (GlcNAc), **92**–**94, 331, 437**–**441**
N-Acetylmuraminic acid (MurNAc), **437**–**441**
N-Acetyl Ser, *33*
N-Acetyl Thr, *33*

Acid base catalysis, **399**–**401**
Acid hydrolysis
  amino acid analysis and, **28**–**31,** 32–33
  partial, *38*–*40,* 63
Acid chlorides, 44
Acid proteases, 251, *253,* **418, 429**–**431, 436**–**437,** *461*
Acrosin, 419
Acrylamide, 286
Actin, 346, 466, *467*
Actinidin, 429
Actinin, *467*
Activated esters, 44
Active sites, enzymes, 248, *257, 272,* **413**–**441, 444**–**451, 454**–**459**
Active site titrants, **393**
Active site-directed irreversible inhibitors, 393–394
Acylation, **88, 94**–**95**
Acyl carrier protein, 443
Acyl enzyme,
  metalloproteases, *461*
  thiol proteases, **418**
  serine proteases, 389, 393, **418**–**428,** *461*
Adenylate cyclase, 453–454
Adenylate kinase, *266,* **407**–**410**
Adenylation, **53**–**55,** 389, 391, **412**–**417,** 470
Adenylylation, **457**–**459**
Adenylyl transferase, 459
ADP, 251, *361*
ADP-ribosylation, 98, 453

Adrenaline, 453
Adrenocorticotropic hormone (ACTH), 82, 83, 122
Affinity chromatography, **346**–**348,** *361*
Affinity labels, **333**–**334,** 393, 421, **426**–**427**
Alanine, *269*
Ala residues, *3, 4, 7, 110, 118,* 154, 181, 185, *186,* 231, **240**–**241,** 256, 260
Albumin, 20, 27, 79, 122, 132, *269,* 347
Alcohol dehydrogenase
  alleles, *126*–*127, 138*
  catalysis, *397*
  evolution, *119*
  hydrodynamic properties, *266*
  isozymes, *138*
  signal peptide, 72
  Zn²⁺, *362*
Aldehydes, 11
Aldolase, **234,** 391–392
Aldose reductase, 130
Alleles, *126, 138*
Allostery, 57, 337, **367**–**382,** *442,* **444**–**452, 456**–**457**
Amadori rearrangement, 464–465
Amidase, 137
Amidation, α-carboxyl, 69, **90**
Amides
  Asn and Gln residues, **9**–**10,** *110*

hydrogen exchange, **282**–**286**
hydrolysis, 9–10, 420
reactions, 35
terminal carboxyl group, 69, **90,** *198*
Amidination, amino groups, 11, *12*
Amino acid residues
  analysis, **28**–**31**
  chemistry, **2**–**20**
  counting, **29**–**31**
  determination, **28**–**43**
  evolutionary variabilities, **110, 118**
  frequency, **4**
  hydrophilicities, **154**
  hydrophobicities, **154**
  mapping positions, **40**–**41**
  mass, **4**
  NMR chemical shifts, **240**
  packing, **231**
  properties, **2**–**20**
  secondary structure tendencies, **186, 256**
  structures, **3**
  volumes, **4, 230, 231**
Amino acids
  activation, **43**–**47, 53**–**57**
  detection, **20**–**23, 29**–**31**
  determination, **28**–**31**
  hydantoins, 32, 34, *35*
  prebiological origin, **107**–**108**
  Schiff base, 401–402
  structures, **2**–**4**
  utilization, **53**–**57,** 463, 465

Amino acyl adenylates, **53–57**, 101, 389, **412–417**

Amino acyl tRNA synthetases, 53–58, **343–346**, 391, **412–417**

a-Amino butyric acid, 64, 101, *187–188*

Amino groups
  acetylation, 11, 33, 35, 69, **87–88, 275**, 453, 472
  pyrrolidone carboxylic acid and, 9, *88*
  counting integral numbers, *31*
  hydrophobicity, 161
  ionization, 6, 10, 12, 144, *326, 395*, 411
  Lys residues, 6, 10–12, 41, 411
  myristoylation, 69, **88**
  posttranslational modification, 69
  reactions, 8, 10–12, 13, **21–23, 31–34**, 41, 274–275, 375
  Schiff base formation, 11, **401–402**, 464–465
  terminal, α, 6, 9, 31–34, 44–47, **87–88**, 259, **274–275**, 411, **434–436**
  ubiquitin and, 471–472

Aminopeptidases, 33, *87, 88*

Amino-terminal residues
  acetylation, 11–12, 35, 69, **87–88**, 453, 472
  degradation and, 472, *473*
  determination, **31–34**
  folding and, *259*
  post-translational modification, **86–88**
  pyrrolidone carboxylic acid, 9, *88*

Amino Tyr residues, 17

Ammonium sulfate, 263–264

Amphipathic, 162, **184**, *186*, 254, 257, 279, *280, 363*

Amphiphilic, 162, **184**, *186*, 254, 257, 277, 279, 280, *363*

Amphitrophic, *278*

Ancestral proteins, 114, **119–121, 127–130**

Anchimeric assistance, 385

Angiotensin, 130, 430, 431

Anhydrides
  amino groups, and, 11, 35, **275**
  carboxyl groups and, 32, 35
  metalloprotease catalysis and, 433
  mixed, 44
  Tyr residues, 16

Anomalous scattering, 205, 209

Antibiotics, 100–102

Antibodies
  antigen binding, 277, *333*, **348–355**, *383*
  binding to other proteins, *333*, **353–355**
  biosynthesis, 68, 351, *383*
  conformational equilibria measured with, **314–315**, 353, *383*
  domain structure, 237, 297, **349–352**
  evolutionary divergence, 122
  F_{ab} fragment, **349–352**
  F_c fragment, **349–352**
  hydrodynamic properties, 27, *266, 269*
  identifying proteins, **59–62**
  internal structural homology, 251, **349–350**
  structural similarity to superoxide dismutase, *259*
  structures, **203**, 253, **348–353**
  unfolding, 297
  Western blotting, 28, 62, 353, *383*

Anticodons, **54–58**

Antifreeze proteins, *168, 326*

Antifreeze solvents, 395

Antigens, 277, *333*, **348–355**, *383*, **406**

Antisense peptides, *169*

α_1-Antitrypsin, 100

Apo proteins, 329

Approximation, 385

Aqueous solutions, **153–156**, *401, 460*

Arabinose-binding protein, **219**, 251

Arginase, *467*

Arginosuccinate lyase, 130

Arginyl tRNA-protein transferase, 87

Arg residues
  addition to amino-terminus, **87**, 472
  addition to carboxyl-terminus, 63
  ADP-ribosylation, **98**, 453
  cleavage at, 13, 38, 63, **81–83**, 419, 422
  diagonal technique, 41
  hydrazine and, 12–13, 32, 35
  ionization, 6, 12
  properties, *3, 4*, **12–13**, *110, 118, 141, 142, 154, 186, 231, 240, 256, 282, 326*

Aromatic rings, **14–17**, *144–145*, 155, 161, **177–179**, 232, 239, *243, 259*, **270–271, 286–287**, 301

Arrhenius plot, 181, **311**, 313

Ascorbic acid, 90, 96

Asn residues
  abbreviation, 4, 115
  ADP-ribosylation, 98
  cleavage at, 36, 38, *40*, **84, 85**
  deamidation, **9–10**, 29, 40, 100, 464, 467, *473*
  determination, 29
  glycosylation, 50, **91–93**
  hydrazine and, 32
  hydroxylation, **96**
  origin, 107
  properties, *3, 4*, **9–10**, 21, *100, 110, 118, 141, 142, 154*, 184, *186*, 216, *231, 240, 256, 282, 327*
  racemization, **9–10**, 465

Asp residues
  abbreviation, 4, 115
  carboxyl proteases and, **418, 430–431**
  catalytic triad, **421–424**, *428*
  cleavage at, 36, **38–40**
  hydroxylation, **96**
  ionization, 6, *395*
  iso, **9–10**, 464
  lysozyme catalysis and, **438–441**
  methylation, 452
  phosphorylation, 97

properties, *3, 4*, **9–10**, *110, 118, 141, 142, 154, 184, 186, 231, 240, 256, 282*
  racemization, **9–10**, 465

Aspartate aminotransferase, 411, 413

Aspartate receptor, *334*

Aspartate transcarbamoylase (ATCase) *131*, 233, 362, **446–452**

Aspartyl proteases, **429–431, 436–437**

Asp-N protease, 38–39

Association constants
  ligand binding, **338–346, 440**
  noncovalent interactions, **155–156**

Asymmetric unit, **204**

Asymmetry
  amino acids, 204
  folded proteins, 204
  polypeptide backbone, 190–191
  quaternary structure, **381–382**, 414

ATP
  binding, **360–361**
  hydrolysis, 67, 75, 324, **406–410**, 468–469
  phosphorylation of proteins, **96–98**

ATPases, *48*, **406–410**, 468–469

Avidin, **341–342**

Azides, 44, *334*

Azurin, 362

Bacteriorhodopsin, *279, 280*

Bacteriochlorophyll protein, 236, *334*

Bandshift electrophoresis, 359, *384*

β-Barrels, 224, *259*, **279**, 334

$(\beta\alpha)_8$-Barrels, **220, 224–225**, *249, 251, 257–258, 259–260, 309, 348*

Bence Jones proteins, 297

β-Bends. *See* Reverse turns

Benzyl halides, 16

Beta sheets. *See* β-Sheets

Bicarbonate, 367

Bicinchonic acid, 23

Bifunctional reagents, *180*, **334**

Binding energies, **337–346, 440**

Biosynthesis of proteins, 18, **49-104**, 107, *442*, 463, 466, 472

Biotin, **341-342**

BiP, 68

Bisubstrate analogues, 403, *406*, **407-409**, **446-452**

Biuret assay, **20-21**

BNPA-skatole, 40

Bohr effect, 371, 375, *380*

Bone, 197

Borate, 12

Borohydride, 11, 19

BPTI (bovine pancreatic trypsin inhibitor)
aromatic rings, **286-287**
bound to trypsin, **331**, 336, 421-422, 434
contact map, **221**
counting Cys residues, **30-31**
diagonal maps, **42**
disulfide bonds, **317-321**
flexibility, *281*
folding pathway, *136*, **317-321**, *328*
hydrodynamic properties, *266*
hydrogen exchange, **283-286**
mutations, *136*
NMR studies, **283-284**
protease inhibitor, 331, 421-422
reduced form, 42
spectral properties, **270-271**
water molecules, internal, **232**

N-Bromosuccinimide, 16

Bromelain, 38

Brønsted plot, **274-275**, **411-413**, *460*

Brownian motion, **265**, *270*

β-Bulge, 223

2,3-Butanedione, 12

Cadmium ions, 18, 367

Caged compounds, **394-396**

Calbindin, *10*

Calcium ions, 156, 293, **361-364**, 452

Calcium-binding proteins, calmodulin, 337, **361-364**, 452-454
domains, 133, **361-364**
enzyme regulation by, 452-454
Gla residues, **95-96**

hydroxylation and, **96**
parvalbumin, 331, **363**
protease, 132, 468
structures, 348, **361-364**
troponin C, **361-364**

Calmodulin
Ca²⁺ binding, 337, **361-364**
kinases and, 97, **452-454**
structure, *198*, 223

Calorimetry, **296-303**, 322, *340*

Calpain, 468

Carbamate, 374-375

Carbamylation, 11

Carbodiimides, **8-9**, 44

Carbon dioxide, 374-375

Carbonic anhydrase
backbone topology, 221
catalysis, 387, 397
evolution, 122
hydrodynamic properties, *266*
mass spectrometry, **37**
Zn²⁺ ion, 362

Carbon monoxide, 364-366

γ-Carboxy Glu residues (Gla), 69, **95-96**, *133*

Carboxyl groups
ionization, 6, **8-9**, 144, *395*
methylation, 453
reactions, **8-9**, 32-33, **34-35**, **44-47**, 54, 98, *326*
terminal, *α*, 6, 9, 32-33, 34-35, 44-47, 54, 98, 470-472

Carboxyl proteases, 81, 251, *253*, **418**, **429-431**, **436-437**, *461*

Carboxymethyl Cys residues, 17, 42-43, 317

Carboxyl-terminal residue
amidation, 69, **90**
determination, **31-35**
folding and, *259*
modifications, **88-91**
removal, **90-91**, **308**

Carboxypeptidase A
active site, **331**, 395, **418**, **432**
catalytic mechanism, **418**, **431-434**, *461*
cis peptide bonds, 221
thermolysin and, *250*, **431-434**
evolution, *250*

hydrodynamic properties, *266*
protein inhibitor, 336
Zn²⁺ ion, **331**, 362, *461*

Carboxypeptidase B, 431, **437**

Carboxypeptidases
carboxyl-terminal residue removal, 32-33, 63
catalytic mechanisms, **431-434**
evolution, *250*
nonglycosylation, 93
posttranslational modifications and, 82
zymogens, **437**

Cartilage, 197

Cascades, 79

Casein, *40*, 97, 122

Catabolite gene activator protein (CAP), **356-360**

Catabolite inactivation, 467

Catalase, 99, 221, 387, 464

Catalysis, enzyme, **385-461**

Catalyst, 386

Catalytic antibodies, **406**

Catalytic triad, **421-428**, 429, **434-436**, *461*

Cataracts, 465

Cathepsins, 430, 469-470

Cavities in proteins, **229-232**, 304-306, *327*

Centrifugation, **24**, 266-268, *326*

Chaperones, **68-69**, 71, 73, **324-325**

Characteristic ratio, 177

Charge transfer-relay system, **421-428**

Charge-transfer complex, 17

Chelate effect, **162-167**

Chemical shifts, NMR, **238-241**

Chemostat, 136

Chemotaxis, *97*, *334*

Chloramphenicol acetyl-transferase, **235**

Chlorophyll, 334

Chloroplasts, 51, **71-74**

N-Chlorosuccinimide, 38

Chondroitin sulfate, 94

Chorionic gonadotropin, 66

Chorismate mutase, 397

Chou-Fasman secondary structure prediction, **255-257**

Chromatography,
affinity, **346-348**, *361*
HPLC, *41*
ion exchange, **317**
paper, **112**
reverse phase, **30**

Chymosin, 430

α-Chymotrypsin
biosynthesis, **79-81**, 325
catalytic mechanism, 411, **418-428**
catalytic triad, **421-428**, **436**
cleavage of polypeptides, 38-39
comparison with chymotrypsinogen, **434-435**
competitive labeling, **274-275**
diffusion coefficient, 269
electrostatic effects, 326, 395
evolution, **247-248**
inhibitor binding, **423**
interaction with denaturants, 295-296
internal structure homology, *253*, *460*
substrate binding, **422-423**
structure, 247, 381, 428, **423**, **435**
unfolding, 299

Chymotrypsinogen
activation, **79-81**, 260, **434-436**
hydrodynamic properties, 27, *266*, 269
structure, 419, **435**

Circular dichroism (CD)
description, **190-191**
secondary structure estimation, 198, **271**
unfolding followed by, **288**, 292

Circular proteins, *167*, *180*

Cis peptide bonds
adjacent to Pro residues, 7-8, *104*, 174, 181, *226*, 312, **324-325**
folded proteins, *176*, *221*
poly(Pro) and, *183*, **188**
protein folding and, *312*, **324-325**
reverse turns and, *226*
structure, **5**, 174

Cis-trans isomerization, **181**, **188**, 196, *198*, **312**

Cistron, **51–53**
Citraconic anhydride, 11, 38
Clathrates, 157–160
Cleland's reagent, *19–20,*
47
Clostripain, 38–39, 63
Clupein, *27*
CNBr, **38–40**, *47*, 63, 308
Cobalt ions, 18
Codons, **54–59**, 112–113,
118
Coenzymes, 338, 401–402
Cofactors, 308–309, 401–
402
Coiled coils
    globular proteins and,
    223, *383*
    structures, **193–196**, *199*
Cold unfolding, **299–300**,
303
Colicin A, 207
Collagen
    amino acid composition,
    43, 196
    biosynthesis, 69, 79, 96,
    103–104, **196–198**
    cross-links, **96**, **196–198**
    degradation, 63, 466
    diffusion, *269*
    evolutionary divergence,
    122
    glycosylation, 93, 96
    hydroxylation, 69, **96**,
    103–104, 196, 325
    internal duplications, 132
    -like structure in globu-
    lar proteins, 223, 351
    structure, **196–198**, 254
Collagenase, 63, 431, 468
Competitive enzyme inhibi-
    tion, **389–391**, 406,
    459–460
Color blindness, *134*
Complement, 79, 108, 133,
350–351
Complementarity determin-
    ing region (CDR), 351
Complementary DNA
    (cDNA), 59
Complementation, **307–309**
Compressibility, *237*, *268*,
281, 285
Concanavalin A
    biosynthesis, **84–86**, *103*
    hydrodynamic proper-
    ties, *266*
    ligand binding, **86**, *336*,
    371
    splicing, **84–86**
    structure, 86, 236

Concentrations of proteins,
    20–23, *264*, 277, 280,
    340
Concerted allosteric model,
    372–381, **444–452**
Configuration, 171
Conformations, **171–173**,
309
Contact maps, 221, 224, **247**
Contact surface, **228–229**
Convergent evolution
    definition, 108
    metalloproteases, *250*,
    431–434
    possible examples, *125*,
    249–251, *259–260*,
    348, 361
    serine proteases, *138*,
    250, 419, 421
Coomassie blue, **22–23**,
    47–48
Cooperativity
    folded protein conforma-
    tions, **301–306**, 316
    helix-coil transition, 185
    ligand binding, **367–382**,
    384
    multiple interactions,
    165–167
    negative, 370, 373, **381–
    382**
    unfolding of proteins,
    289–291, 297, 316,
    320
Copper ions, 18, **20–21**,
    90, **361–362**, 364, 367
Correlation time, 268
Corticotropin, (ACTH), **83**,
    122
Corticotropin-like interme-
    diate lobe peptide
    (CLIP), **83**
COSY spectra, NMR, **241**,
    284
Cotton effect, **190**
Coulomb's law, **142–144**
Crambin, 216, 238, *255*
Creatine aminohydrolase,
    222
Creatine kinase, *397*
Cro repressor, **356–360**
Cross-links
    aging and, 465
    disulfides, 24, 26, 179–
    180, *292*, 306
    fibrous proteins, **196–198**
    insulin folding and, 124
    ligand binding and, **334**
    lysozyme and, **275–276**
    polymers and, 179–180

posttranslational modifi-
    cation, 69, 93–95
quaternary structure de-
    termination, *278*,
    334
stabilizing effect, *167*,
    179–180, 306
Cryoenzymology, *394–396*
Crystal lattice
    description, **202–204**, 237
    diffusion through, 330,
    395
    protein conformation
    and, 217, 262, 281,
    284–285, 372
Crystallins
    evolution, **122**, *130–131*
    internal homology, 251
    longevity, *465*
Crystallization, **202–203**,
    259
Cyanate, 11, *32–33*
Cyanide, 19
Cyanogen bromide (CNBr),
    38–40, *47*, 63, 308
Cyanylation, **38–40**
Cyclic AMP, 97, **452–454**
Cyclic AMP-dependent ki-
    nase, **452–454**
Cyclic dipeptide crystals,
    300–303
Cyclin, 463, *472*
Cyclization, **179–180**, *182*
Cyclohexane, 157–160,
    168, 229
Cyclohexanedione, 12, 41
Cyclophilin, **324**, *333*
Cyclosporin, **324**, *333*
Cyclosporin synthetase, *102*
Cys residues
    absorbance, 20
    acid hydrolysis and, 29
    cleavage at, **38–40**
    counting integral num-
    bers, **29–31**, *48*
    diagonal maps, **41–43**
    disulfide formation, *47*,
    68, **98–99**, *103–*
    104, 179–180, **316–
    321**, **324–325**, *328*
    evolutionary conserva-
    tion, **116–118**, 248
    farnesyl groups and, **90**
    glycosylation of, **91–93**
    heme groups in cy-
    tochromes, **246**, 248
    ionization, 6, 17, *395*
    isolation, **41–43**, *138*
    lipid attachment, **90**, **94–
    95**, *103*, 278

mapping, *41*
metal-ion binding, 18,
    208–210, **251–252**,
    358, **361–362**, 367
oxidation, 16, **18–20**, 29,
    40, 41, *47*, 99–100,
    464
palmitoyl groups and,
    94–95, *103*
posttranslational modifi-
    cation, **18–20**, **90–
    95**, **98–99**, **101–
    102**, *103*
properties, *3*, *4*, **17–20**,
    110, *118*, *141*, *142*,
    154, *186*, *231*, *240*,
    256
thiol proteases and, **418**,
    428–429
Cysteic acid, **18**, 29, 42
Cystine, 18
Cytochrome b_5
    evolution, *122*, 251
    heme group bound, 348,
    362
Cytochrome b_{562}, 348
Cytochrome *c*
    ancestral sequence, **120**
    antibody binding, *333*
    dipole moment, 344, *366*
    electron transfer, 344,
    346
    evolution, **114–125**, 244,
    246–247
    heme group, 73–74,
    348, 362
    hydrodynamic proper-
    ties, *27*, *266*
    oxidase, 72
    sequences, **115**, *117*, **120**
    structural homology,
    244, **246–247**
    structure, 246, *384*
    targeting, 73–74
    Tyr residues, 273
    unfolding, 289, 297, 299,
    308
Cytochrome c_1, 72
Cytochrome c_2, 116, **246**
Cytochrome c_3, 334, 348
Cytochrome c_{550}, 116, **246**
Cytochrome c_{551}, *27*, 116
Cytochrome c_{555}, 116
Cytochrome P450, **218**, 364
Cytochromes
    electron binding, **366–
    367**
    evolution, **114–125**, *367*
    heme group attachment,
    73–74, 348, **362**

Cytochromes *(cont.)*
  membrane-associated, 278
  mitochondrial incorporation, 73–74
  structures, **246**, *384*
Cytoplasm, *264*, 280, *340*
Cytotoxin, 122

Dansyl chloride, **31–33**
Deamidation, **9–10**, 29, 40, 100, 464–465, 467, *473*
Debye units, 145
Debye-Hückel, **156**, 263
Debye-Waller factor, 214
Degradation of proteins, *9–10*, 88, 100, 442, 457, 459, **463–473**
Dehydrobutyrine, 101–102
Dehydrogenases, 250, *348*, **360–361**
Deletions of residues, 111, 248
Denaturants, 24, 26–28, **291–296**
Denaturation, 26–28, 201, 264, 287
Density of proteins, 24, **141**, 266–267
Deoxyribonuclease, 93
Dermatan sulfate, 94
Design of proteins, *172*, **258–259**
Detection of proteins, **20–23**
Detergents, **24–28**, *48*, 162, 263, 277, 278
Diabetes, 465, 467
Diacyl glycerol, 97
Diagonal dot-plot, 109, **111**
Diagonal electrophoresis, **41–43**
Diazo compounds, 9, 430
Dicyclohexylcarbodiimide, 44
Dielectric constant
  definition, 143
  folded proteins, **272–273**, *367*
  water, 143, 156, 264
Diethylpyrocarbonate, 13
Difference Fourier, **330–332**
Diffraction, X ray, **202–216**, **330–332**
Diffusion
  enzyme catalysis limit, 388, 442

ligand binding and, **344–346**, *382*
  membranes and, **280**, 345
  molecules, 261, **265–270**
  polymers, **180–182**
  protein interiors and, 285, **286**
Diffusion coefficient
  definition, **265**
  rates of encounter, *179*, **344–346**
  typical values, **266**, *269*
  unfolded polypeptide chains, **180–182**
Dihedral angles, **174–176**, **183**, **221–223**, 226
Dihydrofolate reductase, *314*, *333*, 403
Dihydroxyphenylalanine decarboxylase, *14*
Diisopropyl fluorophosphate (DFP), **273–274**, **426–427**, 469
Dimethyl sulfoxide, *264*, 395
Diphosphoglycerate (DPG), **332**, 371, 373, 375, 380
Dipole
  cytochrome *c*, 344, *346*
  α-helix, 6, **183–186**, 255, 306, 335–336, 360, 429
  ligand binding and, **335–336**, 344–346, 426, 429
  peptide bond, **6**, **145**, 183
  random coil, *199*
  β-sheet, *198*
  water, 156
Dipole moment, **145**, 344
Dispersion forces, **146–147**
Dissociation constants of ligands, **338–346**
Distance distribution function, 385–386
Distance geometry, **243**
Distance maps, **221**, 224, **247**
Distribution coefficient (K_d), **25–27**
Distribution functions, **177–179**
Disulfide bonds
  chirality, 18–19, 271
  cross-links, 224, 26, 179–180, 292, 306
  determination, **41–43**, 47
  folding intermediates trapped by, **316–321**

formation, **18–20**, *68*, 69, **98–99**, *103–104*, *179–180*, **324–325**, *328*
  immunoglobulins, *349–350*
  interchange, **19–20**, **307**, **316–321**, **324–325**
  mixed, 429
  properties, **18–20**, 327, 464
  redox proteins, 366
  spectral properties, 20, 271
  stability, **163–164**, **316–321**
  stabilizing effect, *179–180*, **306**, **316–321**, *327*
  variation during evolution, 248
Disulfide isomerase, **99**, *103–104*, 307, **324–325**
Dithionitrobenzoic acid (DTNB, Ellman's reagent), 19
Dithiothreitol (DTT, Cleland's reagent), **19–20**, *47*
Divergent evolution, **108–138**, **244–249**, **251–253**, *259*
DNA binding, **165**, 330, *334*, 346, 348, **355–360**, 371, *383*, *384*
DNA polymerase, 114, *417*
Docking protein, 65, 77
Dolichyl phosphate, 91
Domains
  aspartate transcarbamoylase, **449–450**
  carboxyl proteases, 430
  cleavage between, 261, 307
  description, *43*, 51, **219**, *221*, 224
  evolution, 108, *132*, **133**, 251
  exons and, **51**, **133**
  folding and, 68, 231, 254, 291, 297, **321–323**
  glutathione reductase, **335**
  hexokinase, **407–408**
  immunoglobulins, 281, **349–352**
  induced fit and, **407–410**
  kinases, **407–410**

ligand binding, **334–337**, 380, **407–410**
  movements, 262, **269**, *281*, **407–410**
  phosphofructokinase, **445**
  phosphorylase, **455**
  thiol proteases, 429
Dot plot, **109–111**
Double-displacement enzyme mechanism, 389
Double-sieve editing, 344, **417**
DQNMR spectra, 241
DTNB, 19
DTT, **19–20**, *47*
Dyes, 22–23, 279
Dynorphin, 82
Dystrophin, *193*

Eadie-Hofstee plot, **387**, 390
Edman degradation procedure, **33–34**, 38
Effective concentrations
  catalysis and, **398–402**, 406, 407
  definition, **163–167**
  folded proteins, **166–167**, 261, **300–303**, 422
  ligand binding, **333–334**, 406
  multiple interactions and, **165–166**
  unfolded polypeptide chains, **165–166**, **179–180**
EF hand, **361–364**, 468
Einstein-Sutherland equation, 265
Elastase
  catalytic mechanism, **418–428**
  degradation of elastin, 100, 468
  evolution, 247, 248
  hydrodynamic properties, *266*
  kinetics of peptide hydrolysis, **425**
  nonglycosylation, 93
  specificity, 38, **422–423**
  structure, 247, 423
  substrate binding, **422–423**
Elastin, *198*, 468
Electron density map, **207**, **210–215**, **330–332**
Electron microscopy, 23, 196, **201–202**, 208

Electron transfer proteins, 344, *346*, **366–367**

Electrophoresis
  bandshift, 359, *384*
  diagonal, **41–43**
  membrane traversal and, 73
  paper, **41**, *112*
  polyacrylamide (PAGE), **25–28**, *113*, *126–127*, 353, *384*
  SDS, **25–28**, **40**, **47**, 334
  two-dimensional, *127*
  urea gradient, **289–291**

Electrospray ionization mass spectrometry, **36–38**

Electrostatic interactions
  binding rates and, **344–346**, *382*
  coiled coils and, **193–196**
  description, **142–146**
  effective concentrations, 163, 165
  enzyme catalysis, **399–402**, 407, 428, 429, 433
  folded proteins and, *104*, 264, **272–273**, 279, *286*, **293**, **294**, **300–303**, *326*
  α-helices, 183, **185–186**, *198*, 255
  ligand binding and, *104*, **335**, 342, 359, 375, *382*, 436, **456–457**
  membranes and, **76–78**
  polyelectrolytes, **165**, 293, **294**
  water and, **155–156**, 162

Electrostriction, 267

Ellman's reagent (DTNB), **19**

Encounter complex, 345

End-to-end distances, **176–180**

Endocytosis, 70, 420

Endopeptidase, 431

Endoplasmic reticulum (ER), **65–71**, 84, 89, 91–96, 98, 99, *102–103*, **324–325**

Endoproteinase Arg-C, 38–39

Endoproteinase Lys-C, 38–39

Endorphin, 82, 87

Energy minimization, 243

Engineeering proteins, **59–64**

Enhancers, 51

Enkephalins, **82–83**, 431

Enolase, 130, *259–260*

Enterokinase, 79, 434

Enthalpy
  ligand binding, 338–339
  protein folding, **296–303**
  transition state, 180–181, 395
  unfolded protein, 292
  water and, 152, **157–160**

Entropy
  conformational, **172–180**, 309, *327*
  enzyme catalysis and, 385, **398–402**
  intramolecular reactions and, **162–167**
  ligand binding, 155–156, 168, 338, 342, *344*
  protein folding, **296–303**
  transition state, 180–181, 395
  unfolded protein, 292
  water and, 152, **157–160**

Enzyme-activated inhibitors, **393–394**, *396*

Enzymes, **385–460**

Epidermal growth factor (EGF), 96, 108, **133**

Epidermin, **101–102**

Epimerization, 464–465

Epinephrine, 453

Epitopes, *62*, 354, *383*

Epoxides, 9, 430

Equilibrium dialysis, 368

Erabutoxin, *27*

Erythrocruorin, 348

Erythrocytes, 264, 467

Ester formation, **398**

Ester hydrolysis, 420

Esterases, *126–127*

Ethyleneimine, 18

N-Ethylmaleimide, **18**

Eukaryotes, **51–55**, 58, 105, 133

Evolution
  convergence, 108, **249–251**, *259–260*, 348, 361
  description, **105–138**
  divergence, **105–138**
  folded conformations and, **244–253**
  molecular clock, **121–123**
  neutral, *114*, **123–125**, 375
  rates, **121–123**, 470
  selection, **123–125**, 375

simulation, **136–137**

Excluded volume effect, 172, 176, **264**, *265*, 280, *340*

Exons, **50–53**, 118, *127*, 133

Exopeptidases, 32, 431

Extended conformation, 191

Extended X-ray absorption fine structure (EXAFS), 361

Exteriors of proteins, **227–232**, 238

Eye lens proteins, 122, *130–131*, 251, *465*

Factor X, 95, **133**

Factor X_a, 63

FAD, 107, **335**, 360–361 366

F_{ab} fragment, **349–355**

Farnesyl groups, **88–90**, 94–95

Fast atom bombardment mass spectrometry (FAB-MS), 36

Fatty acid synthase, 443

F_c fragment, **349–352**

Ferredoxin
  elongation by gene duplication, **132**, 251
  evolutionary divergence, *122*, *260*, **252**
  internal structural homology, **251**, 252
  iron-sulfur complex, **251**, **361–362**
  sequence, 132
  signal sequence, **72**

Ferritin, **367**

Fibrinogen, *68*, 124, 132, *133*, 269

Fibrinopeptides, 114, **121–124**

Fibroin, silk, **193**

Fibronectin, **133**

Fibrous proteins, 43, **122**, **193–199**, 254

Field desorption mass spectrometry (FDMS), 36

Fingerprints, **112**

Flavin adenine dinucleotide (FAD), 107, **335**, 360–361, 366

Flavin mononucleotide (FMN), 230, 360–361, 366

Flavodoxin, 230

Flexibility of proteins
  allostery and, 372
  crystallographic, 213–215, **236–238**
  degradation and, *473*
  domain movements, 262, 269, *281*, **407–410**
  effective concentrations determined by, **165–167**
  folded proteins, **236–238**, 269, 281, **281–287**
  ligand binding and, **337**, **339–340**, *383*
  NMR, 181–182, 243
  quaternary structure and, 262
  unfolded polypeptides, **173–182**

Fluorescamine, **21–23**, 29

Fluorescence
  aromatic amino acids, **14–17**
  conformation and, 270
  dansyl amino acids, 32, 179
  depolarization, **268–270**, 286
  energy transfer, **177–179**
  fluorescamine, 21–22
  quenching, **286**
  unfolding followed by, **288**, 326

Fluoro-2,4-dinitrobenzene, **31–32**

Fluoromethoxy carbonyl (Fmoc), **45–47**

FMN, 230, 360–361, 366

Folded conformations of proteins, **201–260**

Folding of proteins, 59, **68–69**, *136*, 254, 261, **309–325**, 434, *436*

Folding unit, 219

Folin phenol assay, 21, 23

Footprinting, 359

Formate dehydrogenase, 58

Formic acid, 16, 189

Formylation, 16, **55**, 100

N-Forymylalanine, **423**

N-Formylkynurenine, 16

N-Formylmethionine, **55**

N-Formyltryptophan, **423**

Founder effect, **126–127**

Four-helix bundle, **220**, **258–259**

Fourier analysis, **207**, **210–213**

Free energy
  folding of proteins, **259–303**

Free energy *(cont.)*
  intramolecular interactions, **162–167**
  ligand binding, **337–346, 440**
  perturbation, 306, 340, 410
  reaction profile, **392, 400, 442,** *459*
  relationships, **411–413,** *414–416, 460*
  transfer, **153–154, 159–162,** 295, 301, 343
  transition states, **180–181, 396–406**
  water and, 152, **157–160**
Freely jointed chain, 176
Freezing at reactive centers of enzymes (FARCE), 386
Frictional coefficient, **189–190,** 265–267
Frictional ratio, **189–190, 265–267**
FRODO, *213*
Fusion proteins, 62–63, *103*
F_v fragment, 350–355

Gal4, *76,* **359–360**
Galactosyl transferase, 130
Gastricsin, 430
Gaussian distribution function, **177–179**
Gelatin, 196
Gel filtration, **24–27, 268,** *325–326*
Genes
  conversion, **134**
  cryptic, 137
  division, **132**
  duplication, 113, 120, **127–132,** 134, 136–137, 251–253, 444
  evolution, **105–138**
  expression, **50–53**
  fusion, **132–134**
  isolation, 59
  orthologous, 130–131
  paralogous, 130–131
  pseudo-, 60, *123*
  structure, **50–53**
  transcription, **51–53**
  translation, **53–59**
Genetic code, **57–60,** 107–*108*
Geranylgeranyl groups, **90**
Gla residues, 69, **95–96,** 133
Gln residues
  abbreviation, 4, 115

biosynthesis, 54
cleavage at, 84
deamidation, **9–10,** 29, 40, 100, 464, 467, *473*
hydrazine and, 32
hydrogen exchange, 282
N-hydroxyethyl, **179**
measurement, 29
N-terminal, 9, 88
origin, 107
properties, *3, 4,* **9–10,** 35, *100,* 110, 118, *141, 142, 154, 186,* 216, *231, 240, 256, 327*
Globins
  evolutionary divergence, **121–130,** *137–138,* **244–245,** 375, 380
  heme-binding, 220, 337, 348, **362**
  hybrids, **134–135**
  O_2-binding, 124–125, **245, 364–366, 372–380,** *384*
  structures, **245,** *258*
Globular proteins, **201–260**
γ-Globulins. *See* Immunoglobulins
Glucagon, *122,* 125, 453
Glucocorticoid receptor, 76, 359
Glucose-6-phosphate isomerase, **404**
Glucosylation, 464–465
Glu residues
  abbreviation, 4, 115
  ADP-ribosylation, **98**
  Ca^{2+}- binding and, 95–96
  catalysis and, *395*
  cleavage at, 38
  conversion to Gla, 69, **95–96**
  conversion to Gln, 54
  ionization, 6, **276,** *326, 395*
  lysozyme and, *395,* **438–441**
  metalloproteases and, **431–433,** *461*
  methylation, 452
  properties, *3, 4,* **8–9,** *110, 118, 141, 142, 154, 186, 231, 240, 256, 276, 282, 467*
  pyro-, 88
Glutamate dehydrogenase, *122*
Glutamine synthetase, **457–459,** 465, 467

Glutaminyl tRNA synthetase, *54*
Glutaredoxin, *99*
Glutathione
  biosynthesis, **100–102**
  disulfide, 98–99, *164,* 317, 335
  function, 464
  redox equilibrium in vivo, **98–99,** 324–**325**
  structure, **100**
Glutathione peroxidase, 58
Glutathione reductase, 251, *253,* **335**
Glutathione transferase, 130
Glycan, 89, *92–93*
Glyceraldehyde phosphate dehydrogenase
  affinity chromatography, **347**
  degradation, *467*
  evolutionary divergence, 114, *122*
  NAD binding, *336, 382*
  negative cooperativity, *382*
  thermostability, *304*
Glycine, *269*
Glycogen phosphorylase. *See* Phosphorylase, glycogen
Glycogen synthetase, 454
Glycolysis, **131,** *325,* 443
Glycoproteins
  biosynthesis, 50, **91–94**
  degradation, 467
  structure, 23, *28, 38*
Glycosaminoglycans, 94
Glycosomes, **74–75,** *102–103*
Glycosylation of proteins, N, 50, 62, 68–70, 86, **91–93,** *103,* 325
  nonenzymatic, 464–465
  O, 69, *92–94,* 196
  protein degradation and, 467
Glycosyl-phosphatidylinositol groups, 69, **88–90,** 94
Glyoxysomes, **74–75**
Gly residues
  carboxyl terminal amide and, **90**
  cleavage at, 10, 38, 40, 63, 83–84
  collagen structure and, 196–197

conformational effect of, *175–179,* 185, *186,* **222,** *260,* 306, *327, 328*
deamidation and, **9–10**
myristoylation, 88
properties, 2, *3, 4,* **7,** *110, 118, 141, 142, 154, 186, 231,* **240–241,** *256*
reverse turns and, **226**
silk structure and, 193
Gold, colloidal, 22, 75
Golgi apparatus, **68–71,** 81, 90, 91, 93–95, 130
G proteins, 454, *457*
Gramicidin, **100,** *102*
Granules, secretory, 81, 90
Greek key topology, **227–228**
Growth hormone, *48, 122, 137*
Guanidinium salts
  binding to proteins, 295–296
  denaturing effects, *24,* **27,** 189, 263, **293–296,** 311
  Hofmeister series and, **156**
Guanido group, **11–13**
Guanylation, 11

α-Hairpins, **225–227,** 358, 430
Haldane relationship, **388,** 422
Half-of-sites reactivity, **381–382,** 414
Halophiles, **303–304**
Haptoglobin, 130, *134*
Heat capacity
  definition, **158**
  hydrophobic interaction, **158–162**
  protein folding, 292, **296–303,** 313
  transition state, 180–181, 313
Heat shock proteins, *70,* 325, 469, 472
Heavy-atom derivatives, 17, 18, **208–210**
Helical wheel, **184**
α-Helix
  amphipathic, **184,** *186,* 254, 279, 280, 363
  coiled coils, **193–196,** *199, 383*
  description, **182–186**

α-Helix (cont.)
  design, **258–259**
  dipole, 6, **183–186**, 255,
    306, 335–336, 360,
    429
  DNA binding and, **356–
    360**
  EF hand, **361–364**
  electrostatic interactions,
    183, 185–186, *198*,
    255
  evolutionary variation,
    248
  folded proteins, **221–
    225**, 227–228, **304–
    306**
  formation, **184–186**,
    198, 199
  hemoglobin allostery
    and, **375–380**
  hydrodynamic proper-
    ties, **177**, 189–190,
    198, **268–269**
  hydrogen exchange, 283
  left-handed, 184, *328*
  membrane proteins and,
    278–280
  nucleotide binding and,
    251, **360–361**
  prediction, **255–257**, *260*
  ribonuclease S-peptide,
    307
  Rossmann fold and, **250**,
    348
  spectral properties, **190–
    193**
  stability, **184–186**, 189,
    198, 199, 314
  structure, **182–186**, 220
  tendencies of residues,
    186, 256, *260*
  troponin C, **364**
π-Helix, 183, **187–188**
3₁₀-Helix, 183, **187–188**,
    454
Helix-turn-helix motif,
    355–357, *360*
Hemagglutinin, 223
Heme group
  binding as ligand, **220**,
    246, 334, 335, 337–
    338, 348, **362, 365,
    378**
  electron-binding, 344,
    346, **362, 366–367**
  O₂-binding, **364–366,
    374–380**
Hemerythrin
  O₂-binding, **362**, 364, *365*
  structure, **220, 235**, 253

Hemocyanins, *253*, 364
Hemoglobin
  allosteric properties,
    372–380, 445
  amino acid sequences,
    128
  concentration in red
    blood cell, 264, 340
  degradation, 465, *467–
    468*
  DPG binding, **332**, 371,
    381
  evolution, 111, 114,
    121–130, 134–135,
    244–245, 375, 380
  gene duplication, **127–
    130, 134–135**, *137–
    138*
  haptoglobin binding,
    130
  heme binding, **220**, 348,
    362, 365, 378
  homology, **121–125,
    127–130**, 245
  hybrid chains, **134–135**
  hydrodynamic proper-
    ties, *266*
  Kenya, **134–135**
  Lepore, **134–135**
  linked binding, **371**
  mutants, *103*, **111–113,
    126–127**
  myoglobin and, **127–
    130**, 245
  O₂ binding, **124–125**,
    245, **364–366, 371–
    380**, *384*
  organic phosphate bind-
    ing, **332**, *371*, 381
  quaternary structure,
    233, **372–380**
  sickle-cell, **111–113,
    126–127**, 332
  solubility, **263**
  structure, **220**, 245, **372–
    380**
Hemolysin, *77–78*
Heparan sulfate, 94
Heparin, 94
Heptad repeat, **194–196**,
    199
Heterologous subunit asso-
    ciation, **233–236**
Heterotropic allosteric in-
    teractions, **367, 371**,
    374, 448
Heterozygote, 126
Hexokinase
  catalysis, *397*
  degradation, 467

induced fit, **406–408**, *410*
internal structural ho-
    mology, 251
  ligand binding, **406–408**
  structure, **408**
High potential iron protein,
    362
Hill coefficient, **369–370**,
    380, *444*
Hill plot, **339–340, 369–
    371**
His residues
  ADP-ribosylation, **98**
  catalysis and, 394, *395,
    413*, 418, 434
  catalytic triad and, **421–
    428**
  cleavage at, 36
  diagonal technique, 41
  flexibility, 287
  hydrogen exchange, 13–
    14
  ionization, 6, **13–14**,
    272, 289, **293**, *326*,
    395
  ligation of heme iron
    atom, 246, 248,
    364–366
  metal ion binding, **358,
    361–362**, 431
  methylation, 466
  NMR and, 13–14, **239**,
    240
  origin, 107
  oxidation, 13, 459, 467
  phosphorylation, **97**
  properties, 3, 4, **13–14**,
    16, 21, *110, 118,
    141, 142, 154, 186,
    231, 256*
  protein unfolding and,
    289, **293**
  reactivity in ribonucle-
    ase, **274**
Histones, *27, 104*, 114, *122*
HIV protease, *430–431*
Hofmeister series, **156,
    263–264, 295–296**
HOHAHA NMR spectra,
    241
Holo protein, 329
Homoarginine, 11
Homocitrulline, 11
Homologous proteins
  folding, 321
  ligand binding, **348–353,
    361–364**, 409
  modeling, **257–258**
  sequences, 43, **108–138**,
    461

structures, **244–249,
    251–253**, 256, **409**
Homology, **108**
Homoserine residues, **39–
    40**, *308*
Homotropic allosteric in-
    teractions, **367, 369,
    374**, 448
Hormones
  enzyme regulation and,
    452–454
  evolution, *122*
  precursors, 61, 79, **81–
    83**, 88
  second messengers and,
    97, 361, 452–454
Host-guest method, 185
Hsp 70, 68
Hybrid proteins, **134–135**
Hydration
  amino acid side chains,
    153–155, 160–162
  ions, **155–156**
  nonpolar surfaces, **157–
    162**
  preferential, **264, 293–
    294**
  proteins, **263–264**
Hydrazine, 12–13, *32–33*
Hydrazinolysis, **32–33**
Hydrindantin, 21
Hydrodynamic properties
  diffusion, **180–182, 265–
    266**, 269
  folded proteins, **264–270**
  frictional coefficient,
    189–190, 266
  molecular size and, **176–
    177, 189–190**
  rotation, **268–270**
  sedimentation, 24, **266–
    268**
  unfolded polypeptides,
    189–190
  viscosity, **189–190**
Hydrogen atoms, 216–217
Hydrogen bonds
  amino acid side chains,
    9, 16, 17
  description, **147–149**,
    168
  effective concentrations
    of, **163–167**, 182
  folded proteins, **221–
    228, 229–230, 300–
    307**
  folding and, 316
  α-helix, **182–185**
  hydrogen exchange and,
    283, 285

Hydrogen bonds (cont.)
  ligand binding and, **335**, **342–344**, 351, **356–357**, **414–416**
  random polypeptides, 165, 182
  reverse turns, 225–227
  salt bridges and, **143**
  secondary structure, **183–188**
  β-sheets, **186–187**
  stability of folded proteins, 165–167, *168–169*, 293, **300–307**
  water and, **155–160**, 162, *168–169*, **232**, 295
  zymogen activation and, **436**
Hydrogen exchange, **5–6**, 14, 237, 241, **282–286**, 292, *314, 333*, 467
Hydrogen tunneling, *392*
Hydropathy, 160, *169*
Hydrophobic interaction
  denaturants and, **295–296**
  description, 140, **157–160**, *168–169*
  effective concentration, 163
  heat capacity change, **158–162**
  ligand binding and, **343**, 422
  protein stability and, **300–307**, *326, 328*
  surface area and, **161–162**, 295, 297
  unfolded polypeptide chains, 162–163
Hydrophilicity, **153–154**, *168–169*, 255–256
Hydrophobicity
  amino acid side chains, **154**, 160–162, **254–255**
  definition, **157–160**
  membrane proteins, **67**, 75–78, 276–280
  proteins, 253–255, *259*
  signal peptides, **66–67**
Hydrophobic moment, **162**, 184, *186, 257*
β-Hydroxy-decanoyl dehydrase, **394**
Hydroxylamine, 13, 38, *40*, 389

Hydroxylation, 69, **96**, 196, 325
Hydroxyl groups, **8**, 12, 16, 35, *161, 301, 343*, 417
Hydroxy-Lys (Hyl) residues, 69, **92–95**, 196–197
N-(Hydroxymethyl)glutamine polymers, **179**
Hydroxyphenyl glyoxal, *13*
Hydroxy-Pro (Hyp) residues, 69, *93*, **96**, **196–197**, 466
Hyperreactivity, **273–274**
Hypervariable regions, **350–353**

Ice, 151, *168*, 237, *326*
Icebergs, 354–355
Ile residues
  acid hydrolysis and, 29
  asymmetry, **3**, *7*
  cleavage at, 39
  epimerization, 464, 465
  mass spectrometry, 35
  properties, **3**, *4, 7, 110, 118, 141, 142, 154, 186, 231, 240, 256*
Ile tRNA synthetase, **457**
Imidazole group, 6, **13–14**
Immunogenicity, **60–62**, 351, **353–355**, 406
Immunoglobulin fold, *94, 259*, **350**
Immunoglobulins, see Antibodies
Immunosuppression, 324
Inclusion bodies, 62–63
Indole group, **16–17**
Induced fit, **406–410**, 433, *460–461*
Infrared spectra, **191–193**, 271, 281
Inhibitors, enzyme
  complexes with enzymes, **331**, **393–396**, **403–406**, **421–423**, **426–428**, **430–431**, 434, **436–437**
  covalent, **452–459**
  enzyme rates and, **389–391**, **444–459**
  protein, *81*, 100, *103*, **125**, 218, **421–423**
Inositol hexaphosphate (IHP), *371, 375, 377*
Inositol pentaphosphate (IPP), *371, 375, 377*
Insertions of residues, evolutionary, 111, 248

Insulin
  biosynthesis, 77, **81–83**, 325
  chains, *27*, **81–82**
  evolution, *119, 122*, **124–125**
  folding, **124–125**
  hydrogen exchange, **283**
  nonequivalence of crystal monomers, 381
  proinsulin, 77, **81–82**, **124–125**, 325
  receptor, *97*, 453
  Zn²⁺ ion, 362
Intensity fluctuation spectroscopy, *265, 270*
Intermediate filaments, *195–196*
Intermediates
  accumulation, **290**, **309–310**, *327, 328*, **392–393**, **442–444**, *459*
  covalent, 389, 393, *461*
  enzyme catalysis, 389, **392–396**, 407, 403–406
  metabolic, 443
  protein folding, **290**, **309–323**, *327, 328*
  reaction rates and, 309–310, *459*
  tetrahedral, **405**, 418, **420**, 424, 429, 432, *460*
Interiors of folded proteins, **227–232**, *259*
Intermolecular interactions, **139–162**, 301
Intermolecular reactions, **397–402**
Intervening sequences, **50–53**, 60, 118, 124, 133
Introns, **50–53**, 60, 118, 124, *133*
Intramolecular interactions, **162–167**, 187, **301–303**, **341–344**
Intramolecular reactions, **164**, **333–334**, **397–402**
Invertase, *67*
Iodination, **15–16**, 273
Iodine, 16, **275–276**
Iodoacetamide, reaction with
  Cys thiols, **17–18**, **29–31**
  His residues, **274**
  Met residues, 17, *41*

Iodoacetate, reaction with
  Cys thiols, **17–18**, **29–31**, **41–43**, *47, 317, 328*
  His residues, **274**
Iodosobenzoic acid, 38, 40
Ionization
  accessibility and, **143–144**
  active site groups, **394–395**, **410–411**, 421, 429, 430, **439–441**
  amino acid side chains, **6, 143**
  environment and, *104*, **143–144**, 275
  folded proteins, 230–231, **262–264**, **271–272**, 275, *326*
  polyelectrolytes, 165
  polypeptide backbone, **5–6**
  signal peptides, 66
  unfolding and, 26, **288–289**, 293
Ion pumps, 280
Ions, **156**, 165, **263–264**, 267, 272, 293, **295–296**, **303–304**, 332, **358–364**
Iron ions, 18, 96, **220**, **361–362**, 364, **367**
Iron storage protein, **367**
Iron-sulfur proteins, **132**, **251–252**, 334, 362, **366–367**
IsoAsp residues, **9–10**, 464
Isocitrate dehydrogenase, **457**
Isoelectric point, **262–263**, 293
Isologous subunit association, **233–236**, 350
Isomorphous replacement, **208–210**, 213–215, 330–331
Isotope editing, NMR, **333**
Isotope effect, *391–392*
Isotope exchange, enzymatic, **391–392**
Isotope incorporation, NMR, 239, *243*
Isozymes, *53, 138*, **347**

J-coupling, NMR, **239–241**
Jigsaw puzzle model, **313**

Kallikrein, 419
KDEL signal, **71**
Keratan sulfate, 94

Keratins, 132, **193–196**, 223, 254

Kinases
  cyclic AMP-dependent, 97, 452–454
  evolution, 348
  induced fit, **406–410**
  phosphorylation of proteins, 96–98, **452–453**, *454–457*, 463
  structures, 88, **408, 409**

King & Altman, *388–389*

K_m for substrates, **386–389**, **425**

Knots, absence in proteins, 221, 227, 254

Koshland-Nemethy-Filmer allosteric model, **372, 374**, 380, **381–382**

Kringle domains, **133**, **322–323**

lac repressor, *136*, **356**, *359*

α-Lactalbumin, *27*, **122**, *131*, *292*, *314*

Lactoferrin, 367

β-Lactoglobulin, *348*

Lactate dehydrogenase
  affinity chromatography, **347**
  evolution, *122*, *130–131*
  hydrodynamic properties, *266*
  isozymes, **347**
  nonequivalence of subunits, *381*
  structure, **220**

Lag period, **310–311**, *382*, *460–461*

Lambda repressor, **304–306**, *328*, **356–357**, *360*

Lamin A, *76*

La protease, 469

Laue diffraction, **216**

Leader sequence. *See* Signal peptides

Leghemoglobin, **348**

Lennard-Jones potential, **146**

Lens proteins, *122*, *130–131*, *251*, *465*

Leucine/isoleucine/valine-binding protein, **218–219**

Leu residues
  acid hydrolysis and, 29
  addition to amino-terminus, **87**
  cleavage at, 39

mass spectrometry, 35
properties, *3, 4, 7, 110, 118, 141, 142, 154, 186, 199, 231, 240, 256*

"Leucine zipper," *199, 360, 383*

Leupeptin, 469

Lifson-Roig model, **184–186**

Ligand binding
  affinities, 336, **338–346**, **440**
  description, **329–384**
  protein degradation and, 466
  protein flexibility and, **336–337, 339–340**
  protein modification and, 456
  protein stability and, **295–296, 339–340**, 466
  rates, 336, **344–346**, 359, 365, 370
  refolding and, 323

Light-harvesting chlorophyll binding protein, *72*

Light-harvesting complex, *279*

Linear free-energy relationships, **411–413**, *414–416*, *460*

Lineweaver-Burke plot, **387, 389**

Linked functions
  allostery and, **371, 373**, *384*
  catalysis and, **403**
  description, **317–318**
  folding and, **317–321**
  ionization and, **429**
  ligand binding and, **337**, 366, *384*

Lipases, *138*, *461*

Lipoproteins, *102*, 108, *133*

Lipotropins, *83*, *122*

Liquids, 139, **149–153**, 343

London dispersion forces, **146–147**

lon gene, 469

Loops, **225–227**

Low-density lipoprotein (LDL), 108, *133*

Lowry protein assay, **21**, 23

Lutropin, *122*

Lysosomes
  protein degradation, 430, **469–470**
  protein transfer to, **65–71**, 93, *102–103*

Lysozyme, bacteriophage T4
  catalytic mechanism, *441*
  crystallography, *203, 210, 238*
  homology with mammalian lysozyme, **125, 250**, *441*
  stability, **293, 303, 306**
  unfolding, *316*

Lysozyme, mammalian
  active site, **395, 438**
  antigenic structure, **354**
  biosynthesis, *68*, 71
  catalytic mechanism, *397*, **437–441**
  compressibility, **237**
  evolutionary divergence, *122*, **125, 130–131**, *137*
  folding, **311**, *313–314, 316*
  function, **125**
  hexasaccharide bound, **331, 438**
  homology with α-lactalbumin, **130–131**
  homology with bacteriophage lysozyme, **125, 250**, *441*
  hydration, 238, **263–264**
  hydrodynamic properties, *266, 269*
  hydrogen exchange, *286*
  iodine and, **275–276**
  ligand binding, **331**, *336*, **438**
  structure, **331, 438**
  unfolding, **296–302**, 311, *316*

Lys residues
  acetylation, **11**, 453
  ADP-ribosylation, 98
  carboxl-terminal, **98**
  cleavage at, **38–39**, 81–**83**, 419, 422
  counting, 29, *31*
  diagonal technique, 41
  flexibility, **181**
  hydroxylation, **96**, 196
  phosphorylation, **97**
  properties, *3, 4*, **10–12**, 35, *110, 118, 141, 142, 154, 181, 186, 231, 240, 256, 282*

Schiff base formation, **11**

Lysylendopeptidase, 39

α-Lytic protease, 248, **418–428**, 434, *436*

Magnesium ions, 156, 293, **361–362**

Magnetization transfer, 333

Malaria, **126–127**

Malate dehydrogenase, *75*, 266, 381

Maleic acid, 430, 447

Maleic anhydride, *11–12*, 18, 38, 41

Maltose binding protein, *78*, 325

Manganese ions, 18

Mannose phosphate, **70–71**, 93

Mass
  amino acid residues, **4**
  proteins, **23–28, 36–38, 43**, 229, 266, *326*

Mass spectrometry, 24, **35–38**

Mating factor, yeast, *76*, 103

Maximum parsimony method, 111

β-Meander, **227–228**

Mechanism-based inhibitors, **393–394**, *396*

Melanocyte-stimulating hormone (MSH), **82–83**, 87

Membrane proteins
  amino acid compositions, 43, 67
  bacteriorhodopsin, *279, 280*
  cross-linking, *334*
  diffusion, **280**, *345–346*
  flexibility, **280–281**
  ion pumps, 280
  membrane-spanning segments, 53, 67, 280
  signal peptides, 76
  structures, 254, **276–280**
  sulfation, **95**
  surfaces, 230, 262
  topogenesis, **65–71**, **75–78**
  topography, **75–78**, **277–280**

Membranes,
  anchors, **88–90, 94–95**, 277, 468
  ligand binding, **345–346**
  traversing, **64–78**, *102*, **280**, 324, 443

Membranes (cont.)
structure, *104*, 162, **276–280**

Mercaptoethanol, 19, 22, *47*

p-Mercuribenzoate, **18**

Mercury derivatives, 18, 208, 448

Mesophiles, 303–304

Messenger RNA (mRNA), **50–59**, **61**, 107–108, **113**

Metabolic pathways
evolution, **131–132**
regulation, 325, **442–444**, *459–460*

Metal ions
bound to proteins, *6*, *8*, 293, 338, **361–362**, **367**
catalysis and, 402
heavy atoms, 17, 18, **209–210**
storage, **367**
thiol oxidation and, 18
transport, **367**

Metalloproteases, **418**, **431–434**, 437

Metallothionein, **367**

Methotrexate, 403

N-Methyl acetamide, **153–155**, *169*

Methylene blue, 13

Methylguanidinium formate, **143**

Methyl iodide, 17, 35

Methyl groups, 260, *343*, **239**

O-Methyl isourea, 11

Methylpentanediol (MPD), **294**

met repressor, **233**, 359–360

Met residues
aminopeptidase, *87*
amino-terminal, **55–56**, 62, 69, **86–87**, 472
cleavage at, **38–40**, 63, 308
diagonal techniques, 41
heme groups and, **246**, 248, 366
methylation, 17
oxidation, 17, 40, **99–100**, *461*, 464
properties, *3*, *4*, **17**, 20, 59, *110*, *118*, *141*, *142*, *154*, *186*, *231*, *240*, *256*
removal, **86–87**, 472

sulfone, 17

sulfonium salts, 17, 39

sulfoxide, 17, 20, 39

Michaelis complex, **386–388**, 406, **424**, **432**, *460*

Michaelis-Menten equation, **386–388**

Microbodies, **74–75**

Microfibrils, 196–197

Microsomes, 103

Microtubules, **90–91**

Miller indices, 205

Minimal mutations method, 111

Mitochondria
biosynthesis of proteins, *53*, **71–74**
degradation, 462
evolutionary origin, 51
genetic code, 58–59
import of proteins, **71–74**, 324–325

Models, protein structure, *217*, *254–255*, *257–259*

Moffitt equation, 190

Molecular drive, *127*

Molecular dynamics, **281–282**

Molecular replacement, 210, 330

Molecular surface, **228–229**

Molecular weight (mass)
amino acid residues, **4**
proteins, **23–28**, **36–38**, **43**, 229, 266, *326*

Molten globule state, **292**, **314**, 325

Molybdenum ions, 18

Monoclonal antibodies, 351

Monod-Wyman-Changeux allosteric model, **372–381**, **444–452**

Mononucleotide-binding domains, **249–251**, **361**

Mosaic proteins, **51**, **108**, **132–134**

Muconolactone isomerase, 236

Multicatalytic proteinase, *469*

Multiwavelength anomalous diffraction (MAD), **209–210**

Muscle, 361, 453

Mutant proteins
ATCase, 448, *451–452*
collagens, 196, 198
degradation, 467–468

descriptions, **111–112**, **209–210**
functional properties, **411–413**, *441*
hemoglobin, *102*, **111–114**
proteases, 421, 426, *428*
ribitol dehydrogenase, 136
stabilities, **304–307**, *327–328*
structures, **111–112**, **208–210**, *238*
tryptophan synthase, 136
Tyr tRNA synthetase, **413–416**

Mutation matrix, **109–110**

Mutations
amidase, *137*
effects on protein stability, 136, 196, 198
evolutionary, **106–138**, **244–253**
hemoglobin, *103*, **111–114**
immunoglobulin generation, **351–353**
neutral, 114, **123–125**, 375
protein structure and, *103*, **111–113**, **134–137**
site-directed, **63–64**, *134–135*, 138, 208, **304–307**, 326, 341, 394, **411–416**, *441*, *452*
tryptophan synthase, 136

m value, **291**

MWC allosteric model, **372–381**, **444–452**

Myoglobin
evolution, *113*, *114*, *119*, *122*, **127–130**, 245
flexibility, *232*, **237**, *327*
function, 245
heme binding, 348, **362**, *383*
His residues, *14*, 272
homology with hemoglobin, **127–130**, 245
hydrodynamic properties, *27*, *266*, *269*
ionization, 272, **293**
ligand binding, *327*, *336*, **364–366**, 377, *383*
sequence, **128**
structure, **245**
unfolding, *292*, **293**, *297*, 299, 308, *383*

Myohemerythrin. *See* Hemerythrin

Myosin
degradation, *467*
diffusion, *269*
His residue methylation, 466
structure, *195*

Myristoylation, 84, *88*, 94

NAD (Nicotinamide adenine dinucleotide)
ADP-ribosylation and, **98**
binding domains, **250**, **251**, **360–361**
dehydrogenases and, **220**, **250**, **336**, **347**, **360–361**, **381–382**
origin, 107
protease for apoproteins, 468
redox function, 366

NADP, **335**

Naphthalene, 129

Nascent polypeptide chain
acetylation, **87–88**
conformation, 57, *199*
folding, **68–69**, **98–99**, **323–325**
formation of Gla residues, **95–96**
glycosylation, 93
hydroxylation, 68, 96
signal sequence, **65–68**

Natural selection, 111, **114**, 118, 120, **123–125**

Negative cooperativity, **370**, 373, **381–382**, 451

Neuraminidase, 236

Neuropeptides, **82–83**, 87, 95

Neurophysin, **82–83**

Neurotoxins, *122*

Neutral mutations, **114**, **123–125**, 375

Neutron diffraction
amplitudes, **205**
catalytic triad characterized with, 421, *428*
description, **216–217**
hydrogen exchange and, **216**, 237, **283–285**

Neutron scattering, **149–150**, 152, *201–202*, 278

Newman projection, **172**, 174

Nickel enzymes, *402*

Ninhydrin, **21**, 29, 42

Nitration, **15–17**, 273, 448, 451

Nitrenes, **334**

2-Nitrophenyl phosphate, **394**

p-Nitrophenyl-p'-guanido-benzoate (NPGB), **393**

o-Nitrophenyl sulfenyl chloride, 41

Nitrothiosulfobenzoate (NTSB), **20**

2-Nitro-5-thiocyanobenzoic acid, **39**

NOESY spectra, 241

Noncompetitive enzyme inhibition, **388–391**

Noncrystallographic symmetry, 204, 209

Nonpolar surfaces, **155–156, 157–160**, 165, **238**, 254, **293–303**, *328*

Norleucine, 46

Nuclear magnetic resonance (NMR)
13C, **13–14, 181–182**, *202*, **286–287**, 330, **333**
1H, *14, 198*, 330, **333**, **238–244**, **283–284**, **286–287**, **333**
15N, *202*, 330, **333**, 421

Nuclear Overhauser effect (NOE), **182**, **239–243**, 333

Nuclear proteins, **75–76**

Nucleation, **184–185**, 196, 202, **313**

Nucleic acids, **50–68**, *102*, **107–109**, 165, 330, *334, 335*, 343–344, **355–360**

Nucleoplasmin, *76*

Nucleotide-binding domains, **249–251**, 348, **360–361**, 407

Octanol, 168

Octarellin, *258–259*

Oligosaccharyl transferase, *103–104*

OmpA, *78*

Oncogenes, 452, 466, 472

Opiomelanocortin, **60–61**, **82–83**

Opsins, **134**

Optical rotary dispersion (ORD), **190–191**, 271, 288

Optical transforms, **205–212**

Orbital steering, 385

Organelles, **64–78**

Orientation, 385

Ornithine, 100

Ornithine decarboxylase, 466, *467*

Ornithine transcarbamoylase, *131*

Orthologous genes, **130–131**

Ovalbumin, 27, 66, 68, 93, 130

Ovomucoid, *125*, 218

Oxaloacetate decarboxylase, **404**

Oxazolones, 32

Oxidation
Cys residues, **18–20, 98–100**, 464, 467
His residues, 13, 459, 467
Met residues, 17, **99–100**, *461*, 464, 467
protein aging and, *464–465*, 467
redox proteins, **364–367**
Trp residues, 16

Oxindolealanine, 16, **275–276**

Oxyanion binding site
metalloproteases, 433
serine proteases, **424, 425–428**, 434
thiol proteases, 429

Oxygen
binding to proteins, 336, **364–366, 370, 371, 374–381**
quenching of fluorescence, **286**

Oxytocin, **82–83**

Ozone, 16

Packing of protein interiors, **227, 229–232, 253–255, 279, 300–303**, *327*, 335

Palindromes, **356**, 360

Palmitoyl groups, 69, **94–95**, *103*

Pancreatic polypeptide, 125, **213**

Papain, 297, 349, *395*, **428–429**

Paralogous genes, **130–131**

Paramyosin, *195*

Partial volume
amino acid residues, 4, 141, 153
folded proteins, 24, **266–267**

Partition coefficients, **153–154, 157–162**

Parvalbumin
Ca2+ binding, **331, 363**
evolution, *122*, 251

Patterson map, **208–209**

Penicillopepsin, *431*

Pepsin,
active site, **395, 430**
catalytic mechanism, **429–431**, *461*
cleavage of proteins, 38–39, 349–350
zymogen, 79, 81, **436–437**

Pepsinogen
activation, 79, 81, **436–437**
biosynthesis, 79, 81
hydrodynamic properties, *269*

Pepstatin, **431**, 469

Peptide bond
absorbance, **190–193**
biosynthesis, **50–59**, *199*
cis/trans isomerization, **5, 7–8**, *104*, **174**, *176*, 181, **188**, 196, *198, 221*, **312**, **324–325**
dipole moment, **5–6, 145**, 183, *186*
hydrazinolysis, **32–33**
hydrogen exchange, 6, **282–286**
hydrolysis, acid, 6, **28–31**, 464
hydrolysis, enzymatic, 29, *103*, **417–437**, **468–472**
planarity, 5, 144, **172–174**, 221
properties, **5–6**, **20–21**, 144, 154, 168, **172–176**, *301*
structure, **2–6**
synthesis, **43–47**, **307–308**

Peptide maps, **112**

Peptides
conformations, **189**, *243*, 262, **314–315**, 353, *383*
definition, 4
detection, **20–23**
generation, **38–41**
mass spectrometry, **35–38**
purification, 35, 41
synthesis, **43–47**

unusual, **100–102**

Peptide unit, 174, *176*

Performic acid, 18, 29, **41–43**, 181, 282

Peroxidase, 99, 364

Peroxisomes, **74–75, 102–103**

PEST regions, 467

Phases, crystallographic, **208–217**

Phenolic groups, 6

Phenylglyoxal, *12–13*

Phe residues
addition to amino terminus, **87**
cleavage at, 36, 422
flipping, 239, 243, **286–287**
hydrophilicity, 144, *154*
hydrophobicity, *154, 168*
properties, **3, 4, 14–15**, *110, 118, 141, 142, 186, 231, 240, 256*
spectral properties, **14–16**, 20, 23, 29, **270–271**

Phenylisothiocyanate (PTC), **29–31, 33–34**

Phenylthiocarbamyl, **33–34**

Phenylthiohydantoin (PTH), 34

Phosphatases, 88, **96–98, 452–453**

Phosphate ions, **156**, *382*, **394**

Phosphate transport protein, *337, 382*

Phosphoenolpyruvate carboxykinase, *467*

Phosphofructokinase, 219, **444–446**

Phosphoglucomutase, 406, *410*

Phosphoglycerate kinase, *266*, **288**

Phospholipase A2, 248

Phospholipase C, 90

Phosphonamidates, **405**

N-Phosphonacetyl aspartate (PALA), **446–452**

Phosphorylase, glycogen, *397*, **453–457**

Phosphorylation, **96–98**, 389, **452–457**

Photoactivated reagents, **334, 394–396**

Photooxidation, 13

Photosynthetic reaction center, 230, **278–280**, *367*

Physiological conditions, 1
o-Phthalaldehyde, **22-23**
Phylogenetic trees, **106, 119-121, 127-130,** 136, *137*
Phytochrome, 464
Ping-pong enzyme mechanism, 389, 392, 420
Plasma desorption mass spectrometry (PDMS), 36
Plasmids, **59-64**
Plasmin, 419
Plasminogen, **322**
Plasminogen activator inhibitor-1, *262*
Plastocyanin, *72, 122,* **362,** *367*
Platinum salts, 17, 208
Pleated sheets. See *β*-Sheets
Plectin, *193*
Point accepted mutation (PAM), **109-110**
Polarizability, 145, 426
Poliovirus, **83-84**
Polyacrylamide gel electrophoresis (PAGE), **25-28,** *113, 126-127,* 353, *384*
Poly(Ala), 179
Polyamino acids, **4,** *48,* **173,** 179, *182, 183,* **188,** 189-192, *198, 199*
Poly(benzyl-Glu), *182, 269*
Polyelectrolytes, **165,** *272,* 359
Poly(ethylene glycol), **264,** *325-326*
Poly(Gly), 179, *183,* **188,** 196
Poly(Lys), 187, **189-192,** *199*
Polymers, **2-6,** *173-182,* **264**
Polypeptides, **2-6,** *173-182*
Polypeptide backbone
  biosynthesis, **53-59**
  cleavage, **35-36**
  hydrolysis, acid, 6, **28-31,** 464
  hydrolysis, enzymatic, 18, 29, *103,* **417-437,** **468-472**
  properties, **2-6**
  random, **176-180**
  reactions, **31-41**
  rotations, **173-176,** *181*
  spectral properties, 190-193

synthesis, **43-47**
Poly(Pro) I and II, *183,* **188,** 196, 223
Polyproteins, 69, **82-84,** 430, 471
Poly(S-carboxymethyl-Cys), *187*
Poly(Tyr), 187
Porin, **279-280**
Posttranslational modifications, 35, 62, 69, **78-103, 452-459**
Prealbumin, **233-234,** 236
Precession photographs, **205-207**
Precursors, 50, **79-100**
Predicting protein structure, *384,* **253-258,** **280**
Prefolded state, **314**
Prenyl proteins, 88-90
Preproteins, 50, **61,** *64,* **65-68,** 72
Primary structure
  assembly, **50-59**
  definition, 31, 171, 217
  elucidation, **31-43,** *48,* 213
  evolution, **105-138**
  gene sequence and, *43, 48,* **59-60**
Primordial soup, **107-108**
Procollagen, 69, **96,** 196
Proinsulin, 77, **81-83,** 325
Prokaryotes, 51, **53-55,** 58, **77-78,** 105, 133
Prolactin, *122*
Prolyl peptidases, 38, 39, *188*
Prolyl-4-hydroxylase, **96,** *103-104,* 325
Prolyl peptide isomerase, *104,* **324-325,** 333
Proopiomelanocortin, **60-61, 82-83**
Proparathyrin, *122*
Propinquity, 385
Pro residues
  cis peptide bonds, 7-8, *104,* **174,** *176, 181,* **226-227,** 304, **312, 324-325**
  cleavage at, 38, *40*
  collagen structure and, **196-197**
  conformational effect of, **176-177,** 184, 306, *327*
  folding of proteins and, **311-312, 324-325**

α-helix and, 184, 304
  hydroxylation, **96,** 325
  peptide bond isomerization, 7-8, *104,* **174,** *181,* **311-312, 324-325**
  properties, 2, *3, 4,* **7-8,** 21, 32, 91, *110, 118, 141, 142, 154, 186, 231, 240,* 242, *256,* 467
  reverse turns and, **226-227**
Pro segments, **79-82**
Prosthetic groups
  binding small ligands, **364-367**
  complexes, 330, 335, 337-338
  covalent attachment, 366
  examples, **362,** 366, 402
  removal, 308
Protease inhibitors
  small molecules, **426-427**
  peptide, *336,* 469
  protein, *81,* 100, *103,* **125,** 130, *138,* **218,** *262,* 307-308, **331,** *336,* **421-423,** 426, *428, 437*
Proteases
  ATP-dependent, **468-469,** 471
  carboxyl, **418, 429-431, 436-437,** *461*
  catalytic mechanisms, **417-437**
  cofactor-specific, 468
  degradation of proteins and, 29, **468-472**
  evolution, *125, 138,* **247-248,** 250, 419, 421
  folded proteins cleaved by, 261, 277, **307-309**
  fragmentation of polypeptides, **38-40**
  lysosomal, **469-470**
  metallo, **418, 431-434,** **437**
  peptide bond synthesis and, **45-47,** 307-308
  posttranslational modifications and, **65-69,** 74
  serine, **247-248,** 250, *273-274,* **322-323,** **418-428**

Streptomyces, 248, *336*
thiol, 132, *138,* **418, 428-429,** *461,* 468
trypsin family, **79-81,** 130, **247-248, 418-428**
Proteasome, 469, 471
Protein disulfide isomerase, **99,** *103-104,* 307, **324-325**
Protein phosphatase 2A, 132
Proteins, definition, **4-5**
Proteoglycans, **92, 94,** 197
Proteolytic enzymes. See Proteases
Proteolytic processing, 69, 74, 75, **78-86,** 95, 196
Prothrombin. See Thrombin
Protomers, 233
Proximity, **163-167,** 385, *402,* 406
Pseudogenes, 60
Pseudopepsin, 436
Pseudosubstrate, 453
Psychrophiles, 303
Pyridoxal phosphate catalyst, **401-402**
  protease for apoproteins, 468
  Schiff base with amines, **11, 401-402**
Pyroglutaminyl amino peptidase, 9
Pyrrolidine ring, 7-8
Pyrrolidone carboxylic acid, 9
Pyruvate kinase, 219

Quasi-equivalent quaternary structure, 236
Quaternary structure
  acquisition, **68-69,** *278,* **321-324,** 443
  allostery and, 337, **367-382**
  aspartate transcarbamoylase, **446-452**
  asymmetry, **381-382,** 414
  cross-linking, *278,* **334**
  definition, 171, 217
  determined crystallographically, **232-236,** *337*
  evolution, 251-253
  glutamine synthetase, **457-459**
  hemoglobin, **372-380**

Quarternary structure (cont.)
membrane proteins, **278–280**
phosphofructokinase, **444–446**
phosphorylase, **455–456**
protein kinases, 454
symmetry, **232–236**, **251–253**, 456

Racemization, 9–10, 402, 464, 465
Radial distribution function, **149–150, 177–179**
Radius of gyration, 176–177, 189–190
Ramachandran plot, **173–176, 183**, 222
Raman spectroscopy, 191–193, 271, 281, *396*
Random flight chain, 176
Random polypeptide chains
conformational equilibria in, **314–315**
denaturants and, 189
description, **173–182**
gel filtration, **25–27**
hydrodynamic properties, **176–180, 189–190**, 198, *269*
NMR properties, **238–240**, *198*
number of conformations, 309
spectral properties, **190–193**
transition to α-helix, **184–186**, *198*
Ras p21 protein, *396*
Receptors, **354–355**
Reciprocal space, 205
Redox potential, 366
Redox proteins, **366–367**
Reentrant surface, **228–229**
Refinement, crystallographic, 209, **214–215**
Reflections, X-ray, **205–217**
Relative rate test, 121
Relaxation time
conformational changes, **181–182**
definition, **181**, 268
rotation of molecules, **268–270**
RELAY NMR spectra, 241
Renin, *431*
Rennin, 430

λ-Repressor, **304–306**, *328*, **356–357**, *360*
Repressors, **356–360**
Retinol-binding protein, *348*
Reverse turns, **225–227**, **255–257**
Rhodanese, 251, *395*
Rhodopsin, *134*
Ribitol dehydrogenase, 136–137
Ribonuclease A
affinity labels, *334*
cleavage, *307–308*
C peptide, *198*
deamidation, *9, 47*
evolution, *119, 122, 137*
folding, **311–312**, *314*
glycosylation, 93
His residues, *14*, **274**
hydrodynamic properties, *27, 266, 269*
neutron diffraction study, *285*
NMR studies, *14*, **181–182**
oxidized form, **181–182**
reduced form, *47, 327*
S-peptide, 307
S-protein, **307**
synthesis, *47, 48*
unfolding, **288, 291, 296**, *297, 299, 326*
Ribonuclease B, 93
Ribonuclease S, **229, 307**, *308*
Ribonuclease T1, *291*
Ribophorins, 67
Ribosomes, **55–59**, 65–67, 107–108, *202, 467*
Ribulose-1,5-bisphosphate carboxylase/oxidase, *72*
RNA polymerase, 51, *467*
RNA world, **107–108**
ROESY NMR spectra, 241
Root effect, 124
Rossman fold, **250, 251**, *348*
Rotamers, 38
Rotation, **181–182**, 261, **268–270**
Rubredoxin, 215, 251, *361*, **362**
R values, **214–215**

Salt bridges
description, **143**, *167–168*
folded conformations and, 165–167, *198*, 275, **293**, 303

trypsin proteases, **434–436**
unfolded polypeptide chains, 182
water and, **155–156**, 165–167
Salts, **156, 263–264, 295–296**, 303–304, *327*
Scatchard plot, 339, 340, 368–370, 387
Schiff base, 11, 389, 392, 393, **401–402**, 464–465
SDS (sodium dodecyl sulfate)
detergent, *48, 199, 277*
electrophoresis, **24–28**, **40**, *47, 334*
SecA, 77
SecB, 77
Secondary structure. *See also* α-Helices, β-Sheets
classification of proteins, **224–225**
definition, 171, 217
description, **182–188**
folded proteins, **221–227**
measurement, 242, 271
occurrence of amino acid residues, 256
packing, 227, **253–254**
prediction, **255–257**
spectral properties, **190–193**
unfolded proteins, 292
variation in related proteins, 248–249
Second messengers, 90, 97, 361, 452–454
Secretion of proteins, 62, **65–71, 77–78, 91–96**
SECSY NMR spectra, 241
Sedimentation, 24, **266–268**, *326*
Seleno-Cys, 58, 60
Seleno-Met, **209–210**
Sephadex, 27
Sequences
alignment, **108–111**
gene, 31, 48, **59–64**, **105–138**
protein
ancestral, **119–121**, 127–130
determination, 23, **28–43**, *48*, **59–62**
evolution, **105–138**
gene sequences and, *48*, **59–62**

nature, **43**
possible number, 43
Sequential allosteric model, **372, 374**, 380–382
Sequential enzyme mechanism, 388
Serine proteases. *See also* Trypsin protease family; Subtilisin
catalytic mechanism, 387, 389, **418–428**
evolution, **247–248**
protein degradation and, 469
Ser residue reactivity, **273–274**
Serpins, 130, *261*
Ser residues
acid hydrolysis and, 29
carbohydrate attachment, 91, **93–94**
catalysis and, **418, 421–428**, *461*
catalytic triad, 138, **421–428**
cleavage at, 39, 40
codons, *57, 137*
phosphorylation, **96–98**, *452–453*, **454–457**
properties, *3, 4, 8, 110, 118, 141, 142, 154, 184, 186, 231, 240, 256, 282, 407, 467*
reactivity, *8*, **101–102**, **273–274**, *407, 467*
serine proteases and, **273–274**, **421–428**
β-Sheets
barrels, **220, 224–225**, 249, *251, 257–260*, *279, 309, 334, 348*
description, **186–187**
dimerization and, **234**; 236, **446**
dipole moment, *198*
evolutionary variation, **248**
folded proteins, **221–225, 227–228, 256**, 279
hydrogen exchange, **283**
immunoglobulin fold, **350**
measurement, 271
model systems, 187, 189, **191–193**
prediction, **255–257**
silk, 193
spectral properties, **190–193**

β-Sheets *(cont.)*
  structure, **186–187,** *198,*
    220, 224, 228, 230,
    234, 250
  twist, 186–187, 223, 227
Sialic acid, 92
Sickle-cell anemia, **112,**
  126–127, 332
Side chains
  analogues, **153–154,** *168*
  conformations, 174, *221,*
    223, *249*
  evolutionary variation,
    110, 118
  hydrophilicities, **154**
  hydrophobicities, **154**
  mobility, 269, 286–287
  NMR chemical shifts, **240**
  packing, **231**
  properties, **6–20**
  secondary structure tend-
    encies, **186, 256**
  surface areas, **142**
  structures, **3**
  volumes, **4,** *141*
Sieve, molecular, **24–25**
Sigma factor, 51, *53*
Signal peptidase, 65–68, 69
Signal peptide peptidase,
  68
Signal peptides, 50, 53, **61,**
  62, 65–74, 78, 79, 89,
  102, 103
Signal recognition particle
  (SRP), **65–67,** 73, 77,
  107, 324
Signal sequences, **65–78,**
  102–103
Silk, *193,* 195
Silver ion, 18, 22
Silver stain, **22–23**
Site-directed mutagenesis,
  63–64, *134–135,* 138,
  208, **304–307,** *326,*
  341, 394, **411–416,**
  441, 452
Skin, 197
Solid phase
  sequencing, 34
  synthesis, **43–47**
Solubilities,
  amino acid residues,
    153–155
  proteins, *156,* 202, **262–**
    264, 465
Solvation
  amino acid residues,
    153–155
  ligands, **335**

nonpolar surfaces, **155–**
  160, 293–303
polypeptide conforma-
  tion and, **189,** 293–
  296
transition states, **400–401**
Solvent
  conformation and, **189,**
    223, 293–296
  observed crystallographi-
    cally, **238**
  organic, *460*
  perturbation spectros-
    copy, **270–271**
  protein crystals, **202–**
    203, 207, 217, **238**
  reaction rates and, **400–**
    401
  surface determination,
    227–232
θ (theta), **189,** 291
Southern bean mosaic
  virus, 236
Soybean trypsin inhibitor,
  421
Space groups, crystallo-
  graphic, **203–207**
Spectral properties
  aromatic amino acids,
    14–17, 270–271
  folded proteins, **270–271**
  Tyr derivatives, **16**
  unfolded proteins, **190–**
    193
Spectrin, *193*
Spliceosome, 52, 107
Splicing
  mRNA, **51–53,** 113
  proteins, **84–86**
Stability of folded conformations, *136,* 160,
  165–167, 292–309,
  339, *473*
Standard state, *338,* 340
Staphylococcal nuclease
  complementation, **308,**
    315
  structure, *202,* **308**
  unfolding, **288–289,** *292,*
    303, 306–307
Statine, **430**
Steady state kinetics, **386–**
  392
Stereo diagrams, **218**
Stereopopulation control,
  385
Stokes radius, **26–27,** *268*
Strain
  conformational, *221*

enzyme catalysis and,
    396, 402, **437–441**
β-Strands, **186–187, 220**
Streptomyces griseus proteases A and B, **418–**
  428
Structural proteins, **193–**
  198, 262
Subdomains, 219
Submaxillary protease, 38–
  39
Substituted-enzyme mecha-
  nism, 389
Substrates, enzyme, 385
Subtilisin
  biosynthesis, 81, 434, *436*
  catalytic mechanism,
    413, **418–428**
  evolutionary conver-
    gence, *138,* 250,
    419, 421
  hydrodynamic proper-
    ties, *266*
  ionization, **272–273**
  protein inhibitor, 421,
    434, *436*
  reactive Ser, *138*
  stability, *304, 461*
  thiol, *429*
  zymogen, 434, *436*
Subunits
  association, 68, **232–**
    236, 321–324
  folding and, **68–69, 321–**
    324
  movements, 262, 359,
    372–382, **444–459**
Succinic anhydride, 29, *326*
Succinimides, **9–10,** 40, *465*
Suicide substrates, **393–**
  394, 396
Sulfate-binding protein,
  335, *382*
Sulfate ion, **156, 263–264,**
  295–296, 332, 335
Sulfation, 69, **95**
Sulfenyl halides, 16, 41
Sulfite, 19
Sulfones, 18, **41–43**
Sulfoxides, 17, *99–100*
Sulfonic acids, 18, 29
Sulfonium salts, 17, **39–41**
Superoxide dismutases
  function, 99, *464*
  hydrodynamic proper-
    ties, *266*
  immunoglobulin fold
    and, *259*
  metal ions bound, **362**

structure, 236, *259*
Supersecondary structures,
  227–228, 258, 348
Suppressors, 58
Surfaces
  amino acid residues,
    141–142
  area measurement, **227–**
    232
  folded proteins, **227–**
    232, *254, 259,* 279
  hydrophobicity and,
    159–162, *259,* **295,**
    297
  nonpolar, **155–162,** 165,
    238, *254,* **293–303,**
    308
Surface tension, **156,** 264,
  293, *295–296*
Svedberg equation, **267**
Symmetry
  crystallographic, **203–**
    209
  internal, **251–253,** *460*
  noncrystallographic, 204
  palindromic, **356**
  quaternary structure,
    233–236, 332, 335,
    368, *374, 375,* 381–
    382, **444–459**
Synchrotron, 216
Synthesis, peptide, **43–47,**
  48

T4 lysozyme. *See* Lyso-
  zyme, bacteriophage
  T4
Tandem mass spectrome-
  try, 37–38
Temperature factor, **214,**
  244, 237, 281
Template, tertiary, **257–258**
Tendamistat, **243–244**
Tendon, 197
Terminal residues
  C, **31–33,** 34–35, **88–**
    91, *259,* 308
  N, 6, 9, 24, **31–35,** 41–
    47, 86–88, *259,*
    274–275, 471–472
Tertiary structure
  definition, 171, 217
  description, **217–221**
  determination, **202–217,**
    238–244
  prediction, **257–258**
  primary structure and,
    244–259

Tetrahedral intermediates, **405, 418, 420, 424,** 429, **432,** *460*

Tetrahydrophthaloyl anhydride, *11–12*

Tetranitromethane, 17, 273

Thalassemia, 126

Thaumatin, *138*

Thermitase, *304*

Thermolysin
active site, **418**
carboxypeptidase A and, 250, **431–434**
catalytic mechanism, **431–434,** *461*
specificity, 38
transition-state analogues, **403–406**

Thermophiles, **137,** *277,* 296, **303–304,** 306, 414, **444–446**

Theta (*θ*) solvent, **189,** 291

Thiocyanate, **34–35,** 156, **296**

Thiohydantoins, 35

Thiol groups, 6, 16, **17–20,** **29–31, 41–43,** 47, *161,* 208, **316–321,** *343,* **361–362,** 367, **428–429,** 464

Thiol-disulfide exchange, 19

Thiol proteases, 132, *138,* **418, 428–429,** *461,* 468

Thiolsubtilisin, 429

Thioredoxin, *99, 312,* 366

Thioredoxin reductase, 325

Thiotemplate mechanism, peptide biosynthesis, **101–102**

Threonine synthase, *131*

Threonine dehydratase, *131*

Threonine synthesis, *132*

Thrombin
catalytic mechanism, **418–428**
cleavage of fibrinogen, 124
evolution, **133,** 247
Gla residues, *95*

Thr residues
acid hydrolysis and, 29
asymmetry, **3, 8**
carbohydrate attachment, 69, **93–94,** 196
cleavage at, 39, 40
epimerization, 464, 465
phosphorylation, **96–98,** *452–453,* **454–457**

properties, *3, 4,* **8,** *110, 118, 141, 142, 154,* 181, 184, *186,* **231,** *240, 256, 282, 467*

Thyroid binding protein, *104*

Thyroglobulin, *27*

Thyrotropin, *122*

Ti protease, 469

TNBS, 11

Tobacco mosaic virus
coat protein disk, 209, 236, 251
diffusion, **268–269**
structure, 23

TOCSY NMR spectra, 241

Tomato bushy stunt virus, 236

Topogenesis, 50, **64–78**

Topology, membrane, **253–255**

Torsion angles
conventions, **174–176**
folded proteins, **221–223**
regular conformations, **182–188**
side chains, **174,** 223

Toxins, *98, 122,* 394

Transferrin, *68,* 132, **367**

Transfer RNA (tRNA), **53–58,** 87, *107–108*

Transglycosylation, 441

Transition state
aromatic ring flips, **287**
binding by enzyme, **396–406,** 407, *460*
carboxyl protease catalysis, **418,** 430–431
definition, 180, *459*
enzyme catalysis and, 391, 393, **396–406,** 411–413
lysozyme catalysis, **437–441**
metalloprotease catalysis, **418, 432**
protein folding, 310, **311–314, 316**
serine proteases, **417–428**
theory, **180–181**

Tyr tRNA synthetase, **415**

Transition-state analogues
description, **403–406,** 407, **460–461**
lysozyme, **439–441**
metalloproteases, **405,** *433*

serine proteases, **426–428**

Transit peptides, **72, 74**

Translation, mRNA, **50–59**

Transpeptidation, 57, 85, 103, *461*

Trifluoroacetic acid (TFA), 29, 34, 46

Trifluorethanol, 189

Trigger factor, *78*

Trinitrobenzene sulfonate (TNBS), 11

Triose phosphate dehydrogenase. *See* Glyceraldehyde phosphate dehydrogenase

Triose phosphate isomerase (TIM)
($βα$)$_8$ barrels, **220, 224,** 249, *251,* **257–258,** *259–260,* 309, 348
catalysis, 397, 407, *410, 445*
evolution, *122*
structure, **220**
transition-state analogue, **404**

tRNA synthetases, **53–58,** **343–346,** 391–392, **412–416**

Tropomyosin, *195–196,* 269, *467*

Troponin, 53, *198,* 223, **361–364,** *467*

trp repressor, **233–234,** 323, **359–360,** 367, *383*

Trp residues
absorbance, **14–17,** 20, *23*
acid hydrolysis and, 29
cleavage at, 36, **38–40,** 422
charge-transfer complexes, 17
electrostatic interactions, 144–145
flexibility in folded proteins, 287
fluorescence, **14,** 270
hydrophobicity, *154, 168*
iodine and, **275–276**
isolation, *41*
measurement, 20, 23, 29, 31
properties, *3, 4,* **14–17,** 59, *110, 118, 141, 142, 186, 231,* **239,** *240, 256, 282*

spectral properties, **14–16,** 20, 23, 29, 31, 270–271

Trypsin
active site titration, **393,** *460*
binding BPTI, 331, 336, 421–422
biosynthesis, **79–81,** *260,* **434–436**
catalytic mechanism, **418–428**
cleavage of polypeptides, 18, 38, 63, 112
evolution, 247–248
hydrodynamic properties, 266
inhibitors, 421–422
structure, 217, **331**
substrate binding, 273–274, **421–428**

Trypsinogen
activation, **79–81,** *260,* **434–436**
BPTI binding, *336,* 434
comparison with trypsin, *260,* 434
evolution, *122*
flexibility, 237, *260,* 434
structure, *260*

Trypsin protease family
biosynthesis, **79–81**
catalytic mechanism, **418–428,** *461*
catalytic triad, **421–428**
convergence with subtilisin, *138,* 250, 419, 421
disulfides, 248
domains, 133
evolution, 130, **133,** 138, **247–248,** 419, 421
internal structural homology, 251, *460*
Ser residue, 273–274
substrate preferences, **422–423**
zymogens, **434–436**

Tryptophan
biosynthesis, **132–134,** **233–234, 359–360,** *442–444*
diffusion, 269

Tryptophan oxygenase, *467*

Tryptophan synthase
function, *443–444*
gene fusion, **133–134**
mutants, *136–137*

Tryptophan synthase *(cont.)*
  prediction of structure, **257–258**, *260*
  structure, *326*, *434–444*
Tubulin, **90–91**
Tunicamycin, 91
Turns, **225–227, 255–257**
Turnover number, 387, 392
Tyrocidines, **100**
Tyrosine aminotransferase, *467*
Tyrosine hydroxylase, *131*
Tyr residues
  adenylylation, **457–459**
  carboxyl terminal, **90–91**
  cleavage at, 36, 38–40, 83–84, 422
  cytochrome *c*, 273
  flipping in folded proteins, 239, 243, **286–287**
  hydrogen exchange, **216**
  interactions, 144–145, *259*
  iodination, 15–16, 273
  ionization, **6**, 16, 286, 293,
  measurement, 20, *23*, 29, *31*
  nitration, **15–17**, 273
  phosphorylation, **96–98**, *452–453*
  properties, *3, 4,* **14–17**, 21, *110, 118, 141, 142, 154, 186, 231, 240, 256*, 216
  spectral properties, **14–17**, 20, *23*, 29, *31*, **270–271**
  structure, *3*, **14**, 215, 216
  sulfation, 69, 95

Ubiquitin, 467–469, **470–472**
Ultracentrifugation, **24, 266–268**, *326*
Ultraviolet absorbance
  amino acid side chains, **14–16**, 29, *23*, 29, *31*
  conformation and, 177–179, **270–271**
  protein concentration measured with, 30, *270*

solvent perturbation, **270–271**
unfolding followed by, **288**
Uncompetitive enzyme inhibition, **389–391**, *459–460*
Unequal crossing-over, **134**
Unfolded proteins
  conformational properties, **287–292, 314–315**, 353
  hydrodynamic properties, **26–28, 189–190**, *269*
  hydrogen exchange, 282–285
  immunochemical properties, **314–315**, 353
  physical properties, 288–292
  refolding of, **309–325**
  spectral properties, 190–193
  statistical properties, **176–180**
Unfolding
  degradation and, 466–468, *473*
  denaturants and, **287–296**
  kinetics, **310–311**
  ligand binding and, 295
  local, 285
  thermodynamics of, **296–300**
  translocation and, *64*
Unit cell, **203–204**
Unperturbed state, 176–177
Urea
  aspartate transcarbamoylase dissociation, 448
  binding to proteins, 295–296
  denaturant, *24*, **156, 189**, 198, 263, **293–296**
  dimerization, *155*
  gradient electrophoresis, **289–291**
Urease, 295, *397*
Uridylylation, 459
Urokinase, **133**

V-8 protease, 38–39
Vacuoles, *467–478*
Val residues, *3, 4, 7*, 29, *110, 118, 141, 142, 154*, 184, *186*, **223**, *231*, 240, 256
Val tRNA synthetase, **417**
Vanadate, 468
Van der Waals interactions
  description, **146–147, 155**
  effective concentrations, 163–164, 182
  folded conformation stability and, 182, **300–307**, *327*
  ligand binding and, 22, *342–343*, 351
  water and, **154–155**, 162
Van der Waals radius, **140–142**
Van der Waals surface, **140–142, 227–232**
Van der Waals volume, **140–142**
Van't Hoff plot, 181, 297, **311**
Vasopressin, **82–83**
Vesicles, 69–70, 81, 468, 470
Virtual bond, 176
Viruses
  biosynthesis, 58, 76, 79, **83–84**
  quasi-equivalence, **236**
Viscosity, intrinsic, **189–190, 288**
Vitamin C, 90, 96
Vitamin K, 95–96
V_{max} of enzymes, **386–390**
Void volume, 25–27
Volume
  activation, **287**
  amino acid residues, **4, 141**, *156*, **230–231, 267–268**
  changes, **237**
  elution, 25–27
  exclusion, 264
  folded proteins, 27, *156, 168*, 229–232, 237, **266–268**
  hydrodynamic, **24–28, 189–190**, 448
  partial, **24, 141**, *168*, 189, **266–268**

transition state, **287**, 316
van der Waals, **140–142**, 267
void, 25–27

Water
  bound to protein surfaces, **238, 263–264**, 267
  freezing, *168, 326*
  hydration of ions, 267
  hydrogen exchange and, 285
  hydrophobicity and, **157–162**
  inside proteins, **229–230**, 232, 263, **445–446**
  ligand binding and, **335**, *337, 338, 342*, **357**, *359, 363, 383*, **430**
  properties, 139, 145, **148–162**, *269*, 461
  protein stability and, 302, *326*
  structure, **150–153**, *168*, 229
  substrate, enzyme, 389, 420, 422, **424**, 431
Western blotting, 28, 62, **353**, *383*

Xenon association, 168
X-linked chronic granulomatosa, *102*
X-ray diffraction, **202–217, 330–332**, 334
X-ray scattering, **149–152**, *201–202*

Zimm-Bragg model, **184–186**, *198*
Zinc fingers, **358–360**, *384*
Zinc ions
  aspartate transcarbamoylase and, **362, 448–449**
  carbonic anhydrase and, **362**
  catalysis and, *402*
  metalloproteases and, **418, 431–434**, *461*
  sites in proteins, *361, 367, 358–360*, **362**, *384*, **449**
Zymogens, **79–81, 434–437**

ISBN 0-7167-7030-x

EAN

9 780716 770305

90000 >